urth Edition
aniels to find

n geography,
as critical
ng exams and

ased around

s to help with

AN INTRODUCTION TO
HUMAN GEOGRAPHY

Fourth Edition

Edited by

PETER DANIELS
School of Geography, Earth and Environmental Sciences,
University of Birmingham

MICHAEL BRADSHAW
Department of Geography, University of Leicester

DENIS SHAW
School of Geography, Earth and Environmental Sciences,
University of Birmingham

JAMES SIDAWAY
Department of Geography, National University of Singapore

PEARSON

Harlow, England • London • New York • Boston • San Francisco • Toronto • Sydney
Auckland • Singapore • Hong Kong • Tokyo • Seoul • Taipei • New Delhi
Cape Town • São Paulo • Mexico City • Madrid • Amsterdam • Munich • Paris • Milan

Pearson Education Limited
Edinburgh Gate
Harlow
Essex CM20 2JE
England

and Associated Companies throughout the world

Visit us on the World Wide Web at:
www.pearson.com/uk

First published 2001
Second Edition published 2005
Third Edition published 2008
Fourth Edition published 2012

ISBN: 978-0-273-74070-4

British Library Cataloguing-in-Publication Data
A catalogue record for this book is available from the British Library

Library of Congress Cataloging-in-Publication Data
An introduction to human geography : issues for the 21st century / edited by Peter Daniels . . . [et al.]. — 4th ed.
  p. cm.
Includes bibliographical references and index.
ISBN 978-0-273-74070-4 (pbk.)
1. Human geography.   I. Daniels, P. W.
GF41.I574 2012
304.2—dc23

2011049058

10 9 8 7 6 5 4 3 2
16 15 14 13

Typeset in 10/12.5pt Minion by 73
Printed and bound by Korotan - Ljubljana, Slovenia

Dedicated to Jasmin Leila Sidaway
(see www.rgs.org/jasminleilaaward)

Brief contents

Contents in detail

SUPPORTING RESOURCES

Visit **www.pearsoned.co.uk/daniels** to find valuable online resources

Companion Website for students

- Online study skills tutorials for students of human geography, offering valuable practical advice on topics such as critical thinking, essay-writing, taking lecture notes, sitting exams and accurate referencing
- Numerous practice essay and exam questions based around chapter topics
- Clickable weblinks to relevant internet resources to help with further research

For instructors

- PowerPoint slides to download including all figures and maps from the text
- Suggested answer outlines to all practice essay questions
- Suggested answer outlines to all practice exam questions
- Useful seminar activities based on the human geography study skills tutorials

For more information please contact your local Pearson Education sales representative or visit **www.pearsoned.co.uk/daniels**

Guide to Spotlight boxes

		Page number	Historical perspectives	Models	Social and cultural themes	Globalization	Population and migration	Resource management	Geopolitics and the nation-state	Spatial inequalities	Economy and development	Cities and urban spaces	Agriculture and rural spaces	Money and space	Environmental issues
I.1	James' wanderings	12	●												
1.1	Bands	28	●		●										
1.2	Tribes	29	●		●										
1.3	Chiefdoms	30	●		●										
1.4	Regulated (pre-modern) states	30	●						●						
1.5	Market-based states	30	●												
2.1	Characteristics of states with a mature capitalist economy and society	44	●	●					●		●				
2.2	Characteristics of the period of merchant capitalism in Europe	46	●	●							●				
2.3	Proto-industrialization	55	●	●							●				
2.4	The factory system of production	55	●	●						●	●				
3.1	Fordist or 'organized' capitalism	71		●	●						●				
3.2	Marxism	73		●	●										
3.3	Globalised or 'disorganized' capitalism	79	●	●							●				
6.1	Sustainable development	145	●					●		●	●				●
6.2	Carbon offsetting	155				●		●							●
6.3	Environmental justice	157			●					●					●
7.1	How many hungry people are there in the world?	164			●		●	●			●				●
7.2	Meat and the environment	171						●					●		●
7.3	The Alliance for a Green Revolution in Africa (AGRA)	172			●										●
9.1	Urban myths	202				●				●		●			
9.2	Relational approaches and global cities	209				●						●			
11.1	The hidden others of the countryside	249			●					●			●		
12.1	Climate porn	270													●
13.1	Cultures: a summary	273			●										
14.1	Perfect competition	295		●										●	
14.2	Transnational corporations (TNCs)	297		●	●					●				●	
14.3	Fair trade?	304			●						●				
14.4	Comparative advantage	305		●						●	●				

Guide to Spotlight boxes (cont.)

		Page number	Historical perspectives	Models	Social and cultural themes	Globalization	Population and migration	Resource management	Geopolitics and the nation-state	Spatial inequalities	Economy and development	Cities and urban spaces	Agriculture and rural spaces	Money and space	Environmental issues
14.5	The importance of a 'new creative class'?	309			●	●					●				
15.1	Follow the thing: Papaya	321				●		●			●		●		
15.2	Towards conventionalization? Fair trade and organic foods	323		●		●		●			●		●		
15.3	Is the future hydroponic? Thanet Earth	330											●		●
16.1	From Fordism to post-Fordism?	337	●	●		●					●				
16.2	Conceptualizing global divisions of labour?	342		●		●				●	●				
16.3	Types of clusters in the global economy?	347		●		●				●	●				
17.1	Services and the production process	361		●		●					●				
18.1	The functions of money	380												●	
18.2	The circulation of money as capital	388		●		●								●	
18.3	Rotating Savings and Credit Associations (ROSCAs)	390		●	●									●	
18.4	Properties of monetary networks	391		●										●	
19.1	Consumerism as the end of culture?	401			●	●									
19.2	Jonah Peretti's email exchange with Nike iD	412			●						●				
20.1	The Carter Doctrine: contexts and consequences	433	●						●						
20.2	Feminist geopolitics	435			●				●						
20.3	The EU as a geopolitical actor	438							●						
23.1	United Nations	488				●			●						
23.2	Governmentality	490							●						
23.3	Immanuel Kant	493	●												
23.4	Defining the Arctic	494	●						●						

Guide to Thematic case studies

	Page number	Historical perspectives	Models	Social and cultural themes	Globalization	Population and migration	Resource management	Geopolitics and the nation-state	Spatial inequalities	Economy and development	Cities and urban spaces	Agriculture and rural spaces	Money and space	Environmental issues
1.1 The Roman Empire at its zenith (first to fourth centuries AD)	38	●						●	●					
1.2 European feudalism	39	●	●	●										
2.1 Summary of the Laws of the Indies by Philip II of Spain, 1573	52	●								●	●	●		
4.1 United Nation's Millennium Development Goal number 4: reduce by two-thirds the mortality rate among children under 5	96				●		●							
4.2 HIV/AIDS	98				●		●							
4.3 Malaria	101				●		●							
4.4 Future migration challenges: 'environmental refugees'	108						●							●
5.1 A nuclear renaissance?	121							●		●				●
5.2 Energy security	131							●		●	●			●
5.3 Resource abundance: blessing or curse?	136				●			●		●	●			
6.1 Old-growth logging in Tasmania, Australia	142							●						●
6.2 Low-impact development in Britain	146				●			●						●
6.3 Implementing a convention? The obstacles faced in tackling climate change	149					●		●		●	●			●
6.4 Consumption and the environment	154				●			●		●	●			●
6.5 Wilderness and conservation debates in Australia	158				●		●	●				●		●
7.1 The Bengal famine of 1942–43	164	●										●		
7.2 The green revolution	170	●										●		●
7.3 Small rural farms and food security	174									●		●		
8.1 The Millennium Development Goals	179			●	●									●
8.2 The 'strategic interests' of foreign aid	181							●		●				
8.3 Hurricane Katrina and the USA as 'Third World'	182			●					●	●				
8.4 The (re)emerging powers and the rise of new 'Southern' donors	183							●		●				
8.5 Bandung, non-alignment and the 'Third World'	185	●						●		●				
8.6 States and the geopolitics of development	191	●			●			●	●	●				
8.7 The World Bank and the IMF	193			●	●			●	●	●			●	

Guide to Thematic case studies (cont.)

		Page number	Historical perspectives	Models	Social and cultural themes	Globalization	Population and migration	Resource management	Geopolitics and the nation-state	Spatial inequalities	Economy and development	Cities and urban spaces	Agriculture and rural spaces	Money and space	Environmental issues
8.8	Climate change and development	196									●		●		●
9.2	Measuring the connectivity of cities	210				●					●	●			
10.1	Geographies of institutional disinvestment	223			●						●	●	●	●	
10.2	Community opposition to asylum seekers	229			●		●						●		
12.1	The limits to growth: a radical critique	259	●				●	●			●				●
12.2	Contesting wilderness	261						●					●		●
12.3	Desertification	264						●					●		●
12.4	Gay animals	266			●										
13.1	The globalization of culture: some examples	276			●	●				●					
13.2	Culture and social networking	277			●										
13.3	Bounded or hybrid national culture?	282			●	●			●						
13.4	Hybridity/diaspora – some examples	283	●		●		●								
14.1	Models of economic location	293	●	●						●	●				
14.2	An oligopoly: Nestlé	303				●					●				
14.3	Facebook versus Google?	310			●						●				
17.1	ICT and the establishment of data-processing factories in the Caribbean and call centres in India	369			●						●	●			
17.2	Boeing's new spatial division of expertise	371				●					●	●			
20.1	The continuing Cold War?	436	●						●						
21.1	The spaces of political protest	445			●				●						
21.2	Policing space	449			●						●		●		
21.3	'Ethnic' territorial cleansing	451	●		●				●						
21.4	Marking territory	454			●				●				●		
21.5	Playgrounds and lecture halls	460			●										

Guide to Regional case studies

	Page number	Historical perspectives	Models	Social and cultural themes	Globalization	Population and migration	Resource management	Geopolitics and the nation-state	Spatial inequalities	Economy and development	Cities and urban spaces	Agriculture and rural spaces	Money and space	Environmental issues
9.1 The slow death and tenuous rebirth of Motor City	207			●		●					●			
10.3 Ethnic division in apartheid South Africa	231	●		●		●			●		●			
10.4 Gated communities in Uruguay	235			●					●		●			
11.1 Rural depopulation in the Pacific Islands: Niue	245			●		●			●			●		
11.2 Productivist agriculture in the Hereford cider industry	252									●		●		
18.1 From Bangkok to Whitley Bay . . .	386			●					●	●			●	
18.2 Grameen Bank	394		●							●			●	
19.1 Consuming McDonald's in East Asia	403			●	●									
19.2 Western goods and cultural indigenization: the case of 'Les Sapeurs'	404			●	●									
19.3 The shopping mall experience: Woodfield Mall, Schaumburg, Ilinois	407			●								●		
22.1 Nations without states: the Kurdish case	470	●						●						
22.2 Zomia: highland resistance to nation-building and state making	472			●				●						

Contributors

Heather Barrett Principal Lecturer in Human Geography, Institute of Science and the Environment, University of Worcester. Urban geography; urban morphology, planning and conservation; pedagogic research; employability, use of video.

Michael Bradshaw Professor of Human Geography, Department of Geography, the University of Leicester. Economic geography of Russia; resource geography and globalization and energy security.

John R. Bryson Professor of Enterprise and Economic Geography, School of Geography, Earth and Environmental Sciences, University of Birmingham. Expertise-intensive industries; industrial design and competitiveness; spatial divisions of expertise; sustainability and innovation.

Neil M. Coe Professor of Economic Geography, School of Environment and Development, University of Manchester. Global production networks and local economic development; geographies of local and transnational labour markets; geographies of innovation; institutional and network approaches to economic development.

Ian Cook Associate Professor of Geography, University of Exeter. Cultural-economic geography; specializing in 'follow the thing' research, new media and (public) pedagogy.

Philip Crang Professor of Cultural Geography, Department of Geography, Royal Holloway, University of London. Cultural and economic geography, especially commodity culture.

Peter W. Daniels Emeritus Professor of Geography, School of Geography, Earth and Environmental Sciences, University of Birmingham. Geography of service industries; services and globalization; trade, foreign direct investment and internationalization in services.

Shari Daya Lecturer in Human Geography, Department of Environmental and Geographical Science, University of Cape Town. Culture; identity; modernity; geographies of production and consumption.

Klaus Dodds Professor of Geopolitics at Royal Holloway, University of London and editor of *The Geographical Journal*. Geopolitics and international relations; Arctic and Antarctic; geography and popular culture.

James Evans Senior Lecturer in Geography, School of Environment and Development, University of Manchester. Ecological planning; local environmental governance; urban sustainability and the aesthetics of environmental science.

Carl Grundy-Warr Senior Lecturer in Geography, Department of Geography, National University of Singapore. Political geography; geopolitics; political geographies of forced displacement; political geographies and ecologies of environmental resource governance.

Tim Hall Director of Studies/Principal lecturer, Department of Natural and Social Sciences, University of Gloucestershire. Urban geography; representations of the city; economic geography of organized crime; pedagogic research.

Phil Hubbard Professor of Urban Studies, School of Social Policy, Sociology and Social Research, University of Kent. Geographies of sexuality, gender and whiteness; urban consumption; geographies of higher education.

Brian Ilbery Professor of Rural Studies, Countryside and Community Research Institute, University of Gloucestershire. Agricultural change and policy; food supply chains and alternative food networks; property rights, tenancy and farm diversification.

Andrew Jones Professor of Economic Geography, Department of Geography, Environment and Development Studies, Birkbeck College, University of London. Globalization; transnational firms; business services; economic practices; knowledge economy; international voluntary work.

Virginie Mamadouh Associate Professor of Political and Cultural Geography, University of Amsterdam, The Netherlands. Geopolitics and critical geopolitics; political geographies of European integration; multilingualism; new media and transnationalism.

Damian Maye Senior Research Fellow, Countryside and Community Research Institute, University of Gloucestershire. Agri-food restructuring; alternative food networks; farm tenancy; geographies of micro-brewing; biosecurity and disease.

Cheryl McEwan Reader in Human Geography, Department of Geography, Durham University. Feminist and cultural geographies; post-colonial theory and development; geographies of transformation in South Africa.

Warwick E. Murray Professor of Human Geography and Development Studies, School of Geography, Environment and Earth Sciences, Victoria University of Wellington, New Zealand. Development geography; globalization; rural change; Latin America and Oceania.

Jenny Pickerill Senior Lecturer in Environmental Geography, Department of Geography, University of Leicester. Environmental and social justice; geographies of resistance and activism; Australian indigenous politics; green building solutions.

Jane Pollard Professor of Economic Geography, School of Geography, Politics and Sociology, Newcastle University. Geographies of money and finance; political economy and regional economic development.

Marcus Power Reader in Human Geography, Department of Geography, University of Durham. Geographies of development and post-colonialism; visuality and critical geopolitics; post-socialism and Lusophone Southern Africa.

Bill Pritchard   Associate Professor of Economic Geography, School of Geosciences, University of Sydney, Australia. Geography of food and agriculture; global trade rules in agriculture; food security and rural development.

John Round   Lecturer in Human Geography, School of Geography, Earth and Environmental Sciences, University of Birmingham. Geographies of everyday life in post-Soviet regions; coping tactics as a response to economic marginalization; state–society relations.

Denis Shaw   Honorary Senior Research Fellow, School of Geography, Earth and Environmental Sciences, University of Birmingham. Historical and contemporary geographical change in Russia; history of geographical thought.

James D. Sidaway   Professor of Political Geography, Department of Geography, National University of Singapore.

Economic and political geography; history of geographic thought.

Terry Slater   Honorary Senior Research Fellow, School of Geography, Earth and Environmental Sciences, University of Birmingham. Historical geography; urban morphology; topographical development of medieval towns; urban conservation.

David Storey   Senior Lecturer in Geography, Department of Applied Sciences, Geography and Archaeology, University of Worcester. Nationalism, territory and place; sport and national identity; rural change and development; heritage and rural place promotion.

Chih Yuan Woon   Assistant Professor, Department of Geography, National University of Singapore. Geopolitics; security narratives in Southeast Asia; Chinese strategies.

Guided tour

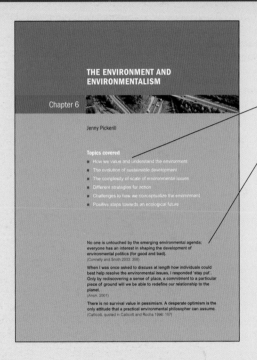

Topics covered introduce, at a glance, the key ideas and perspectives of the chapter.

Memorable **quotes** from important commentators past and present help contextualize and bring the subject to life.

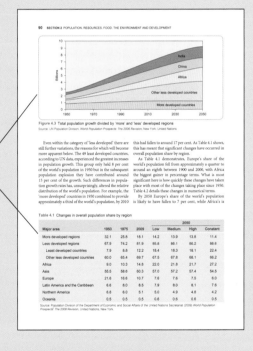

Figures and **tables** graphically illustrate the main points, concepts and processes in the text, to help your learning.

Regional case studies provide you with an insight into a real-life issue specific to a region.

A wide range of **maps**, both ancient and modern, create a clear visual representation of worldwide geographical data.

Snappy **Spotlight boxes** improve your understanding by expanding on the central concepts of the chapter.

Key terms are highlighted throughout the text and are then collated to form a full Glossary at the back of the book. This helps you with unfamiliar concepts or jargon and is particularly helpful for revision purposes.

Thematic case studies, often accompanied by thought-provoking images, investigate crucial geographical themes and illuminate how they can be applied to real life.

Learning outcomes provide a key checklist of learning milestones, to test your understanding of the chapter.

The **Further reading** section helps direct your independent study to the most appropriate texts, journals and websites.

Acknowledgements

As ever, the on-going success of this project relies on the goodwill and enthusiasm of the contributors. We would also like to thank the panel of reviewers who assessed the third edition and made suggestions on the ways in which it might be improved. This edition therefore includes some new chapters that reflect the feedback from the panel and which continues to ensure a lively collection of contributions; from those who have contributed from the outset, from those who joined as new contributors to the third edition, and from those who are new to the fourth edition. Without exception, all have enthusiastically responded to the editors' edict that they should seek to adopt an accessible writing style that will engage readers and encourage them to make connections between many of the issues discussed in the book and their day-to-day. This is achieved by using contemporary/everyday examples that makes the material more meaningful and less abstract. The importance of taking a global perspective is retained while also encouraging the contributors to include as wide a range of examples as possible, especially from Europe. We have also sought to ensure good integration between the case studies/artwork and the text so that the reader can see why a particular feature is situated at a particular point in a chapter, what it is there to illustrate, how can it provoke readers to think through the issue, and how it relates to the main narrative. It has also been very important to ensure that the fourth edition continues to offer readers the cutting edge perspectives on the latest debates, issues and controversies that were a feature of the third edition. The contributors have risen to all these challenge and we would like to acknowledge their constructive response to the dialogue that this has necessitated along the way.

The editors and contributors are indebted to Rufus Curnow at Pearson who initiated the discussions that led to the decision during 2010 to prepare a fourth edition. Just as the process was gathering momentum he handed over the role of Senior Commissioning Editor to Patrick Bond who has since provided invaluable support and encouragement during the process of pulling together the final manuscript and liaising with the contributors over contracts and queries associated with the artwork and illustrations. In relation to the latter, we are also very grateful for the patience shown by Helen Leech, as she wrestled with the complexities of melding some of the pre-existing artwork and illustrations with the ideas for adjustments made by the contributors as well as requests for completely new items. Elizabeth Wright subsequently took over from Helen Leech and has patiently managed all the toing and froing involved with copy editor and the inevitable queries that arise at the page proof stage of production. We had interesting exchanges of ideas with the designer, Maggie Wells, as we attempted to identify a new and striking cover design; her willingness to contemplate all our suggestions was commendable. There are other members of the editorial and production staff at Pearson that the Editors do not meet, such as those tasked with preparing the artwork or undertaking the copy editing and proof reading; we owe them all a debt of gratitude for seeing this project through to completion.

There are also a number of individuals whose efforts deserve particular mention and thanks. Ann Anckorn, Drawing Office Technician, in the School of Geography, Earth and Environmental Sciences at the University of Birmingham helped with redrawing some of the artwork and converting some statistical material into appropriate charts. Mike Bradshaw wishes to acknowledge the support of the Leverhulme Trust in the form of a Major Research Fellowship, which has allowed him to devote more time to studying the global energy dilemmas first introduced in his chapter in this text. James Sidaway would like to thank his colleagues at the Department of Geography, Planning and International Development Studies at the University of Amsterdam, where he was based during the preparation of this edition.

For all of the editors, production of the fourth edition has taken place at a time when we have all been facing numerous challenges and demands on our time, whether as a result of moving to pastures new, making the transition to formal retirement, or simply coping with the ever-changing landscape of higher education. However, we have continued to work closely together as a team; a task made easier by the artifacts of information and communications technology even though we have become geographically more dispersed. Sticking to the task has also been made easier by the continuing enthusiasm shown by all the contributors and the team at Pearson; this has ensured that we have produced a text that we can all be proud of. The most important motivation, of course, is our hope that the ideas, perspectives and challenges discussed in this book will encourage readers to connect with human geography; after all, the vitality of the discipline depends on these students being enthused and critically, as well as creatively, engaging with human geography within and beyond the lecture and seminar room.

PWD
MJB
DJBS
JDS

Birmingham, Leicester and Amsterdam,
December 2011

The editors and publishers would like to thank the following reviewers of the third edition for their invaluable input into the shaping of ideas for this fourth edition:

Dr Simon Arlidge, University of the West of England
Dr David Atkinson, University of Hull
Dr Christopher Bear, University of Aberystwyth
Johan Berg, Stockholm University, Sweden
Dr. Colm Breathnach, University of Strathclyde
Dr Stuart Dawley, University of Newcastle upon Tyne
Professor David Gilbert, Royal Holloway, University of London
Dr Lisa Lau, Keele University
Dr Eric Olund, University of Sheffield
Dr Samarthia Thankappan, University of York

and to the following reviewers of the second edition for their input to shaping the third edition:

Professor Cara Aitchison, University of the West of England
Dr Nicola Ansell, Brunel University
Dr Freerk J. Boedeltje, Radboud University Nijmegen, The Netherlands
Dr Luke Desforges, University of Sheffield
Professor Isabel Dyck, Queen Mary, University of London
Dr James Faulconbridge, Lancaster University
Dr Beth Greenhough, Queen Mary, University of London
Associate Professor Michael Haldrup, Roskilde University, Denmark
Associate Professor Brita Hermelin, Stockholm University, Sweden
Dr Simon Leonard, University of Portsmouth
Professor Anders Lundberg, University of Bergen, Norway

Associate Professor Per Gunnar Røe, University of Oslo, Norway
Dr Stephen A. Royle, Queen's University Belfast
Dr Sam Scott, University of Liverpool
Dr Bettina Van Hoven, University of Groningen, The Netherlands
Professor Neil Wrigley, University of Southampton

The editors and publishers would also like gratefully to acknowledge the contribution of Ian Cook of the University of Exeter (with the assistance of Suzy Reimer, David Bell and Paul Cloke) to the third edition.

Publisher's acknowledgements

We are grateful to the following for permission to reproduce copyright material:

Cartoons

Cartoon 8.3 from The World Bank and the State: a Recipe for Change? By Nicholas Hildyard, © Bretton Woods Project, www.brettonwoodsproject.org

Figures

Figure 2.2 from *Descrição da Fortaleza de Sofala e das mais da India*, Fundacão Oriente, Lisbon (Carneiro, A de M 1990); Figure 2.6 from *Yorkshire Textile Mills 1770–1930*, HMSO (Giles, C. and Goodall, I.H. 1992) p. 102, MD94/04156, © Crown copyright. NMR; Figure 2.7 from *Yorkshire Textile Mills 1770–1930*, HMSO (Giles, C. and Goodall, I.H. 1992) p. 102, Courtesy of the Ironbridge Gorge Museum Trust; Figure 4.4 Demko, G.J., *et al.*, *Population Geography: A Reader*, © 1970 The McGraw-Hill Companies, Inc.; Figure 4.5 after *Human Population: Fundamentals of Growth: Future growth*, Population Reference Bureau (2007) Available at http://www.prb.org/Educators/ TeachersGuides/HumanPopulation/FutureGrowth/ TeachersGuide.aspx?p=1; Figure 4.11 after *Education and HIV/AIDS: A Window of Hope*, World Bank (Bundy, D. and Gotur, M. 2002) p. 10, Figure 1.3: HIV/ AIDS and education: the consequences, © International Bank for Reconstruction and Development/ World Bank; Figure 4.14 from The Determinants and Implications of an Aging Population In Russia, *The Rand Corporation, Conference Proceedings, CF-124*, Figure 6.2 (Vassin, S.A. 1996), RAND Corporation used with permission; Figure 5.1 from *Natural Resources: Allocation, Economics and Policy*, 2 ed.,

Routledge (Rees, J. 1985) p. 15; Figure 5.2 from *Global Change and Challenge: Geography in the 1990s*, Routledge (Rees, J. in Bennett, R. and Estall, R. (eds) 1991); Figure 5.3 from *Natural Resources: Allocation, Economics and Policy*, 2 ed., Routledge (Rees, J. 1985) p. 20; Figure 5.4 reproduced by permission of SAGE Publications, London, Los Angeles, New Delhi and Singapore, from Jones, G. and Hollier, G., *Resources, Society and Environment Management* (© SAGE Publications, 1997); Figure 5.9 from *BP Statistical Review of World Energy June 2011*, BP (2011) p. 16; Figure 5.11 from *Environmental Resources*, Mather, A.S. and Chapman, K., Pearson Education Limited © Pearson Education Limited 1995; Figure 15.1 from Opening spaces through relocalization: locating potential resistance in the weaknesses of the global food system, *Sociologia Ruralis*, 42, pp. 347–69 (Hendrickson, M. K. and Heffernan, W.D. 2002); Figure 15.2 from Forging linkages in the commodity chain: the case of the Chilean salmon farming industry, 1987–2001, *Sociologia Ruralis*, 43, pp. 108–27 (Phyne, J. and Mansilla, J. 2003); Figure 15.3 first published in *Environment and Planning A*, 2005, Ilbery and Maye, 'Alternative (shorter) food supply chains and specialist livestock products in the Scottish-English borders'; Figure 16.1 reproduced by permission of SAGE Publications, London, Los Angeles, New Delhi and Singapore, from Dickon, P., *Global Shift: Mapping the Changing Contours of the World Economy* (© SAGE Publications, 2007); Figure 16.3 adapted from Foster, W., Cheng, Z., Dedrick, J. and Kraemer, K.L. (2006) *Technology and organizational factors in the notebook industry supply chain*, Figure 2, The Personal Computer Industry Center Publication, UC Irvine; Figure 16.7 from *Commodity Chains and Global Capitalism*, Gereffi, G. Copyright © (1994) by Gary Gereffi and Miguel Korzeniewicz. Reproduced with permission of ABC-CLIO, LLC.; Figure 16.8 New models of regulating production networks adapted from Sum, N-L. and Ngai, P. (2005) Globalization and paradoxes of ethical transnational production: code of conduct in a Chinese workplace, *Competition and Change*, 9, 181–200, Figure 1 © Maney Publishing; Figure 17.1 from *Service Worlds: People, Organizations, Technologies*, Cengage (Bryson, J.R., Daniels, P.W. and Warf, B. 2004); Figure 17.2 from Mahajan, S. (2005) *Input-Output Analysis: 2005*, p. 23, Office for National Statistics, London. Office for National Statistics licensed under the Open Government Licence v.1.0.; Figure 17.3 Transnational corporations and spatial divisions of 'service' expertise as a competitive strategy: The example of 3M and Boeing,

The Service Industries Journal, 28 (3), pp. 1–17 (Bryson, J. R. and Rusten, G. 2008), reprinted by permission of the publisher (Taylor & Francis Ltd, http://www.tandf.co.uk/journals).

Maps

Map 1.1 adapted from *Changing the Face of the Earth: Culture, Environment, History*, 2nd ed., Blackwell (Simmons, I. G. 1996) p. 48; Map 1.2 after Sherratt, A., *Cambridge Encyclopedia Archaeology* (1980), Cambridge University Press; Map 2.3 after *Atlas of the British Empire*, (Bayley, C. 1989), © Octopus Publishing Group; Map 2.5 after *The Hamlyn Historical Atlas*, Hamlyn, London, (Moore, R.I. (ed) 1981); Map 2.8 after Pounds, N.J.G., *An Historical Geography of Europe* (1990), Cambridge University Press; Map 4.10 from http://www.unaids.org/documents/20101123_2010_HIV_Prevalence_Map_em.pdf; Map 5.7 and 5.8 from *BP Statistical Review of World Energy June 2011*, BP (2011) pp. 21, 31; Map 14.1 from Maplecroft (2010); Map 20.1 after *The Fate of the Forest: Developers, Destroyers and Defenders of the Amazon*, Verso (Hecht S. and Cockburn A. 1989) p. 127; Map 20.2 adapted from *Political Geography*, Glassner, M.I. Copyright © 1993 John Wiley & Sons, Inc. Reproduced with permission of John Wiley & Sons, Inc.; Map 20.4 from Arquivo Histórico Militar (Galvão, H. 1934) Archival Ref: PT/AHM/DIV/3/47/AV2/2325; Map 20.6 after Political geography and panregions, *Geographical Review*, 80, pp. 1–20 (O'Loughlin J. and van der Wusten, H. 1990).

Tables

Table 2.1 from R. I. Moore (ed), *The Hamlyn Historical Atlas*, Copyright © Octopus Publishing Ltd., 1981; Table 5.1 adapted from *Global Change and Challenge: Geography in the 1990s*, Routledge (Rees, J. in Bennett, R. and Estall, R. (eds) 1991) pp. 5–26; Table 5.2 from *BP Statistical Review of World Energy June 2011*, BP (2011) p. 41; Table 5.3 from The World Bank: *The Little Green Data Book 2010*; Table 5.4 and 5.5 from *Key World Energy Statistics* © OECD/International Energy Agency 2010, page 11, page 13; Table 6.1 and 6.3 from Carter, N., *The Politics of the Environment: Ideas, Activism, Policy* (2001), Cambridge University Press; Table 6.2 from *The shallow and the deep, long-range ecology movement*, Inquiry 16 (Naess, A. 1973) pp. 95–100, Oslo; Table 6.4 from Robert Garner, *Environmental Politics*, published 2000, (Macmillan) reproduced with permission of Palgrave Macmillan; Table 10.1 from *Dangerous Disorder:*

Riots and Violent Disturbances in Thirteen Areas of Britain, 1991–92 by Anne Power and Rebecca Tunstall, published in 1997 by the Joseph Rowntree Foundation. Reproduced by permission of the Joseph Rowntree Foundation; Table 10.2 from The changing dynamics of community opposition to human service facilities in *Journal of the American Planning Association*, Taylor & Francis Ltd., (Takahashi, L.M. and Dear, M.J. 1997), reprinted by permission of the publisher (Taylor & Francis Ltd., http://www.informaworld. com); Table 14.1 from OECD (2009) Figures 2009; Table 14.3 from *World Tourism Highlights, 2010 Edition*, United Nation World Tourism Organization (2010) © UNWTO, 9284400112; Table 14.4 from *Major FDI Indicators*, UNCTAD (2010); Table 14.5 from Z/ Yen Limited, *The Competitive Position of London as a Global Financial Centre*, Corporation of London (November 2005); Table 14.6 from Internet usage, by world region, 2010, www.internetworldstats.com/stats. htm, Copyright © 2000–2011, Miniwatts Marketing Group. All rights reserved.; Table 16.1 reproduced by permission of SAGE Publications, London, Los Angeles, New Delhi and Singapore, from Dickon, P., *Global Shift: Mapping the Changing Contours of the World Economy* (© SAGE Publications, 2007); Table 16.3 adapted from *The growing power of retailers in producer-driven commodity chains: a 'retail revolution' in the US automobile industry?* (Kessler, J. and Appelbaum, R. 1998) unpublished manuscript, Department of Sociology, University of California at Santa Barbara, USA; Table 17.1 from *Input-Output Analysis: 2006*, ONS (Mahajan, S. 2006) p. 37, Office for National Statistics licensed under the Open Government Licence v.1.0.; Table 17.2 from A 'Second' global shift? The offshoring or global sourcing of corporate services and the rise of distanciated emotional labour, *Geografiska Annaler* (Bryson, J.R. 2007); Table 18.1 adapted from World Federation of Exchanges, www.world-exchanges.org/files/ statistics/excel/EQUITY609.xls

Photos

Alamy Images: Alan Payton 124, Alex Segre 335, Cro Magnon 13, David R Frazier Photolibrary, Inc 477, Frankie Angel 398, Glow Images 458, Images of Birmingham 409, ImageState 116, Jim West 432t, 474, Justin Kase 447, Mar Photographics 214, Marvin Dembinsky Photo Association 206, Megan Whittaker 486, Nathan Griffith 443, Photofusion Picture Library 448, Realimage 134, Richard Levine 173, Ryan Rodrick Beiler 463, Superstock 498, Thibaut Petit-Bara 454, Vario Image GmbH & Co KG 461; **Ancient Art and Architecture Collection:** 28bl; **Art Directors and TRIP Photo Library:** Graham Grieves 467; **BAA Aviation Photo Library:** 479; **Banksy:** 263; **Bridgeman Art Library Ltd:** National Gallery, London 262; **Corbis:** Angelo Homak 66, Benjamin Rondel 274, Bettmann 380, 432bl, 490, 4332br, Bill Ross 480, Claro Cortes IV/Reuters 400, Dave G. Houser 284, Gillian Darley/Ediface 70, Ira Nowinski 430, Michael Setboun 436b, Reuters 75, Sherwin Castro/Reuters 369, Stefano Bianchetti 34br, Walter Hodges 298; **T. Paul Daniels:** 299; **Denis Shaw:** 63, 211, 227; **G.P. Dowling:** 406; **Mary Evans Picture Library:** 67; **Ronald Grant Archive:** 269; **Robert Harding World Imagery:** Neil Emmerson 74; **image courtesy of NASA Earth Observatory:** 6, 11; **Image courtesy of The Advertising Archives:** 324, 397; **Getty Images:** 445, 446, 488, 503, Bloomberg 489, Chris Donio 456, De Agnostini 493, Evening Standard/Stringer 118, Fox Photos 133, Hulton Archive/Stringer 72, National Geographic Society 403, Robert Stiggins/Express 399, Superstock Superstock/© DACS 434, Topical Press Agency 365; **Joanne O'Brien Photography:** 279; **Paul Lunnon:** 258, Sean Sprague 352; **Michael Holford Photo Library:** 28cl; **Warwick E. Murray:** 238cl, 238b, 243, 244, 249, 250, 251, 238cl, 238b, 243, 244, 249, 250, 251; **Nottingham Evening Post:** 229; **Oxfam:** 193; **Panos Pictures:** Ami Vitale 113, Chris Stowers 353, Martin Adler 191, Stefan Boness 401; **PhotoDisc:** 258cl; **Photofusion Picture Library:** Paul Risdale 80; **Jenny Pickerill:** 142, 145, 153, 157, 160, 142, 145, 153, 157, 160; **Press Association Images:** Dita Alangkara 119, Eugene Hoshiko 312, John Froschauer 371, Luke Palmisano 384, REMY DE LA MAUVINIERE 225; **Reuters:** Eduardo Munoz 165, Toby Melville 487; **Rex Features:** Everett Collection 436t, Sipa Press 420; **Science Photo Library Ltd:** RIA Novosti 120; **Shutterstock.com:** Andy Z 212, Gr8 360; **Terry Slater:** 51, 61tl, 61tr, 61bl, 61br, 51, 61tl, 61tr, 61bl, 61br; **STILL Pictures The Whole Earth Photo Library:** Blickweinkel/Luftbild 34tl, Sebastian Bloesch/Das Fotoarchiv 25; **The Hereford Mappa Mundi Trust:** 40; **The Kobal Collection:** MGM/ Filmways 497b, Paramount 497t; **Tim Hall:** 215; **Time Magazine:** 433; **TopFoto:** 435; **Vintage Ads:** 77; **www.blackgoldmovie.com/dvd:** 411 All other images © Pearson Education

In some instances we have been unable to trace the owners of copyright material, and we would appreciate any information that would enable us to do so.

GEOGRAPHY: FINDING YOUR WAY IN THE WORLD

Introduction

Peter Daniels

Michael Bradshaw

Denis Shaw

James Sidaway

Geography is indispensable to survival. All animals, including American students who consistently fail their geography tests, must be competent applied geographers. How else do they get around, find food and mate, avoid dangerous places? (Yi-Fu Tuan 2002: 123)

We routinely make sense of places, spaces, and landscapes in our everyday lives – in different ways and for different purposes – and these 'popular geographies' are as important to the conduct of social life as are our understandings of (say) biography and history. (Gregory 1994: 11)

Whether it be global pandemics or climate change, food production or the survival of indigenous peoples, the activities of volcanoes or the depletion of glaciers, geography is the most all-encompassing of subjects. It is about the world around us. Whether we are travellers or explorers or adventurers or scientists or farmers or architects or teachers or pupils, we are all affected by this most exacting and rewarding of subjects. (Palin 2010: 255)

A discipline concerned to appreciate the earth's diversity requires a pragmatic approach, responding to changes in its subject matter both substantively and methodologically. Human geography has done this in recent decades in part because of its size – which reflects its success as an academic enterprise. Student demand sustains a substantial disciplinary presence in many universities, and the scholars appointed to teach them have developed a wide range of research and teaching expertise, based on a plethora of multi- and inter-disciplinary interactions. (Johnston 2011: 105–6)

This book is the fourth edition of *An Introduction to Human Geography*. The first edition was published in 2001. The task of compiling the first edition therefore dates back to 1997–98 so that it is now approaching fifteen years since we wrote the first introduction. While the fourth edition incorporates many revisions and further changes to the structure and contents, we have retained our original goal, which was to provide an introduction to human geography through a survey of the field, focusing upon contemporary *issues* and adopting contemporary approaches. What introductory textbooks in human geography choose to include and foreground (and what is excluded or neglected) was a few years ago the subject of heated debate in a leading disciplinary journal (*Transactions of the Institute of British Geographers*). According to one of the protagonists, new textbooks (in part via their influence on a prospective new generation of geographers) become part of what shapes the dominant themes for research and scholarship in a discipline. Textbooks are thereby implicated 'in strategies to mobilize support for a particular set of disciplinary practices' (Johnston 2007: 437).

Textbooks also reflect *where* they are written and where they are read. For example, most textbooks written in North America devote a significant number of words to explaining what geography is and what constitutes a geographical approach. Early on in one of the most widely used American human geography introductions, Marston *et al.* (2011: 2) note that:

> The power of geography comes from its integrative approach, which addresses global connections, historical trends, and systemic political-economic and sociocultural relations by drawing on the intellectual tradition in both the natural and social sciences.

The reason why introductory textbooks in the United States need to explicitly consider definitions of human geography and devote space to explaining what a geographical approach amounts to is that many of the students taking a module in introductory human geography are not geography majors, that is they will not go on to specialize in geography, and they will usually have experienced limited or no exposure to it as a discipline at high school. Geography is seldom taught in the American 'K–12' (Kindergarten to pre-University) school system. The institutional setting within which the present textbook has been put together is different. The editors and most of the contributors currently teach in British or continental European universities or in countries where the influence of a British style of educational system is more evident. Many of their students have chosen to specialize in geography at a school or college and have made a choice to read for a degree in geography. Consequently, students from these states often have ideas of what the subject matter of the discipline is (although this may turn out to be rather different from much of what they will subsequently encounter at university, see Bonnett 2003 and Stannard 2003). Allowing for this context, part of the rationale for many introductory human geography modules is to explore how university-level geography differs from students' experiences of the subject so far. In such modules students are challenged not to take the world for granted and to understand that human geography is not a direct reflection of a straightforward reality that is 'out there', but a 'social construction'. In other words, although certain experiences, beliefs and value systems are shared, these also vary and interpretations of the world differ from different vantage points in time and place. Moreover, environments and nature so altered by humans and their meaning and representation are always shaped by (indeed they can only be seen and understood via) cultures. These vary in space and time. There is a world of difference between how the 'environment' might be seen for example by an oil prospector, and someone who lives on the land being surveyed and views it as their home, or an environmental activist pointing to the destructive ecological affects of oil extraction and the use of fossil fuels. The very idea that there is a *single* notion of what human geography is about (that is valid for all times, places and perspectives) becomes problematic. As Chris Gibson (2009: 219) has noted:

> Human geography exists as a formal academic discipline in tens, if not a hundred different countries and languages with diverse meanings and contexts. Only a fraction of these stories have been properly documented.

This begs the question, when defining the field, of whose human geography, where and when?

Moreover, human geography and the world that it seeks to interpret and represent are dynamic. In the years since we embarked on the first edition of this book a great deal has changed in the world; the spectacular and violent events of 11 September 2001 in the USA, for example, continue to resonate. Tracing these could begin with the landscape of Manhattan, where the 'twin towers' once stood, but it would include too the enlarged prison in Guantánamo, Cuba (where some of those accused of terrorism are still detained by the US government) to Afghanistan and Iraq, and most recently to the compound in northern Pakistan where Osama Bin Laden was killed by US military forces, to the nature and extent of airport security procedures and surveillance almost everywhere. At the same time, global climate change and environmental catastrophes, such as the Deep Water Horizon disaster in the Gulf of Mexico or the nuclear accident in Japan, together with sustained high energy prices, have moved green issues to the centre of political debates in many countries and in international organizations (see Chapters 5, 6 and 12). And dramatic natural disasters, such as the Indian Ocean tsunami of 24 December 2004, the Kashmir earthquake of 8 October 2005, Hurricane Katrina (New Orleans) on 25 August 2005, floods in Pakistan in the monsoon of 2010 and most recently the Great Eastern Japan earthquake and tsunami on 11 March 2011, have not only caused enormous loss of life and dislocation, but also have many longer-term economic, environmental, social and political impacts. In the aftermath of the explosions and radioactive contamination at the Fukushima nuclear power station (triggered by that earthquake in Japan), the long-standing political campaign in Germany to shut down nuclear power there won a victory when the government of the European Union's largest economy agreed a timetable to close all German nuclear power stations amidst rhetoric about future green technologies.

At the same time, ongoing processes of migration, urbanization and economic transformations (such as the fast pace of development in parts of China and India) are producing new spaces, connections and flows that require new maps and geographical narratives. New divisions are being created, old conflicts revived and on first appearances, the world might appear to be more fragmented and contested than it was at the start of the twentieth century. The aftermath of the 2008 financial crisis, which is still being felt in many economies across the world, seems to have accelerated the shift in the centre of geo-economic power further towards Asia.

Equally, however, economic and cultural globalization continues: some music, fashions and products can be found almost anywhere; though not everywhere! They are often merging and mixing, reworking older patterns. Many economies across the world are so deeply interconnected and intertwined that we can begin to wonder whether there is such a thing as a 'national' economy (see Section 4) or, more accurately, to regard it as an artifact of twentieth-century economics and politics, rather than a helpful way of understanding contemporary economic geography. Certainly, nowhere was immune from the effects of the global financial crisis, though the extent and duration of the impact differed greatly. Prime Minister Putin famously declared that Russia would not suffer from the crisis, yet in the end it was one of the hardest hit among the G20 economies. Now more than half of the world's population lives in cities, which are the focal point of the networks of connectivity and the largest of which are in what are known as 'developing (or 'less economically developed' or sometimes 'Third World') countries' (see Chapter 8 on these categories and Chapters 4, 9 and 10 on cities and population).

As we write, many of these cities in the Arab world have become the sites of massive demonstrations, conflicts and even revolution as popular uprisings have challenged the established political order. In a recent report the World Bank (2011) suggests that the world economy is in the midst of a transformative change, in large part as a result of the rise of a number of dynamic emerging market economies in Asia and South America. Yet, as some of the other chapters that follow will detail, these processes have many deep historical roots and antecedents.

Changing worlds: changing human geographies

Geographical knowledge is not – and should not attempt to be – static and detached from what is going on in the world, but is rather dynamic and profoundly influenced by events, struggles and politics beyond university life.

(Blunt and Wills 2000: x–xi)

The root of the word 'geography' combines *geo* (earth) and *graphy* (writing). To engage in geography is to write about the earth (which includes its lands and seas, resources, places and peoples) or, more widely perhaps, to represent the earth in text (which includes maps: some of the most complexly crafted of all texts). Of course, many other branches of knowledge such as history, anthropology, sociology, oceanography, politics and geology (from which geography draws) and many other sciences are also in some way or other about the world. Traditionally, what has been distinctive about human geography is that it puts an emphasis on people in places and spaces, on landscapes modified by human interventions and on complex spatial connections. Today, human geography embraces issues such as spaces of identity, consumption and sexuality, which are fundamental parts of the human geographies of our interconnected, contradictory, complex, conflict-ridden and fantastically diverse planet. Such topics coexist with (and complicate) longer-established geographical issues associated with, for example, resources, population, cities or the changing distribution of economic activity. Indeed, we would argue that some of the most interesting, but tricky, challenges for human geography arise from its ambition to incorporate human experiences, achievements and transformations in its efforts to understand and to influence the issues that it is addressing. In other words, human geography is not just about 'out there', it is also about 'in here'. Thus, you all create your own personal geographies derived, for example, from where you live, where you work or when you travel. Moreover, through such myriad social–cultural, political and economic geographies, we inhabit a world, as John Pickles (2004: 5) points out,

> that has, in large part, been made as a geo-coded world; a world where boundary objects have been inscribed, literally written on the surface of the earth and coded by layer upon layer of lines drawn on paper.

Those geo-codes have increasingly become digital. They are stored, transformed, transmitted and negotiated electronically; as in the signal that your mobile phone is transmitting regarding your current location (unless you have turned it off) or the data about you that is in archives, online or within that phone (be it on or off).

Such complexity, connections and challenges (as well as diverse 'geo-codes') are evident when we consider 'globalization'. Since the early 1990s, the latter has been something often considered to be a defining characteristic of the times in which we live. When we edited the first edition, we were conscious of, and regularly argued amongst ourselves about what the term 'globalization' takes for granted, what kind of stances we should encourage the contributors to take towards it, and how far we could agree once we saw the arguments emerging in the chapters and subsections. Such discussions were repeated and finessed in preparing subsequent editions. Indeed, as many of the chapters have spelt out, there is certainly a significant sense in which a combination of technical, political, cultural and economic transformations throughout the twentieth century enhanced the sense of global interconnection and this is continuing in the twenty-first century. Take the case of this book. The copy that you are reading may well have been printed far from where you picked it up. Or you may be reading it as an e-book. Either way, it was published by a multinational company, whose ownership and 'home' location

may not be immediately evident. The shares of the publisher (Pearson) are traded in London, part of the daily turnover on the world's largest stock exchange. In addition to textbooks such as this, Pearson publishes the daily *Financial Times*, the weekly *Economist*, and Penguin books. The company found itself in the headlines in March 2011, when it was revealed that the holdings of the Libyan Investment Trust, who held 3.2 per cent of Pearson's shares, had been suspended, along with the freezing of other assets belonging to the Libyan regime (and the Qaddafi family who formed its core), following the outbreak of a civil war there – in which Britain and other outside powers had taken sides with the anti-Qaddafi rebels. In turn, that stake reflects the flows of money associated with geographies of resources (oil and gas) and the modern financial system that are considered in later chapters here. But, in turn, Libyan economic and political stances cannot be understood without reference to colonial histories (the then Ottoman Turkish lands that today comprise Libya were invaded by Italy in 1911, in a bloody war that cost thousands of Arab lives), nationalism, revolution and geopolitics that led the post-colonial Libyan state (established in the early 1950s in the debris of Italian fascist imperialism and the Second World War) into violent conflict with the West, some of its neighbours (whose boundaries themselves were drawn by competing European colonialisms) as well as more conservative Arab regimes.

In addition to such flows of finance and relations of ownership, power and knowledge, the writing and production of the book involves technologies such as desktop and laptop computers (themselves usually manufactured in Asia, but whose components use raw materials more often from Africa), and software written and distributed by some of the world's largest US-headquartered transnational companies (Apple and Microsoft). From the very start of work on the first edition back in 1997, numerous drafts of the chapters and sections were emailed between authors and editors, sometimes between offices next door to each other, or sometimes across Europe. In turn, the second edition was prepared with one of the editors based in Singapore and the three others in the English Midlands, and contributors in a range of locations. And the most convenient place for the editors to meet to finalize outstanding matters for that edition turned out to be in Philadelphia, whilst we were all attending the centenary convention of the Association of American Geographers in March 2004. Subsequent editions were finessed at meetings in Birmingham, but with follow-up communications and drafts circulated by email. During the compilation of this edition, two of the editors met in Amsterdam in May 2011, emailing sections of this introduction between us whilst working in the same office; but doing so using email servers based in Amsterdam and Leicester. One of us subsequently saved a back-up of the draft and later downloaded it elsewhere, by storing it in a Gmail account (the location of whose server is not readily discernable), a service offered by a company (Google) established in 1998, some time *after* work began on what became the first edition of this book. Just a few decades ago the writing and production process would have been very different. There would have been much less dependence on electronic communication, more extensive use of surface or air mail to exchange pieces of the manuscript, and the production of the book, copy-editing, proofreading and printing largely completed locally. Can we therefore see this book as both an artifact and representation of globalization?

To describe contemporary technological, economic, cultural or political tendencies as 'globalization' is, we agreed, to invoke a certain *geographical imagination*, a vision of the growing significance of a global scale of action, of the world as a single place. Yet, we know that many people and places remain relatively marginalized from these supposedly hypermobile ways of living and working. And some people and places benefit from them more than others. Of course, such power and inequality are not themselves new. Consider the profits and consequences of the slave trade, for example. Cheap sugar and cotton for northern hemisphere markets and manufacturers in Europe and America enabled new links, markets and economies. But those humans traded as unpaid workers (slaves) and objects of exploitation that did the work did not see it that way (if they lived long at all). Today, a lowly paid worker (or self-employed prospector) in a diamond mine might experience the global trade in these minerals rather differently to the companies that dominate the jewellery business. Certainly the benefits of globalization are uneven. This is not new.

As an alternative to the term 'globalization', we might use other terms: 'imperialism', 'power' or 'capitalism', for example. Each carries particular connotations. In this way, 'globalization' serves as a particular concept that is used to make sense of the world and one on which a certain geographical imagination, of an increasingly connected and 'shrinking' world for example, is emphasized. Yet just because something

is *imagined* and interpreted in particular ways and by reference to particular geographies, it does not make it any less *real* to those caught up in it. Globalization is seen by some as a broadly positive force, breaking down barriers, making capitalism more efficient and spreading its benefits throughout the world. For others it is a more negative process enabling another round of exploitation, often with the further destruction of local cultures and identities and further commodifying life and nature. Everything is for sale, everything has a price. Obviously, globalization has been contested, in terms of both the meanings attributed to it and the evaluation of its consequences. Judge for yourself, but as you do so, do not make the mistake of assuming that everyone everywhere shares your vantage point and experiences of the world. And having boomed as a term and way of talking about the world in the 1990s, recent years have seen use of the term begin to decline. Although based on a selective sample of books published in English since 1900, data provided by Google's scanning of books indicates that the number of those using the term 'globalization' peaked in the mid-2000s after a steep rise from almost zero in the 1980s (see Figure 1).

Some events and moments have had global coverage. More than a decade ago, the coming of the millennium (and perhaps soon after what became known as '9/11') were represented to us (and lived and made around the world) as an enormous global collective likened by some to the moment when the American Apollo 8 spacecraft in 1968 captured the first image of the earth as a whole from space, or the reception of photographs of the earth from space by subsequent NASA missions (see Plate 1). These images have since circulated widely and are credited with reshaping human perceptions of the planet (see Cosgrove 2001).

Yet the way that some events become 'significant' or 'global' reflects where they happen and who they affect. The spectacular losses of thousands of lives in Manhattan on 9/11 (11 September 2001), for example, became a global media event and subject of debate in the way that the death of more than a million from 1998 to 2003 in the Democratic Republic of Congo through war never did. The idea that this book is one published in the twenty-first century, has roots in a Western (Christian) timetable. According to the calendars of China, Islam and Judaism it is not the twenty-first century, and Islamic, Hindu, Jewish and Chinese 'New Years' do not coincide with the first of January. Furthermore, the calendar that produced 2000 as a millennium year

Figure 1 Use of the term 'globalization', 1900–2009, based on a survey of digitized material in Google books.

Source: http://ngrams.googlelabs.com/graph?content=globalization&year_start=1900&year_end=2008&corpus=0&smoothing=1

Plate 1 'The Blue Marble'. Photograph taken by a crew member aboard Apollo 17 on 7 December 1972.

(NASA)

of its identity and knowledge to ideas, people and artifacts originating outside what have been considered its geographical limits. Many observers subsequently claimed that a new age (and perhaps the geopolitics of a new century) began on sunny New York, Pennsylvanian and Washington DC mornings in September 2001, when the drama of hijacked civilian aircraft being used as weapons rapidly became a global media event. Yet the many causes and consequences of '9/11' are far from new. As geographers you will be able to explore their links to flows of resources, power and geopolitics and to the technologies, cultures and economies of mobility that enabled the use of airliners as weapons to strike symbolic targets. Moreover, how the events and meanings of '9/11' have been represented, experienced and understood in different places vary enormously.

Human geography is caught up in challenging but promising processes of outgrowing its modern roots as a Western science tied up with imperial knowledges of the 'rest' of the world. As it does so, assumptions, categories and taken-for-granted maps will come under scrutiny and be challenged. This book has set out some of the ways in which this is taking place as human geographers rethink what it means to represent the complexity, diversity and (uneven) interconnectedness of spaces, places and landscapes. As readers think more about this, it may be instructive to compare this book – whose contributors are mostly based in British universities – with three other introductory human geography texts (Le Heron *et al.* 1999; Wait *et al.* 2000; Marston *et al.* 2011) written (and, like *An Introduction to Human Geography*, published by multinational information, media and publishing companies) from Australia,

was established first in the regions that we now refer to as Europe and the Middle East and then refined and disseminated over what we now call 'centuries'. In England the calendar now used was not established as official until 1751. And arcane controversies over whether 2000 or 2001 marked a 'new century' were in part rooted in the fact that European mathematics had no concept of zero, until it was imported, via the Arab–Islamic civilization, from India to Europe. So whilst the year 2000 is a Western construct, it turns out to rely on the idea of zero whose origins are elsewhere. One of the lessons here is that the West owes much

New Zealand and the United States respectively. Consider how the contents and style of this text, like those others, is marked by *where* it was written. Geography, as a subject, has both history and its own geography; it has varied in space and time. As a student of the subject, you might want to venture online, or into the recess of a library, to discover past textbooks, such as Haggett (1972) which forty years ago was a ground-breaking text. It is important to appreciate this – and the longer-term – intellectual heritages of geography.

The origins and evolution of academic geography

A brief glance at the available disciplinary histories of geography gives an indication of a long association between geography and the militarized attempts to claim territory on behalf of a particular imperial project.

(Nayek and Jeffrey 2011: 5)

As a distinct subject (with students reading for a degree in it) geography has been present in European universities since the mid- to late-nineteenth century. Before then, however, geographical knowledge was studied and taught in many universities (not only those in Europe, but in the great centres of learning in the predominantly Islamic world, such as Baghdad and Cairo), as part of a variety of programmes of study – sometimes alongside mathematics and geometry for example, or as part of (or alongside) natural history, astronomy or cosmography (see Withers and Mayhew 2002). In the second half of the nineteenth century universities were reorganized around modern disciplines, increasingly with discrete departments and separate degree schemes. This happened as the old curriculum – based on classical learning in philosophy and sciences – began to break down with the rise of science, commerce and new capitalist rationalities. The fact that the late nineteenth century was also a time when many European states (chief amongst them Britain, France and the Netherlands) were engaged in overseas colonization and empire-building and all were keen to foster their sense of national identity and territorial coherence gave geography a new practical relevance. Children needed to be taught, it was argued, about their nation and its place in the world and their teachers thus required a degree in geography. At the same time, geographical knowledge had direct strategic and military relevance (this was the moment too of the birth of geopolitics, as detailed in Chapter 20) and commercial and imperial

relevance, such as in schemes to exploit the perceived agricultural potential of colonies in Africa and Asia, for example. Nineteenth-century ideas about the relations between climate, environment and 'race', and (to use the language of that time) 'civilization' and progress, were caught up with the emergence of the discipline, but so too were the impacts of Darwin's ideas (which influenced physical geography too: in the conception of the way that landforms evolve).

The early years of the modern discipline were therefore inescapably tied up with nationalism and empires. This continued into the early twentieth century, with a growing number of geography departments being established at universities in the USA and Canada, in many Latin American countries, in the European colonies and dependencies (in places such as Singapore, Christchurch, Rangoon, Hong Kong and Jakarta) and in Japan. In some instances, such as Russia, the earliest departments were organized in the 1880s and this was tied to nation-building rather than overseas expansion. Such was also the case in Germany and in Scandinavia. By the time of the Second World War, geography was relatively well established – and the practical knowledge (of military 'intelligence', for example) it yielded in wartime helped to consolidate the discipline's place in universities in many countries. After 1945, it faced new challenges as the range of other disciplines expanded, but (with some exceptions such as the closure of a few departments in the United States), it benefited from the first big post-war expansion in the number both of universities and of students in the 1960s. Human geography increasingly reoriented itself to the technological and scientific spirit of the times, fed by a new phase of military competition in the Cold War (on this, see Barnes and Farish 2006, and see Johnston and Sidaway 2004 for a history of 'Anglo-American' human geography since 1945 and also Hubbard *et al.* 2002: 22–56 for a brief history and Nayek and Jeffrey 2011 for a longer account of geographic thought).

In an age of formal decolonization, where the old imperial disciplinary role was waning, human geography also found new fields of study and outlets for its graduates (such as conservation, development and planning: but also many others, by virtues of the broader skills they would acquire within a geography degree). There were, however, fierce debates about the appropriate focus (for example over the status of regional geography) and methods (such as the role of statistical analysis), which meant that what undergraduate students were exposed to (and thus had to learn) to pass a geography degree continued to change (though unevenly, depending on where they studied).

By the 1960s and 1970s, the world was changing and so was human geography. At first it led geographers to adopt statistical techniques, seeking to render the discipline more scientific. But other, more radical social and political changes (think of the hippies, the rise of feminism and gay liberation, or the movements against the American war in Vietnam and the wider spirit of revolution that came to the fore in the late 1960s) kindled interest in the underlying economic causes of inequality, turmoil and conflict in capitalist societies. An early 1990s textbook introducing the ensuing theoretical debates in human geography, noted how, by the 1970s, 'human geography as an academic discipline had just entered into a period of considerable turmoil' (Cloke *et al.* 1991: viii). They go on to note how:

> One of the most obvious characteristics of contemporary human geography is its diversity of approach. Within human geography today there is an unprecedented liveliness to the engagement with issues of method and theory. Rarely, if ever before, has the subject seen such a plurality of research methodologies and encompassed such a broad sweep of topics of investigation. (Cloke *et al.* 1991: 1)

This liveliness has continued, reflecting both changes in the world (including political, economic and cultural shifts) and accompanying technical developments, theoretical exuberance and shifting funding arrangements for, and ways of, running universities. Cloke *et al.* (1991) looked back to what one particularly creative geographer had once termed a geographical imagination. They thus cited David Harvey (1973: 24):

> This imagination enables the individual to recognize the role of space and place in his own biography . . . to relate to the spaces he sees around him, and to recognize how transactions between individuals and between organizations are affected by the spaces that separate them. It allows him to recognize the relationship which exists between him and his neighbourhood, his territory, or, to use the language of the street gangs, his 'turf'. . . It allows him to fashion and use space creatively and to appreciate the meaning of the spatial forms created by others.

Nearly 40 years later, another leading geographer claimed that:

> it is worth affirming the importance of the geographical imagination, as a matter of both practical wisdom and scholarly reflection, and not least for its pleasure and enchantment, for people's love of learning about the world and their place within it. (Daniels 2011: 186)

Today though, the taken-for-granted 'him' of Harvey's (1973) quote would be qualified with 'him or her': and with this perhaps recognition of significant gender differences in experiences and assumptions. For example, where do men and women experience space differently and how does this relate to spaces of power, sexuality, work and reproduction? Such sensitivity has evolved out of the ways that by the 1970s geographers, David Harvey amongst them, but later joined by others from many different backgrounds, started to ask more difficult questions about inequality, power, exploitation and difference. In turn, capitalism entered a phase of heightened restructuring (involving recessions, new technologies and new forms and places of production and regulation, as detailed in Section 4). Reflecting the times, human geographers became more concerned with inequality, economic and political crises and contradictions. Feminist, humanistic, ecological and other critiques also started to impact on human geography and feed into re-evaluations both of its history (the way that early twentieth-century imperial geography was shaped by racism and sexism, for example) and contemporary contents. Moreover, the boundaries between many of human geography's sub-disciplines, such as urban, political, historical or cultural geography, became more blurred. However, along the way, the modelling, data processing and visualizing capabilities of geographic information science continued to be refined (Fairbairn and Dorling 1997; Fotheringham *et al.* 2000; Schuurman 2004) and the Internet, digitisation and mobile telecommunications produced new capacities for communication and altered relations and perceptions of proximity and distance (see www.zooknic.com for work on the geographies of the Internet). For some, outside the discipline, the decline of the Cold War in the late 1980s brought the 'end of history'. For others, the development of information and telecommunications technology and the globalization of the economy were creating a 'borderless world'; some even proclaimed the 'end of geography'.

For the most part, the evolution of human geography has been driven by the adoption of ideas, concepts and theories from other disciplines. The challenge for human geographers has been to apply these to a geographical context. Economic geography, for example, initially took ideas from economics and business studies and applied them to help explain how economic activity is distributed across space. The geography of economic activities was seen as the outcome of this process with the job of the geographer being to map and describe that pattern of economic activity. This often seemed to relegate the economic geographer (and geographers more generally) to the role of suppliers of information for others to decipher.

In recent decades, however, partly in the context of all the talk of globalization, many other social scientists started to reconsider geography. The status of the subject was enhanced. At the same time, reorganization in many universities led to a fresh phase of adjustments and some reworking of disciplinary boundaries. In many places geography found itself being merged with environmental sciences (or sometimes other social sciences) leading to concerns about its longer-term identity and coherence. New theories, new methods and approaches meant that geography (as you will discover in the course of a degree in it) as a university subject was very different from the descriptive subject shy of theory and with few methods other than illustrative maps, figures and text that was associated with geography in the first half of the twentieth century. Of course, geographers have long claimed that *geography matters* for all the social and physical sciences. But for some reason this is often acknowledged only when internationally renowned economists, political scientists and sociologists actually say that geography must be included in their analyses and interpretations. But as geographers we should beware; other disciplines tend to adopt their own 'take' on geography, one that is sometimes antiquated and simplistic – reinventing industrial location models in the name of 'geographical economics', for example, or attributing to tropical environments a key role in determining global patterns of development. Nevertheless, geographers can take some comfort and sense of possibility from the fact that many other disciplines are recognizing the *spatiality of human society*.

Put simply, the fact that social and economic processes take place across space matters. Indeed the way they do so is vital to how they operate. Human geography is not just about *describing* the spatial manifestations of economy and society: it is about *explaining* how space is configured and shapes economies, societies and social processes. Thus, geography is not a passive outcome; it is a critical component of social and economic processes. More than that, geography in its broader definition provides an interface between the human and the natural worlds. We would argue that geography is a *key* subject for the twenty-first century, in part because many of the challenges that face humanity are at the interface between human societies and natural environments. One of the oldest themes of human geography (and indeed of geography as a whole, including its physical side), that of human–environment relations, has become an urgent agenda for the twenty-first century.

As this textbook has evolved so the environment and environmental issues have become more significant in our discussions. Initially, there was no chapter on the environment, though it was touched on in the chapter on resources and development. In this edition there are two chapters dealing explicitly with environmental issues and the social construction of nature. Furthermore many other chapters discuss issues such as global warming and competition over access to increasingly precious resources such as clean drinking water. In addition, the environmental costs of high levels of mass consumption and the consequences of economic development in the 'emerging economies' are also discussed. In short, this edition reflects the general environmental concerns that we are currently experiencing reflecting fresh debates about how human society and the way we live is threatening the viability of the global ecosystem. This realization presents a major opportunity for geographers who are well equipped to deal with the interrelationships between environment and society.

Human geographers have also become more aware of the ways that knowledge is socially constructed. This is complex, reflecting ideology. But in the simplest terms, the way you see the world is partly a function of who you think you are and, hence, *where* you see yourself as coming from. Human geographers have come to realize that much of the knowledge and understanding that they claimed to be universal is in various ways Eurocentric: it comes from somewhere and any idea that it is universally true (for all and everywhere) might be challenged. Thus, as we have noted, 1 January 2000 was not the beginning of a new century or a new millennium for many of the world's populations; yet, an instant measured on the timescale of a particular religion that, until recently, dominated Western society, was widely taken to be a global event. Similarly, Eurocentrism has been associated with a very 'white' view of a multi-ethnic world. For example, the fairly widespread self-perception that white folk are not really themselves members of a particular 'ethnic' group, except amongst racists and white supremacists. Whereas in global terms, they are arguably a distinctive 'ethnic minority' whose identity has been forged through comparisons and interactions with other people classified as non-white. In turn, however, such ideas about race and ethnicity are rooted in colonial histories (and hierarchies) that are evidently themselves particular (imperial) ways of knowing, ordering and interpreting the world. These resurface in contemporary racism, and have their own historical geography. Ideas about race, which today may seem self-evident and obvious, would not have been present in, for example, the Roman Empire, which had other ways of ordering and stratifying society and demarcating insiders, barbarians, citizens and slaves. All this means that geographers must now reconsider the ways in which assumptions and value judgments shape the way they view the world. They must accept that *all* descriptions of the world are culturally determined, often politically motivated, and can always be contested.

A straw poll would quickly reveal that there remain differences between the authors, and among the editors, over the contents of the book, or how it should have been written. We have already sketched some of the ways that the nature and contents of introductory textbooks vary according to the location of their authors and targeted readers. What you see here is the result of compromise. The same type of book written by a group of geographers in, for example, Australia (see Wait *et al.* 2000), New Zealand (see Le Heron *et al.* 1999), India, Russia or South Africa would undoubtedly be different. At the same time, each reader of this book will interpret its contents in different ways. The position(s) that you occupy in your society and where you are on the globe when you read the book will tend to shape the way you look at the world and/or the way that you interpret its arguments. The majority of the readers will probably reside in the so-called 'Western' world. The first edition was reprinted in India in 2003 as a local edition (authorized for sale in India, Bangladesh, Pakistan, Nepal, Sri Lanka and the Maldives) and a Chinese translation is planned. Those readers will frequently have different worldviews from those of many students in the United Kingdom, for example. For all readers, a key factor shaping their life chances is where they were born (or at least what country they have citizenship of or the right of residence in, or are studying in). So, think about your position; realize that this book presents the views of human geographers working at a particular place and at a particular time. We have set out to challenge readers to think about the ways that human geography interprets the major social, cultural, economic, environmental and related issues that face the world in the early years of the twenty-first century. It is for you to use this book to inform your own insights and opinions as geographers; for some it could be part of a life-long engagement with, and contribution to, human geography.

Geographical approaches

Today, human geography is characterized by a wide variety of approaches; there are many ways of writing geography and approaches to doing so. The idea of a single, all-encompassing geographical approach (which may have been evident at times) is hardly credible now. Nonetheless there are some common themes, topics and approaches. Chris Gibson (2009: 218) observes that:

> While diversity defines contemporary human geography, there are common questions of geographical scale, causality, agency and structure, interrelationships and networks, place and movements.

This book will chart many of these questions, through the subsequent thematic sections and their chapters. As an issue-based approach, designed to introduce readers to the ways that the discipline can inform analysis and critical understanding of a series of concerns, there is not the scope here to say much more about the evolution of debates in human geography, whose histories we have already touched upon. You may further investigate their trajectory through some of the other readings listed at the end of this chapter or track down and read Chris Philo's (2008) fluent summary and follow his 'map' of changing approaches in human geography.

But at this point please accept our word that this story (like Philo's 'map') becomes immensely complex, intertwined and convoluted; reflecting those intellectual trends and social changes mentioned earlier.

But, it is important to appreciate some shared concepts. Most disciplines have their central concepts and ideas that define what it is they study and how they study it. Human geography is often accused of borrowing ideas from elsewhere, rather than generating its own; but it is possible to identify a set of concepts that make it distinct and different. We might disagree on what they might mean (and each is the subject of a vast literature), but ideas of place, space and scale are certainly central to human geography and have been for a long time, as has an interest in landscape, difference, connectivity and unevenness (or areal differentiation, as it was once termed). Much of our discussion so far has highlighted how globalization processes, while promoting processes that lead to homogenization, are throwing differences in access and prosperity into sharp contrast, both within and between countries. We now better understand how people and places are interconnected in complex ways through networks and relations of production, consumption and communication (transportation, media

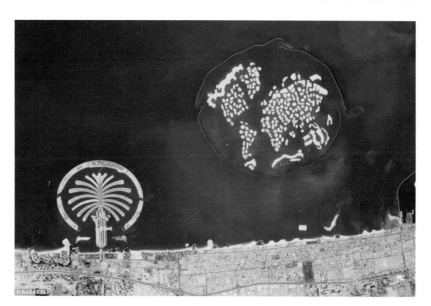

Plate 2 Artificial islands in the shape of 'The World' off the coast of Dubai, United Arab Emirates.

and telecommunications). This in itself is not particularly new, but perhaps it is the spatial reach and density of such interactions that define the world in which we live. The winds of economic, social and cultural change blow across particular places and are reinterpreted locally to reproduce the uniqueness of place, which in turn reworks globalization. In the context of globalization, one of the major challenges for contemporary human geography is to explain how places have succeeded, and in many cases failed, in what Amin and Thrift (1994) have called 'holding down the global'. Put another way, how they have created, and in some cases recreated, a role for themselves in the global economy. In some cases places strike it lucky, they are the right place at the right time; in others they work hard to market a particular role for themselves. For example, places such as Dubai seek to reinvent themselves, and even to reproduce the world (see Plate 2). We don't have to travel to distant places to discover what Massey (1994) termed a 'global sense of place.' Almost any place will be connected with elsewhere. It will have been shaped by and may shape many other places, near and far.

All of the core ideas of human geography mentioned above can be deployed to interrogate and explain the places where we live and the places where we travel. As geographers, we travel the world differently engaging in landscapes and seeking explanations as to why something came to be where it is or how the physical environment shaped the pattern of human occupation. In Spotlight box I.1 one of the editors explains a recent engagement with a particular place – or set of places connected on a path. It is our hope that, as a practicing human geographer, you will also find your own literal

or metaphorical paths through this discipline and be inspired to think geographically. For us, then, human geography is a way of seeing, enquiring and understanding past, present and potential future worlds.

Approach of the book

Writing an introductory textbook is a challenging task. No matter what its contents, or how they are organized, the finished product will not satisfy everyone. The book not only has to engage the attention and interest of the students who will learn with it, but also has to satisfy the instructors who will teach with it. Thus we are caught on the horns of a dilemma: how to make the text as accessible and as user-friendly as possible to students, while not appearing to be too naïve to the instructor assigning it. We also have to acknowledge that each geography programme is different, at least in detail if not in overall objective, and that a textbook fulfilling every need would be very large indeed. Broadly speaking, however, most first-year geography programmes will have a course that offers an introduction to human geography. Such a course is nearly always compulsory. It is often supported with a 'world regional geography' course or a course devoted to topical 'issues in human geography'. Increasingly (though this was always the case in North America) such courses are required to meet the needs of non-geography students who are taking geography as an option or elective. This poses a major challenge for the instructor confronted by a class comprising those interested in human geography but coming to terms with a discipline whose contents,

Spotlight box I.1

James' wanderings

One of us especially likes walking. It is a deeply geographical practice. For (even enabled by the geo-technologies of GPS and mobile Internet) it is easy to get lost, or find oneself in the 'wrong' place, or feel 'out of place', or discover new places. Walking requires an intimate encounter with place, whilst negotiating space and landscape. James has done this often in different cities and even written scholarly geographies about it, linking it to the geopolitics and memories in/of a particular one: the English city of Plymouth (Sidaway, 2009). He is far from alone in this of course and his paper draws on an enormous literature about walking in cities from many sites and a vast range of authors. But sometimes it is nice to get away from the city (even the one he currently lives in: Amsterdam) and walk in more remote places. So in the weeks when the editors were finalizing this edition, he took a short break in northern Spain and walked a section of one of the routes there that is commonly referred to as the Camino de Santiago (Way of St James), mapped in Figure 2. As the account of another who walked this way has noted:

the Camino de Santiago is really a network of routes, many of Roman origin, extending throughout Europe

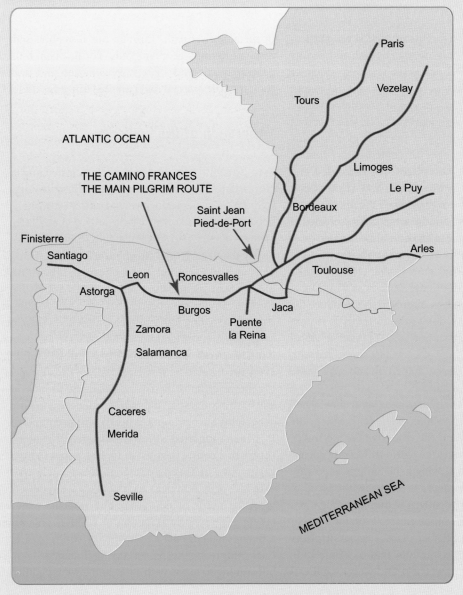

Figure 2
A map of the routes known as Camino de Santiago.

Plate 3 A walker on The Camino de Santiago.
Source: © Cro Magon Alamy Images

that have been used by pilgrims since the eleventh century to reach Santiago de Compostela [whose Cathedral houses the shrine dedicated to the Apostle St James] . . . . The early medieval pilgrimage played an important role . . . . fostered by the reigning political forces.

(Frey 1998: 5)

Frey's book also describes how, with the lessening of Christian influence in the latter half of the twentieth century, the ancient pilgrimage routes declined, before undergoing a rapid renaissance in recent years. Today many thousands walk these routes daily (Plate 1.3). Not all are pilgrims. Some go mostly for the exercise, others for a change of scene, as a holiday, or as a personal or even spiritual journey that may not be grounded in a formal religion. In recent years a Hollywood movie – starring Martin Sheen and directed by Emilio Estevez – further publicized *The Way*. If you don't have the time or energy for the walk, sample excerpts from that Martin Sheen movie (*The Way*) are online. But to walk it is to experience everyday geographies of slow movement – as on any other popular long-distance walking trail, such as those in Appalachia, the UK, Germany, or on even grander scales and higher altitudes in, say, Nepal or Peru.

Long-distance walkers must negotiate and become, sometimes painfully, aware of scale, topography and landscape. When you are walking, distances, pace, elevations and inclines matter fundamentally. Weather, food and water become things that may not be taken for granted, as must what you decide to carry or decide to leave behind (as in other travel of course: including those using sophisticated technologies like aircraft). Economic geographies and cultural geographies are evident. Place names and language differences for starters, or carrying enough money; where to eat and sleep and who gains from all that expenditure? So are political geographies, if you know where to look (which, after reading Section 5 here, you will). But glance up and see an aircraft far overhead – during a long day of remote walking perhaps – and encounter other walkers from across the world, or discuss where you started from with them, and the question of scale becomes more complicated. This is because, when you pause to think about it, any place on the path becomes part of a vast network extending in many directions. To make such a journey also links mind, body, landscape and movement in ways that are hard to capture in words: geographical imaginations may be rich, but some things you can only learn – or fully appreciate – by doing them. Moreover, others walk long distances not for pleasure or spiritual gain, but simply to collect water or earn a living. And other demands, injuries, age or disability, or preferences mean that hiking long distances is not something that everyone can, or might want to, do.

approach and language are very different from what they had expected, other students more interested in physical geography, and non-geographers who are there out of curiosity or requirement. Such a diverse audience also poses a challenge to the textbook writer and those who would assign the book to students. As Noel Castree (2011, 5) has recently noted, 'A century ago, a small number of university geographers in England and elsewhere worked hard to create a subject that is, today, far larger and more buoyant than they could possibly have imagined'. But with this comes the multiple challenges of representing a sprawling and dynamic subject, large classes, more diverse ranges of students with diverse expectations as to what doing a degree is worth, and how and why geography might compete, challenge or be marketable.

When you browse in your university library, at a bookshop or online, you will come across a variety of human geography textbooks. Broadly speaking these are of three types: systematic texts organized by major sub-discipline in human geography, topical texts organized by key topics or issues, and world regional texts using major regions (for example 'Europe' or 'Southeast Asia') to introduce human geography. It will also be evident that these books are clearly aimed at different markets or audiences. Assigning a course text is almost a matter of fact in the United States (where most students will expect a 'course text' and guidelines on exactly what to read), but elsewhere, certainly in the United Kingdom, assigned texts are a more recent phenomenon. Over the past couple of decades, however, an assigned text is more often seen as a necessary *starting point* for an introductory course in human geography almost everywhere. With student budgets often stretched, there is also a good case for purchasing a textbook that will cover more than one course and that will be of lasting use in later years. The benefit of regular updates in the form of new editions is that they will provide access to the most recent literature, although given the time it takes to publish and the limits to the scope that can be given to specific topics, textbooks should *always* be supplemented by other reading.

There is no doubt that introductory textbooks do not always travel well. This is because they are often designed for a particular educational system; but more because they also reflect a specific version of what is important in the world and what constitutes human geography. In addressing this problem, we have sought to produce a textbook that offers an *issue-oriented* introduction to human geography. The major sections focus on important issues facing the world, but they also relate to major sub-disciplines in human geogra-phy. The individual chapters bring together ideas from a variety of fields in human geography. They also incorporate coverage of the globe, drawing ideas from a variety of places to reflect diversity and difference and inequality as well as the interdependent nature of the world. While this integrated approach may not sit well with the traditional ways of producing introductory human geography textbooks, we believe that it acknowledges the complexity of human society as well as the practice of contemporary human geography.

Structure of the book

This book is divided into five sections, supported by an Introduction. Section 1 provides an essential historical context for the four issue-oriented sections that follow. While it is natural for us to proclaim that geography matters, it is also the case that historical geography matters. A cursory examination of an atlas printed at the start of the last century will show you that the world has changed a great deal. Many of the contemporary issues discussed in this book, such as the geographies of development and of inequality, owe much to the way the world looked in 1900, or considerably earlier. A sense of history is critical to an understanding of contemporary human geography. However, if we cast our minds even further back into the records of antiquity we are reminded that societies are transitory; nothing lasts forever. States and civilizations have come and gone and this should teach us that there is little permanent about the current world order. At the same time, what is past, and not always immediately visible, often has profound influence on the world we live in now. Section 1 highlights some of these issues and then focuses on the emergence of capitalism and its relationship with the making of the twentieth-century world that immediately preceded our own.

The remaining four sections comprise groups of chapters that focus on a related set of issues. It makes sense to work with Section 1 first; after that the sections can be utilized in whatever order you see fit. Each section starts with a brief summary of the important issues covered by each of the chapters and there is logic to the order in which the chapters are presented within each section. Section 2 examines the interrelationship between population change, resource production and consumption, global economic development and the environment. Section 3 focuses on social and cultural issues within an urban and rural context and at a global scale and includes a chapter on the social construction of nature and what this means for the analysis and interpretation of how, for example, to manage the global environment in a sustainable way. Section 4 examines the globalization of economic activity, production networks, the emergence of a global financial system and means of exchange (which has played an increasingly visible role in recent decades), and the importance of consumption to an understanding of the geography of economy. Section 5 considers a variety of political geographies at differing scales and how they interface and intertwine. Here too the sense of continuity and change is evident, as in shifting global geopolitical scenarios since the decline of the old superpower Cold War confrontation between the Soviet Union and the USA that had lasted for 50 years from the 1940s to the 1990s and whose legacies remain active.

How to use the book – and going beyond it

Each chapter is designed to introduce a set of specific issues and particular fields in human geography. Each begins with a list of topics covered. The text is broken into manageable sections with supporting materials; important concepts and specific examples are presented in boxes. Each chapter concludes with a list of learning outcomes. These provide a means of assessing whether you have understood the content of the chapter. To assist you in reading beyond the text, each chapter has a set of initial recommended readings, at least some of which you should be able to find in your library (or sometimes via its website, in the case of access to academic journals). Where relevant, we have also included a list of websites that contain information you may find useful in your own research and writing. In addition, at the back of the book there is a glossary of important terms and an extensive bibliography. Such lengthy bibliographies are not always found in an introductory text; however, we want this book to remain an essential reference tool for the duration of your geography programme. Each chapter contains references to the bibliography giving you access to the specialist reading that is essential when you wish to research or otherwise follow through on specific topics in detail. One way of using this book is to read the chapters to support your lectures and seminars, the recommended readings to prepare for examinations, and the references, websites and bibliography to research and write assignments. In other words, reading this textbook is the beginning of your introduction to human geography, not the end.

So, where else do you go from here? As a geography student (or any other kind of student for that matter),

you will be faced with choices, for example between human and physical geography or between more specialized subjects within one of those traditions. It is in the nature of academic disciplines to subdivide and to compartmentalize knowledge. Geography is unusual in that it sits at the intersections between the humanities, social sciences and natural sciences. This means that it internalizes both the disadvantages of disciplinarity and the potential advantages of multidisciplinarity. In other words, as Harrison (2009, 163) notes:

> it incorporates perspectives from a very wide range of diverse subjects from the social sciences (including history, sociology, politics, economics, and psychology) and from the natural sciences (including physics, biology, ecology and geology). To some, this diversity has been regarded as a sign of weakness, suggesting a subject with such a broad range of research fields must treat those in a superficial manner. Others have seen it as a sign of strength, arguing that geography has avoided the intellectual trap of increased specialization and held on to a holistic view.

Much of the promise of geography lies on the margins between the sub-disciplines, between human and physical geography or, for example, between economic and cultural geography. In your studies you should make the most of this promise, seek out the space on the margins and look for the connections between the human and natural environments, the economic, the social and the cultural. Many of the challenges that face humanity also occupy these margins. The emphasis on places and spaces and how they are interconnected, imagined and represented (the hallmarks of human geography) is itself promising. Engaging with such promise will deepen, stretch, enrich and should challenge your understanding of the world.

Geography is a worldly discipline. At once grounded and practical, human geography can sometimes be also deeply theoretical, abstract and philosophical; it mobilizes words, maps and numbers, and articulates social sciences and the humanities, in a symbiotic relationship with the science of physical geography.

All this also means that, not only does a degree in geography provide you with a critical way of seeing the world, it also equips you with a knowledge base and a range of skills that will serve you well in workplaces. As many students have to take on an ever-increasing debt to fund their studies (or have deferred earning or taken time away from other rewarding things in order to study), it is not surprising that they should worry about whether studying geography will lead to a rewarding career, in every sense of the term. As Dr Rita Gardner,

Director of the Royal Geographical Society (with IBG) makes clear:

> There has never been a better or more important time to study geography. With growing interest in issues such as climate change, migration, environmental degradation and social cohesion, geography is one of the most relevant courses you could choose to study. Geographers are also highly employable.
>
> (www.rgs.org/HR/rdonlyres/57E073A9-2201-4F42-82C8-61613068FFCA/0/Why study geography.pdf)

Geography graduates are numerate and literate, effective team workers, must think analytically and critically, and have a wide range of technical skills. These are all qualities that are valued in the workplace, be it in government, the public sector, commerce or civil society. Analysis of a randomly selected sample of the Quarterly Labour Force Survey in the UK (First Quarter 2010) provides evidence of the employability of geographers. Geography graduates had a relatively high employment rate (85 per cent), with 67 per cent of geography graduates being employed in professional and managerial jobs. In the UK the Russell Group of research-intensive universities see geography as a 'facilitating subject' when studied at school as it leads to such a wide range of degree courses at University. This diversity also means that you can also pursue a wide range of careers when you graduate with a geography degree.

Whatever you end up doing and wherever geography takes you intellectually and professionally, we hope that the chapters and arguments brought together in this book will (when supplemented by further reading, critical thought, writing and conversation) encourage readers to continue to engage with this promise inherent in geography. As we noted above, each chapter provides a set of suggested readings and, where relevant, useful websites. This chapter is no exception. Here we consider other resources that provide an introduction to human geography. The readings focus on other introductory texts and the study skills necessary to begin to conduct your own research. The website guide lists places to visit to find out more about geography as a discipline.

Further reading

This first set of readings provides a sample of other textbooks that you can use to supplement your reading. Individual chapters from them are often cited elsewhere as references in this book, but they are also worth browsing to get a sense of the different ways in which human geography is presented.

Cloke, P., Crang, P. and Goodwin, M. (eds) (2005) *Introducing Human Geographies,* Arnold, London. The second edition of an innovative text: 43 inviting short chapters that reflect the diversity of approaches in contemporary human geography. Designed more for the UK market, but well worth delving into wherever you may be. A new edition is in production.

Marston, S.A., Knox, P., Liverman, D., Del Casino, V. and Robbins, P. (2011) *World Regions in Global Context: People, Places, and Environments*, 4th edition, Prentice Hall, Upper Saddle River, NJ. A comprehensive text written principally for the US market with lots of colour pictures and diagrams. Probably the best US textbook. Supported by a website, which is worth browsing, even if you do not have the book.

Le Heron, R., Murphy, L., Forer, P. and Goldstone, M. (1999) *Explorations in Human Geography: Encountering Place,* Oxford University Press, Auckland. This is a stimulating textbook written by a group of New Zealand geographers specifically for New Zealand geography students. It may be difficult to find in the northern hemisphere, but reading it could turn upside down your own taken-for-granted view of the world.

Wait, G., McGuirk, P., Dunn, K. and Hartig, K. (2000) *Introducing Human Geography: Globalization, Difference and Inequality,* Longman, French's Forest, NSW. Like the above, another rewarding Antipodean text; this one is published in New South Wales, Australia.

Research and study guides, readers and further insights into human geography

This section is dedicated to publications that will assist you in conducting geographical research and/or expand your knowledge of debates and approaches in the discipline. The list comprises dictionaries, and books on methodology, approaches and theory in human geography as well as a few (such as autobiographies of influential geographers) that are hard to categorize. You should also seek out the various 'readers' (comprising reprints of influential papers that usually first appeared in journals) that have been produced. Led by publishers, these have proliferated in recent years, along with encyclopedias, dictionaries and companions to the discipline, or for specific sub-disciplines. Some bring together the most influential readings in a particular area of human geography, whilst others are collections of specially commissioned review essays that assess the status of particular areas of research.

Agnew, J., and Duncan, J.S. (eds) (2011) *The Wiley-Blackwell Companion to Human Geography*, Wiley-Blackwell, Oxford. Comprising commissioned original essays this is a rich starting point to get a feel of how the discipline has evolved and what it contains today.

Agnew, J., Livingstone, D. and Rogers, A. (eds) (1996) *Human Geography: An Essential Anthology*, Blackwell, Oxford. A collection of classic geographical writings, and a sober reminder that there is very little that is wholly new.

Atkin, S. and Valentine, G. (eds) (2006) *Approaches to Human Geography,* Sage, London, New Delhi and Thousand Oaks, CA. A good way into the range of theoretical (and methodological) debates in human geography.

Benko, G. and Strohmayer, U. (eds) (2004) *Human Geography: A History for the Twenty-First Century*, Arnold, London. This book sought to address histories of continental European and Anglophone geography and the relations between them. It only scratched the surface.

Blunt, A. and Wills, J. (2000) *Dissident Geographies: An Introduction to Radical Ideas and Practice*, Prentice Hall, Harlow. The geographies explored in the text all share political commitments to critique and challenge prevailing relations of power. From gay and lesbian geographies to geographies of anarchism and anti-racism, this book brings out some of the ways in which the production of geographical knowledge is tied to politics and struggles outside, as well as within, universities.

Castree, N. (2005) *Nature*, Routledge, London and New York. A challenging introduction to how geographers have studied nature and what this tells us about the nature of academic geography.

Claval, P. (1998) *An Introduction to Regional Geography*, Blackwell, Oxford. First published in French in 1993 as *Invitation à la Géographie Régionale*, this book draws on French (and some anglophone) traditions to review the past of, and prospects for, regional geography. An introduction to a significant tradition and work that overlaps with but is also beyond the mainstreams of English-language geography.

Clifford, N. and Valentine, G. (eds) (2010) *Key Methods in Geography*, 2nd edition, Sage, London. An introduction, covering methods used in both human and physical geography, and those they share.

Cloke, P., Crang, P., Goodwin, M., Painter, J. and Philo, C. (2002) *Practising Human Geography*, Sage, London. A more advanced (but very readable and inviting) guide to research methods and writing strategies in human geography. For

details of quantitative techniques or GIS you will need to go elsewhere, but the book is very inspiring on how to use qualitative methods.

Crampton, J.W. (2010) *Mapping: A Critical Introduction to Cartography and GIS*, Wiley-Blackwell, Oxford. One of a growing series of 'Critical Introductions to Geography' and a fantastic and up-to-date guide to maps and mapping.

Cresswell, T. (2004) *Place: A Short Introduction,* Blackwell, Oxford. Explores how human geographers have studied, debated and complicated a concept that may seem self-evident and familiar, but which turns out to be a range of assumptions, ideologies and contests.

Daniels, S. and Lee, R. (eds) (1996) *Exploring Human Geography: A Reader*, Arnold, London. The individual articles were not intended primarily for a student audience, but the collection provides a survey of human geography in the 1980s and 1990s.

Dorling, D. and Fairbairn, D. (1997) *Mapping: Ways of Representing the World*, Prentice Hall, Harlow. Maps have always been associated with geography and geographers. This text will help you to understand how maps express the will to describe, understand and control.

Flowerdew, R. and Martin, D. (1997) *Methods in Human Geography: A Guide for Students Doing a Research Project*, Longman, Harlow. An indispensable guide to a wide variety of the research techniques employed by contemporary human geography.

Gould, P. and Pitts, F.R. (eds) (2002) *Geographical Voices: Fourteen Autobiographical Essays*, Syracuse University Press, Syracuse, NY. A collection of autobiographies reflecting the experiences and insights of some leading geographers of the past 60 years.

Gregory, D., Johnston, R.J., Pratt, G., Watts, M. and Whatmore, S. (eds) (2009) *The Dictionary of Human Geography*, 5th edition, Wiley-Blackwell, Oxford. Over a thousand pages. We think it is an indispensable reference. But, like vintage wines, some prefer earlier editions (each one thicker and weightier than the last). If you want to spend an afternoon learning lots about academic geography, track down older ones and look at how the entries have changed.

Hoggart, K., Lees, L. and Davies, A. (2002) *Researching Human Geography*, Arnold, London. One of a number of books that provide an introduction to research methods in human geography, a good place to look when starting your own dissertation or research project.

Holloway, S.L., Rice, S.P. and Valentine, G. (eds) (2003) *Key Concepts in Geography*, Sage, London. Worth reading alongside this text, especially so if you are taking introductory classes in human and physical geography, since this collection addresses both.

Holt-Jensen, A. (2009) *Geography: History and Concepts* (4th edition, Sage, London). Translated from Norwegian, an introduction to the history, philosophy and methodology of human geography.

Hubbard, P., Kitchin, R., Bartley, B. and Fuller, D. (2002) *Thinking Geographically: Space, Theory and Contemporary Human Geography*, Continuum, London. This book represents an attempt to demystify the nature and role of theory in contemporary human geography.

Hubbard, P., Kitchin, R. and Valentine, G. (eds) (2008) *Key Texts in Human Geography*, Sage, London. A starting point, designed to lead you beyond textbooks into classics in the field. Many of these were not written for a student audience and so this entrée is especially helpful in approaching them.

Johnston, R.J. and Sidaway, J.D. (2004) *Geography and Geographers: Anglo-American Human Geography since 1945*, 6th edition, Arnold, London. First published in 1979, the sixth addition of this book provides an updated survey of the major trends in human geography since 1945 in the English-speaking world. Includes a discussion of geography as a discipline that will help you to better understand what academic geographers do and how the discipline is structured. A seventh edition is planned.

Livingstone, D. (1992) *The Geographical Tradition: Issues in the History of a Contested Enterprise*, Blackwell, Oxford. On publication this immediately became a landmark scholarly study of the long-term history of human geography. It still is.

Moseley, W., Lanegran, D. and Pandit, K. (eds) (2007) *The Introductory Reader in Human Geography*, Blackwell, Oxford. Another reader (collection of papers previously published in academic journals). A good way to get a feel for debates and approaches in a range of sub-disciplines (economic, political, urban and other parts of human geography).

Nayak, A. and Jeffrey, A. (2011) *Geographical Thought: An Introduction to Ideas in Human Geography*, Pearson, Harlow. This does what it says in the title.

Rogers, A. and Viles, H. (eds) (2003) *The Student's Companion to Geography*, 2nd edition, Blackwell, Oxford. The second edition of a useful introduction; the sections contain information on the fields of geography, methods and data sources.

Thrift, N. and Kitchin, R. *et al*. (eds) (2000) *The International Encyclopedia of Human Geography*, Elsevier, Oxford. This A–Z is the most comprehensive survey of human geography ever assembled. An online version is available. If your university has an online copy you can lose yourself inside it for hours and only scratch the surface.

Tuan Yi-Fu (1999) *Who am I? An Autobiography of Emotion, Mind and Spirit*, University of Wisconsin Press, Madison, WI. The autobiography of a Chinese-American who arrived in the United States as a 20-year-old graduate student in the 1950s and went on to become one of America's most original and respected geographers.

Women and Geography Study Group (1997) *Feminist Geographies: Explorations in Diversity and Difference*, Longman, Harlow. Although published in the 1990s, this remains a good way to begin to understand how feminist geographies have reshaped the study of human geography.

Wylie, J. (2007) *Landscape*, Routledge, London. A text that examines how human geographers (and others) have conceptualized and narrated landscape.

As this introduction has sketched, human geography (understood as the subject, and those who study and teach it) has its own geography (understood as locations), in so far as it is unevenly developed (in some countries it is studied as part of the school curriculum, in other places – most significantly perhaps in the USA – it is hardly studied at schools). How and why this has come to be is in turn a complex reflection of wider histories and geographies, including the historical geography of empires (see Chapter 2), which meant that aspects of a British-style educational system were long imposed on many territories in Africa and Asia (and in Ireland). For more systematic reflections on the relative position of geography in the academy in a range of countries, see the set of papers in the *Journal of Geography in Higher Education*, 31(1), compiled by and following a brief introduction by Kong (2007). Browsing other issues of that journal can be instructive, especially since it is full of ideas and guides to research and geographical learning.

This and other material from scholarly journals must usually be downloaded via a computer logged in to a university network; since university libraries pay the subscription costs to publishers of such journals. However, the following is free to all, anywhere with internet access: Barder, H. and Engel-Di Mavro, S. (eds) (2008) *Critical Geographies: A Collection of Readings*, Praxis (e)Press: www.praxis-epress. org. This reprints a collection of classic and inspiring geographies; starting with an 1885 essay by an anarchist Russian geographer on 'What Geography Ought to Be'. We suggest you start here.

But why stop with the Earth? If you read the following book, you will soon see how geography of Mars is, in the end, more about the Earth than Mars. With time, you might begin to agree with us that all geographies of other places tell as much about where they are written as that which they purport to describe:

Lane, M.D.K. (2011) *Geographies of Mars: Seeing and Knowing the Red Planet*, University of Chicago Press, Chicago, IL.

Finally, if Martian geographies are not to your taste, but you are also taking classes in physical geography, see if you can impress those who seek to teach you physical geography, by asking them what they think about:

Clifford, N. (2009) Globalization: a physical geography perspective, *Progress in Human Geography*, 33, 1, 5–16.

Useful websites

The Internet connects you with a wealth of information that is of relevance to an introductory course on human geography. It is impossible to touch on more than a few potential sources here. This section provides access to the sites of some of the major geographical societies in the United Kingdom and North America. All of these sites provide links to societies and organizations across the globe, as well as links to geography-related sites. In addition, many of these sites carry links to electronic versions of their journals (although to access these, you will probably have to go through the website of your university library or be on a computer that is recognized as being connected to the net via a university network) and other publications as well as information and links to resources such as maps, statistics, atlases or photographic collections. Your own department will have a website; visit it and find out what the academics who teach you are also saying about what they do. Visit the sites of other departments across the world and discover how geography is represented and taught in different places. Thus, the Internet provides you with your own virtual guide to human geography. With the partial exception of the Canadian Association of Geographers, the links below focus on the Anglophone world. Use the International Geographical Union website to seek out geographers across the globe.

www.aag.org The website of the Association of American Geographers. Founded in 1904, the AAG is the primary academic geography organization in the United States. This site provides a wealth of information on the Association's activities, as well as access to its two scholarly journals: the *Annals of the Association of American Geographers* and *Professional Geographer*.

www.amergeog.org The website of the American Geographical Society. The Society was established in 1851 and is the oldest professional geographical organization in the United States. The society publishes the *Geographical Review* and *Focus*.

www.cag-acg.ca The website of the Canadian Association of Geographers. Founded in 1951, the CAG-ACG is the primary academic organization in Canada. The association publishes *The Canadian Geographer* and organizes national and regional meetings.

www.geography.org.uk The website of the Geographical Association, which was founded in 1893 and is the national subject teaching organization for all geographers in the United Kingdom. The Association publishes three journals: *Primary Geographer*, *Teaching Geography* and *Geography*.

www.iag.org.au The website of the Institute of Australian Geographers. The IAG was founded in Adelaide in 1958 and is the principal body representing geographers and promoting geography in Australia. The institute publishes the journal *Geographical Studies*.

www.igu.net.org The website of the International Geographical Union, the international umbrella organization for academic geography. This website contains a list of geography departments throughout the world. Equally you could just choose a university and see if it has a geography programme and what you can learn about it online.

www.nationalgeographic.org A public face of geography in the English-speaking world. Home to the *National Geographic* magazine, the site also provides access to the maps and photographs that appear in its magazine, as well as a searchable index.

www.rgs.org The home of the Royal Geographical Society (with the Institute of British Geographers). The RGS was originally founded in 1830 and the RGS-IBG is still the primary academic organization in UK geography. The website provides information on the societies as well as a route to its three key journals: *Area*, the *Geographical Journal* and *Transactions*.

In addition to the websites of the various learned societies, there are dedicated 'web portals' that provide access to a wealth of information to support your geographical studies. The two examples below are UK-based, but their outreach is international and their links may take you to portals or subject gateways based elsewhere.

www.jisc.ac.uk/resourceguides Resource Guide for Geography and the Environment. This website professes to be the essential guide for exam revision and dissertations and theses. It provides access to online catalogues, publications, and subject gateways and data services. It requires a username and password and therefore may need to be accessed via your university library portal.

www.geosource.ac.uk Geosource: Geography and Environment Information on the Internet. This is a free service led by the University of Manchester. It currently contains over 5,000 individually selected information-rich Internet resources covering five subject areas: the environment, general geography and environmental science, geographical techniques and approaches, human geography and physical geography.

As you will probably already know, there are dozens of commercial websites that will help you to try to short-cut and circumvent learning and scholarship, by buying putatively top-grade custom-written essays for almost any assignment. Like other even seedier sides of the Net, these have their own geography, in terms of by whom and *where* they are produced, *where* the money goes, plus the spatiality of the wider economies, trends and technologies that enable them. In the words of a geographical account of these published almost a decade ago:

> The technology of the Internet has connected remote places and facilitated the diffusion of any number of economic activities such as call centers, off-shore banking, and data processing. The Internet adult industry is yet another example of how a combination of regulatory issues, lower costs for content, and low barriers to entry results in a restructuring of production and consumption. While allowing access to a whole new range of people, the Internet is still shaped by existing structures of regulation, power, and hegemony. In short, the space of flows cannot be understood without reference to the 'space of places' to which it connects.
>
> (Zook 2003: 1283)

As for those commercial essay-writing websites, although some pretend otherwise, they rest on a specialist form of fraud known as plagiarism, severely sanctioned in all universities when detected.

For annotated, clickable weblinks and useful tutorials full of practical advice on how to improve your study skills, visit this book's website at **www.pearsoned.co.uk/daniels**

WORLDS IN THE PAST:
CHANGING SCALES OF EXPERIENCE AND PAST WORLDS IN THE PRESENT

Section 1

Edited by Denis Shaw

The world in which we live today has been moulded and is constantly being remoulded by the forces of capitalism (defined in Chapter 2, pp. 44–45). As modern capitalism spread across the globe it inevitably changed the societies which preceded it. But the pre-capitalist societies, while adapting to capitalism, did not necessarily lose all their distinctive features, and capitalism itself was changed in the process of interacting with such societies. The fact that human societies in the modern world differ among themselves in all kinds of ways is only partly the result of the way global capitalism works and of the ways in which societies are now adapting to the various forces affecting them. Societies also differ because of their past histories, histories that stretch back into pre-capitalist times. The world into which modern capitalism spread was itself already enormously varied and had been changing in innumerable ways over thousands of years. This section presents a brief survey of that long history and argues that the human geographer cannot understand the present-day world without knowledge of the worlds which have existed in the past.

Chapter 1 of this section considers what the world was like before capitalism and what if any generalizations can be made about this long period of human history. Chapter 2 analyses the rise and spread of modern capitalism down to the end of the nineteenth century. Its history during this period is intimately linked to the rise of the European powers and to the worldwide spread of their influence. This, of course, is not to deny that capitalism probably sprouted, and then died, in other societies in other periods, or that it might have gained worldwide significance on the basis of a non-European core had circumstances been different. During the twentieth and twenty-first centuries, which are the subject of Chapter 3, capitalism became a truly global phenomenon and radically changed in the process. This chapter thus provides a more immediate background for the sections that follow.

In summary, this part of the book emphasizes the variety and complexity of the world before globalization. And it demonstrates that only by knowing something about that world can we understand globalization itself.

PRE-CAPITALIST WORLDS

Chapter 1

Denis Shaw

Topics covered

- Why geographers need to know about the past
- Ways of classifying past and present societies
- Bands, tribes, chiefdoms, regulated states, market-based states
- Hunting and gathering
- The invention and spread of agriculture
- The rise of cities, states and civilizations
- Medieval feudalism
- Problems of studying past societies

Much of this book is about how capitalism has spread across the globe and even now is remoulding human societies in new and unexpected ways. This chapter is about what the world was like before capitalism. In other words it is about how human beings have lived during most of their existence on this planet. It will also say something about why it is important for human geographers to be aware of that past.

Capitalism has changed the world profoundly and often very quickly. In the first half of the twentieth century it was still possible for geographers and others to travel to regions where its influence was minimal and where people still lived in pre-capitalist societies. Today this is much less true. Even so, the geographer can still find numerous areas where the way of life has many pre-capitalist traits, though perhaps increasingly influenced by the modern world. Let us consider some examples.

About 2 per cent of the population of Australia is composed of Aborigines whose ancestors first settled the continent up to 50,000 years before white Europeans (Broome 1994). Today their way of life ranges from that of suburban professionals to remote out-stations and homelands where people live partly off the land in traditional style. When the British established their first penal colony on Australian soil in 1788, more Aborigines lived on the continent than today, and they were divided into some 500 groups displaying a wide variety of languages, cultures, economies and technologies. All were hunters and gatherers leading a semi-nomadic existence and using some similar tools, like stone core hammers, knives, scrapers and axe heads as well as wooden implements like spears, digging sticks and drinking vessels. However, there was also a variety in technology and economy that reflected the diverse environments in which Aboriginal peoples lived. Whether living on coasts, riverbanks, in woodland or desert regions, they were efficient and skilled at exploiting the flora and fauna of their surroundings. Aboriginal peoples who hunt and fish today are often using skills honed by their ancestors over thousands of years. Their languages and cultures similarly reflect the experiences of many hundreds of generations.

The Inuit or Eskimo peoples living in the Arctic, mainly coastal, territories of Greenland, northern Canada, Alaska and the extreme eastern tip of Russia, are another group who in the past lived almost entirely by hunting and gathering (Sugden 1982). Hunting and gathering still plays an important role in many of the remoter communities. Like the Australian Aborigines, the Inuit showed a remarkable ability to exploit the resources of what, in their case, is an extremely harsh environment. Because of their environment, Inuit communities demonstrated less economic variation than the Aborigines but they were far from uniform. There were marked seasonal movements depending on the character of the local resources as groups moved between exploiting the fish and mammals of the open ocean, hunting seals on the ice, and seeking caribou, musk ox and similar fauna inland. The rich Inuit culture is also varied but by no means as varied as the Aboriginal (there are only two interrelated languages, for example). Considering the vast geographical spaces that the Inuit occupy, the latter fact is remarkable and may reflect a relative ease of communication by sledge and boat as well as the frequency of migrations. Like the Aborigines, the Inuit have had to adapt in the recent period to an intrusive, mainly white culture, but are still sufficiently conscious of their distinctive histories and life-styles to have a strong sense of identity.

The Australian Aborigines and the Inuit are but two of many human groups existing either now or in the recent past who have lived by hunting and gathering (see Figure 1.1). The present-day world also contains other groups whose way of life to a greater or lesser extent reflects pre-capitalist characteristics. Pastoral nomads, for example, live mainly by raising and herding domestic animals (cattle, sheep, goats, camels, yak, reindeer and others), that provide them with food, clothing and other necessities. Pastoralists are particularly found in marginal (semi-arid, sub-arctic, sub-alpine) lands in parts of Eurasia and Africa (see Plate 1.1). Another example is the many peasant farmers still to be found in parts of sub-Saharan Africa, southern Asia and elsewhere. Some still grow much of their own food and market relationships have yet to become of central importance to their lives (Wolf 1966). To the extent that such people eat and otherwise depend on what they grow and raise rather than on what they can sell or earn outside the farm, their way of life is similar to that of peasants over the centuries. Of course, nowadays the number of peasants is declining and they are increasingly exposed to the influences of the outside world. But again they remind us of a world before capitalism.

The point, then, is that today's world still contains many societies whose ways of life differ from those lived by most of this book's readers and which recall earlier periods in human history. In order to understand those ways of life and the world as it is now, the geographer must know something about the world before capitalism. Learning about that world helps us to see how capitalism has changed the world and the pluses and minuses of that process.

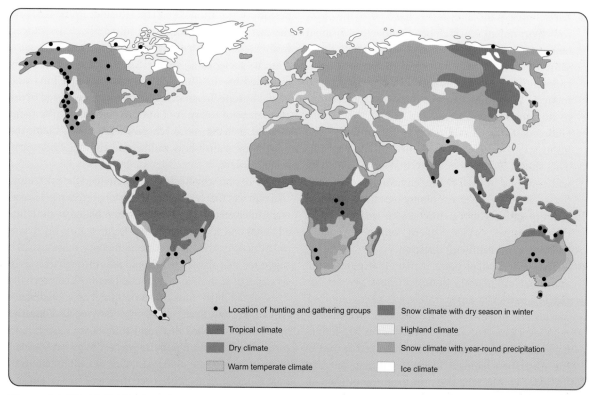

Figure 1.1 World distribution of hunter-gatherers today and in the recent past, with indication of their habitats (based on a classification of climate).

Source: adapted from Simmons (1996: 48)

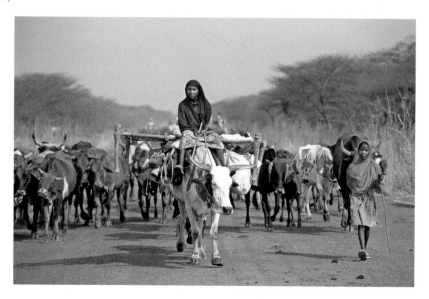

Plate 1.1 Pastoral farmers with their cattle, West Africa.

(© Sebastian Bolesch/Das Fotoarchiv/ Still Pictures)

A number of arguments can be advanced to suggest why it is important to know something about the world before capitalism:

■ Understanding the past helps geographers and others to understand themselves and their societies. Societies and individuals are products of the past, not just of the present. The present cannot be understood in ignorance of the past. Studying the past provides answers not merely to questions about how things were but also about how things are now (and, more tentatively, about how things will be in the future).

■ An example of the latter is the character of the physical environment in which people now live. Human beings have lived on this planet for so many thousands of years that over vast areas the physical environment has been profoundly modified. The world today is the product of thousands of generations of human activity. In fact, can any of the world's landscapes now be said to be truly 'natural'? (The implications of this are explained in Chapter 12.) This shows us that it is likely that human activity will continue to modify our environment into the future. Keeping the environment exactly as it is now is a very unlikely option. But how far we can control those changes, or keep them within acceptable limits, is a more difficult question.

■ A further important point is that Aborigines, Inuit and other **indigenous peoples** (those peoples native to a particular territory that was later colonized, particularly by Europeans) have recently been asserting their rights. For example, Australian Aborigines have asserted their right to be recognized as part of the Australian national identity, pointing out that the story of Australia does not begin with Captain Cook's landing in 1788, as the whites have so often assumed, since the Aborigines were in Australia long before. Indigenous peoples have also been claiming rights to local resources long ago usurped by outsiders and are trying to protect their distinctive cultures. Such peoples often feel they are the victims of the past. By studying that past and trying to understand the variety of cultures and ways of life that exist in the world and how they came to be, geographers are more likely to understand and respect such feelings.

Thus, while human geography is primarily about the present, it cannot afford to ignore the past. Some of the above points will be illustrated in the following pages.

1.1 Making sense of the past

Despite the spread of capitalism and globalization, the world of today remains immensely varied and complex. The same is true of the world in the past. The question for the geographer is how to make sense of this complex past; how to make it amenable to geographical analysis and understanding.

Different scholars have tried different ways of answering this question, but all suffer from shortcomings. For example, the economist W.W. Rostow, in his famous theory of economic growth typical of capitalist development, simply described the pre-industrial

period as 'traditional society' (Rostow 1971: 4–6). Yet the fact is that pre-industrial human societies ranged from the smallest communities of hunters and gatherers to societies as sophisticated and geographically extended as the Roman Empire, ancient China and feudal Europe. Inasmuch as such societies existed in the past, and not today, it is not possible to observe them directly. But the more scholars learn about them, the more they realize how unlike one another these early societies were.

One reason why Rostow oversimplified the pre-industrial past was that he was not much interested in it. He was really interested in modernity, as exemplified by the United States and its Western allies. Furthermore, his theory can be described as 'unilinear evolutionary' inasmuch as it suggested that all societies should pass through a series of set stages of economic development before finally arriving at the age of 'high mass consumption'. The United States (where Rostow lived) and other 'developed' countries have already reached this age. Evolutionary models have certainly long been a popular way for Western societies to try to make sense of the past. One of the attractions of such models has no doubt been their tendency to suggest that the West is the most 'developed' and thus most 'progressive' part of the world and, by implication, the best (and a model for development: see Chapter 8, pp. 184–6). But quite apart from the questionable assumption that all societies should seek to imitate the West, this raises important issues about the meaning of 'progress' and 'development', two terms that certainly carry very positive connotations in the West. For example, while those parts of the world usually deemed most developed and wealthiest certainly use most energy (see Chapter 5) and enjoy access to more material goods, they also make huge demands on the environment, perhaps ultimately to everyone's undoing. Thus, what is progress in this context? Similarly, it has often been noted that the price people tend to pay for more 'development' and wealth is that they have to work harder and for more hours. Unless hard work is regarded as a virtue in itself, this again raises questions about the meaning of progress. The point here is not to disparage all forms of evolutionary theory or idea (for example, to deny that human societies have, by and large, become more complex and spatially extended through time; see Dodgshon 1987). Rather, it is to suggest that what is most recent or new or complex is not necessarily best. Thus, in what follows, words like 'modern' do not imply 'better', nor do 'ancient' or 'primitive' imply 'worse'.

An alternative method of trying to make sense of the past, and one that does not have the unilinear evolutionary structure of, say, Rostow's theory, is to classify human societies into a series of types. Classification can be described as a way of simplifying a complex world by grouping together phenomena that are regarded as having some common feature or property, particularly where the latter is deemed especially significant. Thus Marxists have commonly grouped societies in accordance with their prevailing mode of production, or in other words with the way in which material production is related to social structure (Hindess and Hirst 1975). As a geographer, primarily interested in spatial structure, Robert Sack classified societies in accordance with their use of territoriality (Sack 1986): a theme taken up in Chapter 21. The point is that there is no right or wrong mode of classification. It all depends on what the purpose of the classification is. Most social theorists have wanted to claim that their mode of classifying societies is particularly significant for understanding how and why societies differ. The problem is that there is a large measure of disagreement about what the best mode is, and each has its shortcomings.

1.2 A classification of human societies

As an example of the kinds of difficulties that face any attempt to classify past and present human societies, this chapter will focus on one mode of classification that has had widespread appeal not only for human geographers but also for anthropologists and other scholars. This is the mode of classifying human societies into bands, tribes, chiefdoms and states (regulated and market-based) (see Bobek 1962; Service 1971; Dodgshon 1987). Spotlight boxes 1.1 to 1.5 describe, in simplified form, the major characteristics of each type of human society according to this classification. It will quickly be seen that the classification describes a range of societies from the simplest and most primitive to the capitalist societies of today. What it does not suggest is the circumstances under which one kind of society may change into, or be succeeded by, another.

One reason why this mode of classification has had such appeal is because it suggests that over the course of human history larger and more complex societies have tended to appear. It also relates size and complexity to the way societies occupy space and their relationships to their physical environment. However, it is worth reiterating that this is only one way of classifying human societies. Indeed some scholars reject the whole idea of classifying human societies because they feel it leads to misapprehensions about how societies are formed and survive, and even to dangerous notions of unilinear evolution. It thus needs to be remembered that this scheme is suggestive only, and does not necessarily describe the way any given society has actually changed through time. In particular, it does not stipulate that societies must develop through a series of set stages, or that one type of society necessarily leads to the next.

Subsequent sections will outline some of the social processes that might have helped to produce, change and perhaps destroy the different kinds of human society described in Spotlight boxes 1.1 to 1.5. In so doing, some of the limitations of the classification system used will become apparent. Before proceeding, however, it is worth making one important point. Our knowledge of prehistoric societies (societies which have left no written records) derives mainly from archaeological data, or material left by such societies and the traces of their environmental impacts. Methods for analysing such data have become ever more sophisticated, but the data themselves obviously become scarcer the older they are. Moreover, there are many kinds of questions, for example about the way early people viewed the world around them, which cannot be answered directly from the archaeological evidence. New discoveries are always liable to change our understanding of past societies, and especially of the earliest ones. Our knowledge of the latter societies, and of the dates attaching to them, must therefore be regarded as especially tentative.

1.3 Hunting and gathering

The time when the first human-type species (hominids) first appeared on earth is very uncertain, but a date of at least two and a half million years ago is often given. Homo erectus, forerunner of the modern human species, is believed to have appeared around 1.9 million years ago, and the modern species (Homo sapiens sapiens) at least 40,000 years ago (but see Foley 1995 for a contrary view). Since agriculture appears to have arisen less than 10,000 years ago, hunting and gathering in bands have been the basic occupations for much of humanity's existence (Spotlight box 1.1).

As noted already, only a limited amount is known about early human societies (see Plate 1.2). In addition to archaeological evidence, much has to be inferred, for example from hunter-gatherers who are still in existence or who existed until quite recently. Needless to say, such inference can be dangerous because modern hunter-gatherers have probably been influenced by other, more 'advanced', societies that now share their world. Their way of life is thus unlikely to be identical with those who existed in the remote past.

Spotlight box 1.1

Bands

- Bands are societies of hunter-gatherers. Most of human history has been lived in this way.
- Bands live by hunting wild animals and gathering food from the surrounding flora.
- Because of the pressures they place on the surrounding environment, bands are almost always nomadic.

- Bands are the smallest and simplest of human groups.
- Bands rarely exceed 500 people.
- For most of the time band members generally live in smaller groups of 25–100 people, which allows for efficient exploitation of the environment.
- Bands usually have a minimum of social differentiation and functional specialization.

The distribution of hunter-gatherers in the recent past shows that they have existed in every major climatic zone and have practised enormously varied economies, depending on the nature and potentials of the local environment (Figure 1.1). This fact alone suggests that a simplistic evolutionary model of human societies will fail to do justice to their complex histories. Some generalizations do seem possible, however – for example,

HARPONS : 1, Pis de la Vache; 2, Lacave; 3– Abri Mi
5 à 7, Sainte Eulalie.

Plate 1.2 Artefacts from early human sites: flints and bone harpoons.

(Michael Holford Photo Library)

that in recent times groups living in the highest latitudes (above 60 degrees) have relied largely on hunting, and those below 39 degrees on plant collecting. Fishing seems to have played an important role for many groups living in intermediate latitudes (Lee 1968).

Hunter-gatherers have lived in this world for such a long period that today's environment is the way it is partly as a result of their activities. For example, any notion that early human beings lived in complete harmony with their environments or always had 'sustainable' economies is simply untrue. Thus hunter-gatherers are credited with the extinction of many animals. An outstanding case is North America where perhaps two-thirds of the large mammal fauna living there just before the arrival of the ancestors of modern native Americans (via the Bering land bridge, possibly about 12,000 years ago) subsequently disappeared. The most likely explanation is the effects of hunting. The discovery of fire, its use by hominids long predating the appearance of Homo sapiens sapiens, was also an important instrument whereby human beings changed their environments. Some scholars have argued that the present-day appearance of such major biomes as the African savannas and the prairies of North America is the result of the human use of fire over thousands of years (Simmons 1996). More recently, contacts between hunter-gatherers and agricultural and industrial societies frequently had even more far-reaching environmental effects. At the same time, other scholars have asserted that the relations between such groups and their physical environments were rarely if ever exploitative in the modern sense. Hunter-gatherers lived in an intimate relationship with their environment and frequently regarded it, or aspects of it, as sacred. Some scholars have thus described their relationship as 'organic' in a way which has been forgotten by our modern societies, much to the detriment of today's global environment (Merchant 2005).

Scholars have wondered about processes whereby bands evolved into tribes, or hunter-gathering economies into agricultural ones (Harris 1996a, b). It has been pointed out, for example, that where the natural environment was especially favourable, the packing of bands tended to be denser, leading to less mobility and perhaps to semi-permanent settlement and closer interaction between bands. There may also have been a much greater degree of manipulation of the environment and its resources than might be expected in a pure hunter-gatherer economy. Altogether the life of bands was much more variable and a good deal less stable than the above classification scheme suggests. The boundary between band and tribe is thus blurred.

1.4 Human settlement and agriculture

Just as the boundary line between band and tribe may not always be easy to define in practice, the same can be said of the boundary between tribe and chiefdom. Some scholars, for example, doubt whether the category 'tribe' (Spotlight box 1.2) is particularly helpful, implying as it does the lack of a hierarchical social structure (Friedman and Rowlands 1977). Such scholars tend to believe that the appearance of agriculture will have sparked off competition for access to the best land, or at least rules whereby such land was allocated and inherited, and that some people and groups will inevitably have lost out. They therefore see a social hierarchy (characteristic of chiefdoms and states – see Spotlight boxes 1.3 and 1.4) beginning to emerge even as agriculture and the process of permanent settlement began.

Be that as it may, it is now widely accepted that the traditional picture of agriculture being invented in a small number of 'hearths' and then spreading across the globe is far too simple. Bands of hunters and gatherers are known to develop an often intimate knowledge of their local environments, and no doubt human beings understood much about the factors influencing plant and animal development long before they began to practise full-blown agriculture. David Harris (1989, 1996a, b) has argued that the development of agriculture was a long drawn-out affair and that there must have been much trial and error before it finally emerged in some places in a fully recognizable form. Some prehistoric northern Australian Aborigines, for example, knew about agriculture, but never adopted it.

Something can, however, be said about when, and perhaps why, agriculture appeared. The archaeological evidence makes it possible to detect and date the remains of domesticated varieties of plants and animals. According to Simmons (1996: 93) there were probably three foci for the initial surges of domestication: around 7000 BC (to use the Christian chronology) in south-west Asia, around 6000 BC in south-east Asia, and around 5000 BC in Meso-America. Different species of plants were associated with each, for example: wheat, barley and oats in south-west Asia; rice in south-east Asia; and maize, squash and gourds in Meso-America. Domesticated animals also began to appear about the same time (though the dog, domesticated from the wolf, appeared before agriculture). Sheep, goats, pigs and cattle are all associated with south-west Asia (Simmons 1996: 87–134), though pigs may also have been domesticated independently in eastern Asia.

As to why agriculture was adopted, scholars differ. The simple answer, that it was more efficient than hunting and gathering, hardly applies to naturally productive environments where it clearly involved much greater effort relative to the return achieved. Some

Spotlight box 1.2

Tribes

- Tribes appeared with the invention and spread of agriculture.
- Agriculture usually demands considerable investment of human effort into a relatively small area. This reduces the propensity to migrate (but pastoral nomads are an exception).
- The appearance of agriculture usually allows more people to live in a smaller space. Agricultural societies thus tend to be bigger than bands, with higher population densities.
- Settlements tend to become more fixed than in bands.
- The greater spatial fixity of society encourages intermarriage and the development of extended family networks. Tribes thus develop a sense of kinship and of common descent.
- Socially, tribes are relatively egalitarian, at least as between kinship groups.

Spotlight box 1.3

Chiefdoms

- Chiefdoms arose with the emergence of 'ranked and stratified' societies.
- 'Ranked' societies are societies where groups and individuals have, on a relatively permanent basis, different degrees of status and power.
- 'Stratified' societies are societies where groups and individuals have, on a relatively permanent basis, different degrees of material wealth.
- A chiefdom implies the presence of a permanent ruling group and/or individual, though kinship

remains the chief bond between ruler and subject.

- Chiefdoms imply a greater degree of centralization and control within society than in a tribe (shown, for example, in the imposition of taxes and tribute on the ordinary subjects). Spatially this might be reflected in the greater importance accorded to a central settlement (proto-city) where the ruler resides.
- The first chiefdoms began to appear about 3000 BC in Europe, but earlier elsewhere.

Spotlight box 1.4

Regulated (pre-modern) states

- Whereas the chiefdom is organized around the principle of kinship, the state is organized on the basis of territory (viz. one is subject to the state if one resides in the territory of that state).
- States, whilst also based on social inequality, tend to be larger and administratively more complex than chiefdoms. This implies greater functional specialization and greater probability of the rise of urban forms.

- Pre-modern states existed in a world before modern capitalism. The market was not central to their functioning and was often controlled ('regulated') in various ways. Thus economic relationships were generally subordinated to political, social and religious considerations.
- Other forms of social relationship, for instance landholding, also often tended to be controlled (regulated) rather than being determined by the market.

Spotlight box 1.5

Market-based states

- Market-based states are modern states whose development is closely linked to that of capitalism and thus of the world economy.

- The market is of key importance to their functioning.
- In the last two centuries these states have been associated with colonial and national projects (see Chapters 2, 3 and 22).

scholars favour population pressure. Others, however, argue that this view is over-deterministic and point out that even hunter-gatherers had the means of controlling population growth. Such scholars tend to favour more complex explanations that embrace cultural preferences and choices as well as environmental and population pressures (Maisels 1993: 25–31).

Just as the factors that led up to the appearance and spread of agriculture are by no means straightforward,

the same is also true of its further development. No simple evolutionary model can account for how agriculture has changed and developed throughout the world. Everywhere traditional agriculturalists showed a remarkable ability to adapt their practices to local environmental conditions. Eventually they largely displaced the hunter-gatherers (Figure 1.2) and gave rise to an enormous variety of agricultural systems (Figure 1.3).

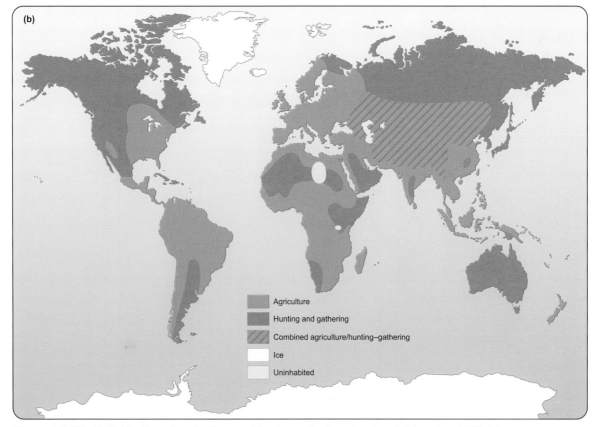

Figure 1.2 World distribution of agriculture and hunter-gathering about AD 1 (a) and AD 1500 (b).

Source: Sherratt (1980: 97, 117)

Figure 1.3 Agricultural regions of the world in the first half of the twentieth century.

Source: after Whittlesey (1936)

Flat polar quartic equal-area projection

1 Nomadic herding
2 Livestock
3 Primitive subsistence agriculture
4a Intensive subsistence, wet rice dominant
4b Intensive subsistence, wet rice *not* dominant
5 Plantations and small farms
6 Mediterranean agriculture
7 Commercial grain farming
8 Crop and livestock farming
9 Commercial dairy farming
10 Commercial gardening and fruit
11 Little or no agriculture

0 — 2500 miles
0 — 4000 km

Agriculture's effects on the landscape were much greater than the effects of hunting and gathering. It changed the vegetation cover of wide areas as forests and grasslands disappeared under fields, practices like terracing and irrigation were introduced, and the effects on soil composition abetted erosion in some places. The grazing of animals and other forms of resource use frequently changed the species composition of forests, grasslands and other areas. Agricultural systems, crops and domesticated animals spread way beyond their initial locations. An outstanding example is the cultivation of rice, which may have started in east Asia around 5000 BC, if not before, and spread as far as Egypt and Sicily by Roman times. Such developments long preceded the European overseas expansion beginning in the fifteenth century AD, which was to have even more far-reaching consequences both for the geography of agriculture and for the environment (see Chapter 2, pp. 47–53).

There is no doubt that the spread of agriculture (whether as a result of migration or through cultural diffusion) had far-reaching effects on human society, encouraging permanent settlement and the emergence of tribal systems (though Crone, 1986, questions the nature of the link between agriculture and tribes). Tribal members, living permanently side by side, almost inevitably developed similar cultural traits such as common languages. As suggested in Spotlight box 1.2, tribes also typically develop myths of common ancestry (indeed, this seems to be part of their definition). In the late nineteenth and early twentieth centuries, a period much influenced by ideas based on Darwinian evolution as well as by myths of European racial and cultural superiority bolstered by European overseas imperial expansion, it became fashionable, on very limited evidence, for scholars to imagine that modern ethnic differences have a biological basis and that certain 'races' or ethnic groups are inherently superior to others. In Britain, for example, many historians and archaeologists regarded the modern English as deriving almost entirely from Anglo-Saxon invaders with both racial and cultural characteristics which equipped them to dominate the 'inferior' Celtic peoples. Similar ideas, of course, reached their ultimate absurdity in the racist theories propounded by Adolf Hitler and the Nazis. Nowadays, with the availability of new scientific approaches like molecular archaeology and DNA studies, we know that the origins of the English and of other peoples are much more complex than was once thought and that 'different' peoples are in fact remarkably similar in their capabilities (Miles

2006: 18–31; Oppenheimer 2007). Different human groups, settling in the same region, may gradually have adopted similar languages and cultural patterns, thus slowly fusing into one people or ethnic group even though they were not originally closely interrelated. In this way the study of even the remote past can cast light on issues which cause contention today and, as in this case, help to discredit theories based on crude racist prejudices and a misreading of history.

1.5 Cities and civilization

The invention of agriculture was only one of the events that moved human societies along a road that ultimately led to the kinds of societies which predominate in the world today. Others included the appearance of cities, states and civilizations. Cities and states usually go together. Maisels (1993: 302), for example, argues that the functional specialization associated with cities allowed rulers to rule more effectively and also to put a greater social distance between them and their subjects. The gathering of specialists around the person of a state's ruler fostered urban life, commerce and the many other things (such as writing, technology, religion and speculative thought) with which is associated the word 'civilization'. Indeed, the very words 'city' and 'civilization' have a common Latin root (civitas, meaning 'state').

As is the case with the appearance of agriculture, the processes that eventually gave rise to cities and civilizations are far from straightforward. It is clear that the development of a social hierarchy was a necessary precursor, but what produced that? Different scholars have debated the merits of alternative potential causes such as land shortages, inheritance rules, or the emergence of particularly strong or charismatic individuals. But while these and other factors may have produced **chiefdoms** (see Spotlight box 1.3), something more seems to be required to explain the rise of cities, states and civilizations. As Maisels writes, chiefs enjoy only 'hegemony' over their peoples, and are bound by rules of kinship which imply mutual obligations. Chiefdoms were notoriously unstable. Kings and other state rulers, by contrast, exercise 'sovereignty' (ultimate power) over a **territory** and its peoples (Maisels 1993: 199). Thus scholars have cast around for particular factors which enabled individual leaders to have their power recognized as legitimate by subject peoples and persuaded the latter to pay over the tribute, food surpluses and services which cities and rulers required to maintain themselves in existence. Such particular factors might have included

Plate 1.3 A medieval town, Altstadt, in Germany, showing evidence of planning.
(Blickwinkel/Luftbild)

religious sanctions, military skills or organizing power. The problem is to show how the kind of social inequality that might be produced by such processes then transforms itself into the stable and legitimate power that a ruling group exercises in the state. Some scholars, therefore, have proposed that states came into being as a result of different processes working together in a contingent way rather than being caused deterministically by a single overriding process (Maisels 1993: 199–220).

Whatever may be the explanation for the initial development of states and associated cities, they gave rise to major changes in human societies (see Spotlight box 1.4). Rather than being organized around the principle of kinship like the tribe and the chiefdom, for example, states were organized on the territorial principle. This means that the rulers of states exercised their rule over defined territories and their inhabitants, no matter who the latter happened to be. This proved an extremely powerful way of organizing human societies. States thus appeared in different parts of the globe. The first organized states are believed to have developed in Sumer in present-day Iraq around 3000 BC. According to Maisels, these took the form of city-states. In China the first state, associated with the Shang dynasty, arose about 1500 BC, and by about 1000 BC the Maya civilization had appeared in Meso-America. In Europe the first states are commonly taken to be the Greek city-states that began to develop in eastern Greece and the neighbouring islands about the eighth century BC. Whether the state arose independently in different places across the globe, or whether the idea of the state was somehow diffused from one or two initial foci, is uncertain.

Early states took many different forms. All faced difficulty in enforcing and maintaining their territoriality in the context of poor communications, and different states approached this problem in different ways. The Greek city-states, for example, typically occupied only small areas and restricted their population sizes. When population in any one state exceeded a certain threshold, that state would found a self-governing colony elsewhere on the sea coast. Travel by sea was generally easier than overland travel and thus settlements could keep in touch (see Plate 1.4). In this way the Greeks eventually colonized much of the eastern Mediterranean coast (Figure 1.4). The huge Roman Empire, by contrast, could not rely solely on sea travel. This was a much more centralized polity and only by having superb military organization, an elaborate

Plate 1.4 A Classical Greek ship of the kind which plied the Mediterranean in Antiquity.

Figure 1.4 The Greek colonization of the ancient Mediterranean and Black Sea *c.* 550 BC.

Source: after Pounds (1947: 49)

system of roads and fortifications, good administration, and a state-controlled religion centred on the person of the emperor, could it survive for several centuries (Figure 1.5 and Case study 1.1). The Mongol empire was also very extensive, controlling much of Asia and a good slice of central and eastern Europe in the thirteenth and fourteenth centuries AD. The Mongols were pastoral nomads whose formidable warrior skills were based on their horsemanship and their ability to outmanoeuvre their enemies. But nomadic empires had a tendency to contract as quickly as they arose. It was difficult for the emperor to maintain control over his swift-moving armies of mounted warriors. Finally, medieval Europe's answer to the problem of territoriality was to decentralize political power through kings to territorial lords and minor nobles (see Case study 1.2). In return for their rights to their land and other privileges, lesser nobles owed their lords a duty of military support in time of need, which meant having to appear with an army of knights, retainers and other soldiers when ordered to do so. The lords in turn owed loyalty to their overlords, and so on in a hierarchy that ended with kings or other rulers. Medieval European states were thus quite decentralized and unstable, as regional lords and their underlings were often tempted to rebel (Figure 1.6). Only after centuries of struggle between rulers, lords and ordinary subjects did the European states of today finally appear.

What all this tells us is that the present-day pattern of states is contingent on and the outcome of quite random historical processes. Had other processes taken place, the outcomes might have been totally different. In other words, there was no historical inevitability about the emergence of states like the United Kingdom, France, China or Nigeria, whatever present-day nationalists might like to think. Often enough state-building has been a murky and messy business in which many innocent people have suffered, and it usually lacked the glorious victories and heroic deeds so often celebrated in modern national mythologies. And if the existence of modern states is the product of historical chance, why should things be any different in the future?

1.6 Pre-capitalist societies

It should be clear by now that pre-capitalist societies were by no means simple. They differed among themselves in all kinds of ways. At the same time, when compared with present-day capitalist societies, they appear to have had a number of distinguishing features. This section will consider some of the more obvious ones.

Geographers and others have sometimes been tempted to think that past societies, lacking modern communications, were essentially small in scale and localized, with life based around the community. It is hoped that enough has been said already to suggest how over-simplified this view is. Huge empires had to be administered and defended, great cities like classical Rome had to be fed. Classical Rome imported some 17 million bushels of wheat each year from Egypt, North Africa and Sicily

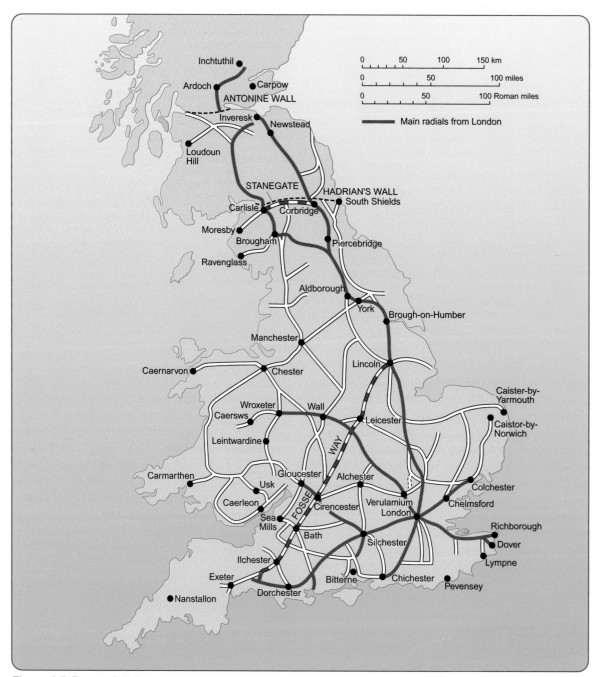

Figure 1.5 Roman Britain.

Source: © Oxford University Press 1984. Redrawn from map VI from Roman Britain in *Oxford History of England*, edited by Peter Salway (1984), by permission of Oxford University Press

(Simmons 1996: 109). In the Roman Empire and other societies, high-value goods often travelled much greater distances. Even the Inuit, living the lives of hunters and gatherers, sometimes travelled hundreds of miles each year on hunting trips, and their geographical knowledge frequently covered an even wider span. Enough has been said above about migrations and the diffusion of artefacts and practices across great distances to suggest

how space could be overcome even if individuals were immobile, at least by today's standards.

There was, however, no avoiding the problems of communication in pre-modern societies. Most people in agricultural societies lived lives which were bound to their villages and the surrounding regions. Their work consisted essentially in winning the means of subsistence for themselves and their families from their

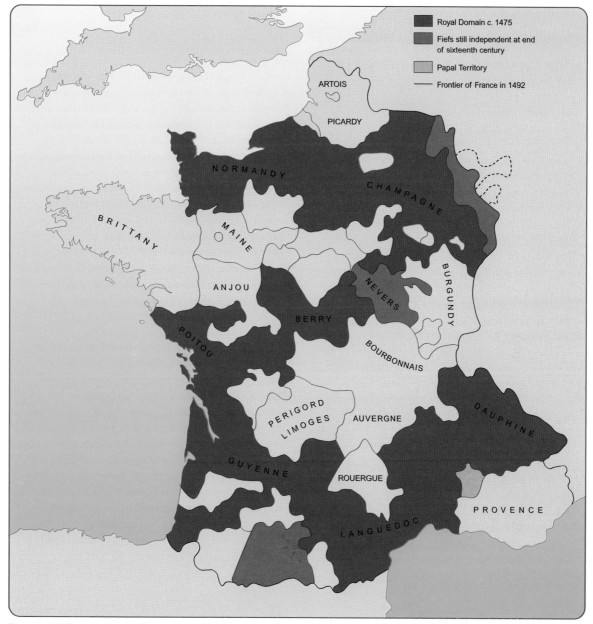

Figure 1.6 France towards the end of the feudal period, showing the political fragmentation typical of many parts of Europe prior to the appearance of modern states.

Source: based on *The Times Atlas of World History* (1989: 15.1). © Collins Bartholomew Ltd 1989, reproduced by permission of HarperCollins Publishers

environment, as well as paying the taxes and meeting the other demands which were made upon them. From the time of the appearance of states down until the twentieth century the lives of most people were lived in the context of such **peasant** communities. As mentioned earlier, such communities are still characteristic of parts of the world today, though increasingly eroded by commercialism and growing contacts with modernity.

Scholars researching traditional peasant societies have frequently pointed to the many cultural differences between such rural dwellers and the cities that ruled over them and which they might occasionally visit. A case in point is religion. Religion has been a fundamental factor in social organization and in outlook in virtually every human society down to the twentieth century. Early states were generally associated with an official religion that legitimized the established order and which was often practised in cities but was rarely fully understood by the peasants, most of whom were probably illiterate (see Plate 1.5). Even where the peasants officially followed

Thematic case study 1.1

The Roman Empire at its zenith (first to fourth centuries AD)

The Roman Empire was one of the greatest achievements of state-building in the pre-modern period and is testimony to the way in which huge distances could be crossed and enormous territories controlled in spite of the absence of modern industrial technologies. At its greatest extent the empire's east–west axis stretched from the Caucasus to Cape Finisterre in north-west Spain, a distance of about 2,800 miles. North–south it reached from Hadrian's Wall, near the present-day Anglo-Scottish border, to the fringes of the Sahara in North Africa, some 1,600 miles. The empire was characterized by two major languages (Latin and Greek), low tariff barriers, a common currency, a common code of laws, and the basis of a common system of education and culture for the Romanized elite. Rome was a truly 'universal state', which lasted in its mature form for half a millennium, and in its eastern (Byzantine) manifestation until the capture of Constantinople by the Turks in 1453.

Like all early states, Rome had problems in enforcing its territoriality – the sheer scale of the empire meant that there were always difficulties in maintaining control over its far-flung provinces. That the empire was able to endure for so long (albeit with many vicissitudes) was a tribute to its highly developed capacity for organization and an astuteness in gaining and retaining local loyalties. The acquisition of new territories proceeded by a mixture of conquest, colonization by Roman and Latin colonists, the establishment of vassal kingdoms and powers, alliances (often forced)

and other means. The Roman army, reformed by the Emperor Augustus after 13 BC, was a masterpiece of military organization superior to any of Rome's enemies in this period. The seas and waterways were patrolled by the navy, protecting merchant shipping. Towns and military bases were interconnected by a network of fine roads, and an official transport system (the *cursus publicus*) with inns and posting stations at regular intervals speeded official communications. There were even roadbooks (*itineraria*) for the guidance of the many travellers, including some tourists. Fortified lines, of which Hadrian's Wall is the most celebrated, protected the most vulnerable frontiers. But the empire was not held together only by military force. It was also by judicious extensions of Roman citizenship and other privileges to local elites, and by fostering helpful religious practices like emperor worship, that local loyalties were ensured.

The empire's economic and cultural achievements were many. Long-distance trade in such items as grain, metals and luxury goods was important, for example, though there were many hindrances to commerce and most people depended on local agriculture and manufacture. It must also be remembered that there was a heavy dependence on slavery. In the end the Romans could never be assured of political stability, even in their heartlands. Reliance on the army contributed to the empire's eventual undoing – generals and their armies competed for political power, the economy was undermined by overtaxation and other factors, and growing military weakness invited rebellion, and invasion by 'barbarians'.

---

the same religion as the elite, as in medieval Europe, they almost always interpreted its teachings in their own ways, mixing them with their own superstitions and 'pagan' beliefs. The life and outlook of the ruler and the elite, if not of all cities, were entirely alien to them. And equally their way of life was alien to the cities.

For this reason, it would be completely misleading to think of the subjects of pre-capitalist states as 'citizens' in the modern sense. Traditional rulers knew little of the countryside, where most of their subjects lived, and cared even less (except, perhaps, where they had landed estates, and even then the running of the estate could be left to officials). What mattered to rulers was law and order, and extracting the taxes and tribute the

state needed to maintain itself in existence. The idea that rulers should care for the welfare of their ordinary subjects, let alone consult them about their policies, is very modern indeed.

It is difficult for most people living in a world dominated by capitalist relationships to conceive of societies where this was not the case. Before capitalism, most people were engaged in subsistence activities: hunting, gathering, farming, fishing or whatever. They might have to pay taxes or tribute, and they might trade some of what they gathered or produced, but market trade, generally organized by a small minority of merchants and traders, was often quite marginal to them. Where market trade did exist in early societies, it was frequently hedged

Thematic case study 1.2

European feudalism

- Feudal society was hierarchical – ranked and stratified.
- The majority of the population were subsistence farmers, engaged in extensive forms of agriculture (for example, by cultivating strips in open fields shared with others, raising some livestock) and exploiting local resources like pastureland, woodland and fish.
- Many of the rural population were obligated to their lords and landholders, for example not to move away from their villages, to work the lord's land, to pay the tribute and taxes the lord and the state required, to serve the lord militarily and to perform other services.
- Lords held their land conditionally from their overlord and ultimately from the king or other ruler. Land could not easily be sold. Lords exercised jurisdiction over those living on their estates.
- Lords owed allegiance to their overlords (and they in turn to their overlords and so on), for example

in supporting them with their military forces and performing other services as required.
- Feudal states were quite decentralized, parcelled out into lordships and they in turn into jurisdictions of various kinds.
- Towns were generally small and few and far between (see Plate 1.3). Most were centres of trade and crafts. Their residents (merchants, burgesses) often enjoyed freedoms and privileges ('liberties') denied to most rural dwellers.
- Merchants and itinerant traders joined together networks of local and long-distance trade, linking local fairs and markets with towns, ports and so on. Craft production and trade were generally subject to a variety of restrictions and controls, and often played only a minor role in the lives of rural dwellers.
- The medieval world-view was strongly influenced by the Church, whose priests and officials were among the few educated people and were universally present.

about by laws and restrictions of various kinds. There might be laws against the taking of interest on loans, for example, regulations on price, or restrictions on where and when trade might take place or who might engage in it. Only with the rise of modern capitalism did the market become central to the way societies functioned.

It is very difficult to generalize about pre-capitalist societies and almost any generalization is open to objection. Case study 1.2 describes the main features of the feudal society that existed in medieval Europe which, according to many scholars, was the seedbed for the development of world capitalism (see Chapter 2). Even here, however, one must be careful – the features described would be more or less true, depending on the date and the place being considered.

1.7 The heritage of the past

Nowadays the past, or what is commonly referred to as heritage, is big business. Medieval cities and cathedrals, historic villages and battlefields, monuments, gardens and ruined temples are objects sought out by international tourists and cultural visitors, whilst tourist providers and also those charged with the conservation of the

past compete to sell the past to visitors. What cannot be viewed out of doors is gathered together into museums which become ever more elaborate and ever more tied to the leisure industry rather than to the educational role which defined them in the past. Cities which are in the business of selling themselves to international investors and global companies use their cultural resources, including heritage, to do so. Altogether, it seems, our age is obsessed with the past, and nostalgia has become part of almost everyone's life.

Not surprisingly, perhaps, the present-day significance of heritage has not escaped the notice of human geographers and other students of society (Graham *et al.* 2000; Johnson 2003). Geographers have been concerned to know what it is about the past which appears to attract so much public interest and how people understand those facets of the past which they encounter. More specifically, since the past is no longer with us and cannot be encountered directly, scholars have begun to ask about which particular past is being viewed under the guise of heritage, and about the ways in which the past is manipulated for public consumption. For the fact is that there are many pasts, depending on the viewpoints of those who wish to describe them and on the messages which those in control of heritage wish to purvey. Whether it be national governments, wishing

Plate 1.5 The celebrated Mappa Mundi or world map, dating from *c.*1300 AD, in Hereford Cathedral, England. Constructed to show Jerusalem in the centre of the world, this map typifies the medieval Christian view of a world explicable only in religious terms and symbols.

(By permission of the Hereford Mappa Mundi Trust)

to convey some lessons about supposed national 'glories' in the past, museum curators trying to explain particular historical events or epochs, or the guardians of historic sites aiming to entertain and titillate tourists, the scope for historical distortion (even if it means only telling one side of the story) is immense. History, in other words, is all around us, and history is controversial. Only by knowing something about the past can the geographer hope to avoid ignorance about the present.

Learning outcomes

Having read this chapter, you should begin to appreciate that:

- The surviving evidence necessarily limits knowledge of past societies, since they cannot be observed directly. In general, the further back in time past societies existed, the more uncertain our knowledge about them becomes.

- People in the past lived very different lives from those of most people alive today. But it is important not to oversimplify or overgeneralize about past societies that varied enormously in their structures and ways of life and in their degree of complexity.

- Classifying societies into types is a way of making sense of the past and of trying to understand the geographical characteristics of varied societies. But all classifications have their shortcomings.

- Overcoming the friction of space was a severe problem in pre-capitalist societies. But it is important not to exaggerate its effects – people often travelled, and often migrated, long distances even in 'primitive' societies, ideas and artefacts travelled vast distances, and empires were successfully established (see Figure 1.7).

CHAPTER 1 PRE-CAPITALIST WORLDS

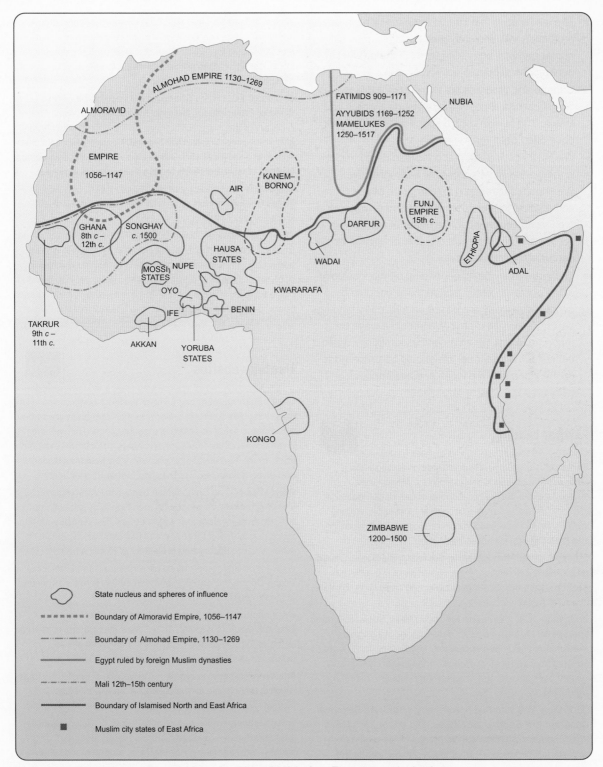

Figure 1.7 The great empires of Africa (AD 900–1500) before European colonization.

Source: based on *The Times Atlas of World History* (1989: 136–7). © Collins Bartholomew Ltd 1989, reproduced by permission of HarperCollins Publishers

- The landscapes that can be seen today have been profoundly influenced by past societies, including hunter-gatherer ones. Our world is the product of generations of human activity.

- Change and progress are two different things. Social change almost always means winners and losers. The human record suggests that the latter have often been the majority.

- We must not assume that our present-day society is the world's most successful or progressive. We should be prepared to learn from others and to allow for the possibility that other ways of organizing society might be better at coping with some problems (for example, that of environmental degradation) than our own.

- There are many pasts depending on the viewpoint of the observer. Many assumptions adopted by people alive today are based on misunderstandings or distortions of the past. We should always be prepared to test our ideas and assumptions against the historical evidence. We should be prepared to do the same with the ideas and assumptions of others.

Further reading

Benton, T. (ed.) (2010) *Understanding Heritage and Memory*, Manchester University Press, Manchester. A set of essays exploring the contentious nature of heritage and memory, including the idea of cultural landscape, arguing that it is never possible to impose a single set of values or interpretations on acts of remembering.

Crone, P. (1989) *Pre-Industrial Societies*, Blackwell, Oxford. An excellent introduction to the character of pre-industrial societies, from the invention of cities and civilization. Written for undergraduates.

Dodgshon, R.A. (1987) *The European Past: Social Evolution and Spatial Order*, Macmillan, Basingstoke. An original examination of European prehistory and history, taken from a geographical perspective. Based on the social classification system highlighted in this chapter. Very detailed, but repays careful study.

Dodgshon, R.A. and Butlin, R.A. (eds) (1990) *An Historical Geography of England and Wales*, 2nd edition, Academic Press, London. An historical geography of the region from prehistoric times to 1939. Chapters 4 and 5 illustrate the issue of feudalism.

Jones, R. (2007) *People/States/Territories: The Political Geographies of British State Formation*, Blackwell, Oxford. People, power and places in the making of the British state.

Jones, R. and Phillips, R. (2005) Unsettling geographical horizons: explaining pre-modern and non-European imperialism, *Annals of the Association of American Geographers*, 95(1), 141–161. This article argues for longer term historical perspectives in human geography.

Simmons, I.G. (1996) *Changing the Face of the Earth: Culture, Environment, History*, 2nd edition, Blackwell, Oxford. A history of the impact of humans on the natural environment from the earliest times to the present. Contains excellent chapters on pre-capitalist societies, especially from the environmental point of view, but tends to be a little technical in places.

The Times Archaeology of the World (1999), new edition, Times Books, London. A lavishly illustrated survey of archaeology.

The Times Atlas of World History (1999), new edition, Times Books, London. A highly acclaimed survey of world history beginning with human origins. This is in fact more a history book with maps than a true atlas and fails to provide detailed maps despite its large format. But it is lavishly illustrated and contains a wealth of factual and explanatory material. Other similar atlases are available.

Useful websites

www.britishmuseum.org   The British Museum, London. The museum holds a unique collection of art and antiquities from ancient and living societies across the globe. The website gives details of the museum's holdings and current exhibitions, hosts discussions on prehistory and past civilizations, and gives guidance on how to find further information.

www.si.edu   The Smithsonian Institution, Washington, DC. The Smithsonian is a focus for many kinds of scientific and cultural endeavour in the United States. The website contains much that is useful for readers of this chapter, especially in relation to the history and prehistory of North America.

www.jorvik-viking-centre.co.uk   The website of the Jorvik Viking Centre in York, England, famed for its reconstruction of life in York in the Viking period.

www.channel4.com/history/microsites/T/timeteam   Channel 4 (UK)'s popular 'Time Team' programme is devoted to the archaeological investigation of a variety of sites, particularly in Britain.

http://english-heritage.org.uk   The website of English Heritage, which is responsible for the upkeep of many prehistoric and historic sites in England, many of which are open to the public.

www.besthistorysites.net   Entitled 'Best of History Websites', a comprehensive guide to history-oriented resources online, starting with prehistory. For teachers, students and others.

http://whc.unesco.org/   The official world heritage site of UNESCO, which designates and seeks to protect 'World Heritage', defined as 'cultural and natural heritage around the world considered to be of outstanding value to humanity'.

For annotated, clickable weblinks and useful tutorials full of practical advice on how to improve your study skills, visit this book's website at **www.pearsoned.co.uk/daniels**

THE RISE AND SPREAD OF CAPITALISM

Chapter 2

Terry Slater

Topics covered

- Definitions of capitalism
- The cyclical nature of capitalism
- The transition from feudalism to capitalism
- The beginnings of European imperialism
- Colonial commerce
- Transatlantic migrations and the slave trade
- European colonial empires and racism
- Industrial and agricultural transformations
- New modes of transportation
- Industrial urbanism

Between 1500 and 1900 there was a fundamental change in the way an increasingly large part of the Western world was organized. Beginning in particular parts of England, and spreading to other parts of Europe, the capitalist economic system was to change, or influence substantially, not only the economies, but also the political, social and cultural dimensions of newly powerful nation-states. By 1900, capitalism was the dominant socio-economic system over a large part of the world, a by-product of the colonial empires of those nation-states.

2.1 What is capitalism?

A number of writers have provided theoretical frameworks for understanding the workings and ramifications of the capitalist system. The first was devised by the Scottish philosopher, historian and father of classical economics, Adam Smith, in his book *An Inquiry into the Nature and Causes of the Wealth of Nations* (1776). Like all theoretical models, Smith's is a simplification of reality, but it introduced a terminology and series of conceptualizations that are still familiar to us today. Smith assumed the model was driven by people's selfish desires for gain and self-interest. Thus, production takes place to generate profit; surplus profits are accumulated as capital; and the basic rule of the system is 'accumulate or perish'. Integral to the system is the determination of prices which, in a free market, said Smith, are determined by the

supply of, and demand for, the factors of production. In order to maximize their profits, industrialists will always seek to minimize their costs of production, including the wages that they pay their workforce, so as to outcompete other producers by having lower prices. One way of doing this was through the division of labour, dividing manufacturing tasks into simple, repetitive operations that could be performed by unskilled, and therefore cheap, labour. Another important way in which manufacturers could seek competitive advantage over their fellows was to be first in the use of new technology. The significance of the invention and adoption of new ways of doing things, whether through science and technology, or ideas and organization, has often been crucial to firms, industries and regions in getting ahead of competitors. This is still very clearly so today in technological 'hotspots' like the Milano and Stuttgart regions of the European space economy. Most scholars agree that the capitalist system began to cohere into an integrated whole in a particular place (England), at a particular time (the seventeenth century), and that it then took another long period (until the mid-nineteenth century) before it matured in the states of north-west Europe and in North America (Green and Sutcliffe 1987: 6–7) (Spotlight box 2.1). We do need to note, however, that recently this viewpoint has been challenged by some historians who believe that it is too Eurocentric and that greater attention should be given to developments in China and the East more generally (Section 2.2.3).

Spotlight box 2.1

Characteristics of states with a mature capitalist economy and society

- The majority of the population in these countries were wage labourers forming a 'working class' who sold their labour power for wages in cash or kind so as to purchase food and other commodities to survive.
- The majority of these wage labourers were male and worked outside the home; females increasingly 'worked' in the home, both in the domestic care of their families and by 'selling' surplus labour time for minimal wages.

- Wage labourers were increasingly closely supervised by managers of the production processes.
- The means of production of wage labourers was owned by capitalists whose aim was to make a profit on their investment.
- There was an increasing disparity between the 'profit' of capitalists and the 'wages' of workers with an increasing propensity for conflict between the two.
- The vast majority of goods and services, including fixed property like land, were distributed through monetary exchange.

2.1.1 Cyclical characteristics of economic development

Smith went on to model the workings of a national economy over a year (macroeconomics) to elucidate the way in which capitalism works in a series of interconnected cycles. These cycles have been studied subsequently by other economists, most notably the Russian N. Kondratieff (1925), whose name is now used for the roughly 50-year-long cycles of boom and depression that have characterized the capitalist world since the mid-eighteenth century.

In seeking the explanation for these cycles of growth and stagnation, Schumpeter (1939) argued that technical innovation leading to the development of new industries was the key to understanding the growth phase of the cycle. Because technical innovation is spatially uneven, then so too has been the geography of economic development under capitalism. The question then arises as to why the geography of innovation is uneven. Recent research has suggested that the socio-institutional structures of regions of innovation, or of those lacking innovation, are the key to explanation: in other words whether educational, governmental and social organizations encourage or discourage enterprise and change. Others have criticized the technological determinism of these theories (Mahon 1987).

2.2 Other perspectives, other stories

2.2.1 Marxism

The viewpoints of orthodox economics are not the only interpretation of the transformation of large parts of the world over the past three hundred years. Karl Marx (1867) took a very different perspective in his analysis of capitalism in the mid-nineteenth century, *Das Kapital* (see Spotlight box 3.2). He proposed that profit arises out of the way in which capitalists (the bourgeoisie) dominate labour (the workers) in an unequal class-based relationship. This unequal class relationship was perceived by Marx to lead inevitably to class conflict. In his later writing he sketched out ways in which labour could gain control of the means of production and thereby 'throw off their chains'. This was to have dramatic long-term consequences for the socio-political organization of the world through most of the twentieth century between the Russian Revolution and the upheavals of 1989 (see Chapter 3, pp. 73–5). We need to note, however, that Marx's writing is dominated by an historical, but not a geographical, perspective: it traces class relationships in time but not in space. Marxist interpretations of capitalist development have therefore always had difficulty in conceptualizing uneven spatial development. It is also highly Eurocentric in its perspective.

2.2.2 World systems theory

Another way of conceptualizing the nature of these changes is to be found in the writings of Immanuel Wallerstein (1979), whose 'world-systems theory' provides a threefold categorization of historical socio-economic systems. Wallerstein proceeded to delimit the spatial characteristics of the capitalist world economy into core, periphery and semi-periphery. These terms are not used in the everyday sense but signify areas in which particular processes operate. For Wallerstein, core processes are those characterized by relatively high wages, advanced technology and diversified production; periphery processes are characterized by low wages, simple technology and limited production. Between the two is the semi-periphery: areas that exploit the periphery but which are exploited by the core, so that they exhibit a mixture of both core and peripheral processes (Taylor 1989: 16–17). In the period between the seventeenth and the end of the nineteenth century some regions of the world were in the core (north-west Europe, for example) or periphery (central Africa) throughout the period. Other regions moved from the semi-periphery into either the core (the southern states of the USA, for example) or the periphery (the states emerging from the ruins of the Ottoman/Turkish empire). Wallerstein's theory has been very influential in explanations of the growth of globalization, but it is an explanation only at the 'structural' level. It says little about the complexities of the social and economic networks that enable the system to work, or of the resistance by groups or individuals trying to change it (Ogborn 1999).

2.2.3 Eurocentrism

Over the past decade a small group of scholars (mostly historians) have claimed that most of the theories discussed above are 'Eurocentric'. That is they tell the story of these three centuries as being about the 'rise of the West' owing to its developing humanism, scientific rationalism and democratic politics. The story of the East is relegated to a few footnotes because it was perceived as passive, unchanging and despotic. This new group of scholars, beginning perhaps with Edward Said (1978) and his concept of 'Orientalism', have tried to show that, in fact, the story of East and West is fundamentally intertwined and that capitalism would not have emerged without the import of eastern technology, ideas and resources along long-established long-distance trade routes developed by eastern merchants and connecting to Europe through the Middle East. Eurocentrism was deeply embedded in the West (at least from medieval times onwards) and also led to a distinctive, socially constructed European 'identity' that was Christian and 'civilized' and could be set against the so-called 'primitive' peoples of North America, Africa and Australasia who could be exploited as European states began to develop their empires (Hobson 2004).

2.3 The transition from feudalism to capitalism

The transformation of the European economy into a capitalist one was a long drawn-out process and historians continue to argue the precise causes of the changes that took place. Writers using an avowedly Marxist frame of reference (Dobb 1946; Kaye 1984) have made many of the most significant contributions in this debate. To them, the development of a class of wage labourers is crucial, together with the assumption of political power by the new class of capitalists. Geographers have been particularly concerned to trace the outworking of these historical processes of social and economic transformation in the particular space economies of local regions (Gregory 1982; Langton 1984; Stobart and Raven 2004).

The disintegration of feudalism (see Case study 1.2), with its carefully controlled market system, was considerably speeded by the after-effects of the Black Death in mid-fourteenth-century Europe. The European population was reduced by between one-third and a half, and towns were especially hard hit. The resulting labour shortage meant that enterprise and new ways of doing things were more likely to be rewarded, since the feudal elite required more revenue to maintain their power. At the same time, the slackening of social and cultural controls meant that new ways of thinking could flourish, especially in the growing towns.

By the later fifteenth century, larger European towns were dominated by what has come to be called merchant capitalism. Merchants were both the providers of capital and principal traders in a regionally specialized, complex, Europe-wide trading nexus based not on luxury goods as in the medieval period, but on bulky staples like grain and timber, and on an increasing array of manufactured products. Towns that flourished economically in fifteenth- and sixteenth-century Europe were those whose inhabitants were able to copy and manufacture imported products more cheaply than the exporting regions, and develop a constant stream of new, innovative, marketable products (Knox and Agnew 1994: 154–61) (Spotlight box 2.2). For most commentators, the key question then becomes: why

Spotlight box 2.2

Characteristics of the period of merchant capitalism in Europe

- Increasing numbers of people sold their labour for money wages. They ceased to work on their own land. This led to increased consumer demand for food, clothing and household goods; in many places there was a notable rise in living standards.
- More producers of both agricultural and craft-manufactured products began to accumulate capital as they produced for this growing market. These prosperous yeoman farmers and manufacturers were the foundation of a new class of capitalists.
- The removal of feudal market restrictions and controls by guilds on production led merchants to invest in the reorganization of production on a capitalist basis.

- Technical innovations transformed industries; the most significant was probably the development of the printing press by Gutenberg, in Mainz, enabling knowledge to be diffused cheaply and rapidly. However, we need to note that paper-making, block printing, and the movable metal-type press all originated earlier in China/Korea.
- The rediscovery of Classical knowledge led to new ways of seeing the world and, consequently, to its rebuilding to reflect these new images, especially in towns.
- Ultimately, these changes led to religious upheaval and political revolution whereby the new capitalist class became dominant in the governance of nation-states.

Figure 2.1 The palace, gardens, park and town of Karlsruhe, Germany, founded in 1715 by the Margrave Karl Wilhelm of Baden.

did these processes of transformation coalesce first in sixteenth- and seventeenth-century England?

We can begin to answer this question by suggesting that, since land was one of the key factors of production, critical in the sixteenth century was the enormous transfer of land in England from conservative ecclesiastical ownership to secular ownership. This was the consequence of the dissolution of the monasteries by Henry VIII in the later 1530s. Even the monastic buildings could be adapted by capitalist manufacturers, as the Wiltshire clothier William Stumpe showed at Malmesbury Abbey where he installed 300 weavers in the 1540s (Chandler 1993: 487–9). Land was also being transformed in the sixteenth and seventeenth centuries through the process of enclosure. This enabled livestock to be raised more efficiently for the urban meat markets and experimentation in new agricultural techniques to be undertaken by yeoman farmers (Butlin 1993: 178–9).

In the seventeenth century it was the social, religious and political transformation of England that was critical, according to Dobb (1946). The struggles between Crown and Parliament during the Civil War and Commonwealth (1642–60) are seen as a conflict between landowners, and capitalist yeomen and manufacturers. Though this struggle was continued through the Restoration (1660) and Glorious Revolution (1688), by the end of the seventeenth century the capitalist bourgeoisie were politically predominant and were able to transform the state to their own advantage.

Elsewhere in Europe landowners remained preeminent over a predominantly peasant agricultural workforce and this period is known as the 'Age of Absolutism'. Landowners, especially the rulers of small states, amassed enormous wealth from their control of land. This wealth was expended on increasingly spectacular landscapes of display: huge palaces, filled with works of art and every luxury, set in carefully designed, and often very intricate, landscapes reliant on vast inputs of labour for their maintenance (Figure 2.1). King Louis XIV of France had set the model at his palace of Versailles, outside Paris, using the resources of a much larger state. Such conspicuous displays of wealth and privilege by the few led, ultimately, to a much bloodier revolution than in England. The French Revolution saw the slaughter of not simply the royal house, but of the landowners, intellectuals and the bourgeoisie, since this was the first of the workers' revolutions, with its cry of '*Liberté, egalité, fraternité*'. However, it led rapidly to the totalitarian militarism of Napoleon.

2.4 An expanding world

As the European economic and cultural world was gradually transformed, European travellers began to voyage beyond the shores of Europe. They travelled overland into

Figure 2.2 The Portuguese fortress and trading station of Sofala, in modern Mozambique, in 1558. For a time it was the administrative centre for all Portuguese trading with India.

Source: Carneiro (1990)

Asia, drawn both by a thirst for greater knowledge and, more significantly, by the rewards that could be reaped from direct exploitation of scarce commodities for European consumption. These included spices, sugar, silk, muslins, porcelain and the like, which had previously been traded through the eastern Mediterranean. In the late fifteenth century, however, this region was coming under Turkish (Ottoman) control. The Portuguese, funded by Italian bankers, were the first Europeans to navigate round Africa into the Indian Ocean, using the existing Arab-dominated sea-based trading networks (Arabs had already rounded the Cape in the other direction), whilst soon after (in 1492), Christopher Columbus, again funded by Italian (Genoese) merchant capital, sailed westwards across the Atlantic to open the Americas to European exploitation using African labour.

For Spain and Portugal, this is the heroic 'Age of the Navigators', as they moved rapidly to take control of these new maritime trade routes looking out to the Atlantic. The merchants of Lisbon and Seville grew rich on this trade. For the rest of the world it marks the beginning of the European colonial empires that were to dominate and control almost every aspect of life for the next four hundred years. The Portuguese established a far-flung network of small fortified port bases around Africa and the Indian Ocean, many of them based beside existing Arab ports, through which they were able to link western Europe into the trade in gold, spices and textiles (Figure 2.2). Equally significant was their use of African slaves to produce sugar on plantation farms in their Brazilian territory. The Spanish founded a more militaristic, oppressive and exploitative empire in central and south America which saw the decimation of local populations through warfare and the catastrophic effects of European diseases, especially smallpox and measles. It also brought enormous inflows of gold and silver bullion into Spain and, ultimately, onwards into the European economy. Blaut (1993) sees this as the event that more than any other kick-started the European capitalist world system.

Fairly quickly, control of the products of this new Western maritime world economy moved from south-west to north-west Europe. The Dutch began to compete successfully with the Portuguese in the Indian Ocean and the centres of European capital moved from Lisbon and Seville first to Antwerp by the mid-sixteenth century, and then to Amsterdam at the end of the century. The merchants of these ports successfully integrated the Atlantic trade with that flowing from the Baltic and from central Europe via the Rhine valley. It was at Antwerp that the first stock exchange was established, enabling capital raised in one business to fund investment in another. Amsterdam's rise was based on having greater military and naval power than its neighbours and using it to enforce the privileges of a distinctive trading monopoly: the state-licensed joint-stock charter company (the Dutch East India Company) (Dodgshon 1998: 74–83).

At the same time, French and English fishermen had followed John Cabot's exploratory voyage across the North Atlantic to begin to exploit the rich fishing grounds off Newfoundland and, subsequently, to trade with native Americans for fur pelts. From these small beginnings French and English merchant influence began to expand through the seventeenth century in the Indian Ocean (using similar company trading monopolies to the Dutch); in the Caribbean, where sugar plantations and slave labour were copied from the Portuguese; and in North America. They superseded the Dutch in the early eighteenth century through superiority of arms, but it was not until the later eighteenth century that the British emerged dominant in the developing Western world economy. Consequently the primary city of the European capitalist system moved once more, from Amsterdam to London (Dodgshon 1998).

The colonialist trading systems that developed in the period up to 1770 have often been described in simple dualist, oppositional, terms (core–periphery; dominant–subordinate; metropolitan–colonial, for example). Blaut (1993) is especially critical of the Eurocentric character of this type of argument. More recently, a more complex, multi-layered interpretation has been posited. We need to note, first, that different parts of the developing world economy operated in somewhat different ways, were based on different product interactions between raw materials, manufacture and consumption, involved complex transportation flows, and required different politico-military frameworks to make them function properly. Meinig's (1986) model of the Anglo-French mid-eighteenth-century North Atlantic economy uncovers some of these complexities in terms of the commercial, political and social systems that were required for the geography of colonial capitalism, and Ogborn (1999) has added details of networks of individuals and patterns of resistance (Figure 2.3). The North Atlantic 'triangular trade' is certainly the best-known of these colonial trading systems.

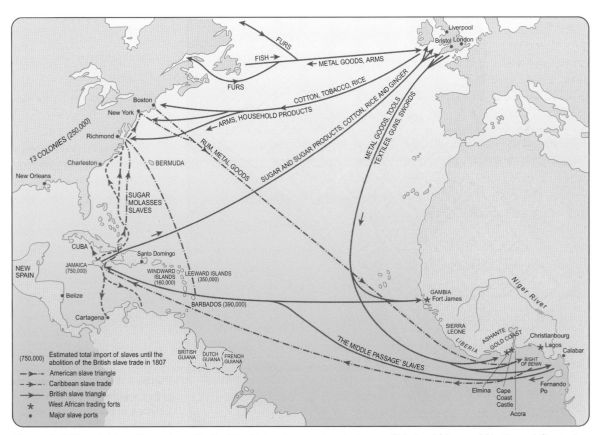

Figure 2.3 The eighteenth-century North Atlantic trading system between Britain, Africa and North and Central America.

Source: Bayley (1989: 46–7)

2.4.1 Colonial commerce

In Meinig's North Atlantic commercial system, London and Paris acted as the source of finance, commercial intelligence and marketing (London's Royal Exchange was founded in 1566). The expanding industries of particular regions of western Europe were growing, in part because they supplied a developing colonial market with manufactured goods such as textiles, tools, armaments and household necessities. These goods were assembled in Atlantic ports such as Bristol, Liverpool, Bordeaux and Le Havre. After crossing the Atlantic they were stored, disassembled and distributed from colonial ports such as New York, Charleston and New Orleans (Figure 2.4, Plate 2.1) by major traders. They sold on to local traders towards the colonial frontier. At the frontier a barter system of exchange was as likely to be found as a money economy. Nonetheless, fur, fish, timber and agricultural produce flowed in the opposite direction for storage and transhipment back to Europe (see Figure 2.3).

In North America, tobacco, rice, sugar, rum and cotton were the major export products. All were produced on the plantation system using slave labour. By 1750, some 50,000—60,000 people were being forcibly transported across the Atlantic each year from all parts of the western African seaboard. Many died en route or shortly afterwards but, altogether, some 3.8 million Africans had been transported by 1750, just over half to Latin America, the remainder to English, French and Dutch North America and the West Indies. These staggering figures meant that African-Americans were easily the largest new culture group to be established in both North and South America in the colonial period (Meinig 1986: 226–31).

Figure 2.4 Plan of the French colonial entrepôt city of New Orleans at the head of the Mississippi delta in 1720, published in London in 1759 by Thomas Jefferys.

Plate 2.1 Madame John's Legacy, c.1788. One of the oldest houses in New Orleans, it was built during the Spanish colonial period (1765–1803). Commercial functions occupy the brick-built ground floor, residence the balcony floor.

(Author)

The African abolitionist Olaudah Equiano, who gained his freedom in England, wrote forcefully from his own experience about the horrors of the 'middle passage' voyage from West Africa to the West Indies. He said:

> The stench of the hold while we were on the coast was intolerably loathsome . . . but now that the whole ship's cargo were confined together, it became absolutely pestilential . . . The air soon became unfit for respiration . . . and brought on a sickness among the slaves, of which many died . . . This wretched situation was aggravated by the galling of the chains . . . and the filth of the necessary tubs (toilets) into which the children often fell and were almost suffocated. The shrieks of the women and the groans of the dying, rendered the whole scene of horror almost inconceivable.
>
> (1789)

By 1750, the trade in slaves was dominated by large European companies using specially constructed and fitted ships, but many smaller traders from the Atlantic ports would sail first to African trading stations to exchange manufactured products, including guns, tools and chains, for slaves. The slaves would then be taken to the West Indies or Charleston, with ships returning to Liverpool or Bristol with a cargo of sugar, tobacco, rum or rice, and sometimes slaves such as Equiano, too. Most of the latter were destined for domestic service in aristocratic households (see Figure 2.3).

2.4.2 Colonial society

The slave trade was perhaps the most significant aspect of the cultural transformation effected by the eighteenth-century capitalist world economy in the Americas and, it should not be forgotten, in Africa. One of the distinctive features of this transformation in the Americas was the 'othering' of Africans by Europeans. They were regarded as in every way inferior to Europeans; they were legally and socially defined as different; and they therefore lived their lives separately from white settlers. The social consequences of this were to reverberate through the next 250 years, into our own times, where spatially segregated city neighbourhoods based on ethnicity are still a commonplace in North American cities, whilst from the 1950s West Indian migration to Britain saw similar ethnic areas develop in many British cities. Where Africans were in the majority in the eighteenth-century colonies, as they were in the West Indies, and in the Carolinas and Louisiana, this enabled distinctive African-American cultures to develop within the confines of political and economic enslavement.

There were, of course, other trans-Atlantic migrations throughout the time period covered in this chapter. The earliest included those seeking to escape religious persecution in Europe, or to establish religious utopias in the new continent. The Puritan 'Founding Fathers' of New England were in this category and, right through to the Mormon migration in the early nine-teenth century, such groups continued to leave Europe's shores for what was perceived to be a better new world. Religious faith also inspired another group of settlers who came to convert the native people of the Ameri-cas to Christianity. The competition in Europe between Catholic and Protestant versions of the faith – between Reformation and Counter-Reformation – was exported to the colonies since both native Americans and African slaves were perceived as being in need of conversion. Consequently, Jesuit monasteries or Franciscan friaries often stand at the heart of Spanish colonial towns and cities, whilst their dedicatory saints gave name to many of today's largest cities: San Francisco, Los Angeles and San Diego, for example (Conzen 1990).

By the beginning of the eighteenth century, sig-nificant migrations of Scots, Irish, French, Germans, Dutch, Swiss and Moravians had taken place, but the dominant migrant group were the English. Between 1700 and 1775 the population of the eastern colo-nies grew tenfold to some 2.5 million and, by 1820, the population of the United States surpassed that of Britain (Lemon 1987: 121). The losers were native Americans who were dispossessed of their lands east of the Appalachians and often taken into slavery.

2.4.3 Colonial politics

There were major differences in the way European states governed their colonial empires. The Spanish developed a highly centralized system in which all as-pects of policy were laid down in Madrid. The Laws of the Indies, which laid down in meticulous detail how colonization should take place and how new towns should be planned, are symptomatic of this centrali-zation (Nostrand 1987) (Case study 2.1). The French governed through a similar centralized structure of military, administrative and ecclesiastical strands of authority linked to government departments and the Crown in Paris. The English, in contrast, allowed each colony to be virtually self-governing and the links to London were many and various. English colonial plan-tation owners and merchants were not averse to threat-ening a governor if their capitalist trading relations were endangered. It should occasion no surprise that it was the English colonies that eventually first rebelled and fought for their independence. Overlapping these colonial territories were the older territorial patterns of native groupings, those of the Iroquois and Creeks remaining well into the eighteenth century (Meinig 1986: 262).

Thematic case study 2.1

Summary of the Laws of the Indies by Philip II of Spain, 1573

Laws 1–14

Establish that exploration cannot take place without royal permission; that governors should learn all they can about their territory; they should consult with and negotiate with local 'Indians'. 'Discoverers' should take possession of land and should name rivers, hills and settlements.

Laws 15–31

Lay down that Spaniards should treat 'Indians' in a friendly manner; they should help priests in their work of conversion.

Laws 32–42

Concern the type of region that should be explored, pacified, settled and brought under Spanish mandate; the siting of towns in such regions, and the establishment of local government.

Laws 43–109

Establish the legal and taxation regimes for new colonies and lay down that each town should have at least 30 families. The ideal size of house plots, farms, herds, and commons is established.

Laws 110–128

Provide details of the plans of new towns, including the size of the plaza, the orientation and breadth of the streets, the location of churches and public buildings, and the allocation of house plots by lot.

Laws 129–135

Describe provision for common land and farms and the character of domestic buildings.

Laws 136–148

Lay down relationships with local 'Indians' and the way in which they should be converted to Christianity so that they 'can live civilly'.

Source: Crouch *et al.* (1982: 6–19)

2.5 Imperialism and racism

A second phase of colonial expansion, through the nineteenth century, was geographically distinct in that it was focused on Asia and, especially after 1880, on Africa. It is also marked by the development of a much more virulent racism in the 'othering' of all non-Europeans. The consolidation of European states in the nineteenth century, especially of Germany and Italy, and the increasing economic competition they offered to the eighteenth-century leaders of the world economy, Britain and France, led them to seek new opportunities through the colonial exploitation of hitherto economically peripheral lands. Britain seized the Dutch Cape Colony at the tip of southern Africa in 1806 to prevent its use by Napoleon's navy. By 1820, 4,000 British settler graziers were in conflict with African pastoralists, despite the fact that the low-wage labour of those same Africans was essential to maintain the farms (Lester 1999). At much the same time, the first British farmers were beginning to establish themselves in Australia as its period as a convict colony began to draw to a close. The recent novel by Kate Grenville, *The Secret River* (2006), gives a perceptive, imaginative account of the struggle for land between farmers and Aborigines at this time and the consequences for both cultures of that struggle. Meanwhile, in Britain itself, Christian-inspired reformers succeeded in bringing an end to the slave trade in 1807, and then to slavery in 1833. Sadly, the invective of both populist and bourgeois writing opposed to abolition characterized Africans, Australian Aborigines, Indians and native Americans as not only unchristian and therefore 'uncivilized' but also as less than human; inferior to Europeans in every way; both savage and yet child-like, and therefore needing to be controlled and disciplined. By the middle of the nineteenth century, these attitudes were widespread amongst all classes in Britain and were fuelled further by popular antagonism to events such as the Indian Mutiny (1857), the Zulu rebellion (1879) and the Maori wars (1860s). By the late nineteenth century, Darwin's ideas on evolution were being misused to give a scientific veneer of respectability to this 'othering' of non-Europeans, to the extent that Australian Aborigines were being shot as 'vermin' rather than seen as fellow humans.

The mid-nineteenth century marked both the deliverance of one continent from colonialism to self-government, as South America broke free from Spanish and Portuguese control, and the colonizing of another, as European powers raced to divide Africa amongst themselves. The interior of Africa remained largely unknown to Europeans until the 1860s except for Christian missionaries. But by the end of the century, borders had been delineated and fought over, natural resources were being rapaciously exploited, and the polities, cultures and economies of native peoples had been disrupted or destroyed (Figure 2.5). Again, this period of history reverberates down to the beginning of the twenty-first century because the boundaries drawn to the satisfaction of European soldiers, administrators and traders did not correspond to African ethnic, cultural and political divisions. In many respects the origins of the genocides of Biafra (1960s' Nigeria) and Rwanda (1990s), and the long civil war in Sudan, which is still ongoing in Darfur, can be laid at the door of the 'scramble for Africa' in the later nineteenth century.

2.6 Industrialization

2.6.1 Proto-industrialization

The early phases of capitalist industrial development in Europe, up to about 1770, are categorized in various ways. To some writers this is the period of merchant capitalism (see Spotlight box 2.2), to others this is the period of proto-industrialization (Spotlight box 2.3). Both terms have generated a substantive academic debate as to precisely what processes of transformation were involved.

Textile manufacture is the most studied production process of this period. Its transformation began in several parts of Europe as early as the fifteenth century. By the early eighteenth century, more and more of the production processes were being mechanized and industrialists were concentrating their machinery into single, multi-floor buildings. By the 1770s, the mechanization of spinning had been completed and weaving followed in the 1820s. The logic of this transformation was completed by steam power and the factory was established as both a building (Figure 2.6) and a system of production (Spotlight box 2.4). Different sectors of the textile industry went through this transformation at different times, and some sectors declined absolutely as capital was diverted to the most prosperous sectors. The textile industry, more than any other, had a relentless tendency to geographical specialization and concentration with the result that many towns and regions were dangerously dependent on a single industry (Laxton 1986). By the 1770s in Britain, west Yorkshire was beginning to dominate light worsted woollen cloth, Lancashire and the north Midlands cotton cloth production, the Welsh borders flannel, and the east Midlands hosiery.

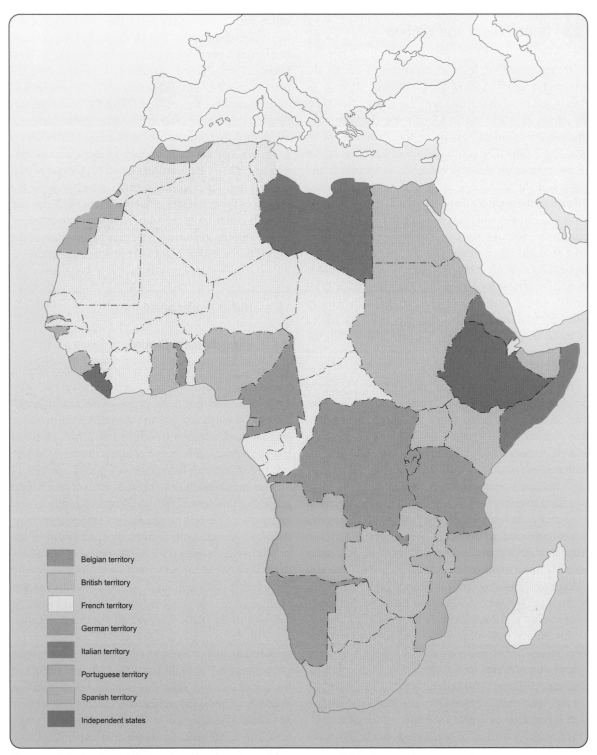

Figure 2.5 Africa in 1914 after the European imperial contest to divide the continent.

Source: Moore (1981: 74)

Figure 2.6 Starkey's woollen textile mill, Huddersfield, c.1850. Though this mill was steam-powered, by mid-century its powerlooms were still housed in multi-storey buildings rather than a single-storey shed, since the site was a restricted one.

Source: Giles and Goodall (1992: 102) © Crown Copyright. NMR

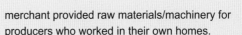

Spotlight box 2.3

Proto-industrialization

- Many early industrial processes were located in the countryside.
- Most industries were characterized by some type of 'domestic' production system whereby the capitalist/ merchant provided raw materials/machinery for producers who worked in their own homes.
- Producers generally worked in family units and were paid piece rates according to their production each week or month.
- Machinery was frequently water-powered.

Spotlight box 2.4

The factory system of production

Under the factory system the capitalist:

- Had complete control of the production process from receipt of raw materials to finished product.
- Had control of the labour force on whom a new, disciplined time geography could be imposed (early factories often worked day and night).
- Could apply capital to the development of new machinery to simplify processes and reduce labour costs through mechanization; and,
- By dividing the tasks of the production process, could reduce labour costs further by employing women and children.

I began work at the mill in Bradford when I was nine years old . . . we began at six in the morning and worked until nine at night. When business was brisk, we began at five and worked until ten in the evening (Hannah Brown, interviewed in 1832).

Legislation dating from this time remains effective in controlling the age at which children and young people can work and their hours of employment.

The complex interaction of resources, labour skills, technological innovation, and capital circulation in particular regional economies, which increased production and lowered prices, meant that new markets had to be developed and transport improved. Market opportunities were to be found both in the growing urban markets at home, and among the colonial populations overseas. Thus, cotton textiles were first imported to Britain from India; the techniques of production were learnt; manufacture commenced; factory production lowered costs; cotton textiles were exported to India; there they undercut and destroyed the Indian industry.

A key transformation of the transport infrastructure, so essential to industrialization, was the improvement of links with growing towns, and with the ports. Canals and turnpike (toll) roads, which were the principal innovations before the 1820s, also lowered the costs of raw materials, especially of coal. This is graphically illustrated by the way in which urban populations celebrated the opening of new canals, not for their industrial potential, but because they dramatically reduced the cost of domestic heating. In Britain, the construction of better roads, and of the canal network, mostly between 1760 and 1815 (the growth phase of the first Kondratieff cycle) was financed by private capital. The state was involved only in providing the necessary legislation to enable construction to take place and fees or tolls to be levied. Thus Josiah Wedgwood (china and pottery), Abraham Darby (iron manufacturer) (Figure 2.7) and Matthew Boulton (jewellery and silver plate), as well as the Duke of Bridgewater (landowner and coal owner), all invested some of their capital in canal companies to improve the distribution of their products (Freeman 1986). A similar pattern is observed in New England. In France and Germany, by contrast, new roads were seen to have a military function in the centralization of state control and were therefore largely financed by the state through taxation.

Figure 2.7 The world's first iron bridge, constructed across the River Severn, east Shropshire, in 1788 by Abraham Darby III's Coalbrookedale Company, was part of the transport infrastructure improvements financed by local industrialists.

Source: The Ironbridge Gorge Museum

2.6.2 Agricultural change

It was not just industries such as textiles, metal manufacturing and pottery that were transformed by early capitalist production: society needed food production to keep pace with a rising population that was increasingly employed full-time in manufacturing. Consequently, the eighteenth century saw the transformation of agriculture into increasingly capitalistic modes. In parts of Europe, especially in Britain, self-sufficient peasant farming began to come under pressure from landowners who used state legislation to remove common rights to land in favour of private ownership through enclosure. Ownership was increasingly concentrated into fewer hands, and farms were consolidated from scattered strips to single-block holdings, again especially in Britain and, rather later, in Scandinavia and northern Germany. Elsewhere in Europe this process was delayed until the twentieth century. In northern and western Europe, landowners invested in transforming production for the new urban markets, using technology and new techniques and crops first developed in the Low Countries in the seventeenth century. In eastern Europe, the semi-periphery, this period saw a reversion to serfdom and near-feudal relations of production. Capital was also deployed to increase the area of land under intensive, rather than extensive, production. Fenland and coastal marshes were drained; heathland soils improved for cultivation,

and the moorland edge of improved pastures pushed higher up hill and mountain slopes.

In the nineteenth century, European agriculture continued to increase production but was unable to keep pace with the demand from the growing urban industrial population. In 1840, Britain repealed the Corn Laws, which had protected local grain producers, and opened its markets to colonial and American producers. New methods of extensive grain farming were used in mid-western North America, and in the new British colonies in Australasia and South Africa. The invention of the steam ship, allowing more rapid transport, and of refrigeration and meat canning from the 1870s, led to similarly extensive ranch grazing of sheep and cattle in these countries. Together with the Pampas grasslands of Argentina, all became an important part of the semi-periphery of Britain's global economy. This export-dominated agriculture was reliant on overseas investment by British capital in new railway networks to transport these products to slaughterhouses and industrial packing plants at the port cities whilst, in the USA, Chicago was growing rapidly on the same economic foundations (Cronon 1991; Miller 1997). It was accompanied, too, by new waves of migration to the colonies and the mid-west of the USA (Table 2.1), especially from Ireland and Scotland, where the Great Famine and the Highland Clearances, respectively, were other manifestations of capitalist agricultural

Table 2.1 European emigration 1881–1910

Source countries	Destination countries	Numbers of migrants
Great Britain	N. America, S. Africa, Australasia	7,144,000
Italy	USA and N. Africa	6,187,000
Germany	North and South America	2,143,000
Austria/Hungary	North and South America	1,799,000
Russia	USA	1,680,000
Scandinavia	North America	1,535,000
Spain	Central and S. America; N. Africa	1,472,000
Ireland	USA	1,414,000
Portugal	South America	775,000
S.E. Europe	USA	c.465,000
France	Central and S. America	223,000
Low Countries	USA	171,000

Source: Moore (1981: 57)

change in the British Isles. The United States absorbed enormous numbers of European migrants throughout the nineteenth century. Irish migrants predominated in the 1840s and 1850s, followed by Scots and English, Scandinavians and Germans through the second half of the century, with Russians, Italians and south-central Europeans from the 1880s onwards (Table 2.1) (Ward 1987).

This later colonization, and the internal colonization of the United States mid-west in particular, were grandly characterized as a moving frontier of settlement by the American historian, F.J. Turner, writing in 1894, who suggested that it also transformed the 'character' of frontier settlers, making them self-reliant, opportunistic, individualistic and democratic. Recent commentators have noted that Turner's hypothesis says little about the continued sweeping aside of the rights of the native populations of these lands who fought a long and ultimately unsuccessful attritional battle to retain it for their own use (the land was regarded as 'open' or 'free' for white settlement); nor does it give any credit for the settlement process to women, who provided both the domestic labour and, often, especially in the initial phase of settlement, much of the farm labour too. The advancing frontier of capitalist agriculture was as much small-scale, incremental and domestic as it was wide-sweeping and large-scale.

2.6.3 Factories and industrial production

Some commentators have seen the 'Industrial Revolution' in Britain as a short period, between 1770 and 1830, of rapid transformation whereby the country's economy moved from an agricultural to an industrial manufacturing basis. More recent writers see the transformation as much more drawn out, extending from the seventeenth century and into the early twentieth century. They also see it as much more regionally diversified and geographically uneven, with particular regions specializing in particular products which transformed their production systems at different times. Most attention has been given to the textile and iron and steel industries, both of which are characterized by large factories requiring massive capital investment in buildings and machinery, driven by steam power, with manufacturing processes increasingly vertically integrated, and employing a large, and increasingly disciplined labour force (see Figure 2.6). The same forces were at work in New England. By 1855, there were 52 cotton mills in Lowell, Massachusetts, employing more than 13,000 people, two-thirds of them women (Groves 1987). In Europe, Lille (France), Ghent (Belgium) and the Wupper valley

of Germany developed as centres of cotton textile manufacture (Pounds 1990: 402).

Many other industries were developed in the same way as new markets arose on a national and international basis and production expanded. Food and drink, for example, is rarely thought of as a factory-based industry, but brewing, once the prerogative of almost every village inn, became increasingly an urban, large-scale industry with regional markets for its beers. In England, Burton-upon-Trent developed rapidly as the 'brewing capital' of Britain, with huge factory-scale breweries, on the basis of its colonial contracts to supply bottled 'India Pale Ale' to troops and civilians stationed in the Indian sub-continent as a substitute for local water. For other industries, however, including many manufacturing industries, 'factories' remained small-scale workshops employing fewer than 50 people (the Sheffield cutlery industry in northern England, and wire-drawing in the eastern Ruhr in Germany are good examples). Indeed, many industries remained almost domestic in scale well into the twentieth century (for example, the Birmingham jewellery and Coventry watch industries, both in central England).

The nineteenth century is characterized not only by this enormous variety in the scale of production, but also by an equally enormous variety of manufactured products. The tag 'The Workshop of the World' is as true of nineteenth-century Britain in general as it is of several of the country's manufacturing cities. In other British, European and American regions there was an increasing specialization so that local economies were dependent on maintaining market advantage on a single industry: the Polish textile city of Lódź and Lancashire's cotton mill towns were of this kind. Such places also produced distinctive working-class cultures, which, in the case of Lódź, was also predominantly Jewish (Koter 1990). Again, in those same mill towns much factory employment was female (see the example of Lowell above), whereas in heavy industry centres, such as ship-building in Newcastle and Sunderland, in the north-east of England, waged employment was male-dominated and women's roles were primarily domestic. E.P. Thompson, in his classic Marxist interpretation of *The Making of the English Working Class* (1963), is especially sensitive to the variety of experiences of working people in this period of aggressive industrial capitalism. There are, however, some common features of nineteenth-century industrial development in Britain and, subsequently, in Europe and North America. First was the fact that the power supply for the majority of industries came to be dominated by steam. There was therefore a move by many industries to the coalfield regions to reduce the cost of fuel (Figure 2.8). The development of the Ruhr region of Germany (Pounds 1990: 412–29), and of western Pennsylvania in the USA (Meyer 1990: 256–8), are good examples of this process. The reason for this change in fuel supply was the invention and commercial development of the rotative steam engine by James Watt and Matthew Boulton in 1769–75.

Secondly this, in its turn, was fundamental to the invention and development of railways that transformed the transport costs of both raw materials and finished products almost as dramatically as it reduced

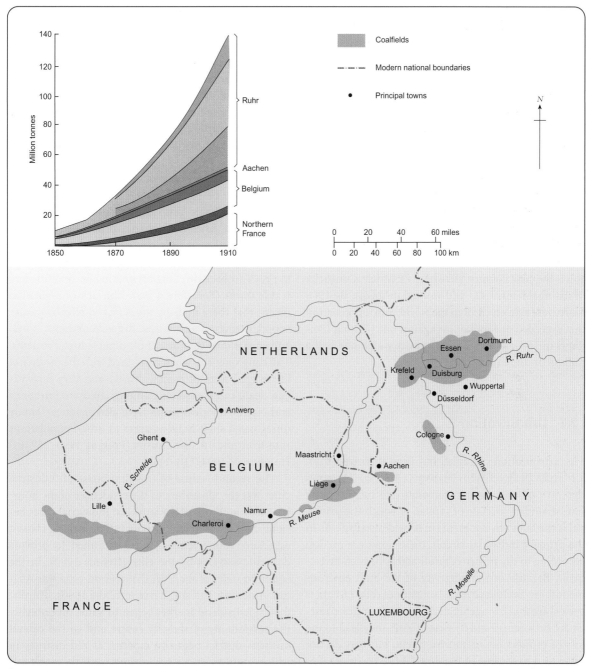

Figure 2.8 The coalfields of north-west Europe saw a concentration of new industrial enterprises after the 1780s as steam-powered machinery became normal in textile manufacturing, iron and steel making and metal goods manufacturing.

Source: Pounds (1990: 44)

the friction of distance. The Western and, increasingly, the colonial worlds saw a new time geography from the 1840s onwards, whereby journeys previously requiring several days to accomplish could be undertaken in a few hours. This was especially important in the industrialization of North America with its vast transcontinental distances (the first transcontinental line was completed in 1869). It was also significant in Britain's imperial control of India, and of Russia's colonization of the interior of Siberia and central Asia. In all three cases the railways enabled industrial-scale exploitation of natural resources (from timber to metalliferous ores) to take place, the profits from which could, in turn, be invested in further industrial, urban or railway development. The rapid expansion of San Francisco following the California gold and Nevada silver discoveries in 1849 and 1859 respectively is a good example of this (Walker 1996). Railways were also major industries in their own right, of course; the Indian railways employed three-quarters of a million people by the 1920s, for example.

Third was the fact that, as capitalists sought economies of scale, the transformations of the nineteenth century were largely urban. By 1851, Britain had become the first country anywhere in which a majority of its inhabitants lived and worked in towns. By 1900 there were sixteen cities in the world with a population that exceeded one million, as against one (Peking [Beijing]) in 1800, and another 27 places which had half a million or more (Lawton 1989).

2.7 Urbanization

The urbanization of capital has been theorized from a Marxist perspective by Harvey (1985a, b). He notes that, as capitalists over-accumulate, the surplus flows into secondary circuits of capital, of which the principal is the built environment (in its widest sense). As in the primary industrial circuit, these flows are cyclical according to the perceived profitability of such investment. Because of the longevity of the built environment these 'building cycles' (sometimes called Kuznets cycles) are much longer than the business cycles, somewhere between 15 and 25 years. Crises may lead to the loss of profitability of an investment, but they do not often lead to the loss of building fabric. Western cities are therefore made in the image of past capitalist decisions and subsequent adaptations to fit new circumstances. Harvey goes on to suggest ways in which class

struggle is written into the landscape of the Western city (1985a: 27–31), which he explores in more detail in an analysis of Paris in the third quarter of the nineteenth century. More recently, Dodgshon (1998: 148–61) has built on these arguments to show how the built environment of towns and cities is a major source of inertia in capitalist economies and societies.

There are perhaps two major phases of development in Western industrial cities before 1900, and the beginnings of a third. The first phase, which began in Britain in the 1770s, was characterized by the need to provide homes for workers once the production process was separated from home. Initially, such housing was provided by a process of 'densification' whereby more and more living spaces were crammed into the existing built-up area. Gardens and yards were used for increasingly high-density basic accommodation, much of which was multi-occupied. By the 1770s in larger British cities this process could go no further and housing began to be provided on the city fringe where it proved a profitable 'crop' for small landowners. Such housing was brick-built, basic, unplanned and unregulated, and regionally distinctive in its plan forms (Plate 2.2(a)). The most notorious of these forms was the back-to-back dwelling originating in the Yorkshire textile towns (Beresford 1988). The lack of regulation meant that these housing areas lacked effective water supplies, sewage and waste disposal systems, social and educational facilities, and connected road systems. The consequences, when combined with the poverty of low wages, were ill health, disease and very high urban death rates, especially among children. European and North American cities were also characterized by unregulated slum housing areas as urban populations rose from the 1820s onwards (Pounds 1990: 368–91; Homburger 1994: 110–11).

The second feature of this first phase of industrial cities was the separation of the residences of the bourgeoisie from those of the workers. Industrial cities became class-divided cities, though the working classes, especially young females, were recruited to provide domestic services by the elite (Dennis 1984). In the southern states of the USA that servant class was African-American and still slaves until the 1860s. Thirdly, the capitalist land market began to value more highly the accessibility of the city centre and so the central business district began to emerge, as did districts in which industry and warehousing were the predominant land use, often close to the port facilities of canal, river frontage or harbour. One of the classic portrayals of this phase of urban development in Britain is

Plate 2.2 English nineteenth-century industrial housing. (a) Back-to-back courtyard house, Birmingham *c*.1840. The door on the right leads on to a court of 12 identical houses fronting a narrow courtyard. Such houses were found in the Midlands in the period 1770–1860. (b) Tunnel-back bye-law houses, Teignmouth, Devon *c*.1880. The distinctive rear elevation resulted from bye-laws derived from the 1875 Public Health Act. These houses are found in all cities in the period 1875–1914. (c) Harborne Tenants Association, Birmingham, 1908. These houses derive from the garden city movement. Experimental estates of the Edwardian period provided the model for twentieth-century municipal housing. (d) Improved industrial dwellings, east London *c*.1885. Industrial Dwellings Associations experimented with tenement blocks with open galleried staircases.

(All photographs the author)

Engels' (1845) *The Condition of the Working Class in England*. Warehouse districts, of course, have recently been 'revalued' in many cities to become today's loft apartments or high-tech offices (Zukin 1989).

The second phase of development, from 1840 to the 1890s, was one of increasing regulation in towns and cities. New forms of local government and the collection of statistical information led to building regulations; the provision of sewerage and better water supplies; the provision of new cemeteries and parks; and better urban transport, especially tramways. All this led in turn to the increasing suburbanization of the better-paid elements of the urban workforce in the last quarter of the nineteenth century. These transformations occurred in most European core economies, and in North American cities, but with temporal variations. Thus, in England, the characteristic late-nineteenth-century dwelling was a suburban, brick-built, tunnel-back, terraced house (Plate 2.2(b)). In North America, excepting only the largest cities, the vast majority of urban housing was timber-built; in much of the rest of Europe, from Stockholm to Naples (as well as in Scotland), the apartment or tenement block was the norm (Plate 2.2(d)). The North American city, from the 1840s, was also characterized by the ghettoization of immigrant populations: first Irish and Germans, then freed slaves from the southern states after the Civil War, then Italians, and finally, from the 1890s, Jewish migrants from the Russian empire of eastern Europe (Ward 1987) (Plate 2.3(d)).

Central business districts were modified by two other major transformations in the capitalist system in this second phase. First, banking services were increasingly required to enable surplus capital to be safely stored, efficiently invested and recycled towards new opportunities; such banks were but one aspect of developing office quarters, including, for example, insurance offices, land agents and other property services, and legal services. Many of these well-constructed buildings have been found alternative uses today, as cafes, pubs or homes, because they have become valued parts of city conservation areas. This tertiary employment sector grew rapidly in the second half of the nineteenth century and 'the office' developed as both a specialized building type and a means of production (Daniels 1975). It was increasingly characterized by female employees using typewriters (the key technological invention) supervised by male managers. In North America the building type showed an increasing propensity to both large floor areas and height at the end of the century. The skyscraper was born in Chicago and relied on the development of steel-frame construction methods and the invention of the elevator (Gad and Holdsworth 1987) (Plate 2.3(a) and (b)). Second was the growth of consumption. This led to the development of a variety

of new retail spaces in city centres, including department stores and arcades (Plate 2.3(c)), as well as an enormous growth in the variety of specialized shops.

The incipient third phase of urban development reflected a minority interest in improving the living conditions of workers by industrialists. Model settlements have a history that dates to the beginning of the factory system. Modern commentators note both the generally higher standard of accommodation in these places, and the capitalist control of the home life as well as the working life of their workers. At the century's end the experimental garden suburb settlements at Bournville and Port Sunlight in England (Plate 2.2(c)) were widely admired and imitated in Germany (Margarethenhöhe in Essen, for example) and the USA (Radburn, New Jersey). At the same time, the first experiments in the large-scale provision of social housing were underway in London using apartment-block housing. These two themes, the garden suburb and social housing, were to combine powerfully in the development of modern planning to shape the urban environment in the first half of the twentieth century in Europe and its empires, but not in North America.

2.8 Conclusion

Within three centuries, Western capitalism had utilized its ill-gotten gains from its first colonial adventures to develop a series of specialized industrial regions supplying worldwide markets. Food production, industrial manufacture, service provision, consumption and transportation were all radically transformed by the capitalist enterprise. Cultures, societies and governments were necessarily impelled to change, too. States became more centralized and powerful. The difference between rich and poor both at the level of individuals and between countries became more marked. For some individuals and groups capitalism brought prosperity, improved living conditions and greater freedom; for others it brought destruction of local cultures, impoverishment, degradation and slavery. At the end of the nineteenth century the South African Boer War gave the first glimpse of the industrialization of warfare which was to scar the twentieth century so deeply.

(a)

(b)

(c)

(d)

Plate 2.3 American nineteenth-century city buildings. (a) Reliance office building, Chicago; architect: Daniel H. Burnham & Co (1895). This is the nearest that a steel-framed building got to being sheathed almost entirely in glass in the nineteenth century. (b) Flat Iron office building, New York; architect: Daniel H. Burnham (1901–2). This famous skyscraper of 22 storeys is in the style of an elongated Renaissance palazzo. The steel frame is clad in limestone and terra cotta. (c) Carson Pirie Scott & Co store, Chicago: architect: Louis H. Sullivan (1899–1904). Department stores were an important retail innovation in large city centres. (d) New York tenement building; 97 Orchard Street (1863–4), now the Lower East Side Tenement Museum. There were two shops in the sub-basement with four three-room apartments on each of the five floors. Seventy-two mostly German immigrants lived there in 1870; in 1900 there were 111 mostly Russian Jewish residents.

(All photographs the author)

Learning outcomes

Having read this chapter, you should know that:

■ Scholars have interpreted and theorized the development of capitalism in different ways.

■ The European colonialist enterprise was critical in the evolution of capitalism.

■ Industrial capitalism transformed all sectors of the economy, including the built environment and social relations between individuals and classes.

■ Industrial capitalism was essentially urban.

Further reading

Blaut, J.M. (1993) *The Colonizer's Model of the World: Geographical Diffusionism and Eurocentric History*, Guilford Press, New York and London. This is a very readable and thought-provoking polemic of post-colonial writing, providing an alternative explanation of the 'success' of European capitalism.

Dodgshon, R.A. (1998) *Society in Time and Space: A Geographical Perspective on Change*, Cambridge University Press, Cambridge. A more advanced text which provides much more detailed arguments and evidence for the themes of this chapter.

Hobson, John M. (2004) *The Eastern Origins of Western Civilisation*, Cambridge University Press, Cambridge. Challenges the Eurocentric interpretation of this period of the world's developmental history.

Knox, P., Agnew, J. and McCartney, L. (2008) *The Geography of the World Economy*, 5th edition, Edward

Arnold, London. A well-written and popular textbook which will give readers another perspective on this period and much else besides.

Langton, J. and Morris, R.J. (eds) (1986) *Atlas of Industrializing Britain, 1780–1914*, Methuen, London and New York. Historical atlases usually treat this period very well with innovative cartography and thought-provoking texts. This one deals with Britain in considerable detail.

Mitchell, R.D. and Groves, P.A. (eds) (1987) *North America: The Historical Geography of a Changing Continent*, Hutchinson, London. There are a number of good historical geographies of North America. This one is well written, thoughtful and copiously illustrated.

Ogborn, M. (2007) *Indian Ink: Script and Print in the Making of the English East India Company,* University of Chicago Press, Chicago, IL. This book explores the geographies of power and knowledge in the rise of the British Empire in India.

Watts, M. (2001) *Late Victorian Holocausts: El Niño Famines and the Making of the Third World,* Verso, London. An account of the interactions between imperialism, ecology and famine in the nineteenth century. It is worth reading in tandem with Crosby, A.W. (1986) *Ecological Imperialism: The Biological Expansion of Europe, 900–1900.* Cambridge University Press, Cambridge.

Useful websites

www.besthistorysites.net   Entitled 'Best of History Websites', a comprehensive guide to history-oriented resources online. For teachers, students and others.

For annotated, clickable weblinks and useful tutorials full of practical advice on how to improve your study skills, visit this book's website at **www.pearsoned.co.uk/daniels**

THE MAKING OF THE TWENTIETH- AND TWENTY-FIRST-CENTURY WORLD

Chapter 3

Denis Shaw

Topics covered

- The second industrial revolution

- Fordism – new patterns of production and consumption in the twentieth century

- Fordist capitalism

- Challenges to liberal capitalism: Nazism, communism

- The end of European imperialism; informal imperialism

- Globalized capitalism

- The world in the early twenty-first century

The twentieth century could be said to have been the period when capitalism finally triumphed over most of the globe. But it was neither a straightforward triumph nor an unchallenged one, and capitalism itself was changed in the process. This chapter is concerned with the various spaces created by and in response to twentieth- and twenty-first-century capitalism – spaces of resistance and reinterpretation as well as spaces of adaptation and acceptance. The patchy and unequal world in which we now live reflects the erratic and conflict-laden nature of the processes which have produced it, and we need to know something about those processes to understand the world as it is now.

3.1 The changing capitalism of the early twentieth century

In the autumn of the year 1933 the writer and journalist J.B. Priestley set out by bus on a trip that was to take him the length and breadth of England and which he later described in his *English Journey* (Priestley 1937). As he left London by the Great West Road, Priestley noted how the road 'looked odd. Being new, it did not look English. We might have suddenly rolled into California.' What struck Priestley as particularly odd was 'the line of new factories on each side' of the road. 'Years of the West Riding', he explained (he was born and raised in Bradford), 'have fixed forever my idea of what a proper factory looks like: a grim, blackened rectangle with a tall chimney at one corner. These decorative little buildings, all glass and concrete and chromium

plate, seem to my barbaric mind to be merely playing at being factories.' Armed with a copy of the now celebrated textbook, *The British Isles: Geographic and Economic Survey* by Stamp and Beaver (1933), which was later to be used by generations of geography undergraduates (including the present writer), Priestley went on to make some astute geographical and social points about these factories: 'Actually, I know, they are tangible evidence, most cunningly arranged to take the eye, to prove that the new industries have moved south. You notice them decorating all the western borders of London. At night they look as exciting as Blackpool. But while these new industries look so much prettier than the old, which I remember only too well, they also look far less substantial. Potato crisps, scent, tooth pastes, bathing costumes, fire extinguishers; those are the concerns behind these pleasing facades' (Priestley 1937: 3–5).

In these few words, Priestley summarized some of the major ways in which the industrial world of the twentieth century was to differ from its nineteenth-century predecessor. The fact that he was making his journey by road was itself significant; 30 years before he would have had to go by rail. The new factories he observed were the products of the technological changes that had been transforming industrial capitalism since the late nineteenth century, and many had clearly developed to serve an expanding consumer market (see Plates 3.1 and 3.2). And the location of the factories by the new arterial highway leading westwards out of London was the result not only of a revolution in transport and communications but also of the locational freedom deriving from the availability of electricity

Plate 3.1 The Hoover factory, Perivale, west London. This splendid example of Art Deco architecture, designed by Gil Wallace in 1932, reflects the new consumer industries which were being established in the years after the First World War.

(© Angelo Hornak/CORBIS)

Low furniture does not hinder the Hoover. The handle can be swung down to the ground if necessary.

Plate 3.2 Using a vacuum cleaner or Hoover, one of the new consumer appliances which were becoming widely available by the 1930s. Photos and advertisements of the period almost always depicted women performing such domestic tasks.

(Mary Evans Picture Library)

and other fuels. The textile industries that Priestley remembered from his childhood in Bradford were tied to the coalfields; the newer industries that were appearing in London by the 1930s no longer needed coal, and were much cleaner and brighter in consequence.

Of course, what Priestley saw along the Great West Road was by no means representative of all the technological changes that had been affecting the industrial economies of Britain and other countries for the previous few decades. What had been happening in these countries was that a whole series of new industrial branches had been developing to supplement, and eventually to eclipse, the traditional activities based on coal, iron and textiles. Not all of these were as pleasing to the eye as those observed by Priestley. In the second half of the nineteenth century, for example, the metallurgical industries had been transformed as a result of a series of inventions allowing the production of cheap steel. Next came the rapid development and proliferation of different branches of the chemical industry (alkalis, dyestuffs, pharmaceuticals, explosives, lacquers, photographic plates and film, celluloid, artificial fibres, plastics). The electricity industry, which began to flourish by the end of the nineteenth century, was dependent on earlier inventions, like the steam turbine. About the same time came the rise of the motor industry, which was in turn associated with other industries like oil and rubber. When Priestley set out on his journey, society was already beginning to adjust to the impact of the many new activities catering to the consumer (most notably, domestic appliances – see Plate 3.2) and to new means of transportation (car, bus, aircraft). Of course the full impact of such developments was to come later, after the Second World War,

while some technologies, like regular TV broadcasting, the jet engine, nuclear power and the microchip, still lay in the future.

So profound were the technological and accompanying social changes that affected industrial capitalism from the late nineteenth century that some historians have described them as a 'second industrial revolution' (Landes 1969: 4). But it is important to remember that the older industries – coal, textiles, railways, some forms of engineering – did not die immediately or indeed quickly. One of the features of the changing industrial geography of the late nineteenth and early twentieth centuries was that countries like the USA and Germany, whose industrialization came later than Britain's, now began to forge ahead on the basis of the newer industries described above. Britain remained overdependent on the older and less dynamic branches (Figure 3.1).

For the first half of the twentieth century, the industrial changes described above only directly affected certain parts of the world, notably western Europe,

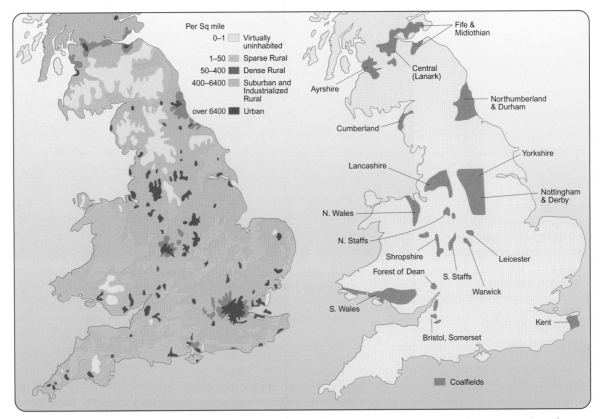

Figure 3.1 The geography of the UK population and coalfields compared. With the significant exception of London, there is a strong correlation between the geography of population and that of coalfields, reflecting both the early stage at which Britain industrialized and urbanized, and the country's long dependence on nineteenth-century industries.

Sources: population, based on Mitchell (1962); coalfields, Stamp and Beaver (1963: 286)

North America, Japan, and by the 1930s the Soviet Union and some other areas. Much of what was later to become known as the Third World, or the developing world, was still agricultural. Yet, in continuation of earlier processes (see Chapter 2, pp. 47–51, 57–8) many colonies and other regions were now being organized commercially to supply the industrial countries with raw materials and tropical products – for example, bananas and sugar from Central America and the Caribbean, Brazilian coffee, Indian tea, Malaysian rubber. They were thus being tied in to the capitalist world economy. Gradually certain of these countries began to adopt the technologies of the industrialized world – the rise of the Indian textile industry is one example – but only later in the twentieth century did industrialization become more widespread.

Thus the foundations of what is now known as a global economy were already being laid in Priestley's day, or even earlier. By the beginning of the twentieth century capitalism had become a world phenomenon, tying far-flung countries together by means of international trade and fostering international capital flows through major financial centres like London. The first multinational corporations were already appearing. All this was aided and abetted by the new systems of communication and transport – telegraph, telephone, radio (from the 1900s), fast steamships, aircraft – which were beginning to span the globe. Of course, none of this bears comparison with the forces of globalization which were to become so significant later in the century. Yet the world was already becoming a smaller place (see Plate 3.7). J.B. Priestley himself suggested this when he compared the Great West Road to 'California'. What might this mean to the average English reader in the 1930s? The answer is – a great deal. The Hollywood film industry was in its heyday and the people, homes and landscapes it portrayed were being viewed, and copied, the world over.

3.2 Fordist capitalism

It would be a mistake to suppose that the advance of capitalism in the twentieth century was a story of unmitigated triumph. On the contrary, its fluctuations and misfortunes were such that one historian felt constrained to call the period the 'Age of Extremes' (Hobsbawm 1995). The first half of the century – Hobsbawm's 'Age of Catastrophe' – was particularly disturbed, with two world wars (1914–18 and 1939–45) and a deep world economic depression (1929–33).

By contrast, the years between 1945 and 1973 were ones of growing prosperity across much of the globe (Hobsbawm's 'Golden Age') to be followed once more by a disturbed period in the wake of the shock oil price rises of the 1970s.

It was Karl Marx who originally emphasized the unplanned, competitive and even chaotic nature of capitalism's development. But one of the features of the twentieth century has been the attempt, by both government and private agency, to regulate and even to control it. The reasons for this phenomenon are many, but they are no doubt linked both to the precipitate nature of technological change during this period and to the severe fluctuations mentioned above. Attempts to 'organize' capitalism have taken a number of different forms. For example, already in the late nineteenth century, and especially in the USA and Germany, there were moves towards the formation of inter-firm agreements, cartels and larger companies and corporations. Large corporations could more easily marshal the huge capital resources which modern industry requires, and also influence their markets more effectively. As noted already, the first multinational corporations appeared at this time, those based in the USA being most notable. Nineteenth-century examples include the German electrical firm, Siemens, and the US Singer sewing machine company. Twentieth-century examples include US-based Hoover, Ford, Coca-Cola, Pepsi Cola, Nabisco (Shredded Wheat), and Kellogg's, all of which invested in Britain in the 1930s. Most of the major international oil firms also date from this period.

A further way in which capitalism became more 'organized' was the phenomenon of mass production, linked especially to the growing consumer market. Here two Americans are regarded as particularly significant. F.W. Taylor (1856–1915) is especially associated with time-and-motion studies, whereby complex tasks on the factory floor could be completed more efficiently and productivity increased. The other is Henry Ford (1863–1947) who organized car production in his Dearborn, Michigan plant using modern methods like the assembly line and interchangeable parts. The result of his centralized approach was a significant reduction in the cost of producing cars, which could now be manufactured on a mass basis. The ensuing mass production methods and associated patterns of mass consumption are frequently referred to as **Fordism**, hence the capitalism of this period is often referred to as *Fordist capitalism*.

Governments were also affected by the desire to 'organize' capitalism and to tackle the many problems

Plate 3.3 A suburban house built in the Modern style in the 1930s, Essex, England. The suburbanization of the inter-war period was a product of the affluence and increased mobility of some middle-class groups during this period. Suburbanization was to become even more prominent after the Second World War.

(© Gillian Darley; Edifice/CORBIS)

to which it seemed to give rise. At the international level, the USA, which emerged after the Second World War as the undisputed leader of world capitalism, took the lead in establishing a series of institutions like the World Bank, the International Monetary Fund (IMF) and the General Agreement on Tariffs and Trade (GATT) to ease international monetary payments, promote trade and encourage economic development. At the national level, numerous countries, especially some of the west European ones, pursued democratic agenda of various kinds, such as attempts to construct 'welfare states' to tackle such social problems as unemployment, ill health, old age and social inequity. Arguably the experience of government planning and controls in wartime (and possibly also a fear of communism – see below) helped pave the way for the optimistic belief in planning and large-scale social engineering that characterized the post-war period. In western Europe especially this was the era of bold experiments in new town and city development, slum clearance and ambitious social housing schemes, regional planning, and extensive controls over land use (Hall 2002). No doubt the success of these schemes was dependent on the spreading affluence that accompanied Hobsbawm's 'Golden Age'. Across western Europe and North America a tide of suburbanization signalled not only a growing ability to own one's own home in a desirable location but also the availability of the social and physical infrastructure, the private cars and the many new consumer products

which now made such a goal possible for many (see Plate 3.3).

The writers Scott Lash and John Urry have described the era ushered in by the methods of Ford and Taylor but reaching its apogee during the period 1945–73 as 'organized capitalism' (Lash and Urry 1987). Some of its principal features (which were especially characteristic of the developed world) are described in accordance with their views in Spotlight box 3.1.

Needless to say, such generalizations would be more or less true, depending on the time and location being considered. Capitalism had different histories in different places, and the exact form it took had much to do with the long-term evolution of each society affected by it.

3.3 Challenges to liberal capitalism: Nazism, communism

Come, bombs, and blow to smithereens,
Those air-conditioned, bright canteens,
Tinned fruit, tinned meat, tinned milk, tinned beans,
Tinned minds, tinned breath.
Mess up the mess they call a town –
A house for ninety-seven down
And once a week a half-a-crown
For twenty years

From 'Slough' by John Betjeman (1937)[1]

[1]Poem: 'Slough' from *Collected Poems* by John Betjeman © 1955, 1958, 1962, 1964, 1968, 1970, 1979, 1981, 1982, 2001. Reproduced by permission of John Murray (Publishers)

Spotlight box 3.1

Fordist or 'organized' capitalism

- Extractive and manufacturing industries are the dominant economic sectors.
- There is an accent on economies of scale, leading to the importance of large industrial plants. Such plants may structure entire regional economies around themselves. Examples might include (in the United Kingdom): the West Midlands, based on cars and engineering; Lancashire, based on cotton textiles; the North-East, based on mining, shipbuilding and heavy engineering; and (in the USA): Detroit, based on cars and engineering; and Philadelphia, based on textiles and port-related activities.
- Manufacturing plants are controlled centrally by big industrial corporations – there is an emphasis on mass production and standardization.

- There is state regulation of the economy to overcome problems generated by the market, for example regional unemployment problems.
- Big industrial cities are the spatial expression of large-scale manufacture.
- There is state-controlled welfare provision to even out social inequalities, address unemployment problems and raise health standards.
- In culture and social provision, there is an accent on mass provision, for example in housing, consumer goods, TV programming and newspapers. The emphasis on mass coverage and standardization leaves relatively little choice, reflecting a modernist perspective.

Source: after Lash and Urry (1987)

John Betjeman's famous fulminations against the town of Slough, situated just west of London and experiencing developments similar to those observed by J.B. Priestley a few years earlier, are in fact a hymn against modernity. Betjeman was railing against many of the social repercussions of the profound twentieth-century changes noted earlier in this chapter. In this he was by no means alone. Similar attitudes have been characteristic of many commentators living in countries where modernity (here defined as the spectrum of economic, social, political and cultural changes associated with twentieth-century capitalism) has brought problems of adaptation.

It has already been noted that the twentieth century cannot be described as a century of uninterrupted progress for capitalism, particularly in its Western 'liberal' form. Capitalism has been subjected to a series of challenges and political struggles that greatly affected the course of twentieth-century history, and in various parts of the world there have been attempts to create spaces in which alternatives to liberal capitalism can flourish. Interestingly enough, in terms of Wallerstein's world systems theory discussed in Chapter 2 (see p. 45) most of these attempts have been associated with countries outside the core, or with those like Germany after its defeat in the First World War struggling to rejoin the core states. Particularly important for the political geography of the twentieth century were the attempts

by Marxists and others to reject the capitalist development model entirely and to reconstruct society on a new basis. This issue will be discussed below. First, however, attention will be paid to another twentieth-century movement which challenged Western-style capitalism without discarding it entirely: Nazism, together with its close relative, Fascism.

Although they had nineteenth-century antecedents, both Fascism and Nazism were essentially products of the inter-war years. Fascism, under its leader Benito Mussolini, ruled in Italy from 1922 until its final defeat in the Second World War in 1945. Nazism under Adolf Hitler ruled Germany from 1933 until it too was defeated in 1945. Various Fascist or neo-Fascist groups ruled or were active elsewhere in Europe, and in some other regions, during this period, and to a lesser extent since.

There is no doubt that Nazism was by far the most influential Fascist movement after 1933, and so the following brief remarks will be devoted to it (Kershaw 2000). One of the problems of discussing Nazism or other forms of Fascism is their lack of a consistent ideology or philosophy. However, certain general points can be made. Like certain Western intellectuals and others, the Nazis were moved by a dislike of facets of capitalist modernity, such as commercialism, materialism, individualism, threats to the traditional family like the rise of female employment (see Plate 3.4),

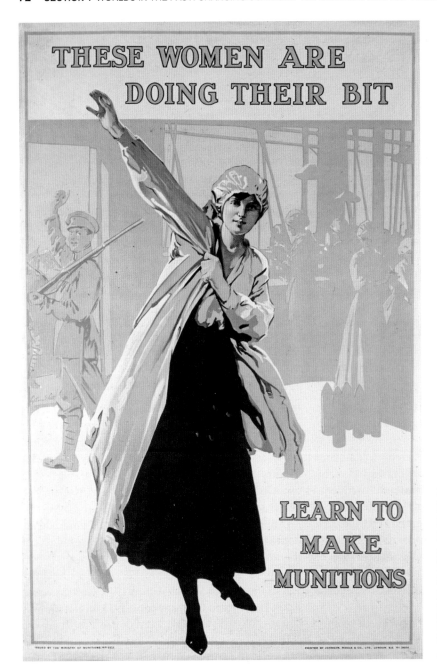

Plate 3.4 Women workers in armaments plants during the First World War. Although Nazis and some other political groups were later to oppose female employment, the labour shortage during the First World War gave women job opportunities which were increasingly taken as the twentieth century advanced.

(Hulton Archive/Stringer/Getty Images)

and similar tendencies that they associated with the 'decadent' Western democracies. They also despised Western-style parliamentary democracy, with its plurality of political parties and class divisions. In its place, they advocated the concept of a single national community, a *Volksgemeinschaft*, headed by a single Leader or Führer (Hitler), who was regarded as representative of, and chosen by, the people (this was the Nazi concept of 'democracy'). The Leader's power was absolute.

Such a creed, however, seemed exceedingly unlikely to come to power in Germany in the early 1930s had it not been for the extreme circumstances reigning there. One was a general sense of resentment at Germany's defeat in the First World War and subsequent national humiliations. As extreme nationalists, the Nazis promised to avenge this defeat. Another was the dire straits to which many of the middle class had been reduced by post-First World War inflation and the Depression that began in 1929. There was also the fear of the many strikes and disorders perpetrated by communists and other left-wing groups (as well as by the Nazis), that were, of course, encouraged by the selfsame economic difficulties.

One of the oddities of Nazism was that it was, at one and the same time, both reactionary and modernizing (Herf 1984). On the one hand, the Nazis looked backwards to an imagined heroic and rustic Germany of the past, to Nordic myths, Germanic towns and landscapes (see Rollins 1995; Hagen 2004) and happy peasants tilling the fields in traditional costume. They tried to bolster the 'traditional' family: women were to remain at home and raise children for the fatherland. There were even attempts to build villages and garden settlements to reflect such ideals. On the other hand, the Nazis were also modernizers who built the autobahns, fostered industry and spent vast sums on the military, organizing their capitalist economy to these ends. Their aim was to turn Germany into a superpower, able to dominate the European continent and regions beyond (see Chapter 20, pp. 427–30).

Nazism's central and most notorious feature was its racism. Building on common European assumptions of racial and cultural superiority (prejudices which have by no means disappeared, even today), they taught that the Germans and related Aryan **races** were equipped by nature to dominate the globe. All non-Aryan peoples were regarded as inferior, particularly the Jews who, because of their culture, religion and cosmopolitan ways, seemed to represent all that the Nazis feared and hated. As time went on, it became clear that the Nazis meant to exterminate the Jews (they killed six million of them) as well as others (gypsies, homosexuals, the insane, certain religious groups) who could have no place in the world they intended to reconstruct. The Nazi death camp at Auschwitz in Poland, which has been researched by geographers (Charlesworth *et al.* 2006), is kept as a memorial to the huge numbers who were murdered and as a solemn warning to today's world of the horrific consequences of racial prejudice.

Perhaps the most significant challenge to liberal capitalism in the twentieth century, however, came from Marxist-style communism (Calvocoressi 1991). By the 1960s and 1970s up to one-third of humanity was living under communist governments which explicitly rejected capitalism as an acceptable way of organizing society. The reasons for that rejection and why it largely failed must now be considered.

An outline of some of the principal features of Marxism is given in Spotlight box 3.2. An important point is that Marx's teachings failed to change those societies at which they were initially aimed – the industrial societies of Western Europe and North America. Marx himself had expected that communism would find support among the growing industrial working classes of countries like Germany and Britain, where factories were bringing such groups together in increasing numbers. However, it was by no means obvious, as the years passed, that the workers of those countries were necessarily being increasingly exploited, as Marx seems to have expected (later, these countries were accused by Marxists of exporting exploitation to colonies and

Spotlight box 3.2

Marxism

- Marxism, which derives from the teachings of Karl Marx (1818–83), is related to other forms of socialism which seek to moderate or reform the injustices of capitalism.
- Unlike some other forms of socialism, Marxism regards capitalism as an innately unjust and exploitative system.
- According to Marxism, capitalism divides society into antagonistic classes: those who own the main sources of wealth (the capitalists) and those who must live by selling their labour to the capitalists (the proletariat).

- Marxism teaches that the capitalists are forced by the very nature of capitalism to maximize the profits they obtain at the expense of the workers.
- Marx thought that eventually capitalism would become so exploitative and prone to crises that its downfall was inevitable.
- It would then be replaced by a much more just, classless society (socialism, gradually maturing into communism).
- In the meantime Marx exhorted the world's workers (especially the industrial workers) to organize politically to hasten capitalism's downfall.

other less developed regions). Instead, Marxism triumphed in Russia (in 1917), in what was in fact the least industrialized of Europe's great powers. Thus, whereas in terms of Wallerstein's world systems theory Marxism was expected to find favour in the core countries, in fact it initially triumphed in a semi-peripheral one. That it did so changed the character of Marxism, which was now faced with the challenge of building socialism in a peasant society, and in virtual isolation from the rest of the world.

What happened after 1917 in Russia (or the Soviet Union as it was now to be called) was of profound importance for the other countries which later adopted communist systems, if only because Russia was the pioneer. What happened there began to assume something of the character of orthodoxy (Hosking 1992). In view of the difficulties they faced, and in all likelihood because of their own inclinations, the Bolsheviks (as the Russian or Soviet communists were called) adopted a highly centralized political system that brooked no opposition and entailed the destruction of the previous ruling and capitalist class. Eventually, from the late 1920s, they implemented a fully centrally planned economic system. This involved the abolition of private enterprise and virtually all forms of market relations, and the collectivization of agriculture. The purpose of this extraordinary economic system was both to speed the process of economic development and to build up the country's military resources. As the 1930s advanced, it became clear that what had earlier seemed a rather vague threat from the outside capitalist world was beginning to take a concrete and menacing form in the guise of Nazi Germany. The centrally planned or command economy (which involved much suffering on the part of the Soviet people) eventually proved equal to this challenge. In the ensuing war with Germany (1941–5), the Soviet Union emerged victorious, but only after sustaining enormous losses.

Soviet victory in the Second World War greatly enhanced that country's prestige. Moreover, because the Soviet armies were now in occupation of much of central and eastern Europe, they were able to ensure that regimes friendly to the Soviet Union (that is, communist regimes) would assume power in those regions. Communism soon spread into other countries, notably China (in 1949), south-east Asia and beyond: in other words, into Wallerstein's 'periphery'. All these countries initially followed

Plate 3.5 Central Moscow, showing one of the open spaces created or enhanced by the Soviet dictator Joseph Stalin for official communist demonstrations and ceremonial display.

(Neil Emmerson/Robert Harding)

the Soviet development model, but soon found that it was necessary to adapt it to their own needs. In the meantime the spread of communism, and Soviet ambitions, excited the suspicions of the capitalist West. From the 1950s, therefore, the world was split into two armed camps, both equipped with nuclear weapons. The ensuing confrontation, known as the **Cold War**, profoundly influenced both sides and encouraged their militarization (see Chapter 20, pp. 427–36). However, many countries, especially in the developing world, tried to avoid taking sides, while China, though communist, began to pursue its own version of communism outside the Soviet sphere.

While the actual form that communism took in the Soviet Union and other countries may have been a modification of Marx's own ideas, it did represent a radical departure from the capitalist development model. Not only was the command economy a very different, state-centred approach to economic development (and one which was copied in many parts of the developing world after the Second World War, with varying degrees of success), it regarded itself, and was regarded, as a threat to the whole idea of capitalism (Nove 1987). Internally, whilst it led to attempts to reconstruct society on a different basis, it did in fact give rise to new forms of spatial inequality (Bater 1986). Interestingly enough it also had certain spatial features in common with Nazism and Fascism, such as an emphasis on creating urban spaces specifically for the purpose of mass ceremonial and display (Plate 3.5). The creation of 'spaces of terror', such as concentration camps for the incarceration or elimination of those deemed unacceptable to the regime, was also a feature of the two systems (Moran 2004; Pallot 2005).

In the end, communism failed to prove itself a successful, viable challenger to capitalism. Especially from the 1970s, the Soviet Union and its eastern European allies fell behind their capitalist rivals in terms of productivity, flexibility and innovation. Whether this was because of problems inherent to command economies as such, or whether it has more to do with mistakes made by the various political leaderships, is hard to say. Whatever the reasons, by the end of the 1980s practically the whole of the communist world was in a state of economic and political crisis. The subsequent fall of communism in eastern Europe, and the splitting of the Soviet Union into 15 separate states in 1991, signalled the end of the Cold War. Since then the post-communist states, plus China which continues to pursue its own version of socialism 'with Chinese characteristics',

Plate 3.6 The fall of the Berlin wall in 1989 signalled the end of communism in the Soviet bloc and the removal of a major twentieth-century challenge to world capitalism.

(© Reuters/Corbis)

have been struggling to adapt to the market economy and in other ways to cope with the problems of post-communist transformation (Gwynne, Klak and Shaw 2003: 59–72, 101–8).

3.4 The end of imperialism?

Father, Mother, and Me
Sister and Auntie say
All the people like us are We,
And everyone else is They.
And They live over the sea,
While We live over the way,
But – would you believe it? – They look upon We
As only a sort of They! –

From 'We and They' by Rudyard Kipling (1912: 763–4)[2]

---

[2]Poem: 'We and They' by Rudyard Kipling, with permission of A.P. Watt Ltd on behalf of Gráinne Yeats

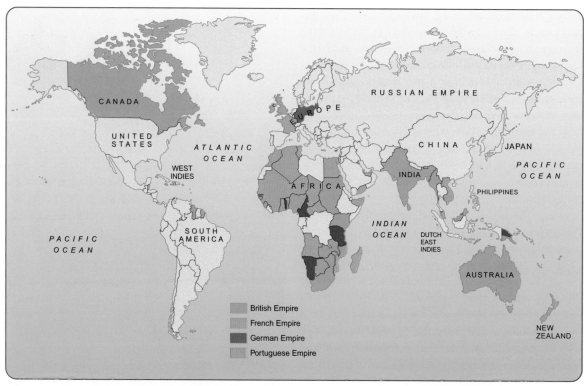

Figure 3.2 The world in 1914 showing British, French, German and Portuguese empires. Note, too, the Russian empire, Dutch East Indies and other colonial territories under Belgian, Italian and Spanish rule.

At the beginning of the twentieth century Britain and several other European powers sat proudly at the centre of a series of empires that spanned the globe (Figure 3.2). As noted in Chapter 2, these empires were the products of a long period of European exploration, settlement, economic exploitation and imperial rivalry. Something of the complacency and condescension with which Europeans commonly regarded their empires at this period is nicely captured by Rudyard Kipling in his comic poem 'We and They', quoted above.

However, the future of European imperialism was already being questioned even before the First World War. The English liberal J.A. Hobson, and later the Russian revolutionary V.I. Lenin, popularized the idea that Europe's overseas colonies were being economically organized and exploited mainly for the benefit of the European 'mother countries', forming an undeveloped periphery to the European core. Lenin taught that imperialism was an inevitable consequence of capitalism – its 'highest stage'. In the meantime various rumblings of discontent were being felt in various parts of the European empires.

But it was the three great episodes of the first half of the twentieth century – the First World War, the Great Depression (1929–33) and the Second World War – that fatally undermined European imperialism. Nationalism, which had had such an impact on the political geography of Europe (see Chapter 22), had also influenced the colonial world where 'national liberation movements' began to demand independence for their countries. Starting with the independence of India in 1947, the next quarter of a century witnessed the break-up of all the European empires (though the Soviet Union, successor to the old Russian empire, finally disappeared only in 1991). Many new, independent states appeared on the map of Africa, Asia and other regions, though not, unfortunately, without considerable turmoil in some cases. The world's political geography was transformed.

In giving (or being forced to give) independence to their colonies, the former imperial powers hoped that they would adopt European-type political systems and capitalist economic systems, partly because these appeared the best basis for future development, and partly because they seemed a reasonable way of upholding European influence. The elites who were now to hold power in the new states were often sympathetic to these aims, since they had frequently been educated by Europeans and wished to see their countries modernized along European or Western lines. However, the adoption of Western models was not always a suitable response to their problems. For example, many of the new states attempted to copy the European idea of the nation-state, hoping to unite their peoples around a common sense of national identity. But the old colonial boundaries had generally been drawn up to suit imperial convenience rather than that of local communities. Thus these boundaries, now the boundaries of independent states, frequently grouped peoples into one state who had no common culture or history while dividing others who did so. This contributed little to the political stability or unity of many new states. Another problem was the feeling among many citizens of the new states that the capitalist economy was responsible for the under-development of their countries. This set off a search for socialist or communist alternatives, much to the annoyance of the West, and encouraged debates about the

meaning of 'development' (see Chapter 8, pp. 184–7). Finally, fundamental questions were frequently posed about how far modernization, as generally understood, was compatible with the traditions of the former colonial peoples. The Islamic revival, discussed below, can be seen as one response to this dilemma.

Just as imperialism had a profound impact on the political and social geographies of enormous areas in Africa, Asia and the Americas, it equally affected the imperial countries themselves. One of the most important manifestations of this in the twentieth century was the flow of migrants from colonies and former colonies, especially in the tropics and sub-tropics, to take up jobs in the former imperial states. This naturally had a far-reaching cultural impact in cities and regions in Western Europe. In fact the whole experience of imperialism led to the mixing of peoples and cultures on a grand scale. It also led to widespread questioning of long-held assumptions about European (and often male) cultural superiority (see Chapter 2, pp. 50–5).

Many scholars have argued that the end of European colonialism did not mean the end of exploitative relationships between the core countries of the world economy (including the former imperial powers) and what was now increasingly referred to as the 'Third World'. According to such thinkers, the formal imperialism of the colonial era had merely been replaced by a more 'informal' alternative, but the basic situation of the core exercising hegemony over the periphery had not really changed (Frank 1969; Wallerstein 1980). From the 1970s, however, some fundamental changes seemed to affect the world economy which, in the opinion of certain scholars, demanded that international relationships be viewed in a new way. Lash and Urry have described these changes as 'the end of organized capitalism' (Lash and Urry 1987).

3.5 Globalized capitalism

In 1960 the industrialized areas of Western Europe and North America produced almost 80 per cent of the world's industrial output. Even Japan accounted for only around 4 per cent. Much of the Third World remained agricultural. Only after this time did industrialization spread beyond its traditional centres (which since the 1930s had included the Soviet Union). Meanwhile many of the older industrial countries began to lose industries, even some of those which had arrived with the twentieth century.

Plate 3.7 Imperial Airways flies to Australia. Imperial Airways was the major British airline company which operated at ever more ambitious international and intercontinental scales during the 1930s.

(Vintage Ad Gallery)

Of course, some areas of the 'Third World' benefited far more from this industrial spread than did others. Most spectacularly, the newly industrialized countries of east Asia, Brazil, Mexico and certain others soon seemed set to join the industrial core. Yet others, like certain Middle Eastern states, earned huge revenues from their energy exports. But there remained many areas which missed out on the new developments (even so the latter part of the twentieth century was a period of unprecedented population growth and urbanization across much of the developing world – see Chapters 4 and 9). The term 'Third World', used to group together countries with such disparate economic characters, seemed increasingly redundant, and the world as a whole seemed as unequal as ever.

How is one to explain the changes affecting the world's economic geography in the last third of the twentieth century and at the beginning of the twenty-first? Geographers and others argued that such changes are part of the process of globalization. Speedier communications meant that the world was becoming a much smaller place as the twentieth century drew towards its close. Capitalism itself was now a truly global phenomenon as markets were internationalized and finance became fully mobile. Before 1960, despite the importance of international trade, the world economy was structured around individual states. After 1960, the world economy became in effect transnational as the boundaries of individual states became ever less important to its functioning. Thus this period witnessed the rise to global importance of the transnational corporations, commercial conglomerates which became major players on a world scale (see Chapters 14 and 16). Because of the wealth and political influence they wielded, such huge companies became increasingly free to switch their operations from country to country as economic circumstances dictated. States, which had previously seemed unchallenged within their own frontiers, found it ever more difficult to control their own economies and began to bid against one another to attract foot-loose investment and the favours of the transnational corporations. Many industries and services now began to locate in parts of the Third World, where costs were cheaper, whilst core industrial countries began to experience deindustrialization and a switch into services and 'control' functions (the headquarters of the transnational corporations still tended to be located in the traditional core countries). A further important result of the development of information technology was that production became much more flexible than before and more geared up to highly specialized markets, changing fashion and the whim of the individual consumer. This new, more flexible approach to production is sometimes known as **post-Fordism**.

Later chapters of this book will explore some of the specific economic and social implications of these changes. Meanwhile, it is worth stressing here how unsettling such developments have been, especially for the core countries. The years of growing affluence after 1945 were succeeded, from the mid-1970s, by a period of greater uncertainty as deindustrialization set in (see Plate 3.8), unemployment rose, insecurity became more widespread, and social inequality became more apparent. By the century's end it seemed that Western-style capitalism and democracy were increasingly threatened and if anything the feeling of threat and uncertainty has grown into the new century. The last part of this chapter will briefly survey some of the factors which lie behind that uncertainty.

Just as Lash and Urry used the term 'organized capitalism' to describe the years down to about 1973 when Fordism reached its apogee, so they have described the last years of the twentieth century as those of 'disorganized capitalism'. Some of its more prominent features are listed in Spotlight box 3.3, once again paying particular attention to how the changes have affected the core countries of the world economy.

3.6 The world in the early twenty-first century

In 1992, some three years after the fall of the Berlin Wall, an event which is regarded as marking the end of the Cold War (see Plate 3.6), the American political scientist Francis Fukuyama published a book entitled *The End of History and the Last Man* (Fukuyama 1992). Here Fukuyama made some startling claims regarding what he saw as the recent global triumph of 'liberal democracy'. The end of the Cold War, Fukuyama asserted, meant that liberal democracy had now 'conquered rival ideologies like hereditary monarchy, fascism and . . . communism' and that this triumph marked 'the end point of mankind's ideological evolution'. He went on to announce that 'while earlier forms of government were characterized by grave defects and irrationalities that led to their eventual collapse, liberal democracy [is] arguably free from such internal contradictions'. Hence 'the ideal of liberal democracy could not be improved on'. A close reading of Fukuyama's text reveals that what he had in mind by 'liberal democracy' was not only a specific political system but

Spotlight box 3.3

Globalized or 'disorganized' capitalism

- The onset of 'disorganized capitalism' is marked by a decline in the relative importance of extractive and manufacturing industries.
- There is a relative increase in the importance of service and consumer industries, especially in employment.
- The use of flexible technologies encourages a reduction in the average size of manufacturing plants with more accent on labour-saving investments and more flexible employment processes, all induced by competition.
- Because of the need for flexibility and cost-cutting, industrial firms tend to 'hive off' many of the services and supporting activities they need to other firms and organizations. There are thus more opportunities for small firms, changing the traditionally specialized nature of the regional economy.
- Regional economies are also affected by the greater emphasis on non-standardized production – traditional regional specializations become less marked.

- The global economy reduces the effectiveness of state attempts at economic regulation – from the state's point of view, the economy becomes less predictable.
- Rising costs, demands for reduced taxation, and growing social inequality challenge the idea of a centralized welfare state.
- Smaller, more footloose industries, the rise of services, better communications and other factors reduce the traditional importance of big, industrial cities by comparison with small towns and rural areas.
- There is a rise in importance of the educated social strata needed to work in the new administrative, control, service and related activities – the so-called 'service class' – with more sophisticated and individualised tastes in consumption and other areas. The age of mass cultural provision is replaced by greater cultural fragmentation and pluralism (sometimes referred to as post-modernism). There is a commensurate decline in faith in large-scale planning and similar activities associated with modernism.

Source: after Lash and Urry (1987)

also a way of organizing the economy, namely through 'free markets'. Just how 'free' markets were meant to be he did not say. But in the opinion of many, what Fukuyama was celebrating at the dawn of the new millennium was in effect the final triumph of Western-style capitalism, and of the Western way of life, over the rest of the world – an awesome claim indeed.

Now in the second decade of the new millennium, and 20 years since Fukuyama published his book, it is probably true to say that relatively few people are as optimistic about the state of the world as he evidently was. In fact the future seems as uncertain as ever and 'liberal democracy' (whatever that might mean in practice) is faced with new challenges, some of which were perhaps only dimly discernible when Fukuyama wrote. Understanding these challenges forms the substance of the chapters which follow. Here we can only summarize the most significant among them.

3.6.1 Economic challenges

Fukuyama published his book at a time when most Western states, and many others as well, had strongly espoused the economic doctrine known as **neoliberalism**. This was a reaction to the disappointing economic record of the 1960s and 1970s and promised, through freeing up markets, privatizing as much of the economy and associated activity as possible, and 'rolling back the state', to achieve high and sustainable economic growth (see Peck 2010). In the event the actual record of economic achievement was much more modest than had been hoped for (Harvey 2006: 30, 42; Dicken 2011: 525) and there was much volatility on the world market. Furthermore, the removal or absence of regulation over much economic activity seemed to encourage financial speculation and even corruption in many places, culminating in a major international banking crisis in 2008 and precipitating economic recession and a crisis in the Eurozone in 2011. As a result many countries found themselves deeply in debt and were forced to make major cuts in public spending which especially affected the poorer elements in society. Whilst, as we have seen, this was not the first occasion that Western-style capitalism has been seized with crisis, it has caused many to question whether the way in which the world economy has been run over the recent period has been sensible

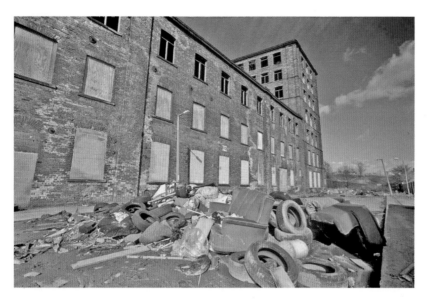

Plate 3.8 Deindustrialization in the 1980s afflicted many parts of the world in the wake of the 1970s' oil price shocks and of major changes in the world economy.

(Paul Ridsdale/Photofusion)

and whether some alternative policy should now be adopted, particularly as some countries like China have pursued rather different (and more state-centred) economic policies and done particularly well. Indeed, some economic analysts have suggested that the West can no longer assume economic dominance over the rest of the world with rising economic powers like China and India now entering the fray and that its days as the centre of the global economy may be numbered. Some of these issues are discussed further in Chapters 14 and 16.

There is an additional reason why many people have begun to question the wisdom of a neoliberal approach to economic policy and this is the growing inequalities that it seems to engender. For example, many countries that have adopted neoliberal economic and social policies have experienced the emergence of huge gaps between rich and poor. Thus, at one end of the scale, many now enjoy previously unimaginable income levels, which benefit among other things from low levels of taxation and which allow them to escape the consequences of the reductions in public expenditure which are the hallmark of neoliberalism. In the UK and other countries the public concern expressed at the levels of payments and bonuses being made to bankers and others whose activities are widely regarded as having precipitated the 2008 banking crisis (which resulted in huge investments of public money to rescue the faltering banks) is symptomatic of a broader dissatisfaction. And, at the other end, many of the poor, who are most dependent on public welfare and public expenditure of various kinds (whose reduction is partly associated with the neoliberal

policy of lowering taxes) suffer deprivation as well as the unemployment and other problems associated with a volatile economy (see Chapter 10). Many indeed wonder whether such levels of inequality are compatible with the maintenance of free and democratic polities. Meanwhile, at an international scale, many scholars regard neoliberal attitudes as being partly responsible for the growing economic inequalities and other disparities between countries in different parts of the world (Stiglitz and Charlton 2005; see Chapter 8). According to such scholars, the factors encouraging this include the inappropriate ways in which the rules governing international trade have been applied to developing countries, and the hypocritical way in which many Western countries, having espoused free trade principles as applied to other countries, actively seek to avoid their full application in the case of their own. In consequence, whilst Western countries pump huge amounts of aid into poorer countries, this is more than counteracted by the trade policies they pursue.

3.6.2 Geopolitical challenges

At the time Fukuyama published his book it seemed safe to assume that, with the collapse of communism and the Soviet Union, the United States would be left as the only remaining superpower, well placed to dominate the rest of the globe. Thus there dawned the vision of a world organized in accordance with a *Pax Americana*, whereby the benefits of American democracy and of the Western way of life more generally might spread to the rest of the globe. Again, however, in the second decade of the twenty-first century, this vision now seems too simplistic, for a number of reasons. One, mentioned above, has been the phenomenal rise of China and of other parts of East Asia in the recent period (Dicken 2011). Economic power, of course, invariably means political power, and in this situation the United States' geopolitical dominance is no longer assured (see Chapters 20 and 23), especially since China is now the world's second biggest economy. A symptom of the changing situation lies in the growing competition and disagreement between the USA and China in such areas as trade relations, currency policy and industrial policy as well as disputes over such matters as human rights and public access to the internet. Thus the blithe assumption that the USA can order the world according to its wishes and that the rest of the world will gladly adopt American values is questionable. This is quite apart from the problem of the numerous poor and peripheral states across the world (sometimes referred to as 'failed states') which neither the USA nor China nor anyone else seems able to control.

Given the rising significance of China and other parts of Asia in the recent period, it has become fashionable among certain scholars to attempt to explain this phenomenon from a long-term, historical perspective (see, for example, Frank 1998; Darwin 2007). Andre Gunder Frank's book, *ReOrient* (Frank 1998), is a case in point. Contrary to the ideas of Wallenstein and others (see Chapter 2, p. 45), his thesis is that by the year 1500 there was already a single global world economy in existence centred on Asia whose roots went back many thousands of years. In this global system Europe was both relatively and absolutely marginal and continued to be so until about 1800 when it assumed dominance over Asia, partly because of its successful conquest of the Americas. Thus, according to Frank, there is nothing particularly surprising about Asia's recent resurgence. Hence Frank is challenging the rather Eurocentric perspective adopted in the first three chapters of this book as well, it might be said, as the perspective adopted by the overwhelming majority of Western historians and other scholars.

3.6.3 Environmental challenges

In the opinion of many people the most urgent problem facing the world today is that of environmental change, and in particular climate change. As we saw in Chapter 1, there is nothing new about the idea of human beings changing the environments in which they live. Indeed they have probably done so ever since they first appeared on earth. But what has become particularly worrying in the most recent period is the notion that such changes are now becoming unsustainable and are beginning to threaten not only our way of life but also, perhaps, our very existence. Concern about such matters goes back many years, to the nineteenth century at least, but became more pressing from about the 1960s when some of the environmental consequences of continuous economic growth became more noticeable. Thus concern began to be expressed about the possibility of the world running out of its most significant resources, especially energy resources. This worry has hardly diminished through time with the search for oil and gas now impinging on the Polar and sub-Polar regions, some of the most ecologically fragile regions on earth. Likewise there is concern over the loss of biodiversity, associated among other things with the pressure to intensify agriculture and to expand food production to feed the world's rapidly growing population (itself a problem of the utmost importance – see Chapters 4 and 15). Needless to say, the impoverishment of the web of life, or the biosphere as it is often called, which sustains life on earth could have catastrophic consequences for everyone. The issue of climate change has come on to the agenda more recently as data have become available showing that the continuous emission of carbon dioxide and other greenhouse gases into the atmosphere, which is a hallmark of industrial societies, may already have led to marked (and in some opinions unstoppable) climate change. Hurricanes, bush fires, floods, droughts, heat waves are, in the opinion of some, symptoms of climate change. In the more extreme scenarios, the complete melting of the polar icecaps, the flooding of much of the earth's low-lying (and most densely populated) regions, and catastrophic changes elsewhere, are predicted.

There is no need to emphasize the point that the more dramatic predictions regarding the dire consequences of continuous and unsustainable environmental change threaten the entire basis of our modern way of life. Either we must change our whole mode of life, it seems, or hope, through further technological development for example, to solve some of the most pressing problems. But even then it is difficult to envisage the model of continuous economic growth in the way we understand it now being sustainable in the longer term, at least on this planet. Not surprisingly, given the rhetoric surrounding environmental issues (and, perhaps, the vested interests who might be threatened if the more dramatic scenarios are taken seriously), not a few people discount such claims, questioning some of the evidence supporting climate change and other predictions. The whole question is therefore highly controversial. Some of the ramifications of the environmental debates, and their political implications, are explored in Chapters 4–7.

3.6.4 Value and identity issues

On the first anniversary of the 9/11 attack on the World Trade Center in New York, US President George W. Bush made the following declaration:

> We will use our position of unparalleled strength and influence to build an atmosphere of international order and openness in which progress and liberty can flourish in many nations. A peaceful world of growing freedom serves American long-term interests, reflects American ideals and unites America's allies . . . We seek a just peace where repression, resentment and poverty are replaced with the hope of democracy, development, free markets and free trade . . .
>
> (quoted in Harvey 2006: 11).

In this declaration, President Bush clearly implied that such 'ideals' are not only American but also need adopting by the rest of the world.

But what exactly is 'freedom' in a world in which, for example, the USA has felt able to imprison so-called 'terrorists' in its Guatanamo Bay prison camp without trial and allegedly even used torture on them? Or what is 'democracy' in a world where the USA and its allies back numerous repressive regimes in their own security interests, or where politicians, having made certain promises to their electorates, appear to feel free to do the exact opposite once they are elected as well as engaging in corrupt practices in certain cases? No wonder that such vaunted 'values' are being questioned in many quarters rather than simply being accepted as presented.

In fact the whole area of values, what they mean and whose interests they serve, seems to be an increasingly contested one. This is linked to the issue of identity. Perhaps linked to feelings of social injustice and/or loss of certainty and personal autonomy in an increasingly globalized world, more and more people appear to feel the need to assert their distinctiveness and their freedom either as individuals or groups. 'New movements' like feminism, gay rights and animal rights are a case in point. Another is the recent rise of fundamentalist religious movements among many established faiths, whether it be among Christians, Muslims, Hindus

or whatever (sometimes, it must be said, in reaction against some of the 'new movements' mentioned above, and sometimes against the materialism and secularism which are such a feature of modern life). Rising nationalism also seems to be a problem in many parts of the world. Thus whereas Fukuyama airily assumed that, in western Europe at least, nationalism was being 'tamed', the demands for regional or ethnic autonomy or recognition seem by no means to be abating. In January 2011, for example, the upper house of the Spanish parliament agreed that senators could debate in any one of five languages even though they all speak Castilian Spanish perfectly well – and that at considerable public expense for simultaneous translation facilities. Elsewhere in the world, nationalism has often been even more of an issue.

Unfortunately the debates over issues of values and identity have sometimes given rise to disorder and even outright violence. Some of these points are explored further in Chapters 13 and 22.

3.6.5 Security issues

The explosion of a terrorist bomb at Moscow's Domodedovo airport in January 2011, killing more than 30 people, is a horrific reminder of a problem which has affected numerous nations over recent years. The Moscow bomb has been linked to the North Caucasus where a war for secession has been going on since 1994. The Chechen fighters, who have spearheaded this bid for independence, are clearly nationalists. The Russian authorities also emphasize their Muslim credentials, thus pointing to their links with terrorist groups fighting in Afghanistan and active elsewhere in the world. The fact is of course the Islamic revival, which is associated with the desire to return to traditional Muslim values on the part of many who reject Western values based on secularism and materialism (and thus Western-style capitalism), is a far more broadly based movement than this misleading attempt to link it with terrorism implies (Park 1994; Esposito 1995). In the opinion of the majority of scholars, Islam and the adoption of terrorist tactics are incompatible.

Clearly, however, the dangers and apparent frequency of terrorist acts are a sharp reminder of the limits to the power of states in an increasingly globalized world. They are a challenge to the democratic freedoms enjoyed in many states (despite the caveats about the meaning of the term 'democracy' outlined above), not least because they encourage governments to adopt more and more stringent security measures.

But terrorism in the above sense is by no means the only security threat to menace states and communities as the twenty-first century advances. One might mention the spread of weapons of mass destruction, and the possibility of cyber-terrorism, as two which are particularly worrying (see Chapter 20). Many states and citizens are also concerned about the issue of international migration as a security concern. Encouraged by such events as wars and disturbances, ecological catastrophes and economic hardship, such movements are often most unfortunately targeted by right-wing political elements eager to warn of the danger of national cultures being 'swamped' and of the need to defend national identity.

3.7 Conclusion

The twentieth century was a period of rapid economic and social change over most parts of the world. It was also a period when the world seemed to become smaller and most regions were gradually drawn into an ever more embracing global system. Yet it would be a mistake to imagine that this was an uncontested process, or one that threatens to bring about a global uniformity. The legacy of the twentieth century for the twenty-first is a world that is both dynamic and uneven, and therefore very uncertain. The rest of this book tries to grapple with this uncertainty.

Learning outcomes

Having read this chapter, you should know that:

■ Capitalism is inherently dynamic and unstable. It was so throughout the twentieth century, and is likely to continue to be so in the future.

■ The concepts of Fordist or 'organized' and Globalised or 'disorganized' capitalism are ways of trying to make sense of the changes which affected capitalist societies during the twentieth century.

■ 'Fordism' and 'post-Fordism' (concerning production and consumption) and 'modernism' and 'post-modernism' (concerning culture) similarly try to make sense of twentieth and twenty-first century change. No concepts, however, can do justice to the complexity of change during this period.

■ Western-style liberal capitalism is only one variant of capitalism. It has been challenged in various ways in the twentieth century, some of which have had long-term consequences for different parts of the globe. Future challenges might prove more successful than past ones.

■ The twentieth century has been an era of nation-states and of nationalism. Towards the end of the twentieth and into the twenty-first centuries the role of the nation-state seems increasingly challenged by globalization. Nationalism, however, may yet flourish as a response to globalization.

■ Modernity may not lead to the disappearance of traditional cultural practices. The Islamic revival,

and similar religious revivals across the world, are a case in point.

■ Despite the view of some that globalization is leading to the emergence of a global culture, it might actually increase the differences between places and hence the importance of geography.

Further reading

Dicken, P. (2011) *Global Shift: Mapping the Changing Contours of the World Economy*, 6th edition, Sage, London. A comprehensive overview of developments in the global economy up to the present time. Chapter 2 considers the longer-term development of that economy with particular emphasis on the period since 1960. It is therefore particularly relevant to the second half of the present chapter.

Frank, A. G. (1998) *ReOrient: Global Economy in the Asian Age*, University of California Press, Berkeley, CA. A rather idiosyncratic book which seeks to challenge the Eurocentric assumption of many scholars that the global economy (and other signifiers of modernity, including 'capitalism') had their origins in Europe and then spread to the rest of the world. Frank argues that the rise of Asia preceded that of Europe with the latter only becoming predominant after about 1800. Thus the recent rise of Asia is only a return to its former preeminence.

Gwynne, R.N, Klak, T. and Shaw, D.J.B. (2003) *Alternative Capitalisms: Geographies of Emerging Regions*, Arnold, London. Examines the effects of globalization and recent economic and political transformations in the world's 'emerging regions', with particular reference to Latin America and the Caribbean, East Central Europe and the former Soviet Union, and East Asia.

Hall, P. (2002) *Cities of Tomorrow*, 3rd edition, Blackwell, Oxford. A splendid account of urban development in the late nineteenth and twentieth centuries in different parts of the world. The accent is on urban planning and design, but there are many social insights.

Hobsbawm, E. (1995) *Age of Extremes: The Short Twentieth Century 1914–1991*, Abacus, London. What quickly became a classic account of twentieth-century history, written by a doyen of British Marxist historians. Any chapter is worth reading, but geography students will find those dealing with economic, social and cultural change especially revealing.

Useful websites

www.si.edu   The Smithsonian Institution, Washington, DC. The Smithsonian is a focus for many kinds of scientific and cultural endeavour in the United States. The website is a very useful source for twentieth-century history and developments, with particular emphasis on the United States.

www.besthistorysites.net   Entitled 'Best of History Websites', a comprehensive guide to history-oriented resources online. For teachers, students and others.

www.stanford.edu/group/spatialhistory/cgi-bin/site/index. php Website of the Spatial History Project at Stanford University, USA. Examines some of the issues raised in this chapter, including that of Holocaust Geographies.

For annotated, clickable weblinks and useful tutorials full of practical advice on how to improve your study skills, visit this book's website at **www.pearsoned.co.uk/daniels**

POPULATION, RESOURCES, FOOD, THE ENVIRONMENT AND DEVELOPMENT

Section 2

Edited by Michael Bradshaw

Writing in 2009, John Beddington, the Chief Scientific Adviser to the UK government, noted that by 2030 the world would need to produce 50 per cent more food and energy, together with 30 per cent more fresh water, while at the same time mitigating and adapting to climate change. He warned that this combination of challenges threatened to create a 'perfect storm' of global events. As Lester Brown (2011: ix) notes in the preface to his book *World on the Edge,* Jonathon Porritt, then Chair of the UK Sustainable Communities Development Commission, which has since been axed by the UK's Coalition Government, endorsed Beddington's analysis, but suggested that the storm was more likely to strike closer to 2020 than 2030 (Porritt 2009). Recent events might suggest that Porritt is right: the summer of 2010 brought widespread drought and in mid-2011 we face high energy costs, rising food and raw material costs and, in the aftermath of a global economic crisis, we seem to be making limited progress on addressing the challenge of climate change. While this section does not deal explicitly with the issues of climate change and water

security, which are normally covered elsewhere in a geography curriculum, it does deal with the issues of population, food, energy, environment and development. Thus, it provides an essential background to understanding many of the issues that contribute to the 'perfect storm' scenario.

This section makes clear both the scale and the complexity of delivering the more sustainable approach to development needed in the face of storm warnings. The chapters highlight the diversity of problems that exist at the global, interstate and local level. In this era of heightened globalization, developments in one locale, region or state can have global impact. For example, the so-called 'developed world' has achieved high levels of consumption, but in doing so has exhausted indigenous sources of energy, raw material and food supply, making consumers increasingly reliant on global supply chains. The same regions now face high production costs, a shortage of skilled labour and an increasingly elderly and dependent population. Elsewhere in the majority of the world there are regions and

→ states that suffer from an absence of economic development and an inability to compete in the global marketplace for much needed resources: a problem often compounded by the fact that they have to export food and natural resources to generate income to support the trappings of statehood. All too often, local political elites misappropriate the national surplus to support their own livings standards and safeguard their control over power. Events in the Arab world in the spring of 2011 reflect in part the peoples' frustration with this situation. At the same time, despite the recent global economic crisis, the continuing rapid growth of new centres of demand in the emerging economies of, for example, Brazil, China and India, highlights the dynamic nature of the global system and the fact that new sources of demand for resources are placing additional stress on the planet's ability to support global capitalism. The analysis presented in this section suggests that there can be no doubt that the storm clouds are gathering and that urgent action is needed to change the relationship between environment, society and development.

DEMOGRAPHIC TRANSFORMATIONS

Chapter 4

John Round

Topics covered

- The geographies of population growth

- Changes in life expectancy and fertility

- The impact of disease and economic change on demographic futures

- Population ageing and the role of migration in ageing societies

4.1 Introduction

Demographic transformations influence almost every topic discussed in this book. Changes in population structure impact on the environment, influence economic development, alter government strategies, stimulate migration and, among many other examples, can lead to social upheaval. While it is clear that the world's population is growing rapidly within this growth there are many geographies for this chapter to explore. To begin we first look at global population growth over the past few centuries, highlighting how the world's population has risen rapidly since the end of the Second World War. After detailing some future population predictions, and showing how tiny variations in birth/death rates can alter the projections, the chapter then investigates case study examples where factors such as disease, economic upheaval and conversely economic stability have rapidly altered population structures. The illustrations include the impact of HIV/AIDS on sub-Saharan Africa, Russia's post-Soviet demographic crisis and Western Europe's ageing population. These examples have led demographers to reassess the traditional **demographic transition model**, which suggested that societies reach a plateau where death rates are slightly below birth rates ensuring gradual population growth. Of course demography is not just concerned with overall population numbers. Demographic challenges such as increasing life expectancy, declining family sizes, reducing fertility levels, increased levels of migration and rapidly ageing populations are some of the issues that countries must face. First, by way of introduction, we examine the geographies of population growth over the last two millennia.

4.2 Geographies of population growth

It is impossible to state accurately what the world's population is. Although most countries conduct regular population censuses these are not an exact science. The data generated by these surveys might be unreliable because, for example, statistical methods prove to be imperfect, households might wish to hide the number of people living there for tax reasons, or governments might inflate numbers for political reasons. For example, the United Kingdom 2001 census showed that there were 800,000 fewer young men living in the country than was previously thought (www.statistics.gov.uk). Initially migration was thought to be the cause but eventually 190,000 men were added to the total figure. Such variations have serious implications for government policy. Many

local authorities argued that the census underestimated their populations, which would have led to them losing government funding. Nottingham City Council, for example, argued that the census failed to include 16,000 people, which would have cost them over seven million pounds in funding per year (http://news.bbc.co.uk). For the 2011 UK census many local authorities took out adverts asking people to complete the census accurately to make sure that the authority received the correct level of government support. In other regions of the world census data can be used for more political means. For example, the Russian census of 2002 declared that almost 1.1 million people lived in the region of Chechnya. Many observers argued that after over a decade of war and outward migration the Russian government was inflating the figure to try and show that life was returning to normal there. The Danish Refugee Council, having surveyed the region in the same month as the official census, suggested the figure was nearer to 700,000 (www.drc.dk). Similarly, it was more symbolic than based on fact that a baby born in Sarajevo in October 1999 was declared by the United Nations as the sixth billionth person on earth. What is irrefutable, however, is that over the last 50 years a population explosion has taken place. It is estimated that at the start of the twentieth century there were approximately 1.3 billion people on earth, by the century's end the world's total population had grown by approximately 4.7 billion people. Even more amazingly the vast majority of this increase has taken place in the last 50 years, as the estimated population in 1950 was just over two billion (Figure 4.1). In October 2011, the United Nations symbolically declared that the world's seventh billion person had just been born in the Philippines.

Between 1900 and 1920 the world's average population growth rate was approximately 0.5 per cent per year. The growth rate then doubled during the 1920s, remaining constant until the end of the 1940s. During the 1950s the rate rose rapidly to 1.8 per cent, rising again to 2 per cent the following decade. Obviously such increases are compound (the yearly increases are in addition to previous growth), meaning that a growth rate of 2 per cent allows the world's population to double every 35 years. From this peak, the growth rate then fell to 1.7 per cent in the 1980s to the current average of approximately a 1.2 per cent increase per year. This, estimates the United Nations population division (www.un.org/popin), sees an annual population growth of 79 million people per year down from a peak of 83 million a year in the 1980s.

One of the most challenging problems that demographers face is trying to project what future population growth rates will be. It is now expected that growth rates will

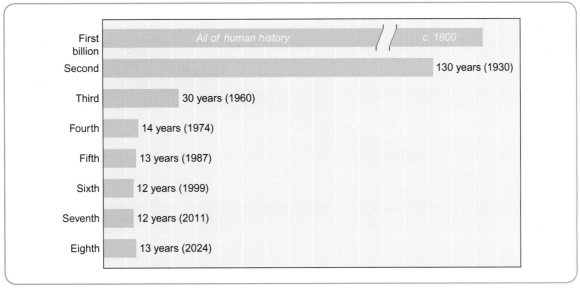

Figure 4.1 Total population growth over time.

Source: graphic from 2010 World Population Data Sheet, Washington, D.C., Population Reference Bureau

continue to decline, with the United Nations predicting a fall below 1 per cent by around 2020, continuing to decline to 0.3 per cent by 2050 (which will be the lowest levels since the mid-eighteenth century). However, even these more modest growth rates will see the world's population grow by over 35 million a year until 2050. This would then give an overall population of 8.9 billion, 2 billion more than present. Given the compound nature of population growth, as we shall see below, it takes only small changes in fertility/death rates to produce large changes in population totals over a long period. Thus, as Figure 4.2 shows, the UN produces a range of predictions based on different **fertility rates** for 2050, ranging from 7.3 billion, and already declining, to 10.7 billion.

While the world's population is growing rapidly this overall figure tells us little of the geographies that exist within it. The most striking aspect of the geography of population change is how population size in the '**more developed countries**' has remained relatively stable since the late 1950s, at around 1 billion. In contrast the total population of the '**less developed countries**' has increased rapidly from approximately 1 billion to 5 billion since 1950. It must be noted here that terms such as 'more developed' or 'less developed' are extremely problematic, as is discussed in Chapter 9, but they are used reservedly here, as they are the terms employed by the United Nations when discussing demographic issues. This difference in growth rates is shown starkly by Figure 4.3, which shows that the rates of population growth in less developed regions will continue to be far above those in more developed regions for the foreseeable future.

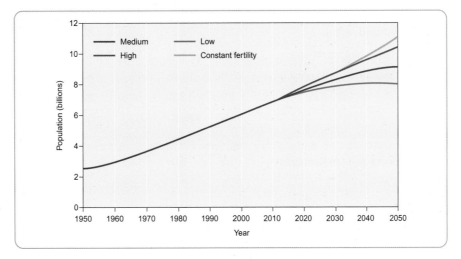

Figure 4.2 How changes in fertility rates might impact on future population growth.

Source: UN Population Division, *World Population Prospects: The 2008 Revision*, New York: United Nations

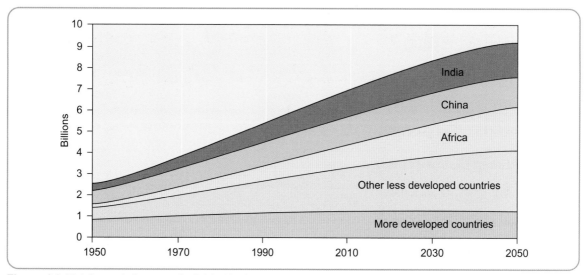

Figure 4.3 Total population growth divided by 'more' and 'less' developed regions

Source: UN Population Division, *World Population Prospects: The 2006 Revision*, New York: United Nations

Even within the category of 'less developed' there are still further variations, the reasons for which will become more apparent below. The 49 least developed countries, according to UN data, experienced the greatest increases in population growth. This group only held 8 per cent of the world's population in 1950 but in the subsequent population explosion they have contributed around 15 per cent of the growth. Such differences in population growth rates has, unsurprisingly, altered the relative distribution of the world's population. For example, the 'more developed' countries in 1950 combined to provide approximately a third of the world's population, by 2010

this had fallen to around 17 per cent. As Table 4.1 shows, this has meant that significant changes have occurred in overall population share by region.

As Table 4.1 demonstrates, Europe's share of the world's population fell from approximately a quarter to around an eighth between 1900 and 2000, with Africa the biggest gainer in percentage terms. What is most significant here is how quickly these changes have taken place with most of the changes taking place since 1950. Table 4.2 details these changes in numerical terms.

By 2050 Europe's share of the world's population is likely to have fallen to 7 per cent, while Africa's is

Table 4.1 Changes in overall population share by region

Major area	1950	1975	2009	2050 Low	2050 Medium	2050 High	2050 Constant
More developed regions	32.1	25.8	18.1	14.2	13.9	13.8	11.4
Less developed regions	67.9	74.2	81.9	85.8	86.1	86.2	88.6
Least developed countries	7.9	8.8	12.2	18.4	18.3	18.1	22.4
Other less developed countries	60.0	65.4	69.7	67.5	67.8	68.1	66.2
Africa	9.0	10.3	14.8	22.0	21.8	21.7	27.2
Asia	55.5	58.6	60.3	57.0	57.2	57.4	54.5
Europe	21.6	16.6	10.7	7.6	7.6	7.5	6.0
Latin America and the Caribbean	6.6	8.0	8.5	7.9	8.0	8.1	7.6
Northern America	6.8	6.0	5.1	5.0	4.9	4.8	4.2
Oceania	0.5	0.5	0.5	0.6	0.6	0.6	0.5

Source: Population Division of the Department of Economic and Social Affairs of the United Nations Secretariat (2009) *World Population Prospects: The 2008 Revision*, United Nations, New York.

Table 4.2 Population growth, 1950–2050, by development categories and major areas

Major area	Population in 2010 (millions)						Population in 2050 (millions)					
	0–14	15–24	25–59	60+	80+	Total	0–14	15–24	25–59	60+	80+	Total
World	1862	1218	3070	759	106	6909	1797	1209	4136	2008	395	9150
More developed regions	204	159	605	269	53	1237	197	134	528	416	121	1275
Less developed regions	1657	1059	2465	490	52	5671	1601	1074	3607	1592	274	7875
Least developed countries	341	173	297	44	3	855	452	288	747	185	18	1672
Other less developed countries	1316	887	2168	446	49	4817	1149	786	2860	1407	256	6202
Africa	416	209	353	55	4	1033	546	348	892	213	21	1998
Asia	1092	756	1906	414	47	4167	937	645	2413	1236	228	5231
Europe	113	93	366	161	31	733	104	70	281	236	66	691
Latin America and the Caribbean	163	105	261	59	9	589	124	87	331	186	40	729
Northern America	70	49	168	65	13	352	76	52	196	125	36	448
Oceania	9	5	16	6	1	36	10	7	23	12	3	51
Percentage distribution by age group												
World	27	18	44	11	2	100	20	13	45	22	4	100
More developed regions	17	13	49	22	4	100	15	11	41	33	9	100
Less developed regions	29	19	43	9	1	100	20	14	46	20	3	100
Least developed countries	40	20	35	5	0	100	27	17	45	11	1	100
Other less developed countries	27	18	45	9	1	100	19	13	46	23	4	100
Africa	40	20	34	5	0	100	27	17	45	11	1	100
Asia	26	18	46	10	1	100	18	12	46	24	4	100
Europe	15	13	50	22	4	100	15	10	41	34	10	100
Latin America and the Caribbean	28	18	44	10	1	100	17	12	45	26	5	100
Northern America	20	14	48	18	4	100	17	12	44	28	8	100
Oceania	24	15	45	15	3	100	19	13	44	24	7	100

Note: Refers only to countries or areas with 100,000 persons or more in 2009.

Source: Population Division of the Department of Economic and Social Affairs of the United Nations Secretariat (2009) *World Population Prospects: The 2008 Revision*, New York: United Nations expected to be nearly three times this. Over the period 2003–2050 Africa's population is projected to grow by just under one billion, representing over one-third of global growth (36.4 per cent) compared to only one-sixth (16.6 per cent) for 1950–2003. Meanwhile, Asia seems destined to remain the main contributor in absolute terms, adding a further 1.4 billion up to 2050, but this is a marked reduction in the pace of growth compared to its 2.4 billion gain between 1950 and 2003.

Rapid changes can also be seen at a country level. In 1950 only 8 countries, using current boundaries, had a population of over 50 million. By 2006 this figure has grown to 22. China has seen its population rise from approximately 550 million in 1950 to over 1.3 billion in 2001, whereas Europe's population has risen from around the same figure to 735 million in the same period. This is despite attempts to dampen population growth through government policies such as the 'one child per family' campaign. This policy made it very difficult for families to have more than one child as any extra would incur great education, health care and taxation costs. Since the mid-1990s, birth rates have fallen steadily but they are still more than double death rates, producing a

significant growth rate. Given the compound effects of previous population growth this will mean that China's population will continue to grow for the foreseeable future. While it is currently the world's most populated country the declining growth rate means that it is likely that India will overtake it in the next 40 years. In its 2001 census India was said to have a population of just over a billion people. This is a staggering rise of over 21 per cent from the 1991 census. In 2009 the World Bank estimated that the population had grown further to 1.15 billion and the Indian government predicts that by 2025 the population will rise to 1.4 billion, a further increase of 36 per cent. It is estimated the total population will increase by a further 400 million by 2050.

4.3 Geographies of changing birth and death rates

To explain changes in population growth rates demographers often use the demographic transition model. As Figure 4.4 shows, this model has five stages to

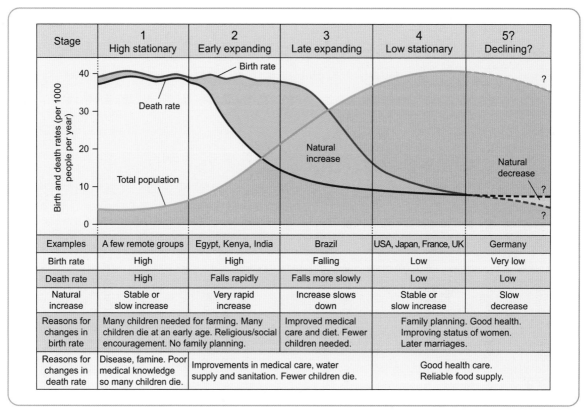

Figure 4.4 The demographic transition model.

Source: *Population Geography: A Reader*, McGraw-Hill (Chung R. (author), Demko G.J., Rose H.M., and Schnell G.A. (eds) 1970)

differentiate between the stages which most countries have passed through in their demographic history (see Lee 2003 for an overview of its development).

Stage one is characterized by rudimentary health care, with death from what are now easily treatable diseases commonplace, leading to very high crude death rates (CDR – deaths per 1000 people). Birth rates (CBR – births per 1000 people) are also high as large families were the norm, partly to ensure that parents will have support when they can no longer work. With birth and death rates relatively equal, overall population size does not increase dramatically during this stage.

As health care and sanitation systems develop, death rates fall as fewer people die from easily preventable diseases and countries move into stage two of the model. However, there is usually a period during which birth rates remain high. This might be for cultural reasons or because economic growth sees an increasing demand for family labour. This gap between death and birth rates sees an increasing population. As death rates can fall rapidly this can often cause a rapid rise in population growth. Figure 4.5 demonstrates how this occurs

From these figures we can see that, in less developed countries, from the middle of the previous century death rates fell dramatically as health care began to improve. However, birth rates, although declining, did not converge with death rates, resulting in a rapid increase in the rate of natural population growth. More developed countries have entered the third stage of the demographic transition model where birth and death rates converge. Birth rates have fallen rapidly for a combination of reasons:

- Increased economic prosperity and the development of pension systems means that many parents no longer have to rely on their children for support during old age.

- Increased urbanization means that families do not have the space to have many children.

- Increased social and economic choices for women, these include improving opportunities in the workplace, much greater levels of participation in higher education than in previous generations and the availability of reliable contraception.

- Cultural attitudes towards large families have changed so that rather than being the norm, having many children is seen as unusual and an increasing number of women are deciding not to have children at all.

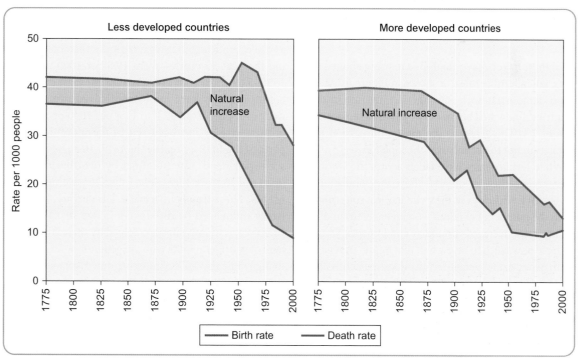

Figure 4.5 How changes to birth and death rates can lead to changes in natural population growth.

Source: Population Reference Bureau (2007)

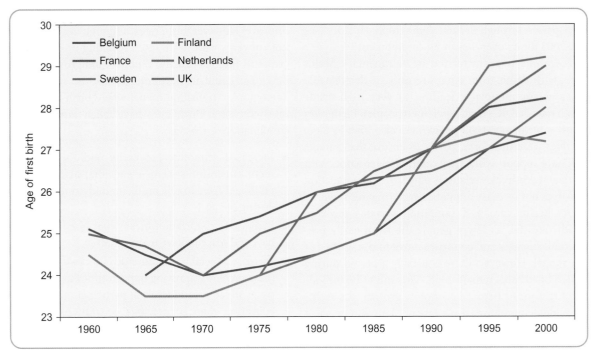

Figure 4.6 Increase in age of first birth in selected West European countries.

Source: data from Eurostat.org

■ Increasing levels of divorce and single-parent families means that large families are less likely.

■ A rapid increase in the average age of first birth (see
Figure 4.6). If the first birth is delayed then a large
family is less likely.

All of these factors have combined to bring about a dramatic fall in fertility rates, as shown in Figure 4.7. As the
graph demonstrates, the average fertility rate in these
countries is now well below the population replacement
level of 2.1. This figure is the minimum needed for a
population not to decline over the long term.

While fertility rates have fallen across the world, in
many regions they are still well above the population replacement level; however, the trend in most countries is a
fall in the number of children women have. In the 1960s
fertility rates averaged around 6 children per woman,
but have now fallen to below half that figure. In Asia and
South America the average fertility rate is currently about
2.5. In sub-Saharan Africa the majority of countries currently have fertility rates of above 4, but many are beginning to decline. The United Nations predicts that by
2050 over three-quarters of LDR countries will have fertility rates below the population replacement level.

As Figure 4.4 demonstrates, countries in the third
stage of the demographic transition model have

experienced a fall in death rates that has now levelled
out. In the countries within this group the majority of
deaths now occur in the over-60 age group as infant
mortality is greatly reduced and deaths from easily
treatable diseases are uncommon. Across the world as
a whole, crude death rates have more than halved during the past 50 years – from 19.8 per 1000 people in
1950–5 to 8.5 in 2005–10. Even for the least developed
countries, rates have more than halved over this period, dropping from 27.9 to 11.4 per 1000 people. The
greatest change has occurred in the more developed
parts of the LDRs, because the LDRs in aggregate have
seen their death rate plummet by two-thirds, ending
up at only 8.8 per 1,000, lower than the average for the
MDRs. These trends in crude death rate are, however,
partly influenced by differences in age structure. The
LDRs' current low rates are helped by their large numbers of younger people who are generally at least risk of
dying, while the MDRs' rates are now rising because of
their increasing proportions of older people.

In terms of life expectancy, a measure that is not
affected by age structure and that provides a direct indicator of life-chances, we can see that many regions
have experienced improvements over the last 50 years,
(see Figure 4.8). For MDRs, between 1950–5 and
1975–80, combined male and female life expectancy

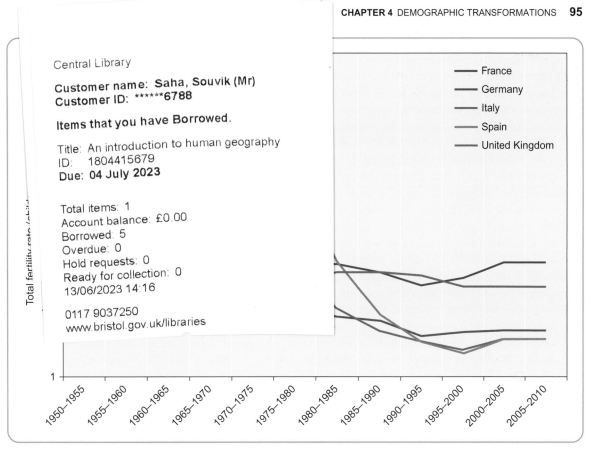

Figure 4.7 Fertility rates in selected West European countries.

Source: UN Population Division, *World Population Prospects: The 2006 Revision,* New York: United Nations

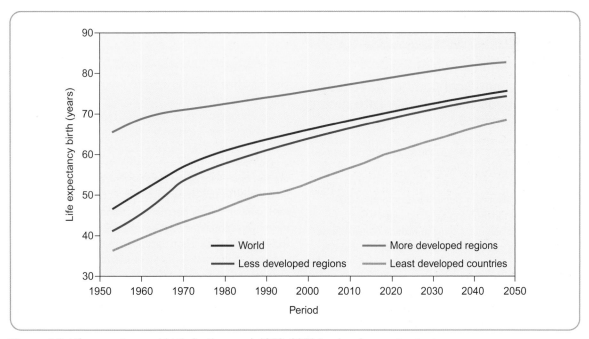

Figure 4.8 Life expectancy at birth (both sexes) 1950–2050 by development category.

Source: UN Population Division, *World Population Prospects: The 2008 Revision,* New York: United Nations rose from 66.1 to 72.3 years. This represents an increase of 1 year for every 4 years of elapsed time. In the last 25 years increases have occurred at a slower pace, with the average only increasing by 3.5 years during this period. This is partly explained by the decline in life expectancy in former communist countries, the reasons for which are explored below, where men can expect to live for only 64 years compared to 74 years in western Europe. In LDRs, life expectancy has increased rapidly but again the rate of growth is slowing. Here, in the 25 years from 1950, life expectancy rocketed from 41 to almost 57 years, but over the next quarter of a century average life expectancy advanced by a further 7 years only. The earlier progress can be attributed to the rapid reduction in malaria, principally due to the use of techniques developed to protect soldiers in the early 1940s fighting in south-east Asia, and to sustained programmes of curbing infectious and water-borne diseases like smallpox, tuberculosis and cholera. Within the space of 20 years, thanks to the import of mortality control measures from the MDRs, many countries were able to pass through an epidemiological transition that had taken over a century for many European countries. Even so, life expectancy remains stubbornly low in the least developed countries in general and in sub-Saharan Africa in particular. This can often be attributed to armed conflict, such as the genocide in Rwanda, and the impact of HIV/AIDs has also caused life expectancy to plummet in many African countries.

The other main obstacle to life expectancy improving in LDRs is the persistently high infant mortality rate (IMR). Whereas the IMR for the period 2005–10 is just over 4 per 1,000 live births in North America, most of Europe and Australia/New Zealand, it averages 22 in Latin America, 56 in south-central Asia and 80 in sub-Saharan Africa. It is one of the United Nation's Millennium Development Goals to reduce by two-thirds the mortality rate among children under five (www.undp. org). See Case study 4.1.

Of course the demographic transition model can only help explain previous population trends and cannot predict future ones. Figure 4.9 shows the classification of countries according to their progression through the model.

As noted above, unexpected events such as the spread of HIV/AIDS (Case study 4.2) or the collapse of the Soviet Union can rapidly alter population trends. Therefore, the chapter now explores the impact of these events in sub-Saharan Africa and Russia respectively. Following this, the consequences of entering stage three of the model, where birth and death rates converge, is examined through an analysis of the problems facing an ageing western Europe.

Thematic case study 4.1

United Nation's Millennium Development Goal number 4: reduce by two-thirds the mortality rate among children under 5.

As discussed in Chapter 8, the Millennium Development Goals are eight development objectives to be achieved by 2015. They were adopted by 189 nations at the UN Millennium Summit in September 2000 (see www.undp.org/mdg for a full overview of the goals). Goal number 4 aims to reduce by two-thirds the mortality rate among children under 5. The UN states that:

> One of the darkest characteristics of poverty is that it seems to prey on the vulnerable and defenceless. In low-income countries, one out of every 10 children dies before the age of five. In wealthier nations, this number is only one out of 143.

In more stark terms this translates to 11 million children dying under the age of 5 every year, more than 1200 per hour, most from easily treatable diseases. Progress is to be measured by the following indicators: under-5 mortality rate; infant mortality rate and the proportion of 1-year-old children immunized against measles. UN data shows, however, that in the first four years of the scheme little progress had been made in sub-Saharan Africa. Few countries have seen infant mortality decrease significantly and some have even seen rates increase. Some countries have experienced a rise in immunization rates but this is not universal to the region (for an excellent database on Millennium Development Goal data go to http://unstats. un.org/unsd/mdg). Without rapid increases in funding to prevent easily preventable child deaths, the UN estimates that in some parts of sub-Saharan Africa the goals will not be reached until 2115.

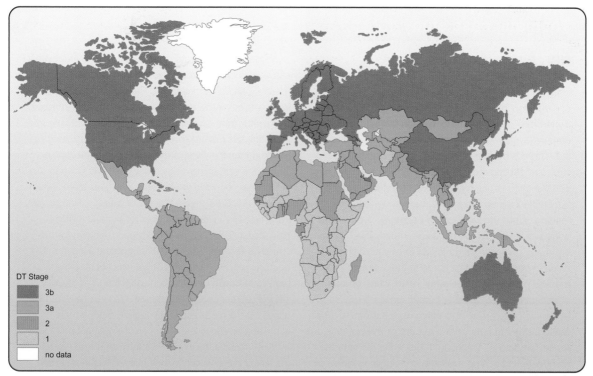

Figure 4.9 Countries classified on the basis of stage reached in the demographic transition by 2000–2005.

Source: data from UN (2003)

4.4 The geographies of HIV/AIDS in sub-Saharan Africa

It is estimated that in 2009 approximately two-thirds of the total number of people infected with HIV/AIDS live in sub-Saharan Africa, equating to approximately 25 million people. The United Nations UNAIDS organization (www.unaids.org) suggests that in 2009 around 1.9 million people were infected with HIV in the region, which is more than the rest of the world combined. Of the 2.9 million people who died of AIDS in 2006, 2.1 million were from sub-Saharan Africa. Figure 4.10 shows how the region suffers disproportionately from the epidemic.

The disease has spread rapidly throughout sub-Saharan Africa. In 1990 only 1 per cent of women attending ante-natal services in South Africa were infected, by 2000 this figure had risen to 22.4 per cent. By 2005, 30.2 per cent (attendees' blood samples are routinely tested for HIV and this provides one of the more reliable sources of infection rates) were infected. In other words, almost a third of the women attending ante-natal services in South Africa are infected with this incurable disease. Why are infection rates so high in the region when compared to the rest of the world? There are many theories, some of the ones highlighted by UNAIDS include:

■ A traditional lack of discussion of sexual health matters.

■ The low social and socio-economic status of women.

■ **Poverty**, which in some cases leads to women undertaking commercial sex work.

■ Migrant labour patterns, which means that family life is disrupted, often leading to an increase in sexual partners.

■ A reported cultural resistance to the use of condoms in many regions.

■ In certain regions cultural attitudes towards relationships, i.e. multiple partners and polygamy, can increase risk of infection.

■ Established epidemics of other sexually transmitted diseases.

In contrast to many other regions of the world, the disease is predominant amongst young women. Across the region this group is four times more likely to be infected than young men. Women also endure further difficulties, as they are often also caring for

Thematic case study 4.2

HIV/AIDS

HIV (Human Immunodeficiency Virus) is a virus that can develop into AIDS (Acquired Immunodeficiency Syndrome). This syndrome leads to the failure of the immune system, making minor infections life threatening. There is currently no cure for this disease. The virus is most commonly passed on by sexual contact and the sharing of infected needles and can also be transmitted during pregnancy and through breast-feeding. While the scale of the pandemic is clear, the origins of HIV/AIDS are not easily identifiable. There are many competing theories as to why and where the disease began, but it first came to the attention of the medical profession in the early 1980s when groups of patients started to suffer from diseases rarely seen within their age group and which were extremely resist-ant to treatment (see www.avert.org or www.unaids.org for further details). By the mid-1980s the disease was named, tests became available and public awareness campaigns on how to avoid infection began. Despite this, the number of people becoming infected grew rapidly. One of the main causes of this is the fact that HIV is asymptomatic and can be carried for many years before AIDS develops, therefore a carrier can infect many other people before they become ill themselves.

While there is no cure, drugs, known as antiretrovirals, can slow down the progression of the disease. This can greatly improve the health and life expectancy of people living with HIV/AIDS and they can block the transmission of the virus from mother to child. They are not a cure and if treatment is missed or a patient develops resistance to the drug then the disease reverts to its normal course. Due to the cost of the drugs many people in low and mid-dle income countries are not able to access them. In late 2006 UNAIDS estimated that out of over 7 million people living with HIV in these countries less than 2 million had access to antiretroviral medication, leading the G8 countries to pledge that by 2010 universal access would be achieved. However, this target was not reached with the World Health Organization estimating that globally still only 25 per cent of people with the virus have regular access to the medication they require. Furthermore, in sub-Saharan Africa less than half of expectant mothers with HIV receive the medication they require to reduce the risk of transmission to their child. While there is still the will to increase access to medication, as UNAIDS highlights, the current economic downturn means that even though the cost of the drugs continues to fall many governments are struggling to purchase them in sufficient quantities or to maintain the health care infrastructure needed for their effective delivery.

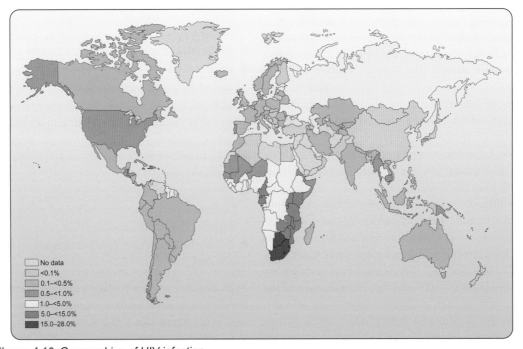

Figure 4.10 Geographies of HIV infection.

Source: Joint United Nations Program on HIV/AIDS (2010) UNAIDS Report on the Global AIDS Epidemic 2010, Geneva, UNAIDS

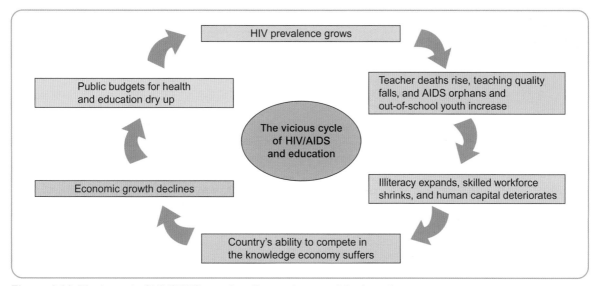

Figure 4.11 The impact of HIV/AIDS on education systems and the broader economy.

Source: Bundy, D and Manorama, G (2002) Education and HIV/AIDS: A Window of Hope. Washington, D.C., World Bank

family members who are ill from the disease. With infection rates reaching 50 per cent in some urban areas it is no surprise that it is estimated that, by 2009, 16 million children had been orphaned by the disease, 90 per cent of whom lived in sub-Saharan Africa. Child care is then often passed on to elderly family members who themselves are often suffering from ill health. UNAIDS suggests that there are now few families in sub-Saharan Africa who are not impacted upon by the disease. However, in many regions there is a great stigma attached to infection. Therefore, even when ill, many people are still reluctant to be tested; this obviously denies them appropriate treatment. Given the prevalence of infection within the working-age group, national economies suffer greatly, not just because of the cost of trying to treat the disease but also from lost tax revenues, skill shortages and general uncertainty. Figure 4.11 shows the impact of the disease on the education sector and how this impacts upon the whole economy.

The disease also has a dramatic impact on life expectancy. As discussed earlier, during the 1970s and 1980s improving levels of health care meant that fewer people were dying from easily preventable diseases, leading to rising life expectancies. For example, life expectancy in Botswana rose to approximately 65 by the mid 1980s and it was predicted that it would have reached 69 by 2011, but for HIV/AIDS. Instead, in 2011 male life expectancy was 55. In Swaziland life expectancy is now

46, without HIV/AIDs it is estimated that it would be 64. In 2006 there were 23 births/1000 population compared to 29 deaths/1000 population, demonstrating a return to the early stages of the demographic transition model. Figure 4.12 highlights this dramatic decline in life expectancies.

As the graph shows, towards the end of the last decade life expectancy began to increase slowly as the use of antiretrovirals and education levels about the transmission of HIV increased.

There is also a clear political economy to the disease in sub-Saharan Africa. Many governments still deny the severity of the disease, and many dispute the causes of the epidemic. The South African President, Thabo Mbeki, publicly disputed whether the HIV virus leads to the development of AIDS (McGreal 2006). He has also argued that in South Africa far fewer people have died of the disease than is officially reported. Such statements lead to confusion amongst the general public and hamper attempts to educate people as to how to avoid contracting the disease. His successor Jacob Zuma stated in a court case that he was protected from HIV/AIDS as he had a shower after unprotected sex with someone infected with the disease. Furthermore, within the region there is a lack of a coordinated response to the disease. However, it is argued that in many areas Western ideals are imposed as a solution to the problem. For example, some religious groups suggest abstinence rather than condom use as a solution, which

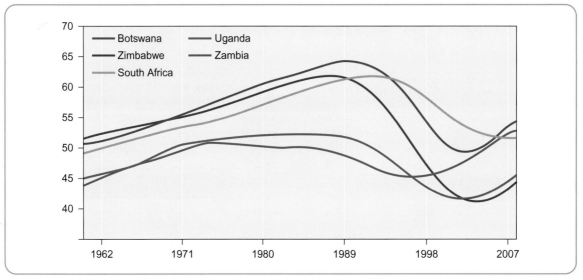

Figure 4.12 Changing male life expectancy in five sub-Saharan countries.

Source: World Development Indicators 2011. Washington, D.C., World Bank

does not reflect the realities of the situation. On a global scale there is much debate about the cost of anti-viral drugs which can increase dramatically the life expectancy of those living with AIDS. Many sub-Saharan states cannot afford the high cost of these drugs, partially because of the economic impact of the disease, as discussed above. Therefore pressure is put upon Western drug companies to make treatment more freely available. Yet there is resistance to providing cheap drugs, given the high cost of developing them, and there are also concerns as to whether they will be distributed correctly. In order to develop approaches which have some chance of success, UNAIDS suggests the following areas must be focused upon:

■ HIV/AIDS can no longer be seen as an 'individual' disease and needs to be tackled on a national level.

■ Educational and preventative measures must take into account social and cultural norms.

■ The double burden of women looking after their family while sick themselves must be considered.

■ The socio-economic structure of the disease must be considered – young women must be targeted for education and empowerment.

■ Interested parties must increase levels of cooperation.

■ Western states need to be prepared to support, both financially and politically, the restructuring of healthcare and education systems.

When looking at the geographies of infection within sub-Saharan Africa, it is clear that some countries did implement the above suggestions at an early stage of the disease. For example, Senegal has stopped infection rates at between 1 and 2 per cent and Uganda has cut HIV/AIDS prevalence from double-digit figures to around 7 per cent. This demonstrates that countries with the political will and the careful use of resources can contain the spread of HIV/AIDS. The first case of HIV/AIDS in Uganda was detected in 1984 and by 1986 a national committee was set up, which the President personally led, to examine responses to the disease. Remember that in 2006 the South African president was still disputing the cause of the disease. In 1991 a public awareness campaign was launched across all sectors of the media with pop songs, billboards and radio messages all popularizing HIV/AIDS protection messages. As a result, HIV/AIDS rates peaked in 1993 and now 98 per cent of Ugandans know that the disease is mainly transmitted through unprotected sex. As the United Nations points out, if such measures are not adopted by all sub-Saharan countries, then the epidemic will continue to decimate generations and will provide a further barrier to sub-Saharan Africa's political and economic development. However, there is some cause for optimism as infections in sub-Saharan Africa fell from 2.8 million in 2006 to 1.9 million in 2008, suggesting that the measures are having some impact.

Thematic case study 4.3

Malaria

While HIV/AIDS might receive more global publicity than malaria, it is an extremely serious health threat in sub-Saharan Africa, for which there is no vaccine. While after the Second World War great efforts were made to eradicate the disease, it proved resistant in many regions of Africa and Asia. Since the turn of the century, attempts to contain the disease in sub-Saharan Africa have faltered for a number of reasons. Firstly, it is argued, civil wars have seen health systems break down, making preventative measures much harder to implement and treatments near impossible to distribute effectively. Secondly, the spread of HIV/AIDS has caused an increase in malaria deaths as the infection exacerbates malaria fevers. Thirdly, and the main reason, drug-resistant strains of the infection have mutated. While it is hard to accurately count the number of people who die from the disease, it is estimated that it is over 1 millions deaths per year, the vast majority occurring amongst African children. The geographer Danny Dorling has produced the following map (Figure 4.13), based on World Health Organization data, which demonstrates how infection rates are endured disproportionately in sub-Saharan Africa.

The recent upsurge in infect...
international aid community incre...
find efficient preventative measure...
One of the most effective measures...
treated nets. However, they need to be...
treated with insecticide, and insects can...
resistant, the cost of which is prohibitive to...
households. As with HIV/AIDS treatments, the drugs used to treat malaria are expensive as combination treatments are needed. Furthermore, they are only really effective if diagnosis is made early, something that often does not happen. Attempts are been made to develop vaccines, but it is proving extremely complex and conservative estimates suggest it will be at least a decade until a breakthrough is made. Perhaps the most intriguing measure is the genetic sequencing of the parasite with the aim of introducing a version that is unable to carry the disease. Unfortunately, it is estimated that it could be up to 20 years until effective breakthroughs are made. While a cure is not likely in the near future, at least the issue is now firmly on the international development agenda, with increasing resources devoted to attempting to develop effective means of preventing the spread of the disease and to find viable treatments.

Figure 4.13 Global distribution of malaria cases

Source: www.worldmapper.org. See this website for further cartograms

4.5 Russia's demographic crisis

If the present tendency continues [population decline] there will be a threat to the survival of the nation . . . our demographic situation today is an alarming one.

(President Putin, in an address to the nation in 2000)

The demographic crisis that Russia is currently enduring is unparalleled outside of wartime in the global North. Since the collapse of the Soviet Union in 1991, Russia has seen death rates soar, fertility rates decline and life expectancy fall, which, combined, saw the country's population fall by just under a million people in 2005. In the period since the collapse of the Soviet Union, Russia's population has fallen by approximately

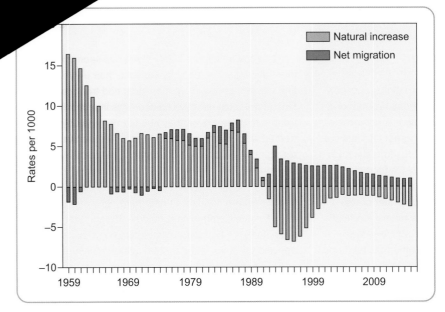

Figure 4.14 Components of population change, Russia, 1959–2015.

Source: RAND Corporation used with permission

2.5 million people. However, the true scale of the problem has been masked by high levels of inward migration as Russians returned to their homeland from the now former Soviet states. Over 5.5 million people made this journey suggesting that the rate of natural decrease was over 8 million people. As Figure 4.14 shows, Russia's population was growing in the late 1980s and while migration partially compensated for the rapid decline in natural population growth in the 1990s this to has now slowed. As the figure shows, the overall population decrease at the turn of the century reached over 800,000 people per year (data from Marquez 2006).

Life expectancy fell for men from 63.8 in 1990 to 57.6 in 1994, the steepest fall ever recorded outside of wartime in the global North, and this figure only recovered to around 59 by 2005. By 2010 it had risen to 62 but this is still the lowest level of life expectancy in the 'developed world'. Meanwhile the life expectancy for women fell from 74.3 in 1990 to 71.2 in 1994, rising to around 74 in 2010. The disparity in life expectancy between men and women is the highest in the world. What is most startling about these figures is the speed with which the demographic situation deteriorated in Russia.

Why has this happened? There are many theories, but we must first look at the many demographic shocks that the country has experienced over the last 100 years as these laid the foundations for the current problems. The Russian revolution in 1917 led to many years of civil war and famine and many millions died prematurely. Not long after this situation had

stabilized, the Second World War broke out. Russia experienced some of the heaviest losses with recent estimates suggesting that over 14 million Russians were killed during this period. The end of the war did not bring stability to the country as Stalin ushered in a period of the 'great terror' with up to 2.5 million people exiled to prison camps in the Russian far north. It is estimated that during Stalin's reign, between 1927 and 1953, at least 2.8 million people died in these camps (Applebaum 2003). It was not until Stalin's death that Russia's demographic situation became comparable to that in Western Europe. Russia soon completed the first stage of the **'epidemiological transition model'**, the 'sanitary revolution', meaning that people should no longer die of easily preventable causes due to the creation of a countrywide health system. Even then there were still great demographic imbalances, as more men died in the above events than women. The legacy of this is that in 2002 there were 67 million men compared to 77 million women in Russia. However, Russia was, by this point, beginning to move through the demographic transition model as birth rates began to rise while death rates fell.

In the 1960s and 70s life expectancy did not converge with levels in Western Europe, mainly for cultural reasons. Traditionally Russia has had some of the world's highest alcohol consumption rates, and a limited range of food products meant that many people led unhealthy lifestyles. In 1985 President Gorbachev tried to tackle this by introducing limits on when, where and how much alcohol could be sold. While very

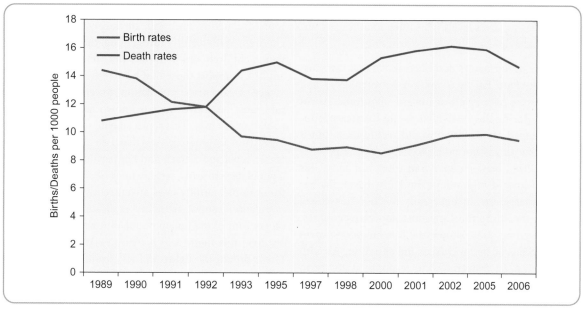

Figure 4.15 How changes to death and birth rates impact upon overall population change in Russia.

Source: UN Population Division, *World Population Prospects: The 2008 Revision*, New York: United Nations

unpopular, these measures did see some increase in life expectancy. However, these gains were short-lived as the collapse of the Soviet Union in 1991 meant such reforms were abandoned. As noted above, this period ushered in great demographic uncertainty and Russia diverted from the path suggested by the demographic transition model. As with other countries, Russia had seen birth and death rates converge, which would have led to a stable population. As Figure 4.15 shows, however, the situation changed dramatically after 1992. Note how this differs from Figure 4.4 above.

Since 1992 the gap between birth and death rates led to a rapid decline in the country's population. There are many theories as to why this demographic crisis has developed and rather than one single cause it is arguably a mixture of several factors.

Historical lag This argues that the current situation is a continuation of the decline in health of the Russian population since the 1960s and was so rapid due to the suppressed deaths delayed by Gorbachev's anti-alcohol campaigns of the mid 1980s. In other words, those who would have died from alcohol abuse in the 1980s, but were not able to do so because of the restrictions, were now able to do so. Furthermore, while impressive in terms of the number of doctors and hospitable beds, the Russian health-care system had not adopted all of the most recent medical advances. While there is no doubt that these issues did contribute to the problem

they do not really explain why the crisis has lasted for so long.

Poverty Current estimates suggest that between 20 and 30 million Russians can be classed as poor, as for many the real worth of their incomes has declined rapidly after the collapse of the Soviet Union. However, the collapse did not bring starvation as people/households developed alternative survival methods, such as undertaking informal work or growing their own food, for example. Given that most people can earn enough money to buy enough food for their household this can only be part of the problem. Furthermore, real incomes have increased since the middle of the previous decade but life expectancy has only just started to rise slowly.

Access to services While Russia still has a free health-care system that has in theory enough doctors and nurses, the system is starved of funding. This means that the wages for many medical staff are only just above poverty levels and equipment is often poor. Many people say that they have to pay to obtain free medicines and high poverty levels mean that many cannot afford all of the medicines that they need.

Cultural issues Many middle-aged men have found it difficult to adapt to the post-Soviet era. They often have had to find work outside of their professions and they feel that the skills they developed in the Soviet era are no longer required. Often this leads to frustration and increasing stress levels. This often manifests in unhealthy behaviour, such as binge drinking.

It is a mix of the factors described above that saw Russia's population fall by 7 million between 1991 and 2008, despite high levels of inward migration. It was assumed during the 1990s that as the Russian economy developed the situation would rectify itself, but it is clear that this is not going to happen. The population stabilized between 2008 and 2009, but again this is due to inward migration. The United Nations has forecast that, given current trends, by 2050 Russia will fall from the 7th to the 14th most populated country. As the opening quote shows, the Russian government is now aware of the gravity of the situation. As with other European countries, one response is to try to stimulate the birth rate by offering incentives to families to have children. Families with more than one child are given support for education and health care, but poverty means that many people feel that they cannot have large families. Unlike western European states, the Russian government is not looking to encourage inwards migration to cover shortages in the workforce. Given the length and gravity of the situation it is clear that it is going to be extremely difficult to reverse the demographic problems Russia is enduring.

4.6 Western Europe: the challenges of an ageing population

It is usual for a region that has entered the final stage of the demographic transition model, where birth and death rates converge, to experience population ageing. This is when the average age of a population rises as people live longer and fewer people are born. In this section we will look at why populations age and the long-term challenges that this presents to society. According to UN calculations, between 1950 and 2000 the average age of the world's population rose from 23.6 to 26.4 years and is expected to rise steeply by 2050 to 36.8 years. While less developed regions are experiencing the fastest rates of growth the 'oldest' countries are still to be found in the more developed regions (Figure 4.16). Europe will experience the highest percentage share of people over 65. This figure is significant as it is assumed that people over this age will not be working and that many will be dependant on state pensions for their income. Also this group will be

less likely to be paying tax, will use health-care systems more frequently and will, obviously, not be having children. With fertility rates falling, there will be fewer young children and then over time fewer working-age adults to support those that have retired.

For a country's population to grow, its fertility rate must be above what is known as the population **replacement rate**. This replacement rate is based on the idea that if a woman has less than two children, then she is not 'replacing' herself and her partner. Therefore, for a population to grow, fertility must be above 2.1. In Italy, for example, fertility rates are currently 1.4 and thus its population cannot naturally replicate itself. In 2000, 18 per cent of the Italian population were aged over 65. By 2050 this figure will increase to 35.9 per cent. For Great Britain the figure will rise from 15.8 to 26.7 per cent, with the European Union average at 22.4 per cent. In the same period the number of 20–24-year-olds in Spain will fall by more than 60 per cent.

Such trends are now posing major challenges to European Union countries. It is estimated that by 2030 that there will be 20.8 million (6.8 per cent) fewer people of working age within the region. For every two people of working age there will be one person post-retirement. In Italy, with Europe's most rapidly ageing population, there will only be 1.45 working-age people per person post-retirement age, whereas in the United States of America the ratio will be 3.2 to 1. This will mean that tax revenues will fall and more people will be relying on the state for support. It is a concern of many governments that with fewer tax payers it will be impossible to provide current levels of support. Even in 2005 spending on the post-working age sector is a large percentage of overall social-protection spending. In Italy and Greece the figure is over 60 per cent. As the European Commission has stated, current levels of social provision are not sustainable, as the population ages, unless taxes are raised to prohibitive levels (EU 2006).

Such trends increase individual vulnerability as societies age. The creation of a post-war welfare state in Britain was based on a 'cradle to the grave' principle. People worked in the knowledge that their state pension would provide enough income for a satisfactory standard of living and that they would be able to turn to the health-care system for free assistance in times of need. Given the growing disparity between the tax base and those reliant on state incomes such a framework is no longer viable. It is estimated that currently more than 12 million people do not have personal or company pensions and will have to rely on the state

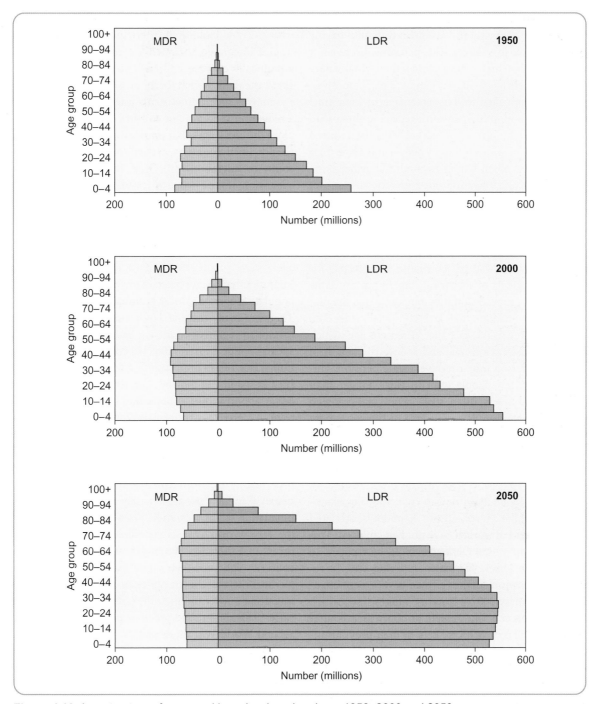

Figure 4.16 Age structure of more and less developed regions, 1950, 2000 and 2050.

Source: data from UN (2003). Note: the 2050 data are based on medium-variant projections

for their retirement income. The British government has recently made it clear that young people should no longer expect to be able to 'get by' on a state pension alone when they retire as the state will not be in a position to pay such pension levels. One of the state's responses was to raise the retirement age of women to

the same as men, 65, and to set out a framework for the gradual increasing of this age. By 2044 the retirement age for both men and women will be 68. These changes met with little protest. There has been more angst, however, about changes to private pension schemes. Organizations that have previously offered generous pension schemes to workers are realising that they are going to have great difficulty paying them as people live longer. This pensions discrepancy has meant that many have closed their final salary schemes (where by paying a set amount each month an employee was guaranteed a percentage of their final salary as their monthly pension), replacing them with schemes linked to the stock market. This has reduced the projected pensions for many people and made their future income much harder to predict. Furthermore, the financial crisis has impacted greatly on pension funds, ensuring that further reforms will take place.

Other European governments are now beginning to react to this problem. The task is made extremely difficult by the issue's extremely emotive nature. Many people would baulk at paying higher taxes to pay for the increasing cost of societal ageing, but at the same time, as the European Commission points out, the majority wish to maintain their levels of social protection. One solution is to increase the retirement age, which would increase tax revenue and reduce the length of time that a person might receive a state pension. Understandably, few people wish to work longer than they expected to and, often, strike action is threatened at even the mention of an increase in the retirement age. Another option is to try and stimulate birth rates so that more young people enter the economy in the future. Governments are therefore resorting to incentive schemes to try and arrest the declining birth rates experienced by many European countries. For example, Germany has a birth rate of 1.37 and it is estimated that 30 per cent of women have not had children. This led *Bild*, a top-selling German tabloid newspaper, to exclaim 'Baby shock: Germans are dying out'. In response to this problem the German government introduced in January 2007 a 'parents' money' law, which means that parents who remain at home to look after their child will receive 67 per cent of their previous income for a year (up to €1800 per month). Those on a low income will receive 12 monthly payments of €450. This is extremely generous when compared to the British system wherein a new mother receives 90 per cent of her previous pay for 6 weeks and then a flat €150 for 26 weeks only. Men can take 2 weeks paid paternity leave at €150 per week for 2 weeks. In Norway mothers receive 10 months at full pay and men *have* to take 4 weeks paternity leave at full pay, helping the country to the third highest birth rate in Europe. Poland and Italy have recently introduced an incentive of a one-off payment for each child born. There is no guarantee that such schemes will work, and it will be a long time

before this cohort reaches working age, many countries are now looking at international migration as a way of increasing their work force. However, this is a politically charged approach as many people worry about the damage that foreign workers will do to their economy – such as increasing unemployment rates or a reduction in wages due to extra competition. To explore the challenges that international migration poses to individual countries, we will now examine the increase in migration into Great Britain following the expansion of the European Union in 2004.

4.7 Migration and population change

Migration has always been an important component of population change at local, regional and national scales. There are many reasons why people decide to migrate to another region or country. These can be push factors such as political instability, civil conflict, forced migration, a lack of job opportunities, famine or pull factors such as stability in the new country, higher pay rates or better opportunities. From around 1800, Europe experienced high rates of out migration, mainly due to push factors such as famine, in Ireland, or poverty, in Italy, for example. Up to the beginning of the First World War it is estimated that over 55 million people left Europe, mainly for North America. In this period approximately 9 million people left Italy and Norway saw its population fall by a quarter. These flows reduced after the First World War as economic growth in Europe meant that there were fewer push factors and recipient countries tightened their migration criteria. For example, in 1924 the United States of America introduced migration quotas and then the Great Depression of the early 1930s diminished previous pull factors. After the Second World War there was a significant baby boom in western Europe, but this did not provide a rapid enough response to the need for labour as the region's economies grew. Thus, rather than seeing mass emigration the region proactively encouraged immigration.

This was usually along colonial lines as infrastructures were in place to facilitate migration. For example, the British government advertised for workers in the West Indies to migrate to fill shortages in transport and hospitals. The further tightening of migration controls in the USA made Britain a more attractive destination and by the late 1950s up to 50,000 people a year were making the journey. Inward migration was also stimulated by India gaining independence from colonial rule as religious tension meant that many people were displaced from their home region. The majority of migrants who came to Britain came from three regions: Gujarat, Punjab and Mirpur (in Pakistan after India was divided into two states after independence). None of these regions were particularly poor but they did have a tradition of international trade and migrants had previously settled in Britain. The formation of diasporas (communities of migrants) enabled further migration as new arrivals had a starting point from which to build their new lives, such as family or community support. Migration streams into Great Britain peaked in 1960–62, as legislation was subsequently introduced that made it harder to enter the country. This was passed against a backdrop of fear that increasing immigration would erode British values and would lead to violence between ethnic groups. The epitome of this was Enoch Powell's 'Rivers of Blood' speech, in which he claimed the country would be 'overrun' if migration was not halted. Although this speech saw him expelled from his shadow cabinet post he did receive popular support and it was no surprise that migrants often suffered from racism and struggled to find well-paying work. When Ugandan Indians were expelled en mass by the dictator Idi Imin many wished to move to Britain due to the diaspora that already existed. Fearing this, Leicester City Council took out adverts in the Ugandan press telling migrants not to move there, stating that they would be unable to help them if they did. Despite this many did move to the region and the city is now recognized across the world as a model for multicultural relations.

With birth rates falling in many Western countries, cross-border migration is now again a topic of much debate, as it constitutes a higher percentage of population growth than was previously the case. Furthermore, the overall number of international migrants is steadily growing. In 2010, the UN estimated that the number of international migrants was over 213 million, 60 million more than in 1990. In MDR countries, on average 10 per cent of the population are classed as migrants and there is a net flow of over two million people between LDRs and MDRs. Castles and Miller (2003) argue that the forms of migration are changing and that we are entering an 'age of migration'. They identify four main ways in which migration is changing:

1. Migration is becoming more 'global', in the sense that more and more countries are being affected at the same time and that the diversity of areas of origin is also increasing.

2. Migration is increasing, with the number of movements growing in volume in all major regions at the present time.

3. Migration is becoming more differentiated, with no one type of movement dominating a country's flows, but instead with combinations of permanent settlers, refugees, skilled labour, economic migrants, students, retirees, arranged brides, and so on.

4. Migration is increasingly feminized, with women not only moving to join earlier male migrants but now playing a much fuller part in their own right, notably being labour migrants themselves as well as often being dominant in refugee flows.

In Europe, over the last two decades, the biggest impact on international migration has been the collapse of the Soviet Union and the subsequent accession of many east European countries to the European Union in 2004 and in 2007. During the Soviet period it was extremely difficult for east Europeans to travel to Western Europe. When the Soviet system began to collapse, signified by the fall of the Berlin Wall in 1989, people in the region were now free to leave their countries. However, many west European countries imposed strict visa regimes to stop mass migration from the region. This meant that people could visit Western Europe for a vacation but were unable to migrate to look for work. This situation changed when the EU was expanded eastwards in May 2004. This meant that over 75 million people joined the European Union, many of them from countries where incomes were much lower than in Western Europe. Prior to this expansion EU members had the right to move freely between countries to look for work. However, existing member states were worried that when the east European countries joined and were given the same migration rights then too many people would come to Western Europe looking for work. As a result, many countries enforced legislation that meant that they did not fully open their borders to citizens from the new member states. This caused great anger in the so-called 'accession states' as such restrictions were not included in the negotiations that took place prior to entry into the EU. Britain, Ireland and Sweden were only countries not to impose numerical restrictions on immigration.

Prior to the expansion taking place, the opening of Britain's borders caused much concern. There were fears that too many people would move in looking for work, putting social welfare systems under great strain, and that unemployment rates would rise rapidly. The UK government certainly underestimated the number of people that wanted to move from eastern Europe.

Initial estimates suggested that approximately 15,000 people per year would migrate to look for work. However, between May 2004 (when accession took place) and June 2006, almost 450,000 people applied to the Worker Registration Scheme (WRS), which the British government put in place to monitor labour migration. Between 2006 and the end of 2010 a further 700,000 people registered under the scheme. The scheme will close at the end of April 2011 as member states are only allowed to place restrictions on movement of labour for seven years after a country accedes. It is extremely difficult to calculate the actual number of migrants who have remained, and if a migrant registers, leaves the UK and then returns to register again in the same year they are counted twice. The true figure is probably higher than the above as people would have worked without registering in order to avoid paying taxes and the scheme did not include self-employed workers. However, it is clear that far more migration has taken place than was expected, leading to much controversy over whether it benefits the country.

In Britain the centre-right press often publishes stories criticizing government migration policy, arguing that unfair pressure is being put on health care, education and other welfare systems and that migrants are forcing down wage rates and keeping unemployment figures high. However, others argue that in fact migrants are predominately undertaking low-paid, low-skilled work, which employers were previously finding difficult to recruit to. Government data suggests that between 2004 and 2010 approximately 80 per cent of migrants were aged between 18 and 24 and that very few brought dependants under the age of 17 with them. This suggests that the majority of the migration is young people looking for work, new experiences and to improve their English and work skills. Amongst people born in EU accession countries the economic inactivity rate, in 2009, was 13.8 per cent as opposed to the national average of 22.5 per cent. To date there is little evidence of 'benefit tourism', as to claim income support a migrant must have worked continuously for the previous 12 months. The regulations will change in May 2011,

Thematic case study 4.4

Future migration challenges: 'environmental refugees'

An issue of growing concern is the increasing amount of migration caused by environmental issues. Although the term 'environmental refugee' was first used in the early 1970s, it was not until the UN published a paper entitled 'Environmental Refugees' in 1985 that the term fully entered the development policy debate. Jacobson (1988) developed this idea further, identifying three different forms of forced environmental migration:

■ Those displaced due to an environmental disaster, such as a flood or earthquake.
■ People forced to migrate because their local environmental conditions put their health in danger.
■ Groups forced to resettle because declining land quality, such as desertification, means that they can no longer sustain their livelihoods.

While there is considerable controversy about the use of the term and the problem of calculating how many people could be included in this category, it is clear that migration forced by environmental problems is a growing issue. While there is still no legal definition of the term, a report by the Environmental

Justice Foundation (EJF) (2008, 4) notes that 'The most widely cited figure for the number of "environmental refugees" is 200 million people who could be forced from their homes by 2050, of which 150 million would be 'climate refugees'. They state that in Africa an estimated 10 million people have been displaced over the last 20 years because of desertification and that globally 26 million have been displaced because of climate change. In 2008, Antonio Guterres, the UN High Commissioner for Refugees, said:

> Climate change is today one of the main drivers of forced displacement, both directly through impact on environment – not allowing people to live any more in the areas where they were traditionally living – and as a trigger of extreme poverty and conflict www.ejfoundation.org.

As the EJF say, the lack of a legal framework makes responses difficult and the only sustainable approach is cooperation between organization and government bodies at all levels. One of the biggest concerns, they argue, is rising sea levels, which will see many countries lose land. This has led the Maldivian government to consider buying land in other countries to provide a living space for its population who will be displaced by rising sea levels.

for the same reasons as above, but applicants will still have to show that they reside in the UK, and thus there is little expectation of a large increase in claimants.

Many migrants were helped by existing migrant diasporas, but the majority of migration has occurred to places where work is available. This has seen the creation of new communities that can be seen by the opening of shops selling Polish food and goods, or adverts for banking in Lithuanian, for example. This is not to deny that migration is causing pressures in some areas. For example, it is estimated that in Southampton up to 10 per cent of the current population are migrants from the new accession states. This inevitably means that demands on state services will increase. At the same time this does not mean that such issues are insurmountable. However, despite the fact that increased migration has clearly not led to a rapid rise in unemployment rates, migration is still an extremely emotive issue. In response to public pressure, the British government put restrictions on the number of Romanian and Bulgarian workers allowed to move to Great Britain after their countries joined the European Union in 2007. The new Coalition government has already said that for future accession states a 7-year ban on their people working in the UK will be imposed, with the Immigration Minister stating that 'we will not make the mistakes of the past'.

4.8 Conclusion

Over the course of human history the world has undergone many different demographic transformations, but the pace of change has increased over the last 50 years, at times quite dramatically. The world's population has grown by over 5 billion during this period, with the vast majority of the growth occurring within less developed regions, which has placed great strain on environmental, food, housing and welfare infrastructures. Conversely, many countries in the more economically developed regions are experiencing natural population decline as fertility rates fall, posing new challenges to governments as their ageing populations make maintaining existing levels of welfare increasingly difficult. In such countries, overall population levels would be falling were it not for inflows of migrants; however, this is also leading to increasing social tensions and pressures on resources.

As noted above, predicting future population levels is extremely difficult. Even small changes in fertility rates can cause large differences in the medium term. Furthermore, as the three case studies have shown, unexpected events can lead to very rapid changes in population levels. In sub-Saharan Africa, HIV/AIDS, while not causing the population to fall, is extremely damaging to the future of the region and has caused life expectancy to plummet. Russia's post-Soviet demographic crisis is another example of unexpected population change that will have a long-lasting impact on the country. Without knowing where such impacts will occur next, or their scale, it is difficult to state with any confidence the rate of population growth over the next 50 years.

Notwithstanding these caveats, it is clear that the world's population will continue to grow and it is likely that the overall population will reach over 8 billion by 2025, barring a catastrophe such as global warfare or disease pandemic, which will pose many questions for governments. Given the difficulties in trying to ensure an adequate diet and housing for the current population, further growth will only exacerbate these pressures.

Learning outcomes

Having read this chapter, you should be able to:

- Recognize the dynamic nature of populations, not only in terms of size but also with respect to the geographies which exist within the overall growth.

- Explore how sudden economic changes or disease alters the course taken by countries through the demographic transition process.

- Appreciate that the completion of the demographic transition no longer appears to herald the end of transformations in the population.

- Acknowledge the increasing importance of migration as a component of national population change alongside fertility and mortality.

- Assess the significant problems that ageing populations will have in terms of social and economic development.

Further reading

Bailey, A.J. (2011) Population geographies and climate change, *Progress in Human Geography*, 35(5), 686–95.

Bailey, A.J. (2010) Population geographies, gender, and the mitigation-development nexus, *Progress in Human*

Geography, (3), 375–86. The first of these articles connects the themes of this chapter with those of Chapters 5 and 6, the second with those of Chapters 7 and 8.

Castles, S. and Miller, M.J. (2009) *The Age of Migration: International Population Movements in the Modern World*, Palgrave Macmillan, Basingstoke. This book, now in its fourth edition, provides a comprehensive survey of past trends and current developments in international migration. See the companion website at: www.age-of-migration.com/UK/index.asp

Harper, S. (2005) *Ageing Societies*, Hodder Arnold, London. This book explores the issues of population in both developing and mature countries. It examines the necessity for people to extend their working lives, the changes ageing societies have on families, the challenges it poses for state social provision and the impact these issues will have on developing regions.

Hugo, G. (2007) *Population geography*. *Progress in Human Geography*, **31**(1), 77–88. One of a series of articles looking at developments in population geography in the Southern Hemisphere. This article looks at population vulnerability and migration.

Jones, H. (1990) *Population Geography*, Paul Chapman, London. This is a clear introduction to geographical perspectives on population. Look especially at the chapters on population growth and regulation, international variations in mortality, fertility in developed countries, fertility in less developed countries, and international migration.

National Research Council (2000) *Beyond Six Billion: Forecasting the World's Population*, National Academy Press, Washington, DC. This presents the findings of a US National Academy of Sciences panel on population projections. It contains detailed examinations of transitional and post-transitional fertility, mortality and life expectancy, and international migration, together with assessments of the accuracy of past projections and of the uncertainties in current population forecasts.

UNAIDS (2010) *UNAIDS 2011–2015 Strategy: Getting to Zero*. Available to download at www.unaids.org/en/media/unaids/contentassets/documents/unaidspublication/2010/JC2034_UNAIDS_Strategy_en.pdf. This report gives a comprehensive overview of the challenges that countries and international organizations face when confronting the epidemic. It also details many success stories and the development of good practice as well as providing country and regional statistics.

United Nations (2010) *World Population to 2300*. Available to download at www.un.org/esa/population/publications/.../WorldPop2300final.pdf. The report provides an overview of global demographic trends and explores in detail population size and growth, urbanization and city growth, population ageing, fertility and contraception, mortality and international migration.

Useful websites

www.un.org/esa/population/unpop.htm Provides access to extracts from the United Nations' population-related publications and statistics.

www.census.gov/ipc/www/idb US Census Bureau (online) International Data Base, United States Bureau of the Census, Washington, DC. This is a computerized source of demographic and socio-economic statistics for 227 countries and areas of the world from 1950 to the present, and projected demographic data to 2050, freely available on the World Wide Web.

www.prb.org Home page of the Population Reference Bureau, based in Washington, DC. This site provides news updates on all aspects of world and US population, plus information about its own publications (including its monthly *Population Today and Population Bulletin*) and links to other websites.

www.populationaction.org Home of Population Action International, an organization dedicated to advancing policies and programmes that slow population growth in order to improve the quality of life for all people.

http://popindex.princeton.edu The primary reference tool to the world's population literature, maintained by the Office of Population Research at Princeton University, with a searchable and browsable database containing over 46,000 abstracts as published in *Population Index 1986–2000*.

www.unaids.org Provides a great deal of statistical data on the epidemic and access to publications.

www.un.org/millenniumgoals Provides a background to the Millennium Development Goals and has numerous publications and reports.

For annotated, clickable weblinks and useful tutorials full of practical advice on how to improve your study skills, visit this book's website at **www.pearsoned.co.uk/daniels**

RESOURCES, ENERGY AND DEVELOPMENT

Chapter 5

Michael Bradshaw

Topics covered

- The nature of natural resources and ways of defining and classifying them

- The factors that influence resource availability

- Types of renewable energy

- The changing geography of oil and gas production and consumption

- Global energy dilemmas

Resources are defined by *society*, not by nature.
(Adapted from Rees 1985:11)

Resources are not, they become; they are not static but expand and contract in response to human wants and actions.
Zimmerman (Peach and Constantin 1972:16)

The large-scale exploitation of the planet's resource base is one of the defining characteristics of the age in which we live (McNeill 2000). Levels of resource consumption are a major marker between the so-called 'developed' (North) and 'developing' worlds (South) (see Chapter 8 for further discussion). According to the World Bank (2010a), in 2008 the 'high-income countries' accounted for 16.0 per cent of the world's population, but produced 71.5 per cent of global Gross Domestic Product (GDP); in 2007 they consumed 48.2 per cent of the world's commercial energy; and in 2006 they accounted for 44.4 per cent of the world's CO_2 emissions. In the early twenty-first century we are, fortunately, increasingly aware of the finite nature of the planet's resource base; recognising that long before its resources physically run out we will have to find alternative ways of doing things to avoid destroying the ecosystem (see Chapter 6). The explosion at the Deepwater Horizon drilling platform in April 2010, and the subsequent oil leak that poured oil into the Gulf of Mexico for three months, provides a stark reminder that the exploitation of natural resources usually results in environmental degradation. There are also many tragic reminders of the human cost of resource extraction, such as the plight of the 33 Chilean miners trapped underground for 69 days or the deaths of more than 2600 coal miners, in China, in 2009 alone. Indeed, all energy sources carry risks, as the crisis in March 2011 at the Fukushima Daiichi nuclear power station in Japan following the devastating earthquake and tsunami showed. This chapter explores the nature of natural resources and considers the relationship between resource production and consumption, economic development and climate change (see Gautier 2008 for a discussion of climate change and the role of energy). The chapter is divided into three sections, each with a distinct task. The first section examines the nature of 'natural resource', it evaluates the various ways of classifying resources and analyses the diverse factors that influence their availability. The second section analyses the specific case of energy resources, and reviews the various types of renewable energy and the changing geographies of oil and gas production and consumption. The third section considers the interrelationship between energy consumption, economic development and climate change. The chapter concludes by discussing the energy dilemmas – the interrelationship between energy security, economic development and climate change – that currently face the different regions of the world.

5.1 Natural resources

The Earth holds a finite stock of resources, the **resource base**. However, what human societies consider a resource has varied through time and across space. As both Zimmerman and Rees acknowledge above, something is a resource because human society attaches value to it. Many of the things we value today as 'resources' held no value in the past and may hold no value in the future. Followers of science fiction, such as *Star Trek*, see a future based on resources and technologies that have yet to be created.

5.1.1 Defining resources

In his seminal work *World Resources and Industries*, Zimmerman (Peach and Constantin 1972: 9) states: 'The word "resource" does not refer to a substance, but a function that a thing or a substance may perform, or to an operation in which it might take place.' He also notes that 'resources are . . . as dynamic as civilization itself.' In fact, one can suggest that each major human civilization was sustained by a particular set of resources and technologies for their exploitation (Smil 1994; Simmons 1996); the archaeological record includes the *Stone Age*, the *Bronze Age* and the *Iron Age*. In reference to the Industrial Revolution of the nineteenth century, Simmons (1996: 208) states that 'industrialization based on fossil fuel energy represents a turning point in the history of human–nature relations.' If the nineteenth century was based on the era of coal, then the twentieth century will been seen as the era of oil (Hall *et al.* 2003). Clearly, the changing notion of what constitutes a 'resource' is an important factor in shaping the relationship between human societies and the natural environment. As Blunden (1995: 164) observed, 'Because definition as a resource depends on usefulness to human society, natural materials may be required as resources by societies in some times and places but not in others.' While it is still the case that people in many parts of the world live in what Mather and Chapman (1995: 139) call 'low-energy societies' (dependent upon plants, animals and human labour), one of the defining characteristics of our age, and a consequence of the processes of globalization, is that the majority of societies now have some shared notion of what sorts of things constitute resources. The Inuit living in northern Canada are as dependent on gasoline for their skidoos as the commuters for their cars in Los Angeles or

London are. Furthermore, echoing Lenin's famous edict that 'Communism is electrification plus Soviet labour power', the provision of electricity is central to the modernization process in all types of societies. There are still 1.4 billion people around the world that lack access to electricity (IEA, UNDP and UNIDO 2010) and it is seen as a key building block for the achievement of the Millennium Development Goals (MDG).

If the notion of resource is dynamic and intimately linked to the evolution of human society, it follows that so-called 'technological progress' both creates and destroys resource value. As new technologies emerge, dependent on particular resources, so old technologies and their associated resources become redundant. For example, today flint has limited value as a building material, yet in the Stone Age it was an essential resource for making tools. Because such resources had 'use value' they were also objects of trade. Other resources that were valued in the Bronze Age,

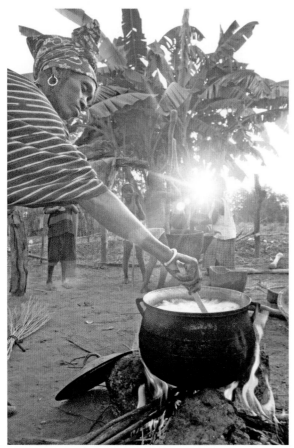

Plate 5.1 The use of firewood for cooking in a 'low-energy' consumption society.

(Ami Vitale/Panos Pictures)

such as copper and zinc, are still valued today as new ways of using them are discovered. New technologies also create new demands and concerns, such as the availability of so-called rare earth metals used in many 'green' technologies or lithium which is needed for rechargeable batteries. The use value of resources also means that control over their supply is an important part of political and economic power. In theory at least, resource-rich regions are able to exploit this natural advantage in their dealings with resource-poor regions. Consequently, gaining control over particular resources has been at the heart of many wars and much conflict (LeBillion 2007). The European colonization of the 'developing world' was motivated by a desire to discover and control new sources of resources. It is also no surprise that even today five of the world's ten largest corporations are oil and gas companies – Shell, Exxon Mobil, BP, Sinopec (China Petroleum and Petrochemical Company) and CNPC (China National Petroleum Company) (Dicken 2011: 243–69). Our discussion so far has implied that all resources share the same characteristics. However, we know that there are many different types of natural resource and the various ways of classifying them is the subject of the next section.

5.1.2 Classifying resources

Natural resources are commonly divided into two types: **non-renewable** or **stock resources** and **renewable** or **flow resources** (Figure 5.1). Stock resources are those, mainly mineral, that have taken millions of years to form and so their availability is finite. Hence, we also refer to them as non-renewable as there is no possibility of their being replenished on a timescale of relevance to human society. Within the category of stock resources, it is useful to distinguish between those that are consumed by use, such as fuel minerals, those that are theoretically recoverable and those that are recyclable, such as aluminium. A further characteristic of stock resources is that they tend to be highly localized, that is they are found in relative abundance only in specific places, such as ore deposits and coalfields. Some stock resources are more abundant than others and their relative scarcity affects their value. For example, aggregate minerals, like sand and gravel, are relatively abundant, while precious metals, such as gold and silver, are relatively rare.

Renewable or flow resources are those that are naturally renewed within a sufficiently short time-span to

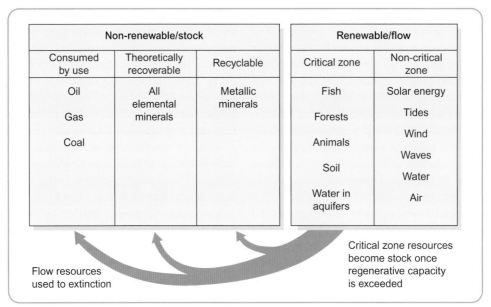

Figure 5.1 A classification of resource types.

Source: J. Rees, *Natural Resources: Allocation, Economics and Policy*, 2nd edn, Routledge, 1985, p. 15

be of use to human society. Again we can distinguish between different types of stock resource. Figure 5.1 divides flow resources into 'critical zone' and 'non-critical zone' resources. The distinction here is between those flow resources whose continued availability is dependent upon management by society (critical zone) and those that will continue to be available independent of the actions of society (non-critical zone). As indicated by the arrows, it is possible for critical zone flow resources to become stock resources if they are mismanaged and their regenerative capacity is exceeded. Thus, resource management aims to ensure that exploitation of a particular renewable resource

does not damage its capacity to replace itself. In recognition of the increasing challenge, and the numerous failures, to manage renewable resources, Rees (1991: 8) has developed an alternative to the conventional two-part typology of natural resources: 'All resources are renewable on some timescale . . . what matters for the sustainability of future supplies is the relative rates of replenishment and use . . . it seems better . . . to think in terms of a "resource continuum" than the conventional two-part typology' (Figure 5.2). Water might once have been considered a relatively abundant resource, but the combined impact of population growth and industrialization is placing increasing stress on

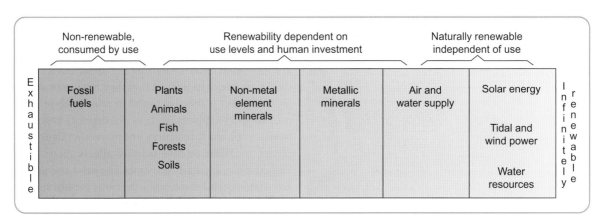

Figure 5.2 The resource continuum.

Source: J. Rees, Resources and environment: scarcity and sustainability, in Bennett and Estall (eds) *Global and Challenge: Geography in the 1990s*, Routledge, 1991, p. 9

the planet's finite supply of fresh water. Millennium Development Goal 7 (see www.developmentgoals.org) is to ensure environmental sustainability, and a target has been set to halve by 2015 the proportion of people without sustainable access to safe drinking water and basic sanitation. A 2010 assessment on the progress of the Millennium Development Goals (United Nations, 58–62) reports that 'if current trends continue, the world will meet or even exceed the MDG drinking water target by 2015'. However, a large rural–urban divide still exists and 'globally, eight out of 10 people who are still without access to an improved drinking water live in rural areas'. By comparison, little or no progress has been made on the sanitation goal, and here the situation is much worse in urban than rural areas. The resource continuum recognizes that access to (clean) air and water supply now have scarcity value and should be considered as finite.

In our discussion so far we have highlighted the complexity of the notion of natural resource and considered the various ways of classifying resources, but what factors affect the availability of particular resources?

5.1.3 Resource availability

As was noted earlier, at any moment in time there is a finite stock of natural resources on the planet, the resource base. Each stock resource has its own resource base, the total quantity of a substance or property on the planet, for example, the total amount of oil geologically present today. However, that total resource base is not the amount available for human exploitation. Figure 5.3 illustrates the relationship between the resource base and the various sub-divisions of resource availability. The term proven reserve is applied to those deposits that have already been discovered and are known to be economically extractable under current demand, price and technological conditions. However, the extent of a proven reserve is dynamic and dependent upon a host

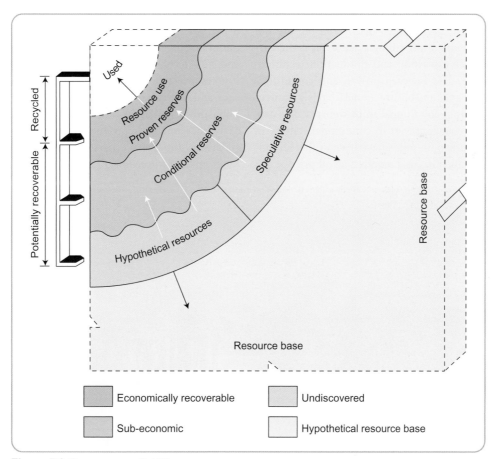

Figure 5.3 Resource availability.

Source: J. Rees, *Natural Resources: Allocation, Economics and Policy*, 2nd edn, Routledge, 1985, p. 20

of interlinked factors. These include the availability of the technology and skills to exploit the resource, the level of demand, the cost of production and processing, the price it can command in the marketplace, the availability and price of substitutes, and the environmental and social costs of developing the resource. The extent to which each of these factors influences resource development also varies across space and time. Today we have the technological ability to recover resources in geological and environmental conditions that were previously uneconomic. For example, the Deepwater Horizon drilling rig in the Gulf of Mexico that was destroyed by an explosion in April 2010 was designed to operate in waters up to 2,400 metres (8,000 feet), to a maximum drill depth of 9,100 metres (30,000 feet). It was pushing the limits of offshore production into ever-deeper waters and deeper fields. The recent growth of shale gas production in the United States relies on new techniques of horizontal drilling and fracturing that release the natural gas trapped in shale deposits far below ground at a competitive cost, resulting in a dramatic surge in natural gas production in the USA.

The category of **conditional reserve** refers to deposits that have already been discovered but are not economic to work at prevailing price levels using currently available extraction and production technologies. The boundary between proven and conditional reserves is dynamic and bi-directional (that is, resources that change from conditional to proven reserves can, if conditions change, revert to conditional status). In the past, Canadian oil sands have been at this boundary. High production costs required an oil price of $65 or higher to be economic, when the price fell below this level investment in oil sands production fell substantially. However, a combination of technologically driven cost reduction and the prospects of sustained high prices has pushed up oil sands production again, despite its negative environmental impacts. The two remaining categories of resource are not readily available to society. **Hypothetical resources** are those that we may expect to find in the future in areas that have only been partially surveyed and developed. Thus, the US Geological Survey has suggested that the Arctic accounts for 13 per cent of the world's undiscovered oil and 30 per cent of its undiscovered gas. There are also hypothetical resources in regions such as Antarctica, but the international community has placed a moratorium on resource development, for the time being at least. **Speculative resources** are those that might be found in unexplored areas that are thought to have favourable geological conditions. Finally, there remains a large part of the earth about which we have no information on its potential resource base. The strength of this classification is that it stresses the highly dynamic nature of the concept of resource reserve. Its weakness is that it lends itself to the idea that we will never run out of resources; there will always be more to discover and new technology will continue to make new resources available for exploitation. Even if that was the case, and clearly it is not, the planet now faces the additional problem that the consumption of resources, such as hydrocarbon fuels, produces wastes that threaten the stability of the global ecosystem. Thus, there is a need to rethink the whole notion of resource scarcity to take

Plate 5.2 As offshore technology advances, so more remote resources can be exploited.

(ImageState/Alamy)

into account the ecological cost of our current 'fossil fuels society'.

Given the highly dynamic nature of resources, at any moment in time, it is very difficult to estimate the level of resource availability. It is even more difficult to speculate about future levels of production and consumption and, thus, the possibility of resource scarcity. Nonetheless, the supporters of 'peak oil' maintain that the world will soon reach the point of maximum oil production (Deffeyes 2001), which is highly contested, not just because extrapolation on the basis of current trends is often misleading, but also because there are a whole variety of factors that can promote resource scarcity (Table 5.1). The resource crisis of the 1970s was motivated by geopolitics, but by increasing the price of energy, it brought about other forms of scarcity. For example, the increased price of oil had a major negative impact upon those countries in the 'developing world' that had embarked upon industrialization and had become increasingly dependent upon imported oil. However, it also shocked the 'developed world' into the realization that energy resources were finite, energy conservation was worthwhile and that alternative sources of energy were required. The 'developed world', and particularly the United States, then recognized the strategic importance of securing 'energy independence', but it remains an elusive goal.

In July 2008 the oil price peaked at $144 a barrel, only to fall below $40 by November 2009 as a global economic crisis took hold. Despite continued geopolitical tensions in the Middle East and North Africa, we still have a relative abundance of oil, in large part because the global economic recession reduced demand. But there is no scope for complacency; the rapid growth of energy demand in the emerging economies, especially China and India, is causing concern that high energy prices will return. This coincides with increasing societal concerns about the ecological consequences of increased energy consumption. Between 1970 and 2004 the largest growth of greenhouse gases came from the energy supply sector, with an increase of 145 per cent (the IPCC Fourth Assessment Report is available at www.ipcc.ch – see Working Group Three 2007: 3). The prospect of increasing demand as a result of economic development in the 'global south', combined with the challenges of climate change, has led to the idea of a 'New Energy Paradigm' that combines traditional concerns about security of energy supply at reasonable prices with the need to devise energy policies that reduce carbon emissions (Helm 2007).

Table 5.1 The dimensions of resource scarcity

Type of scarcity	Concern
Physical scarcity	• Exhaustion of minerals and energy. • Human populations exceed the food production capacity of the land. • Depletion of renewable resources such as fish, soils or timber. • Growing demand for water for human use threatens aquatic ecosystems and the ability of river systems to replenish themselves.
Geopolitical scarcity	• Use of minerals exports as a political weapon. • Shift in the location of low-cost minerals sources to 'hostile' or unstable blocs of nations.
Economic scarcity	• Demand at current price levels exceeds the quantity supplied, resulting in shortages. • Needs exceed the ability of individuals or countries to pay for resource supplies (resource poverty). • Rich economies can always outbid the poor for essential resources, creating unequal patterns of resource use. • Economic exhaustion or falling demand for specific minerals or renewable resources causes economic and social disruption in producer regions or in nations dependent on them.
Renewable and environmental resource scarcity	• Distribution of essential biogeographical cycles (e.g. the carbon cycle and the greenhouse effect) threatening sustainability of life on earth. • Pollution loads exceeding the 'absorptive' capacity, causing economic health and amenity problems. • Loss of plant and animal species (biodiversity) and landscape values, with wide, but poorly understood, long-term consequences.

Source: Adapted from J. Rees (1991) Resources and environment: scarcity and sustainability, in Bennett and Estall (eds) *Global Change and Challenge: Geography in the 1990s*, Routledge, London, p. 6

Plate 5.3 The energy crisis in 1973 brought queues at the petrol pumps in many Western countries and a three-day week in the UK.

(Evening Standard/Stringer/Getty Images)

5.2 Fuelling the planet

Of all the different types of resource that we have discussed so far, in recent history it is those that provide energy services (we transform energy resources to obtain services such as heating, light, transportation, etc.) that have been the most sought after. For most of human history societies have utilized renewable sources of energy; flow resources such as wood that can be depleted, sustained or increased by human activity; and continuous resources, such as water to drive watermills or the wind to turn windmills, that are available irrespective of human activity (Figure 5.4). They have also used draught animals, horses, buffalo, etc., but these need to be fed and housed. The Industrial Revolution changed the way in which certain parts of the world powered their economies, leaving the remainder dependent upon renewable resources.

5.2.1 The dominance of fossil fuels

Since the invention of the steam engine about 200 years ago, much of human society has become ever more dependent upon the exploitation of non-renewable energy resources (see Smil 2010 for an introduction to the energy system and energy transitions). According to BP (2010), in 2009 nearly 88 per cent of the primary commercial energy consumed on the planet came from the three fossil fuels (coal, oil and gas). These resources were formed from the decomposition of organic materials millions of years ago and have been transformed by heat and pressure into coal, oil and natural gas. The advantage of fossil fuels as a source of energy is that they are readily accessible, a highly concentrated source of energy, easy to convert using proven technologies, and cost-efficient in production and use. Unfortunately, the combustion of fossil fuels is also the single largest source of greenhouse gases (GHG).

Oil is the most versatile fossil fuel while coal, even though it is relatively bulky, is still used in many industrial processes, as a source of heat, and the most important fuel for electricity generation. In recent years, particularly in western Europe, it has been unable to compete with the versatility of oil or with cleaner natural gas. The great advantage of oil is its transportability over long distances in large volumes using pipelines,

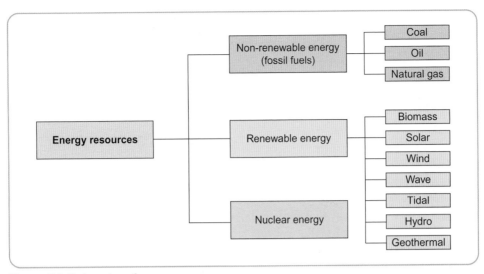

Figure 5.4 Categories of energy resources.

Source: Jones and Hollier (1997: 173)

ocean tankers, or smaller rail and road tankers. It has spawned a huge petrochemical industry producing a vast range of products, including plastics and textiles, that have substituted products based on renewable resources, such as wood and cotton, for those based on non-renewable resources. In addition, oil has created many resources, such as jet engine fuel, for which commercially viable substitutes have yet to be developed. Consequently, advanced industrial societies are totally dependent upon the continued supply of oil at a reasonable price, which explains the importance that governments attach to energy security.

In recent years, natural gas has become an increasingly important source of energy and a feedstock for petrochemicals. In many advanced industrialized economies it has replaced coal, and even oil, as the favoured resource for electricity generation and domestic heating. However, the transport flexibility of natural gas is inferior to oil; pipeline transport is most economic but the infrastructure investment needed requires sizeable reserves, a secure source of production (security of supply) and stable markets (security of demand). Liquefied natural gas (LNG), produced via cooling and high pressure, is an option but the process requires substantial investment in plant and consumes large quantities of energy. Nevertheless, LNG production is becoming an increasingly important fuel, particularly in the Asia-Pacific region, while the European

Plate 5.4 Special LNG tankers are used to distribute liquefied natural gas to markets worldwide.

(AP Photo/Dita Alangkara)

Union sees LNG supplies as a means of compensating for declines in domestic production and diversifying its sources of supply.

5.2.2 Alternative sources of energy

Given the finite nature of hydrocarbon fuels, the problems of scarcity and the environmental consequences of burning fossil fuels, it is no surprise that alternative sources of energy have been sought. There are two alternatives (see Figure 5.4): first, developing new ways of harnessing renewable energy sources, and second, developing nuclear power. At present all commercial nuclear power stations use so-called fission reactors that release energy, mainly in the form of heat, which is then used to generate electricity. In the 1950s, nuclear energy was heralded as a cheap and clean alternative to fossil fuels, but safety issues discouraged the expected expansion of the nuclear industry (see Case study 5.1).

Today, since it does not generate CO_2, the nuclear option for electricity production is back on the agenda even though many of the concerns about high costs, safety, the disposal of radioactive waste and the decommissioning of power stations remain. The disaster at Fukushima Daiichi power plant in Japan has reopened the debate and we can expect a reassessment of the role that nuclear power can play in meeting the growing demand for energy and mitigating climate change.

The other alternative to fossil fuels is the development of 'new' renewable sources of energy (Evans 2007: 81–114). Many of these renewable sources involve applying new technologies to historic sources of energy supply and scaling them up to provide an alternative source of electricity generation and fuel. While no single source is likely to provide an alternative to fossil fuels, collectively they can start to reduce our reliance on hydrocarbons. A recent review by the Swedish Academy of Sciences (Destouni and Frank 2010: 19)

Plate 5.5 For more than two decades the Chernobyl disaster spelled the end for nuclear power as a clean and safe alternative source of energy.

(Ria Novosti/Science Photo Library)

Thematic case study 5.1

A nuclear renaissance?

During the 1950s the peaceful use of nuclear power was heralded as the great hope for the world's energy needs. Harnessing the power of the atom promised a cheap and clean source of electricity. The energy crisis of the 1970s increased interest in a nuclear solution. By developing a nuclear power industry, the developed world hoped that it could reduce its reliance upon OPEC-supplied oil. The 1970s and 1980s saw a rapid expansion in the generating capacity of the world's nuclear power plants. Between 1960 and 1990 total global electrical generating capacity had grown from zero to 328 gigawatts. In the last decade or so growth has stalled (in 2008 the total global capacity was 372.5 gigawatts) with the real possibility of a decline in total capacity. What went wrong and have conditions changed such that a return to nuclear power is now back on the agenda?

In short, as the nuclear power industry developed, expanded and matured it became increasingly apparent that the supposed benefits were far outweighed by the environmental problems it posed. Reddish and Rand (1996: 79) listed four criticisms levelled at nuclear power by environmental groups and these remain the major concerns today:

1. Abnormal radiation levels from normal operations will cause cell damage, malignant cancers, genetic diseases, etc.
2. Reactor operations and transport of irradiated fuels cannot be guaranteed safe against catastrophic accidents.
3. Radioactive waste has to be disposed of safely and reactors have to be decommissioned, all of which is extremely costly.
4. The radioactive materials needed to fuel the reactors need to be kept safe and secure from theft or misappropriation, to make nuclear weapons (the so-called proliferation problem).

Over the years public trust in nuclear power has been significantly eroded by a number of serious accidents, such as those at Windscale, Cumbria (United Kingdom) in 1957, Three Mile Island (USA) in 1979, Chernobyl in the Ukraine in 1986 and now Fukushima in 2011. Furthermore, as the true cost of building, maintaining and decommissioning nuclear power plants has become apparent many have questioned the economics of nuclear power relative to other sources of energy.

Despite these problems, in 2008 nuclear power accounted for 15 per cent of global electricity generation, but until recently it has been very much an option for the developed world, partly because of cost and technology, but also because of concerns over the proliferation of nuclear weapons. In 1996, OECD member states accounted for 86.6 per cent of total nuclear power production, but by 2005 that share had fallen slightly to 83.8 per cent (IEA 2006: 437). In 2008 the top three countries in terms of nuclear capacity were the United States (27.1 per cent), France (17.0 per cent) and Japan (12.4 per cent). In the same year, nuclear power accounted for 76 per cent of total domestic electricity generation in France and 42 per cent in Sweden. Sweden had decided to phase out its nuclear power stations and switch to conventional thermal power stations, but in 2009 it reversed this policy. Until recently, the high cost of constructing nuclear power stations, rising public opinion against them and the increasing availability of cheaper alternatives, such as LNG, meant that substantial expansion seemed unlikely. However, current concerns about climate change and the associated desire to reduce CO_2 emissions, plus the tight supply of oil and gas and associated high-energy prices have changed government attitudes towards nuclear power. Hedberg *at al.* (2010: 3) report that at present 55 new reactors are being built in 30 countries, 137 are planned and a further 295 are proposed for the future; there are currently nearly 440 fission reactors in operation in 30 countries (for further details visit www .world-nuclear.org).

For the emerging markets of China, India and Russia the expansion of nuclear power is seen as a key element of policies to meet the energy demand associated with rapid economic growth. Russia, for example, has a programme to increase the share of nuclear power in electricity generation from 15 per cent now to 25 per cent by 2020, it also exporting nuclear power stations to emerging markets. China has set a target to build 40 gigawatts of nuclear generating capacity by 2020 and India has a target of 40 gigawatts by 2030. Many OECD countries are also reconsidering the nuclear option because, first, it addresses the need to reduce greenhouse gas emissions and second, increased reliance on imported oil and gas is raising concerns about energy security. In particular, following supply disruptions in 2006 and 2009, the European Union is concerned about its increasing reliance on supplies of natural gas from Russia. Given the high costs and long lead times to build nuclear power plants, a decision needs to be made now to avoid energy shortages in the future. The environmental movement remains vehemently against nuclear power as it argues that none of the long-standing problems have been adequately addressed. Thus, the nuclear debate is being revisited in the United Kingdom and elsewhere in the OECD, as the disadvantages of the nuclear option are set against the possible benefits it might bring in terms of addressing climate change and promoting energy security (see Case study 5.2). The UK government is now committed to building a new generation of ten nuclear power stations, all on the site of existing soon-to-be-decommissioned nuclear stations. The United States' Energy Information Administration (EIA 2010: 4) predicts that electricity generation from

nuclear power will increase from 2.6 trillion kilowatt-hours in 2007, to 3.6 trillion kilowatt-hours in 2020 and 4.5 trillion kilowatt-hours by 2035. The strongest growth will be in non-OECD Asia, where nuclear power electricity generation is projected to grow at an annual rate of 7.7 per cent a year from 2007 to 2035. However, the disaster at Fukushima will undoubtedly result in increased public concern about the safety of nuclear power and the lessons learnt may increase the costs of nuclear power further, calling into doubt its viability. The problem is that if nuclear power falls from favour again it is likely to result in even higher levels of fossil fuel consumption as power companies turn to coal and gas as a low-cost alternative. Alternatively, it could spur investment in renewable energy, but that seems unlikely.

maintains that by 2050, renewable energy can be expected to provide up to 35 per cent of global energy supply and nearly half of electricity production, which still means 65 per cent of global energy will come from fossil fuels and nuclear power.

■ *Biomass energy* was the very first form of energy used by humans and involves the burning of plant and animal residue to produce heat. In much of the world this is still the traditional source of energy. The IEA, UNDP and UNIDO (2010: 20) maintains that there are currently 2.7 billion people in developing countries who rely for cooking primarily on biomass used in inefficient devices. In industrially developed economies, wood waste from pulp and paper mills is used to generate heat and electricity, and in some households wood-burning fires are used for space heating, although often more as a lifestyle choice than out of necessity. More recently, the use of biomass to produce alternative fuels, so-called biofuels, has gained popularity in the developed world; but it is increasingly apparent that if not properly managed biofuels production creates as many problems as it solves. The land being taken up to produce crops for biofuel production is either a natural habitat supporting biodiversity or agricultural land that could be used for food production. At the same time, increased demand for crops such as corn is increasing the cost of animal fodder and basic foodstuffs. Further, much of this production requires the use of fertilizers and pesticides that are hydrocarbon-based and consume considerable amounts of energy when they are produced. The net result is that many 'first

generation' biofuels may actually consume more energy to produce than they generate when used to run a vehicle. So-called 'second generation' biofuels aim to use non-food feedstock, organic waste and algae and may become a viable alternative to fossil fuels. Finally, on a small scale, gas from landfill sites and from the anaerobic digestion of vegetable waste is also being developed to produce methane.

■ *Solar energy* is ultimately the source of all energy, with the exception of nuclear power. Solar energy usually involves capturing the sun's rays in order to produce thermal energy to warm buildings, heat hot water or generate steam to produce electricity. Alternatively, photovoltaic cells turn the sun's rays directly into electricity. As the efficiency of solar panels increases and their cost declines, so solar energy is becoming a viable small-scale local solution at the household or building level. However, it is subject to daily or seasonal variations and is weather-dependent, but it can provide an efficient supplemental source of energy for heating domestic hot water, and so on (visit the website of the International Solar Energy Society at www.ises.org). There are now some very ambitious large-scale projects aimed at harnessing the sun's energy and transmitting it via high-voltage transmission lines to centres of demand. The so-called 'Desertec Concept' is based on the proposition that 'Within 6 hours the deserts receive more energy from the sun than humankind consumes in a year' (www.desertec.org) and it proposes the development of a network of concentrating solar power installations in North Africa to generate electricity for export to Europe. As the technology improves, energy costs increase and the costs of solar energy decrease, it will increasingly become part of the low-carbon energy solution.

■ *Wind power* is of historical significance and has now made a return in the form of 'wind farms'. Improved design has increased the efficiency of modern windmills and they are becoming an increasingly common sight in exposed coastal and upland locations in the developed world. Wind farms can generate electricity for local use or for the power grid. However, some would argue that modern wind farms lack the aesthetic qualities of traditional windmills and therefore many see them as a source of both visual and noise pollution. In Europe, Germany and Denmark have led the way in wind power. In September 2010, Vattenfall opened the Thanet Offshore Wind Farm, which has 100 wind turbines and covers an area of 35 square kilometres. It is the largest operational offshore wind farm anywhere in the world and has a capacity of 300 megawatts, which has boosted UK offshore wind capacity by 30 per cent, producing on average enough electricity to supply more than 200,000 homes. The UK government has plans for further offshore wind farms to be built off the coast in the Thames Estuary, the Greater Wash and the North-west, and the energy from them is expected to power 15 per cent of UK households. In terms of total installed wind capacity in 2009, the USA comes first with 22.3 per cent, followed by Germany at 16.3 per cent and China at 5.9 per cent (for more information on wind power see www.windpower.org). But it is China that leads the way in terms of new installed capacity. In 2009 alone China installed 13.8 gigawatts of new wind power capacity (compared to 6.3 gigawatts in 2008). However, despite this impressive level of activity, wind power still only accounts for less than 1 per cent of China's total electricity demand (World Wind Energy Council 2010: 23). China badly needs new sources of energy that will not aggravate already substantial energy-related environmental problems and it has ambitious plans for renewable energy. It also sees the supply of equipment, such as wind turbines, as a new growth area for industry for both domestic consumption and export.

■ *Hydropower generation* also pre-dates the industrial revolution, but dams and hydroelectric schemes have now replaced the watermills that powered the early machines. In many regions of the world hydroelectricity has become an important source of energy, but, in all but a few cases, it cannot offer a large-scale solution to a country's energy needs. The building of dams to store water and build a 'head' to drive the turbines that generate the electricity floods large areas of land; it also starves river systems of flow and damages deltaic and riverine environments. The construction of Three Gorges Dam on the Yangtze River in China involved planning the displacement of 1.2 million people and the creation of a reservoir occupying 111,000 hectares. Other forms of waterpower, such as wave and tidal power, are also being developed. At present, there is only one large-scale commercial tidal power station in the world, the La Rance Tidal Power Plant in Brittany, France. The UK government was considering building a 10-mile long barrage across the mouth of the River Severn between England and Wales to exploit the Bristol Channel's 14-metre tidal range, the second highest in the world. It was claimed that the project would generate 5–6 per cent of the electricity needed by England and Wales. However, after a 2-year feasibility study the government decided in late 2010 not to develop the scheme, but to keep the option open for future consideration. A cost in excess of £30 billion made the project high cost and high risk in comparison to other ways of generating electricity. New forms of wave power and tidal power are being developed, such as the Pelamis Wave Energy Converter, but these are still at the developmental stage.

■ The final form of renewable energy is *geothermal energy*. As the name suggests, this form of energy uses naturally occurring heat in the earth's crust and is only really viable in areas of volcanic activity. There are two forms of geothermal energy. One is wet-rock geothermal energy where steam or hot water is trapped from boreholes or surface vents and used to heat buildings or generate electricity. The other is hot–dry geothermal energy, which is accessed by boreholes drilled into hot dry rocks. Water is then forced down and the steam used to generate electricity. At present geothermal energy accounts for about 10 per cent of New Zealand's electricity production. In global terms, it is the United States that produces the most electricity with geothermal power. In 2009, US geothermal power plants produced 15.2 billion kilowatt-hours, or 0.4 per cent of total US electricity generation, and seven states have geothermal power plants. At a very local scale there is growing interest in using ground-source heat pumps to heat individual buildings. This technology does not require a geothermal source, but makes use of renewable energy stored in the ground.

Plate 5.6 Thanet Offshore Wind Farm.

(www.vattenfall.co.uk/en/thanet-offshore-wind-farm.htm)

In nearly all of the above cases, renewable energy is already providing localized, small-scale solutions to energy needs but, at present, does not provide a viable alternative to fossil fuels. The challenge is to 'scale up' renewable energy production to provide a low-carbon alternative. This can take the form of large-scale centralised systems, like an offshore wind farm or a concentrating solar array, or decentralised generation within individual households and buildings. At present, most of these alternatives require considerable capital investment and access to advanced technology and they do not, therefore, offer an obvious solution to the fast-growing energy needs of the 'developing world'. That said, the 'developed world' can afford to develop renewable energy technologies and considerable progress has been made. The European Union aims to increase their contribution from 7 per cent in 2007 to 20 per cent by 2020 (Commission of the European Communities 2007: 13). The UK target is 15 per cent by 2020. If fossil fuel prices remain high or increase further and/or a tax is placed on carbon emissions, this will stimulate investment in renewable energy making it a viable alternative for those that can afford it. But, to provide a more global solution, those technologies, and the finance to purchase them, will need to be made available to the regions experiencing the most rapid increases in energy demand, i.e. the global south.

5.2.3 The changing energy mix

The balance between the various sources of energy is known as the energy mix. During the last century it changed substantially, but a substantial regional variation remains, as shown in Table 5.2. Simply because it accounts for most of the world's energy consumption, the evolution of the energy mix tends to be linked to the development of the industrialized world where coal was by far the most important energy resource at the start of the twentieth century. The oil industry was still in its early stages; the first modern oil well was drilled in the US state of Pennsylvania in 1859. Natural gas was not recognized as a resource at all. Hence, the dominance of coal was not threatened. Advances in transportation technology, such as the invention of the jet engine, and the emergence of a petrochemicals industry triggered a rapid increase in the demand for oil in the boom years after the Second World War. In 1950 coal accounted for 61 per cent of the world's commercial energy consumption with only 27 per cent coming from oil; by 1970 it was 30 per cent from coal, 44 per cent from oil and 20 per cent from natural gas. Even so, the absolute level of global coal production continues to increase at the same time as economic growth in the developed world is increasingly underpinned by oil and, more recently, by natural gas. The recent 'dash-for-gas' in the developed world is driven by a desire to reduce dependence on oil imported from the Middle East and the fact that electricity generation using natural gas generates less greenhouse gases. Thus, in the past hundred years, the energy mix in the industrialized world has undergone two major transitions, from coal to oil and from oil to gas.

The energy mix in the developing world is somewhat different (see Table 5.2). For the most part it

Table 5.2 Regional variations in energy balance, 2010 (per cent of total consumption)*

	Oil	Natural gas	Coal	Nuclear	HEP	Renewables	Per cent of total
European Union	38.2	25.6	15.6	12.0	4.8	3.9	14.4
OECD	38.0	25.1	19.8	9.4	5.6	2.2	46.4
NON-OECD	29.8	22.7	38.1	1.6	7.2	0.6	53.6
North America	37.5	27.7	20.1	7.7	5.4	1.6	23.1
Europe & Eurasia	31.1	34.4	16.4	9.2	6.6	2.3	24.8
S. & C. America	46.1	21.7	3.9	0.8	25.7	1.8	5.1
Africa	41.7	25.4	25.6	0.8	6.2	0.3	3.1
Middle East	51.4	46.9	1.3	–	0.4	0.1	5.8
Asia-Pacific	27.7	11.2	52.1	2.9	5.4	0.7	38.1
World	33.6	23.8	29.6	5.2	6.5	1.3	100.0

*Includes commercially traded fuels only. Therefore excludes wood, peat and animal waste as fuels. Also excluded is wind, geothermal and solar power.

Source: BP *Statistical Review of World Energy* (2011: 41). (Also available at: www.bp.com)

remains more reliant upon oil for its commercial energy and in some regions coal remains the dominant resource. For example, China in 2009 accounted for 46.9 per cent of global coal consumption that generated 70.6 per cent of its commercial energy. Over the past few years China has met its growing energy needs by burning even more coal, raising obvious concerns about climate change. Elsewhere, dependence upon oil has made countries in the developing world particularly vulnerable to sudden periodic increases in oil price over the past three decades, but they lack the capital and technology to develop their own energy potential or to diversify their energy mix. However, as it depletes its own sources of oil and gas the developed world has become increasingly dependent upon supplies from the developing world. Historically, the international oil companies (IOCs) were the key actors delivering the energy resources of the developing world, but increasingly the governments of the oil-producing states have seized control over their oil and gas industries. Now IOCs, such as BP, ExxonMobil, ChevronTexaco and Shell, work in partnership with regimes whose human rights records are condemned by governments and NGOs in the developed world – the most high-profile case being Shell in Nigeria. Furthermore, the most attractive oilfields are reserved for the national oil companies (NOCs) of the reserve-holding states. The age of 'easy oil' is over for

IOCs and they must now make do with access to the most technologically and environmentally challenging fields.

5.2.4 The changing geography of energy production and consumption

Whereas in 1950 North America, Europe and the Soviet Union accounted for nearly 90 per cent of world energy demand, by 1990 their share had fallen to two-thirds (Jones and Hollier 1997: 181). In 1997 the member states of the Organization for Economic Cooperation and Development (OECD) accounted for 58.4 per cent of global energy consumption but by 2006 this had fallen to 50.1 per cent and by 2010 to 46.4 per cent (Table 5.2). The redistribution of energy consumption has not been due to decline in the developed world, although during the 1990s this was the case in the so-called 'transition economies', but to increased demand in the developing world (Table 5.3). There are substantial variations in terms of level of per capita energy use and energy intensity, with high-income countries having by far the highest level of energy use. The high usage levels in the middle-income countries and in Europe and Eurasia reflect the energy legacies of the post-socialist states in Central Europe and the former Soviet Union while in the developing world China and India are the new major emitters. The past decade has

Table 5.3 Energy consumption and efficiency

	GDP per unit of energy use (2005 ppp $/kg oil equivalent)	Energy use per capita (kg oil equivalent)	CO_2 emissions per unit of GDP (2005 ppp $/kg oil equivalent)
World	5.4	1,819	0.5
Low income	3.2	423	0.4
Middle income	4.4	1,242	0.6
Lower middle income	3.9	1,013	0.8
Upper middle income	5.2	2,130	0.5
Low & middle income	4.3	1,127	0.6
High income	6.5	5,521	0.4
East Asia & Pacific	3.6	1,295	0.9
Europe & Central Asia	3.7	2,948	0.7
Latin America & Caribbean	7.5	1,273	0.3
Middle East & N. Africa	5.0	1,275	0.6
South Asia	5.0	484	0.5
Sub-Saharan Africa	3.8	662	0.5
Euro Area	8.2	3,789	0.3
China	3.4	1,295	1.0
United States	5.5	7,766	0.4
India	4.9	529	0.8

Source: World Bank (2010b) *The Little Green Data Book 2010*. (Also available at: www.worldbank.org)

seen a global shift in the energy system. Population growth and economic development in countries like Brazil, India and China is driving up energy demand, while the developed economies are seeking reduced levels of energy consumption. According to BP, during 2008 China's energy consumption grew by more than 8.7 per cent, or some 19.5 per cent of global energy production, while global energy consumption actually fell by 1.1 per cent. This differential growth rate, together with the decline in indigenous production in the developed world, has major implications for global geopolitics (Bradshaw 2009).

There are three elements to the geography of energy production: the global distribution of reserves and production, the distribution of consumption, and the resultant pattern of trade between energy-surplus and energy-deficit regions. As noted earlier, the potential to produce coal, oil and natural gas is geologically determined, but the actual exploitation of a deposit depends on numerous factors. Reserve to production ratios (how long current reserves would last at current rates of production) for the three fossil fuels show that coal is the most abundant fossil fuel, followed by natural gas and then oil. World reserves of coal in late 2009 would last 119 years at current rates of consumption, current oil reserves would last 45.7 years, and natural gas reserves 62.8 years (BP 2010). While the OECD countries have more than sufficient coal to meet their needs, it is these economies that have most reduced their reliance upon coal, making the developed world dependent upon the very fuels for which it has the lowest reserve to production ratios. This means that the developed world is likely to become ever more dependent upon imports of energy from the developing world. Take oil reserves: the Middle East still accounts for the majority of the world's oil – 56 per cent in 2009, with a reserve to production ratio of 84.8 years. Natural gas reserves are dominated by the Middle East and the former Soviet Union where three countries, Iran, Russia and Qatar, account for 53 per cent of global reserves (BP 2010: 22). Thus, the majority of the world's oil and gas reserves are in regions that are presently perceived to be politically and/or economically unstable by the major importing states.

The diagrams shown in Figures 5.5 and 5.6 compare the share of world production and consumption of oil and gas by major region. Asia-Pacific, Europe and North America are the major deficit regions, Africa, the Middle East, the former Soviet Union and South and Central America are the major surplus regions. Despite being the second largest producer in 1997, the United States still imports almost as much oil as it produces (Table 5.4). Likewise, North Sea production falls well short of meeting Europe's demand for oil and is now past its peak production. The result is a movement of oil production dominated by Middle Eastern supply to Europe and North America and substantial movements to Singapore and Japan (Figure 5.7).

The major exporters and importers of natural gas are listed in Table 5.5, while Figure 5.8 shows the movements of natural gas. The less 'transportable' nature of natural gas is clearly shown in Figure 5.8; for the most part there is a balance within each region, excepting Africa and the former Soviet Union (Russia and Turkmenistan), which have a surplus of natural gas, and Europe, which has a deficit. Natural gas supply into Europe is via pipeline from West Siberia (Russia) and North Africa or as LNG from North Africa. At present most of this LNG originates from the Asia-Pacific region, but this is changing as supply from the Middle East (Qatar) to Asia and from the west coast of Africa to Europe increases. The recent rapid growth of unconventional gas production, which has seen the US pass Russia as the world's largest gas producer, has dramatically reduced the anticipated need for the US to import gas.

These geographies of production and consumption arise from a complex interaction of economic and

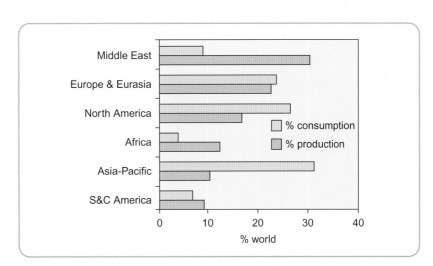

Figure 5.5 Distribution of oil production and consumption in 2010.

Source: *BP Statistical Review of World Energy* (2011: 8, 11). (Also available at: www.bp.com)

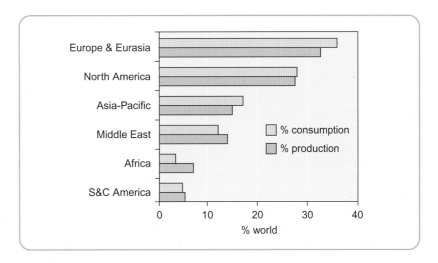

Figure 5.6 Distribution of natural gas production and consumption in 2010.

Source: *BP Statistical Review of World Energy* (2011: 24, 27). (Also available at: www.bp.com)

Table 5.4 Producers, exporters and importers of crude oil (million tonnes)

Producers in 2009	Mt	% of world	Exporters in 2008	Mt	Importers in 2008	Mt
Russia	494	12.9	Saudi Arabia	355	United States	564
Saudi Arabia	452	11.8	Russia	241	Japan	199
United States	320	8.3	Iran	120	China	175
Iran	206	5.4	UAE	108	India	128
China	194	5.0	Nigeria	102	Korea	116
Canada	152	4.0	Angola	92	Germany	105
Mexico	146	4.1	Norway	90	Italy	88
Venezuela	126	3.3	Kuwait	89	France	83
Kuwait	124	3.2	Iraq	88	Spain	61
UAE	120	3.1	Venezuela	74	Netherlands	57
Rest of World	1,509	39.2	Rest of World	593	Rest of World	514
World	3,843	100	World	1,952	World	2,090

Source: OECD/IEA (2010) *Key World Energy Statistics 2010*. (Also available at: www.iea.org)

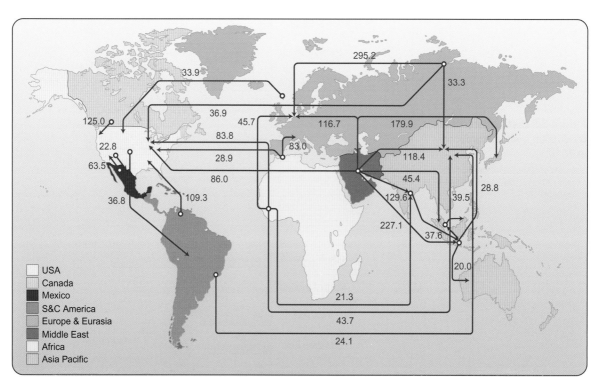

Figure 5.7 Major oil trade movements in 2010 (million tonnes).

Source: *BP Statistical Review of World Energy* (2011: 21). (Also available at: www.bp.com)

Table 5.5 Producers, exporters and importers[1] of natural gas (billion cubic metres)

Producers in 2009	BCM	% of world	Exporters in 2009	BCM	Importers in 2009	BCM
United States	594	19.2	Russia	160	Japan	93
Russia	589	19.0	Norway	100	Germany	83
Canada	159	5.1	Canada	76	United States	76
Iran	144	4.6	Qatar	67	Italy	69
Norway	106	3.4	Algeria	55	France	45
China	90	2.9	Indonesia	36	Ukraine	38
Qatar	89	2.9	Netherlands	30	Turkey	38
Algeria	81	2.6	Turkmenistan	27	Spain	34
Netherlands	79	2.5	Malaysia	24	Korea	33
Indonesia	76	2.5	Trinidad & Tobago	21	United Kingdom	29
Rest of World	1,094	35.3	Rest of World	140	Rest of World	214
World	3,101	100.0	World	736	World	749

[1]Exports and imports include pipeline gas and LNG

Source: OECD/IEA (2010) *Key World Energy Statistics 2010.* (Also available at: www.iea.org)

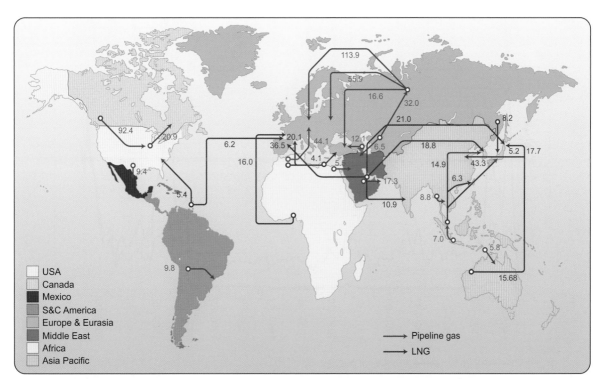

Figure 5.8 Major trade movements of natural gas in 2010 (billion cubic metres).

Source: *BP Statistical Review of World Energy* (2011: 31). (Also available at: www.bp.com)

political factors. Not surprisingly, there are substantial regional differences in the cost of oil production. The record-low oil prices of the late 1990s focused attention upon the actual cost of producing oil and Cambridge Energy Research Associates put the cost of production in the Middle East as low as US$2/barrel (*Economist*, 4 March 1999: 29). That same report estimated the cost of production in Indonesia at US$6, Venezuela and Nigeria at US$7, Mexico at US$10, the United States and the North Sea at US$11 and Russia at US$14. The current period of high oil price is promoting exploration activity in high-cost locations, such as the deep water offshore in the Gulf of Mexico and off the coast of Brazil. This prompts two questions: why develop high-cost fields when the Middle East has massive reserves and very low production costs, and what will happen to the high-cost producers if oil prices fall? The latter may

seem a remote prospect at present, but ultimately climate change policies will reduce demand for fossil fuels, either by replacing them with low-carbon alternatives or pricing them out of the market by taxing carbon.

The answers lie in the realms of geopolitics as much as economics (see Klare 2008). Prior to the first oil shock of 1973–4 the industrialized world had become increasingly dependent upon supplies of cheap oil from the Middle East (Odell 1989). When the Middle Eastern oil producers formed OPEC (the Organization of Petroleum Exporting Countries, see www.opec.org) and used the oil embargo to punish the industrialized West for supporting Israel in the Arab–Israeli war the price of oil rapidly escalated. The first oil shock was followed by another in 1979–80 following the Iranian Revolution and a mini-shock in 1990–1 because of the Gulf War (see Figure 5.9). Following the second shock

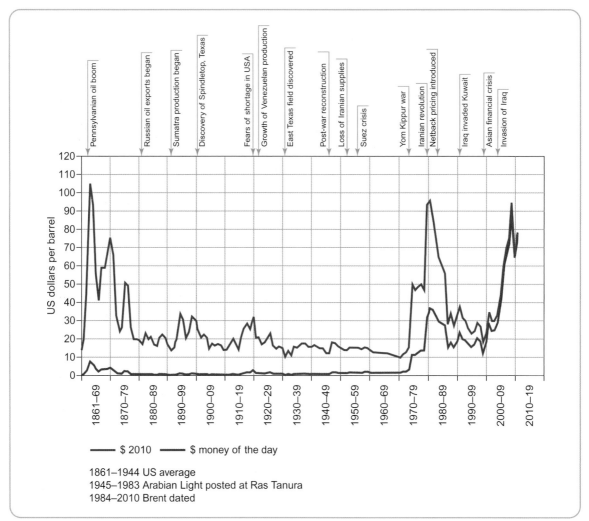

Figure 5.9 Historical trends in the world crude oil prices, 1861–2010.

Source: *BP Statistical Review of World Energy* (2011: 16). (Also available at: www.bp.com)

prices started to decline. There were at least three reasons: first, high energy costs promoted conservation; second, the high price of oil and the actions of OPEC promoted production by high-cost non-OPEC producers; and third, recession and economic restructuring reduced the growth of demand in the developed world. As a response to very low prices, OPEC members agreed to cut back production which, combined with global economic recovery, saw prices recover by 1999. However, subsequent events dramatically illustrate the volatile nature of oil and gas markets.

In the aftermath of 11 September 2001, a global economic downturn kept oil prices low. Then the global geopolitical situation changed dramatically with the war in Afghanistan and then the second Iraq War. But despite this, a surge in demand from emerging markets, principally China and India,

plus tight supply as a result of underinvestment in the previous decade and continued instability in the Middle East all contributed to a short-term 'supply gap'. By mid-2008 a combination of tight markets and financial speculation in oil futures pushed the price to a record $147 a barrel. Then the collapse of Lehman Brothers triggered a global financial crisis, demand fell and by the end of the year the price was below $40 a barrel. It has since recovered and is now fluctuating at a relatively high $70–80 a barrel (see Figure 5.9). There are now concerns that as economic growth returns surging demand, particularly in the emerging economies, will not be met by sufficient production increases, resulting in an 'oil crunch' and high prices, which might trigger further economic recession. The current uncertainty highlights the complex relationship between energy

Thematic case study 5.2

Energy security

At present, the United States – with something less than 5 per cent of the world's total population – consumes about 25 per cent of the world's total supply of oil. In 2025, if current trends persist, we will be consuming half as much petroleum again as we do today; however, domestic production will be no greater than it is today, and so the entire increase in consumption – approximately 10 million barrels of oil a day – will have to be supplied by foreign producers. And because we can't really control what goes on in those countries, we become hostage to their capacity to ensure an uninterrupted flow of petroleum.

Klare (2004: 11)

A generally accepted definition of 'energy security' is the reliable supply of energy at reasonable prices. As we know from discussions earlier in this chapter, resource scarcity is seldom about the physical shortage of a particular resource: this is particularly true of oil and gas. When the first edition of this book was in preparation we were in a period of very low oil prices; today we face sustained higher oil prices. Furthermore, there seems to be a growing consensus that in the longer term we have to reduce our reliance on the consumption of hydrocarbons to avoid the more catastrophic consequences of climate change.

Despite the global economic crisis, we seem to be experiencing a period of pressure on global oil supplies because of increased demand from growing economies, such as China and India, combined with supply interruptions caused by hurricanes and oil leaks in the Gulf of Mexico, labour unrest in Nigeria, declining production in long-established fields and delays in bringing on new fields. Thus, there is increased competition for access to supplies of oil and gas. Furthermore, an increasing share of the oil and gas consumed by the developed world (OECD) is imported from non-OECD and OPEC suppliers, particularly in the Middle East. In such a context, it seems that it is the business of national governments to devise policies to promote energy security. These may take the form of seeking to promote energy conservation, promoting other sources of energy supply, promoting cooperation with, and investing in, oil- and gas-producing states (often via support for the activities of state and private oil companies) and *in extremis* taking military action to protect the supply of oil. Many would argue that since the so-called 'Carter Doctrine' of 1979, when President Jimmy Carter, in the wake of the Soviet invasion of Afghanistan (see Spotlight box 20.1, p. 433), 'declared that any move by a hostile power to acquire control over the Gulf region would be regarded "as an assault on the vital interests of the United States of America", which would be opposed "by any means including military force'" (quoted from Rutledge 2006: 48), US foreign policy in the Middle East has been all about oil (Harvey 2003; Klare 2004). Both President Bush and Prime Minister Blair denied that the invasion of Iraq was motivated by a desire to secure Iraqi oil production; even if it was, it failed as Iraqi production is now lower than before the war and this is helping to keep the oil price up. Similar accusations are now being made about military intervention against Libya. Whatever the motivations for military intervention in the Middle East and North Africa, it is a fact that the United States, Europe and Asia are dependent on supplies of oil from that region. It is also true that Europe has sought to reduce this reliance by importing oil and gas from Russia and that this is now also proving a source of geopolitical concern as Russia is using its perceived status as an 'energy superpower' to pursue its own political agenda.

The combination of concerns about the geopolitical dimension of energy security, becoming increasing reliant on unreliable and potentially hostile sources of supply, and about climate change, the need to reduce levels of hydrocarbon consumption, have created a new energy paradigm (Helm 2007) and have also prompted a need to rethink the concept of energy security (Yergin 2006). As Daniel Yergin (2006: 70) puts it in a highly influential article in the journal *Foreign Affairs*: 'a wider approach [to energy security] is now required that takes into account the rapid evolution of global energy trade, supply-chain vulnerabilities, terrorism, and the integration of major new economies in the world

market.' According to Yergin, this new approach or framework must abide by the following principles:

■ Diversification of supply to reduce the impact of a disruption in supply from one source.
■ A 'security margin' in the energy supply system that provides a buffer against shocks and facilitates recovery after disruptions.
■ Recognition of the 'reality of integration', that there is only one global oil complex and a worldwide system that moves 86 million barrels of oil every day. For all consumers stability resides in the stability of this market.
■ The importance of information that underpins well-functioning markets.

In addition to these principles, Yergin (2006: 70) maintains that there is a need to recognize the globalization of the energy security system, which means that the developed world must engage with the likes of China and India, and a need to expand the concept of energy security to include the entire energy supply chain and infrastructure. In the longer term, it may be the case that the solution lies in developing an alternative to today's hydrocarbon economy; however, for the next 30 years or more we will depend on oil and gas for the majority of our energy needs and the competition for increasingly scarce resources means that concerns about energy security can only grow. Next time you switch on the light think about the complex set of issues that now lies behind keeping the lights on!

consumption and economic development and the volatility of energy prices.

5.3 Energy and development

The experience of the developed world suggests that in the initial phases of industrialization there is a direct link between increased energy consumption and economic development. That is, as industrial activity grows it consumes more and more energy. However, in the past 40 years it has become increasingly apparent that in the developed world the relationship between economic development and energy consumption has changed. This is largely due to the processes of de-industrialization and economic restructuring that have seen the less energy-intensive service sector replace

heavy industry and, to a lesser degree, manufacturing as the major generators of wealth. At the same time, conservation and technological change has made industry more energy-efficient. Similarly, a change in transportation technology has increased energy efficiency As a result, today's post-industrial societies have decoupled the link between economic growth and increased energy consumption. That said, it is still the case that the developed world consumes a large amount the world's energy; although non-OECD primary energy consumption surpassed that of the OECD for the first time in 2008.

5.3.1 Energy consumption, economic development and climate change

The relationship between energy consumption and economic development is usually depicted as a scatter plot showing the level of per capita energy consumption on one axis and GNP per capita on the other (Figure 5.10). The chart shows that the higher the level of GNP per capita the higher the level of energy consumption. Obvious outliers are easily explained away: some countries, such as the Soviet-type economies that had a bias towards heavy industry and were notoriously wasteful in their use of energy traditionally had higher levels of energy consumption than GNP. Others, such as Japan, have low levels of energy consumption but high GNP because they introduced energy-saving technologies at an early stage and moved energy-intensive industry offshore. Finally, there are some countries, such as Canada, where climatic extremes (both hot and cold) require a large amount of energy to be used for heating or for air conditioning. There are also other factors, for example the more urbanised a country the higher its level

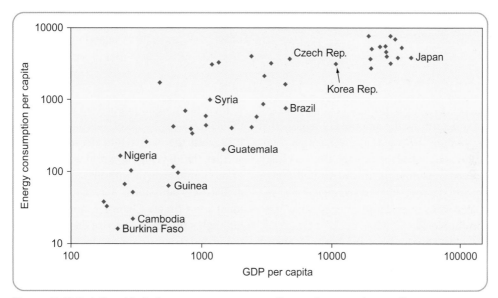

Figure 5.10 Relationship between energy consumption and economic growth.

Plate 5.7 When 'muck was money'.

(FOX PHOTOS/Getty Images)

Plate 5.8 Not a smokestack in sight on this new industrial estate.

(Realimage/Alamy)

of per capita energy consumption. Finally, the size of a country may determine the amounts of energy that are required to move between places. Such an argument could be marshalled to explain the very high levels of energy consumption in the United States, as could climatic factors, but the average American household also consumes more energy than anywhere else in the world.

On the face of it the relationship between energy consumption and economic development seems unproblematic. However, as Chapter 8 reveals, the whole question of what stands for 'development' is itself problematic and contested. The measurements used in Figure 5.10 systematically understate the relative position of the developing world. First, because the use of commercial energy as a measure of energy consumption ignores the role of non-commercial biomass energy sources, it therefore only measures the 'modern' sectors of the economy. Second, the monetary measure of GNP per capita inadequately measures the 'human condition' in many of the world's poorest countries. That said, a more representative measure would only move them slightly 'up the curve' and it would not modify the clear relationship between industrialization and energy consumption. The Stern Review (2006: xi) into the economics of climate change notes: 'CO_2 emissions per head have been strongly correlated with GDP per head. As a result, since 1950, North America and Europe have produced around 70 per cent of all the CO_2 emissions due to energy production, while developing countries have accounted for less than one-quarter. Most future emissions growth will come from today's developing countries, because of their more rapid population and GDP growth and their increasing share of energy-intensive industries.' It suggests that we need to 'decarbonise' development so that the developing world can enjoy the benefits of economic growth without substantially increasing global CO_2 emissions. A recent report on 'Development and Climate Change' echoes this sentiment and highlights the importance of ensuring that the developing countries do not get 'locked in' to the development of high-carbon energy infrastructures (World Bank 2010c: 190).

The historical relationship between industrialization and energy consumption implies that all the countries of the developing world will eventually follow an energy and development trajectory similar to the developed world. Thus, the patterns of energy and development can be equated to a stage model of the kind proposed by W.W. Rostow (Figure 5.11). This model examines the change in the energy ratio (the relationship between the rate of change in energy consumption and the rate of change in economic growth) over time as an economy develops (Mather and Chapman 1995: 154). In the pre-industrial phase the energy ratio is less than one, as these are 'low-energy' societies mainly dependent on subsistence agriculture. As discussed earlier (also see Chapter 2), the industrial revolution resulted in a change in the relationship between society and resource consumption; harnessing new sources of energy was at the heart of this process. As economies industrialize, the energy ratio begins to exceed one. In the early phases of industrialization energy efficiency is low and economies are dominated by energy-intensive industries. As the industrial economy

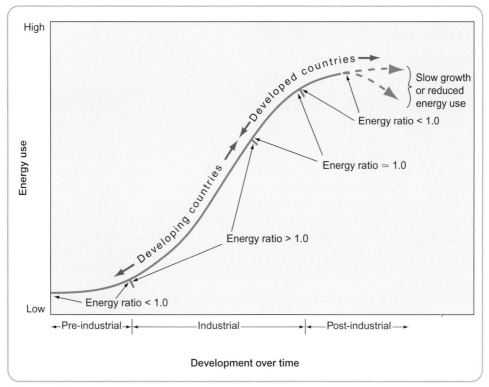

Figure 5.11 Energy ratios and economic development.

Source: *Environmental Resources*, Mather, A.S. and Chapman, K., Pearson Education Limited © Prentice Hall (1995)

matures so the energy ratio declines due to increased efficiency, increased energy costs, or the actual decline in the level of economic activity. This stage is typified by the developed world during the 1970s. Eventually, the economy moves into a post-industrial phase and the ratio falls below one. If the model seems very familiar, it is much like that for demographic transition (see Chapter 4) because it describes the evolution of the energy economy as it actually occurred in western Europe and the United States.

What evidence is there that the rest of the world will inevitably pass through the same stages? It could justifiably be argued that the newly industrialized countries (NICs) of Asia have followed this pattern of industrialization; however, the time taken has been compressed as the NICs have tried to diversify their economies and promote service sector growth without the advantage of substantial indigenous energy resources, a fact that has made them vulnerable to the effects of energy price volatility. At present it is the rapidly growing energy demands of the emerging economies such as China, India and Brazil that is driving the growth in global energy demand and sustaining energy prices. With that has come increased CO_2 emissions, making the energy needs of these emerging economies a major obstacle in global climate change negotiations. Meanwhile, as noted earlier, a much bigger question mark hangs over the developing world. Is it inevitable that it will have to industrialize to improve living standards? If so, then access to energy resources and/or energy-saving and low-carbon technologies will have to be a central component of the development agenda in the twenty-first century. Furthermore, as we have seen in this chapter, many of those same countries are also likely to be the major suppliers of oil and gas and other minerals and there is plenty of evidence to suggest that such resource-based development often fails to provide a sustainable basis for improving living standards (see Case study 5.3) and can generate conflict and instability. The first decade of the twenty-first century suggests that the global shift in economic growth and energy demand will generate a whole series of new challenges to the existing world order.

Thematic case study 5.3

Resource abundance: blessing or curse?

From all that has been said in this chapter, one would assume that having a relative abundance of natural resources would convey an advantage in terms of prospects for economic development. However, the reality is somewhat different. There is now substantial evidence to suggest that resource-abundant economies have actually performed worse in terms of rates of economic development than resource-poor economies. The geographer Richard Auty (1993, 2001) was the first to term this phenomenon the 'resource curse thesis'. Sachs and Warner (2001) have weighed up the evidence and conclude that there is sufficient evidence to suggest that 'high resource intensity tends to correlate with slow growth'. In other words, economies that have a high degree of dependence on the resource sector tend to grow more slowly than resource-poor economies. While the 'resource curse thesis' is now generally accepted, there are exceptions to the rule, those usually cited being Botswana, Chile and Malaysia, and there is no single explanation for this underperformance. Sachs and Warner suggest that most forms of explanation follow some form of 'crowding out' logic, whereby the dominant resource sector inhibits the development of the non-resource sector. Resource economies are particularly susceptible to fluctuations in income because of the volatility of resource prices, added to which, as we know from earlier discussions, the resource base itself is often soon depleted. Once the 'resource boom' is over, the economy is not sufficiently developed or diversified to sustain living standards and a period of 'bust' often follows. Such a 'boom and bust' cycle is by no means inevitable: effective government policy can use resource income to promote a more diversified economy that can sustain living standards once the boom has passed. A review of the literature (Stevens 2003) suggests that there are number of dimensions to the 'resource curse' (though Stevens favours the term 'resource impact'):

■ In many resource-rich economies there has been a failure to save income during boom periods to cover periods of bust (some states have created so-called 'stabilization funds' to save for a rainy day), plus a tendency to spend income on consumption

(usually through increased imports) and on prestige projects.

■ There is also a failure to redeploy income from the resource sector to promote a more sustainable pattern of economic development. In some instances there is also a tendency to use resource income to subsidize and protect the activity of inefficient producers in the non-resource sector. Later, when the resource income dwindles and the economy is opened up to international competition, these inefficient producers then fail.

■ In many instances, often as a consequence of a colonial heritage, the resource economy remains relatively isolated from the rest of the economy. This minimizes the multiplier impact of large-scale resource-based investment projects, beyond their payment of taxes. One solution is to impose a 'local content' requirement on the resources companies, forcing them to use local suppliers of goods and services. However, given the relative lack of 'economic development' in many resource economies it is often difficult to source goods and services locally.

■ The increase in export income associated with a resource boom tends to result in a strengthening of the value of the domestic currency of the resource economy. This can have the effect of making the cost of domestic production in the agricultural and manufacturing sectors higher than the cost of imports. This is known as 'Dutch Disease' following the experience of the Netherlands; the net result is a decline in the competitiveness of the non-resource sector, which aggravates the problems discussed above.

■ Finally, there is increasing evidence that suggests that a sudden influx of resource income tends to promote crime and corruption, armed conflict and an abuse of human rights. Such problems not only have a direct impact on the welfare of individuals, but also promote increased social inequality and undermine the effectiveness of the state (see Ross 2001; Renner 2002; Le Billon 2007).

The continuing underperformance of resource-rich economies has led the World Bank and the major resource companies to reassess the impacts of resource development. The World Bank through its Extractive Industries Review (www.eireview.org) is reconsidering whether it should be promoting resource-based development as a means of improving living standards and promoting sustainable development. The EIR's final report, called *Striking a Better Balance*, concludes: 'the Extractive Industries Review believes that there is still a role for the World Bank Group in the oil, gas and mining sectors – but only if its intervention allows extractive industries to contribute to poverty alleviation through sustainable development and that can only happen if the right conditions are in place'. The EIR was prompted by the World Bank's decision to finance the construction of the Baku-Tbilisi-Ceyhan (BTC) pipeline from landlocked Azerbaijan to the Mediterranean coast of Turkey. The collapse of the Soviet Union has prompted a dash to gain access to new energy resources in the Caspian and Central Asia, and more recently in Siberia and the Russian Far East (Kleveman 2003). The BTC pipeline also prompted the UK government to champion the Extractive Industries Transparency Initiative (http://eiti.org/), which is 'a coalition of governments, companies, civil society groups, investors and international organizations. The EITI now includes 23 countries and supports improved governance in resource rich countries through the full publication and verification of company payments and government revenues from oil, gas and mining.' High oil and gas prices and concerns about energy security have also promoted increased oil and gas activity off the shores of West Africa and elsewhere (Gary and Karl 2003; Rowell *et al.* 2005; Ghazvinian 2007) as the developed and emerging economies (China in particular) seek to gain control of sources of energy (Bradshaw 2009). Sustained high energy prices are resulting in a new influx of resource revenues and the expansion of oil and gas production into new areas will create a new set of resource-abundant economies and regions; NGOs such as Revenue Watch (www.revenuewatch.org) are now closely monitoring the flow of revenues arising from these new resource projects (see Caspian Revenue Watch 2003). The dangers of the resource curse are now well recognized, but only time will tell if the international community can assist the newly resource-rich economies to avoid the pitfalls of the past.

5.4 Conclusions: global energy dilemmas

This chapter has considered the nature of resources through a detailed study of the relationship between energy and economic development. We conclude by considering the world's energy dilemmas. It is not just a matter of whether or not there is access to sufficient energy resources to meet demand, which is increasingly in question. Rather, it is also a question as to whether the global ecosystem can absorb the consequences of continued increases in energy production and consumption. Put simply, at the global scale and to avoid catastrophic climate change, can an energy system that delivers secure and affordable energy services that are also environmentally benign be created? The exact nature of the energy dilemma varies across the globe, so there is no single energy problem, nor a universal solution (for further discussion of global energy dilemmas see Bradshaw 2010; 2012).

In the developed market economies (including the Asian NICs), the energy dilemma relates to the economic and political costs of geopolitical scarcity and the environmental impacts of high levels of GHG (greenhouse gas) emissions (see Table 5.1). Increasingly, these economies are finding themselves dependent upon what they perceive to be 'hostile' and 'unstable' sources of energy supply. The response has been to develop high-cost resources close to home, to promote increased efficiency and conservation and to seek low-carbon alternatives, such as renewable energy. However, providing energy security and reducing GHG emissions is bound to increase the cost of energy services. The developed world can afford to seek technological solutions to its energy dilemma; furthermore, very substantial reductions in fossil fuel consumption and GHG emissions are essential to create 'headroom' for the developing world.

In the transition economies of Central Europe and the former Soviet Union the energy dilemmas are somewhat different. Many 'smokestack' industries have been closed in the last two decades but many of the states in this region must still secure access to energy resources at market prices and liberalise their energy systems. At the same time, the resource-rich states, such as Azerbaijan, Kazakhstan, Russia and Uzbekistan, see energy export revenues as the cornerstone of their economic transformation. An inherited reliance upon Russian energy imports is a source of insecurity and diversifying their sources of energy supply, especially after the Russia–Ukraine gas disputes in 2006 and 2009, is now a priority.

The resource-abundant economies of the emerging and developing world face particular challenges as a consequence of their resource wealth. For many, but not all, access to revenue from oil exports has proved a mixed blessing. However, it would be wrong to label all energy-exporting economies as victims of the resource curse. There is a world of difference between the oil and gas exporting states of the former Soviet Union, the Middle East and West Africa. They all have their peculiarities, but all are vulnerable to the volatility of energy prices. Thus, for the present at least, their energy dilemma relates to the challenge of using energy exports to finance a more sustainable model of economic development. In the longer run, these states must deal with the fact that climate change policies will reduce the demand for their principal source of revenue, which makes economic diversification all the more imperative.

Over the last decade or so a group of rapidly industrialising economies has emerged and their rate of economic development, population growth and urbanization is driving rapid growth in demand for energy services. Although economies such as China and India possess substantial energy resources of their own, they tend to be reliant upon coal; at the same time, the rate of demand growth has outstripped the capacity of their domestic energy industries to match demand. The net result has been a rapid increase in energy imports, particularly oil and increasingly gas. This growth in energy consumption is also associated with increasing GHG emissions. As discussed earlier, energy- and carbon-intensive types of industrial activity dominate their current economic structures, and they have yet to decouple economic growth from growing energy demand. Somewhat paradoxically, because of their very large populations, their per capita levels of energy consumption and CO_2 emissions are still low. As individual incomes increase there is huge potential for consumption-driven energy demand to increase, through car ownership, for example. Thus, the energy dilemmas for these fast-growing economies relate, on the one hand, to securing the necessary energy to fuel continued economic development; while, on the other hand, ensuring that increased energy consumption does not result in further environmental degradation and rapid growth of GHG emissions.

In the developing world a dualistic pattern of resource consumption seems to have developed in relation to energy. On the one hand, there remains a 'low-energy' rural society dependent upon biomass energy and subsistence agriculture, in which the majority of people exist. Here population pressure is leading to environmental problems as forests are stripped for fuel wood and land ploughed up or over-grazed. On the other hand, there exists a growing 'high-energy' society linked to increasing commercial agricultural (see the next chapter for more on this), industrial activity and the spread of urbanization. Much of this activity is aimed at supplying natural resources and agricultural products to the developed world. The two sectors of society combine to create an increasing demand for energy services, and increased pressure on the environment. The balance between the sectors also varies greatly among the countries of the developing world. Paradoxically, solutions to the problem of rural poverty will inevitably be increased demands for energy services, though the absolute levels of per capita energy consumption will remain low.

Given that the majority of the world's population lives in the emerging and developing worlds that are also experiencing population increase and rapid urbanization, the growth in energy demand in global south will be the major factor contributing to increases in GHG emissions over the coming decades. More than a decade ago, in its *Global Environment Outlook 2000*, the United Nations Environment Programme (UNEP 1999: 2) concluded that: 'A tenfold reduction in resource consumption in the industrialized countries is a necessary long-term target if adequate resources are to be released for the needs of developing countries.' It is this increasing energy demand in the developing world that now raises the spectre of scarcity in any number of forms. To address the many global energy dilemmas, the developed world must address its own energy profligacy and at the same time help the developing world to shape an entirely new relationship between energy and development.

Learning outcomes

Having read this chapter, you should understand that:

■ The notion of what is a resource is socially constructed and varies through time and across space.

■ There is a variety of different ways of classifying natural resources.

■ There is a complex set of factors that influence the availability of natural resources.

■ There are substantial regional variations in energy production and consumption and a growing mismatch between the countries that consume most of the world's energy and those that produce it.

■ There is a complex and increasingly challenging relationship between energy consumption, economic development and climate.

■ Different types of resource dilemmas confront the different regions of the world.

Further reading

Bridge, G. (2011) Resource geographies 1. Making carbon economies, old and new, *Progress in Human Geography*, **35**(6), 820–34. The first in a series of planned reports on resource geographies to appear in this flagship journal signify a renewed interest in resource issues.

Dicken, P. (2011) *Global Shift: Mapping the Changing Contours of the Global Economy,* 6th edition, Sage, London. The latest edition of this excellent textbook has a new chapter on the extractive industries.

Gautier, C. (2008) *Oil, Water and Climate: An Introduction*, Cambridge University Press, Cambridge. An introduction to the essential background on climate change needed to understand the relationships between energy, population and environment.

Jones, G. and Hollier, G. (**1997**) *Resources, Society and Environmental Management*, Paul Chapman, London. Now quite dated, but provides a reference source for the early discussion on the nature of natural resources and ways of classifying them.

Klare, M. (2008) *Rising Powers Shrinking Planet: The New Geopolitics of Energy*, Metropolitan Books, New York. Provides an accessible analysis of the geopolitical consequences of the current global energy shift.

Robbins, P., Hinstz, J. and Moore, S.A. (2010) *Environment and Society*, Wiley Blackwell, Oxford. Has lots of material relevant to this chapter and others in this section.

Smil, V. (2010) *Energy Transitions: History, Requirements, Prospects*, Praeger, Santa Barbara, CA. Provides essential background on the nature of the energy system and places the current energy dilemma in historical perspective.

Useful websites

There is a huge amount of material available on the Web that relates to the issues discussed in this chapter. Some of the key reference sources are listed below; additional websites have been referenced in the text.

www.bp.com Home of the BP *Statistical Review of World Energy*. This website, which is updated on an annual basis, has a wealth of statistical information on the energy sector, much of which is downloadable in Excel format or as PowerPoint slides.

www.iea.org The official site of the International Energy Agency. The site contains information on the Agency's operations and publications, as well as statistics on energy production and consumption. Also has links to other international organizations.

www.eia.doe.org The official site of the US government's Energy Information Administration. In addition to information on US energy matters, it contains reports on individual countries and a massive list of links to other US government agencies, international agencies, foreign governments and commercial company sites.

www.ipcc.ch The official site of the Intergovernmental Panel on Climate Change (IPCC), established by the UN Environment Programme (UNEP) and the World Metrological Organization (WMO) to provide the world with a clear scientific view on the current state of knowledge on climate change and its potential environmental and socio-economic impacts. The site hosts all of the IPCC's Assessment Reports and supporting documentation.

www.worldbank.org The official site of the World Bank. The site contains information on World Bank operations and publications, as well as downloadable statistics and briefing documents.

www.wri.org The World Resources Institute is an independent centre for policy research and technical assistance on global environmental and development issues. Its website provides information on its own activities and publications, some of which are downloadable, as well as links to other organizations in the same area. The Institute, together with the UNEP, UNDP and World Bank, produces the biennial *Resource Report,* which is an indispensable reference guide to the state of the earth's resources.

For annotated, clickable weblinks and useful tutorials full of practical advice on how to improve your study skills, visit this book's website at **www.pearsoned.co.uk/daniels**

THE ENVIRONMENT AND ENVIRONMENTALISM

Chapter 6

Jenny Pickerill

Topics covered

- How we value and understand the environment

- The evolution of sustainable development

- The complexity of scale of environmental issues

- Different strategies for action

- Challenges to how we conceptualize the environment

- Positive steps towards an ecological future

No one is untouched by the emerging environmental agenda; everyone has an interest in shaping the development of environmental politics (for good and bad).
(Connelly and Smith 2003: 358)

When I was once asked to discuss at length how individuals could best help resolve the environmental issues, I responded 'stay put'. Only by rediscovering a sense of place, a commitment to a particular piece of ground will we be able to redefine our relationship to the planet.
(Anon. 2001)

There is no survival value in pessimism. A desperate optimism is the only attitude that a practical environmental philosopher can assume.
(Callicott, quoted in Callicott and Rocha 1996: 157)

6.1 What kind of world do you want?

What kind of world do you want? If we don't act now, we won't have a choice. Do you want a world where climate change continues to devastate our environment?

(Friends of the Earth flyer, June 2006)

There is a need for utopianism when thinking about the environment. Environmentalists are often accused of being unrealistic and demanding the impossible. Yet we seldom ask ourselves the question 'What kind of world would we like?' If we considered it a little more we might be taking bigger steps towards curbing our environmental impact and seeking environmental justice. Environmental issues affect all places and societies – the urban and the rural, the rich and the poor – nobody is unaffected. The environment matters and our impact upon it is obvious, from the air that we breathe to the energy that we use.

The rise of environmental activism has highlighted the importance of environmental protection with some notable successes. Britain's road-building programme of the 1990s was curtailed, and the logging of old-growth forests has been all but stopped in Western Australia. Britain has the 'oldest, strongest, best-organized and most widely supported environmental lobby in the world' (McCormick 1991: 34), and there are examples worldwide of people taking a stand and making changes to protect the environment. However, 'there is no doubt that environmental issues have had a big impact on contemporary politics, and yet the frequency with which governments adopt a business-as-usual response to environmental problems raises the cynical thought that perhaps nothing much has really changed' (Carter 2001: 2). Thus, despite high levels of environmental concern and an active environmental movement, we are faced with a wide array of environmental problems – from global climate change to river pollution, from water supply issues to nuclear power. To understand why this is, we need to explore: how we value the environment and the different philosophical approaches to

environmental issues; the complexity of scale and responsibility; different strategies for action; what a focus on 'the environment' ignores; and finally what positive steps we should take.

6.2 Valuing the environment

We each value the environment in different ways and for different ends. Historically we have always debated what parts of our human and non-human world have what values. These values are used to answer ethical questions about the environment, for example whether we should eat meat, or whether it is justifiable to cut down a tree in order to burn it to heat a house. Understanding values helps us understand why some people care greatly about the environment and others less so.

We can use three categories of value when talking about the environment. As summarized in Table 6.1 these are intrinsic, inherent and instrumental values. An object, or living being, has an **instrumental value** when we view it as a resource, or as of use to us for a specific end. For example, a forest has an instrumental value if we view it just as a source of wood for fuel. At the other end of the continuum an element has **intrinsic value** if we view it as important in and of itself, with no reference to how we might use it or to how it might make us feel. For example, we view human life as intrinsically valuable.

In between these, **inherent value** (sometimes referred to as 'weak' instrumental value) refers to how we value an element beyond its use as a resource but still relate it to how it makes us feel. For example, we might value a woodland because we enjoying walking through it and for its role in atmospheric stability. In all of these discussions of value we must remember that 'humans will always be the distributors of value, but they are not necessarily the only bearers of value' (Carter 2001: 36).

We can understand this better if we consider the rule of no-substitution. If something is instrumentally valuable, its value will disappear if there is a

Table 6.1 Defining value

Instrumental value	Value which something has for someone as a means to an end
Inherent value	Value something has for someone, but not as a means to a further end
Intrinsic value	Simply the value something has. No appeal need be made to those for whom it has value

Source: Carter (2001: 15)

(preferable) substitute. For example, if we could solve eyesight problems with a pill, we would no longer need opticians. But if something is inherently valuable for itself, and not as an end, it is not substitutable in this way. Hinchliffe and Belshaw (2003) use the example of the wild tiger to illustrate the consequences of these values. The wild tiger population is threatened by the international trade in tiger parts, often for use in Chinese medicine. One solution is to establish tiger farms. However, if we find tiger farms acceptable then there is room for instrumental valuations. If tigers' inherent value is paramount, then the notion of farming becomes objectionable. The implications of such values are also explored in Case study 6.1.

We can also consider the values we give the environment by understanding the terms 'anthropocentrism' and 'ecocentrism'. The anthropocentric approach argues that only humans have intrinsic value and the environment is only of instrumental value. In other words the environment is only useful as a resource for humans. In this approach we do not have any responsibility to ensure the environmental sustainability of our actions. Ecocentrism critiques anthropocentrism and suggests that non-human entities have intrinsic value. However, for many the situation is not this black and white. There is a large intermediate area within which some non-human entities have value, others do not,

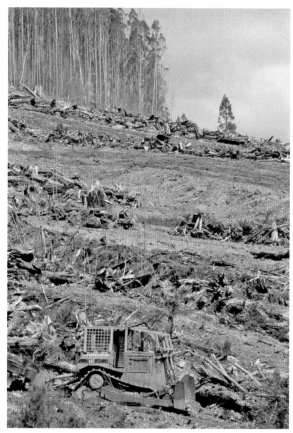

Plate 6.1 Logging in Styx Valley, Tasmania, Australia
(Jenny Pickerill)

Thematic case study 6.1

Old-growth logging in Tasmania, Australia

The logging of old-growth forests is a particularly controversial issue in Australia, with views sharply polarized (Hutton and Connors 1999). Old-growth forests have had only minimal human interference, take hundreds of years to grow and contain the tallest plants on earth. They are often considered to be of high conservation value.

The state government of Tasmania has shown little inclination to curb the logging of old-growth forests. This is despite a consultation process (Tasmania Together) where the population voted to end deforestation. In May 2003 the government lifted a ban on logging in the Tarkine – the single greatest stretch of temperate rainforest in Australia. The forests of the south-east and north-east are also threatened by expanding logging operations. Every year approximately

15,000 hectares of old-growth forest are logged in Tasmania.

In a country which draws much of its wealth from its natural resources (forests, minerals, uranium, fishing and agriculture), logging is often justified by the need to provide jobs in rural areas and to provide much-needed export revenue. Loggers argue that Australia is limited in its resources and consequently needs to make the most profitable use of them; that they have historically 'managed' forests through selective logging, consequently arguing that all forests in Tasmania have been influenced by humans for generations.

Environmentalists have argued that logging is simply a waste of resources. Logging in Tasmania is heavily subsidized and uneconomical (not only in the price obtained for products but in its negative impact upon tourism). Most of the wood is sold as woodchips to be made into paper. Moreover it is unsustainable and irreversible and many of the trees destroyed would take over 300 years to regrow, e.g. the Houn Pine in the south-west of the island takes 3,000 years to grow to full size. Old-growth forests cannot simply be replaced; their destruction results in the loss of biodiversity of a barely understood and fragile ecosystem. Deforestation does not only have localized effects but also plays a key role in CO_2 absorption and can upset hydrological systems beyond the area of logging. Environmentalists also believe that the largest trees are important in and of themselves. The Wilderness Society, a key environmental group in the Australian logging debate says, 'our forests are places that inspire and rejuvenate us all – nurturing us as well as the delicate ferns and mosses on the forest floor, or connecting us to timeless grandeur as we touch the tallest flowering plants on Earth' (2006).

The campaigns to save the old-growth forests have ranged from small-scale local protests to garnering international support. The Styx Valley in the south-east is an example of an iconic campaign. It contains the world's tallest hardwood trees – *Eucalyptus regnans*. Many are taller than a 25-storey building. The forest is also home to many native species of wildlife, including the majestic Wedge-tailed Eagle and the Yellow-tailed Black Cockatoo. While a few of the largest trees in the valley have been protected, between 300 and 600 hectares are being logged each year. Furthermore the main form of logging is destructive clear felling and burning (not selective logging).

Action on a local level has involved occupying threatened trees and targeting the woodchip mills. Regionally, environmentalists have been using their research to argue against clear fell logging, and to prove the environmental implications of logging. Environmentalists have appealed to the federal government and the Australian populace for support through media campaigns and advertising. In 2010 there was an historic breakthrough in the 30-year-long campaign against old-growth logging, when the environmentalists came to an agreement with timber communities, forest unions and the industry in a jointly signed Statement of Principles on how a sustainable timber industry could be created while also protecting the remaining old-growth forests. The agreement would protect key old-growth forests like the Tarkine and Styx Valley amongst others. The main logging company – Gunns – promised to end its operations in the state. While it is too early to say whether this agreement will hold, and the timeframe of ending logging is a little unclear, this is the first time that all the protagonists have come to an agreement, and it is likely to signal the end of old-growth logging in Tasmania.

Analysis of the conflicts about protection of forests enables us to understand how the environment is valued differently within a community, in other words why some argue for its protection and others choose its destruction. The state government and logging companies clearly view the environment as a resource for use by humans, thus they believe the environment has an instrumental value. They will only be interested in managing its sustainability if that will add value to its use and exploitation in years to come. For environmentalists it is more complex. On one level, many acknowledge the instrumental value of forests (for making furniture, for example) but think the current methods of forestry are wasteful. But in addition environmentalists believe the environment has an inherent and an intrinsic value which is not related to human need. In this way many believe it should be protected from all but recreation, and left untouched for future generations to enjoy.

Q How do you value forests like those in Tasmania and how does that affect your perspective as a geographer?

but where humans are valued as *more* important than the environment. Here the environment has some value beyond human use. Thus there is a continuum between these approaches and in 1973 Arne Naess distinguished between 'deep' and 'shallow' ecology, identifying a continuum between the two (Naess 1973). Shallow ecology is an anthropocentric position. Deep ecology, as defined by Naess, is a form of ecocentrism but one where everything has intrinsic value (see Table 6.2 for the eight basic principles of deep ecology). The approach asks us to consider humans not only as part of nature but also of *equal* value to other non-human entities. This is a holistic vision whereby all human and non-human entities are interconnected and interdependent. A holistic perspective enables us to understand that if we upset one element it will have an impact on all other elements.

What are the consequences of valuing non-human entities in this way? If we were to adopt such an approach it would require us to make major and radical changes to the way we live. No longer could we see the non-human world as merely providing resources for our existence: rather, deep ecology asks us to radically reduce our population, restrict economic growth,

Table 6.2 The eight basic principles of deep ecology

1	The well-being and flourishing of human and non-human life on Earth have value in themselves (synonyms: intrinsic value, inherent value). These values are independent of the usefulness of the non-human world for human purposes.
2	Richness and diversity of life forms contribute to the realisation of these values and are also values in themselves.
3	Humans have no right to reduce this richness and diversity except to satisfy vital needs.
4	The flourishing of human life and cultures is compatible with a substantial decrease of the human population. The flourishing of non-human life requires such a decrease.
5	Present human interference with the non-human world is excessive and the situation is rapidly worsening.
6	Policies must therefore be changed. These policies affect basic economic, technological and ideological structures. The resulting state of affairs will be deeply different from the present.
7	The ideological change is mainly that of appreciating life quality (dwelling in situations on inherent value) rather than adhering to an increasingly higher standard of living. There will be a profound awareness of the difference between big and great.
8	Those who subscribe to the foregoing points have an obligation directly or indirectly to try and implement the necessary changes.

Source: Naess (1973: 95–100)

and to live a simpler life by giving up excesses and 'to touch the Earth lightly'. There is a strong element of self-sacrifice to this in order to ensure survival of the non-human world. Naess has been criticized for asking humans to make such sacrifices when we would not ask animals to do the same. However, Naess 'accepts that human beings . . . are sometimes justified in lending greater moral weight to human well-being than to that of non-human life' (Cooper 2001: 214). What deep ecology really challenges us to do is to value non-human life as we would humans and thus truly consider how our daily actions could be different. Although some of the sacrifices might appear difficult and controversial (such as radically reducing populations), Naess clearly illustrates the extent of actions required to put into practice a deep concern for the environment.

An examination of values has illustrated some of the potential conflicts in how we view the environment. Environmentalists often argue that we need to rethink humans' relationship with the non-human world and they use values to argue that non-human elements hold greater value than often assigned them.

6.3 Limits to growth and the challenge of capitalism

This rejection by environmentalists of the anthropocentric approach whereby the environment is viewed instrumentally, and that instead it has an intrinsic value, is a cornerstone of modern environmental thought. A second key insight is that there are limits to growth. The earth is a finite system and thus eventually growth will have significant impacts upon the ability of the environment to sustain future generations. This notion was introduced by Meadows *et al.* in 1972 with their publication of *Limits to Growth*. Using computer models they sought to project trends in resource use and population growth into the future. They predicted natural limits on future growth and a doomsday scenario of environmental catastrophe. Although we now know these predictions to be overly pessimistic the idea of limits and finitude (scarcity rather than abundance) struck a chord.

However, it also became apparent, for people in the minority world at least, that it was possible to maintain economic growth and reduce resource consumption per unit of wealth created. This argument saw the emergence of an alternative view to ecocatastrophism: a view that would grow into the concept of sustainable development.

Connelly and Smith (2003) identify two broad interpretations of what an environmentally sustainable society would look like: a deep green approach, and the more reformist ecological modernization. The deep green interpretation takes many of Naess' principles and seeks to put them into practice. It is a radical approach, a strong interpretation of sustainable development that calls for change and restructuring at all levels of society in political, economic and social arenas.

Spotlight box 6.1

Sustainable development

The 1987 Brundtland Report adopted the position that it was possible to pursue economic growth without compromising the environment, and it is generally accepted as introducing the concept of sustainable development (the seeds of which had been sown some 15 years earlier). It also provided the first, and still widely used, definition of sustainable development: 'development which meets the needs of the present without compromising the ability of future generations to meet their own needs'.

However, its meaning has now become highly contested and vague, leading to weak interpretations. For example, how can present and future needs be determined, and whose needs does it refer to? Many attempts have been made to refine the concept, which have led to numerous, often contradictory, definitions. The result is that 'sustainable development' is often little more than a soundbite, signalling recognition that the environment matters but failing to provide substance.

The real success of the Brundtland Report was the emphasis it placed on the relationships between economic growth, social conditions and environmental degradation, an emphasis that placed sustainable development firmly on the global political agenda, and that over the next five years would see a concerted effort by UN agencies and many governments towards a global summit on environment and development. In 1992 – after five years of negotiations and preparatory meetings – the UN Conference on Environment and Development (also known as the Earth Summit) was held in Rio de Janeiro. However, the tension between the need for economic growth and environmental protection continues. A group of powerful majority world countries formed a group at the 2009 Copenhagen Climate negotiations. BASIC (Brazil, South Africa, India and China) assert that there should be a differentiation of responsibility for carbon emissions. In other words, that industrialised countries, particularly the USA, should have to reduce their emissions more and more quickly than others. Thus, they continue to argue that they should be allowed to employ a less robust version of sustainable development than the minority world.

Q Is it morally justifiable and politically plausible for the minority world to try to dictate the terms of sustainable development and good environmental management to the rest of the world?

In essence it is a critique of everything about the way we currently live today, a critique of contemporary industrial society. Accepting that there are limits to growth means changing our notions of what makes a good life and learning to live more simply (thus slowing growth): for example, consuming less overall but buying more locally produced goods, producing less waste, travelling less and being conscious of our impact on the environment in our daily lives. The result is an emphasis on 'localization', living small-scale,

Plate 6.2 Low impact house: Tony's Roundhouse, Brithdir Mawr, Wales.

(Jenny Pickerill)

Thematic case study 6.2

Low-impact development in Britain

Low-impact development (LID) is a form of low-impact living, a deep green vision where we minimise our environmental impact in all aspects of our daily lives. LID is a radical form of housing and livelihood that works in harmony with the landscape and natural world around it. Simon Fairlie, a pioneer of LID in Britain argues that 'a low impact development is one that, through its low negative environmental impact, either enhances or does not significantly diminish environmental quality' (1996). It is also an example of a mixture of mitigation and adaptation to climate change – building homes that will reduce carbon usage and cope with a changed climate.

A LID has a low visual impact by blending with its surroundings, is built from local, recycled or natural materials, is small-scale and environmentally efficient. They are autonomous in the sense that they generate their own energy (through wind, solar or water power) and deal with their waste (through recycling and composting), thus they are not connected to mains supplies. In addition LID is about more than just the construction of environmentally friendly dwellings: it is about the *way* we live. Implicit in the LID approach is an emphasis on minimizing vehicle use, creating livelihoods largely from the land on which we live, reducing consumption and purchasing other items locally. Moreover, LID can positively enhance biodiversity and is reversible.

Why do we need such a radical approach? In Britain the built environment is responsible for nearly half of the country's carbon emissions (Goodier and Pan 2010). The Government has attempted to make changes to the source of these emissions. Britain has committed to zero carbon house building by 2020 – zero net emissions of carbon dioxide from all energy use in the home. In 2009 the *UK Low Carbon Transition Plan* was introduced, which outlines strategies for changes to the country's energy, transport, employment, skills and training through to housing. The 2012 Green Deal (Department of Energy and Climate Change 2010) will also provide grants for environmental technological improvements to existing houses. While this is a welcome move, for many environmentalists it is too little too late. It also fails to take account of energy used in construction materials and the build process, and of the ways people live and consume. It is a technological 'fix' approach to the problem which

requires very little of the inhabitants – apply the technology and simply carry on as before – or of society. Moreover, to improve the energy efficiency of a standard house can be costly and thus off-putting, whereas a LID is often low-cost *by design* because of its focus on appropriate scale and materials.

In Britain there is a growing number of LIDs. Hockerton Housing Project is a LID in Nottinghamshire which is an earth-sheltered, self-sufficient terrace of five single-storey houses. Built into the side of a hill with large windows to the south, they rely on passive solar heating (and no other form of heating) and generate all their electricity through wind turbines and photovoltaic systems. The houses are highly energy-efficient and waste is disposed off through a reed-bed system into a large lake. They also grow their own vegetables and generate income through education. Another example is Tony Wrench's roundhouse at Brithdir Mawr, south-west Wales (Plate 6.2). This is built from home-grown timber, straw and earth. It is built into the earth with a turf roof so that, from behind, you can hardly see it. It is highly insulated, uses recycled windows and uses passive solar heating and a wood burner for heating.

Until recently it has been very difficult to get planning permission in Britain for LIDs. Many communities have secretly constructed their dwellings or fought long retrospective planning battles to keep their homes. However, on 13 July 2006, eight years after the discovery of roundhouses at Brithdir Mawr and the ensuing planning controversy, Pembrokeshire County Council and Pembrokeshire National Park adopted an innovative policy on low-impact development. This policy explicitly allows rural greenfield development under a number of conditions aimed to ensure developments remain low impact. Under the guidelines, development must be highly sustainable, using local, renewable, recycled and/or natural materials and built to high standards of eco-design, with the emphasis on 'low impact', including visual effects. Moreover, the proposal must offer positive environmental, social and/or economic contributions with public benefit. These could include services to the community, economic diversification for the area, opening paths for walkers and improved biodiversity. Furthermore, residents must prove a need to be on the land (rather than clustered within existing rural settlements). Thus residents' livelihoods will be 'tied directly to the land on which it is located', through, for example, agriculture, forestry or horticulture and these land-based activities will substantially meet household needs. In 2009 the first large-scale new eco-village was approved under this policy. Lammas eco-village (www.lammas.org.uk) is a low-cost nine house community on 76 acres of mixed pasture and woodland next to the village of Glandwr,

Pembrokeshire. They are building straw-bale, timber-frame and cob houses on an old farm site.

Q What measures should we put in place to ensure our buildings, and the way we live in them, have minimal environmental impact?

perhaps more communally to share the load of everyday tasks, in energy-efficient buildings, and often with an emphasis on producing for your own needs, for example, through growing your own vegetables. See Case study 6.2 for a British example of low-impact living.

The ecological modernization interpretation is far less radical and has been adopted by many more in mainstream politics. Supporters of ecological modernization criticize the deep green vision as naive and utopian with an unnecessary and ineffective focus on localization. In essence this approach argues that economic growth does not need to be slowed to ensure environmental protection. It does not undermine the limits to growth thesis entirely, but suggests that we can 'reorient' economic growth and use technological solutions for environmental problems, thus overcoming the environmental impact of growth. It is a very weak interpretation of sustainable development that allows for a reformist response where the current dominance of free-market capitalism is not challenged; 'the future of environmental politics on this reading is a technical question of how to make capitalism more environmentally sensitive' (Connelly and Smith 2003: 359).

6.4 The complexity of scale and responsibility

The scale of response to environmental issues is a key point of debate. As we have seen, deep greens argue that localization is one of the most important elements to enacting environmental sustainability. If we adopt a more localized lifestyle then we can minimize our environmental impact. However, our environmental problems are far more complex, and global, than the notion of 'localization' suggests. To understand this we need to explore the evolution of environmental concern from one of local problems to global environmental issues.

There is no clear beginning to environmental concern. Plato, Lucretius and Caesar all noted the

problems of soil erosion. We now know that many of today's most pressing environmental issues were first identified in the nineteenth and early twentieth centuries. For example, in his 1869 book *Man and Nature*, George Perkins Marsh presented evidence for consequences of tropical deforestation. Carter (2001) has categorized the evolution of environmental issues (Table 6.3) into three key periods. Pre-1960s a great deal of environmental concern was focused upon small-scale localized problems such as local pollution and wildlife protection. In the 1960s concern grew about the inter-relation between human actions and more complex environmental implications. For example, *Silent Spring*, written by the biologist Rachel Carson in 1962, distilled the essence of the broad concerns of the era, and is generally regarded as a milestone in modern environmentalism. Carson pointed out the long-term ecological consequences of intensive agriculture, in particular the

Table 6.3 Evolution of environmental issues

First generation: Preservation and conservation (pre-1960s)	• Protection of wildlife and habitats • Soil erosion • Local pollution
Second generation: 'Modern environmentalism' (from 1960s)	• Population growth • Technology • Desertification • Pesticides • Resource depletion • Pollution abatement
Third generation: Global issues (late 1970s onwards)	• Acid rain • Ozone depletion • Rainforest destruction • Climate change • Loss of biodiversity • Genetically modified organisms

Source: Carter (2001: 4)

ways in which synthetic pesticides (notably **DDT**) persisted in the food chain and poisoned birds and mammals (including humans). At the same time there were growing concerns about **overpopulation**. Paul Ehrlich, Garrett Hardin and Barry Commoner were considered **neo-Malthusian** because they saw environmental problems as a consequence of population growth, following the arguments set out by Thomas Malthus in the late eighteenth century. The misguided simplicity of neo-Malthusian explanations of environmental problems is shown in studies that have tried to explain land degradation in terms of population dynamics (Boserüp 1990). Not only is the evidence contradictory, for example population growth in northern China has been linked to increased land degradation (Takeuchi *et al.* 1995), whereas in northern Yemen (Carapico 1985) and Kenya's Machakos Hills (Tiffin *et al.* 1994) it has been related to population decline, but also it is clear from these studies that linking land degradation to one explanatory parameter – population size – masks the complexity behind population dynamics and labour–land relationships (Millington *et al.* 1989).

By the early 1970s there was recognition not just of the interconnectedness of environmental issues, and of the problems of cumulative damage, but also that these were global problems that needed internationally coordinated responses. Environmental scientists identify two types of global environmental change. On the one hand, systemic change occurs when there is a direct impact on a physically interconnected, global system, i.e. the atmosphere or the oceans: for example, the effects that increased emissions of greenhouse gases have on global climate. Cumulative change, on the other hand, occurs when many discrete events become significant because their distribution is global or because, added together, their impact is felt across a large proportion of the globe. For example, across the Americas, Africa and Asia-Pacific, the relatively small amounts of tropical forest lost in each area being converted from forest to farmland, or being lost to a mining operation, add up to a global-scale problem affecting humid tropical forests, one of the most biodiverse biomes found on earth: forests of vital significance to the global water and carbon balances. Recognition of this duality is important in formulating policies and strategies. Those actors addressing systemic issues need to be global, with all countries agreeing and adhering to a particular policy or strategy. Those focusing on cumulative issues require both global approaches (e.g. in the fields of data exchange and comparative research) and local approaches (to reduce the impacts of a problem because

slightly different causes exist in different geographical locations) (Middleton 2003).

In 1992, after five years of negotiations and preparatory meetings, the UN Conference on Environment and Development (also known as the Earth Summit) was held in Rio de Janeiro. It brought together 178 government delegations, over 120 heads of government, over 50,000 **NGO** representatives and a press corps of around 5,000. A rift opened up between the representatives of the majority world and the minority world countries, the former being worried that environmental concerns would be used by the minority world to 'limit [their] development for the sake of the Earth'. Shifting tensions between geopolitical blocs developed over issues such as biodiversity, climate change and desertification. Whatever its failings, the key point is that the Earth Summit resulted in a number of specific agreements that now form the basis of a global strategy for sustainable development and it had a wider, less quantifiable impact on environmental issues worldwide, particularly in that:

■ it elevated the importance of the environment on the political agendas of most countries (most countries now participate in the conventions that were agreed to at Rio, and many have used these conventions to pass new environmental legislation);

■ the multilateral negotiations led to some issues gaining internationally important status (perhaps most notably biodiversity loss, which had only been recognized by scientists some 12 years earlier), and which changed long-standing, often ineffective attitudes to nature conservation;

■ legislation at the international scale was negotiated;

■ there was recognition that many stakeholders had legitimate voices in environmental policy-making (the recognition of the stakeholder principle was vital in allowing actors outside government to become effective in bringing about changes in environment and development, a point that proponents of the bottom-up approach to development had long argued for).

The Rio Conference was itself a milestone as it heralded an era of global environmental policy-making and legislation. Each of the three main conventions agreed at Rio has been followed by annual conferences of the parties (COPs) (i.e. the countries that have signed a convention) at which progress towards ratification and implementation of the conventions has been discussed. Whilst the Convention to Combat Desertification and the Convention on Biological

Diversity proved relatively straightforward to ratify and implement, the Framework Convention on Climate Change and its most important protocol – the Kyoto Protocol – have proved difficult to ratify (see Case study 6.3). Moreover, discussions post-Kyoto, for example at the Copenhagen Climate Conference in November 2009, have proved disappointing – resulting only in weak non-binding accord. The intention was that a new stronger agreement on carbon emissions reduction would have been established at this meeting. Much of the blame for the failure to achieve this was accredited to minority world countries unprepared to accommodate the demands of majority world countries who argued that they should be allowed higher emission rates.

Thematic case study 6.3

Implementing a convention? The obstacles faced in tackling climate change

Although there is residual controversy surrounding the issue of climate change, three basic propositions can be accepted:

1. There is a natural greenhouse effect that warms the earth's surface by some 33°C. Without this human habitation would be impossible.
2. Since the onset of industrialization, some 150 years ago, the amount of CO_2 released into the atmosphere has increased markedly. Carbon dioxide is a greenhouse gas – that is, it traps the long-wave radiation emitted by the warming of the earth's surface by the sun and, in doing so, heats up the atmosphere. Industrialization and fossil fuel energy has also increased the amounts of other greenhouse gases, e.g. methane and nitrous oxide, and has introduced new compounds such as chlorofluorocarbons (CFCs), which promote climate change and, in some instances, deplete the ozone layer.
3. Land-use change, particularly deforestation, has reduced the earth's ability to absorb CO_2 through the process of photosynthesis.

These processes are promoting an increase in the level of greenhouse gases in the atmosphere at a time when the earth is experiencing a natural warming trend. While natural processes account for some of the warming trend, it is now accepted that elevated levels of CO_2 in the atmosphere are a major cause of climate change. The single largest anthropogenic source of CO_2 is the burning of fossil fuels (for details on climate change see Hulme 2009).

Solutions to the problem of climate change lie in reducing fossil fuel combustion in transportation, industry and power generation. The question is how?

The United Nations Framework Convention on Climate Change (UNFCC) was signed at Rio in 1992 (visit the UNFCC website at www.unfcc.int for further details). The ultimate objective of the UNFCC was to 'stabilize greenhouse gas concentrations in the atmosphere at a level that would prevent dangerous anthropogenic interference with the climate system. Such a level should be achieved within a time-frame sufficient to allow ecosystems to adapt naturally to climate change, to ensure that food production is not threatened and to enable economic development to proceed in a sustainable manner'. Conferences of the parties (COP) are attended by all the states that have ratified or acceded to the Convention (over 194 by 2010) every two years. The third COP held in Kyoto, Japan, in 1997 saw agreement on the 'Kyoto Protocols', the first actions to address greenhouse gas emissions. The breakthrough at Kyoto was due to a flexible approach that allowed individual countries to set their own targets, while aiming at a 5 per cent global reduction in CO_2 over 1990 levels. The target date for individual countries was set as some time between 2008 and 2012. In addition, the minority world countries agreed to reduce emissions in 2000 to their 1990 level, a target few have met. The European Union has an emission reductions target of 28 per cent, the USA 27 per cent and Japan 26 per cent. The transition economies of Central Europe and the former Soviet Union were granted concessions; for example, Russia has a zero-change target. Countries seeking admission into the European Union – such as Poland – accepted reductions. Not surprisingly, a complex set of rules and regulations surrounds the targets, making it possible to trade 'emissions reductions' between states. Thus, for example, if Japan helped Russia to reduce its emissions below the 1990 level, that reduction would be set against Japan's target.

For the protocol to come into force, the countries responsible for 55 per cent of the greenhouse gas emissions in 1990 have to ratify it, this was reached in May 2002. A total of 187 countries had ratified the Kyoto Protocols by November 2009, their total emissions amounting to 63.9 per cent of 1990 emissions (http://unfccc.int/files/essential_background/kyoto_protocol/application/pdf/kpstats.pdf). However, the USA has refused to ratify, and in 2010 it was the second largest single emitter of CO_2, second only to China.

In Britain there has been mixed leadership in the two decades since the Rio Conference, but by 2010 major changes for the environmental good were beginning to be made. Britain met its Kyoto pledge, exceeding it by almost 50 per cent, and in 2008 the introduction of the Climate Change Act meant that Britain had a legal framework to ensure a cut in its carbon emissions, the first such legislation in the world. It specifically aimed to ensure an 80 per cent reduction in CO_2 (using the 1990 baseline) by 2050. In 2010 the new Coalition Government cancelled the problematic Heathrow airport expansions plans for a third runway, as well as refusing permission for additional runways at Gatwick and Stansted. It also set out proposals to change and increase the duty paid on flights. This government is also an advocate of encouraging behavioural change through so-called 'nudge' politics. Its first environment-related initiative is the launch of the Green Deal (from 2012), where homeowners will be able to install energy-efficiency measures (such as insulation) with the costs coming out of the savings on their energy bills, removing the need to pay upfront. Whether these changes will be enough remains to be seen. David Cameron, Prime Minister at the time, said in 2010 that he doubted further international agreement on emissions reduction was likely to be possible and blamed the USA and China for obstructing the discussion process.

At the same time as the minority world committed itself to emissions reductions, the majority world was persuaded to accept the concept of the **clean development mechanism (CDM)**. The majority countries feel that emissions reductions in the minority world should be on such a scale that would compensate for increased future emissions because of economic growth in the majority world. CDM is aimed at enabling economic growth without a commensurate increase in greenhouse gas emissions. However, this will require transfer of considerable amounts of capital and technology. In other words, the minority world will have to help the majority world implement the CDM. Thus the CDM is a mechanism through which minority world countries are able to offset their emissions by investing in emission reductions wherever it is cheapest globally, effectively earning credits towards their own reduction targets by assisting in the implementation of CDM in the majority world. O'Riordan (2000: 202) notes that 'the CDM mechanism denies the right of Third World nations to select their own CO_2 future. This is "ecological colonialism" by another name. While the minority world has access to the capital and technology to achieve the Kyoto targets, it still requires the political will. In the majority world, political will amounts to little without the material requirements to balance the economic growth with environmental needs.' There are now significant concerns as to the efficacy of this mechanism. The Copenhagen Accord in 2009 has done little to quell these fears, and is widely regarded as too weak and vague to really tackle these key concerns. Although there were similar criticisms of the outcomes of the Cancun climate talks in 2010, progress was made in establishing a new Green Climate Fund. This is intended to raise and distribute £64 billion a year by 2020 to assist in technology transfer and low-carbon development for the poorer countries.

There is an alternative on the table to this clean development mechanism: this is known as contraction and convergence (C&C). At COP9 in Milan many representatives admitted privately that 'C&C [was] what [they] had been waiting for' (Pearce 2003: 6). Contraction means that greenhouse gas emissions would be reduced globally, resulting in dramatic cuts during the next half century. Convergence would see each country's emissions reduced, until by 2050, according to authorities in the United Kingdom and Germany, everybody in the world would have an equal right to pollute – the amount being 0.3 tonnes C per person. Carbon trading permits would help the heaviest polluters to reduce their emissions rapidly. C&C also overcomes the USA's objection to Kyoto, which is that their Asian economic competitors such as China have no emissions reductions targets. However, as the failure of the talks at Copenhagen illustrate, there remains opposition and barriers to any international emissions reduction agreement. This has been compounded further by the financial crises in the minority world since 2008, reducing the ability of such countries to invest in environmental projects.

Q What will it take for all countries and elites everywhere to cast aside their domestic concerns and privileges in favour of the global and collective good?

Although a few governments have found it difficult to ratify conventions and make efforts to achieve targets, the implementation of sustainable development actions by local administrations – **Local Agenda 21s** – has been far more successful. The International Council for Local Environment Initiatives (2002) reported that 6,416 LA21s were underway or committed to in 113 countries, that national campaigns were underway in 18 countries, and that formal stakeholder groups had been established in 73 per cent of administrative units with LA21s.

The problem of scale can make it hard to determine how to take the best action. Not only do we need to unpack the causes of environmental change and identify likely solutions, but also 'those making decisions concerning global environmental problems need to consider the ramifications for national economies and local populations' (Harris 2004: 13). In other words, we (individuals, business, nations, etc.) need to take responsibility for our environmental impacts *and* understand the consequences of mitigating these problems for other people (local populations, the economy, etc.). The problem of climate change is thus 'a local and social problem as much as it is a global environmental problem' (Barnett 2007: 1361). For example, if we want to reduce the negative consequences of driving cars we reduce individual use, encourage companies to design more fuel-efficient vehicles, and encourage governments to increase the price of petrol through tax rises, subsidize eco-alternatives and introduce road pricing. However, in practice, governments of countries like the USA have argued in the past that it would damage their economy, making them less globally competitive (leading to job losses). It was only in 2007 that US President George Bush began to make pledges to tackle climate change. In his State of the Union Address, January 2007, he argued that the USA must 'reduce gasoline usage in the United States by 20 per cent in the next 10 years . . . increase the supply of alternative fuels . . . reform and modernize fuel economy standards for cars' (Bush 2007). Barack Obama, taking office in the USA in 2008, has set more aggressive targets for emissions reductions, but by 2010 his proposed Climate Change Bill had become stuck in political quagmire and looked unlikely to be approved by Senate. It is too early to judge the success of these initiatives and China and India argue that they should be allowed to benefit from economic development through the use of fossil fuels, just as the USA already has (see Chapter 5).

All these complexities are shaped by how we value the environment, at what scale we view these problems, and thus whose responsibility we believe it is to make changes. It is clear, then, that we all need to take responsibility for our actions but that a solely localized approach by some of us will allow the harmful environmental consequences of others to go unchecked. Multiscalar action is required, from local to international, involving all of us in the processes of change. But how have people sought to make such change happen and how does it work in practice?

6.5 Strategies for change

There are a wide variety of ways in which action is taken, through international agreements, the legislature (i.e. formal party politics), lifestyle changes, community living, and through direct action (Dobson 2001). We can broadly conceive of these as a continuum between reformist and radical approaches (see Table 6.4). A reformist approach advocates working within the current political and economic system. It reflects the shallow green perspective that it is possible to modify our actions to reduce environmental impacts. The radical approach advocates major changes to how we act, which cannot be accommodated within the current system: a deeper green world-view; 'the most radical [green aim] seeks nothing less than a non-violent revolution to overthrow our whole polluting, plundering and materialistic industrial society and, in its place, to create a new economic and social order which will allow human beings to live in harmony with the planet. In those terms, the Green movement lays claim to being the most radical and important political and cultural force since the birth of socialism' (Porritt and Winner 1988: 9). Both approaches are part of the broad and diverse environmental movements. We can use some examples to explore these strategies in practice: in national party politics (the German Green Party), political lobbying (Friends of the Earth UK), direct action (Earth First!), individual actions and finally business responses.

The West German Green Party (now the German Green Party) – Die Grünen – made its entrance into formal politics in 1983 with 5.6 per cent of the national vote and 27 seats in Federal Parliament (Doyle 2005). The party emerged from a vibrant environmental movement that was particularly active on anti-nuclear issues. When it was first launched the party was closely aligned to those seeking radical change. Petra Kelly, an early leader of the party, described Die Grünen as

Table 6.4 Reformist and radical approaches to environmentalism

Reformist	Radical
1. Modified sustainable economic growth/ ecological modernisation.	1. Limits to, and undesirability of, economic growth.
2. Large role for technological development as a provider of solutions for environmental problems.	2. A distrust of scientific and technological fixes.
3. Environmental solutions can co-exist with existing social and political structures.	3. Radical social and political change necessary: either authoritarian or decentralized and democratic political organization.
4. Anthropocentrism and a commitment to intragenerational and intergenerational equity.	4. Intrinsic value of nature or, at least, a weaker version of anthropocentrizm; a commitment to social justice within human society and between humans and non-human nature.

Source: from Robert Garner, *Environmental Politics*, published 2000, (Macmillan) reproduced with permission of Palgrave Macmillan

'half party and half local action group – we shall go on being an anti-party party' (Kelly 1984: 21). Thus the party sought to be part of formal politics but simultaneously subvert the process by seeking to practise more **participatory democracy** and pushing for fundamental change. They did this through the legislation they sought, funding for environmental groups and organizations, and structurally by rotating their representatives sitting in parliament by limiting the length of time to two years, limiting the length of office of their leaders and practising participatory decision-making processes.

However, over time 'the party has been colonized by the demands and temptations of parliamentary activity' (Dobson 2001: 127). The party has dropped many of its earlier commitments (environmental sustainability, disarmament, social justice and democracy) and abandoned 'their experimental attempt at institutionalizing direct democratic structures within the framework of **representative democracy**' (Poguntke 1993: 395). The radical faction lost out to those in the party who believed in reformist change. This approach led to the party forging coalitions with other parties, eventually forming an alliance with the Social Democrats to form the national government in 1998, consequently weakening their stance on war and nuclear power, and taking a stance *against* those undertaking environmental direct action (Doyle 2005). The German experience is a cautionary tale about the potential for environmental action to take place through formal political arenas. However, just as a party might become institutionalized, this acts as a stimulus for other non-political party action; 'the very fact that social movements tend to be absorbed by the system provides a motive for the radicalisation of other groups' (Rucht and Roose 1999: 91). So although forming a political party might not be the most effective strategy for action (and in countries like Britain and Australia without proportional representation, it is almost impossible for Green Parties to gain any form of power [Caroline Lucas became the first and so far only Green Party MP in Britain in May 2010]), in practice few environmentalists follow such a singular path: rather they will be involved in and support other forms of action simultaneously.

Environmental NGOs and related groups serve two key roles in society: they often identify environmental issues and, importantly, they seek to translate environmental concern into practical strategies for change. In 1971 Friends of the Earth (FOE) was launched in London; it has become one of the biggest and most influential of the British environmental NGOs, and is part of an international FOE network. Its formation reflected a frustration with the staid conservation movement's lack of action against those perpetrators of environmental degradation (Rawcliffe 1998). They have a five-pronged strategy: to lobby those in political power and industry, often using legislative activity to press for change; to generate scientific research and publish it in accessible formats; to employ the media to attract attention to particular issues; to mobilize the public through local groups; and to coordinate and cooperate with other groups to run large-scale campaigns.

By the early 1990s however, FOE had evolved from a radical group to one that seemed less keen to press for far-reaching change. In effect it had become reformist – pragmatically seeking inclusion of environmental concern within the current political and social system. Many saw this reformist approach as being ineffectual, slow and hierarchical. Drawing inspiration from other movements and other countries, smaller, more radical environmental groups (or informal networks) began to proliferate. The visions they espouse are often idealistic and the steps to realizing their dreams often challenging, such as dramatically reducing consumption, abandoning the use of cars, or changing our eating habits. They pose a systematic challenge to existing societal practices, often rejecting 'representative' democracy and formal politics and promoting grassroots participation in environmental decision-making and a 'do-it-yourself' (DIY) approach. The tactics employed reflect their belief in the need for radical change and often involve the use of non-violent direct action. Direct action is 'intended to have an immediate effect on a situation, as distinct from political activity which might have a roundabout effect through representatives, or demonstrative activity whose effect was to get publicity' (Rooum 1995: 27). Examples include road blockades and office occupations. Earth First! (EF!) formed in the United States in 1979 and spread to the United Kingdom in 1991. Based upon anarchist ideology, it is a network of autonomous groups that eschew formal membership and hierarchy (and thus leadership) and espouse consensus decision-making structures. It is based on the belief that actions speak louder than words, and has become synonymous with (often illegal) non-violent direct action and the DIY approach: one of their slogans is 'If not you, then who?' (Wall 1999; Doherty 2002). EF! has been influential in British anti-roads protests, e.g. Twyford Down (1992), Newbury (1995–6) and A30 Devon at Allercombe and Fairmile (1994–7). They used protest camps with tree-sits, tunnels and lock-ons with the aims of (i) physically preventing road construction; (ii) generating media publicity; (iii) educating the public; and (iv) acting as a catalyst for mass mobilization. The defining characteristics of such activism are the continuously evolving creativity in tactics coupled with a broad concern for a variety of issues, for example 'they see exploitation of the Third World, the global poor, women, animals and the environment as a product of hierarchy, patriarchy, anthropocentrism, racism and, most prominently, capitalist economic relations' (Seel and Plows 2000: 114).

Such actions have been influenced by many others and replicated worldwide. In Australia, environmentalists have employed direct action since the late 1970s. Actions at Terania Creek (New South Wales), Daintree (Queensland) and surrounding the Franklin Dam (Tasmania) all focused on preventing the logging of old-growth trees. Activists set up protest camps and used tree-sits and roadblocks to deter environmental destruction (Cohen 1997; Doyle 2000). Such tactics are still employed today at anti-logging protests at East Gippsland (Victoria). India has an even more extensive history of *satyagraha* (non-violent action) inspired by Mahatma Gandhi. In the early 1970s, villagers (mostly women) from Uttar Pradesh physically began to try to prevent local logging – resulting in the Chipko Andalan movement (McCormick 1989). More recently, villagers threatened with losing their land, homes and livelihoods by the construction of the Narmada Dam have committed to drowning rather than forced resettlement (Shiva 2002). Such examples clearly illustrate the implicit link between environmental concern and social justice.

A key tenet of many environmental approaches is the need for individuals to take action. This is both because it is relatively easy for individuals to make changes to their lifestyles *and* because it is only with all individuals making changes that bigger changes can occur: 'the only possible building blocks of a Greener future are individuals moving towards a Greener way

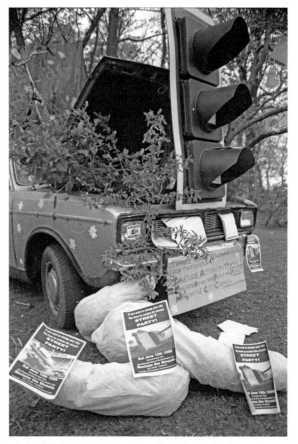

Plate 6.3 An anti-car stunt calling for radical environmental activism.

(Jenny Pickerill)

of life *themselves* and joining together with others who are doing the same' (Bunyard and Morgan-Grenville 1987: 336). Thus individuals are encouraged to 're-duce, reuse, recycle'. The three R's are deliberately in the order of magnitude of impact. Thus, for example, reducing our consumption (see Case study 6.4) will have a far greater environmental benefit than recy-cling, which is time, energy and financially inefficient. However, unfortunately at present, 'it seems unlikely that a massive number of individuals will experience the conversion that will lead to the necessary changes in their daily behaviour' (Dobson 2001: 131).

Thematic case study 6.4

Consumption and the environment

> Most of the luxuries, and many of the so-called comforts of life, are not only indispensable, but positive hindrances to the elevation of mankind. With respect to luxuries and comforts, the wisest have ever lived a more simple and meagre life than the poor . . . simplicity, simplicity, simplicity!
>
> (Henry David Thoreau, *Walden*, 1854)

As Thoreau illustrates, environmentalists have long argued for the need to consume less. There is an explicit link between economic growth, increased consumption, and negative environmental impact, and 'if the global population adopts the lifestyle currently seen in the minority countries, the environmental ef-fects would be enormous' (Harris 2004: 271). Indeed the British Coalition Government's response to recent economic problems has been to emphasise the need for further consumption in the belief that economic growth is driven by constantly increasing levels of consumption. Although environmentalists argue for a need to limit economic growth per se there is a particu-lar focus on consumption, because if we were able to reduce demand for environmentally destructive goods it follows that production would also fall. Moreover, reducing consumption can be a personal act, available to all, and is part of the broader green vision of a life where we spend leisure time absorbed by non-material pleasures such as socializing, exercise, reading, etc., rather than shopping.

Like Thoreau, British environmentalist Jonathan Porritt argues for a life of 'voluntary simplicity' (1984: 204). But what would a simple life look like, and how can we define what are justifiable needs and excessive luxuries? Although we can agree on what constitutes basic needs (shelter, food, clothing), there is disagree-ment as to what other items would be part of a greener vision. For example, are cars allowable? If so, what type of car? This ambiguity has led to the rise of green consumerism – purchasing goods with less environ-mental impact. In recent years there has been rapid growth in this sector, for example, eco-friendly washing liquids, organic food, or 'natural' shampoos. However, as Dobson notes, 'there is nothing inherently green . . . in green consumerism . . . none of it – absent other strategies – can bring about the radical changes envis-aged by ecologism' (2001: 131–2). The trouble is that green consumerism is not about buying *less*, just pur-chasing different products: 'The underlying aim of this green consumerism is to reform rather than fundamen-tally restructure our patterns of consumption' (Porritt and Winner 1988: 199). A similar argument can be made for why recycling in and of itself is limited. Recycling is not about *reducing* what waste we produce in the first place, just dealing with our waste more appropriately.

Green consumerism and the notion of 'sustainable consumption' are becoming mainstream practices partly because they do not require us to make major changes in our lifestyles (Burgess *et al.* 2003). In con-trast, smaller-scale models of production and consump-tion, such as local organic food networks, or LETS and Time Bank schemes (which use alternative currencies to encourage non-monetary transactions) are being advo-cated as more environmentally sensitive and sustainable (Seyfang 2009). Moreover, there is now an annual in-ternational 'Buy Nothing Day' which encourages people to 'host swap shops – free concerts and shows – credit card cut-up tables – create a shopping free zone with carpet and chairs – conga against consumerism – protest against sweatshops . . . There is only one rule – BUY NOTHING!' (www.buynothingday.co.uk).

For more information on consumption see the report by the Sustainable Development Commission (2006), 'I will if you will – towards sustainable con-sumption,' at www.sd-commission.org.uk.

Q 'If something cannot be manufactured, built or grown without causing irreparable ecological damage, can't we strive to create something to take its place, or simply decide to do without it?' (Tokar 1994: 80). What could you do without in your consumption and what would make you willing to give it up?

Finally, businesses can play a key role in strategies for environmental change, both good and bad (Doyle and McEachern 2008). Businesses have sought to develop technologies that reduce environmental impacts (wind farms are one example), establish ethical banking practices (such as the Co-operative Bank), and practise Corporate Social Responsibility throughout their operations. Globally the United Nations Global Compact attempts to encourage such action by asking business to sign up to ten universally accepted principles in the areas of human rights, labour, environment and anti-corruption. These are models that attempt to marry capitalist approaches with softer environmental costs. However, some companies have lobbied hard against any climate change mitigation strategies that they perceive would impact upon their profits. For the automotive industry in the USA this has included lobbying against tax increases on gasoline or for producing smaller more fuel-efficient vehicles. Chrysler's chief economist Van Jolissaint has even got as far as questioning the existence of climate change and the 'quasi-hysterical' policies of the Europeans (Schifferes 2007). Those that market themselves as being 'green' also face accusations of greenwashing – appearing green while changing none of their fundamental approaches, for example see Spotlight box 6.2. Moreover, businesses' entry into green markets can lead to a dilution of the original principles and eventually subvert the environmental benefits. For example, a core element of organic food production is to reduce pollution. This principle is enhanced when people purchase organic food locally (for example, through a box scheme from their local farm). However, when supermarkets saw organics as a market opportunity, they began to import organically produced food from Africa and Latin America, undermining many of its environmental benefits (Hughes 2005; 2007). Here supermarkets failed to take account of the importance of reducing the 'air miles' that food has travelled as part of the measure of the environmental impact of food consumption.

The dilemma remains how best to enact change, be it through NGOs that are able to lever access to the halls of power but in doing so can appear elitist, or through radical direct action which encourages grassroots participation and individual responsibility but which in turn can be small in scale and effect. For now, the answer appears to lie in the vibrancy of environmental movements composed of a variety of contesting, challenging and supportive groups, incorporating international NGOs alongside more radical small-scale groups.

What can one person, one household, one community or one administrative unit do when governments of some of the world's most economically powerful states cannot be brought onside? The environmentalist's answer is to be a good global citizen, to put pressure on administrations, governments, corporations and supranational organizations through political and non-political means, and to educate others. We have gone some way down this route in the past four decades, but significant problems remain.

Spotlight box 6.2

Carbon offsetting

Carbon offsetting is the practice of financially contributing to projects which lead to a reduction in carbon dioxide emissions. This contribution is then counted as a reduction to the purchaser's carbon footprint. In other words it is offset against their actual carbon emissions. It is an indirect form of reducing carbon usage. Companies increasingly seek to buy carbon credits through funding energy efficiency, reforestation or renewable power projects, often in majority world countries. In 2005 HSBC, one of the world's biggest banks, asserted that it had gone 'carbon neutral'. It achieved this by buying carbon offset credit from renewable energy projects in New Zealand, Australia, India and Germany. Individuals are also offsetting some of their most carbon intensive activities, such as air travel, through contributing to tree planting programmes.

Critics of carbon offsetting argue that it is a poor substitute for actually reducing our carbon emissions and is another way that companies and individuals can buy their way out of responsibility for the environment. There is significant debate as to its validity and effectiveness, with various different forms of measurement being used and the time-frame of, for example, forest growth still yet to be taken properly into account.

6.6 What is missing from our focus on 'the environment'?

One of those problems is the way in which we conceptualize the environment. Explanations by scientists of the human causes of environmental degradation have, until recently, almost entirely focused on direct causes. Middleton (2008), for example, develops a fourfold typology of society–environment relationships (see Table 6.5). However, we are beginning to acknowledge the complex web of relationships between humans and the environment, and its relation to poverty, development, injustice and democracy.

Since Rio the emphasis has changed to one that seeks to integrate environment and development more closely and for the environmental agenda to be led by concerns about poverty. These new lines of thinking are underpinned by the World Bank's Poverty Alleviation Strategies, the Millennium Development Goals which were agreed by the UN General Assembly in September 2000 (www.un.org/millenniumgoals), and as a general response to lack of progress on poverty and debt in the Third World (see Chapter 8 for more discussion on this theme). For many in the majority world the links between poverty, development and the environment are explicit. It is not a simple case of poverty leading to environmental damage: there are countless examples of poor people being key actors in environmental campaigns in India, Indonesia, the Philippines and across South America. Certain forms of development have significant environmental consequences – such as dam building, fossil fuel power generation and natural resource extraction (see Chapter 5 for further information). But we should understand these developments within the context of a history of colonial exploitation and contemporary processes of competitive economic globalization.

Table 6.5 A typology of society–environment relationships

Type of relationship	Examples
Deliberate environmental manipulation	1. Diversion of watercourses for irrigation which leads to changes in river flows. Positive benefits include flood control and water availability for agricultural production over a long period, e.g. the Indus Irrigation Project in Pakistan. 2. Construction of large dams which leads to forced relocation of people, loss of productive land and wetland habitats, and changes in water resources and ecology downstream, e.g. Three Gorges Project in China.
Unanticipated impacts of activities that have been carried out to improve society	1. Irrigation in the former Soviet Union led to the desiccation of the Aral Sea because of diversion of too much water from the rivers that flowed into it. This led to the loss of the Aral Sea commercial fishery and health problems related to wind erosion of agrochemicals trapped in former lake-bed sediments. 2. Expansion of cereal cultivation onto semi-arid grasslands has led to unexpected episodes of severe soil erosion (e.g. the US 'dustbowl' in the 1930s).
Complex problems in which cause and effect are not obvious. These are often characterized by human activities and natural forces combining synergistically	This applies to climate change, where there is still debate as to what proportion of elevated temperatures is due to natural climate change as against the effects of burning fossil fuels.
Linkages characterized by overlap and interaction	Contemporary deforestation is causing changes to people's livelihoods, losses of biodiversity, changes in the regional water balance and changes in global climate. All of these impacts overlap at scales ranging from local to global.

Source: after Middleton (2003)

Moreover, **ecological modernization** has failed to consider international justice (Connelly and Smith 2003). Where industrialized nations have sought to reduce their environmental damage they have in many cases simply exported their environmental problems to less industrialized countries: for example, the export of harmful computer wastes to China (Shabi 2002). This process has also occurred internally, see Spotlight box 6.3. Furthermore, environmental protection has resulted in the dispossession of Indigenous peoples from their land to make way for national parks (see Case study 6.5). This practice results from two mistaken beliefs: first, that humans will only ever have a detrimental impact on their surroundings and second, that preservation of an environment is more important than human ties to a particular place, and in some cases, human survival.

Thus the industrialized world's approach to 'the environment' often fails to understand majority world experiences, the need to incorporate humans *into* our understandings of the environment, and to acknowledge that some people have limited environmental choices. Recent work has sought to change how we view the environment, suggesting that the division between human and non-human is an artificial and unhelpful one (see Chapter 12). Environmental concerns have been recast to be ones of social survival. For example, the impact of sea-level rises on Pacific Islanders will be so catastrophic that it will potentially wipe out entire nations. Rather than being seen as a hindrance to wildlife preservation, Indigenous people's environmental knowledge and management skills are slowly beginning to be valued. Moreover, their often alternative ways of viewing the environment, in the case of Indigenous Australians for example, as being a melding of culture, land and sky, challenge our

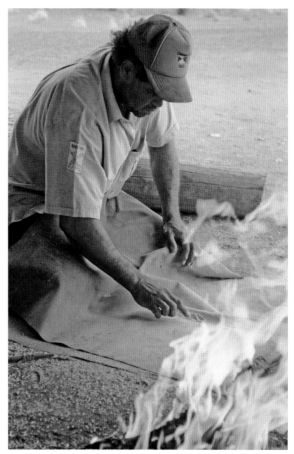

Plate 6.4 Teaching Indigenous knowledge about the land, Australia.

(Jenny Pickerill)

conceptualizations. The need to understand how we can trigger behavioural changes and the psychology behind environmental concern are becoming priority areas for environmental research (Hulme 2004).

Spotlight box 6.3

Environmental justice

The environmental justice movement called for acknowledgement that it is often the already marginalized who suffer most from environmental degradation. Robert Bullard in his book *Dumping in Dixie* (1990) argued that toxic dump sites were more likely to be near neighbourhoods of black Americans than whites – race being a major factor in the quality of a person's environment. This is a form of environmental racism and environmental injustice.

It is also a critique of environmentalists' focus on wilderness preservation rather than the plight of the underprivileged, and 'has dragged other parts of the US environmental movement into a political place where the social ramifications of being green have to be confronted' (Doyle and McEachern 2001: 69). Understanding environmental problems as routed in an inequality is helpful in understanding some of the contemporary issues in dealing with climate change. Parks and Roberts (2008) argue that 'global inequality may be a central impediment to interstate cooperation on climate change policy' (p. 621) and therefore it is this injustice that we need to tackle first and foremost.

Thematic case study 6.5

Wilderness and conservation debates in Australia

Environmental campaigns worldwide are often framed as conserving 'wilderness' and preventing exploitation of natural resources. Organizations lobbying for protection of vast tracts of Alaska, Canada and Australia employ an emotive language using words such as 'pristine', 'untouched', 'undisturbed', 'intact expanse' and 'wild frontiers' to rally support. Photographs of lands empty of people or any human structures often accompany these words. Whether directly, through hunting for example, or indirectly through their development activities, people are usually held responsible for losses of wildlife. Until the 1980s, conservationists' solutions relied on controlling people's actions through strategies like hunting bans and establishing protected areas. These strategies – collectively known as protectionism – are overtly neo-Malthusian and have dominated nature conservation. This approach has proved successful in arguing for the need to place legal boundaries around tracts of land through World Heritage status or as National Parks.

Despite this, there is, increasingly, recognition not only that Indigenous people have historic rights of possession to some of that land, but also that Indigenous environmental knowledge and land management practices can be beneficial for conservation outcomes. This recognition and the legal changes that have accompanied it (such as the development of Native Title in Australia) have forced environmental and conservation groups to reconsider their approach. Simultaneously a global movement has emerged, focused on environmental justice and environmental racism. Yet this operates in tandem and separate from those groups concerned with biodiversity and wildlife preservation. Many environmental groups have been slow to change their language or practices when it comes to viewing areas as 'wilderness' devoid of people, cautious of meddling with their model of success and entering the complex and contested arena of Indigenous politics.

There are many reasons to urge such groups to change: such practices are a continuation of an unjust colonial system complete with racism and stereotypes; they not only fail Indigenous populations but lead to the loss of valuable environmental knowledge and academics have long called for the need for a more nuanced understanding of nature – not viewing it as a separate entity from social and cultural influences

(Castree 2003). Moreover, while progress at the global level through UN resolutions to ensure that Indigenous people are not forcibly removed from their lands has been slow, small changes can occur at a local scale with positive effects for Indigenous communities. In other words, environmental groups have the power and capacity to markedly improve the situation for the betterment of conservation outcomes *and* Indigenous people.

Language is a key articulation of power in understandings of the 'environment'. The term 'wilderness' and its relation 'wild' is highly problematic when talking of any landscape, but especially so in Australia, a land inhabited by the Indigenous population for tens of thousands of years. Yet these terms are still employed. Such a romanticization of the environment draws upon the writings of Henry David Thoreau, John Muir and Miles Dunphy, and is used to sell Australia and garner support for environmental protection.

Indigenous people have been highly critical of the use of such a word. It has obvious colonial and racist connotations. This narrow vision, propagated by environmental groups, of what 'wilderness' entails is dualistic. Indigenous Australians argue that there can be no such division between the environment and culture and 'land is a much more energetic configuration of earth and air, water and minerals, animals and plants, as well as people, than a surface area contained by lines on a map' (Whatmore 2002: 71). It is precisely because of these interrelations between environment and culture that Indigenous people need to live on their land in order to care for their country; 'the land needs the people and the people need the land . . . these are important cultural environments that require people to manage them according to tradition and culture, to maintain . . . and to encourage species' (Damian Britnell, Chief Executive Officer, Bamanga Bubu Ngadimunku Inc., Mossman Gorge Aboriginal Community).

The Indigenous critique of 'wilderness' and approach to 'country', especially calls to live on land, is controversial. However, the majority of environmental groups in Australia now have (albeit only very recently) a vague policy of supporting Indigenous rights. Some groups have attempted to go further and alter their language and practices to implement this policy. The Wilderness Society has shifted away from large-scale agreements and broad alliances with Indigenous groups to grounded community-level cooperation. This approach appears more successful. Although they have not discarded the contentious goal of creating National Parks, they are working more in collaboration with traditional owners, supporting a class of National Park which is owned as Aboriginal Land but managed under the Nature Conservation Act as a National Park, supporting their attempts to gain Aboriginal Freehold land, and to establish Indigenous Protected Areas (Schneiders 2006: 27). They are developing 'tenure-blind' conservation measures: less concerned with who owns the land and more focused on good land-management practices.

The Australian Conservation Foundation has taken a different path. They have instigated a northern Australia Programme which aims to be a long-term proactive approach towards creating a bi-cultural organization. It is focused on coming up with joint goals rather than finding campaign partners for issues they think are important. To this end ACF has even opposed the creation of new National Parks when traditional owners have objected. In terms of language, ACF avoids the words 'wilderness' and 'biodiversity' altogether and instead talks about 'nature' and 'culture'.

Both these organizations have made progress in modifying their language and practices in response to the charge that they were party to 'a form of ecological imperialism' (Langton 1998: 18). However, two key problems remain. First, the majority of Indigenous engagement by these two groups has been confined to northern Australia. A view persists that there is no need to engage with Indigenous politics further south, perpetuating the myth that only those Indigenous people who have a more apparent and historic (according to non-Indigenous adjudicators) connection to their homeland need consultation. Second, although the language has changed, the underlying premise of why a landscape is of enough value to protect has not altered. Both still employ a biophysical-based and scientific method to determine value. We are a long way from bridging the gap between 'environment' and 'culture' and widening the ways in which we value landscape.

Source: adapted from Pickerill (2008)

Q How would an Indigenous Australian approach change how we looked after the environment?

6.7 Environmental futures

Society is taking action: policies are being designed and implemented at global, regional and local scales. But there are issues that still dog progress:

- Remediation and restoration of many types of environmental damage takes years.
- The costs of environmental remediation and protecting environments are often enormous.
- Our incomplete knowledge about many aspects of the environment, and how environment–society linkages work, can act as a barrier to finding effective solutions.
- The speed of economic and cultural change is rapid, the desire and drive for economic development strong; in contrast, the negotiation and implementation of global and regional environmental policies are slow.

Despite these obstacles it is up to us to decide what kind of world we want and to work towards it. We have explored a range of ways in which we understand and value the environment, and different ways in which we could 'reconceive the environment', and strategies, for change. In summary, there are four key steps we can take towards reducing our environmental damage.

First, we need to understand our environmental impacts better: globally, regionally, locally and individually. At an individual level we can use the concept of an ecological footprint to evaluate our resource use. Second, we can *participate* in environmental decision-making. Connelly and Smith (2003: 361), through their concept of ecological democratization, argue that the only way to achieve environmentally sustainable practices is through 'a commitment to both justice and participation'. They advocate extensive citizen participation and development of democratic institutions to tackle environmental issues at all scales. In many ways this is an extension of many pre-existing local initiatives. It is also a step beyond advocating education. As Pepper notes, 'people will not change their values just through being "taught" different ones. What, then, is the real way forward, if it is not to be solely or even largely through education? It must be through seeking *reform at the material base of society, concurrent with educational change*' (1984: 224, emphasis in original). Third, we need to recognize and make use of innovative grassroots solutions. Seyfang and Smith (2007) argue that it is the grassroots-community level where

Plate 6.5 A Reclaim the Streets demonstration in Hull, May 1999, which illegally occupied the streets for a day in protest at a car-obsessed culture.

(Jenny Pickerill)

significant innovation occurs that can offer us inspirational visions of alternative futures. We have already seen the challenge posed by low-impact development. Some of these innovations are small-scale in their approach, others, such as permaculture, are now being adopted globally. Fourth, and least radically, we need to improve our management of the current situation. We have existing environmental issues that we need to mitigate immediately, even while we begin to explore alternative futures. For example, in Britain there are growing numbers of car users. A 2006 government report, The Eddington Transport Study, supported road pricing, rather than more road building,

in order to overcome existing congestion problems, and that 'all transport users should meet all their external economic, social and environmental costs' (see www.hm-treasury.gov.uk/independent_reviews/eddington_transport_study/eddington_index.cfm). This approach has led to the introduction of the congestion charge in London. However, plans to implement similar schemes in Manchester and Edinburgh have been publically rejected. We have to decide if road pricing is an appropriate management strategy, or whether advocating having a travel allowance, akin to rationing, would be more efficient. Given these choices, which pathways forward are you going to take?

Learning outcomes

Having read this chapter, you should understand:

■ How we value human and non-human life/the environment.

■ The evolution of environmental concern.

■ The complexity of scale when looking for solutions to environmental problems.

■ The possibilities and limitations of different strategies for action.

■ How we can begin to make progressive steps towards reducing our environmental impacts.

Further reading

Bingham, N., Blowers, A. and Belshaw, C. (2003) *Contested Environments*, Open University Press, Milton Keynes. This book explores why environmental issues are so often controversial, and uses concepts of value, power and action to unpack environmental concerns.

Doyle, T. (2005) *Environmental Movements in Majority and Minority Worlds: A Global Perspective*, Rutgers University Press, London. A good introduction to a variety of environmental movements in the USA, Australia, Britain, Germany, Philippines and India, incorporating protection of forests, rivers and wilderness areas, and movements against mining, road building and nuclear power.

Dryzek, J. S. and Schlosberg, D. (eds) (2004) *Debating the Earth: The Environmental Politics Reader*, Oxford University Press, Oxford. A useful reader incorporating a wide variety of contemporary and historic authors on key themes of environmental politics such as democracy, justice, wilderness and the place of people.

Harris, F. (ed.) (2004) *Global Environmental Issues*, John Wiley and Sons, Chichester. Taking a global perspective this book examines environmental problems as complex issues with a network of human and biophysical causes.

Hulme, M (2009) *Why We Disagree About Climate Change: Understanding Controversy, Inaction and Opportunity*, Cambridge University Press, Cambridge. This book seeks to reconsider how we perceive climate change and views it as an environmental, cultural and political phenomenon rather than simply a threat. It combines a scientific approach with social and political standpoints.

Middleton, N. J. (2008) *The Global Casino: An Introduction to Environmental Issues*, 4th edition, Arnold, London. An up-to-date introduction to a wide range of environmental issues in 22 chapters – from tropical deforestation to climate change, from urban environments to food production, as well as chapters on sustainable development and war. It provides an excellent starting point for delving deeper into individual issues.

Palmer, J.A. (ed.) (2001) *Fifty Key Thinkers on the Environment*, Routledge, London. A useful summary of key environmental thinkers.

Peet, R., Robbins, P. and Watts, M.J. (eds) (2011) *Global Political Ecology*, Routledge, London. This book includes a large variety of case examples in exploring the relationships between environmental issues, economics, poverty and politics.

Peterson Del Mar, D. (2006) *Environmentalism*, Longman, New York. A short thought-provoking book which examines the complex relationship between prosperity and environmentalism.

Robbins, P., Hintz, J. and Moore, S.A. (2010) *Environment and Society: A Critical Introduction*, Wiley-Blackwell, London. An accessible introduction to the relationship between society and environmental challenges, usefully illustrated with a broad range of examples.

Useful websites

www.climatecrisis.net Provides additional information, science and 'take action' suggestions for Al Gore's film *An Inconvenient Truth* about climate change.

www.ipcc.ch Provides information from the Intergovernmental Panel on Climate Change on the science of climate change and future change scenarios.

www.lowimpact.org Low Impact Living Initiative. A non-profit organization whose mission is to help people reduce their impact on the environment, improve their quality of life, gain new skills, live in a healthier and more satisfying way, have fun and save money.

www.wilderness.org.au The Wilderness Society Australia. Contains information on old-growth logging in Australia and the wide variety of other environmental campaigns with which this Australian environmental advocacy organization is involved.

www.foe.co.uk Friends of the Earth is an international environmental organization which campaigns for change on climate change, food and biodiversity.

www.unep.org/GEO/geo4 *GEO-4: Global Environment Outlook 2007*, Earthscan, London. The third book in a series of three-yearly syntheses of the global environment produced by the United Nations Environment Programme.

www.zerofootprintoffsets.com/calculator.aspx Zeroprint Carbon Calculator: enables you to estimate your carbon footprint.

For annotated, clickable weblinks and useful tutorials full of practical advice on how to improve your study skills, visit this book's website at **www.pearsoned.co.uk/daniels**

FOOD SECURITY

Chapter 7

Bill Pritchard

Topics covered

- Famines
- Undernourishment
- Entitlements
- Green revolution
- Genetically modified foods and world hunger
- Food-aid dependency
- The Right to Food

Starvation is the characteristic of some people not *having* enough food to eat. It is not the characteristic of there *being* not enough food to eat. While the latter can be a cause of the former, it is but one of many *possible* causes.

(Opening words to Amartya Sen's *Poverty and Famines: An Essay on Entitlement and Deprivation*, 1981)

7.1 Introduction

'Food security' is a term that is used in many different ways. The purpose of this chapter is to discuss this concept in a manner that is consistent with its usage at the international level by entities such as the UN Food and Agriculture Organization. We begin by placing contemporary debates about food security into an historical perspective.

7.2 Hunger in human history

Widespread hunger has stalked humankind perennially. Although this threat might seem far removed from the quotidian concerns of most people living in the global North during the early twenty-first century, the challenge to feed and sustain the human population has been a pre-eminent social, political and economic struggle across history, and remains central to the planet's circumstances today.

In western Europe, battles over food and hunger were seminal to the cycles of political evolution and revolution which forged nation states. Rousseau's apocryphal description of a French princess (not Marie Antoinette, as commonly attributed) asking why the peasants would prefer bread to cake ['*Qu'ils mangent de la brioche!*' ('popularly translated as 'Let them eat cake!')] is a testament to the symbolic and material power of food in political and social life. Throughout the period right up until the twentieth century, periodic civil strife over food was a defining element of European life. In the British Isles, there was ongoing hunger and malnourishment for the working and agrarian classes throughout the onset of the industrial period, which erupted episodically into food riots, such as those affecting England and Wales in 1772–73. During the nineteenth century, a key element in the debate over the passage of the Corn Laws (which allowed cheaper, imported food into Britain) was the need to feed (and, thus, politically placate) the fast-growing urban working class. All the time, however, the threat of widespread hunger was ever-present. In Ireland, as is well known, the collapse in potato production because of *phytophthora infestans* (potato blight) instigated famine conditions over the period 1845–52. Perhaps less well known, similar conditions also afflicted the Scottish Highlands over this same period, with a level of death and out-migration which was of almost equal magnitude to the Irish famine.

As European peoples spread across the world, these challenges travelled with them. The American tradition of Thanksgiving traces its origins to the near-starvation of the Pilgrim settlers in 1620–21. Following a successful harvest that kept the settlers alive, a Thanksgiving feast was held. Comparably, in Australia, the First Fleet settlers of 1788 faced imminent starvation within two years of arrival due to the expiration of their rations and farming failures.

Amongst European peoples worldwide, the politics of hunger continued to hold important sway right up until the end of the Second World War, often associated with interruptions to seasonal harvests brought on by climate or war. Thus, in 1815 when the far-distant Tambora volcano erupted in the Dutch East Indies (now Indonesia) and sent a shower of ash into the upper atmosphere, a resultant global cooling of temperatures decimated food harvests in the northern hemisphere the following summer, creating a hunger crisis in cities on America's Atlantic seaboard. The author Simon Winchester describes this event as 'the last great subsistence crisis of the Western world' (2003: 295). Food insecurity remained a present threat in urban America through the nineteenth century, including New York's 'bread riots' in 1837, and food riots in many cities during the Civil War (1861–65). In Europe, the wars of the nineteenth and twentieth centuries were often accompanied by food shortages and hunger. Perhaps the last ever full-scale famine in western Europe occurred in 1944–45 in the western Netherlands, when Dutch resistance to Nazi rule caused disruptions to food transports and an estimated 18,000 people died of hunger (van der Zee 1982). More recently (though perhaps less dramatically), the end of the Cold War saw heightened food insecurity in the former Soviet Union over several years as its economic system restructured to the logics of capitalism. Yet for all the importance of hunger within the record of Western civilization, non-European populations have been the overwhelming bearers of hunger across human history. An East India Company trader visiting (what is now known as) the state of Gujarat in India during the great famine of 1630–32 recorded that: 'from Suratt [sic] to this place [Agra] our noses were never free of the stinck [sic] of corpses' and 'the [roads] were so full of dead bodyes [sic] that we could hardlie [sic] pass them without treading on or going over some' (cited in Keay 1993: 115). The strengthening of British colonial rule in ensuing centuries exacerbated hunger vulnerabilities. Local politico-administrative systems were recomposed to fit within

Thematic case study 7.1

The Bengal famine of 1942–43

Estimates of deaths from the 1942–43 Bengal famine remain disputed. The Famine Commission set up by the British authorities reported 1.5 million deaths. Amartya Sen has suggested the death toll was closer to 3 million, and some other Indian sources suggest it was higher still. Ó Gráda (2007: 19) adopts

a 'mid-range' estimate of 2.1 million deaths. For the Famine Commission estimate, see p. 68 in Weigold, A. (1999) Famine management: the Bengal famine (1942–44) revisited, *South Asia*, **22**(1), 63-77. For Amartya Sen's discussion of the death toll, see p. 52 in Sen, A. (1981) *Poverty and Famines: An Essay on Entitlement and Deprivation*, Oxford University Press, Oxford.

colonial logics, which had as their core the extraction of profit to service the Empire. A series of famines in India during the nineteenth century saw the deaths of an estimated 30–40 million people. The last major famine under British rule occurred in Bengal, in 1942–43, when a combination of war-induced decisions to restrict the transportation of rice and cyclonic devastation resulted in the starvation deaths of between 1.5 million and 4 million persons (see Case study 7.1).

Histories of China, other parts of Asia and Africa reap a similar story. In Ethiopia, for example, widespread crop failures and the related onset of famine occurred with regularity once or twice a decade from recorded history many centuries ago until 1984–85, when a severe famine impelled global action. Though the short-term and long-term implications of the global community's responses in 1984–85 (the 'Live Aid/

Band Aid' initiative of Sir Bob Geldof) remain grist for ongoing controversy and debate, the debilitating famine cycle in Ethiopia was broken in the two decades thereafter. The economic historian Cormac Ó Gráda (2007: 5–6) observes that 'By the 1990s, famine-induced deaths were confined to poverty-stricken and often war-torn pockets of the globe'.

7.3 The present scale and geography of global hunger

Notwithstanding this receding of large-scale repeat famine cycles, the estimated number of persons living in chronic hunger has actually increased during recent years. A key turning point in this recent history

Spotlight box 7.1

How many hungry people are there in the world?

As illustrated in Figure 7.2, the FAO estimates that there are almost one billion undernourished people on the planet. In all probability, however, this is a lower bound estimate of global hunger. The methodology by which these data are collected (via national household surveys and modelling the effects of changes in incomes and food prices) tends to assume that if a household has the capability to acquire enough food to meet the daily minimum energy requirements of its members, then no-one in

that household is hungry. But, as is well known, food is not always distributed equitably within households (in many societies there is a prevalence of men being fed first and women often going hungry). Also, just because a household *can* feed itself doesn't mean it *does* feed itself; in many households across the globe, money that should be used for food is used for alcohol, drugs and gambling, for instance. Indeed, one of the world's leading food researchers, Per Pinstrup-Andersen, cites an estimate of two billion iron deficient people on the planet as evidence to suggest that global hunger is far greater than official estimates suggest (Pinstrup-Andersen 2009: 6).

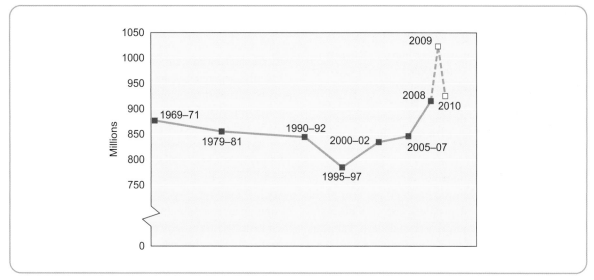

Figure 7.1 Number of undernourished people in the world, 1969–71 to 2010

Source: adapted from FAO (2010) *State of Food Insecurity*, p. 9

was 2008, when there was a confluence of rapidly ris-ing prices for food (Figure 7.1) and energy (see Chap-ter 5). These trends sparked a dramatic increase in global hunger. In 2008, it was conservatively estimated (although see Spotlight box 7.1) that the number of people living in chronic hunger was estimated to have exceeded one billion people, for the first time in hu-man history. These events triggered food riots across 60 countries, which saw at least one government (in Haiti) fall (Plate 7.1). Significantly, the events of 2008 arguably constituted the first-ever *global* food crisis. Because of the highly integrated international

character of food commodity markets, ripple effects from one country to the next occurred in a rapid and coordinated fashion. With hunger a matter of global political and economic concern, international policy-makers were shaken from their complacent assump-tions about the world community slowly 'winning the war' on malnutrition. Even though the measured glo-bal number of undernourished persons fell sharply in 2009 (in synch with falling food prices), the general consensus is that such a decline (welcome though it is) does not necessarily indicate a turning point for the better. As suggested by the UN's Food and Agriculture

Plate 7.1 Food riots in Haiti in 2008.

Organization (FAO) in its 2010 *State of Food Insecurity* report:

> The number and proportion of hungry people in the world are declining as the global economy recovers and food prices remain below their peak levels, but hunger remains higher than before the food price and economic crises, making it more difficult to meet the internationally agreed hunger-reduction targets. (FAO 2010: 8)

Furthermore, global hunger is dispersed across the world in a highly uneven way, with 40 per cent of the world's undernourished people living in China and India. A further four countries (Bangladesh, Pakistan, Democratic Republic of the Congo, and Indonesia) account for a further 26 per cent (FAO 2010: 10). Moreover, with the exception of China, all of these countries are failing to meet the Millennium Development Goal (MDG) for reducing hunger (see Figure 7.2 and Case study 8.1). (The MDG for hunger aims to reduce by half [between 1990 and 2015] the proportion of people in the world suffering from hunger – measured statistically as (i) the prevalence of under-weight children under 5 years of age, and (ii) the proportion of the population below minimum dietary energy consumption.) Clearly, this map paints a depressing picture of global food security; many of the world's countries with the highest rates of food insecurity are also those currently displaying worst progress in generating improvements to these situations. Moreover, it needs remembering that hunger and undernourishment also exist in the developed world. The recent global recession has exacerbated the extent of hunger: in 2008, 14.6 per cent of the US population was food insecure at some point in the year, including 5.7 per cent with 'very low food security', meaning 'that the food intake of one or more household members was reduced and their eating patterns were disrupted at times during the year because the household lacked money and other resources for food' (Nord *et al.* 2009: 1).

The rapid changes in the incidence of global undernourishment in recent years have triggered renewed debate amongst researchers and policy-makers on the causes and manifestations of hunger. Central to these current debates is the newly articulated concept of **food security**. To a large degree, this concept has become the foundation stone for contemporary strategic interventions of international organizations (such

Progress achieved (1990–92 to 2005–07)

- ■ Already met MDG 1 or very close to meeting the target
- ▨ Progress sufficient to reach MDG 1 if prevailing trends persist
- □ Progress insufficient to reach MDG 1 if prevailing trends persist
- ■ No progress or deterioration
- □ Not relevant – prevalence of hunger was below 5% in 1990
- ▨ Missing or insufficient data

Figure 7.2 Progress towards Millennium Development Goal 1: hunger target.

Source: www.fao.org/fileadmin/templates/es/Hunger_Portal/MDG_Progress_Map.pdf

as the FAO) and non-governmental organizations (NGOs) (such as Oxfam, ActionAid and Red Cross/Red Crescent). In the following section, the concept of food security is defined, and afterwards, its application to key issues relating to global hunger is discussed.

7.4 Defining food security

Until the mid-1990s, the prevailing perspective in international debates on food and hunger tended to prioritize production-centric explanations with the emphasis on maintaining food stocks in the context of an unpredictable world. Hence, in the case of the Irish famine of 1845–52, cited above, famine was seen to be caused by the unexpected threat of potato blight. In Ethiopia or the Indian subcontinent, famine vulnerability was tied to the unpredictability of the monsoon. Clearly, the implicit suggestion within these explanations was that such problems were premised on catastrophic and unavoidable circumstances – 'acts of God', as an insurance policy might say. Hence, in 1974, the World Food Conference defined food security as: 'Availability at all times of adequate world food supplies of basic foodstuffs to sustain a steady expansion of food consumption and to offset fluctuations in production and prices' (cited in FAO 2006: 1).

Fundamental aspects of this perspective remain important within present thinking. Rarely, if ever, are rapid descents into widespread hunger *not* associated with a food production shock of some kind (climatic, environmental, biological or human-induced, e.g. war). Such cataclysms obviously impact on people's abilities to feed themselves and, from a policy perspective, require an immediate response in the form of food aid. Nevertheless, since the mid-1990s there has been an increasing recognition of the need to build more sophisticated conceptual understandings of what causes hunger. Instead of seeing particular biological, climatic or environmental contexts as catastrophic events that define these problems, the emergent approach is to understand food production shocks as *triggers* that reverberate (in socially uneven ways) through populations. This approach refocuses the attention of researchers and policy-makers to the broader systems that connect people to the food system. This perspective goes under the banner of the food security approach.

The chief and most influential proponent of such an altered perspective has been Amartya Sen, an Indian economist and philosopher. Sen was awarded the Nobel Prize in Economics in 1998 for a body of work that included landmark analyses of the causes of famine. Sen addressed the causes of hunger and famine from a radically different set of premises from that which prevailed in mainstream thinking. He saw that famines in different circumstances worldwide each could be understood as having a common socio-politico-economic themes. Whereas dominant contemporary explanations about these episodes looked to a fall in food production as the root causes of hunger, Sen documented a more complex reality. Hence in the Irish famine, the potato crop failed but Ireland remained a large exporter of foodstuffs across to England. In Ethiopia in the famine of 1972–73, food was trucked from famine areas to the capital of Addis Ababa. The great Bengal famines of 1942–43 and 1972–73 were associated with 'moderate' falls in production being translated into severe market shortfalls due to hoarding, speculation and 'administrative chaos' (Sen 1981: 76). Extending this further, Sen's approach asked why some 'poor years' of agricultural production produced widespread hunger or famine, whilst others did not. Clearly, other factors came into play. China's disastrous 'Great Leap Forward' of 1957 caused famine not just because of crop failures, but also through ill-conceived central state planning. Famine in Cambodia during the late 1970s was instigated by the brutal maladministration of the militaristic Khmer Rouge. The 1984–85 famine in Ethiopia was associated not only with drought, but in a context where the national government diverted a massive share of scarce national resources to the military. Famine in North Korea in 1998 was associated with crop failures in the context of an autocratic regime that prioritized national security and protection of the interests of the elite class.

From this analysis, Sen proposed the concept of *entitlements* as a way of understanding hunger. This concept sought to explain hunger and famine by way of asking questions about the social, cultural and economic frameworks that bestowed rights to ownership within populations. As Sen contended, even the seemingly straightforward notion of possessing a loaf of bread (and thus forestalling one's hunger) assumes a chain of entitlement relations:

> I own this loaf of bread. Why is this ownership accepted? Because I got it by exchange through paying some money I owned. Why is my ownership of that money accepted? Because I got it by selling a bamboo umbrella owned by me. Why is my ownership of the bamboo umbrella accepted? Because I made it with my own labour using some bamboo from my land. Why is my ownership of the land accepted? Because I

inherited it from my father. Why is his ownership of that land accepted? And so on. Each link in this chain of entitlement relations 'legitimizes' one set of ownership by reference to another, or to some basic entitlement in the form of enjoying the fruits of one's own labour. (Sen 1981: 1–2)

Thus, in Sen's writings on this subject, he asserted that the ability of a person to avoid starvation depended on her/his entitlements, which, in turn, was constructed from her/his *ownership bundle* (the combination of labour powers, resources and assets s/he can use to acquire food) and the *exchange entitlement regime* s/he faced (the rights to resources s/he can access to transfer an ownership bundle into food). Or, in slightly more simplified language: 'Entitlements are defined as the set of all commodity bundles over which a person can establish command given the legal, political, economic and social arrangements of the community in which they live (including traditional rights such as access to common resources)' (FAO 2006: 1). Applying this framework to Ireland in 1845–52, Sen contended that famine was caused because the pre-eminent asset within the ownership bundle of the rural Irish poor (their control over small plots of land on which to grow potatoes, their staple food) became suddenly worthless with the onset of potato blight, and the population had few other rights over resources that could replace this loss (piecemeal social safety net efforts, such as the workhouse system, were associated with poor management and/or were hotbeds for infectious diseases [Ó Gráda 2007:17]). Hence, although potato blight was the trigger, it was this combination of ownership bundles and exchange entitlements amongst the poor that provided the ultimate cause of the Irish famine.

Sen's work was an important precedent for the ways that policymakers approached these issues at the 1996 World Food Summit (WFS), in Rome. This Summit is popularly credited with launching the concept of food security into mainstream food policy discourse. As defined at the WFS:

> Food security exists when all people, at all times, have physical, social and economic access to sufficient, safe and nutritious food that meets their dietary needs and food preferences for an active and healthy life. (FAO 1996)

Expressed in this way, the concept serves a dual purpose. At one level, it encapsulates a *normative goal*; an aspiration for the world community to attain this outcome for 'all people, at all times'. But also, it

represents a *definitional signpost*. The clear focus of the quote above is that 'food security' is defined in terms of the extent to which a food system satisfies individuals' nutritional, livelihood and social circumstances. Notably, unlike the 1974 World Food Conference definition, cited earlier, this does not equate 'food security' with maximizing food production, or having large stocks of food in reserve. These supply-side factors are obviously important, but the over-riding message in this definition is that the attainment of food security is primarily about the social, economic, cultural and political circumstances that either enable or restrict the provisioning of food to needy populations. Thus, the WFS definition of food security can be seen as a political enactment of Sen's notion of entitlements. Ultimately, what is important for addressing hunger is *not only* the ability to produce food, but a social-political-economic regime that enables people in need to gain that food.

In order to give this definition operational relevance, however, it needed elaboration. Accordingly, in the immediate years after the 1996 WFS, a FAO Working Group developed a framework for understanding the processes that created particular patterns of food security/insecurity. This framework – which is shown diagrammatically in Figure 7.3 – conceives food security as being an outcome from three dimensions of the food system:

- *Availability*: The supply-side factors that shape the availability of sufficient quantities of food of appropriate quality.
- *Access*: The political, social, cultural and economic processes that connect supply-side processes to individuals.
- *Utilization*: The elements of clean water, sanitation and health care that ensure that food that is made available and is accessible (i.e. the two categories above) generates nutritional well-being for consumers.

Reflecting the importance of these three dimensions within contemporary debates on food security, the remainder of this chapter uses this framework as an organizing device to bring into focus a series of issues relating to food security in the contemporary world. We focus on the first two categories – availability and access – because the third relates primarily to nutrition and dietary science and, as such, is somewhat outside the scope of the current chapter.

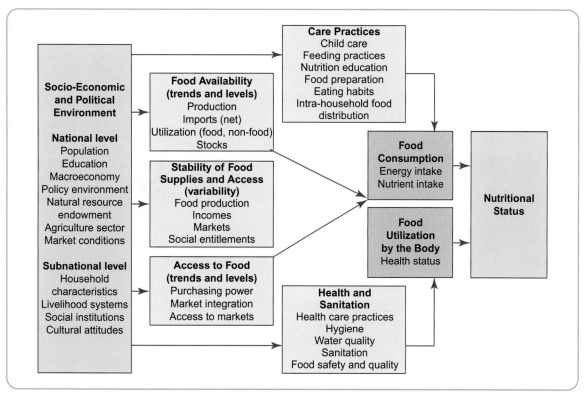

Figure 7.3 Conceptual framework of possible causes of low food consumption and poor nutritional status.

Source: adapted from FAO (2008) *Climate Change and Food Security: A Framework Document*, p. 4

7.5 Food availability

As noted above, the concept of food availability refers to the overall capacity of the planet to produce enough food to eliminate hunger. (Whether that food gets to the hungry, of course, is another matter – to be discussed in Section 7.6). By-and-large, the past century has witnessed unparalleled progress in this regard. Increases in the quantum of food being produced on the planet dramatically exceeded population growth, so that, during the twentieth century, food became cheaper. Figure 7.4 illustrates this trend. (The international wheat price is a good indicator of overall food prices.)

A myriad of factors were behind these production gains: more land put to agricultural use, improvements in water infrastructure and technologies; developments in post-harvest technologies and storage facilities; increased capitalization and economies of scale in farming (allowing machinery such as tractors and combine harvesters to be employed); innovations in livestock breeding and husbandry; and intensive

protein production (fish, poultry and pork, in particular). However, two key factors stood out. The first was in the early decades of the twentieth century, when developments in plant genetics converged with the invention of synthetic, nitrogen-based fertilizers to enable agribusiness firms to match particular seed varieties with specific chemical inputs (see Goodman and Redclift 1991: 95–100 and Pritchard 1998: 66). These innovations generated profound boosts to agricultural productivity, especially in the 'new world' agricultural heartlands of the USA, Canada, Argentina and Australia. The second key shift occurred in the 1960s and 1970s, with the development of **green revolution** varieties of staple cereal crops, notably, rice and wheat (see Case study 7.2). Notwithstanding ongoing controversy about the social and environmental impacts of the green revolution, discussed in Case study 7.2, with the passage of time its legacy remains indisputable. It provided a boost to world food output precisely at a time when the global population was increasing very rapidly, thus forestalling worst-case eventualities of potential famine and undernourishment.

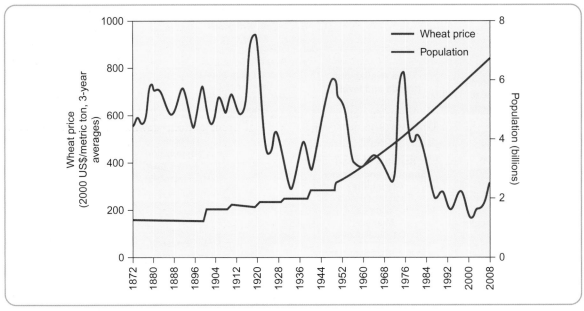

Figure 7.4 Wheat prices and population 1872–2008.

Source: adapted from Von Braun (2008) *Food and Financial Crises: Implications for Agriculture and the Poor*, p. 2

Yet in meeting this challenge, new questions have been opened about 'what kind of food system' the world has created. During the last decade, in particular, the ethical basis of food production (in terms of animal welfare, social issues and environmental sustainability) has come under intensified scrutiny. Many have asserted that technology has bolstered global food production, but at what cost? The intensification of protein production systems has been associated with factory farm regimes that raise questions about animal rights and welfare. Renowned philosophers, such as Peter Singer, contend that there has been a moral

Thematic case study 7.2

The green revolution

The start of the 'green revolution' is generally traced to the plant-breeding research of Norman Borlaug in Mexico, commencing in the late 1940s. By the early 1960s, Borlaug and his colleagues had developed and globally distributed a series of high-yield 'dwarf wheat' varieties with dramatically increased potentials. Also around this time, a series of parallel developments in rice research (undertaken mainly in the Philippines) saw new high-yield varieties developed. These innovations became the centrepieces for major international development assistance efforts during the 1960s and 1970s. Proponents argued that because of their undeniable capacities to boost global cereals production, they held the keys to eliminating world hunger. Additionally, these agendas enmeshed with the Cold War politics of the time – by assisting the rural poor, it

was argued, Communist sympathies would be muted. Thus, as one historian has contended: 'American plant-breeding science thus became part of the cold war's defence of capitalist political economies' (Perkins 1997: 239). However, there were (and remain) many dissenting voices about the social and environmental implications of green revolution interventions. It put many rural societies onto a pathway of increasing inequality, because richer farmers were best able to exploit these technologies and, once they had captured the benefits, used the returns to buy more land (Atkins and Bowler 2001: 221). Ecologically, green revolution varieties were associated with increased pesticide use (often in poorly regulated contexts – thus leading to health and environmental problems), intensified water extraction and the replacement of multi- for mono-cropped agricultural systems, thus reducing local agro-biodiversity (see Shiva 1991).

vacuum at the heart of agendas to optimize global food production (see Singer 2009). Moreover, the boost to global agricultural production has occurred hand-in-hand with increased dependence on international food trade (see Fold and Pritchard 2005), raising questions about the labour standards and livelihood conditions that underpin global agriculture and food production. These concerns include such matters as low wages, hazardous working conditions (food workers are often exposed to dangerous agro-chemicals, especially in the developing world) and the exploitation of smallholder producers by commodity traders. Ensuing public debates have given rise to an expansion of fair trade initiatives, corporate social responsibility agendas, and various types of codes of conduct; however, legitimate questions still remain about the social ethics of the global food system.

Perhaps the harshest criticisms of the contemporary global food system, however, relate to its impacts on the environment. The huge boost to global agricultural production in the twentieth century depended on a massive injection of energy inputs into agriculture, with little regard to the ecological limits of natural resources (see Spotlight box 6.1). Thus, in the Indian State of Punjab – the nation's breadbasket and a key site for the green revolution – the demands from water-thirsty farming systems is seeing the water table fall at an average rate of 55 cm per year (Aggarwal *et al.* 2009). This issue shows the inevitable connections between food, water and energy security. Unless new solutions are proffered, the ability of Punjab's farmers to feed the nation will be seriously compromised,

and thus India's poor record on food security (see Figure 7.2) will be harmed further. The problem of ecological limits is more severe still, however, with regards to animal protein. Marine fish stocks appear to be in rapid decline, with more than half of the world's catch coming from less than 7 per cent of the oceans, 'in areas characterized by an increasing amount of habitat damage from bottom trawling, pollution and dead zones, invasive species infestations and vulnerability to climate change' (Nelleman *et al.* 2009: 23). If the world's population is to continue to eat seafood at its current rate, consumption will need to be sourced increasingly from aquaculture. But the rub, here, is that aquaculture systems are more resource-intensive than marine capture, because they depend on fish being fed from manufactured foodstuffs. Hence, it seems that the only viable way to sustain the global appetite for seafood is to push production into more energy- and resource-intensive (i.e. unsustainable) ways. Comparable arguments can also be made with regards to meat production (see Spotlight box 7.2 and Case study 6.4).

So does the continued operation of the current global food system imply that to feed the world's population in the twenty-first century, the only option is to further degrade the global environment? As the journal *Nature* editorialized in 2010, the challenge of feeding the world was easy . . . 'Easy, that is, if the world brings into play swathes of extra land, spreads still more fertilizers and pesticides, and further depletes already scarce groundwater supplies' (*Nature* 2010: 531). According to many, a key element in resolving the Hobson's choice of food versus the environment is to embrace genetically

Spotlight box 7.2

Meat and the environment

Increased consumption of meat is a central feature of the current global food system. As populations within developing countries become wealthier, their consumption of meat tends to rise dramatically. Researchers worldwide agree that the effects on the global environment of this consumption are overwhelmingly negative: '[T]he livestock sector is by far the single largest anthropogenic user of land. The total area occupied by grazing is equivalent to 26% of the ice-free terrestrial surface of the planet. In addition, the total area dedicated to feedcrop production [i.e. crops grown to feed animals for meat production] amounts to 33% of total arable land' (Steinfeld *et al.*

2006: xxi). Hence, the expansion of grazing lands for meat production (and the additional land used for feedcrops that go into meat production) is a key factor driving global deforestation and reduced biodiversity. Further to these points, the feeding of crops to animals for meat production represents a diversion of output away from humans. On average, it takes 3 kg of grain to produce 1 kg of meat. It is estimated that if global per capita meat consumption could be stabilized at year 2000 levels, then by the year 2050 an extra 400 million tonnes of cereal would be available for human consumption: 'enough to cover the annual calorie need for 1.2 billion people' (Nelleman *et al.* 2009: 26–7). See also Jarosz (2009) and D'Silva and Webster (2010).

modified foods (GMFs). In this chapter we cannot do justice to the debate on GMFs, which is complex and multi-faceted. However, in terms of the debate on food security, it can briefly be observed that advocacy of genetic modification needs to tackle the same kind of charges that confronted the green revolution; namely, it intensifies the high input–high output technological treadmill of farming, with potentially highly uneven social impacts (richer farmers are, in general, better placed than smallholders to access genetically modified materials). Thus, whilst genetic modification may assist farmers to generate higher yields and/or enable lower pesticide use, its wider social and environmental implications remain an ongoing matter of contention.

In the early twenty-first century, sub-Saharan Africa is the cradle for debate on what role genetic modification could and should play in agricultural development. By and large, the green revolution of the 1960s and 1970s exerted relatively little influence on sub-Saharan African agriculture. Thus, in 2006, an agreement to fund a coordinated plan to develop a 'uniquely African' green revolution was established, under the name the Alliance for a Green Revolution in Africa (AGRA). The founding donors for AGRA were the Rockefeller Foundation and the Bill & Melinda Gates Foundation, with partnership support from a wide range of international development and agricultural research organizations. AGRA affirms that it has learnt from mistakes of the original green revolution, and professes to implement initiatives that assist the interests of smallholders (as opposed to large farms), but some NGOs have been critical of its activities (see Spotlight box 7.3). At the

time of writing, AGRA says it 'is not funding the development of new varieties through the use of genetic engineering' (AGRA 2010); however, this does not preclude future involvement in these activities, 'when it is the most appropriate tool to address an important need of small-scale farmers and when it is consistent with government policy'.

7.6 Food access

The debates discussed above frame the question of whether there is global capacity to feed the world's population. However, for food production to actually reach the mouths of the needy requires a further set of considerations to come into view. As the scientific journal *Nature* (2010: 532) acknowledges: 'Nor are science and technology by themselves a panacea for world hunger. Poverty, not lack of food production, is the root cause'. This set of issues – which invoke the concept of *food access* – raises questions about the entitlement mechanisms that connect people with food.

In a general sense, people gain access to food in three ways. First, people can grow or procure their own food directly, either by cultivating crops, tending livestock or by hunting, fishing or foraging. Of course, for most of human history this has been the dominant food access regime for most people on the planet. Even today, a very large proportion of the world's population relies heavily on such mechanisms to ensure their day-to-day survival. Second, people can buy food through the use

Spotlight box 7.3

The Alliance for a Green Revolution in Africa (AGRA)

The NGO ActionAid (2009) has published a broad critique of AGRA. It argues that: (i) too much faith is placed on technological solutions, rather than seeking to establish conducive and stable political, economic and social frameworks to underpin these investments; (ii) the program's enthusiasm for private sector-led solutions blinds it to the potential for smallholders to become dependent on technologies and seeds owned by large corporate interests; (iii) 'large private consulting firms' have played too great a role in project developments, rather than African governments;

(iv) not enough attention has been placed on smallholders' 'access to land, water, infrastructure, information and credit' (p. 12). In sum, ActionAid advocate a food sovereignty approach, which suggests that the best livelihood course for smallholders is to extract themselves from the potentially exploitative effects of involvement in the global economy. The NGO contends that the 'overall goal [of a restructured AGRA] should be to reverse the current dependence on global agricultural technological and commodity markets, which submerge local knowledge and technologies and deepen the extraction of farming surpluses (through the unfair pricing of food and inputs)' (p. 22).

of money. (This is akin to Amartya Sen's example of the umbrella seller, quoted above.) Most of the readers of this book, one would imagine, obtain the vast majority (if not all) of their food through this method. Third, people can gain access to food through disbursement programs (such as food banks, food aid and food voucher welfare programs) operated by governments, international organizations (such as the UN's Food World Program) or charities/NGOs.

These three general mechanisms of food access do not operate separately to one another. Households across the world have diverse, multi-dimensional modes of food access. The food security of a household in a coastal village of India might depend on fish being caught daily by one household member, fruits and vegetables to be bought at a local market using the cash income earned by another household member, and rice and cooking oil being acquired at a public distribution outlet operated as part of a government food program.

Comparably, in the developed world, self-sufficient production and food disbursement programs play an important (indeed, possibly increasing) role for many households. Urban agriculture schemes based around community gardens and individual allotments have grown steadily during the past decade, and media reports have suggested that this trend has accelerated since the onset of economic recession in 2008 (Plate 7.2). The symbolic development of a kitchen garden on the White House lawn by first lady Michelle Obama tapped this zeitgeist. Likewise, the recession has also spawned increased demand for emergency food disbursements, across both sides of the Atlantic. In the US, the non-profit organization 'Feeding America' provided food to 37 million people during 2009, which was an increase of 46% from just before the recession, in 2006 (Feeding America 2010). Recipients of such food disbursements frequently don't rely wholly on these programs, but move in and out of dependence in line with changes to their individual and/or household situations. Comparably, it is rarely the case that individuals/households depend entirely on production from agricultural allotments or community gardens; for most, these provide an ancillary coping strategy in times of financial stress or are part of a lifestyle philosophy to reduce engagement with the corporate food system.

Focusing on the developing world (where these issues tend to have more pressing livelihood implications), two vital debates are pertinent to contemporary discussions of how food access arrangements impact on food security outcomes. The first relates to the future of small farms; whether food security for the rural poor in developing countries is ultimately best served by seeking to maintain smallholder, self-sufficient production. The second relates to national food policies; the extent and character of governments' interventions in food markets to secure food security for the poor (a debate about food security via disbursements).

Plate 7.2 A community garden/urban agricultural allotment.

(Alamy Images)

7.6.1 Food access via self-production or the market? Debate on the future of small farms

The future of small farms in the developing world is currently a topic of major discussion. In essence, what is at stake is the question of what works best for hungry people: a food entitlement regime based on people growing their own food, or one in which they buy food at local markets, shops and supermarkets.

The economist Paul Collier kick-started the most recent instalment of this debate with an influential article published in the journal *Foreign Affairs*, in the midst of the 2008 food crisis. Collier's chief charge was that 'the world needs more commercial agriculture, not less' (2008: 68), and that '[T]he first giant that must be slain is the middle- and upper-class love affair with peasant agriculture' (2008: 71). Collier asserted that foreign aid projects and 'rural-friendly' international financial programs from the West were inspired by a wrong-headed, allegedly 'romantic' attachment to support small-scale agriculture. In Collier's vision, food security is best attained by encouraging small farmers to abandon their plots of land, and then to aggregate those holdings into large-scale farms. Former peasants could then work on those farms, or move to cities where incomes are generally higher (see UNDP 2009). In 2009, the World Bank's *World Development Report* chimed in to support key aspects of this argument, arguing that the welfare of the world's hungry and poor would be best served by policies that encouraged the movement of people from rural to urban areas (a so-called 'three D's' strategy: encouraging population *density* (i.e. population growth in cities), *distance* (encouraging people to be mobile, in search of jobs) and *division* (removing barriers like tariffs and some restrictions on migration that inhibit the flow of goods and people)). The assumption of these visions is that (i) large-scale agriculture enables food to be produced

at lower costs, and (ii) the higher wages paid in cities places more money in the pockets of the poor; a tandem process that generates quantum improvements in food security. Among advocates of this position, Brazil is seen as a poster-child; over the past decade, the Brazilian government has sought to address poverty and hunger through encouraging the development of large-scale agriculture to generate export income and jobs.

Yet many researchers question the assumptions that lie behind these visions. Encouraging rural smallholders to divest their land and move to cities makes sense only if countries are at a stage of development where sufficient numbers of city-based jobs are available. If this isn't the case, rural migrants add to the ranks of squatter-settlements with highly insecure food and economic prospects. A more reasonable approach, according to a number of leading geographers (see Wiggins *et al.* 2010) is to establish more gradualist, 'multi-track' strategies of rural poverty-reduction. According to the proponents of these views, smallholders' control over land has foundational importance in allowing own-production for basic sustenance. Hence, food security strategies should aim to bolster the livelihoods of the poor through constructing viable rural economies, rather than advocating abandonment of rural settlements for an uncertain urban life (see Case study 7.3).

7.6.2 Food aid and provisioning programs

In many developing country contexts, food aid and food provisioning schemes provide important complements to peoples' access to food via own-production or the market. Usually this involves food programs (operated by national governments or international agencies, for example the World Food Program) warehousing and distributing major food staples (rice, wheat, etc.) to populations in need.

Such disbursements are crucial in preventing starvation deaths in times of crisis, but, long-term, have

---

Thematic case study 7.3

Small rural farms and food security

In many parts of the developing world, the majority of the food insecure population live in rural areas (FAO 2005: 5). Many of these people exist on small plots of land that are sub-economic (that is, on their own, they can't produce enough food or income to sustain

a household). Thus, many agricultural smallholders in the developing world suffer chronic hunger, despite having the ability to grow their food. A good summary of the debate about the future of these populations can be found in a special issue of the journal *World Development* dedicated to this problem (Wiggins *et al.* 2010).

been criticized because of the potentially adverse implications of creating cycles of food-aid dependency. The crux of this criticism is the alleged *displacement effects* of food aid. This arises in contexts where the food security of rural populations is dependent upon the sale of agricultural outputs to cities and towns. In such cases, periodic large disbursements of food aid can swamp local markets, thus driving down prices and taking away the incomes of farmers who rely on market sales. Closely related to these criticisms is the charge that food aid disbursements can often be *donor-driven* (meaning that they serve the interests of the senders, not the recipients). For example, the decision by the US government to heavily subsidize its agricultural sector has gone hand-in-hand with policies to procure large amounts of staple food crops for international food aid. As many critics have argued, therefore, US policies to disburse food aid can be seen as strategies that help prop up the incomes of American farmers, as much as they are policies aiming to assist hungry populations. Notwithstanding these criticisms, however, an exhaustive study of these issues by the FAO (2006) concluded that the adversities of food-aid dependence tended to arise mainly when food aid was poorly managed: 'On balance, the report finds that food aid can support food security both in emergencies and in cases of chronic hunger *if it is properly managed*' (p. vii).

In many developing countries, international food aid however plays a secondary role to food provisioning policies operated by national governments. India is a case in point. The national government operates the Food Corporation of India (which purchases and stores key food items from farmers) and the Public Distribution System (which distributes food to the poor at highly subsidized prices through a nationwide network of 'Fair Price' shops). In theory, these activities should ensure a basic level of nutritional subsistence to the poor. However, well-documented problems of maladministration and corruption have ensured that these arrangements work less than perfectly (see Programme Evaluation Organisation 2005; Ram *et al.* 2009). Hence, recent years have witnessed a keen public debate in India on how to restructure these schemes to ensure that people in need receive their full food entitlements. Technological innovations will play a key part in future arrangements, with bureaucrats seeking to use smart card and electronic monitoring systems to cut down on system 'leakage'.

Finally, in this regard, is debate over the *right to food*. The above discussion implicitly assumes that the task of getting food to people in need relies on the goodwill of the international community and/ or national governments. However, it is increasingly recognized that governments have human rights obligations to feed their citizens. In 1948, the UN Declaration of Human Rights recognized peoples' Right to Food, but this was not clearly incorporated into any international legal framework until 1996 (when the Right to Food was adopted by the World Food Summit) and then 2000, when the UN Commission on Human Rights appointed (for the first time) a Special Rapporteur on the Right to Food (a person with overarching authority to report on international progress on the Right to Food). These initiatives triggered a shift in many countries to a rights-based approach to food, that is, the recognition by governments in many countries that their people had a legal right to be provisioned with food, which could be justiciable (that is, heard in court). Thus, in India, in 2010 the national government agreed to implement Food Security legislation that would have the purpose of enshrining people's legal *right* to food.

7.7 Conclusion

The issue of food security is extensive and complex. This chapter has sketched key themes in relation to this concept, but many more have remained outside its purview. For example, a crucial area for ongoing consideration is the question of how food security and climate change connect to one another. During the past few years, there has been widespread discussion of these issues using the notion of *food system stability,* the extent to which food systems are resilient to environmental and social change of various kinds.

The vital message of this chapter is that the *food security* needs to be understood as a social outcome. It exists when all people, at all times, have physical, social and economic access to sufficient, safe and nutritious food. Thus, the concept of food security should not be confused with debates on the technologies to maximize food production, or on whether any country (whether the UK, Japan or Botswana) is reliant on food imports. What matters are the socio-cultural, political, environmental and economic systems that link a food system to a population. To think about food security, thus, is to think about the environmental sustainability and social justice of the planet.

Learning outcomes

Having read this chapter, you should understand that:

- Food security is defined socially. It concerns not simply the amount of food that is produced, but the extent to which a food system is able to satisfy the nutritional requirements of everyone in a population. Therefore, food security research must focus on the three core elements of food access, availability and utilization.

- During the past half century, the world has been able to successfully address problems of famine, but levels of chronic malnutrition remain of major concern. Increases to global food prices since 2007–08 have exacerbated these problems.

- Proposals to increase world food supply through technologies such as genetic modification should be assessed in terms of how they address the needs of the most vulnerable and whose interests they serve.

- A key question in the world today is whether the long-term food security needs of the planet are better served by a shift to large-scale agriculture, or the maintenance of smallholder systems of farm production.

- Food security is a human right, and recent approaches to this issue have stressed peoples' Right to Food.

Further reading

The World Bank's 2009 *World Development Report* prompted considerable (mainly critical) debate by geographers. This included the following article:

Jarsoz, L. (2009) Energy, climate change, meat and markets: mapping the coordinates of the current world food crisis, *Geography Compass*, 3(6), 2065–2083. This article relates to issues considered in Spotlight box 7.2 and other chapters in this section.

FAO (various years) *The State of Food Insecurity,* FAO, Rome. This is an annual report which provides the most up-to-date assessment of global conditions with regards to food security. Each year, a different aspect of this issue is highlighted in the report.

Shiva, V. (1991). The Green Revolution in the Punjab. *The Ecologist,* 21(2): 57–60. A famous article which provided a critical perspective on the Green Revolution and provided the foundation for alternative viewpoints on technology and agriculture.

Wiggins, S., Kirsten, J and Llambi, L. (eds) (2010) The Future of Small Farms: Special Issue, *World Development,* 38(10): 1341-1526. Articles in this special issue of *World Development* summarize key aspects of debate around the fate of smallholders in ensuring global food security.

Sen, A. (1981) *Poverty and Famines: An Essay on Entitlement and Deprivation,* OUP, Oxford. This book changed the way that leading policy-makers began to view the issue of food security. It was a major element in justifying Sen's subsequent award of the Nobel Prize for Economics.

Useful websites

www.fao.org The Food and Agriculture Organization of the UN. This is the pre-eminent global body charged with responsibility for international coordination of food and agricultural efforts.

IFPRI www.ifpri.org The International Food Policy Research Institute is a policy and research arm of the Consultative Group on International Agricultural Research. It provides extensive information on international food prices and under-nourishment.

www.gatesfoundation.org/Pages/home.aspx The Bill & Melinda Gates Foundation funds extensive research and extension activities in the areas of food, agriculture and hunger.

www.foodfirst.org Food First is a US-based Non-Government Organization with a long track record in monitoring global food and hunger.

www.mssrf.org The M.S. Swaminathan Research Foundation is a leading Indian Non-Government Organization in the area of food policy.

WORLDS APART: GLOBAL DIFFERENCE AND INEQUALITY

Chapter 8

Marcus Power

Topics covered

- The importance of the 'three worlds' schema
- Development as knowledge and power
- The view from 'the South' and the view from 'below'
- Alternative geographies of global development and inequality

What is the geography of the Third World? Certain common features come to mind: poverty, famine, environmental disaster and degradation, political instability, regional inequalities and so on. A powerful and negative image is created that has coherence, resolution and definition. But behind this tragic stereotype there is an alternative geography, one which demonstrates that the introduction of development into the countries of the Third World has been a protracted, painstaking and fiercely contested process.

(Bell 1994: 175)

Perhaps one of the most interesting features of the word 'development' is that it has produced a bewildering array of labels for people and places that are not considered 'developed'. As Morag Bell suggests in the above quote, the space of the Third World has lots of tragic stereotypes of famine, poverty, drought, etc., to animate it and make it seem coherent and defined. Learning to understand and appreciate social or economic differences relative to the area of the world in which we live is a difficult and value-laden process. The term 'development' is notoriously hard to define and often refers simply to 'good change', a positive word that in everyday parlance is practically synonymous with 'progress' and is typically viewed in terms of increased living standards, better health and well-being and other forms of common good that are seen to benefit society at large. Conditions of poverty clearly vary between different areas, as does the way poverty is experienced, so simplistic stereotypes of 'Third World poverty' are of little use. As Hart (2001) suggests, a distinction can usefully be made between 'big D' Development and 'little d' development. The former refers to a post-Second World War project of intervention in the 'Third World' that emerged during the time of decolonization and the Cold War. The phrase 'little d' development, however, points to the development of capitalism as a 'geographically uneven, profoundly contradictory set of historical processes' (Hart 2001: 650). D/development can thus be viewed simultaneously as both a project and a process.

In recent years there has been an upsurge in the categorization and labelling of non-Western peoples and spaces as demands for aid effectiveness and concerns with inequality grow. Thus countries are variously seen as belonging to the 'Third World' or the 'global South', to be somehow 'lesser developed', 'underdeveloped' or even 'backward' in some way. Constructing categories of peoples and places is often seen as an efficient way to manage resource allocation and for tracking whether disadvantaged people are 'benefiting' from 'development' (Eyben et al. 2006). In some ways the lack of an agreed set of international development indicators and measures or of common systems of data collection tells its own story of the failure of international development since 1945 (see Case study 8.1). A major problem with the geographies of development produced from these statistics is that they have allowed some observers to label whole areas as 'Third World' or 'lesser developed' as if the same

could be said of all its constituents (Wood 1985). As Morag Bell puts it:

> In emphasising what people are deprived of (as is implied by poverty), they [statistics] impose a negative uniformity upon non-western societies. As objects of study poor people become categories and are labelled as a homogeneous group.
>
> (Bell 1994: 184)

Thus people's life stories and multiple identities are reduced by imposing labels such as 'poor' or 'refugee', they can be 'top down' and can lead to particular interpretations of the underlying problem. Further, labelling whole regions and spaces as 'developing' or 'developed' reduces and overlooks the political, economic, social and cultural diversity of the places and communities included within these gross generalizations, simplifications and aggregations (Wood 1985). Labels can stigmatize people and places but they can also be resisted and can serve to mobilize people (e.g. through labels like 'citizen' or 'disabled'). This chapter seeks to explore the notion of 'Third World' development and outlines some of the ways in which this has been theorised and practised. In order to do this, the chapter examines the historical and geopolitical dimensions of development. Rather than seeing poor people as 'objects of study' we also need to attend to the ways in which poor people themselves understand and make sense of their experiences of poverty.

Moving beyond the labelling of 'Third World' peoples and places as a homogeneous group, it is important to grasp how places and peoples are spatially and socially *differentiated* through development and inequality, experiencing progress and 'good change' in a variety of ways. We need to encourage a diversity of ways of enframing the problems of 'development' and resist the temptation for neat categories and easy quantification. The key question here is how to listen to and understand the differences that define and distinguish how development is lived in different sorts of places and amongst different groups of people. Do we always need to rely on quantification and statistics to understand these issues or might we look to learn from other kinds of stories and narratives? Thus what constitutes 'knowledge' in development policy and practice is often defined in relation to quantitative measures, statistical data, formal academic research and Western 'science' but we can also learn a great deal from fictional writing, literary perspectives and the use of audio-visual media in order to reveal a

Thematic case study 8.1

The Millennium Development Goals

One area where a consensus has been established in recent years by the international community is in the establishment of the Millennium Development Goals (MDGs). These are eight international development goals that all 192 United Nations member states and at least 23 international organizations have agreed to achieve by the year 2015. They include (1) eradicating extreme poverty and hunger; (2) achieving universal primary education; (3) promoting gender equality and empowering women; (4) reducing child mortality; (5) improving maternal health; (6) combating HIV/AIDS, malaria and other diseases; (7) ensuring environmental sustainability; and (8) developing a global partnership for development. Together with these eight goals are 21 targets and a series of measurable indicators for each target. To date, progress towards meeting the goals has been uneven with some countries doing well, others less so, and a few doing badly. Their success or failure will have immense consequences, not only for the world's poor, but also for the credibility of collective action by the international community. Targets that will be missed include those for sanitation, which lags behind water in both funding and attention, whilst progress towards achieving an improvement in maternal health has been dismal (Shepherd 2008). Questions of gender equality are fundamental to the achievement of all the MDGs (Jones *et al.* 2008), yet most of the goals are 'gender

blind'. The achievement of the goals also requires continued economic growth. There is a widespread perception of the MDGs as a Northern project – despite their adoption by the UN member states. MDG7 on environmental sustainability is one example. Mired in controversy as a result of contested definitions, ill-defined targets and indicators that are based on Northern concerns, such as rainforests, MDG7 has not resulted in strong environmental strategies in the South. The perception that the MDGs are rooted in a Northern agenda feeds a further perception: 'if they [the North] want them, let them pay for them' – a perception that the MDGs are to be achieved through aid alone. In fact, the necessary public expenditures will only be made consistently over long periods of time if they are underpinned by a social compact between state and citizen. As we head towards 2015 increasing global uncertainties (such as the global economic crisis and climate change) have led some observers to question and rethink the MDG approach to development policy. This 'after 2015' debate is about questioning the value of an MDG-type, target-based approach to international development, about progress so far on poverty reduction, about looking to an uncertain future and exploring what kind of system is needed after the MDG deadline has passed (IDS 2009). A major conference was held at UN headquarters in New York on 20–22 September 2010 to review progress to date, with five years left to the 2015 deadline (see also Chapter 5 for a discussion of energy services and the MDGs).

different side to experiences of development (Lewis *et al.* 2005). This chapter is divided into four further sections. The first discusses two of the most important conceptual perspectives that have been formulated on the relations between development and inequality: the *modernization* and *dependency* schools. Although there have been many other different strands of development thinking, both these perspectives have been widely influential and remain relevant to an understanding of theory and practice today. Section 8.3 outlines the need to view development historically and formulate a sense of how it has been redefined through time. How have historical forces shaped our understanding of the geography of development and in what ways are the legacies of the past important

to understanding contemporary global economic difference and inequality? The next section focuses attention on people and places and explores how the subjects of development experience developmental interventions. Three major 'brokers of development' (non-governmental organizations (NGOs), states and international institutions) dominate the literature on development, but how do these agents of the development process understand the needs and aspirations of intended beneficiaries? The final section of the chapter poses questions about the political, cultural and economic implications of globalization for the future of development theory and practice and examines possible alternatives to development based around local cultures and knowledges.

8.1 Geography and 'Third World' development

By whom is development being done? To whom? These are potent questions. They should particularly be asked when 'solutions' are put forward that begin 'We should . . .', without making clear who 'We' are and what interests 'We' represent.

(Allen and Thomas 2000: 4)

The principal problem with labelling vast regions as 'lesser developed', 'developing' or 'Third World' is that a set of negative assumptions are often made and a 'negative uniformity' is imposed. Thus the picture of unevenness and injustice in the contemporary world that comes to us through these labels is not always a sharp, coherent and precise one and often this unevenness is not effectively conveyed in the statistical measures that are taken as indices of what constitutes 'development'. Crucial then to the construction of a 'Third World' is a process of setting worlds apart and a politics of labelling that it is necessary to understand and be critically aware of. All too often the 'developing world' has been defined as a 'problem' for Western governments that can only be resolved with the intervention of Western 'experts', donors, technology, expertise or ideology (see Case study 8.2).

Definitions of the 'Third World' have been contested, as have the origins of the phrase (Mountjoy 1976; O'Connor 1976; Auty 1979; Pletsch 1981; Wolfe-Phillips 1987), yet the concept of 'three worlds' can hardly be said to convey a precise meaning. The geographical boundaries of the 'Third World' are thus incredibly vague and open-ended (Chaliand and Rageau 1985; Griffiths 1994) and the idea has different meanings in different cultures. The 'three worlds' schema is very much a Cold War conceptualization of space and is strongly associated with the global social and political conflict between capitalism and communism, between the USA and USSR, in the second half of the twentieth century. The term posited a first world of advanced capitalism in Europe, the USA, Australia and Japan, a second world of the socialist bloc (China's position within this has been much debated) and a third world made up of the countries that remained when the supposedly 'significant' spaces of the world had been accounted for. This extremely rudimentary schema of classification has provided many observers with a convenient shorthand expression and is undoubtedly linked to the formation of 'mental maps' and 'imagined

geographies' of inequality (Bell 1994). These 'mental maps' and 'imagined geographies' are often created where people have no direct experience of encountering the regions or peoples they are describing or referring to. In an interesting study of student travellers to the 'Third World', Desforges (1998) illustrates the links between travel and identity for many students who sought to 'collect places' that offered authentic individual knowledges and personal experiences that could be gained through travel. This involved a framing of 'Third World' others and places as different and the assumption that it was possible to 'collect' experiences of Third World places.

Given that definitions of development have often been contested, these 'imagined geographies' of the 'Third World' frequently overstate a set of (tragic) stereotypical and supposedly 'common' features such as overpopulation, environmental degradation or political chaos. In discussing unevenness and inequality, therefore, a representation or an imagination of the people and places subject to the development process is constructed which is only partial and provisional. An 'alternative geography of the Third World' is proposed here, based on a recognition that development has been a 'protracted, painstaking and fiercely contested process' (Bell 1994: 175). This chapter tries to offer such a critical 'alternative geography' that aims to challenge some of the 'representations of peoples and places upon which grand [development] theories are based' (Bell 1994: 193–4).

It is extremely difficult to generalize about the people and places of the 'Third World' since relations between different regions of the world are imagined and simplified by geographical expressions like 'North' and 'South', 'rich' and 'poor', 'developed' and 'underdeveloped', 'First World' and 'Third World'. These terms have to be approached with some caution. Becoming particularly popular after the publication of the Brandt Report in 1980 (Brandt 1980), the idea of 'North' and 'South' is widely seen as preferable to the 'three worlds' scheme. However, no definitions were mentioned in the report and some of the boundaries it delineated between 'North' and 'South' were considered erroneous. Subscribers to this, and the three worlds scheme, have been criticized for the simplicity of these divisions and their failure to recognize diversity and difference within these spaces; the world does not consist of a series of individual national economies in the way often suggested in United Nations and World Bank reports. Many commentators have argued instead that there is a single global economic system with ever-closer degrees

Thematic case study 8.2

The 'strategic interests' of foreign aid

In many ways, examining the case for aid can tell us a great deal about the history of development theory and practice and provides a useful opening on to wider discussions of North–South relations. In 2009, total net official development assistance (ODA) from members of the OECD's Development Assistance Committee (DAC) rose slightly in real terms (+0.7 per cent) to US$119.6 billion, representing 0.31 per cent of DAC members' combined gross national income (OECD 2010). This is well below the UN target of 0.7 per cent, which has been reached and exceeded by only five countries (Denmark, Luxembourg, the Netherlands, Sweden and Norway). Around two-thirds of this is government-to-government or *bilateral aid* and the remainder is *multilateral aid* disbursed by agencies like the World Bank group, the UN bodies and the European Union. Critics of foreign aid have argued that it has been less effective than private investments and commercial loans in stimulating long-term economic growth. Other critics point to the dubious Cold War record of foreign aid and its subsidizing of autocratic regimes and inflammation of regional conflicts. Thus, many critics of foreign aid have sought to highlight the 'strategic interests' at work in its distribution as well as the inequality and unpredictability of aid provision. During the 'War on Terror', for example, there is evidence that the USA has used foreign aid donation to secure support from a number of states in the South such as Pakistan and Angola. Strangely, except for a brief period during the mid-1970s, anti-poverty measures have not been an important focal point of foreign aid, whilst aid has led to many reversals as well as to advances. Seen as a (simultaneous) remedy to problems of growth, governance, poverty and inequality,

it has become (not unlike the idea of development) overburdened with expectations (Sogge 2002) and an overambitious enterprise (Rist 1997). A 2005 report by ActionAid claimed that only 39 per cent of total overseas development assistance is 'real aid' and that the remainder is 'phantom aid', made up of overpriced technical assistance, double-counted debt relief and poorly targeted and uncoordinated aid (ActionAid International 2005). Given that, on average, aid constitutes less than 1 per cent of government spending in OECD countries, the Millennium Development Goals would seem to be unattainable. Although aid from OECD donors has increased in recent years, much of this is in the form of debt relief, or humanitarian or military aid, which does little to advance long-term development. There is evidence, however, that aid may increasingly be used to help countries of the global South deal with the implications of climate change. The Copenhagen Accord, the main outcome of the UN climate change conference held in Copenhagen in December 2009, set aside funding for the development and transfer of low-carbon technologies to developing countries. The Accord commits 'developed' countries to provide US $30 billion in short-term financing between 2010 and 2012 and to mobilize US$100 billion by 2020 (some of which will come from multilateral sources). This funding will be used for mitigation, adaptation, technology transfer and capacity building in developing countries.

Not surprisingly (given the size of its economy) the United States remains the largest DAC donor, providing US$28.67 billion in 2009, yet this constitutes only 0.20 per cent of US gross national income. India (8th) and China (15th) were amongst the top 20 recipients of ODA in 2009, yet these countries are also increasingly active as donors of aid themselves, along with other 'rising powers' like Brazil, South Africa and Nigeria.

of interdependency (Allen 1995). Our task is to think critically about the uneven and unequal geography of the world economy and the ways in which the global capitalist economy *actively produces* inequality and uneven development. We also need to remember that the pursuit of development is not exclusive to the so-called 'Third World' and that issues of poverty and inequality are also highly pertinent in the supposedly 'developed' 'First World' (see Case study 8.3). The onset of a global economic downturn in 2007 has implications for the livelihoods of almost everyone in an increasingly

interconnected world and there has been evidence of rising levels of unemployment, poverty and homelessness in many Western countries.

For some observers, a big part of the economic development and wealth of the rich countries is wealth that has been directly imported from the poor countries. The key question here then is: to what extent does the world economic system *generate* inequality and run on inequality? Where does this leave the primary liberal idea of 'development for all'? Is it potentially fraudulent and misleading to hold up an image of

Thematic case study 8.3

Hurricane Katrina and the USA as 'Third World'

One of the costliest and deadliest hurricanes in the history of the USA, Hurricane Katrina, devastated much of the north central coast of the United States in 2005, killing 1,836 people and causing US$81.2 billion worth of damage. Over 60 countries pledged monetary dona-tions or other assistance to the USA, including Kuwait (US$500 million), Qatar (US$100 million), South Korea (US$30 million), China and Pakistan (both US$5 mil-lion) and Bangladesh (US$1 million). Cuba and Venezuela were the first countries to offer aid, in the form of US$1 million, 1,100 doctors and 26.4 metric tonnes of medicine but both offers were refused by the US government (British Broadcasting Corporation 2005b). Media reporters embedded in the mayhem caused by the Hurricane reported that a 'third world country' had suddenly appeared on the Gulf Coast or noted that the scenes in cities like New Orleans resembled 'third world refugee camps'. The *U.S. News & World Report* referred to 'the Third World images of death and devastation reeled across the nation's TV screens', whilst on *Fox News*, anchor Shepard Smith lamented that things would not be the same anymore, noting that Americans would 'forever be scarred by Third World horrors unthinkable in this nation until

now'. As Brooks (2005) has argued, however, the 'Third World' did not somehow 'sneak into' the USA with this hurricane, it had been there for a long time already, given that (at a conservative estimate) around 13 per cent of Americans live in poverty (around 18 per cent of children). According to Brooks, 'when you break down key economic development indicators by income group and race you find that conditions for poor Americans rival those in developing countries'. Just as it has often taken a mass famine or tsunami to generate media attention for the 'Third World' in Western countries like the USA, so it has taken the destruction of a US city for the media to notice the 'Third World' within the United States. It was almost as if by displacing disasters and human suffering to the 'Third World', this natural disaster was seen as not really happening in the United States (Wa Ngugi 2005). The disaster illustrates some of the ways in which the media shape and construct an imaginary 'Third World' but also how the concept of 'three worlds' can produce a kind of distancing where problems of poverty, hunger and forced migration are seen to be worlds away. Hurricane Katrina also cruelly exposed the myth of the United States as a fully 'developed' country and highlighted the inequality and poverty that exists within US cities (see also Chapter 7 for a discussion of food insecurity in the USA).

wealthy Western living conditions as something that is available to all and a realistic or attainable objective? If the global economy is a pyramid and everyone is en-couraged to stand on top, how is this to be arranged? Although there are some notable exceptions (such as the recent growth seen in Brazil, India and China – see Case study 8.4) development equality – 'catching up' with the wealthy through economic activity – increasingly seems less likely given the current uncer-tainty and instability in international financial, currency and commodity markets and falling levels of trade and investment.

'Development' has served in part as a kind of 'light-house' (Sachs 1992) or as a 'lodestar' (Wallerstein 1991) into which several different movements, governments and institutions have invested faith and meaning. Like the term 'Third World', development partly represents a geographical imagining, a representation of a better world,

and a belief in the idea of correctable inequalities/ injustices between nations, states and regions and within the existing global economic orders. Escobar (1995) argues that the idea of the 'Third World' also gives enormous power to Western development insti-tutions to shape popular perceptions of Africa, Asia or Latin America. The Third World thus is defined by and becomes intelligible through the languages and repre-sentations of the agencies and institutions of global development.

In a famous speech in 1949, US President Harry Truman spoke of the emergence of an 'underdeveloped' world that presented a 'handicap and threat both to them and the more prosperous areas' (Truman 1949). Born of the ashes of colonialism and of the fires of anti-communism in the United States, 'development' occu-pied centre stage as an ideal in the global arena after the Second World War and in particular during the Cold

Thematic case study 8.4

The (re)emerging powers and the rise of new 'Southern' donors

The United States emerged preeminent after the Second World War and built a post-war international order around a range of governance institutions, including the United Nations, the International Monetary Fund (IMF), the World Bank, the General Agreement on Tariffs and Trade (GATT) and regional security alliances. The end of the Cold War consolidated this American-led global institutional order but in recent years a group of fast-growing non-Western countries – led by China and India – are rising up the ranks of the world system as the boundaries used to differentiate rich and poor, 'first' and 'third' world countries are becoming more fluid. In the coming decades these (re)emerging powers will have economies that will rival the United States and Europe and they are already becoming an international economic force. According to the *Economist*, 'developing countries' now produce half of the global gross domestic product (GDP) (*The Economist* 2006). Additionally these (re)emerging powers hold most of the world's financial reserves and are placing significant new demands on energy and raw materials. The rise of countries like China, India and Brazil thus has potentially far-reaching implications for power and for the post-war institutions of governance in world politics but also for the international landscape of development cooperation. At the 2006 annual meeting of the IMF and World Bank in Singapore, representatives from several of these rising countries noted that their economies were stronger than several of the advanced economies – and they insisted on gaining a greater voice over the setting of policies. Their collective size and impact on trade, finance, energy, and the environment will make them important players in the years to come. Further, many of these rising powers are emerging as aid donors themselves, providing development assistance to a range of other non-Western partners and often heralding this as 'South–South cooperation'. These new 'Southern' donors question the very idea of development (cooperation) as a Western concept but also the development paradigm as a whole (Six 2009).

War. Truman went on to explain the need for 'modern, scientific and technical knowledge' as a pathway to overcoming this 'handicap' of underdevelopment and announced the beginning of a 'bold new program' within the 'developed world' to resolve inequality and remedy impoverishment in 'backward' areas (Sachs 1992). Since then 'development' has become

> an amoeba-like concept, shapeless but ineradicable [which] spreads everywhere because it connotes the best of intentions [creating] a common ground in which right and left, elites and grass-roots fight their battles.
>
> (Sachs 1992: 4)

Ideological interpretations of development have varied but they have usually always claimed 'the best of intentions'. The idea of development is often discussed in relation to 'developing countries' (itself a value-laden expression) but given its 'amoeba like' nature, development has taken numerous forms in numerous locations. A sense of development being simultaneously important in both 'developing' and 'developed' societies was recently formulated in a Human Development Index (HDI) (UNDP 1998). Again assuming that 'development' can be depicted in a single measure, the UNDP Index combines data (by country) on life expectancy, literacy, income, environmental quality and political freedom. Development is defined here as a series of numerical stages with a linear progression indicated (statistically) as degrees of departure from Western norms and ideals. Human development is represented as a series of league tables that measure progress with the unquestioned underlying assumption that every country can reach the highest levels of performance set in 'the West' or have access to the 'purchasing power' of the 'upper income' world. As in the case of GNP per capita, the HDI does in many ways (albeit at a very crude level) point to growing gaps between different areas of the world. Its significance is undermined by the failure of its 'generalized' indicators. However effective statistical measurement of development can become, it is necessary to be critical of the way in which statistics are collated and definitions formulated and to be careful in the assumptions and prescriptions made as a result. It is through the amassing of statistics and surveillance by global development agencies that the image of an underdeveloped, primordial, traditional and war-ravaged 'Third World' is (re)produced (Ahluwalia 2001).

8.2 Conceptualizing development

The idea of progress has become an article of faith in definitions of 'development', which have themselves taken on a 'quasi-mystical connotation' in the sense that seeking to become 'developed' has been constructed as an objective that is unquestionably positive and beyond reproach. 'Development thinking' (Hettne 1995), or the sum total of ideas about development theory, ideology and strategy, has however often been caught in a 'Western' perception of reality. Conceptualizing development is thus partly about the negotiation of what constitutes progress and improvement and the definition of what constitutes 'appropriate' intervention in the affairs of 'poor' or 'lesser developed' countries. Since all-encompassing definitions have been contested and controversial, little consensus exists today but some core conceptions have emerged, many of which have continued relevance in the contemporary world. Although there are many different strands of development thinking to explore, the modernization and dependency approaches have been two of the most influential in the twentieth century. In discussing these different conceptions it is important to think about where and when they have emerged (Slater 1993). Most reflect some of the priorities of development thinking characteristic of their era. The formation of development theories therefore depends on different perceptions of 'development challenges' at different times (see Case study 8.5).

8.2.1˜The modernization school: an anti-communist manifesto

With the end of the Second World War and after the United Nations was established, conceptualizations of development received a decisive stimulus. Between 1945 and 1981, UN membership rose from 51 to 156 nation-states (Berger 2001). With many new states being formed after the end of colonialism and in the context of the Cold War between the USA and the USSR, theorizing development became a more complicated and contested enterprise. In some ways the task of development was seen by some as seeking to provide 'an ethos and system of values which can compete successfully with the

attraction exercised by Communism' (Watnick 1952–3: 37). In this regard many observers called for the modernization of 'underdeveloped areas' and painted a picture (following President Truman) of 'underdeveloped peoples' confined to 'backwardness' but torn between the appeal of communism and the prospect of Western modernization. This was an essential characteristic of the *modernization school*, which was often dualistic, opposing 'traditional' to 'modern' life-styles, 'indigenous' to 'westernized', as if no country or citizen could belong to both categories.

Hirschmann (1958) was a key proponent of modernization ideas, voicing the optimistic view that the forces of concentration ('polarization') will 'trickle down' from the core to the periphery at national, regional and global scales. Hirschmann explained how development economics might 'slay the dragon of backwardness' (quoted in Rist 1997: 219). Friedmann's (1966) core–periphery model adopts many of the same assumptions about the polarization of development in 'transitional societies' and the 'trickle-down' effect of development. This model suggests that a number of stages exist in the national development of countries or cities leading to a final stage that represents the culmination of the development process. Geographers at the time sought to contribute to the mapping of modernization geography, seeking to look at how progress trickled down along urban hierarchies, through transport systems or with the introduction of modern technology. The message from these modernization geographies was that underdeveloped countries could move briskly into the modern tempo of life within a few years, whilst the state (concerned with macro issues and the national economy) would be the key monitor and broker of development.

One of the modernization theorists who identified stages of growth in the development process was the American economic historian W.W. Rostow (1960). Again the focus was on a top-down 'trickling' of capitalist development from urban-industrial areas to other regions (Stöhr and Taylor 1981). Rostow (1960) predicted that nations would 'take off' into development, having gone through five stages, which he likened to the stages an aeroplane goes through before take-off, from taxiing on the runway to mid-flight. Ranging from stage one, 'traditional society', to stage five, the 'age of high mass consumption', the theory takes our faith in the capitalist system for granted since Rostow assumes that all countries will be in a position to 'take off' into development. Further, in common with other

Thematic case study 8.5

Bandung, non-alignment and the 'Third World'

The Bandung conference was a meeting of representatives from 29 African and Asian nations, held in Bandung (Indonesia) in 1955, which aimed to promote economic and political cooperation within the 'Third World' and to oppose colonialism. The conference was sponsored by Burma, India, Indonesia, Ceylon (Sri Lanka) and Pakistan and tried to cut through the layers of social, political and economic difference that separated nations of the 'Third World' in order to think about the possibility of *common agendas and actions*. The aims of the 29 nations that attended included a desire to promote goodwill and cooperation among Third World nations and to explore and advance their mutual as well as common interests. Bandung was in many ways the 'launching pad for Third World demands' where countries distanced themselves from the 'big powers seeking to lay down the law' (Rist 1997: 86). It was not hard for countries with shared histories of colonial exploitation to find something in common, since the 'agenda and subject matter had been written for centuries in the blood and bones of participants' (Wright 1995: 14). In his opening speech to the conference on 18 April 1955, President Sukarno of Indonesia urged participants to remember that they were all united by a common 'detestation' of colonialism and racism (Sukarno 1955: 1) and pointed out that colonialism was not dead or in the past but also had its 'modern' (neo-colonial) forms. The conference was especially successful in hastening the arrival of new international institutions explicitly dealing with 'development' (Rist 1997). Samir Amin coined the term 'Bandung Era' in a discussion of the national approaches to development that emerged from this period (Amin 1994; Berger 1994, 2001). The Non-Aligned Movement (NAM) was then founded (among 29 states most of which were former colonies) in 1961 on a principle of non-alignment with either the USA or the USSR in the Cold War. Three years after the NAM was set up, the 'Group of 77' was then established (mostly made up of the attendees of previous conferences) to press for changes from UN agencies in the way they viewed 'developing countries'. The G-77 has since expanded to include 131 member countries and has been very active in promoting the collective interests of its members and enhancing their joint negotiating capacity within the UN. The G-77 recently played an important role in the negotiations during the UN conference on climate change in Copenhagen in December 2009 and within this China, India, Brazil and South Africa, the so-called BASIC bloc of nations, took centre stage in brokering the final agreement. The NAM has also continued to function as a space through which nations of the so-called 'Third World' meet and discuss common issues and objectives and now has some 115 member nations. The organization has been restructured since the end of the Cold War and continues its commitment to founding principles whilst also now incorporating a focus on reforming the international economy and cooperating with NGOs, the United Nations, the European Union and the G-8. Other related fora have also become increasingly important, such as the G-20, which was first established in 1999, in the wake of the 1997 Asian Financial Crisis, to bring together major 'developed' and emerging economies to stabilise the global financial market. In recent years the G-20 has been seeking to promote international economic cooperation to tackle the financial and economic crisis that spread across the globe in 2008. China, India, Brazil and South Africa are all members along with the USA, the UK, France, Russia and Germany.

modernization approaches, Rostow's model devalues and misinterprets 'traditional societies', which represent the 'lowest' form or stage of development. The advanced state of modernization is always represented as 'Western modernization'; traditional societies seem like distant, poor relations.

Geographies of inequality and development cannot be neatly summarized as a set of prescriptive stages. Urban areas, for example, are themselves subject to uneven development and inequality. Modernization approaches have also largely failed to address the importance of gender, assuming that men and women occupied equal positions in terms of power relations and decision-making. Further, the 'trickle-down' effect has often failed to materialize among those who have been the subjects of modernization projects. This approach seemed to suggest that development could be mimicked, copied and replicated and that underdeveloped countries should try to reproduce the development paths of richer 'developed' nations like Britain or the USA. As Gunder Frank has argued, this approach can also be understood as reflections of the 'Sinatra Doctrine':

> Do it my way, what is good for General Motors is good for the country, and what is good for the United States is good for the world, and especially for those who wish to 'develop like we did'.

(Frank 1997: 13)

In the modernization schema, it is implied that there is nothing before the beginning of development in a 'developing country' that is worth retaining or recalling, only a series of deficiencies, absences, weaknesses and incapacities (Abrahamsen 2000; Andreasson 2005). The approach was also in a sense very much based around a 'top–down' rather than a 'bottom–up' approach, implying that the process could be brokered by states or development institutions rather than emerge from the 'grass-roots' struggles of 'Third World peoples' as had been called for in some more radical approaches. In terms of common criticisms, the division of the world into modern and traditional has often been seen as problematic (Pletsch 1981). Modern societies were much more fractured and were divided by ethnicity, class and politics and were not as united and responsive to the blueprints of planners as was often assumed. The scale of modernization programmes was also often a problem in that they assumed that 'big is beautiful' (involving large dam-building and irrigation projects, for example). Furthermore, the school and its practitioners often depoliticized development, making few if any references to history and culture. Rostow was able to say little about the 'final stage' of his organic model of development since his underlying principle was that growth had no limits (Rist 1997) and instead he simply discussed a number of very generalized scenarios. Like so many theorizations of development that followed, it ended with a creed, a set of principles about what was to be done, and heavily invested faith in the goals of mass consumption and westernization.

8.2.2 The dependency school: beyond 'core' and 'periphery'?

One of the major weaknesses of the modernization approach was that the notion of a 'trickle-down' diffusion of development implied that there were precise demarcations available of where the 'core' ends and the 'periphery' begins when this has never really been the case. Radical **dependency approaches** that emerged in the 1960s and 1970s challenged this notion of positive core–periphery relations, identifying instead

exploitation between 'satellites' and 'metropoles'. Whilst the modernization approach was taken up by many international institutions and bilateral donors, dependency approaches comprised all those opposed to US post-war imperialism and allied in some way to the movement of 'third worldism' (see Case study 8.5). The dependency school is most commonly associated with Latin America, but also emerged in Africa, the Caribbean and the Middle East. Drawing upon Marx's writings about the unevenness of capitalist development, dependency scholars such as Celso Furtado (1964) and Milton Santos (1974) drew attention to the mode of incorporation of each country into the world capitalist system, identifying this as a key cause of exploitation. The dependency school is most commonly associated with the work of André Gunder Frank who published a number of seminal pieces on 'The development of underdevelopment' in the late 1960s (Frank 1966). This thesis is relatively uncomplicated in that it views development and underdevelopment as opposite sides of the same coin; the development of one area often necessitates the underdevelopment of another. Dependency on a metropolitan 'core' (e.g. Europe, North America) increases the 'underdevelopment' of satellites in the 'periphery' (e.g. Latin America, Africa). Unlike many modernization approaches, dependency theorists (*dependistas*) sought to view development in historical context, and argued, for example, that colonialism helped to put in place a set of dependent relations between core and periphery.

These peripheral satellites, it was argued, were encouraged to produce what they did not consume (e.g. primary products, including natural resources) and consume what they did not produce (e.g. manufactured or industrial goods). Thus, rather than likening the process of economic growth to an aeroplane setting off into a blue sky of urban-based, Western life-styles and consumption, the dependency school was arguing that many 'underdeveloped' areas had been stalled on the runway by unequal relations and a history of colonialism, denying them a chance of ever being airborne, 'modern' or 'industrialized'. A dependence on natural resources can lead to slower growth and patterns of development that are geographically uneven (see also Chapter 5 for a discussion of the 'resource curse' thesis).

In many ways, dependency approaches were so directly opposed to modernization approaches (almost point by point) that eventually both 'seemed to checkmate each other' (Schuurman 2001: 6). The dependency school even appeared to preserve the dualistic and binary classification of the world into 'developed'/'underdeveloped', First World and Third World, core and periphery, and also lacked a clear statement of what 'development' actually is. Key criticisms directed at the *dependistas* were that the theory represents a form of 'economic determinism' and also overlooks social and cultural variation within developed and underdeveloped regions. The dependency framework seemed to leave the simplistic impression of an 'evil genie who organizes the system, loading the dice and making sure the same people win all the time' (Rist 1997: 122). Another point of contention was that the dependency theorists seemed to be calling for a de-linking from the world capitalist economy at a time when it was undergoing further globalization and economic integration. Elements of the dependency writings were, however, quite thought-provoking and remain relevant, particularly their contention that the obstacles to development equality were *structural*, arising not from a lack of will or poor weather conditions but from entrenched patterns of global inequality and 'dependent' relationships. A brief comparison of the two schools of development thinking is useful here, particularly given that dependency was primarily a critique of the classical traditions of modernization. According to Bell (1994) both perspectives deal in dualistic either/or scenarios:

> these contrasting schools of thought share a common starting point – poverty is viewed in deprivationist terms while their different views of development have a positive connotation.

> (Bell 1994: 184)

Concepts of development like the modernization and dependency perspectives are far from being static, uniform or unified, however. Neither represents a singular commonly agreed approach and both are complex and varied. In some ways it is the ability of each school of thought to capture and accommodate difference and diversity between states and within regions that is most important here. As we shall see in the following section, in order to do this it is important to further consider theories and practices of development in their specific historical and geographical contexts.

8.3 Development practice (1): the historical geography of development

Whilst both modernization and dependency approaches alluded to the importance of 'tradition', many early writings about development lacked a sense of

historical perspective (Rist 1997). As Crush (1995) points out, development is primarily 'forward looking', imagining a better world, and does not always examine issues of historical and geographical context. It is particularly important to examine the significance of Empire in the making of international development. Between 1800 and 1878, European rule, including former colonies in North and South America, increased from 35 per cent to 67 per cent of the earth's land surface, adding another 18 per cent between 1875 and 1914, the period of 'formal colonialism' (Hoogevelt 1997: 18). In the last three decades of the nineteenth century, European states thus added 10 million square miles of territory and 150 million people to their areas of control or 'one fifth of the earth's land surface and one tenth of its people' (Peet with Hartwick 1999: 105). Colonialism has been variously interpreted as an economic process of unequal exchange, as a political process aimed at administration and subordination of indigenous peoples, and as a cultural process of imposing European superiority (see Chapters 2 and 20). According to the dependency theorists it was in this period that the periphery was brought into an expanding network of economic exchanges with the core of the world system. A new sense of responsibility for distant human suffering also first emerged during this time as the societies of Europe and North America became entwined within global networks of exchange and exploitation in the late eighteenth and early nineteenth centuries (Haskell 1985a, 1985b). Thus the origins of a humanitarian concern to come to the aid of 'distant others' lay partly in response to the practices of slavery in the transatlantic world and to the expansion of colonial settlement in the 'age of empire':

> Not only did colonisation carry a metropolitan sense of responsibility into new Asian, North American, African and Australasian terrains, it also prompted humanitarians to formulate new antidotes, new 'cures' for the ills of the world.

> (Lester 2002: 278)

These new antidotes, cures and remedies were to have enduring significance for the shaping of twentieth-century global development theory and practice, which also often carried an implicit 'metropolitan sense of responsibility'. Colonial development was also associated with an unconditional belief in the concept of progress and the 'makeability' of society, being heavily conditioned by the dominance of the evolutionary thinking that was popular in Europe at the time. Imperialism was viewed as a cultural and economic necessity where colonies were regarded as the national 'property' of the metropolitan countries and thus needed to be 'developed' using the latest methods and ideas. With this came a missionary zeal to 'civilize' and modernize the colonized and their ways of life. An important contention here then is that colonialism 'conditioned' the meanings and practices of development in a number of important ways.

After 1945 and under US President Truman, 'underdevelopment' became the incomplete and 'embryonic' form of development and the gap was seen as bridgeable only through an acceleration of growth (Rist 1997). As we shall see, there was an important sense in which, after 1945, 'development' was seen as a process that was entrusted to particular kinds of agents who performed the acts and power of development upon distant others. Globally, development would have its 'trustees', guiding 'civilized' nations that had the 'capacity' and the knowledge or expertise to organize land, labour and capital in the South on behalf of others. This is an important idea then that quite a paternal and parental style of relationship was established through the imperial encounter between colonizer and colonized in ways which continue to have a bearing on the definition of North–South partnerships in the 'post-colonial' world. Additionally, what is also relevant here is that many 'post-colonial' states continue to maintain important political, cultural and economic ties with their former colonial rulers (see Figure 8.1).

Decolonization is thus simultaneously an ideological, material and spatial process, just as complicated as colonization (Pieterse and Parekh 1995). Colonialism put in place important political and economic relations but the cultural legacies of colonialism bequeathed deep social and cultural divisions in many societies. In the process of decolonization, 'development' became an overarching objective for many nationalist movements and the independent states they tried to form. Although experiments with development were tried in many colonies, the idea of development was invested with the hopes and dreams of many newly emerging states who wanted to address these inequalities and divisions in their societies (Rahnema 1997). An important issue here concerns the extent to which colonial state machineries were reworked and transformed after independence (Power 2003). The colonial state had rested on force for its legitimacy, a legitimacy that was thus highly superficial. Colonial states also had a role in creating political and economic communities, defining the rules of the game and the boundaries of community whilst creating power structures to dominate them. The colonial

state was also the dominant economic actor, creating a currency, levying taxes, introducing crops, developing markets, controlling labour and production. Above all, colonial state administrations sought the integration of the colonial economy into the wider economies of empire, to make linkages with the metropole and to establish flows of peoples and resources. After the formal end of colonialism, new states have had to formulate alternative methods of garnering legitimacy for their authority (i.e. other than the use of force preferred by the colonists).

As the European capitalist system expanded and became ever more global in its reach, the structures of economic, social and political life that existed in colonies (before colonialism) were often radically remade.

The historical process by which 'gaps' began to emerge between 'North' and 'South' has been interpreted in a variety of ways, but a key question has been: to what extent did European expansion and colonialism 'underdevelop' (Frank 1966) large areas of the world? The impact of imperial expansion was not uniform; the geographical patterns of expansion varied, as did the motivations for it. Hall (1992) argues that an important divide was put in place between 'the West' and 'the rest' as a direct result of this imperial expansion, reminding us that 'the West' is much more of an idea than a geographical reality. By accelerating contact between cultures and economies 'the West' was presented as 'the best' and most advanced or 'civilized' of all humanity. Many accounts of the history of European expansion are thus dominated by the presumed supremacy of 'Europe' and 'the West' with only limited references to the complex histories and cultures of the areas that were colonized. Similar criticisms have been made of some development theories (e.g. modernization) that have also often been based around 'Western' philosophies, experiences and histories (Hettne 1995; Power 2003).

Many recent histories of development have dated its beginnings as an area of theory and state practice to President Truman's speech of 1949. The idea of development is, however, much older than this (Cowen and Shenton 1996) and has much more diverse geographical origins. Development was not a simple 'gift' following contact with Europeans but predates the 'age of discovery' (1400–1550) and the 'age of empire' (1875–1914) (see Section 1). It is also worth remembering that Indigenous peoples in Africa, the Americas, Asia and Australasia had highly developed and sophisticated cultures and technologies prior to colonization (Dickenson et al. 1996).

1945–1955
1955–1965
1965–1975

Figure 8.1 Decolonization and the proliferation of independent states, 1945–1975.

8.4 Development practice (2): subjects of development

In addition to thinking historically about the production of global difference and inequality, it is also necessary to listen to what we might call the *other side of the development story*, or the views of women, the views of people in the South, the views of people that are excluded socially, and the view from 'below'. People are central to the development process and an integral element in all development strategies. However, their needs have often been ignored and there has been a general failure to consider the different implications of development policies for individuals, households and communities. Development strategies have also been driven by economic objectives, relegating provision for people's basic needs, their empowerment or efforts to raise their living standards to second place (Potter *et al.* 1999). Development strategies often assume that 'people' and 'communities' are homogeneous and passive rather than differentiated and dynamic. Furthermore, the knowledge and skills of people in particular communities have often been overlooked in favour of broader and less sensitive approaches to 'modernization' that bypass local cultural traditions and histories. Three major agents of development have dominated the literature on development: nation-states, international development institutions and NGOs (as Case study 8.6 illustrates, geopolitical factors are also driving the contemporary development agenda). It is difficult, however, to view any one of these three 'brokers of development' in isolation since, for example, nation-states are the primary focus of the United Nations while NGOs often try to work at the interface between national governments and international governmental institutions.

Many international development organizations have often made erroneous assumptions about the distribution of power and decision-making within the nations and communities where they operate. Consider the level of the household, for example. It is taken for granted that an equal division of labour exists between men and women within households and communities. As we have already noted, national statistics often conceal significant variations within and between specific regions, between rural and urban areas or between men and women. These aggregations of household labour and resources often fail to capture the changing and important contributions of household members such as children. The contributions of women to the productive and reproductive activities of the household are still widely overlooked in 'official'

national development statistics (Momsen 1991; Østergaard 1992). Nonetheless, the household has been and is likely to remain the basic living unit in many parts of the global South, the key to controlling and organizing production, consumption and decision-making. Household sizes and structures differ considerably from those of the global North, often including an extended family of grandparents, aunts, uncles and other more distant relatives. Unequal access to resources means that all these households do not have the same level of access to income, assets (e.g. land) or local and national power structures. It is important for international development organizations to understand the different roles and responsibilities of men and women within these diverse households. The contribution of women in particular to household income and welfare needs to be much more carefully understood (Moser 1993; Kabeer 1994; Sweetman 2000; Perry and Schneck 2001).

In development policy and practice it has often been assumed that the household in the 'global South' corresponds to the Western model of the nuclear household, yet these households are far from homogeneous in their structure and in the amount of control over decision-making and resources that women enjoy. The contribution of women to the development process, of their knowledge, time, labour and capital, has often been underestimated by development 'experts'. Mohanty (1991) rejects the idea of the stereotypical 'Third World woman' as the passive recipient of social services that has often been constructed by development planners and theorists alike. Similarly, the visibility of women's productive, reproductive and community-management roles in the recording of national development statistics has been widely criticized by gender and development (GAD) researchers (Rathgeber 1990; Moser 1993; Visvanathan *et al.* 1997; Momsen, 2009). Women were often not defined as 'productive' contributors to their societies, as it was abstractly assumed that women were largely confined to the private, reproductive or domestic sphere (Marchand and Parpart 1995). In agriculture, for example, access to credit, rural extension services, land ownership and farm technology has often been denied to women, despite the fact that women produce more than half the world's food (Koczberski 1998). In this regard it is important to note that many countries in the South have attempted to create indigenous versions of European welfare states, which has proven economically very difficult for many states, leading to the uneven development of citizenship benefits amongst different socio-economic groups and between men and women (see Chapter 20). Again, however, we see

Thematic case study 8.6

States and the geopolitics of development

At the time of the terrorist attacks in Washington, DC and New York City in 2001, world leaders made numerous comments about the links between international poverty and terrorism, suggesting that this was to become *the* key issue of the twenty-first century. The ensuing *Global War on Terrorism* led by states defining themselves as the 'free' or 'civilized' world began in Afghanistan and has shaped contemporary development thinking and practice in many important ways. These recent events highlight the need to understand more fully the geopolitics of international development and in particular the way in which many states in the South are seen to lack necessary degrees of 'order', hierarchy and societal organization. In thinking about geopolitics it is important to explore the rights and abilities of each state to control the territory encompassed by its boundaries or to gain international recognition for its territorial authority. This also means thinking about the way in which foreign aid is used by Western donors to solicit support for the war on terrorism from states

in poorer countries. The state in Angola, for example, was offered a wide range of incentives in 2003 in order to gain Angolan support for the USA when the African country had a temporary seat on the UN Security Council. Foreign aid has sometimes been seen as a kind of political narcotic, fostering addictive behaviour among states that receive it and thus come to depend on it. States are thought to exhibit the symptoms of dependence – a short-run 'fix' or benefit from aid, but external support sometimes does lasting damage to the country. During the Cold War, the USSR in particular played on its ambiguous position as both 'inside' the international system and 'outside' it, presenting itself as the natural 'midwife' for completing the independence of newborn states (Laïdi 1988). With the accelerating pace of decolonization and the creation of independent states in the South, geopolitical questions were addressed from a set of new or 'Third World' perspectives, new forms of North–South geopolitics which had emerged from the legacies of colonialism and the perception that 'underdeveloped' countries had distinct geopolitical considerations from those of Western societies (see also Chapter 20).

the problems created by trying to draw on a universal set of standards based on the experiences of Western liberal democracies and the problems created by ignoring or overlooking the particularity and diversity of peoples and places. Development planners have often neglected the importance of women's 'double day'

(Moser 1993) or the limitations placed on their time by the need to combine reproductive (e.g. childcare) with productive (e.g. farming) activities. In addition to these two important roles, women often participate in the management of their communities. They organize political, social and economic structures and play

Plate 8.1 The environmental pollution left by Shell's oil refineries in Nigeria has had particular implications for the people of Ogoniland who have seen very little of the profits generated by oil reinvested in their state. This farmer has lost all of his productive farmland to an oil spill. Some observers have argued that Shell's relationship with the Ogoni people and the Nigerian state is one of 'neocolonialism'.

(Martin Adler/Panos Pictures)

a crucial role in the management of the environment and natural resources (Dankelman and Davidson 1988; Sondheimer 1991; Braidotti *et al.* 1994; Shiva and Moser 1995; Buckingham-Hatfield 2000).

Although the importance of women's participation in development planning has increasingly been recognized by international institutions, the capacity of national governments to respond effectively to the needs of their societies at the 'grass roots' has been severely curtailed by indebtedness, reduced commodity prices, dependence on international donors, recession and protectionism in the global economy. The state has typically been seen as the primary agent of development, particularly since the 1960s when many countries attempted ambitious large-scale modernization projects with the backing of international development institutions and commercial lenders. The formation of new states, particularly in sub-Saharan Africa, followed at a time when development ideals were becoming increasingly popular worldwide. The 'age of great optimism' imbued these emerging states with a sense of the importance of development in the post-colonial era. Again, as with so many other aspects of 'development', there are no universal definitions of the state and states do not behave in any predetermined way. States in the South have become subject to enormous pressures from the societies within their borders, becoming sites of protest and contention when their activities impinge on the lives of their subjects and citizens.

Ideology also has an important bearing on the way the role of the state in development is viewed. For example, reflecting the World Bank's 'neoliberal' framework, the annual World Development Reports (WDRs) often view an effective state almost exclusively 'through the lens of economic efficiency', with states seen as providing either 'barriers' or 'lubricants' to free-market economic reforms (see Case study 8.7). Neoliberalism promotes and normalizes an economic growth-first strategy where the social and welfare concerns of citizens are seen as less of a priority. Neoliberal development models also naturalize an image of international markets as fair and efficient, privileging 'lean government', deregulation and the removal of state subsidies. More importantly, in many ways neoliberalism forecloses alternative paths of development, narrowing the ideological space in which it is possible to think outside the 'development box' (Power 2003). Behind neoliberalism is the core assumption that the economy should dictate its rules to society rather than the other way around.

The Bank's neoliberal vision of development has emerged clearly in debates about globalization and the world economy. Globalization is a key means through which many people seek to understand transitions amongst human societies at the start of the third millennium (see Section 4). The Bank attaches a great deal of importance to the impacts of globalization and the increased interconnectedness of the world economy and the ease of travel and communication that characterise the contemporary world. In some ways, however, globalization has been 'fetishized' by the Bank and other international development agencies to the extent that as an idea it now has 'an existence independent of the will of human beings, inevitable and irresistible' (Marcuse 2000: 1). There has also been a tendency to speak of the 'end of geography', envisioning a time when all places will have similar social and cultural characteristics as global corporations and media agencies disseminate similar products and images across the globe (Holloway and Hubbard 2001). It is necessary, however, to avoid the view that globalization is a relatively new and recent phenomenon, led by the 'unstoppable' juggernaut of global changes in information, knowledge and technology. What is often not clear in these debates is what role there is for the people of the South to shape their own destinies and the specific impacts on places, territories and communities.

Not surprisingly (given its economistic focus) the World Bank often 'locks society out' (Hildyard 1997:43) in its definition of the state (Figure 8.2). Although the

Figure 8.2 This illustration depicts the idea that World Bank 'experts' view the role of the state in development as comprising a simple formula or recipe, with several distinct 'ingredients'.

Source: *The World Bank and the State: A Recipe for Change?* by Nicolas Hildyard, Bretton Woods Project, www.brettonwoodsproject.org

Thematic case study 8.7

The World Bank and the IMF

Both the International Monetary Fund (IMF) and the World Bank are based in Washington, DC. The Fund was set up in the 1940s as a result of the Bretton Woods conference to maintain currency stability and develop world trade by establishing a multilateral system of payments between countries based on fixed exchange rates and complete convertibility from one currency to another. The IMF is not controlled by the people or the governments of low-income countries and is not like other UN agencies where every government has a vote (Wood and Welch 1998). Instead the fund is run by a select band of high-income countries that fund its operations (the USA has an 18 per cent share of votes and has de facto control of the organization). The World Bank is now increasingly involved in macroeconomic issues and is always chaired by an American, while the President of the IMF is usually European. The latter is dominated by OECD countries and in particular by the USA and is organized in quite a similar way to the Bank.

The International Bank for Reconstruction and Development (better known as the World Bank) was founded as a means of reviving war-damaged European economies, a mandate that was later extended to 'developing nations'. The Bank is funded by dues from members and by money borrowed in international markets. It makes loans to member nations at rates below those of commercial banks to finance development 'infrastructure' projects (e.g. power plants, roads, hydro-dams) and to help countries 'adjust' their economies to globalization. Like the IMF, the Bank was an early and enthusiastic supporter of the neoliberal agenda. Both agencies make heavy use of intrusive conditionalities in their funding schemes where lending is only made if recipients meet certain specified conditions, even though the World Bank recently acknowledged that its capacity building efforts with states in the South (which incorporate several lending conditions) have had only a limited impact. The 2000/2001 World Development Report (WDR) prepared by the Bank was entitled *Attacking World Poverty* (World Bank 2001a) and promised 'the most detailed-ever investigation of global poverty' (World Bank 2001b). The result of more than two years of research, the WDR drew on a 'background study' that sought the 'personal accounts' of more than 60,000 men

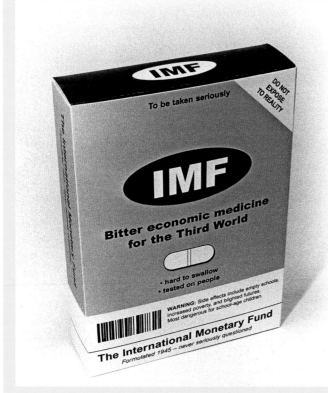

Plate 8.2 The IMF and its hard-to-swallow 'bitter economic medicine' has had major economic implications for 'Third World' countries depicted in this Oxfam campaign. Its side effects include empty schools, increased poverty and 'blighted futures'.

(Oxfam GB)

and women living in poverty in 60 countries around the world. This approach allowed the Bank to claim that their vision of development is based upon 'the testimony of poor people themselves'. The 2010 WDR (World Bank 2010c) focuses on development and climate change and argues that 'developing countries' can shift to low-carbon paths to development whilst continuing to reduce poverty but only with technical and financial assistance from high-income countries. In 2009 total World Bank lending amounted to some US$46.9 billion of which Latin America and the Caribbean received the highest share (US$13.8 billion), followed by Europe and Central Asia (US$9 billion), and East Asia and the Pacific, which received US$6.9 billion (World Bank 2009). The Bank spends significant sums on employing consultants that offer technical expertise but there is little effort made to measure the effectiveness of this spending. Five hundred days of technical assistance costs roughly the same as employing 5000 teachers but which does more to cut poverty?

WDR calls for greater participation in decision-making in order to bring 'government closer to the people', a number of key policy areas are seen as 'off-limits'. Despite its dismal environmental record (where lending policies have often financed ecological destruction), the World Bank has sought to appease its critics by including environmental NGOs (non-governmental organizations) in its operations in recent years. Thus the Bank has co-opted its environmental critics in order to 'green' itself and illustrate its commitment to environmental conservation (Sheehan 2000). According to Chossudovsky (1997) the World Bank has claimed to have shifted policy focus towards 'sustainable development' and 'poverty alleviation' in recent years, but neo-liberal policy prescriptions continue their dominance within the international development industry whilst the Bank has been able to maintain many aspects of its core policy focus by co-opting critics and cloaking itself in the guise of green environmentalism.

The term 'empowerment' is often used by the Bank and other sections of the 'development industry' (Ferguson 1990) to describe their commitment to popular participation in decision-making for development. Facilitating 'people's participation' is now on the agenda of most international development institutions. Again, however, some development agencies have manipulated this agenda in order to present themselves in more positive ways, as 'listening' to the voice of the people and as facilitating popular and local 'ownership' of development plans and policies (Cooke and Kothari 2001). For Rahnema (1997), the increasingly widespread acceptance of the idea of participation suggests that it has been severely diluted and has lost some of its more radical potential:

> governments and development institutions are no longer scared by the outcome of people's participation . . . there is little evidence to indicate that the participatory approach, as it evolved, did, as a rule, succeed in bringing about new forms of people's power.

> (Rahnema 1997: 118–23)

Thus many commentators have become somewhat disillusioned with the development concepts of 'participation' and 'empowerment', arguing that development planners and politicians have often tried to manipulate experiences of participation to suit their own ends and that empowerment has been stronger on rhetoric than in reality, suggesting that advances in widening political participation have been slow. Empowerment speaks of making people the agents of their own development while doing little about the causes of inequality, and without transforming the nature of ties with international development institutions and their ideologies.

The ideals of empowerment and popular participation and their increasing acceptance in development circles is partly a consequence of the rising influence of NGOs which have often been highly localized and focused on particular issues. NGOs have been defined as a 'residual' category that includes a wide variety of formal associations that are not government-led, even though definitions vary enormously and many NGOs do receive finance and support from their home governments. They can operate at a local, national or international level, but there is often little overall coordination between different NGOs even though they usually work on similar issues, sometimes in the same country or city. The role of NGOs as agents of development has grown vastly in the past 30 years. NGOs have a particular geography of power, with command and coordination functions usually located in the North (decision-making and fundraising) and network power in the South (the necessity to be represented in capital cities and 'on the ground' in provinces). A key issue that has been raised in critiques of NGO activity is the closeness of their contact with local cultures and knowledges. Although many have argued that the work of NGOs is often poorly monitored and sometimes imposes foreign solutions in delicate situations, many observers have argued that they can provide an effective link between state and society, often working in 'very practical ways' (Edwards and Hulme 1992: 14). The relationship between the state and NGOs is crucial to the prospects of inclusive democratic development. Particularly in situations of political instability or conflict, NGOs can provide critical linkages in the absence of effective or disputed state administration.

8.5 Conclusions: geography, unevenness and inequality

> Radical development geography must adopt a framework that is liberated from the tyranny of dualism and allows for changes in the world economy and variations between states – one that accommodates alternative historical and contemporary geographies, cutting across the North–South divide. (Bell 1994: 188)

Development has nearly always been seen as something that is possible, if only people or countries follow through a series of stages or carry out a series of prescribed remedies and recipes. It is also often assumed that an organic and inherently positive and 'natural' process of evolution is somehow at work, one that is both progressive and forward-moving: 'from little acorns do giant oaks grow'. Development also often means simply 'more': whatever we might have some of today we might or should have more of tomorrow (Wallerstein 1994). After nearly six decades of global development the promise that tomorrow will bring something 'more' is wearing very thin for many poor people.

As Radcliffe (1999: 84) has argued, '[r]ethinking development means reconsidering the categories we use in development geography, and unpacking the power relations that shape them'. This question of the power of development and the unpacking of the power relations that are inscribed in development practices is a critical one. A rethinking of development is also necessary because the lines that have so far divided North and South are now present within *every nation-state* and are making ever less appropriate the conventional language used to interpret the geography of development in the world economy (rich/poor, North/South, First World/Third World, developed/developing) (Power 2003). It is also necessary to (re)formulate a view of development that focuses on the relationships to households and communities and not just (as often happens) on 'formal' institutions like the state, transnational corporations, the international development agencies or NGOs.

At the very core of development studies and development geography there has been a normative preoccupation with the poor (saying what 'ought' to be *done*) and with the 'marginalized and exploited people of the South' (Schuurman 2001: 9). Perhaps in addition to this focus on the poor we should also ask questions about the rich, about wealth and consumption? The notion of a 'shrinking world' has been influential in discussions of the impacts of globalization, but patterns of global inequality have not shrunk after 50 years of development theory and practice. In addition, although development thinking has itself been far from monolithic, many conceptions had the habit of 'robbing people of different cultures of the opportunity to define the terms of their social life' (Esteva 1992: 9). It is crucial then to avoid painting a picture of development (and its subjects) as monolithic and to avoid repeating this habit of robbing people of their cultural and historical difference and diversity. The *Dictionary of Human Geography* (Johnston *et al.* 2000) defines development as a process of 'becoming' and as a 'potential state of being', but also mentions how:

> The achievement of a state of development would enable individuals to make their own histories and geographies *under conditions of their own choosing*.

> (Johnston et al. 2000: 103; emphasis added)

International development institutions have often overlooked local perspectives and knowledges of development and state approaches have often been heavily centralized and 'top-down'. Escobar (1995), for example, has argued that development has 'created abnormalities' such as poverty, underdevelopment, backwardness and landlessness and then proceeded to address them in a way that denies value or initiative to local cultures. Whilst many sections of the 'development industry' have in recent years come to acknowledge the importance of people power and popular participation in development (Friedmann

1992), 'empowerment' has been co-opted by the mainstream of development theory and practice and has been stripped of its radical potential. UN agencies in particular have been criticized for their neglect of participatory approaches and their excessive bureaucracy, but prospects for reform look bleak. In some ways, the policy prescriptions of macroeconomic global institutions like the World Bank and the IMF claim to be leading the campaign to eradicate world poverty, but a wide variety of 'anti-capitalist' and 'anti-globalization' protests in recent years have shown that these prescriptions are inflexible and ignorant of the particularities of place and country-specific contexts. There have also been transnational movements like the *Jubilee 2000* and *Make Poverty History* campaigns or the series of 'Live 8' music concerts in 2005, which have all sought to put pressure on Western political leaders.

Many recent critiques of development appear disillusioned with the future of the development industry and its capacity to understand and alleviate world poverty. Further, many 'Western' politicians are only gradually beginning to wake up to the realities of these contemporary global inequalities. A number of them see these concerns as those of distant geographies, a world of problems pushed and 'worlded' beyond the universe of immediate moral concern. The effects of poverty and inequality (a bit like those of climate change, see Case study 8.8) are thus regarded as diffuse and long-term (Wade 2001). Additionally, rather than thinking of a single, interconnected world economic system, this impoverishment and inequity is constructed as somehow unique or exclusive to the peoples of the 'Third World'. If our concern then is to build a more radical development geography it needs to be understood that poverty also occurs in

Thematic case study 8.8

Climate change and development

Climate is closely intertwined with development. For one thing, climate is a resource in itself, and it affects the productivity of other critical resources, such as crops and livestock, forests, fisheries and water resources. In addition to natural climate variability, long-term climate trends and climate change are already having a discernible impact on development including direct impacts in relation to climate sensitive activities such as agriculture, and indirect consequences on socio-economic issues such as poverty. The effects of climate change may therefore be especially critical to the achievement of development objectives related to the most vulnerable groups and communities. The adaptive capacity of those affected by climate change ultimately depends on their access to economic, ecological, social and human resources including institutional structures, decision-making processes, information and public awareness (Huq *et al.* 2006). The projected impact of climate change on access to natural resources, heat-related mortality and spread of vector-borne diseases such as malaria, for example, has direct implications for the achievement of several of the Millennium Development Goals (MDGs) (OECD 2006). In certain cases, there are direct trade-offs between development priorities and the actions required

to deal with climate change. Governments and donors confronting immediate challenges, such as poverty and inadequate infrastructure, have few incentives to divert resources to investments that are seen as not paying off until climate change impacts are full-blown. Further, some development practitioners view climate change as a long-term problem that does not compare with more urgent concerns such as food security, HIV/AIDS or pollution (Huq *et al.* 2006). Despite the magnitude of its likely impacts on countries of the global South, until recently climate change has been viewed largely as an environmental concern, of little relevance to development policy-makers or practitioners. These countries are particularly vulnerable because their economies are generally more dependent on climate-sensitive natural resources, and because they are less able to cope with the impacts of climate change (OECD 2006). Similarly, development approaches have been given less attention within the climate change community, who instead favour natural science approaches focused on reducing greenhouse gas emissions. Whilst there exists an established focus on how development can be made climate-friendly, much less attention has been paid to how development can be made more resilient to the impacts of climate change. There is thus a continuing need to bridge the gap between the climate change adaptation and development communities.

'developed' countries and that the aid and 'development' policies of such countries, far from being a part of the solution, may actually be considered a part of the problem. Marginality and deprivation (or for that matter, excessive consumption amongst the affluent) in Europe, North America or Russia and other post-communist 'transition economies' should also be seen foremost as issues of 'development' (Woods 1998; Jones 2000). Furthermore, the rapid economic growth experienced by (re)emerging powers in the South like Brazil, India and China, and their increasingly significant roles in development cooperation and aid disbursement, requires us to adopt a framework for analysis that is 'liberated from the tyranny of dualism' (Bell 1994: 188) and that allows for changes in the world economy and variation within and between states. This also means moving beyond the tragic stereotypes of a single condition of 'Third World poverty' and a single 'geography of the Third World'. This could make a world of difference.

Learning outcomes

Having read this chapter, you should understand:

■ The importance of critical and alternative approaches to the study of global development.

■ The historical forces and pressures that shape development theories and practices.

■ The power and politics of 'labelling' and categorizing poor peoples, places and the spaces of development.

■ The role of some of the key brokers of development such as states, United Nations organizations, communities and NGOs.

■ The differentiation and diversity of peoples, places and spaces of development at the beginning of the twenty-first century.

■ The importance of a people-centred view of development and the need to understand the *other side of the development story*, or the views of women, the views of people in the South, the views of people who are excluded socially, and the view from 'below'.

Further reading

Bryceson, D.F., Gough, K.V., Rigg, J., Agergaard, J. (2009) Critical commentary: The World Development Report 2009, *Urban Studies*, 46: 723–38.

Buckley, R.M. and Buckley, T.D. (2009) Discontent with the World Bank's excursion into economic geography: lions and butterflies once more? *Environment and Planning A,* 41(12): 2806–15.

Chant, S. and McIlwaine, S. (2009) *Geographies of development in the 21st century*, Edward Elgar, Cheltenham. Explores the immense social, cultural, political and economic variations among countries and in different places in the global South, providing an engaging introduction to development.

Desai, V. and Potter, R. (eds) (2008) *The Arnold Companion to Development Studies*, 2nd edition, Arnold, London. Contains short and accessible chapters on a wide variety of development themes. A useful companion text in studying development, with suggestions for further reading on each topic. A third edition is currently in preparation.

Potter, R., Binns, T., Elliott, J.A. and Smith, D. (2008) *Geographies of Development*, 3rd edition, Longman, London. Updated and revised in a third edition, this book offers a wide-ranging discussion of theories of development, urban/rural spaces and the important institutions of global development.

Power, M. (2003) *Rethinking Development Geographies*, Routledge, London. Draws out the spatial dimensions of development and outlines how the discipline of geography has been implicated and involved in the theory and practice of 'development'. The book offers a critical and stimulating introduction to the imperial and geopolitical dimensions of development, looking at cold war and colonial constructions of 'The Tropics' and the 'Third World'.

Rigg, J. (2007) *An Everyday Geography of the Global South,* Routledge, London. Draws on more than 90 case studies from 36 countries across Asia, Africa and Latin America, starting with the lived and 'everyday' experiences of 'ordinary' people as they encounter development.

Rigg, J., Bebbington, A., Gough, K.V., Bryceson, D.F., Agergaard, J., Fold, N. and Tacoll, C. (2009) The World Development Report 2009 reshapes economic geography: geographical reflections, *Transactions of the Institute of British Geographers,* 34: 128–136.

Sidaway, J.D. (2012) Geographies of Development: New Maps, New Visions?, *The Professional Geographer* 64(1): 49–62. This paper reconsiders what remains of the 'Third World'.

Williams, G., Meth, P. and Willis, K. (2009) *Geographies of Developing Areas: The Global South in a Changing World,* Routledge, London. Another rewarding survey.

Willis, K. (2005) *Theories and Practices of Development,* Routledge, London. Explores the development theories behind contemporary debates such as globalization and transnationalism and traces the main definitions of 'development' and 'development theory' over time. It contains student-friendly features, including case studies, with examples, definitions, summary sections, suggestions for further reading, discussion questions and website information.

Useful websites

www.eldis.org The ELDIS Development information research gateway site with a useful country search facility and good web links.

www.guardian.co.uk/global-development A site focusing on global development and tracking progress towards the MDGs.

www.globalissues.org or http://www.focusweb.org The first site contains useful material on global issues and has introductory material on geopolitics and on foreign aid; the second is the home of 'Focus on the Global South'.

www.forumsocialmundial.org.br The website for the World Social Forum.

www.twnside.org.sg The home of the Third World Network (TWN).

www.ids.ac.uk/go/climate-change The Climate Change and Development Centre (CCDC) at The Institute of Development Studies.

www.iied.org International Institution for Environment and Development (IIED).

http://peopleandplanet.org UK students campaigning on world poverty, human rights and the environment.

www.brettonwoodsproject.org The website of the Bretton Woods Project, an organization seeking to monitor and challenge the power of the World Bank and IMF and to seek alternatives.

For annotated, clickable weblinks and useful tutorials full of practical advice on how to improve your study skills, visit this book's website at **www.pearsoned.co.uk/daniels**

SOCIETY, SETTLEMENT AND CULTURE

Edited by Denis Shaw

Geographers are interested in the ways places differ from one another and also in the similarities and interconnections between places. But the complexities of the changes that human societies have experienced since the late twentieth century render the geographer's task of making sense of the world increasingly difficult. Old dichotomies, like that between 'urban' and 'rural', and the lines which geographers have customarily drawn on their maps to divide country from country, region from region and culture from culture, seem to make less and less sense in a globalizing environment. Even so the results of such processes seem not to be a world that is more uniform. While distant places interconnect and interact with increasing intimacy, new spaces appear where people are excluded from such developments, and almost everywhere some individuals who share spaces with others are denied the rights and advantages that the latter enjoy.

It is such complexities, making themselves felt in urban and rural spaces and in the sphere of culture, that form the subject of this section.

The emphasis is on the multifaceted effects of globalization on human societies and the way people think. Thus the world has become more urban – most of the world's population now lives in cities and only three major regions of the world (sub-Saharan Africa, south and much of South-east Asia, and China) remain predominantly rural (though this still represents a substantial proportion of the world's population and seems set to change in the foreseeable future). But the experience of urbanization differs radically between cities, spaces within cities, and individuals. The corollary is that the world has become less rural – rural spaces may be urbanized, linked into global networks, or lose population to the cities – but again those spaces, and the individuals within them, are affected in very different ways by such processes. And the world has become more 'cultural' in the sense that present-day changes are as significant for their cultural effects as for any others. Issues like identity and the meanings and ways of life that are shared by, or that divide, individuals and groups become matters of concern, contestation and even conflict.

Chapter 9 on cities considers how the term 'urban' is to be defined and the implications of urbanism for the ways in which people live and interact. The significant issue of the role cities now play in global and regional economic and power networks is addressed before the chapter turns to examine urban problems like poverty, cultural diversity and environmental sustainability and how these might best be addressed. Chapter 10 on social inequalities argues that all modern societies are riddled with inequalities, some of which are more fundamental than others. The chapter demonstrates that one of the clearest ways in which social disadvantage is reflected is through its geographical expression, particularly in the existence of spaces described as being 'on the margins'. Chapter 11, entitled 'Rural worlds', is similarly concerned with definitions of the 'rural', and analyses the alternative consequences for both rural spaces and individuals of the processes of urbanization and globalization. Chapter 12 on social constructions of nature argues that nature is not some pre-given physical reality, existing totally outside ourselves, but is in fact an idea which is constructed by society. It is never easy to draw a line between things that are deemed 'natural' and things that are 'social', and different ways of conceiving of what is 'natural' can be used to advance particular political agendas. Even science is not immune from this process – as the chapter claims, 'science constructs nature as a mirror to reflect the things that society wants to see in itself' – and therefore we should always be ready to challenge the assumptions lying behind theories about 'nature' and the political stances and policies to which they give rise. Chapter 13 on culture defines the latter as a process and as a system of shared meanings. The argument is that globalization is not producing a single global culture although the idea of a global culture is becoming important. Global cultural processes interact with processes occurring at national and regional levels to produce a complex world in which it is necessary to foster approaches that will welcome and celebrate cultural differences rather than shun and fear them.

This section makes clear the fact that the geography of a globalizing world is as much a geography of exclusion as it is one of interconnection and interlinkage. It shows that it is a dynamic world where social and cultural change is very much the order of the day, and yet it is a world punctuated by continuities. Finally, it suggests that it is a world in which people can make a difference, where they can be more than mere ciphers dominated by impersonal forces like global financial and power networks, and where ideas, attitudes and relationships matter.

CITIES: URBAN WORLDS

Chapter 9

Tim Hall and Heather Barrett

Topics covered

- Sociological and administrative definitions of cities
- Edge cities
- The rural–urban fringe
- Levels of urbanization by world region
- Impacts of economic change on urbanization patterns
- Networks and connectivity between cities
- Connectivity and power
- Researching urban connectivity
- The internal structures of cities
- Postmodern urbanization
- Influences of the state and the planning system on urban form
- Global urban poverty and inequality
- Cultural diversity in cities
- Cities and sustainability

There is a tendency, certainly within the media but also at times within the academic discourse of urban geography, to talk about cities in very unequivocal terms. Talking about cities in this way tends to perpetuate certain *urban myths* (see Spotlight box 9.1). Cities appear to be, for example, spectacular, diverse, connected, fragmented, growing or dying. This is not to suggest that these impressions are untrue, rather that there is a tendency when we talk about the city to focus on these eye-catching aspects and to exaggerate them at the expense of the subtleties and complexities that exist in cities. The tendency to talk about cities in this way is not new. A number of commentators have recognized the existence of urban myths throughout history, tracing the earliest to Ancient Greece and Rome (Williams 1973; Gold and Revill 2004; Rennie Short 2006). It is important not to dismiss this way of thinking about cities entirely. Such myths emphasize the sheer diversity of the urban world and begin to provide a framework for understanding and working through this diversity. However, relying entirely on them as a guide carries with it the danger that much of the detail that makes up the daily 'stuff' of urban life eludes analysis. A review of almost any local newspaper, for example, tells of a range of more mundane concerns that typically constitute the daily life of cities. Gloucester is a small city in the south-west of England. Its newspaper, *The Citizen*, for example, regularly runs stories on issues such as the problems caused by an infestation of seagulls in the city centre, recurrent minor incidents of anti-social behaviour and vandalism, deaths and injuries caused by traffic accidents, financial problems affecting independent

Spotlight box 9.1

Urban myths

Urban myths refer to long-standing cultural attitudes towards the city expressed through imagery. Powerful pro- and anti-urban imagery can be traced back to biblical times; indeed many enduring impressions of the urban can be first found in the bible. Pro-urban myths include seeing the city as civilized, modern, liberating, exciting or perhaps romantic, while anti-urban myths see the city as ugly, alienating, corrupting, dirty or dangerous, amongst other things. Art and literature are good sources for tracing the development of urban myths. We could think, for example, of the negative urban imagery in the fiction of Charles Dickens. Dickens, in a number of his most famous novels, such as *Oliver Twist* (1838) and *Hard Times* (1854), portrayed the city in a very negative light, describing it as dirty, industrial, polluted and crime-ridden. By contrast the American novelist Richard Ford paints a much more positive picture of the city, or at least of small towns; he is less enthusiastic about big cities such as New York. In novels such as *The Sportswriter* (1986) and *Independence Day* (1995) he sees even the most ordinary of suburban landscapes as possessing a rich, epic quality. The passage at the end of *Independence Day* during which the protagonist, Frank Bascombe, cruises around the town of Haddam, New Jersey observing the start of their Independence Day celebrations takes on an almost hypnotic, blissful quality. There are countless other examples of variations on pro- and anti-urban myths from literature as well as other artistic media such as painting, poetry, music and film.

There are many excellent studies of urban myths, the most famous of which is undoubtedly Raymond Williams' (1973) *The Country and the City*, which was a hugely influential scholarly work. More recent studies that build on Williams' work include those by J.R. Gold and G. Revill (2004) and John Rennie Short (2006). Urban myths are significant in that they reflect widely held cultural attitudes towards the urban. They tell us what societies think about cities at different times. Dickens' urban imagery is reflective of, and in part helped reinforce, widespread attitudes towards the city during the Victorian period. Perhaps Richard Ford's reflect attitudes towards the suburbs at the end of the twentieth century? Certainly he writes about environments in which many people wish to live. Often these attitudes towards the city are implicated in wider moral debates. These urban myths are also significant because they are still constantly reproduced. Anti-urban myths have been a staple of recent media coverage of New Orleans after Hurricane Katrina and of the abandonment of neighbourhoods in Detroit and other North American cities, for example (see Case study 9.1). While a bad image is a serious issue for a city to have to deal with, a positive one is a resource that it can exploit.

Q Think of an example of a city that you regard as having a very negative urban image. What is the nature of this image, how did it come about, and is the city doing anything to address this?

retailers, educational successes of local schools, and reports from the city's sports clubs. This daily diet of local news is typical of cities the world over. Such mundane issues though are rarely the substantive concerns of urban geographers who tend to focus on the 'big issues' rather than the minutiae of life in the city:

> Urban geography and urban studies is a discipline populated by Big Things. Cities for a start. They are by definition big. Motorways and mass transportation systems, urban redevelopment projects, and suburban shopping malls are pretty big too. Then there are skyscrapers, mega-projects, new-towns, *edge cities*, again all large, obvious, written across the landscape and close to the heart of urban geography's sense of itself … Given its attraction to the Big, it is perhaps of little surprise that urban geography has not been particularly good at, or indeed often has not been interested in, making sense of many of the smaller elements that make up a city. (Latham 2008: 215–15)

A corollary of this tendency within urban geography to see the city in unequivocal terms has been the tendency to focus on a narrow range of big cities, such as Los Angeles, London, Mexico City and Tokyo, where the processes of urbanization are manifested most clearly and spectacularly. However, it would be wrong to dismiss both the mundane concerns outlined above and smaller, seemingly less interesting cities. These concerns make up the fabric of everyday life for the majority of urban dwellers and, despite their large population sizes, the majority of the world's urban population do not live in large cities but in smaller urban settlements (Bell and Jayne 2006).

> 52 per cent of the world's urban population resides in cities and towns of less than 500,000 people. A similar picture is painted for developed and developing countries, as 54 and 51 per cent of their urban population, respectively, live in such cities. Despite the attention they command, megacities – cities with over 10 million people – are home to only 9 per cent of the world's urban population. (UN HABITAT 2009: 27)

Further, although the issues above may seem trivial or of only limited interest in themselves, they constitute part of much bigger, broader, more significant and wide-ranging urban issues and for this reason deserve our attention. Thus, we need to try to appreciate the fine grain of daily urban life, the big issues and the relations between these. This chapter, then, aims to outline ways to think through the complexity and diversity of

the urban world and to talk about the ways that geographers have tried to reconcile both big issues and more mundane concerns in their work. It will do this first by focusing on three aspects of cities that form enduring themes in urban geography, which can be thought of as the multiple geographies of the urban world, as follows:

- The macro-geography of the urban world: the ways in which processes of contemporary urbanization are changing the distribution of the urban population around the world.
- The host of networks and connections that exist between cities and the effects that these webs of connections have on cities.
- The diversity of the internal worlds of individual cities.

Although these will be considered separately here, it should be remembered that they are interrelated issues. Each of these aspects affects the others. These can be thought of as some of the basic or fundamental questions of urban geography. Geographers have sought to explore these in a variety of ways over time devising often radically different theoretical perspectives on them. As Paddison argues:

> Historically then, the study of cities is identifiable with continuities and discontinuities – continuities in terms of the basic questions cities pose, discontinuities in terms of how they have been studied and theorised. (Paddison 2001: 4)

The discussion of these basic questions is situated within a reflection on some key recent theoretical debates and challenges within urban geography. The chapter concludes by a looking at a number of contemporary issues. While some of these have echoes in earlier phases of urban geography, others have emerged more recently.

9.1 Defining the urban world

Trying to define the urban world is fraught with difficulty. One does not need to be a student of urban geography to recognize characteristics of urban settlements and the differences between cities, towns and villages. However, organizing this understanding systematically and using it to define what and where is urban has proven more problematic. Fundamentally we need to ask if we can recognize any qualities that are uniquely or distinctly urban. Although there are undoubtedly differences between rural and urban, the question here becomes: are these differences of type or of degree? This question formed the basis of a significant strand of sociological debate during the twentieth century.

The search for qualities or processes that are uniquely urban has long been at the heart of sociological enquiry into the city. This originated in the work of Ferdinand Toennies (1887) who distinguished between *Gemeinschaft* (traditional communities) and *Gesellschaft* (modern societies characterized by more instrumental social relations). This spawned a tradition of sociological enquiry that explored the supposed differences between rural and urban societies and the ways of life that characterized them (Savage *et al.* 2003: 109). The most influential of these were the work of Georg Simmel, particularly his essay 'The metropolis and mental life' (1903/2004) and the longer treatment of his ideas in *The Philosophy of Money* (1907/1978), and Louis Wirth's essay 'Urbanism as a way of life' (1938). Influential in their time, they have been the subject of much debate across a number of disciplines concerned with questions of the urban. Broadly speaking, Simmel's perspective was temporal, arguing that the culture of the modern period (roughly from 1850 onwards) has been urban, while Wirth's was spatial, seeking to recognize distinctly urban qualities that differed from those of the rural. Simmel's concern was primarily with the effects of the development of the money economy in the modern period on social relations. He argued that it produced blazé, reserved, instrumental relations between people that are found at their most developed in cities, the centres of the money economy. This he contrasted with the social relations of earlier historical periods. Simmel's work has often been misinterpreted as attributing causality to cities. However, it was money rather than cities per se which, he argued, produced the distinctive social relations that he observed (Byrne 2001; Savage *et al.* 2003; Hubbard 2006). Wirth, by contrast, attempted to show that the characteristics of cities, notably their size, density and heterogeneity, produced social relations of a different type to rural areas.

> Wirth's basic argument was that city life was characterized by isolation and social disorganization, and that this was due to the fact that all cities were large, dense and heterogeneous. (Savage *et al.* 2003: 108)

Although undoubtedly pioneering and influential, these works have failed subsequently to sustain the notion that distinctly urban qualities, processes and ways of life exist. Criticisms, particularly of Wirth's arguments, have included the recognition of many cultural groups within cities, a plethora of subcultures rather than the singular urban culture that Wirth proposed, and the identification of close-knit communities in cities, though ones frequently undergoing profound change (Hoggart 1957; Young and Wilmot 1962), which runs counter to Wirth's assertions of isolation and social disorganization as the key characteristics of urban society (Byrne 2001; Savage *et al.* 2003).

Defining the urban for administrative purposes, the measurement, management and planning of cities, has proven equally problematic as attempts to define it sociologically. Amongst administrative definitions of the urban, population size is perhaps the most basic criterion used to define the nature of settlements. Internationally though there are huge variations in the ways in which this criterion is applied. While in parts of Scandinavia a settlement of 300 can be classed as urban, in Japan only settlements whose population size exceeds 30,000 qualify. Higher population densities are often cited as a characteristic of urban areas but here again there is no natural or universally agreed cut-off point that separates urban from non-urban settlements. Indeed, the extensive depopulation that has recently been observed in some old de-industrialized cities such as Detroit in the USA and across parts of East and Central Europe (Hall and Pfeiffer 2000; Oswalt 2004) has meant that large parts of these long-established urban areas now have very low population densities.

The task of drawing boundaries around cities, suggesting a neat separation from rural areas beyond, is further complicated by the ambiguous nature of the rural–urban fringe. Here is a zone where rural and urban functions intermingle and where often substantial components of the hidden infrastructure of the city, such as sewage works, power plants and server farms, along with major transport and distribution functions such as airports and motorways sit in otherwise rural surroundings. Commentators have long argued over whether these zones should be more accurately defined as rural, urban, transitional zones, or as something altogether quite different and distinct (Gallent *et al.* 2006; Gallent and Anderson 2007). A casual stroll around these areas soon demolishes the impression that there is a neat break between urban and rural.

While the landscapes of the rural–urban fringes of many cities are regarded as somewhat prosaic outcomes of urbanization, the edges of some, particularly American, cities are regarded as more spectacular manifestations of these processes. The processes operating here and their outcomes reveal the inability of a relatively static administrative framework to contain the dynamic and unruly processes of late capitalist urbanization. **Edge cities**, as they are generally referred to, are settlements that contain many of the traditional functions of cities but which have grown rapidly through private development during the last 30 years. They owe their name to their location on the edges of, or at times beyond, existing cities. They seem to be producing new geographies of the urban world, prompting some commentators to suggest that they represent a new **postmodern** phase of urban history very different from that of the nineteenth and twentieth centuries (Dear and Flusty 1998; Soja 1989, 1996, 2000; Dear 2000). We look at some of these new geographies of the urban world in the following section. Edge cities, though, can be thought of as alternatives to the city rather than extensions to it, as in the case of suburbs which constitute the archetypal landscape of earlier rounds of urbanization in many parts of the developed world. Often edge cities straddle administrative boundaries, fragmenting their true form and extent within official statistical returns. The rapidity with which edge cities have colonized rural land in some areas means that where these processes are at their most prevalent, the administrative framework of regions is more a reflection of historical rather than contemporary processes and patterns of urbanization. The amorphous nature of post-modern urbanization has led some to question whether edge cities and associated settlement forms should be thought of as cities at all (Pile 1999: 30). This neatly returns us to the difficult sociological questions of the urban discussed above.

While acknowledging that it is difficult to define exactly what and where is urban, it should not be assumed from this that urban questions are unimportant. 'Cityness', namely the density, not only of people but also of institutions and built forms (Pile 1999), and the intense heterogeneity and juxtapositions of cities, clearly matter (Amin and Thrift 2002: 2). The nature and quality of urban life are different in different places and this is influenced in large part by the surroundings within which urban lives are lived. Living in a suburb of a small European town, for example, is very different to living in the heart of Tokyo. While we should not fall into the trap of fetishising the city, of attaching uniqueness and causality to its qualities, we should not dismiss the differences within and across the urban world. Indeed geography is all about acknowledging these differences and this has emerged as a key aspect

of recent theoretical debates in urban geography. Differences in the nature of places matter and have effects on the creative possibilities open to urban dwellers (Florida 2002) and to the fears and problems they face (Valentine 1989; Patel 2000; Hayward 2004; Aas 2007).

9.2 Multiple geographies of the urban world

9.2.1 Global patterns of urbanization

Current trends in urbanization are changing the macro-geographies of the urban world. By macro-geographies we mean broad patterns in the distribution of the global urban population (see Pacione 2009: Chapter 5). The extent to which different regions of the world are urbanized varies hugely. This is a long-standing historical feature in which generally levels of urbanization correlate with those of development. The most highly urbanized nations, then, have tended to be the most developed. Consequently, until the mid-twentieth century significant urban development was largely concentrated in Europe, North America and Latin America. However, since then, urban growth in these regions has been slow with only modest growth in the proportion of their populations living in towns and cities. In addition, within these regions there has been uneven growth by city size, with the largest growth occurring in smaller urban areas. This reflects the changing economic geographies of key sectors in these regions and the choices of mature urban populations for suburban living and relocation to smaller urban areas and the countryside (Champion 1999).

The fate of large cities in these regions has been mixed. A small number of global or major capital cities, such as London, Paris and New York, have maintained their position or grown in population and influence during this period. This has occurred mainly on the back of the growth of their financial sectors and their primacy within international, national and regional economic systems (Short 2004; Kim 2008). These cities have not been without their problems though. Many have seen significant shrinkage in their manufacturing sectors coupled with problems related to their failure to adequately accommodate, manage and plan for large immigrant populations (Sandercock 1998; 2003; Saunders 2010). The problems of Paris' North African immigrant populations that culminated in civil disturbances in a number of suburban *banlieues* typify this (Wacquant 2007). Whereas bald population statistics suggest overall buoyancy in these cities there has been widespread redrawing of their internal social geographies and this has not been without cost (see Chapter 10).

Second-tier, regional cities, particularly those with a large manufacturing economy, have tended to struggle in the latter half of the twentieth century. While this is the result of a general global shift of manufacturing away from its traditional heartlands (see Chapter 14) the outcomes of this are mediated locally. Since the 1950s many of these cities have suffered significant population loss as their manufacturing economies have contracted. Liverpool, in the North West of England, for example, lost an average of 10,000 people per year during the 1970s. Attempts to regenerate many cities of this type through ambitious property-led, city centre developments seemed to be stemming this decline as, by the late 1990s, many cities such as Liverpool, which had experienced long-term decline, were seeing their populations stabilize and even in some cases grow slightly. These optimistic signs have been tempered somewhat in recent years, however, as the global recession has impacted on the financial, business, travel and property sectors around which these urban renaissances have typically been constructed (Martin 2010). The fortunes of many of these cities seem to be a reflection of broad trends in national or international economics. For some cities, however, there are signs that the process of decline might be more terminal. These include the city of Detroit (see Case study 9.1) and many in Eastern and Central Europe as mentioned above. Here an almost total reliance on industries that have now declined or

relocated globally, a failure to stimulate significant economic growth in Eastern Europe since 1989, and stark inequalities within the deeply racially divided populations of the North American rustbelt, appear to be keys in understanding the problem.

The most significant growth in the urban world in the last 30 years has taken place in those regions with traditionally low percentages of urban populations, namely Africa and Asia. Asia has witnessed particularly rapid urban growth during this period, headed by China and India, something that is predicted to continue into the foreseeable future. Despite both continents experiencing significant urban growth, the nature of growth in Africa and Asia has been quite different. While China, for example, has seen growth in its urban populations concentrated largely in major urban centres, Africa's has predominantly taken place in small and intermediate cities.

These broad changes in the macro-geographies of the urban world are captured clearly in shifts in the number and location of the world's mega-cities (cities with more than 10 million inhabitants) (Figure 9.1). The United Nations predicts that the number of these cities will increase to 27 in 2025 and that the majority of them will be located outside the developed world. The rapid recent urban growth and its concentration in large settlements are particularly apparent here. Cities from this region are predicted increasingly to dominate the mega-city league table, whereas longer-standing mega-cities in the West and Latin America, such as New York–Newark and Mexico City, will slip down the rankings. A key issue related to this that has attracted the attention of urban analysts has been that

Plate 9.1 Abandoned buildings and neighbourhoods in Detroit.

Regional case study 9.1

The slow death and tenuous rebirth of Motor City

Detroit has recently been constructed within the media as an archetype of urban abandonment. Its decline (between 30 and 40 per cent of its metropolitan neighbourhoods are now largely abandoned) has been seen as symptomatic of the end of the industrial age in the USA and a possible harbinger of the future of many formerly prosperous cities in the former industrial heartlands of the global North. The statistics below (Table 9.1) starkly demonstrate the decline of the city since the middle of the twentieth century. What they fail to show, however, is the extent to which this decline is racially divided. The story of Detroit in the twentieth century is one of rapid industrial expansion, racial segregation, selective suburbanization, and subsequently economic collapse. It is a story that leaves many questioning the city's future.

The era of mass production emerged in Detroit in the early twentieth century when Henry Ford's Dearborn plant opened in 1913. It pioneered a method of production that became standard around the industrial world and ushered in the economic and cultural changes that created the era of mass consumption. Much of Detroit's subsequent development was tied up with the automobile industry. The city, as it grew in the early twentieth century, developed around the manufacturing plants of America's three largest automobile companies: Ford, General Motors and Chrysler. Its social geographies were reflective of the strongly racialized organization of labour within the industry. The racial segregation of the city was compounded by the suburbanization of the city in the middle years of the twentieth century that took many of the city's wealthier white residents to new settlements beyond the city limits. This was particularly noticeable following a series of serious civic disturbances in inner city Detroit in 1967 that were quelled by the National Guard leading to a number of deaths. This left predominantly black residents to occupy the progressively degrading inner city suburbs.

The periodic crises that have affected the American automobile industry since the oil price crisis of the mid-1970s have impacted particularly heavily on Detroit. The American car industry has been criticized for a failure to innovate during its years of success which left it vulnerable to changes in demand and competition from abroad. A revival in the fortunes of the city's automobile manufacturers in the late 1990s, based around the growth in popularity of sports utility vehicles (SUVs), was halted and reversed by recent spikes in the price of oil and the demand for smaller, more fuel-efficient cars largely provided by overseas manufacturers. In 2009, General Motors and Chrysler were forced to turn to the US government to avoid catastrophic bankruptcies. In addition to these crises, Detroit has been badly affected by the problems in the US housing market linked to sub-prime mortgage lending. A lack of alternatives in the formal economy has seen the growth of informal and illegal economic actively based around scrapping metal and other goods from abandoned factories and office blocks, illegal gambling, for instance on dog fighting, and drug dealing. The combination of a narrow economic base rooted heavily in automobile manufacture and stark racial division between the inner city and suburbs has underpinned the current crisis facing the city.

There are some tenuous signs though that the city is mapping out an alternative future from amidst its ruins. The abandonment of such large areas of its inner city has opened up opportunities for artists and entrepreneurial urban agriculture projects to make productive use of these spaces. Although it is too early to judge the extent to which these will lead to a reversal of the decline the city has experienced in recent years, they suggest it is too soon to suggest that Detroit will become America's first 'post-city'.

Table 9.1 Demographic data Detroit metropolitan area 1950–2010

Year	Population	Change	Rank (US cities)
1950	1,849,568	–	5
1960	1,670,144	–9.7%	5
1970	1,514,063	–9.3%	5
1980	1,203,368	–20.5%	6
1990	1,027,974	–14.6%	7
2000	951,270	–7.5%	10
2009	821,792 (est)	–13.6%	11 (2007)

Decline 1950–2009: –1,027,776 (55.6%)

Detroit–Warren–Livonia unemployment rate: 14.4% (August 2010)

United States national unemployment rate: 9.6% (September 2010)

Source: United States Census Bureau/Bureau of Labor Statistics

Source: BBC Television (2010) *Requiem for Detroit* (broadcast 13/03/10)

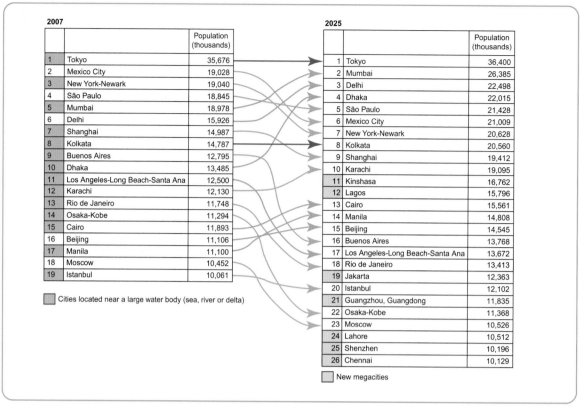

Figure 9.1 Mega-cities, cities with more than 10 million inhabitants, 2007 and 2025.

of power in the urban world. Namely, does this changing geography of mega-cities represent a shift in the distribution of the world's most powerful cities? The issue of power highlights the importance of connectivity in the contemporary urban world.

9.2.2 Global urban connectivity

It is perhaps not surprising that in a global age the issue of connectivity has gained significance within the analysis of cities. Connectivity simply refers to the connections that exist between places, although in actuality the nature of these connections is often highly complex. Commonly places are connected through the movement of goods, people, information, ideas and money. The current age is often represented as one of connection and cities act as vital hubs in global or regional networks of various kinds. Connectivity between cities is not a characteristic of the recent past. Networks of trade, politics, empire and travel, for example, are observable back to ancient history. However, recent technological changes that have led to the development of truly global communications systems have wrought

both qualitative and quantitative discontinuities in the nature of these networks. The speed, capacity, intensity, quantity and quality of the connections between cities has grown significantly since the 1980s (Harvey 1989; Dodge and Kitchen 2000; Hubbard 2006). While not dismissing the importance of historical connections and networks stretching around the world (see Sheppard 2002) the ubiquity and intensity of connection in the current age is unparalleled. Social scientists have been quick to pick up on the significance of this. For example, they have adopted various forms of network or *relational approach* (Dicken *et al.* 2001; Sheppard 2002; Bosco 2006; Murdoch 2006) that have sought to problematize the sedentary tendencies of much traditional social theory. Both have been influential in reshaping urban geography in recent years (see Spotlight box 9.2).

The example of London in Spotlight box 9.2 reminds us that, if the contemporary world is a networked one, then the major nodes in these networks are cities. It is here that all major networks coalesce, be they networks of travel and migration, trade, communication, finance or politics. These networks do not coalesce

Spotlight box 9.2

Relational approaches and global cities

Relational approaches have encouraged scholars of the city not to regard cities as bounded, isolated entities. Rather they have viewed cities as nodes within a variety of networks, many of which are global in their extent. When thinking about the hinterland of cities, for example, relational approaches have not referred to the contiguous spaces surrounding cities, as was the case in the earlier models of Christaller and von Thünen. Rather they have argued that cities' hinterlands potentially encompass spaces that are spatially distant but which are nonetheless closely bound together within networks. This has led some to question the assumption that we can identify a small subset of 'global' cities – cities such as New York, London and Tokyo commonly painted as the control centres of the global economy – and to argue that all cities are global to a greater or lesser extent (Taylor 2004). London may be a more significant node within global networks than cities such as Gloucester, Oaxaca or Hobart, but these cities are certainly not, somehow, 'non-global' (Hubbard 2006: 185).

Relational approaches represent a fundamental rethinking of prevailing conceptions of space. They potentially profoundly affect the ways in which we conceive of cities. Ash Amin discusses this in the context of the representation of London in maps:

> it will take a lot to displace the A–Z or concentric circle image of London by a *relational* map that incorporates the network of sites around the world that pump fresh food into a distribution centre called Covent Garden, that draws neighbourhood boundaries around settlements in postcolonial countries with which social and kinship ties remain strong, that makes us see sites such as Heathrow airport or Kings Cross station as radiations of trails shooting out across the land and far beyond. (Amin 2004: 34; cited in Hubbard 2006: 204)

While many, if not most, cities have hinterlands that are potentially global in their extent, there are significant differences between the specific connections that flow in and out of different cities and hence in the nature of these hinterlands. This reminds us that it is important not to be seduced by the global rhetoric that infuses much of this discourse. It is the materiality of the city, the internal mixes of people, firms and institutions and the interactions that take place within cities that affect the ways in which cities articulate with the outside world (and visa-versa of course) (Thrift 1996). The internal worlds of different cities are highly diverse and it is important to appreciate that this is a component of, and a context within which, the geographies of global networks are formed.

evenly across the urban world, however. Inevitably there is a geography to these networks within which some cities are significantly more important nodes than others. It is here that an important relationship between connectivity and power begins to emerge that destabilizes earlier notions of power associated with cities that were, in large part, related to the size of those cities. For many years it has been taken for granted that the largest cities were the most important. The relational view of cities, though, raises some questions about the validity of this assumption. Rather than simply reading off the power or importance of cities as some function of population size it is now clear that a more accurate and relevant analysis of power in a connected age takes account of the nature and extent of the connections flowing in and out of cities (Case study 9.2). It is possible, for example, for some relatively large cities to be poorly connected, and hence marginal, in the current age despite their size. John Rennie Short (2004; see also Hubbard 2006: 180) cites cities such as Teheran (Iran), Dhaka (Banglagesh), Chongqing (China) and Khartoum (Sudan) as examples of such cities. Dhaka was actually the tenth largest city in the world in 2007 with a population of 13,485,000 and is estimated to become the world's fourth largest city by 2025 (UN-HABITAT 2009). Given the rate of its projected growth it may well exceed this position after 2025. However, despite this it is relatively little connected into global financial networks, nor is it a major global trade hub or centre of global media networks, except those serving the Bangladeshi diaspora. In all the attempts to quantify the

Thematic case study 9.2

Measuring the connectivity of cities

Attempts to quantify the significance and connectivity of cities in the current age have been a major focus of **global cities** research since the mid 1990s. Analysts have employed a variety of sources of data including the distribution of financial and producer service offices, air travel, financial flows and electronic communications data to map the networks and connections that flow between the world's cities. Using these methods it is possible to rank cities according to their connectivity and apparent influence within the global economy. Typically this analysis posits the cities of London, New York and Tokyo as the summit of global city hierarchies with various tiers below encompassing major regional centres, global cities of lesser significance, important national centres with some

strong international links and finally smaller cities that nonetheless occupy significant niches within the global economy (Hubbard 2006: 173–4). This sophisticated analysis has provided a glimpse of the nature, distribution and operation of power in a global age. However, it has drawn some criticism for its focus on data that reflect the processes of Western economic globalization, thus ignoring other potentially important dimensions of contemporary globalization (Short *et al.* 1996; Robinson 2006). Hierarchies of global cities of culture, religion or politics, for example, might look quite different from those economic rankings that have prevailed in analysis to date. This is not to dismiss the importance of this global cities research, rather, to argue that it represents only one take on globalization, albeit arguably the most significant, amongst many.

significance and levels of connectivity of the world's major cities (see Sassen 1994; Smith and Timberlake 1995; Taylor 2004; Derudder and Wilcox 2005) it occupies a very lowly position, if it is present at all.

9.2.3 The internal structure of cities

The trajectories of urban development vary a great deal around the globe. They reflect the particular combinations of local and general factors that come together in different times and at different places. Consequently the internal worlds of cities are similarly complex and heterogeneous. Despite this, geographers have expended a great deal of effort attempting to find patterns and regularities within this complexity and devise models of the internal structure of cities. Some of the most influential work in the subject is reflective of this tradition within urban geography (see Burgess *et al.* 1925; Hoyt 1939; Harris and Ullman 1945; Shevky and Bell 1955; Mann 1965; Soja 1989, 1996, 2000; Dear and Flusty 1998; Dear 2000). Work exploring the internal structure of cities has tended to be characterized by a focus on either the physical morphology of cities, their buildings, streets and spaces, or the complex and dynamic economic and social geographies that overlie this morphology and that are both reflective of it and active in its transformation. The two-way relationship between the city's physical form and its social and

economic geographies is one key to understanding the nature of cities' internal worlds. The focus is on the former here but with a cautionary note not to attribute to it too great a deterministic quality.

While a variety of processes have shaped the geographies of cities it is possible to recognize some broad historical and geographical distinctions within which certain processes have been dominant. The very earliest cities that appeared in places such as Mesopotamia between 5000–6000 BP and later the Indus Valley, Egypt, China and the Mediterranean tended to reflect a combination of local physical and environmental conditions and the religious and cultural beliefs of societies. Later the emergence of mercantile and subsequently industrial capitalism and processes of colonization became both determining influences on the evolving structures of cities and systems through which ideas about urban form were transmitted around the world. More recently many commentators have argued that globalization and post-industrial or postmodern capitalism has become the dominant influence on urban form across the globe. Some have argued that this most recent phase is one of global urban convergence where difference between the internal worlds of cities is being reduced (Plate 9.2), although other authors have recorded evidence of diversity as well as similarity within postmodern cities (Poulsen *et al.* 2002). These forces, certainly since the medieval period, have also been

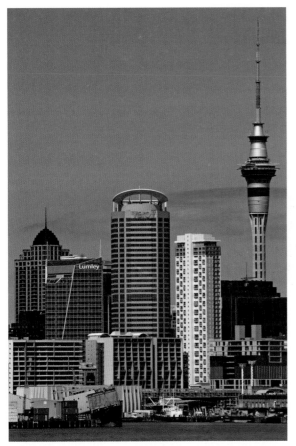

Plate 9.2 Global urban convergence? Auckland, New Zealand.

mediated through the state, which has demonstrated differing degrees of centralization historically and geographically. For example, the cities of communist Eastern Europe and China demonstrated a distinctive 'camel-back' urban form, with a high-density centre surrounded respectively by low density industrial and high-density residential zones, a form resulting from centralist state-control of land markets and planning and not seen in cities of the capitalist world (Pacione 2009: 184). Further, observable differences between the forms of British and American cities during the twentieth century were attributed largely to the history of UK government state intervention in the housing market (Mann 1965; Ward 2004).

Broad categorizations such as these, or typologies that suggest distinctive epochs or categories either in the evolution of urban form or in types of cities, are undoubtedly helpful in that they begin to work through the complexity and variety of cities. However, they do carry with them the risk of overemphasizing either the pervasiveness of processes such as postmodern urbanization, or the distinctiveness of urban epochs. For example, the extent to which it can be said we are currently witnessing a new phase of urbanization producing cities under conditions of postmodern globalization that are distinctively different from those produced under industrial capitalism has been the subject of considerable debate (Harvey 1989). Similarly, too ready a recourse to models of urban form carries with it the risk of masking a patina of local distinctiveness in the internal worlds of cities. It is important to remember that, eye-catching as the epochal processes of urbanization may be, their imprint on cities is always uneven and their effects are mediated locally. Some cities, perhaps smaller cities in particular (Bell and Jayne 2006), might be relatively lightly touched by these processes while elsewhere processes of contemporary urbanization might be transforming the architectural landscape of cities while leaving the underlying structure or plan relatively unaffected. Cities' internal worlds, then, are often palimpsests within which it is possible to discern many phases of partial urban transformations. The universalizing tendencies within urban theory and its masking or disregard of urban cosmopolitanism has recently been the subject of some significant critiques (Hubbard 2006; Robinson 2006).

9.2.4 Postmodern urbanization

The main process of contemporary urbanization that has caught the attention of urban theorists is that of postmodern urbanization. Some have argued that we are witnessing a new phase of postmodern urbanization that is producing cities that are distinctly different in their form and their social and economic characteristics from earlier waves of urban development. This claim actually rests on empirical investigation of only a small number of cities, of which Los Angeles has assumed a primacy (Plate 9.3). A key motif of postmodern urbanization is an apparent fragmentation of both urban form and the social and economic geographies of the city. The postmodern city is said to demonstrate a chaotic urban form characterized by the juxtaposition of spectacular 'fragments' of commercial and residential development and redevelopment set within extensive areas of environmentally and economically degraded and ethnically segregated space (Soja 1989, 1996, 2000; Dear and Flusty 1998; Davis 1990; Dear 2000). These urban geographies, it has been argued, reflect profound changes in the processes that shape cities and their urban landscapes. Rather than the relatively stable phase of industrial capitalism, mediated

Plate 9.3 Los Angeles, constructed as the archetype of post-modern urbanization

Source: Shutterstock

by an engaged but not centralized planning system in much of the global North during the twentieth century, we now have a phase of chaotic capitalism and more emasculated planning systems driven primarily by economic rather than social imperatives (Hall and Hubbard 1996, 1998).

Some of the most detailed explorations of the landscapes of the postmodern city include Mike Davis's and Ed Soja's readings of Los Angeles, a city constructed within urban theory as the archetypal postmodern metropolis (Scott and Soja 1996). Soja has highlighted a number of economic, social and cultural processes that have underpinned Los Angeles' development. Economically he cites the growing significance of new sectors such as producer services, high-technology, cultural entertainment and knowledge-based industries, each of which show distinctive clustering patterns around which new urban forms have emerged. These developments have disrupted the more mono-centric, regular patterns of urban form characteristic of industrial capitalism. Los Angeles is characterized by a number of edge city, or post-suburban, developments, such as those spread across Orange County, which have emerged around these and other new or growing economic sectors. In addition the global connectivity of cities discussed above has seen Los Angeles' functions as a key hub, particularly with links to the Pacific region, underpin growth through an influx of global capital to the city.

Socially Los Angeles has been characterized by heightened levels of, particularly ethnic, segregation and polarization and increasing levels of tension and unrest across its culturally and economically diverse landscapes. Perceptions of Los Angeles as a dystopian realm, while a staple of films such as *Bladerunner* (1982), *Falling Down* (1993), *Escape from LA* (1996) and *Lakeview Terrace* (2008), appear to reflect widely held views of the city by its residents. Increasingly, residential, leisure and commercial landscapes are characterized by 'paranoid' or 'carceral' architecture based on protection, surveillance and the exclusion of perceived threats. Evidence would suggest that, rather than making residents feel safer, such 'fortress landscapes' merely exacerbate levels of segregation between communities and, because they preclude the possibility of cross-cultural mixture and dialogue, heighten perceptions of threat associated with unknown or demonized urban populations (Sandercock 2006). A counterpoint to this dystopianism, and perhaps a response to it, has been the pervasiveness of fantasy and simulation throughout the city's landscape. Originally deriving from the fantasy worlds of expositions, world's fairs and theme parks such as Disneyland (Ley and Olds 1999), they now permeate the city more widely and are found in examples such as themed shopping malls and residential districts (Crawford 2004).

While Los Angeles offers a convincing narrative of postmodern urban development, the question remains, as with all urban models, to what extent does this one case represent a more widespread process or a harbinger of future urbanization? Although scholars have traced the imprint of postmodernity across urban landscapes and forms well beyond Los Angeles (Watson and Gibson 1995; Poulsen *et al.* 2002), there has yet to emerge a consensus around the extent to which it represents an epochal break in the history of urbanization. Tracing its imprint across the urban world remains an ongoing project for urban geographers. Certainly though, the discourse of a uniform postmodernism as an inevitable form and type for cities around the world would appear to be an oversimplification. There is evidence, for example, of significant differences, as well as similarities, between postmodern cities (Poulsen *et al.* 2002), while the interactions of postmodernism with post-colonial urbanization across the global South are as likely to lead to complex patterns of locally mediated outcomes as they are to the emergence of a uniform global postmodern 'blandscape'.

9.3 Contemporary urban issues

The definition and identification of contemporary urban issues varies with scale. The example, mentioned at the start of this chapter, of the issues highlighted by a local newspaper within a single city, points to a range of issues that are primarily reflective of the local urban context. However, the specific minutiae of these issues should not distract us from the fact that they are also reflective of more broadly conceived urban concerns. Examples of minor incidents of anti-social behaviour are constitutive of broader issues of crime, safety and conflict within the city (Valentine 2001). This in turn might speak of the relationships between different groups of people, for example, distinguished on the basis of age, ethnicity or lifestyle. The problems caused by seagulls in the city centre, for example, speak to broader concerns about the relationships between society and nature, the human and the non-human, in cities (Wolch 2002; Hinchcliffe 2007) and perhaps of cities and sustainability. It is to these broadly conceived contemporary urban issues that we now turn. Although different authors might highlight different sets of concerns, we wish to highlight three pressing issues facing the urban world in the early twenty-first century. These are issues of global urban poverty and inequality, cultural diversity within cities, and the relationship between cities and sustainability.

9.3.1 Urban poverty and inequality

There is nothing inherently urban about poverty and inequality (see Chapter 10). However, the urban world is the context within which they are seen at their most intense and perhaps intractable. While poverty here is understood primarily as an economic phenomenon reflecting levels of wealth, development and income its impacts are much wider, with both social and public health implications, for example. Levels of poverty and inequality are both an issue within cities and between cities and are fundamental to many of the problems that they experience.

There are huge global disparities in the equality of income distribution between countries. While the average income of the top 20 per cent in Sierra Leone is roughly 58 times greater than that of the bottom 20 per cent, in Slovakia this figure is only around four times greater (UN Development Programme 2009). Cities tend to reflect these national and regional patterns of wealth and development but it tends to be within cities that the juxtapositions of wealth and poverty are at their most intense and visible. The effects of this inequality are compounded by the failure of governments in countries within which it is at its greatest to intervene effectively and to support their poorest residents. Put simply, where the wealthiest in a country command so much of the money available there is neither sufficient left over to support the poorest, nor often the political will, given the connections that frequently exist between the economic and political elites. This inequality, and the processes that underpin it, produce fractured cities that are very difficult to plan and manage in any meaningful way. They are not good cities.

Manifestations of urban poverty and inequality differ in the cities of the global North and South. In the global South the most archetypal landscapes of poverty are the, often extensive and now long-standing, **informal settlements** (shanty towns or favelas) that ring many cities (Potter and Lloyd-Evans 1998). These are a product of the high levels of rural to urban migration and rapid urbanization in these regions and the inability of formal housing channels to provide enough affordable housing for rapidly urbanizing and largely impoverished populations. The provision of basic services such as clean water and sanitation are often impossible in these settings. While these are typically portrayed through reference to poverty and exclusion, some have seen them as sites of hope that provide self-help solutions or alternative futures where the state, the market and international development policy are not able to provide adequate housing (Keivani and Werna 2001). The modernization of some cities such as Dubai has created huge visual contrasts within their landscapes between the monuments to modernization that have been constructed recently and the camps of the

Plate 9.4 Landscapes of urban inequality. High rise developments and construction workers in Dubai.

construction workers employed to build them (Plate 9.4). The persistence of poverty within many countries of the global South, alongside the embrace of modernization, has produced cities in places such as India whose landscapes, and economic and social geographies can be read as manifestations of ongoing inequality in these countries.

In the global North, postmodern urbanization has been interpreted by many as an urbanization of inequality. The commentaries on Los Angeles discussed above, for example, all highlight the juxtapositions of spectacular wealth and grinding poverty found within it. Inequality lies at the heart of the fear that has been said to permeate residents of postmodern cities which has seen them retreat behind **gated suburbs** and apparently become so reliant on other fortress landscapes. Of widespread concern within debates about contemporary urbanism has been the apparent emergence of an international urban underclass, not only in postmodern cities but also those associated with high levels of immigration and deindustrialization, who have become cut off from routes into the formal economy and who have as a consequence turned to informal and illegal economies, most notably drug dealing, as rational-choice alternatives. The appeal to such evocative labels, which bear the weight of hundreds of years of demonizing the urban poor, carries with it the danger of universalizing conditions that might be particular to only some cities. While groups whose poverty is long-standing and whose ties to the formal economy have been effectively severed do exist,

their experience is not typical of all of the urban poor. Poverty for the majority in the global North is much more mundane than evocations of the underclass would suggest, and involves struggles to make ends meet and struggles over inadequate housing, healthy food, precarious employment and restricted mobility. Certainly the majority would resist having the underclass tag attached to describe themselves and their experiences.

However they may be manifest, poverty and inequality remain some of the most significant and long-standing urban issues. Indeed, many commentators argue that poverty and inequality are increasing as a result of the adoption of a range of **neoliberal** urban policies in cities around the world (Cochrane 2007), which have seen reductions in state support for local populations in the form of cuts to welfare services.

9.3.2 Cultural diversity in cities

The sociologist Sharon Zukin has argued that urban 'publics have become more mobile and diverse' (Zukin 1995: 3). This has resulted in increasing levels of cultural diversity becoming a defining characteristic of contemporary cities. Leonie Sandercock coined the phrase 'mongrel cities' to describe cities in which 'difference, otherness, multiplicity, heterogeneity, diversity and plurality prevail' (2006: 38). This brings with it the potential for innovation, mixing, intercultural dialogue and the emergence of new, hybrid cultural forms. However, it can also pose challenges and give rise to more negative

Plate 9.5 Mongrel cities: The Dominican Day Parade New York City August 2010. Whilst parades like this affirm the cultural diversity of the contemporary city, they do so primarily through celebrations of national identity rather than through inter-cultural dialogue.
(Tim Hall)

outcomes. Sandercock's work starts from the premise that mongrel cities are an undeniable reality and that it is an obligation of both the city-building professionals and citizens to respond to the challenges those cities present.

Many myths and images of multicultural cities are promoted through the marketing campaigns of cities keen to stress the contributions of different groups to their public cultures. By contrast, Richard Sennett (1994) has argued that mongrel cities tend to lack civic culture (Plate 9.5). He argues that these cities are overwhelmingly characterized by tolerance, rather than hostility, between different cultural groups. However, this is a tolerance characterized by indifference towards other cultural groups rather than an active engagement with diversity. It constitutes a form of indifference across which very little inter-cultural dialogue takes place and a form of stasis in which cultural groups are predominantly inward looking. The result is that little common ground emerges between groups, few meaningful cross-cultural institutions develop and little cross-cultural political capacity emerges through which cities can be shaped in desirable ways. The challenge of cultural diversity in cities, then, is how to develop this missing engagement and solidarity between different cultural groups (Colhoun 2002: 108; cited in Sandercock 2006: 39). Sandercock, drawing on the work of Sennett, argues that this involves 'the challenge of living together

not simply in tolerant indifference to each other, but in active engagement' (Sandercock 2006: 40).

Leonie Sandercock, in reviewing the work of social theorists in this area, highlights the contributions of the geographer Ash Amin in seeking to explore routes to more convivial futures for mongrel cities (2002). Amin is particularly critical of several popular policy approaches to developing multiculturalism. Many of these are design-led and revolve around the supposed potentials of the public realm as a resource, shared between cultural groups, through which, with appropriate design, inter-cultural encounter can be fostered (Rogers 1999). Sandercock though, drawing on Amin's findings, offers a sceptical assessment of the utopian potentials of public space.

> The depressing reality, Amin counters, is that far from being spaces where diversity is being negotiated, these spaces tend either to be territorialized by particular groups (whites, youths, skateboarders, Asian families) or they are spaces of transit, with very little contact between strangers. (Sandercock 2006: 44)

Rather than pursuing conventional design-led policies, Amin argued for an alternative focus on the potentials of 'micro-publics', spaces such as workplaces, schools, community centres and sports clubs that provide contexts in which inter-cultural dialogues, although often mundane or 'prosaic negotiations' (Sandercock 2006: 44), can take place. Micro-publics, Amin argues, offer contexts within which cultural differences can be bridged around a recognition of shared interests, presences and common goals.

Whilst keen not to idealize these micro-publics, Amin recognizes that inter-cultural dialogues will not just happen in these spaces. Indeed within many existing micro-publics such dialogues are not typical. Rather these potentials are something that often need to be shaped and fostered. It is easy to be sceptical of the potentials of micro-publics in this regard. Multiculturalism is not necessarily a characteristic of many sports clubs, for example. It is common in many cities globally to see sport organized formally or informally, to some extent, along ethnic and national lines, as with the case of soccer associations for Latin Americans in US cities (Hamilton and Chincilla 2001: 169) and Asian cricket teams and leagues running in parallel to the official league structure in British cities such as London and Birmingham. This is not to say that there is no integration of these communities within mainstream leagues and teams, rather that they are formed as a result of perceptions by some players of exclusion from mainstream clubs, along with the costs of membership. They may also serve as ways of fostering social contact and cultural and national solidarity amongst members. However, given the failure of conventional policy routes towards multiculturalism these settings offer potential alternatives through which to incubate future policy. It would be naïve, though, to pretend that they are any more immune to the politics of indifference, hostility and nationalism than any of the city's other spaces and settings. The challenge, then, is both how desirable dialogues and encounters might be fostered within these micro-publics, and how they might feed into macro-public civic and political movements shaping more convivial futures in mongrel cities.

9.3.3 Cities, sustainability and resilience

The ecological crises of the current age are well known. They are discussed elsewhere in this volume in some detail (Chapters 5, 6 and 7). The challenges posed by dwindling global supplies of accessible oil and climatic chaos are global in their extent and fundamental in their significance to the planet (Atkinson 2007a, 2007b, 2008; Newman *et al.* 2009). The relationships between cities and these ongoing crises are becoming an ever greater concern of urban scholars, as witnessed by the growth of literature on cities and sustainability since

early key publications from the mid 1990s (Haughton and Hunter 1994). Concern here is two-way involving both the ways in which cities pose a threat to the environment and the ways in which the environment poses a threat to cities. The latter is evidenced by the recent growth in the urban resilience literature (Coafee *et al.* 2009; Newman *et al.* 2009). The former concern involves both the mapping of the contributions of cities to these crises and outlining possible solutions to them while the latter is concerned primarily with the ways in which cities might build resilience to crisis into their infrastructures and social and economic structures and practices. There is overlap between these literatures as the latter is concerned both with reactive measures to mitigate the impacts of crises but also the development of long term practices and processes that may contribute to the resolution of ecological crises, or at least their reduction and management.

Without doubt, given that over half of the world's population now lives in cities, the rates of future growth of the world's urban population and the ecological demands made by cities and their populations, the urban world will be the key site across which the impacts of these ecological crises are felt and the key site in any possible resolution of them. It should be remembered that when scholars of the city discuss sustainability, crisis and resilience, their discussions are not restricted to matters of environmental science. Human geographers have key contributions to make to these debates, since inequality as highlighted above, for example, is an inherently unsustainable condition. It compromises resilience as the impacts of ecological crises are felt unevenly across social groups, the poor being less able to insulate themselves from them. Further, it undermines attempts at collective solutions to ongoing problems of environment and sustainability. The debates around the relationships between cities and ecological crises have encompassed many dimensions including consumption patterns and resource use amongst urban populations (Rees and Wackernagel 2008), sustainable urban size and form (Breheny 1995), the roles of neighbourhoods and green spaces in promoting sustainable development (Carley 1999; Chiesura 2003), transport and mobility (Hall 2003) and encounters with nature in the city (Hinchliffe *et al.* 2005; Benton-Short and Short 2008). Just as many sub-disciplines of human geography have been reshaped as they have undergone a cultural turn in recent years, it is increasingly likely that urban geography will be reshaped as it undergoes an ecological turn in the near future. The imperatives for this could not be greater. Atkinson, in a series of essays (2007a, 2007b, 2008), argues that the current crisis will lead to no less than the collapse of modern civilization as we now know it. Ultimately, he predicts:

> It is not at all clear how fast and through what stages the collapse will unfold because there are many variables which will interact differentially and depend crucially on political decisions taken – and possibly major conflicts – along the way; however, we can be sure that in general the decline will be inexorable. By the latter decades of this century, a radically altered world will have emerged, with a greatly reduced population living surrounded by the defunct debris of modernity, comprised of fragmented and largely self-reliant political entities. Our complex, 'globalized' world of megastates and technological hubris will be but a fading memory. The impacts of global warming and other environmental legacies of our age will reduce the options for reconstruction, possibly fatally. (Atkinson 2008: 79)

The challenge facing urban dwellers, urban managers and urban geographers in the future is to try to ensure that Atkinson's and other gloomy prognoses for the future of the urban world are not realized.

9.4 Conclusion

The urban world, as we have seen, is big, dynamic, diverse and complex: a potentially intimidating topic of study. However, it is also a fascinating, rewarding and important topic for which there are economic, social and ecological imperatives that demand it is better understood and managed.

This chapter has attempted to provide a broad framework through which to approach cities. This involves, at its heart, the resolution of the small and immediate concerns of the urban here and now with the broader trends, patterns and issues of which they are part.

Urban geography and urban geographers at all levels have important contributions to make to the future resolution of the urban issues identified at the end of this chapter and many more besides. The twenty-first century will certainly be an urban century, but how urban systems and cities will change, and what these will look like at the end of the century, will continue to pose intriguing questions for urban geographers to address.

Learning outcomes

Having read this chapter, you should be able to:

- Appreciate the nature and diversity of the urban world. You should be able to recognize the characteristics of cities, the nature of urbanization, the diversity of global urban forms and their characteristics.

- Recognize the difficulties inherent in trying to define and represent the urban, either within scholarly discourses or through systems of administration, planning and management.

- Think geographically about cities as socio-spatial entities. Cities are the products of social relations. This is reflected both in the relationships within cities, for example between different groups of people, and between cities. Understanding the nature and effects of these relationships is an important aspect of differentiating cities.

- Appreciate the range of issues that face cities in the present era and the ways in which these can vary between different parts of the world and between different types of city. In addition it is important to critically evaluate the different routes to the resolution of these problems that have been advocated.

- Understand the different ways in which cities have been conceptualized in geographical and sociological thought. Scholars of cities have developed many different takes on the city, approaching it from different angles and studying different aspects of it. Evaluating the strengths and weaknesses of these different approaches is a key skill for the urban geographer.

Further reading

Amin, A. and Thrift, N. (2002) *Cities: Reimagining the Urban,* Polity, Cambridge. An important contribution to recent debates about the city. Challenges the tendencies within urban geography to see the city as a bounded entity and to focus on the apparently 'big' issues at the expense of the everyday experience. Offers fresh ways of looking at the contemporary urban world.

Hall, T. and Barrett, H. (2011) *Urban Geography*, 4th edition, Routledge, Abingdon. An accessible guide to the subject that focuses particularly on the recent history of cities. Contains a number of student-friendly case studies, examples and exercises.

Hubbard, P. (2006) *City*, Routledge, Abingdon. A panoramic sweep over the contemporary urban world viewed through the lens of urban theory from geography and beyond. An important critical synthesis of ways of thinking about and looking at the city. Combines a historical sensitivity with a keen eye on the city as it is today.

Latham, A., McCormack, D., McNamara, K. and McNeill, D. (2009) *Key Concepts in Urban Geography*, Sage, London. A series of engaging, concise essays on a range of key issues and concepts. Critical, occasionally quirky and contemporary in its organization, approach and outlook. Acts as both a key starting point for the novice urban geographer and a handy source of reference.

LeGates, R.T. and Stout, F. (eds) (2007) *The City Reader*, 4th edition, Routledge, Abingdon. A landmark collection of key writings on the city from the nineteenth century to the present day. An authoritative collection that presents key articles along with useful biographical essays on the authors and overview of their career and work. An essential resource for the serious urban geographer.

Pacione, M. (2009) *Urban Geography: A Global Perspective*, 3rd edition, Routledge, London. The most comprehensive urban geography textbook currently available. Covers cities from the developed and developing world through a series of well organized chapters filled with a wealth of references, examples and illustrations.

Robinson, J. (2006) *Ordinary Cities: Between Modernity and Development*, Routledge, Abingdon. Offers a seri- ous challenge to the tendency of traditional urban theory to universalize the experiences of a small number of cities and fail to engage with the complex diversity of the global urban world. Argues from a post-colonial standpoint for a cosmo- politan urban theory that is able to embrace and include global urban diversity. An important critical intervention in thinking about the city.

Browse the journals *Urban Studies* or the *International Journal of Urban and Regional Research* for many urban case studies.

Useful websites

www.unhabitat.org The site of the UN Habitat programme concerned with shaping better futures for cities.

www.lboro.ac.uk/gawc The home of the Global and World Cities Network at the University of Loughborough. A wealth of data and reports on the networks of global cities.

www.city-data.com Rich data archive for US cities.

www.cabe.org.uk/sustainable-places Commission for Architecture and the Built Environment – UK. Site for the promotion of sustainable architecture and design.

maps.google.com Zoomable maps, satellite imagery and street-level panoramas.

www.thisisgloucestershire.co.uk The home of Gloucester's *Citizen* local newspaper mentioned in the chapter. Check out the latest on the seagull infestation and other stories concerning Gloucester's citizens. Most other towns and cities will have their equivalent newspaper/site.

For annotated, clickable weblinks and useful tutorials full of practical advice on how to improve your study skills, visit this book's website at **www.pearsoned.co.uk/daniels**

SOCIAL INEQUALITIES, SPATIAL DIVIDES

Chapter 10

Phil Hubbard

Topics covered

- The socio-spatial dialectic
- The connections between social exclusion, poverty and marginalization
- Stereotyping, representation and the exclusionary urge
- Racial exclusions: places and spaces of ethnicity
- Gentrification as purification strategy

All societies are riddled by inequalities, some of which are more fundamental than others (for example, a person's skin colour, age, or religion typically plays a critical role in shaping their status in contemporary society, whereas their hair colour or shoe size tends to be less significant). Quite why some social differences matter more than others is a key debate in the social sciences – one to which geographers contribute in a variety of ways. One contribution involves forms of 'mapping' that show that different social groups are differently distributed in space. Such mappings typically demonstrate that particular social groups, defined by age, sexuality, class, gender, religion or ethnicity, dominate in particular areas, often to the exclusion of other groups. Census of Population sources, for example, can be mapped to show that the city is far from an undifferentiated social space, but is a complex 'patchwork' in which different social groups take their allotted place: while few residential streets or neighbourhoods are homogeneous in their social make-up, there appear to be strong tendencies towards clustering, segregation and socio-spatial sorting. If anything, these tendencies are becoming more clear-cut over time: in the UK, for example, it is now easier than ever to typify some areas as young or old, black or white, settled or migrant (Dorling and Rees 2004). When other variables are bought into play – such as economic measures of wealth, income and employment – it becomes obvious that such patterns reflect the differential social status of these groups, with those with privileged social identities tending to be found in more valued locations (something reflected in the house prices in different areas).

Through such contributions, geographers have demonstrated there is a definite, if complex, two-way relationship between society and space, with spatial orderings reflecting social orders – and vice versa. This reciprocal relationship – the socio-spatial dialectic – constitutes the primary focus of *social geography,* a sub-branch of the discipline concerned with elucidating the spatial imprints of social relations and, conversely, demonstrating the centrality of space in social life. This chapter explores some of the key concepts in contemporary social geography. Rather than focusing on social elites, it does this by focusing on the geographies of the least powerful and advantaged in society. As this chapter will demonstrate, one of the clearest ways that the social disadvantage of such groups is expressed is through its geographical manifestation in the form of spaces described as being 'on the margins' – e.g. 'slums', 'problem estates' or 'ghettos'. For those who do not live in such spaces (and this inevitably includes most social geographers), these areas can often be spaces of fear and fascination because they represent spatial concentrations of those who fail to match mainstream ideas of how people should live and work. As this chapter will demonstrate, the stigma surrounding these places compounds their poverty and isolation, as once a place is labelled as degraded and deprived, its residents may find it hard to be accepted by mainstream society. Underlining the fact that space plays an *active* role in shaping people's standard of living, this chapter's focus on geographies of segregation and exclusion demonstrates why a sensitivity to place matters, and illustrates some of the ways social geographers contribute to pressing debates on social inequality.

10.1 Poverty and marginalization

While social geography has always had a strong interest in mapping social groups and their location in space, for many social geographers this is a means to an end: namely, highlighting the importance of spatial processes in perpetuating social inequality (Chapter 9). Social geography hence circles around questions of the quality of life that people experience in different spaces. In simple terms, quality of life can be defined as the extent to which our needs and desires (be they social, psychological or physiological) are fulfilled. Though this can only truly be measured through subjective assessments of the extent to which individuals are satisfied with their own life, many studies by geographers have reached consensus that it is easier to attain an acceptable or good quality of life in areas that offer ample economic opportunities, good social facilities, and a quality environment. Given this, we can see that our life chances are closely related to our place of residence. One well-known aspect of this is the so-called 'postcode lottery' around health services in the UK: depending on where you live, you will enjoy very different levels of medical provision through the National Health Service. Where you are born, and where you live, thus has a major bearing on where (and how) you die.

The same is true of prosperity (or lack of it). Geographical studies of wealth and income suggest that we should not just think about places with concentrations of rich people, but should recognize the existence of *rich places*. This distinction is subtle but significant, for it suggests that wealth is amplified in some places. To elaborate: people who are born in places characterized by high concentrations of affluence are more likely to be wealthy themselves because the concentration of money in the area means standards of education, health care and employment are relatively good. While house prices are high, housing standards are good, and may be an important source of inherited wealth. As Dorling (2004) explains, there is a circularity here whereby affluence breeds affluence.

Surprisingly, geographers have said little about geographies of affluence, despite the obvious relationship that exists between income and quality of life (Beaverstock *et al.* 2004). In contrast, much has been said about geographies of poverty, with social geographers often drawn towards this topic because they feel their relatively privileged position might allow them to draw attention to – and perhaps improve – the life chances of those living in poor areas. Poverty can be defined as the condition where individuals or households are unable to afford what might be perceived to be the normal necessities of life. Inevitably, judgments of what we need to consume to survive vary across time and space, meaning geographers are generally more interested in *relative* rather than *absolute* poverty. On a worldwide scale, for example, it is evident that indicators such as average income per capita provide only a very superficial insight into experiences of poverty in different nation-states (see Chapter 8, pp. 180–4). Relative income measures – e.g. the definition of poverty as being 60 per cent of median equivalent income – are somewhat more meaningful, but fail to take into account what that income can secure in terms of goods, services and the necessities of life. Consequently, recent attempts to measure poverty have generally defined people as being in poverty if they lack the financial resources needed to obtain the living conditions that are customary and 'normal' in the society to which they belong. Identifying such thresholds does not suggest that everyone's needs are the same, but does suggest an income level that is socially unacceptable for any individual to live below. For example, the Millennium Poverty Survey suggested that 26 per cent of the British population were living in poverty in 1999 on the basis that their income would not secure at least three of the 35 items considered by at least 50 per cent of the British population as necessary for maintaining an acceptable standard of living. While some of these items may not have been considered a necessity in previous decades (and others remain luxuries that many in the global South can only dream of), comparison with previous studies suggests the incidence of poverty has actually risen sharply. In 1983, 14 per cent of British

households lacked three or more necessities because they could not afford them, a figure that increased to 21 per cent in 1990 and over 25 per cent in 1999 (Pantazis *et al.* 2006; see also Dorling *et al.* 2007).

Increases in costs of living clearly have an impact here, with decreases in poverty in the early 2000s seen to be reversing in more recent years as disposable incomes failed to keep pace with rises in fuel costs, food and housing. In 2010, the minimum weekly income for a single person of working age was suggested to be around £227, rising to £673 for a couple with two dependent children (Davis *et al.* 2010). Roughly translated, such figures suggest that nearly 13 million people in the UK cannot afford what are considered acceptable standards of living. While such deprivation is particularly pronounced among those who are out of work, many of those in the lower strata of the labour market are also living in poverty (the persistence of low-paid, part-time or precarious forms of employment being significant here, alongside the restructuring of welfare provision for those on low incomes). Coupled with the rapidly rising incomes characteristic of managerial and professional occupations (particularly in 'knowledge-rich' sectors such as law and finance), the consequence is an increasingly polarized society. For example, between 1979 and 1991, the average British income grew by 36 per cent, yet for the bottom tenth of the population it dropped by 14 per cent. While the rate of income polarization slowed in the 1990s – partly because of welfare reform – the gap between the 'haves' and 'have-nots' has accelerated in the twenty-first century (Dorling and Rees 2003, 2004). By 2008/09, the average income of the top fifth of households was around 15 times greater than for those in the bottom fifth (£73,800 per year compared with £5,000) (see Figure 10.1) and most measures of income inequality suggest the UK is more unequal than at any previous time in the last 30 years.

While some nations do not exhibit the levels of social polarization apparent in the UK, it is important to note that levels of social inequality evident in many other nations are far in excess of this (see Spicker 2006). Moreover, in each and every case this social inequality is spatially expressed, whether as a division between rich and poor regions, an urban–rural divide or differences in wealth between cities. Yet, some of the sharpest contrasts between wealth and poverty are found *within* cities, and one of the major contributions social geographers have made to debates surrounding poverty is to draw attention to areas of acute urban need (see Chapter 9, pp. 213–14). According to the

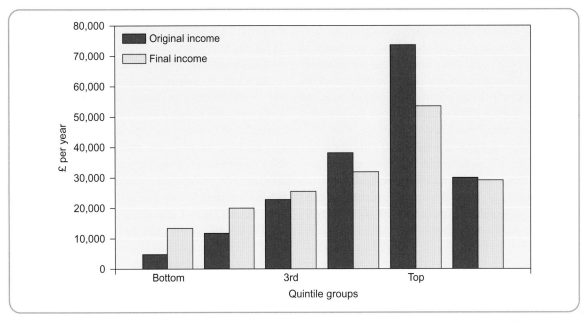

Figure 10.1 Average original and final (i.e. after taxes and benefits) income per household by quintile, UK, 2008–9.

Source: National Statistics Online (2009)

United Nations (2001), over one billion of the world's population lived in 'slum' urban environments without access to safe water, acceptable sanitation and secure, tenured housing of an acceptable standard. Ninety-five per cent of these concentrated slum environments are found in the 'Third World', and it has accordingly become something of a geographical cliché to juxtapose the assuredly affluent 'Westernized' city centres now typical of many of the world's rapidly expanding cities with the 'slum' dwellings that are often just a stone's throw away. For instance, profiling Jakarta in Indonesia (population 9 million), Cybriwsky and Ford (2001) contrast the city's Golden Triangle of prestigious residential districts (e.g. Cikini, Kuningan and Menteng) with the *rumah liar* ('wild houses') characteristic of its sprawling, chaotic *kampungs*. They thus conclude that Jakarta 'has extraordinary contrasts between the worlds of prosperity and poverty, and significant challenges ahead for continued development as a global metropolis' (Cybriwsky and Ford 2001: 209).

Yet such 'extraordinary' contrasts between landscapes of wealth and poverty are not just restricted to mega-cities or those cities living with the legacy of colonial rule: for example, while London remains the most deeply divided city in Britain in terms of income disparities, most British cities display obvious contrasts between the affluent landscapes of the central city and the 'pockets' of deprivation that surround it. Notable here are those peripheral areas of 1960s local authority housing, which, in the 1980s and 1990s, came to be characterized by high rates of economic inactivity. While these are not 'slums' according to the UN's definitions, evidence from 20 of these 'unpopular' housing estates suggests they possess unemployment rates, on average, two and a half times higher than the surrounding urban area (Power and Tunstall 1997), representing pernicious spatial concentrations of relative poverty (Table 10.1). In 2010, more than half of those living in the socially rented housing typical of such estates had less than 60 per cent of median income, and could thus be defined as living in poverty (Department of Work and Pensions 2009).

The particular combination of problems encountered in such deprived estates created national headlines in the 1990s as confrontations between police and local youths (especially joyriders accused of stealing cars) escalated into fully fledged riots. The spatial concentration of unemployment on these estates, especially among young men, has been hypothesized as the most important underlying cause of these disturbances. Figures for the 13 estates affected by serious disorder between 1991 and 1992 suggested that nearly 80 per cent of their occupants were either economically inactive or unemployed

Table 10.1 Social characteristics of 12 'riot' estates and 20 'unpopular' estates from 1991 UK Census of Population

	British average (%)	20 'unpopular' housing estates (%)	12 'riot' housing estates (%)
Unemployment	10	34	31
Economic inactivity	36	44	45
Population under-16	19	31	31
Population under-24	33	46	48
Lone parents	4	18	15
Ethnic minority population	6	26	11
Pupils aged 15+ gaining 5+GCSEs (grade A–E)	43	20	20

Source: A. Power and R. Tunstall, *Dangerous Disorder: Riots and Violent Disturbances in Thirteen Areas of Britain, 1991–92*, Joseph Rowntree Foundation, 1997. Reproduced by permission of the Joseph Rowntree Foundation

(Table 10.1). A simultaneous withdrawal of economic infrastructure – including businesses, shops and banks (Case study 10.1) – meant that there was little money circulating locally. The mythical black market (an informal economy based on petty theft, drug dealing or cash-in-hand trading) did not offer any sort of escape for the majority of residents, given that black markets rarely flourish when there are few people who can afford to buy stolen goods, let alone legitimate ones (Green 1997).

Even so, it has been suggested that the spate of vehicle thefts occurring on such estates was not primarily

Thematic case study 10.1

Geographies of institutional disinvestment

During the late 1990s, debates on poverty began to focus on the role institutional disinvestment played in exacerbating the problems encountered by those living in 'poor places'. One dimension of this was the withdrawal of financial services from poorer areas, with as many as 15 per cent of bank branches in the United Kingdom closed down in the early 1990s, the majority in low-income, inner-city areas (Leyshon and Thrift 1994). Another was the closure of once-vibrant neighbourhood shops as retail chains concentrated on the development of large superstores catering to mobile and affluent consumers (Williams and Hubbard 2001). Given that many of these were located in out-of-town and edge-of-town locations, this left many neighbourhoods described as food 'deserts', where a range of affordable and varied food was available only to those who had private transport:

> Food deserts . . . are those areas of cities where cheap, nutritious food is virtually unobtainable.

> Car-less residents, unable to reach out-of-town supermarkets, depend on the corner shop where prices are high, products are processed and fresh fruit and vegetables are poor or non-existent.
>
> (Laurence 1997)

This has raised serious concerns about the health of those in poor places, with the lack of fresh fruit and vegetables and a reliance on fatty, fast foods implicated in the making of significant health inequalities. Confirming the intensity of food-access problems in some of the large local authority housing estates in British cities, Wrigley *et al.* (2003) noted significant improvements in diet following the construction of a new superstore in the Seacroft district of Leeds, an area that had previously suffered considerable retail disinvestment. This suggests that large-scale corporate intervention can significantly improve retail access and diet for people living in stigmatized neighbourhoods, albeit that small-scale, community-based initiatives might also provide a solution in some cases (Cummins and MacIntyre 1999).

economically motivated, as Campbell's (1993) vivid account of the Blackbird Leys estate, Oxford, suggests:

> The lads had stopped making cars, but that did not stop them stealing them. The collapse in manufacturing work for men was succeeded in Oxford by the rise in car theft, a crime that was emblematic of the Eighties and Nineties. . . . Car crime on the estate was about much more than trade, however. It was about a relationship between young men, power, machinery, speed and transcendence. . . . It was the spectacle of the displays that vexed and humiliated the police, uniting thieves, drivers and audiences in alliance against the authorities.
>
> (Campbell 1993: 32–3)

The idea that car crime embodies a form of resistance suggests joyriders were not motivated simply by greed or want, but by a more complex desire to reassert their masculinity by confronting authority. Such actions can, however, only be understood in the context of the economic transitions that led to significant changes in the relationship between masculinity, work and domesticity on such estates. Here, Nayek (2003) suggests that the disappearance of a once-familiar gender division of labour organized around a male 'breadwinner' (and the expectation of women's domestic role) was the cause of much male anxiety in areas of high unemployment. With the shift to a service-based economy valuing 'feminized' attributes such as keyboard skills and communication proficiency over the robust 'masculine' qualities associated with the culture of manual labour, young men unwilling (or unable) to adapt to demands of service-based work found themselves increasingly cut off from the world of work. Without a regular source of income, or any real incentive to look for work, the traditional relationship between masculinity and spatiality broke down as these young men rarely left the estate for any particular reason. Yet, in the relatively secure space of a stolen car, Massey (1996) argues, masculine identities could be reasserted. Driving with speed – 'on the edge' – thus represented young men's attempt to reaffirm their masculinity, albeit in a manner that created conflict with other estate dwellers (as well as police).

While geographers have stressed that joyriding and other uncivil actions are intimately connected to (gendered) processes of economic restructuring and institutional disinvestment, the media largely ignored such arguments in favour of a rhetoric depicting Britain's peripheral estates as breeding grounds of immorality, characterized by high rates of teenage pregnancy, widespread drug abuse, alcoholism and antisocial behaviour:

> The sense of decline and neglect in many of these areas is palpable: the built environment in many of these areas has now taken on all the classic, ominous characteristics (boarded-up windows, barbed wire surrounds) of the enclaves of high crime and violence associated with Los Angeles and its ghettos in the months leading up to the 1992 riots. Public space is often colonized by young men in baseball caps and cheap khaki.
>
> (Taylor 1997: 6 © Guardian News & Media Ltd 1997)

While there are as many as 2,000 social housing estates in Britain, accounting for around two million people in total (Goodwin 1995), the term 'no-go' estate became a convenient way for the media and public alike to label a wide variety of social environments (only a small number of which experienced unrest). Such labelling exacerbated neighbourhood decline: those who could, moved out, leaving behind only the most vulnerable (Hastings and Dean 2003).

From a parlous position in the 1990s many of Britain's most unpopular estates improved in terms of employment, education and health, thanks principally to substantial government investment (Tunstall and Coulter 2006). Yet some housing estates remained shunned by both state and market, blighted by a debilitating cycle of labelling and marginalization. In the words of Lynsey Hanley (cited in *The Guardian*, 12 August 2011: 32) these estates remained surrounded by an 'invisible forcefield' that asks outsiders why they might want to enter, and insiders why they might want to leave: in the aftermath of the rioting and looting of summer 2011, much was hence made of the deprivation, hopelessness and criminality that is associated with living in places which are prevented from becoming integral parts of the local social fabric. A related example here is provided by the *banlieues* (suburbs) of social housing surrounding Paris and other major French cities. In October 2005, three young men in one of these *banlieues* – Clichy-sous-Bois (NE Paris) – sought to evade a police check (a regular occurrence in the *banlieues*) by hiding in an electricity substation. Two were electrocuted, prompting an anti-police protest that rapidly spread across the country. In over a week of riots, 10,000 cars were torched and 3,000 people arrested. Video footage of burning cars was screened nightly by the world's media, prompting commentators to speculate why such scenes were happening. Whilst some sections of the press suggested these were simply copycat protests in which young men wanted to appear

Plate 10.1 *Banlieue* riots.
(Associated Press/PA Photos)

on television, more trenchant critiques suggested that at their source were structural inequalities that needed to be understood in geographical and historical context. For, as Dikeç (2006a) argues, the *banlieues* are not all deprived (and some are reasonably affluent). Yet they have always had negative connotations, identified as '*les quartiers difficiles*' which contain the social problems that have, by and large, been cast out of the central city. At the turn of the millennium, these areas of social housing possessed unemployment rates twice the national average, with 40 per cent of young people unemployed. The fact that some of these areas of social housing were explicitly designed to house North African immigrants in the 1960s is also highly significant here, given that the *banlieue* has always been understood as a 'problem area' (i.e. designed to solve the problem of housing new migrant labour). Writing before the riots, Tissot and Poupeau (2005: 5) argued that when one talks of poverty in France, one can talk of the 'socially excluded', the 'immigrants', or even of 'the young', but it is easier for reports to dwell on the places where it is found: '*les banlieues*'. Post-millennium, and in a context of considerable Islamaphobia in the West, the problem of the *banlieue* was transformed from one where inhabitants of the *banlieue* were a problem for one another to one where they were imagined as a threat to the city (and the nation) as a whole.

The connections made here between ethnicity and space are significant, and we will return to these subsequently. At this point, it suffices to say that the white-dominated media in France could not accept that the protest was one in which French citizens were justifiably declaring their anger at the deteriorating housing, lack of transport and poor schooling that characterized their *banlieue*. Rather, this was depicted as a riot in which 'outsiders' from the margins sought to attack the centre (Dikeç 2006b). Such arguments support Shields' (1991: 5) view that marginal places 'carry the image and stigma of their marginality which becomes indistinguishable from any empirical identity they might have had'. In this sense, it is clear that media stereotypes play a crucial role in creating and perpetuating social and spatial inequalities.

10.2 Social exclusion and cultural stereotypes

The fact that the media exacerbated the deprivation experienced in the *banlieue* and outer city estates by drawing broad-brushed stereotypes of place suggests that when we explore social inequality, we need to think carefully about the way representations of people and place entwine. This involves consideration of cultural issues (see Chapter 13), and shifts attention from the economic manifestations of inequality to questions of why some groups become identified as threats or problems for mainstream society. No matter how widespread the rhetoric that we are all created equal has been, historically it is difficult to identify a society that has not made distinctions between the social mainstream (or norm) and those Others who seemingly threaten the coherence of society. The notion of a scapegoat is certainly an ancient one, and history can tell of many groups – e.g. Jews, prostitutes, homosexuals – whose lives have been made intolerable because they are seen to pollute the body politic. Such groups have often been cast out, physically, as well as metaphorically, socially excluded on the basis that they disturb social and spatial orders. In situations where they have been allowed to remain, their occupation of space has often been fiercely contested: in many contexts, the socially excluded have been unable or unwilling to occupy the spaces associated with mainstream society, carving out their own geographies on the margins.

Writing in the context of the urban West, Winchester and White (1988) suggest that socially excluded groups include the unemployed, the impoverished elderly, lone-parent families, ethnic minorities, refugees and asylum seekers, the disabled, illegal immigrants, the homeless, sexual minorities, prostitutes, criminals, drug users and students. If you are reading this book, the chances are that you are a student. If so, the identification of students as an excluded group may be causing you some puzzlement. Even though some students are highly indebted, may come from poor backgrounds or have to take part-time jobs to fund their (supposedly) full-time studies, few live in the conditions of poverty described above. In fact, in the UK the average student is from the South-east (the most affluent part of the United Kingdom) and has parents who are part of the professional or managerial class (Dorling 2004). This means most students are able to afford most things identified as adult necessities (and even when they are unable, they may be assisted through hardship grants or student loans). So why is it appropriate to describe students as an excluded group?

The answer is that students are identified as a distinctive and different group because their lifestyles are often perceived to lie outside the norms of society, making them infrequent or unwelcome visitors in many of the spaces that are the loci of mainstream social life. Historically, this is related to popular understandings of students as politically radical, embracing alternative fashion and music in a way that marks them out as part of the 'counter-culture'. In the contemporary context, it is also associated with stereotypes of students which suggest they transgress the established social boundaries between work and play (and day and night) by holding late night parties (at times when other people are sleeping) and sleeping all day (when other people are working). Now, as you read this you will no doubt argue that the reality of the situation is a good deal more complex: not all students enjoy drinking, clubbing and partying to all hours – and those that do are often able to do so while devoting considerable time to their studies. Furthermore, while some students remain deeply committed to issues such as environmentalism, animal rights or the restructuring of Third World debt, stereotypes of student radicalism, drug-taking and 'free love' which were consolidated in the late 1960s are now woefully inaccurate as descriptors of contemporary student cultures.

That said, it is possible that you may have experienced exclusion in your own life as a student. No doubt there are parts of the town or city where you study where you are told students are not welcome and where you may have felt 'out of place'. For example, some pubs or clubs let students know, in subtle or not-so-subtle ways, that they are not welcome. You may even know of instances where students have been verbally abused (or even attacked) just for being students. Thankfully, such instances are rare. In contrast, opposition to studentification – the increased student occupation of the local housing market – is becoming very common. This phenomenon is mainly limited to university towns in the UK where on-campus accommodation is not sufficient to house burgeoning numbers of students and where particular areas become associated with high levels of rental accommodation targeting students (Smith and Holt 2007) (Plate 10.2). Though student occupation can increase house prices (with the buy-to-let market proving extremely lucrative), student housing is often opposed by long-term residents who blame it for neighbourhood decline. In some areas, the consequences of studentification have been reported to be nothing short of catastrophic, with lobby groups arguing for stronger controls on the licensing of student housing to prevent neighbourhoods becoming 'over-run' by young, transient populations, displacing longer-term residents (UK Universities 2005). Highly dramatized media stories evoke the environmental and social transformation occurring in districts undergoing student expansion, illustrated with images of overgrown gardens, rubbish left out and sheets used as makeshift curtains. Additionally, campaign groups opposing student occupation allege students cause noise and nuisance, displace local long-term residents and fail to contribute to community life (Hubbard 2008, 2009).

We can see here that media stereotypes of students and student housing are mutually reinforcing in a variety of ways, and certainly many opponents of student housing allege that students do not look after their house because they are too busy socializing. A counter

Plate 10.2 A studentified landscape of rental housing in multiple occupation, Loughborough, UK.

argument here is that the deleterious state of many student houses is down to the absentee landlords who fail to look after them, and, in any case, not all student households are noisy and disregarding of their neighbours. No matter, for all it takes is one or two instances of antisocial behaviour to reinforce popular stereotypes and instigate campaigns aiming to reduce student occupation in particular areas (such as Storer in Loughborough, Headingley in Leeds, Elvet in Durham and Selly Oak in Birmingham). In some cases, long-term residents even adopt the metaphor of 'studenticide', arguing that studentification needs to be reversed lest it destroys the urban social fabric: surveys suggest homeowners in UK cities now fear student neighbours more than squatters.

This discussion of studentification is highly relevant to our discussion of social exclusion, given it demonstrates that some groups may experience stigmatization irrespective of their income or class. Yet talking about students as a marginal group remains problematic given that students occupy their ambivalent social position for a short time only, and typically move from a marginalized and indebted position to one of relative affluence within a few years as they convert their education and cultural capital into financial rewards. In other instances, socially excluded groups suffer social stigmatization that is more likely to be associated with long-term financial exclusion or deprivation. For example, Gleeson (1998) suggests that people with a physical impairment or mobility problem typically find themselves excluded from workplaces designed around an able-bodied ideal. Here, the portrayal of

the disabled as representing what Shakespeare (1994) describes as the 'imperfect physicality' of human existence is a major factor, with the able-bodied remaining anxious about those who are visibly different.

Nonetheless, our consideration of studentification (and the associated development of student 'ghettos') shows us that highly diverse groups can become known through narrow social stereotypes, with these stereotypes shaping the relationship between them and dominant social groups. This suggests that social exclusion has an important symbolic dimension, being created in the realms of *representation*. Here, the term representation is taken to encompass the wide range of media – such as films, TV, internet sites and newspapers – through which we come to understand the world and our place within it (Woodward 1997). Inevitably, such media provide a partial and simplified view of the lifestyles of heterogeneous social groups. Although most people do not necessarily accept these stereotypes uncritically, they inevitably find themselves drawing on them in their everyday life. What is especially significant about these stereotypes is that they are ideological in nature, in the sense they are generally created by (and in the interests of) dominant social groups – typically white, able-bodied, heterosexual, middle-class men. It is those who do not conform to their view of the world who are society's Others.

A vivid example of how such stereotyped images contribute to social and spatial exclusion is provided by the media reporting of HIV/AIDS that occurred in the mid-1980s. In the ensuing moral panic that followed diagnosis of the virus and its mode of transmission, a sometimes hysterical media began to focus on the groups most readily identified as at risk – non-white ethnic minorities, intravenous-drug users and gay men (Watney 1987). The presence of HIV infection in these groups was generally perceived to be no accident, but portrayed as a condition affecting those whose 'inner essence' diverged from that of normal society (and whose lifestyles were judged to be incompatible with the maintenance of 'family values'). Hence, sensationalist stories of sexual immorality, irresponsibility and wilful hedonism among those groups most affected by the virus were used to create social barriers between the 'healthy' and those regarded as sexually promiscuous, socially irresponsible and unclean. Consequently, this symbolic marking of HIV-infected groups as deviant informed their social exclusion, encouraging widespread discrimination, prejudice and neglect. As Wilton (1996) has shown, this had clear spatial effects, with those infected with HIV exhibiting 'diminishing geographies' as their access to the workplace, the home and the street was subject to increasing constraint. In his analysis, Wilton stresses that these constraints were not a product of the physical onset of AIDS but of the social stigma surrounding it. Symptoms of this stigma have included employers refusing to take on those with HIV, dentists refusing to treat them and, as Wilton relates, community groups opposing the construction of AIDS hospices in their neighbourhood.

Such neighbourhood opposition to community facilities and welfare services is so widespread in Western society, particularly in suburban landscapes, that few stop to question why people might object to such developments in their neighbourhood. Certainly, many NIMBY (Not In My Back Yard) campaigns are fought with reference to detrimental environmental impacts (such as noise or air pollution during construction). Such impacts are, to an extent, quantifiable; what is less measurable is the concern that homeowners have about the arrival of stigmatized populations. However, for Takahashi and Dear (1997), community opposition to facilities for those living with AIDS/HIV is indicative of the more general antipathy displayed towards Other populations. Their survey of homeowners in US cities reveals a 'continuum' of acceptance, where facilities for populations depicted as 'different' (such as homes for the elderly) are regarded more favourably than those for populations stereotyped as 'dangerous' or 'deviant' (such as those living with HIV) (Table 10.2). This frequently results in the concentration of facilities in inner-city areas where home-ownership rates are

Table 10.2 Relative acceptability of human service facilities, based on a US survey of 1,326 respondents

Facility type	Mean acceptability, 1 = low, 6 = high
School	4.75
Day care center	4.69
Nursing home for elderly	4.65
Medical clinic for allergies	4.40
Hospital	4.32
Group home for mentally retarded	3.98
Alcohol rehabilitation center	3.80
Homeless shelter	3.73
Drug treatment center	3.61
Group home for mentally disabled	3.51
Group home for people with depression	3.47
Mental health outpatient facility	3.45
Independent apartment for mentally disabled	3.30
Group home for people living with AIDS	3.20

Source: Takahashi and Dear (1997: 83)

low and community opposition is least vocal (Case study 10.2).

This 'ghettoization' of facilities for those living with HIV, the homeless, asylum seekers and populations dependent on welfare demonstrates that the geographies of marginal groups are, to a lesser or greater extent, the product of dominant imaginary geographies casting minorities as 'folk devils' who need to be located elsewhere. As Sibley (1995: 49) contends, this elsewhere might be nowhere, as when the genocide of gypsies and Jews was undertaken by the Nazis, or it might simply be a space 'out of sight' of mainstream populations (such as the red-light districts that are the focus of sex work in many British cities – see Hubbard 1999). In *Geographies of Exclusion*, Sibley (1995) offers a theoretically informed account of how these imaginary geographies fuel exclusionary practices. Drawing particularly on psychoanalytic ideas about the importance of maintaining self-identity (literally, maintaining the boundaries of the Self), he argues that the urge to exclude threatening Others from one's proximity is connected to ideas about the importance of bodily cleanliness, many of which may be inculcated in early

Thematic case study 10.2

Community opposition to asylum seekers

In accordance with the United Nations' protocol on refugees, the United Kingdom is obliged to offer asylum to those in fear of persecution on the basis of their race, religion or culture. However, in the late 1990s, a mounting backlog of asylum applications, together with media stories of 'bogus' asylum seekers, combined to demonize this group in the public imagination. One strategy adopted by the government as it sought to diffuse concern about the 'wave' of asylum seekers 'flooding' Britain was the dispersal of asylum seekers away from the Channel ports where their visibility had been exploited by right-wing groups intent on exploiting racist fears of difference. As part of this process of planned dispersal, the Home Office proposed the construction of accommodation centres for asylum seekers in a number of rural locales. However, this policy often met with considerable local opposition. For example, in February 2002, residents of Newton in Nottinghamshire learnt that the Home Office was planning to convert nearby RAF Newton into an accommodation centre for asylum seekers. A series of protests organized by the Newton Action Group expressed vehement opposition to the proposal, with campaigners arguing that the development would cause increases in crime, vandalism and social disorder, and that property prices would fall. One resident claimed that parents would be 'terrified of letting their children play out', while another asked 'Can we be assured there are no child abusers, drug addicts or convicted rapists among them?' (see Plate 10.3). The fact that this is by no means an isolated example demonstrates how effective the media have been at exploiting public anxieties about the threat that asylum seekers pose to the sanctity and purity of the nation. As an extremely vulnerable group, asylum seekers have been less able than some to oppose these racist myths, and many have reported harrowing experiences of discrimination. For example, in the Sighthill area of

Glasgow, the arrival of 1,200 Kurdish and Kosovan refugees in 2001 prompted a number of vicious assaults by local white youths who claimed the asylum seekers were being shown favouritism by the local council. Following a series of attacks, it took the murder of 22-year-old Kurdish asylum-seeker Firsat Dag to draw attention to the everyday racism fuelled by a media rhetoric that depicted asylum seekers as 'social security scroungers'.

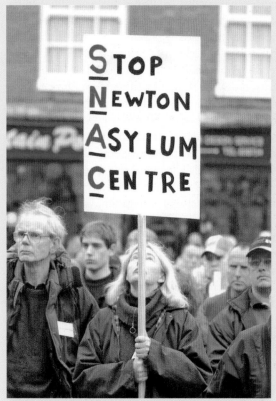

Plate 10.3 Protests against a planned asylum centre in Newton, Nottinghamshire in 2002.
(Nottingham Evening Post)

infancy. Developing psychoanalytical ideas, Sibley details how fears of the Self being defiled are consequently projected (or mapped) on to those individuals and groups depicted as polluting or dirty. In turn, Sibley argues that individuals adopt a series of exclusionary strategies designed to retain 'psychic distance' between themselves and **abject** groups, building

symbolic, psychological and physical boundaries between Self and Other. This engagement with psychoanalytical theory offers geographers an important insight into processes of marginalization, suggesting that the exclusion of Other groups can only be understood with reference to the deep-seated urge people have to *purify* their surroundings.

10.3 Racial exclusions

The discussion above suggests that while society is highly diverse, some social differences are 'amplified' into significance through practices of representation. An example here is the division of society on racial lines (see also Chapter 9, pp. 207, 214–16.) Like class, age, or religion, race is a social construct, an invention of the many discourses and images that suggest that our skin colour matters, with cultural difference following biological differentiation. Race, therefore, is intimately linked to racism. Unpacking the imaginary geographies that cast minorities to the margins, research in social geography has exposed the way that media negativity identifies ethnic minority groups as the source of social problems such as criminality, poor health and environmental decline (see Smith 1989; Anderson 1991; Dunn 2001). Sibley (1999) accordingly argues that the equation of whiteness with purity, order and cleanliness in northern European cultures has been responsible for the creation of negative stereotypes of (for example) Africans, Afro-Caribbeans, people from the Indian subcontinent and Roma gypsies. He contends that each of these groups has consequently been allocated only a marginal role in national cultures because of their 'threatening' nature. In turn, such marginalization fuels spatial segregation, with the creation of ethnic-minority enclaves in the poorest neighbourhoods of many European cities being a product of personal prejudice in the job and housing markets that is fuelled by negative media stereotypes (Wacquant 2010). For example, in the UK there remains talk of white and non-white groups leading 'parallel lives', despite efforts at partnership designed to instill 'community cohesion': evidence suggests that the often-convivial everyday encounters that occur in public and workspaces between non-white and white individuals are not necessarily mirrored in the creation of residentially mixed neighbourhoods (Swanton 2010). In this context, some have suggested ethic segregation in the UK is largely 'voluntary' – and hence self-segregation by both white and non-white groups. While this conclusion is disputed in some quarters (see Peach 1996; Phillips 2006), it suggests that cultural racism and processes of fear, resentment and prejudice may be more significant than institutional or state-sponsored attempts at segregation and bordering (Anderson 2010). In other contexts, this segregation has been institutional and via 'tighter' forms of control: for example, under apartheid in South Africa, the Group Areas Act prevented black people from straying into residential areas declared 'white', leaving a legacy of cities that are sharply divided on ethnic lines (see Case study 10.3).

While segregationist policies have long been illegal in the USA, whiteness remains profoundly encoded in the city–suburb distinction that exists there, with the suburbs seen to provide a refuge for the white middle classes from the dirt, disorder, and, above all, criminality of the coloured inner city (Valentine 2001). Indeed, the imagined association between areas of predominantly black residence and the welfare-dependent 'ghetto' has become so pernicious in the USA that it dominates white assumptions about Afro-American lifestyles, creating stereotypes that often bear little resemblance to black people's urban experiences. Events such as the Rodney King riots of 1992 (prompted by the acquittal of a white police officer for the filmed beating of a black Los Angeles resident) consolidated the reputation of the US ghetto as a crime-ridden environment; although the areas of unrest were typically those with the poorest income levels, high population densities and high school-dropout rates, it was the 'racial character' of these areas that was highlighted in the subsequent reporting (Smith 1995).

This association between US minority populations and specific criminalized places has been cemented over the years in those media that have used 'ghetto' as a coded term for the imagined deviance of black people (especially black men). As McCarthy *et al.* (1997) argue, the American middle classes tend to learn more about black inner cities through 'long-distanced' media images than through personal everyday interactions. Stories of 'black-on-black' violence abound in the media, with black urban males depicted as the main criminal threat in the USA (Wilson 2005). The cultural responses – in the form of black media practices – have been ambivalent in their outcomes, with the gangsta rap that emerged in the 1990s accused of banalizing everyday violence and gun crime. Even 'new wave' black cinema such as *Menace II Society, Boyz 'n the Hood, Jungle Fever* and *Straight Outta Brooklyn*, which sought to show something of the reality of everyday life for Afro-Americans, ultimately served to stereotype such places as landscapes of endemic violence and drug dependence (Benton 1995). In this context, 'urban' cinema, film and literature can be seen as a corroboration of dominant white myths that imagine drugs, guns and criminality to be part of everyday ghetto life. Such white fears and fantasies about the black inner city are acidly invoked in Wolfe's (1988) fictional account of the Wall Street trader, Sherman McCoy, who one night takes the wrong turning on his drive home only to be

Regional case study 10.3

Ethnic division in apartheid South Africa

Apartheid-era South Africa is often cited as representing an extreme example of socio-spatial segregation, with a white minority population passing a series of Acts in the mid-twentieth century designed to circumscribe the mobility of coloured, Indian and African people. Under apartheid laws, non-whites were denied access to major cities except if in possession of a pass that indicated they were gainfully employed in the city. As such, while the ruling white minority depicted non-whites as racially inferior, they recognized their importance as a source of unskilled and semi-skilled labour. Initially, non-white workers were housed in barracks or municipal hostels where they were subject to regimentation designed to curb the 'animal instincts' of natives and instil the importance of 'the proper habits of work and life' (Popke 2001: 740). The understanding that non-whites possessed useful and productive bodies, yet were prone to indolence and incivility, was clearly connected to colonial representations that associated non-whiteness with savagery, animality and nature, characteristics that were repressed in 'civilized' Western cultures (Anderson 2000).

However, the continued growth of non-white 'shanty towns' around the periphery of major cities fuelled white fears of 'swaart gevaar' (black danger), and led to national legislation designed to construct buffer zones between the white city and non-white townships. Both physically and symbolically on the margins of South African cities, the townships thus became abject landscapes, depicted by the white media as spaces of vice, disease and lawlessness. Repeated attempts at creating racially sanitized, planned townships were thus part of an explicit attempt to modernize South African society while maintaining the distance between poorer non-white groups and the affluent white city.

Rather than suggesting that the apartheid city represents a racist distortion of the processes of segregation evident elsewhere in the world, more recent commentaries on the apartheid era suggest it illustrates the racist conceits that underpinned twentieth century modernism (Robinson 1997; Popke 2001). As such, it is possible to identify planning and 'improvement' schemes as underpinned by racist imaginaries throughout the world (and not just in those nations with a colonial history). Indeed, despite the abolition of apartheid Group Area Acts, racist fears of difference continue to fuel socio-spatial exclusions in South Africa, with the growth in gated communities being merely one manifestation of this (Jurgens and Gnad 2002).

confronted with a side of New York that he remains cosseted from in his 'yuppie' housing development:

> At the next corner, he turned – west, he figured – and followed that street a few blocks. There were more low buildings. They might have been garages, they might have been sheds. There were fences with spirals of razor wire on top. But the streets were deserted, which was okay, he told himself. Yet he could feel his heart beating with a nervous twang. Then he turned again. A narrow street with seven or eight storey apartment buildings; no sign of people, not a light in a window. The next block, the same. He turned again, and as he rounded the corner – astonishing. Utterly empty, a vast open terrain . . . here and there were traces of rubble and slag. The earth looked like concrete, except it rolled down this way, and up that . . . the hills and dales of the Bronx . . . reduced to asphalt, concrete and cinders.
>
> (Wolfe 1988: 65)

This depiction of the Bronx as a wasteland inhabited by 'lots of dark faces' again reminds us of the ways

in which fears of difference (in this example, based on skin colour) can provoke anxiety and feed the urge to exclude. From the perspective of the white narrator in Wolfe's book, the urban decay of the Bronx mirrors his perceived view of the city's black population as Other: they are street people rather than the 'air people' who inhabit his social world (adopting Raban's, 1990, memorable description of those who can afford to distance themselves from marginal places).

For many Afro-Americans, ghetto life is similarly typified by anxiety, but an anxiety originating from problems of poverty (in 1990 the average white household income was 1.72 times higher than that of black households, with one-third of black families living below the US poverty line). As the renowned feminist and cultural theorist bell hooks relates, the experience of living in the black ghetto is one that consistently reinforces feelings of subordination and inferiority:

> To be in the margin is to be part of the whole but outside the main body. As black Americans living in a

small Kentucky town, the railroad tracks were a daily reminder of our marginality. Across those tracks were paved streets, stores we could not enter and restaurants we could not eat in, and people we could not look directly in the face. Across these tracks was a world we could work in as maids, as janitors, as prostitutes, as long as it was in a service capacity. We could enter that world, but we could not live there. We always had to return to the margin, to cross the tracks, to shacks and abandoned houses on the edge of town.

(hooks 1984: 9)

Until the 1960s it was not uncommon to find specific ordinances and zoning laws controlling the development of black neighbourhoods. Although such laws have long since been overturned, the legacy of these remain in the racist practices of mortgage financiers, banks and estate agents, who may seek to maintain the blackness or whiteness of particular neighbourhoods in the interests of maintaining house prices (Short 1996). It is thus unsurprising that the Otherness of the Afro-American population is mirrored in what has been termed **hypersegregation**, with black people remaining more spatially concentrated than any other US ethnic minority. However, Dissimilarity Indexes exploring how the Afro-American population is distributed relative to the white population suggest that their segregation has actually declined over the period 1980–2000 (see Table 10.3). Yet the declines remain modest, and some metropolitan areas actually experienced an *increase* in residential segregation over the period 1980–2000 (the seven highest increases all in the southern USA, including Columbus, GA; Goldsboro, NC; Athens, GA; and Danville, VA). Wilson (2005) also notes a deepening of poverty within the most ghettoized neighbourhoods, measured through the ratio of high- to low-income residents, with the ratio of rich to poor in Chicago's Englewood and Woodlawn neighbourhoods increasing by 40 per cent between 1990 and 2000, and 80 per cent in Cleveland Hough and Fairfax in the same period. Workfare programmes fail to alleviate the situation, with the poorest forced into dead-end and demeaning employment for minimal wages.

While the disappearance of well-paid work from US inner cities (and the associated out-migration of the black middle class) has been postulated as the major cause of such concentrated black poverty in the USA,

Table 10.3 The five most segregated cities in the USA, measured by white/Afro-American Dissimilarity Index, 1980–2000

MSA/PMSA name	Year	Total population	Black or Afro-American population	Dissimilarity Index (D)
Detroit, MI	2000	4,441,551	1,045,652	0.846
	1990	4,266,490	942,794	0.874
	1980	4,387,613	892,347	0.874
Milwaukee-Waukesha, WI	2000	1,500,741	245,151	0.818
	1990	1,432,149	197,183	0.826
	1980	1,396,972	150,662	0.839
New York	2000	9,314,235	2,451,277	0.810
	1990	8,546,583	2,249,997	0.813
	1980	8,274,039	1,907,434	0.812
Newark, NJ	2000	2,032,989	473,829	0.801
	1990	1,915,600	424,037	0.825
	1980	1,963,131	409,244	0.827
Chicago, IL	2000	8,272,768	1,602,248	0.797
	1990	7,410,858	1,427,357	0.838
	1980	7,175,269	1,429,217	0.878

1 = complete segregation, 0 = complete integration

Source: US Census Bureau, www.census.gov/hhes

Mohan (2000) argues that such factors alone cannot explain the continuing segregation of Afro-Americans. Instead, he suggests it is necessary to consider the wholesale institutional abandonment that has accompanied the economic stigmatization of Afro-American communities. Coining the phrase 'desertification' to describe the planned shrinkage of essential health, welfare and emergency services in the Bronx in the 1970s and 1980s, Wallace and Wallace (1998, 2000) suggested that such disinvestment was part of a deliberate attempt to cut back on public spending by withdrawing services from the very districts most in need. As they detail them, the long-term consequences have often been catastrophic (including increases in infant mortality rates, low birth weights, cirrhosis, TB and AIDS-related deaths). At the same time, many areas stigmatized as pockets of black poverty have become insurance no-go zones where small businesses find it impossible to get insurance because premiums are so high, and large percentages of the population are denied car insurance because of excessive costs. Moreover, the intensification of the activities of police, courts and prison is recognized to have disproportionately affected such 'imploding ghettos', with the mass incarceration of young, working-class African American men seen by many as a state strategy designed to 'discipline' and manage the socially dispossessed (Waquant 2010). It is for such reasons that some commentators still feel justified in speaking of the 'apartheid' of American cities (Wallace and Wallace 1998).

10.4 Gentrification: reclaiming the margins?

As already noted, all societies possess marginal places, whether these are spaces of poverty, sites of illicit activities or areas occupied by stigmatized populations (and perhaps all three). Often, the boundaries of such places have remained consistent over time, even if the individuals and groups they contain have changed (an example here might be the persistence of the inner city as a disordered urban zone). This is clearly connected to questions of power, with dominant social groups able to physically and symbolically claim the centres, marginalizing the less powerful:

> Centre/periphery distinctions tend frequently to be associated with endurance in time. Those who occupy centres establish themselves as having control over resources which allow them to maintain differentials between themselves and those in peripheral regions. The established may employ a variety of forms of social closure to sustain distance from others who are effectively treated as outsiders.
>
> (Giddens 1991: 131)

In this chapter we have already examined numerous processes which tend to spatially isolate and exclude less powerful groups – e.g. discrimination in the housing market, institutional racism, NIMBY politics and so on. Yet there are significant counter processes, with urban policy in particular often prefigured on the idea that marginal places need to be bought back into the mainstream in the interests of economic development as well as social cohesion. These policies are underpinned by an increasingly complex diversity of agencies and partnerships, blurring the distinction between public and private sectors as property developers and policy-actors work in tandem to regenerate deprived areas and engineer an 'urban renaissance'.

The impact of regeneration is decidedly mixed. Suffice to say, many policies of regeneration attempt to 'take back' the margins from marginal people, rather than empowering them, effectively transforming landscapes of poverty and decay into prestigious (and *ordered*) space through selective investment and redevelopment. Hence, the net result of such improvement policy is often the onset of gentrification. A much-debated phenomenon in social geography, gentrification is the process by which poor neighbourhoods are transformed by an influx of affluent homebuyers and renters. In some instances, this involves individual gentrifiers buying existing properties and transforming them (often with governmental subsidy); in others, it involves the wholesale replacement of low-income housing with prestigious apartments by corporate developers (Davidson and Lees 2010). Either way, the net result is often one of displacement, with former residents frequently unable to afford to live in gentrified areas. Marginalized populations may see their neighbourhoods enjoy a spectacular renaissance, but often do not share in its subsequent prosperity (Lees 2008).

For such reasons, many inner city districts appear to be bifurcating: either they are being abandoned to the urban poor or, conversely, are being gentrified. While the latter might instigate processes of improvement that have long-term benefits for marginal groups (and the jury is still out on 'trickle-down' urban theory), most geographers remain sanguine about their impacts. Gentrification, it seems, can be conceptualized as yet another form of profit-generation for the wealthy, representing a property wager rather than an investment in people. Indeed, one of the leading urban commentators, Neil Smith (1996), argues that gentrification ultimately only occurs if the gap between actual and potential land value is such that property development becomes lucrative. In such circumstances, the processes of disinvestment associated with marginalization become the precursor to a subsequent wave of reinvestment. But for profits to be realized, the marginalized groups and land-uses which occupy these devalued spaces must be driven out, so that potential developers can view them as safe investment opportunities. It is here that the role of the local state is crucial. Smith (2002: 442) points to this when he details how squatters, the homeless, squeegee merchants and 'street people' were ruthlessly dealt with in New York following the election of Mayor Rudolph Giuliani and appointment of Police Commissioner William Bratton. Espousing a rhetoric of Zero Tolerance for miscreants, these individuals were pivotal in identifying the urban disadvantaged as a disorderly population. This urge to tame urban disorder was to trigger notorious police brutality against minorities, justified with reference to the need for improved quality of life, but actually intended, Smith argues, to make particular areas of the city safe for corporate gentrification (and the associated invasion of upper-income groups). One notable example of this process was the wholesale removal of the homeless and their supporters from Tompkins Square Park at a time when developers were seeking to sell the Lower East Side as a space for 'family' residence. The seizure and subsequent 'purification' of the park (and surrounding areas) serves to underline how exclusionary urges may turn a genuinely public space into a space reserved for those who accord with mainstream ideas of living and working (see also Mitchell 1996).

The vicious **urban revanchism** that Smith documents is a symptom of the deep-rooted fear amongst white middle- and upper-class citizens of Other populations, be they the unemployed, sex workers, the homeless and immigrants. It is very common for mainstream middle-class groups to wax lyrical about urban living, and celebrate the diversity of the city, but all too few seem prepared to leave the suburbs unless it is for a similarly homogeneous gentrified inner-city district (Atkinson and Flint 2003). Moreover, these tendencies are not merely confined to the urban West, and it is easy to find instances of marginalized populations being displaced throughout the global South as shanty towns or **informal settlements** are bulldozed to make way for new middle-class developments, gated communities, shopping centres and highway developments

(see Case study 10.4). Though often resisted, the net result is the corporate gentrification of city centres worldwide, to the extent that the CBD of Manila now looks very much the same as the CBD of Sydney (Winchester *et al.* 2003). While such processes of corporate gentrification may ultimately attract mobile consumer capital, and tie Third World cities into a network of world cities, the consequence is of course the production of new landscapes of exclusion.

The prognosis here for marginal groups is not good, as some of the main policies advocated in their name seem destined to serve interests other than their own. But while such practices of regeneration tend to carve up space in favour of powerful middle-class elites, it is dangerous to suggest that poorer groups are unable to resist these processes. For example, some excluded groups may well resist forms of social closure by **transgressing** into the spaces of the powerful, challenging taken-for-granted expectations about where they should locate (Cresswell 1996). Such transgressions may trigger a moral panic (such as that which surrounded asylum seekers in the late 1990s or people living with HIV in the 1980s), encouraging new forms of social and spatial control. This control is often underpinned by the use of police power, as was the case with the French *banlieue* riots, where curfews were rigorously enforced in affected areas. In such cases, practices of spatial ordering and surveillance quickly reassert the social order, and remind people of their place. However, transgressions may also set in motion social changes by challenging assumptions about who or what belongs where. Parades, sit-ins, strikes, squats: all enact an opposition that, however fleeting, may bring about a change in social attitudes. For instance, gay pride marches in Western cities have often drawn attention to homophobia in society, and over time have encouraged the repeal of discriminatory legislation; perhaps less successfully, Muslim-identifying populations have pursued a variety of public actions designed to topple the Islamophobia that is rampant in the Western nations.

Developing these ideas on transgression and resistance, geographers have hence conceptualized places on the margin as sites where the relatively powerless can organize themselves into self-supporting cultures of resistance and cooperation. For example, so-called 'gay ghetto' areas in the USA (e.g. Castro in San Francisco, or West Hollywood in Los Angeles), France (e.g. the Marais, Paris) and South Africa (e.g. De Waterkant, Cape Town) have been transformed from marginal spaces of persecution to relatively affluent centres of gay male cultural life through political organization, creativity and activism, creating new gay identities in the process (see

Regional case study 10.4

Gated communities in Uruguay

One of the most obvious symptoms of social division in our cities is the presence of gated communities (Kenna and Dunn, 2009). These communities currently enjoy a widespread popularity because their design promises security as well as a sense of belonging: residents typically sign up to a legally binding code of conduct that ensures all residents understand their rights and responsibilities as members of a community. Physically separated from the wider city, and often offering a privatized form of social life through the provision of enclosed gardens, gymnasia, shops and facilities, gated communities have been postulated as offering a new form of urbanism that may rekindle sociality and encourage diverse peoples to live together. In most cases, however, gated communities have been found to promote social dissociation and sharpen social segregation, being woven into processes of gentrification and urban revanchism by catering to middle-class tastes (and incomes) (Atkinson and Flint 2004).

In the case of Montevideo (Uruguay), there are still only a handful of gated communities (certainly if compared to its Latin American neighbours). Alvarez-Rivadulla (2007: 51) points out that the majority of these are peripheral to the city centre in the Carrasco suburb, and offer a semi-rural setting in which the urban threats of downtown Montevideo are far removed: 'bird song replaces traffic noise, the grey of the city gives way to shades of green'. As is the case in most gated communities, however, the residents are remarkably homogeneous in terms of age, family status

and class, as well as sharing certain cultural dispositions (such as their enthusiasm for golf). Their social life tends to be strongly family-oriented, and inward- rather than outward-looking. The occupants are manifestly not part of any global elite, but are significantly more affluent than the average for the city. Significantly, as Alvarez-Rivadulla (2007) notes, this is not a group that had ever mixed with less affluent groups (even if it shares the same urban public spaces and neighbourhoods with them). As such, gated communities reinforce already-existing social tendencies, feeding on fears of difference and myths of urban danger.

Alvarez-Rivadulla (2007) concludes that the reasons that the occupants of Montevideo's gated communities move to them are not so different from the motivations listed by suburban dwellers in studies in the UK and the USA: a desire to escape the unpredictability of the city, seek stability and put down roots. Located on the periphery of the city – an area traditionally associated with Montevideo's urban poor – these gated communities have effectively displaced working class and marginal populations in shanty dwellings, and raised land values in the vicinity, exacerbating contrasts between the traditionally affluent neighbourhoods and the neighbourhoods belonging to the urban poor. In a reversal of the trends we noted earlier in the chapter, an archipelago of affluent gated communities can be seen to be emerging within a sea of poverty (as opposed to cities in the USA and the UK, where we noted the existence of islands of deprivation increasingly dissociated from an otherwise mobile and affluent city).

Forest 1995; Visser 2003; Sibalis 2004). Likewise, alternative economies thrive in many racialized areas (particularly in the cultural industries – food, music, fashion, arts and media), allowing 'ethnic entrepreneurs' to bring different values and ideas to the attention of wider audiences, making them more mainstream in the process. This gradual **demarginalization** process may, over time, bring excluded populations into the mainstream.

10.5 Conclusion

Social inequality is just one of many issues explored by social geographers. Nonetheless, it is one of the most

important issues facing society in the twenty-first century. As we have seen, all societies possess their 'wild' and 'untamed' zones – ghettos, slums, *banlieues*, *favelas* – as well as less obvious sites of marginalization (including sites of rural deprivation). While we need to be wary of suggesting these sites share common characteristics (e.g. the multi-ethnic French *banlieue* or British inner-city stands in stark contrast to the ethnic homogeneity of the US 'ghetto'), there are a number of key themes geographers routinely invoke when analysing their geographies: namely, exclusion, residualization, transgression and social control. Taken together, these themes emphasize the fact that space is never an abstract surface on which social life is played out, but is socially produced. As such, research on geographies of residential segregation (e.g. Wacquant 2007), structural changes in urban retailing (e.g. Cummins *et al.* 2008), inequalities in health and well-being (e.g. Shaw *et al.* 2008), or access to education (e.g. Butler and Hamnett 2010) does not just provide valuable insights into the particular combination of problems that beset our most vulnerable populations; it also makes a key contribution to *theoretical* debates about the role of space in constructing socially significant categories of identity and difference (such as class, gender, sexuality and race). In the final analysis, concepts such as exclusion and resistance are not solely of relevance to the issues of social inequality discussed in this chapter: they are fundamental to making sense of the diversity that characterizes the contemporary world.

Learning outcomes

Having read this chapter, you should be able to:

■ Understand that society and space are divided in ways that often segregate stigmatized 'Other' groups.

■ Appreciate the complexity of the processes encouraging this segregation, including people's desire to distance themselves from populations represented by the media as threatening or polluting.

■ Identify the characteristics of spaces of exclusion, which tend to lack economic, social or political infrastructure, and are typified by health problems and a poor quality of life for those who live there.

■ Recognize the importance of spatial stigmatization in exacerbating social divides; once a place has obtained a reputation as a space of exclusion, it is unlikely to attract investment of a type that will benefit its inhabitants.

■ Understand that efforts to reclaim spaces of exclusion in the name of improvement and 'urban renaissance' often instigate gentrification – a process that often triggers new forms of spatial purification and exclusion rather than social mixing.

Further reading

Caldeira, T. (2001) *City of Walls: Crime, Segregation and Citizenship in São Paulo*, University of California Press, Berkeley, CA. A fascinating account of the fear and anxiety that divides this Brazilian city, which makes frequent comparisons with the forms of racism and hate evident in North American cities.

Cloke, P., May, J. and Johnson, S. (2010) *Swept up Lives? Reinvisioning the Homeless City*, Wiley-Blackwell, Oxford. An account of homeless lives in Britain that offers a valuable corrective to accounts of urban revanchism in which homeless people are denied agency and the complex politics of welfare and care are reduced to a single exclusionary urge.

Del Casino, V. (2007) *Social Geography: A Critical Introduction*, Wiley-Backwell, Oxford. A lively and accessible introduction to social geography that focuses on issues of inequality and segregation across the course of people's lives.

Dikeç, M. (2007) *Badlands of the Republic: Space, Politics, and French Urban Policy*, RGS/IBG Book Series, Blackwell, Oxford. Develops ideas about the social and political imaginations that have marginalized French immigrant populations in the *banlieues* (the 'badlands' of his title), offering significant insight into the 2005 urban unrest in French cities.

Dorling, D., Rigby, J., Wheeler. D. *et al.* (2007) *Poverty and Wealth in Britain, 1968–2005*, Joseph Rowntree Foundation, York. A project report that provides a fulsome discussion of different measures of deprivation, distinguishing between 'core poor' and 'breadline poor' groups, and concludes that Britain is becoming less equal over time, with rich and poor living further apart. Available as a free download from: www.jrf.org.uk/publications/poverty-and-wealth-across-britain-1968-2005.

Hamnett, C. (2003) *Unequal City: London in the Global Arena*, Routledge, London. A detailed examination of the economic and social changes that have taken place in London over the past 40 years, noting significant spatial shifts in occupational structure and income. Though generalizing from a world city is dangerous, Hamnett's case study makes a number of important points about the geographical imprints of polarization.

Sibley, D. (1995) *Geographies of Exclusion: Society and Difference in the West*, Routledge, London. Remains a landmark theoretical statement on geographies of exclusion, bringing geographical ideas into dialogue with psychoanalytical literatures. The book is illustrated throughout with examples ranging from the marginalization of gypsies to the exclusion of non-white voices from the spaces of the academy.

Wacquant, L. (2009) *Punishing the Poor: The Neoliberal Government of Social Insecurity*, Duke University Press, Durham. NC. A powerfully argued book that suggests contemporary penal strategies in the USA (and beyond) are responding not so much to fears of crime but social resentments underpinned by racial – and racist – imaginaries. The focus on the 'hyperincarceration' of young, black men from 'ghetto' environments suggests neoliberal policy continues to gain its legitimacy from spatial ordering.

For annotated, clickable weblinks and useful tutorials full of practical advice on how to improve your study skills, visit this book's website at **www.pearsoned.co.uk/daniels**

RURAL WORLDS

Chapter 11

Warwick E. Murray

Topics covered

- Definitions and meanings of 'rural'
- Trajectories and gaps in rural geography as a sub-discipline
- The interlinking of rural change in the global North and South
- Demographic change in rural worlds: depopulation, counterurbanization and rural–urban drift
- Political change in rural worlds: countryside movements and rural resistance
- Cultural change in rural worlds: social constructions of the country, commodification of rural areas, social change
- Economic change in rural worlds: productivist vs. post-productivist agriculture, multifunctional rural areas

Ask somebody what he or she thinks of when you use the term 'rural' and the response you get will be influenced heavily by where that person 'comes from' in both a geographical and sociocultural sense. It is an assumption, but somebody in England might mention green fields, rolling hills, neat hedges, black and white houses, cider apples, cosy pubs and village fetes. Notions of what constitutes 'rural' are likely to be much less pleasant in Tonga, for example. There it might be suggested that poor infrastructure, absence of electricity, grinding poverty, chiefly, and patriarchal governance systems, vulnerable housing and environmental deterioration characterize such areas (see Plates 11.1 and 11.2 for example). This is not to suggest that all is well with the countryside in the rich world (global North) and ill in the poor world (global South); marginalization and poverty are increasingly recognized to be part of rich-world rural areas, just as there are pockets of privilege in rural areas in poorer countries. Furthermore, within and between the global South and global North there are widely different rural geographical imaginations; in the USA the concept of 'rural' tends to be associated with wilderness, whereas in western Europe notions of the rural tend to be more idyllic. In the global South for example, Chilean notions of the rural – of unkempt green fields, horseback riders, unpaved roads, set against the snowy Andes – are very different to what might be

Plate 11.1 'Shire Cottage' – a rural home in Tarrington, Herefordshire, England.

(Warwick E. Murray)

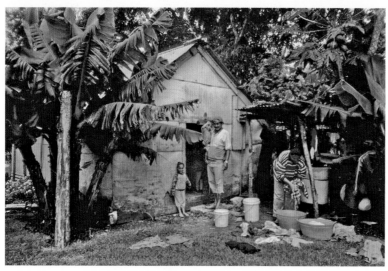

Plate 11.2 A rural home in Tongatapu Island, Tonga.

(Warwick E. Murray)

imagined in sub-Saharan Africa, where drought and desert might be imagined. The point being made here is that imaginaries, together with the realities of the 'rural' are not homogenous across the globe, varying across time and space in ways that alter how the rural is defined, interacted with, inhabited, interpreted and socially reconstructed.

Much recent work in rural geography, in the Anglo/American/Australasian (**AAA**) tradition at least, has sought to show that there are in fact rural worlds within worlds and that peoples' identity and social position play a role in the way rural spaces are conceptualized and experienced. Furthermore, as already hinted at, the experience of the rural varies widely between the global North and global South, though much more geographical research on the latter is required. Although rural spaces differ vastly across the world, arguably they form interlinked parts of one global complex. As such, as globalization unfolds, the social, economic and environmental shifts that are occurring in rural worlds everywhere do not occur in isolation (Murray 2006). This chapter investigates a range of perspectives on rural worlds from different geographical viewpoints. In doing so, it draws on many examples from across the globe to paint a picture of the dynamic and increasingly interwoven spaces we call 'rural'.

From the vantage point of highly urbanized societies that tend to characterize the global North (and Latin America in the global South), it is easy to forget that just under 50 per cent (49.9 per cent in 2009) of the world's population lives in rural areas. In both absolute and relative senses the vast majority of these rural dwellers live in the poor world; in the richer countries in 2009 there were just under 310 million rural inhabitants, representing 25.1 per cent of total population there, whilst in the South there were close to 3.1 billion rural inhabitants, accounting for 55.4 per cent of the total (UN 2010). As such, it could be argued that rural geography, at the global scale, should be principally concerned with issues of development, poverty and inequality in poorer countries. This chapter seeks balance in this respect, building on examples from both the global North and South. The issues at stake in rural worlds across the world are serious and pressing, and concern billions of people. In the rural global South, poverty and deprivation are high and environmental degradation is advancing rapidly. In the rural global North social justice concerns and local environmental issues, though arguably less pressing than those in the South, are also

considerable. The commodification of the countryside in the global North and the use of the rural idyll as a marketing tool to sell 'countryside' products mean that the rural is increasingly important all across society in richer countries. Far from being stagnant or in decline, as is sometimes assumed, the 'rural' is shifting rapidly – and these changes have implications for the whole of human society. This chapter looks first at how we might define the term 'rural'; it then turns to how geographers have grappled with the issues in such places. We then move on to consider changing rural geographies in terms of demographic, political, cultural and economic shifts.

11.1 Words and worlds: what is 'rural'?

What is meant by the term 'rural'? As already discussed above, meanings and definitions vary across cultures and places. At the general level, there are two ways we can approach defining the rural. The first set of approaches might be termed *empirical*, and includes functional approaches including measuring land-use characteristics as well as demographic approaches that involve such things as population density measures. The second set of definitions can be termed *conceptual* and does not use directly quantifiable measures of rurality, drawing instead on social constructs, which have to do with how we *imagine* the countryside.

Governments have their own methodologies for delimiting what is meant by rural, which are generally empirical and are important for planning and development purposes. Using a functional definition assumes that there is a rural/urban dichotomy and that the rural is defined by what it is not – that is, not urban. But this rural/urban dichotomy will vary between places greatly, as both the rural and the urban are variously understood. In the Pacific Island territories, such as Fiji, Samoa or Rapa Nui, many of the areas that are defined as urban in official statistics would likely be seen as rural areas to people from the North; such areas often have lush vegetation, informal communal gardens and relatively low-density populations. Rapa Nui (Easter Island) is defined as 100 per cent urbanized in official statistics, for example, as all of its approximately 3,800 inhabitants are located in the capital Hanga Roa. This low-density village-like settlement would not be considered 'urban' by Western standards, however. In this sense, using functional as well as demographic and other empirical measures is questionable when intended for comparison across societies. The point, again, is that defining the rural is not a precise science.

Human geographers realize that the rural and urban overlap in many ways. It is common to refer to peri-urban, rururban and semi-urban areas. In this regard, some have argued that rural and the urban are best thought of as lying on a rural–urban continuum. Overall, it is now recognized that seeking to define the rural and the urban in opposition to each other (what are sometimes called binaries) is not the most useful approach. Both **political economy** and more recent cultural approaches to rural geography emphasize, in various ways, that rural and urban spaces are constructed by processes that cut across both, and that whilst the impacts of these processes on the ground will not be equal they are responding to common stimuli.

Conceptual definitions of rural are important, and not as obscure as they may appear at first. As we will see, economic opportunities, planning, infrastructure development and many other 'concrete' outcomes are predicated on how individuals, collectives and governments imagine the rural. Under such definitions rural is something in the mind – and thus rural 'social' space need not correspond with rural 'geographic' space. More will be said on this issue when we consider how rural geography has itself changed over recent years.

Cloke (2000: 718) defines the 'rural' as:

> Areas which are dominated by extensive land uses such as agriculture or forestry, or by large open spaces of underdeveloped land, which contain small, lower order settlements demonstrating a strong relationship between buildings and extensive landscape, and which *are perceived as rural by most residents*. (Emphasis added)

The emphasis is added above in order to highlight the fact that rural geographers have come to think of 'rurality' as much as a state of mind as one based on a specific configuration of functions. Cloke goes on to argue that while some rural areas are still defined functionally, in those closer to urban centres 'rural is more of a socially constructed and culturally constructed and therefore contested category' (Cloke 2000: 718). This latter point is most relevant to richer countries where the blurring between urban and rural is increasingly pronounced and where cultural commodification of the countryside and the construction of the rural idyll are widespread. As we will see, rurality in the global South generally has very different connotations.

11.2 Changing rural geographies

In order to understand the geographies of rural space, it is useful to review the evolution of rural geography as a sub-discipline. Rural geography is the study of the relationship between humans and the environment in rural areas, the nature of rural localities, economies, societies, cultures and environments and how this varies across space. Trajectories within the sub-discipline vary markedly across the world, but in the Anglo-American-Australasian (AAA) tradition there has been a relatively common path. Within the context of geography in general, rural geography has, until recently, been somewhat neglected. This may be because in richer countries the rural population is a minority. Also, it could be argued that much Western scholarship carries an urban bias, based on the assumption that it is in such areas that social progress is designed and takes place. The past two decades have seen a rise in interest in rural geography in the AAA tradition and rural areas in the global North are seen as spaces where *general* cultural and economic processes can be researched, understood and interpreted. Indeed, rural studies are increasingly interdisciplinary. Rural geographies of the poor world have been treated very differently and have tended to be conflated with development studies: thus there is very little cultural geography of the rural global South (see Murray 2008).

Early geography paid some attention to rural areas but this declined with the quantitative revolution and the rise of spatial science in the 1960s. Rural geography at this time was essentially agricultural geography, as 'rural' and 'agricultural' were – for good empirical reasons – more explicitly interchangeable at the time. Rural geography made something of a comeback in the 1970s and early 1980s, and during this period the sub-discipline could be described as functionalist, concerned with such issues as rural planning, land use change and urban encroachment. In hindsight, this approach can be seen as 'uncritical' and theory-free. The relevance of such rural geography was questioned in terms of its contribution to broader society in the late 1980s. Some also argued that by focusing on the distinctiveness of the countryside it ignored processes that cut across the increasingly blurred rural–urban divide.

By the early 1990s in the AAA tradition, there was resurgence in rural geography based on concepts from **political economy**. This focused in particular on a critique of the role of the 'restructuring' impacts of **globalization** and – although not referred to as such then – **neoliberalism**. This approach saw the application of concepts from neo-**Marxist** and **world systems** perspectives and was focused on how circuits of capital across the world conditioned, and were conditioned by, the nature of rural space in different locations and how the state intervened in such flows. Some such concepts have been applied to agriculture in the South, particularly commodity chain analysis, although not always under the title of rural 'geography'. Themes considered during this period included diversification, environmental change, international food chains and deregulation, as well as attempts, in later forms at least, to bring together social and political issues with economic concerns (Cloke 2005a). The latter included work on commodification, **gentrification**, accessibility and **counterurbanization**. Much of the work in this area, particularly in **agri-food systems** has continued until the present. Indeed one might argue that in New Zealand and Australia, emphasis has remained on the political economy of agriculture and on the impacts of globalization, and has not shifted to cultural interpretations to the extent that it has in the United Kingdom. Furthermore, across the world traditional land use studies remain popular; Chilean rural geography is dominated by land use analysis, functional definitions, and quantification oriented towards planning and poverty alleviation, for example. In this sense it is more meaningful to talk of rural 'geographies' rather than a singular geography (Roche 2002).

The most recent turn in rural geography has been profoundly influenced by the general cultural turn in human geography which itself was stimulated by the shift to **postmodern** ideas in the social sciences. This shift has by no means been homogenous and there are commentators who have argued that this has made our rural geographies less rather than more relevant (see Wilson and Rigg 2003). There can be little doubt that the new cultural geographies of rural areas are far less applicable to the global South, although the extent to which they have been explored there at all is woeful. The cultural approach, then, seeks to break down old binaries and structures, interrogating rurality as a **social construct**, which is participated in by persons of shared cultural, social and moral values. Crucially, the social space of the rural (which can be imagined and thus be located anywhere) need not necessarily overlap with the geographical space of the rural. Thus the new rural geography engages with ideas that have to do with the social construction of categories, and the way they are then represented and reproduced.

Partly in response to the cultural turn, **political economy** approaches have broadened to include, among other things, studies of those who are 'marginalized', including geographies of rural women, the young, the old and those in poverty, for example. There have also been a number of more contemporary geographies exploring, among other things, rural sexuality, feminism, **travellers** and those who, because of dominant cultural constructions of the countryside (in the United Kingdom at least), have been severely 'othered' (Little and Leyshon 2003). Finally, there has been a shift in the AAA tradition towards the exploration and promotion of 'alternative' rural and agricultural production networks and sustainable rural livelihoods, including rural collectives and organic projects, for example.

Linking the new cultural rural geography and political economy rural studies is an important challenge (Argent 2002; Philips 2002). Furthermore, building in empirical and theoretical input from geographers in areas other than the United Kingdom and the United States is also very important. Most rural dwellers live outside the countries where the rural geography agenda is set, and in conditions that are very different to those found there. In this regard, more research on the rural geography of poorer countries is required (Wilson and Rigg 2003; Murray 2008). Overall, the important thing to remember is that numerous traditions in rural geography coexist and that 'paradigms' or perspectives never neatly succeed one another; communicating across and between these world-views is a central challenge for rural geographers.

11.3 Shifting rural worlds

One of the core themes of this chapter is interaction: within rural worlds, between rural and urban spaces, and across continents as rural areas are bound together in global networks. Notwithstanding this, it is useful to subdivide the subsequent discussion into themes. In what follows we will consider geographical change in rural demographies, polities, cultures and economies. Relative weight is placed on the latter two categories, although this does not mean that they are more significant than other areas. Processes that operate in any one of these spheres overlap in various ways with processes in others, and reference to this will be made as we proceed.

Rather than considering the global North and South separately, examples from each and the links between them are emphasized within this framework. The forces that cut across these boundaries can be summarized as modernization (see modernity in the glossary) (including increased urbanization and industrialization), and its latest incarnates, neoliberalism and globalization. These processes fall unevenly in different places, given local histories and respective positions in the global system. As global divisions of labour spread further across the planet and borders become increasingly permeable, rural and agricultural spaces in very different places form constituent parts of an evolving global network of rural spaces. As such, land use and identity and all other aspects of rural change in the North must be seen as intimately tied to change in the South.

11.3.1 Dynamic rural demographies

Rural demographies across the world are highly dynamic. This demographic change has been conditioned, to varying extents, by generally increasing levels of urbanization, the shifting economic base of the countryside away from agriculture combined with the rolling out of neoliberalism, and shifting cultures in rural areas. These shifts have been configured in very different ways across space but, in general terms, depopulation in the countryside has been an almost universal trend since the industrial revolution in the global North, and since the Second World War in the global South. Table 11.1 shows that whilst absolute world rural population rose from 1.7 billion to over 3.3 billion between 1950 and 2009, in relative terms it declined, falling from 71 per cent to 49.9 per cent of the total global population.

Table 11.1 also shows the breakdown of rural population by region, illustrating that in a proportional sense Africa and Asia are far more 'rural' – with levels of over 60 per cent – than the other global regions. Looking at regional data at this scale often hides important variations – for example, Oceania includes the highly urbanized countries of Australia and New Zealand, as well as relatively rural Pacific Island countries such as Vanuatu, Fiji and Tonga. Overall, however, it is clear that there is a significant North/South divide in terms of the demographic rural/urban dichotomy; in the richer countries there are just over 300 million rural inhabitants, representing approximately 25 per cent of the total population, whilst in the South there are over 3 billion rural inhabitants, accounting for just over 55 per cent of the total population at the present time.

Since the Industrial Revolution the relative loss of population in the countryside in the West has proceeded as national economies became based first on industry and then services, both of which are generally,

Table 11.1 Rural population by region and development grouping, 1950–2030

	1950		1975		2000		2009		2030(*)	
	Population (million)	%	Population (million)	%	Population (million)	%	Population (million)	%	Population (million)	%
Africa	191	85.3	310	74.6	518	63.6	611	60.4	729	49.3
Asia	1162	83.2	1820	76	2313	62.9	2402	58.3	2236	45.9
Europe	271	49.5	232	34.4	206	28.3	201	27.5	152	21.7
Latin America	97	58	125	38.8	129	24.6	121	20.7	113	15.7
North America	62	36.1	64	26.2	66	20.9	63	18.1	53	13.3
Oceania	5	38	6	28.5	9	29.5	11	29.8	11	26.2
Global North	*390*	*47.9*	*350*	*33.1*	*320*	*26.8*	*310*	*25.1*	*240*	*19.2*
Global South	*1400*	*81.9*	*2210*	*73.1*	*2920*	*59.7*	*3059*	*55.4*	*3050*	*43.9*
World Rural	*1790*	*71*	*2560*	*62.8*	*3240*	*53.3*	*3406*	*49.9*	*3290*	*40.1*
World Total	2520		4070		6090		6829		8200	

* = projected

Source: based on data from UN (2006b)

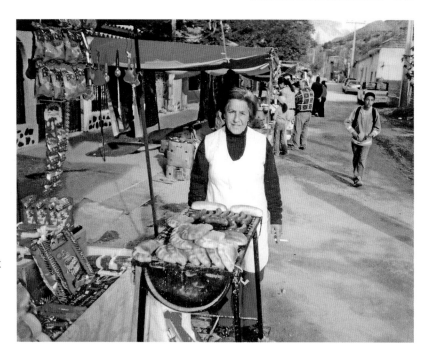

Plate 11.3 The monthly market in Chile Chico, a small isolated rural settlement high in the Chilean Andes on the border of Chile and Argentina.

(Warwick E. Murray)

though not exclusively, concentrated in urban areas. This population loss has commonly been highly selective, often taking the young and sometimes the skilled, leaving depleted labour markets and service provision (see Plate 11.3). Indeed, as noted above, rural populations across the world are both ageing and in relative decline. Rates of urbanization are most rapid by far in the global South, while the level of urbanization remains low compared to the West. The most rapid rates of relative rural population decline, then, are found in poorer regions of the planet, which has significant and generally deleterious consequences for the rural society left behind and implications in terms of how they should be managed.

There have been some exceptions to the rule of relative rural depopulation and one of the most studied demographic processes in the global North has been counterurbanization. In the 1970s in the USA, for example, fuelled by rising net in-migration, smaller towns grew at a more rapid pace than larger ones, reversing the historical trend in urbanization towards large agglomerations. In the 1980s and early 1990s in the United Kingdom smaller settlements and some rural areas experienced growth whilst inner cities recorded net losses. Hereford, an agricultural centre in the west of England located on the Welsh border, became the fastest growing settlement in the country in the late 1980s, for example. The real and perceived diseconomies of scale of large cities as well as changes in the nature of work and private mobility led to this outcome. Cultural

shifts are also present in this demographic change. As people have tired of the costs inherent in urban living, such as congestion, pollution and social distance, some have sought out the idyll of rural life, closer community, and many other stereotypes that abound regarding the countryside in the global North. In New Zealand and Australia the rise of the 'lifestyle block', an acre or two of land in the countryside where the 'good life' can be practised, within commuting distance of a large city, is a growing phenomenon (see Plate 11.4).

The universality of counterurbanization in the West has been called into question and some see it as very time- and place-specific: related to a period of manufacturing and service decentralization in Europe and USA that took place in the 1970s and early 1980s. Countries that have experienced periods of growth in the proportion of the total population living in rural areas include Germany (1970–85), United Kingdom (1950–70), Finland (marginally from 1980 onwards) and Australia (1975–90) (UN 2006b). There has been a clustering of relative rural population gain in Central and Eastern Europe since 1990 in places including Latvia, Estonia, Slovakia, Romania, Czech Republic and also Russia, for example. However, in many Western countries – including Spain, Holland, Belgium, Canada, Sweden and New Zealand, for example – counterurbanization has not occurred at all at a national scale. Outside of Europe there are virtually no examples of relative rural population gain over the past 60 years, except in very small countries or small

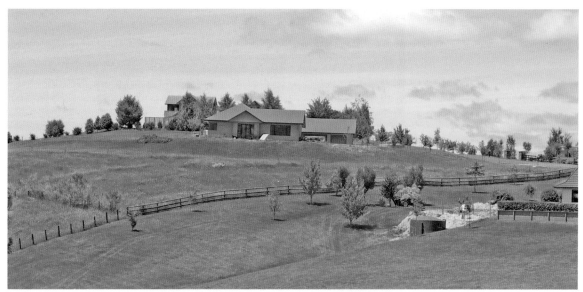

Plate 11.4 A rural 'lifestyle block' near Lake Taupo, Taupo, New Zealand.
(Warwick E. Murray)

island nations. However, the rural–urban drift has slowed in the global North; the rate of decline in rural population has slowed from rates of 1.35 per cent per annum in the period 1950–55 to 0.66 per cent between the years 2005 and 2010 (UN 2009). Rural depopulation is undoubtedly the dominant pattern in richer countries, albeit at lower rates than previously, particularly as inner cities are rejuvenated and people appear to be seeking settlement closer to their place of work. Recently, the rise of Internet communications has led some to suggest that a new process of counterurbanization could well be imminent, as people can increasingly locate at a distance from their place of employment. Whilst possible in theory, such technology does not obviate the essential human need to meet face to face, and to date there is no evidence that cyber-technology is reversing rural depopulation anywhere on the planet.

Although counterurbanization is not a worldwide phenomenon, representing perhaps a fascinating yet ultimately unimportant diversion from the principal trend of rural depopulation, there is no doubt that the social structure of the rural population in the global North is shifting because of demographic change. When one looks below the scale of the whole country, it has often been possible to detect smaller regional units of buoyant rural population growth based on migration patterns. This shift is influenced by urban-to-rural migration among certain groups and changes the nature of rural areas in a variety of ways. Some rural areas, particularly those close to larger urban settlements,

have become effectively dormitory zones. When in-migrant populations in the rural United Kingdom – characterized inherently by high mobility – bypass local services such as small shops, local pubs and post-offices, the impact on the socio-cultural nature and economic base of rural areas can be dramatic. The influx of urban migrants in some areas has occasionally led to tensions between the existent and in-migrant population, in terms of how rural settlements are governed and managed. Through the 1970s and 1980s Welsh nationalists, in response to the perceived invasion, burnt a number of holiday homes purchased by non-Welsh newcomers, for example. Interestingly, however, studies in the United Kingdom have shown that social collectives, such as playgroups and village football teams, have been characterized by high participation by in-migrants, clearly keen to illustrate their commitment to their new home and live out the imagined community spirit that some might seek. Cloke (2005b) argues that differing densities of population and territorial size lead to different impacts with respect to the processes outlined above. In the United Kingdom, for example, rural areas that are subject to immigration effectively become suburbs. This is contrasted to the situation in the USA, Australia and, to an extent, New Zealand, where the migratory 'footprint' of any given city will be relatively dispersed. Again what should be clear from this discussion is that there are variable outcomes across global rural space.

The demographic processes at work and outcomes in the rural areas of the global South are, generally speaking, very different to those outlined in the preceding discussion, although the theme of rural depopulation in terms of the proportion of total population living in rural areas is common. As already noted this process is generally much more rapid in poorer countries in terms of absolute rates of migration. Although rural–urban drift has ebbed and flowed in different places at different times, this shift has become one of the defining features of social change in the South over the past decades, leading to the abandonment of rural areas in extreme cases and the explosion of urban populations in receiving areas. Latin America has seen a particularly marked rural depopulation over the past 50 years in proportional terms, for example, as has the Pacific Island region (see Case study 11.1).

A number of factors explain this rural–urban drift in the South, with the weight of each varying from place to place. Generally speaking, the move abroad of transnational corporations beginning in the 1960s, the increasing mobility of capital from the 1980s onwards, and the concomitant rise in industrial employment, together with development theory and policy

that see such modernization (i.e. industrial urban development) as beneficial and desirable, have combined to precipitate this outcome (Potter *et al.* 2008). The process has been aggravated by the legacy of unequal landholding structures in rural areas, where peasant farmers often exist side by side with large landowners sometimes descended from colonial elites. The commercialization of agriculture and de-agrarianization of national economies at the same time have often led to fewer opportunities for small-scale producers, as larger players have access to the credit and collateral that are required to purchase the technology in order to compete. Thus smaller producers have either lost their land to become temporary labourers on large farms or have migrated to the city – leading to what is known as proletarianization in the case of the former, and de-peasantization in the case of the latter (Murray, 2006). The latter outcome has been the overriding consequence. There are instances when this rural population decline has been stopped or even reversed, as is the case in the Monte Patria *comuna* (district) of northern Chile during the peak years of the grape export boom

Regional case study 11.1

Rural depopulation in the Pacific Islands: Niue

Niue provides a good example of rural depopulation in the global South. Niue is a very small island nation with a population under 2,000. Although Alofi, its capital, is defined as an urban area, functionally and demographically it is rural, being low-density and dominated by subsistence agricultural production. From a peak of 5,000, this rural island has been losing population since the 1960s, a process which accelerated when the former colonial power, New Zealand, offered dual citizenship for Niueans on independence in 1974. Owing to a mixture of natural disasters, environmental catastrophes, and misguided economic development policy, the islanders have moved consistently and in very high relative numbers to the cities of New Zealand, especially Auckland, and to a lesser extent to Australia. This loss has its positive side – remittances sent back from Niuean migrants, often second and third generation, form part of the backbone of the country's economy, with over 50 per cent of GDP accounted for by such flows.

Unlike remittance-based migration from some rural areas to urban areas in the South, this case is not circular in its nature: it is largely permanent. However, this loss of population means that Niue is the most rapidly depopulating rural zone and country in the world. The consequences of this demographic shift have been devastating: services in retail and transport have declined as they become uneconomic; whole settlements have been abandoned leaving empty shells where once were houses; and the country's agriculture suffers from severe labour shortages. After the devastating cyclone Heta in late 2004, some have asked whether the country can survive and at what point it should be declared unviable. The disappearance of the rural society on the island of Niue would deny the world a unique and fascinating culture that contributes to the world's human diversity. A similar story is true across many of the Pacific Islands such as Rapa Nui, Samoa, Tuvalu and Tokelau, as rural population is lost to the process of both intra- and transnational urbanization (Murray and Terry 2004).

in the early 1990s and is now the case in the rapidly expanding wine export regions of the Central Valley (Overton and Murray 2010), but this is not common.

In short, rural areas in poorer countries have become increasingly less attractive economically and socially and a large-scale drift to the cities has occurred. In some cases this has been coupled with severe environmental decline because of over-exploitation, rapid commercialization or climate change. Severe drought has increased over recent years in the sub-Saharan rural zones, for example, leading to a rise in rural–urban drift. Under such conditions, political circumstances, including wars, and neoliberal economic policies such as export orientation, have often aggravated environmental tensions leading to full-scale food shortages and famine. Problems in rural areas in the South are made all the more significant when one considers that despite rapid relative population loss, the largest *absolute* rural populations are found in poor countries and this is set to be the case in the foreseeable future. Table 11.2 shows the largest 10 rural populations by country in 2009, projected into 2030, with the USA being the only country in the global North featuring in the lists.

11.3.2 Dynamic rural polities

Political power in the countryside in both the global North and South is shifting considerably in terms of how it is gained and exercised. As such, the governance of rural spaces is increasingly complex and contested. In Northern countries rural areas have traditionally been over-represented in parliamentary politics, owing in part to the inherited importance of the landed elite in national affairs of state as well as the desire of some democratic governments to escape accusations of urban bias. Within the countryside itself this created space for various countryside interest groups, alliances, political parties and social movements that draw on particular imaginaries of the countryside to make their case. In New Zealand, for example, a small political party – the Outdoor Recreation Party NZ – won 1.3 per cent (approximately 25,000) votes in the 2002 general election on a platform of 'preserving' rural areas for hunting and fishing as well as broader rural-first policies and as a consequence of this success was integrated into a larger party – the United Future – in subsequent elections. The Countryside Alliance in the United Kingdom is another case. With the motto 'love the countryside', the movement was formed in 1997 and grew rapidly in reaction to the single issue of the proposed ban on fox-hunting contained in the Labour Party 2001 election manifesto. The movement soon mushroomed to include a range of disgruntled rural groups who argued that central government was increasingly over-interventionist in rural affairs. The Alliance argues that over 100,000 individuals are members. The movement culminated in a massive march on UK Parliament in October 2002 involving approximately 400,000 people. The imagined conflict between city and country – traditional and modern, and 'authentic' and 'fake' – was invoked in speeches during the march. Many groups that are truly marginalized

Table 11.2 Ten largest rural populations by country, 2005 and 2030 (millions).

Country	Rural population 2009	Country	Rural population 2030 (projected)
India	842	India	859
China	725	China	574
Indonesia	129	Pakistan	124
Bangladesh	118	Bangladesh	124
Pakistan	116	Ethiopia	94
Nigeria	79	Indonesia	84
Ethiopia	69	Nigeria	70
Vietnam	62	Vietnam	63
USA	57	Dem. Rep. of Congo	60
Philippines	47	Uganda	57

Source: UN (2009)

in the countryside, such as travellers for example, were not represented to the same extent as the 'sport' of fox-hunting. This has led some to argue that the Alliance was a smokescreen erected by the elite, utilizing stereotypical interpretations of the countryside, to push a single issue pertinent to them, whilst drawing in support from the rural masses. Casting the particularities of the situation to one side, what is interesting is the way the Alliance used 'imaginaries' of the countryside in order to pursue a specific political point. It is also clear that these imaginaries – of fox-hunting and hounds and red suits on horses – are not necessarily shared, or even considered desirable, by the majority in the countryside in the United Kingdom.

But representations of the rural for political gain are not just made from within the countryside itself: arguments about rural identity are sometimes invoked in order to make broader political gains. Countries in the European Union, especially France, have used arguments concerning the importance of 'rurality' to make the case for the maintenance of subsidies for agriculture, claiming that such funds help sustain the lifestyles and landscapes that make a broader contribution to the character of the European countryside. More recently, European governments have couched this argument in terms of the protection of the 'multifunctionality' of the countryside; that is to say, sustaining agriculture has positive knock-on effects in other areas (such as environmental preservation, for example) that others in broader society also value. Some in the South see this as a form of protectionism, allowing the continuance of subsidized agriculture in richer countries. At the same time that such support persists, global financial institutions insist on the adoption of free-market, non-subsidized, neoliberal policies in poorer countries, a situation which discriminates against millions of rural inhabitants.

In the global South the practice of modernist development has often been described as 'urban biased' (Kay 2001). The persistent neglect of rural development initiatives in favour of industrial urban policy has been written about widely and consistently, most notably in Africa and Latin America. In the latter, rural zones have been neglected by administrations that have been over-centralized, partly a consequence of a legacy of urban-based Hispanic society, and partly the result of the opening of economies to foreign capital which exploits cheap labour pools in urban areas during the recent phase of neoliberalism. There are numerous examples of rural resistance to the neoliberalization of rural areas and its associated urban bias

in places as diverse as Thailand, Fiji and Brazil. One of the most notable examples is that of the Zapatistas in Chiapas, Southern Mexico (Murray 2006). In some countries where agro-exports have risen, population outflows have been stemmed, but the effective control of resources still remains in urban areas, resulting in a process of what could be called internal colonialism. Rural dwellers in the South have to endure the double impact of often over-centralized and urban-biased governments, and the economic control of large corporations based in capital cities or, often, outside of the borders of the country itself.

There are also examples of resistance to 'modernist' rural production methods in the South by populations in the global North. Participation in alternative food networks, organic supply chains, and moral or fair trade linkages as a form of resistance has become a topic of some significance over the recent past. Some have conceptualized alternative projects as just part of neoliberalism, often controlled by retail power, where alternatives are mainstreamed and conventionalized quickly. Yet some fair trade NGOs maintain that consumers can play an important role in terms of caring at distance that can make a difference to rural spaces in poorer countries. McCarthy (2006), however, argues that to date alternative ethical networks have done little to alter existing North–South dynamics and power remains concentrated in the former.

11.3.3 Dynamic rural cultures

Traditional rural cultures are being altered, in some cases irrevocably by the shifting relationship between the urban and the rural, although what is meant by 'traditional' is contested. As noted previously, the increasing intermingling of urban and rural populations and processes raises questions as to what is meant by 'rural' at all. Indeed some have argued that the concept of rurality exists only in the mind and that the 'binary' of rural and urban is a false one. Thus rural *social* space and rural *territorial* space are not necessarily the same thing and 'rurality' is something that has to be deconstructed. Some have even argued that we have reached a 'post-rural' condition in that pure rurality no longer exists. Cultural geographers have been hard at work trying to tease out the way that meanings of rurality are perceived by different actors in the global North and how this affects behaviour, geographical or otherwise.

Postmodernists argue that today's world is a **simulacrum** where symbols or signifiers of an event replace direct experience. Reality is thus experienced through representations of it, creating a hyper-reality that is disembedded from territorial space. It is possible to argue that this has occurred in the case of the perception of the rural by urban consumers who wish to experience the 'rustic' and the relatively 'untouched', which they imagine to be associated with the countryside. This might include: wearing country clothes, driving big four-wheel drive all-terrain vehicles around city streets, designing the interior of their homes like country cottages and so on. It could be argued that many people who live in the rural–urban fringe, extend this hyper-reality to their everyday existence. The idea that we are in a post-rural condition probably takes the argument too far; it certainly does so in the case of the global South where such notions are at best not applicable and, at worst, fanciful, irrelevant and elitist. There can be little doubt, however, that the cultural meanings associated with the rural have caused a shift in the relationship between society and geographical space, and between culture and nature (see Chapter 12).

The foregoing trends have influenced the cultural consumption of the countryside and the allied commodification of the rural experience. In the global North, rural imaginaries are used to sell things. When objects, traits and ideas of a particular culture are brought into capitalist circuits, or when new traits are invented in order to stimulate economic gain, 'cultural commodification' can be said to be occurring. Increasingly, cosmopolitan consumers seek out 'authentic' experiences that will help them differentiate themselves from the masses. There are further examples across the 'cultural' economy – world music, poverty tourism, 'ethnic' food aisles in supermarkets, cultural quarters: Chinatown, Little India and the like – being established in cities as diverse as Singapore, London and Auckland. Although the flow of cultural products from the periphery to the core is predated by imperial and, more recently, the globalized diffusion of Westernization, there can be little doubt that there is a tendency towards the consumption of 'the Other' in richer markets (see Chapter 19).

In the case of rural areas, cultural commodification has been especially important. This is most obvious in the advertising world, which uses rural references in order to sell products that have little or nothing to do with the countryside. For example, the advertising campaign for Speights Beer in New Zealand involves images of typical rural South Island men who are tough in outlook and pioneering in spirit. As pioneers they are spurred on by the taste of Speights beer, which allegedly gives them energy, and which they prefer

above all 'urban' temptations (such as a box at Eden Park rugby ground, the main stadium home of the All Blacks, in one advert). The good-humoured rivalry between the North and South Islands of New Zealand, which is a proxy for urban and rural in some ways, is used to sell a product which 'real men' drink. In reality, although headquarters are in the southern city of Dunedin, most Speights beers are brewed in Auckland, a large urban conglomeration much like any other in the global North.

There is also a trend towards 'place-making' in rural areas across the global North, often for the purposes of creating niche tourism. This involves taking a trait, or inventing one, which is characteristic of a locality and using this for economic gain. Ludlow, in the English county of Shropshire, for example, has reinvented itself as one of the principal gourmet destinations in the country based on the coincidental location there of a number of top restaurateurs in the 1990s. This has created a hub of restaurants, cafés, wine bars and tea shops, which bears little relevance to the 'authentic' history of this small rural town. In New Zealand, Taihape has pronounced itself 'gumboot' (wellington boot) capital of the world, has an annual world gumboot throwing competition, and has erected statues to make this obvious (Plate 11.5)! Towns that have been bypassed by the establishment of large highways turn this to their advantage by advertising the 'best of rural life' on large highway billboards to attract weary travellers, as in the case of Marton, New Zealand. These examples illustrate attempts to use rural imaginaries to create diversified livelihoods in areas where traditional agricultural production patterns are rapidly restructuring. In such efforts culture, invented or otherwise, plays an important economic role.

A phenomenon linked to cultural consumption is the creation of rural spectacle in order to market place (Cloke 2005b). The construction of 'fresh' and 'natural' experiences is important in this respect. Adventure tourism in New Zealand and Australia – including white water rafting, bungee jumping and zorbing, for example – is an interesting example of how rural areas have reconfigured their economic bases. At Kiwi 360 near to Tauranga, New Zealand, it is possible to visit orchards, take part in activities and tour a museum to kiwi fruit production. The multiple connotations of the word 'kiwi' – which refers to a native flightless bird, a fruit, and often used to describe a person of New Zealand citizenship – are clearly used in marketing such rural spectacles. Related to the concept of rural spectacle is the evolution of cultural visits to

Plate 11.5 Taihape, New Zealand – gumboot capital!
(Warwick E. Murray)

rural areas. In New Zealand, in the area around Rotorua (central North Island), there are numerous 'authentic' Maori villages that have been developed for tourists to visit and experience indigenous culture. Today, most Maori live in urban areas and their lifestyle is far removed from the hyper-reality depicted in such locations, but such businesses provide an important source of income for those involved. Cultural commodification has the potential to bring economic

gains but it can also straightjacket places into fossilized cultural representations, leading to the creation of damaging stereotypes. Notwithstanding this, some have argued that such ventures help maintain traits of Maori culture that might otherwise die out, thereby rejuvenating rural indigenous culture.

In the rural global South cultural issues are very different. Rural cultures are, of course, vibrant and enormously varied. In some cases 'national' culture is rooted

Spotlight box 11.1

The hidden others of the countryside

As mentioned previously, researching the 'hidden others' of the countryside in the West has now become more important in the AAA geographical tradition. This research has revealed a darker underside to the rural idyll where people are excluded because of various identity traits or socio-economic characteristics. Work has been undertaken on travellers, marginalized sexualities, the unemployed, the elderly and the poverty-stricken, for example. Poverty in the countryside in the West can be disturbingly high, and this has been particularly the case as the economic base has rapidly restructured over the past two decades under neoliberalism. In New Zealand, for example, the neoliberal restructuring of the 1980s and 1990s led to the closure of many small- and medium-scale dairy reception plants, as the New Zealand Dairy Board (later Fonterra) rationalized its operation

geographically. This, together with other rural economic shifts, led to very high levels of poverty in some rural regions such as Taranaki and the East Coast, rural decline and depopulation, and some unintended consequences such as the closure of rural rugby clubs, further undoing the social and cultural fabric of the countryside (Willis 2001). Despite limited state intervention it has proven impossible for many such areas to regain dynamism, and pockets of poverty, crime and work-poor cultures in small rural towns such as Patea and Waverley remain stubbornly present to this day.

One of the ironies is that many recent countryside movements and alliances, such as those discussed in a preceding section, represent anything but such disadvantaged groups in the case of the global North, although the involvement of all classes in often elitist causes is frequently solicited in order to create the impression of a 'rural consensus'.

Plate 11.6 The annual bareback riding tournament high in the Limarí Valley, Norte Chico, Chile.

(Warwick E. Murray)

in traditional and conservative imaginaries of the rural (see Plate 11.6), although this culture is sometimes fossilized and invented. Notwithstanding this, very little is known of the rural cultural geography of the South, and geography in the North has tended to ignore it. There is a rich tradition of anthropology of rural areas in the Third World, but this tends to focus on the unusual rather the everyday cultures of the rural poor. In some post-colonial countries rural areas are often home to 'indigenous' cultures, or at least they were so before the onset of rural–urban drift. This raises particular issues with respect to rural development planning in poorer counties. In Chile, for example, many Mapuche – once the only unconquered indigenous group in Latin America – live in the south of the country in a marginalized economic and social environment. Those Mapuche that have moved to the capital city have often fared no better in the urban context. Far from the buzzing and primate metropolis of Santiago, the rural Mapuche have, until recently at least, been left behind by the economic and political progress made in the country over the past 15 years. In contrast to the global North, the experience of 'rurality' can be painful and poverty-stricken for such groups (see Spotlight Box 1.1).

Across the Third World the **peasantry** is declining and this is leading to rapid change in the cultural life of rural dwellers. As neoliberalism unfolds, peasant farmers and other small-scale producers are forced from their land and, as discussed previously, leave for the cities. In Fiji, despite leaving their homes, urban migrants from the countryside still retain much of their rural culture – although it is metamorphosizing. Through the concept of *Vanua* indigenous Fijians are linked to the

land; indeed they see themselves as indivisible from the land, and in particular the rural places they hail from. Second- and third-generation urban Fijians who live in the urban agglomeration based in and around Suva (population circa 350,000) might well still consider themselves as 'coming from' Taveuni, Ovalau, Tailevu or some other rural area and they will make this known very early on during conversation. Notwithstanding some continuity, there are also significant changes; the influence of urban Fijians that return to the village, temporarily or otherwise, is contested and often bemoaned by those who stay behind, as they bring with them concepts, such as property rights, individualism and capitalism, which hitherto were alien in such societies.

The marketization of the countryside in ways that increase tourism and so forth are rarely practised in the global South, although it is possible to find exceptions to this rule. Wine tourism in rural Chile, for example, is increasingly popular with the urban middle classes who travel to the countryside to taste and buy wine on estates that recreate the colonial hacienda period as a rural idyll (Overton and Murray 2010). Similarly, there are small rural towns across Africa and Latin America that specialize in handicrafts for tourists. In rural Tonga, women make tapa cloth, beautifully decorated bark cloth which is traditionally used for ornate and ceremonial purposes, for sale at tourist markets in the capital Nuku'alofa. In rural Sarawak (Malaysian Borneo) it is possible for foreign tourists to undertake longhouse visits. However, in general, agricultural production, combined with other diversified – yet often threatened – livelihood forms, are the mainstay of such localities, and culture as a commodity at least plays no role at all.

Plate 11.7 Pluriactivity in rural New Zealand – off this stretch of State Highway 1, north of Levin, you can buy petrol, book a river paddle tour and have a cup of tea in an aeroplane!

(Warwick E. Murray)

11.3.4 Dynamic rural economies

In the past the economy of rural localities was essentially agricultural. The terms 'rural' economy and 'agricultural' economy are no longer interchangeable, however. In this section we look at how the rural and the agricultural have decoupled and what has replaced agricultural livelihoods, at least in part. This shift has arisen as the world economy has 'de-agrarianized', involving a large-scale shift away from agricultural to industrial and service activities. Increasingly, and particularly in the North, there is a tendency towards pluriactivity (diversified livelihoods) in rural areas (see Plate 11.7). However, diversification away from agriculture, viewed at the global scale, has been uneven, leading to the creation of very different rural economic landscapes across the world. In order to understand this transition, below we consider 'productivist' agriculture, as well as newer concepts of 'post-productive' and 'multifunctional' rural spaces.

The contemporary agri-food system can be conceptualized as a complex that stretches across continents, linking producers, consumers, supermarket retailers, and government policy makers among other agents, all of which represent nodes within commodity networks. Driving the agri-food system is agribusiness, which is associated with the industrialization and globalization of agriculture (Le Heron 1993). This has led to a shift in the nature of farming in many parts of the world involving a delocalization of activity and the creation of *long networks*, which see production oriented away from local and national markets and towards the glo-

bal economy. The global agri-food network has not led to the homogenization of rural space. At the same time as we witness the industrialization of agriculture, especially in parts of the global South, we are also seeing the rise of niche, organic and alternative agriculture. Furthermore, in places such as Latin America and South-east Asia where large-scale commercial and export-oriented agriculture is increasingly evident, we see the importance of subsistence farming.

Productivist rural economies

Productivist agriculture can be defined as the highly intensive production of a limited collection of primary commodities (McCarthy 2005). The agricultural sector has industrialized and commercialized, eclipsing non-capitalist agriculture. Agribusiness has expanded through vertical and horizontal integration, forming conglomerates that link the field to the supermarket. This has allowed companies to reduce the costs of transacting, and internalize the risk of the inherently vulnerable business of farming. Consequently, agricultural sectors have increased in terms of ownership, and a new political economy of agriculture based largely on monopoly capitalism and mass production has emerged. Some have referred to this as **Fordist** agriculture. Companies such as Monsanto, Nestlé, Fonterra and Bulmers are major players in this new economy.

Given the liberalization of the world economy from the 1970s onwards, agribusiness TNCs have increasingly invested abroad. Cheap labour and land as well as less strictly enforced environmental regulations provide incentives for agribusiness TNCs to diffuse the production component of their operations to poorer countries. This is especially the case in fresh fruit, horticultural and floricultural sectors, resulting in a proliferation of counter-seasonal exports to the Northern markets. As a network, for example, floriculture now generates over US$50 billion in sales per annum, and involves fresh-bloom supply from over 80 countries, most of which is channelled through the Netherlands to third countries in the rich world. Globalization has drawn Third World rural spaces into global capitalist circuits, yielding profound impacts. Export-oriented agriculture destined for the North has, in the case of Latin America, led to the evolution of monocultural localities which are extremely vulnerable economically, socially and environmentally (Murray 2006).

The evolution of agribusiness has profound implications for local rural socio-economies in the global North as well. It has, for example, led to the relative decline of the family farm (Whatmore 1995). Furthermore, agribusiness has altered the rural landscape, removing hedges and practising other policies intended to capture economies of scale. Rural choices have also been impacted by the consolidation of agribusiness (see Case study 11.2).

Post-productive agriculture and rural worlds

We are witnessing the return, in some places, of smaller-scale agriculture, sometimes as a direct resistance to the globalization of agriculture. Simultaneously we can also observe the various uses of rural spaces intended to diversify livelihoods and add 'off farm' components. These can be termed '**post-productive**' rural landscapes where the goals, aside from maximizing agricultural yields, include cultural commodification, optimizing ecological value, and stimulating niche-based agriculture.

The transition to post-productive agriculture is linked to the perceived costs and falling productivity of the globalized model and the regulatory crisis it is facing. Government intervention is being rolled back, and large-scale production operating at a loss can no longer be tolerated. There is some evidence to suggest that the main bastion of productivist protectionism, the European Union's Common Agricultural Policy, may be dismantled over the next decade or so. In this sense the transition to post-productive agriculture is consistent with neoliberalism and cannot be seen as resistance to it. Indeed, if we view the evolution of this part of the system as the consequence of the farming-out of the labour- and land-intensive parts of the agri-food production complex to the periphery, then the rise of post-productivism forms part of the broader evolution of global capitalism.

In contrast to the above, some commentators see the rise of post-productive agriculture as indicative of resistance to globalization, greater concern for the environment and the search for smaller scale, *short network*, sustainable and alternative food networks. Consumer demand has transformed Western consumption patterns, and food safety and quality issues also motivate elites in the South. This has led to a worldwide boom in organic farming, for example, that has impacted agricultural landscapes in the North and South. Ironically it is potentially easier to establish such sectors in the South. The concept of 'organic' farming in many parts of the Pacific and peripheral Latin America is what has been practised for centuries.

Regional case study 11.2

Productivist agriculture in the Hereford cider industry

In the 1980s rapid growth in the UK cider industry led to a fundamental restructuring of the Herefordshire economy as the largest global producer of cider – HP Bulmers – began an aggressive campaign to dominate local supply networks and to secure the lucrative national market. The company purchased land from small-scale farmers and consolidated orchards in a way that allowed it to take full advantage of mechanized harvesting. At the same time it employed hundreds of medium-sized farmers on two-yearly contracts in order to shore up supply. Bulmers also took over a number of competing medium-sized cider producers including Symond's (producer of Scrumpy Jack). This granted Bulmers a regional bilateral **monopsony** position (single buyer and seller simultaneously) in the county (and for many miles beyond). Small-scale cider producers were out-competed, and the grubbing up of many old varieties of cider apples in place of the new mass-production bush-stock varieties accelerated the decline of small-scale farmhouse cider producers, thereby eroding diversity and choice.

In reality, we face a situation in rural worlds where a mixture of productivist and post-productivist functions exist. At the global scale there is, of course, a division of labour that sees much productivist agriculture located in poorer countries because of favourable locational comparative advantages, such as cheap land and labour, counterseasonality, and lax environmental regulations. Niche agriculture is undoubtedly becoming more common in Northern countries. However, this dichotomy blurs a complex reality. In the United Kingdom post-productive activities in the countryside are advanced, but in places such as New Zealand, Australia and the USA there is only limited evidence of a post-productive shift. New Zealand is considered one of the most cutting-edge of the agricultural economies but it is still heavily productivist, and it appears that because of a boom in demand from China large-scale dairy farming is in fact on the rise as more and more land is converted to dairy farming (see Willis 2005). In the global South the evidence for post-productivism is even scarcer. There are some examples, such as the rise of wine tourism in Chile, but in general we are seeing a shift *towards* productivism as neoliberalism is allowed to penetrate localities in the periphery. The rise of GM and its diffusion is likely to accentuate this trend. We thus have a significant North–South divide in agriculture, and rural geography might do well to revisit older ideas including dependency and structuralism, in order to interpret these differentiated outcomes in the context of one global system.

Wilson and Rigg (2003) have sought to 'test' the applicability of post-productive agriculture outside of the United Kingdom and the European Union. They argue that the concept of post-productivism (see post-productivist transition in the glossary) is flawed because only when it was invented was the term 'productivism' applied to the preceding period (Roche 2005). Taking an anti-modernization viewpoint they argue that the linearity of an assumed development towards post-productivism is unsound. In their view, 'multifunctionalism' is a much better way of characterizing what is occurring across the world – and perhaps always has been.

Multifunctional rural worlds?

Post-productivism has been heavily criticized; use of the term is problematic because it makes assumptions about what is meant by 'productivist'. The nature of productivist agriculture has varied, and still does vary across the world; it persists in some localities whilst in others it has never arrived, and probably never will. In places such as the Pacific Islands, subsistence agriculture plays a very important role in agricultural production and rural lives, and commercial networks bypass millions of rural dwellers. Although modernist discourses might portray this as backward, the socio-cultural and nutritional role of subsistence and semi-subsistence is critical. Yet agriculture of this nature is being squeezed by globalization.

In recognizing the problems associated with the concept of post-productivism, the term 'multifunctional' has become employed in Northern rural studies. Wilson argues that the productivist debate has been 'conceptualized from a UK-centric perspective that has largely failed to discuss whether the concept has applicability in Europe or beyond' (2001: 77). Instead he suggests that the use of 'multifunctional agricultural regimes' is a better way to characterize the shifts that are occurring in European rural space. Multifunctionalism refers to the notion that 'rural landscapes typically produce a range of commodity and non-commodity use values simultaneously and that policy ought to try and recognize and protect the entire range of values' (McCarthy 2005: 774). It is argued that this concept is more useful than post-productivism as it offers a positive characterization: it recognizes the continued importance of commodity production and is sensitive to geographical difference.

Multifunctionalism implies that the countryside across the world is used for both productive and post-productive purposes and that the combination of uses, including commodification and conservation, varies from place to place. Whilst this formulation reflects complex rural worlds much more accurately than previous constructions, it is also vulnerable to criticism. In some ways the concept says everything and nothing at the same time. Furthermore, commentators from the South have argued that its use in policy formation in the European Union, for example, is a disguised form of protectionism. As noted previously governments, particularly in France, have argued that subsidies to agriculture create positive social externalities by supporting other, often less tangible, functions of the countryside which help maintain 'rurality'. Meanwhile, poorer countries are expected to open their markets and perform the productivist element of the global agri-food complex. Indeed, it could be argued that there is a move away from multifunctionalism in the South, as monocultural neoliberalized sectors replace formerly diversified livelihoods.

11.4 Conclusion

'Rural' is a relative term; it shifts across time and space. It is also an enormously diverse category and therefore difficult to make generalizations about. It is increasingly difficult to define rural spaces, be they social or geographical. In both the South and North rural spaces are increasingly politicized, often in ways that move beyond 'normal' electoral politics as new political movements abound. Rural identities have conditioned this political change and have themselves shifted in response to new social, economic and cultural configurations. Environments are threatened in all worlds because of neoliberalism and the rapid shift in the economic exploitation of rural areas. Certain constructions of rural culture are used to sell the countryside and to represent its interests while 'other' cultures remain marginalized or are disappearing. In short, rural spaces are increasingly dynamic demographically, politically, culturally, environmentally and economically. Rural geography has had to move swiftly in an attempt to capture the complex changes.

This general shift, at the global level, has been influenced by the unfolding of neoliberalism and globalization. At the world scale, whilst it is true that rural spaces in the global North are becoming increasingly multifunctional and are shifting away from 'Fordist' agriculture, in the South we are seeing an erosion of multifunctionality and pluriactivity. This is certainly the case where neoliberal agriculture penetrates and the diversified livelihood-sustaining activities that have often been built up over decades and centuries are eroded. These are dynamic times for rural spaces and populations all across the world, as the insecurities, socio-economic and environmental, associated with the acceleration of globalization abound. This is especially the case in the South where neoliberal governance runs unrestrained. Although rapid advances have been made in rural geography in and of the West, it is imperative that geographers turn attention to rural spaces of the South, which have so often borne the brunt of the evolution of global capitalism. Furthermore, by shifting focus in this way geographers can contribute to a more balanced and democratic human geography that is relevant to rural society as a whole.

Learning outcomes

Having read this chapter you should be able to:

- Discuss the contested definitions of 'rural' from both a functional and conceptual point of view.

- Account for the evolution of dominant trends in rural geography in the AAA (Anglo-American-Australasian) tradition and how these have influenced interpretations of rural processes.

- Appreciate the impact of neoliberalism and globalization on rural spaces and how change in the global South and global North is linked.

- Consider the evidence for rural depopulation across the world and outline the arguments for counter-urbanization in richer countries.

- Understand shifting rural polities and the nature of the alliances and resistance movements that have evolved over the recent past across the world.

- Interpret the commodification of the 'rural' in richer countries and the contribution of culture to the economic dynamism of the countryside in the global North.

- Comprehend the concepts of 'productivist', 'post-productivist' and 'multifunctional' rural economies, and debate the shifts between the three in various parts of the world.

- Recognize that more research on issues in the rural global South is required if a more democratic rural geography is to be attained.

Further reading

Challies, E.R.T. (2008) Commodity chains, rural development and the global agri-food system, *Geography Compass*, 2(2), 375–94. This a review of the state of the art with respect to commodity chain analysis and how this links to global geographies of rural change.

Cloke, P.J., Marsden, T. and Mooney, P. (eds) (2005) *Handbook of Rural Studies*, Sage, London. This is a very broad-ranging and complete edited collection of viewpoints from some of the top writers in the field of rural studies that is particularly strong on cultural perspectives.

McCarthy, J. (2007) Rural geography – globalizing the countryside, *Progress in Human Geography*, 31(1), 1–6.

An excellent stock-take of work in rural geography that seeks to explore the 'globalized' nature of rural areas across the world.

Munton, R.M. (2008) (ed.) *The Rural – Critical Essays in Human Geography*, Wiley-Blackwell, London. A useful way to follow up arguments here through a collection of rural geography essays from the AAA perspective, focusing mainly on the British case.

Murray, W.E. (2008) Neoliberalism, rural geography and the global South, *Human Geography,* **1**(1), 33–8. This paper presents an argument for taking the rural geographies of the global South more seriously and highlights the importance of viewing rural change in any location as part of a global whole

Roche, M. (2002) Rural geography: searching rural geographies*, Progress in Human Geography*, 26(6), 823–9. This review article provides an excellent stock-take and discussion of the main trends and major scholarly articles in rural geography through the 1980s and 1990s.

Wilson, G. and Rigg, J. (2003) 'Post-productivist' agricultural regimes and the South: discordant concepts?, *Progress in Human Geography*, 27(6), 681–707. This article is an excellent critical review of the non-applicablility of post-productivist rural geography concepts to agriculture in developing countries.

Woods, M. (2005) *Rural Geography,* Sage, London. Although focused on western cases, this is a landmark text and a good way to learn more.

For annotated, clickable weblinks and useful tutorials full of practical advice on how to improve your study skills, visit this book's website at **www.pearsoned.co.uk/daniels**

SOCIAL CONSTRUCTIONS OF NATURE

Chapter 12

James Evans

Topics covered

- The concept that nature is not a pre-given physical reality, but an idea that is constructed by society

- The politics of representing things, behaviours and landscapes as natural

- Environmental myths

- Science and the construction of human nature

- Nature and environmental change in the media

12.1 Questioning nature

The one thing that is not natural is nature.

(Soper 1995: 7)

What is nature? There is an easy answer. It is the birds and the bees, the plants and the landscapes around us – from the familiar things in our lives, like pet cats and dogs and the park at the end of the road, to more distant things that we see on television, like Amazonian rainforests and Giant Pandas. Nature is the set of things that are *separate* from humans. We have the social world on the one side, with its politics, injustices and cultural achievements, and on the other we have the natural world, a pre-given set of biological entities – the domain of natural scientists – that have no politics and no culture . . . things that simply *are*.

Albert Einstein once quipped that 'the environment is everything that isn't me'. But is it really this simple? Can we draw lines between ourselves and the surrounding world so easily? The microbes that occupy our gut to the food like fruit and vegetables that we eat every day to sustain and reproduce ourselves suggest that it is hard to draw clear lines between our own bodies and the surrounding environment. Food itself is far from natural. We use fertilizers, pesticides and mechanized agriculture to give nature a helping hand. Biologically, food represents the culmination of thousands of years of selective breeding to create more productive plants and animal. In more recent times humans have created new organisms using genetic modification in laboratories. Bruce McKibben (1999) has taken this argument to its logical conclusion, arguing that it is impossible to find a part of the planet that has not been affected by humans in some way – even the remotest inland Antarctic ice sheets have trapped CO_2 in them from industrial emissions. This observation leads him to argue that 'nature', in its traditional sense as something separate from and untouched by human society, has now *ended*.

It is similarly hard to draw a line between human society and nature at larger scales. Since our ancestor *Homo erectus* began using primitive tools some 1.9 million years ago, the history of human society has been one of environmental transformation, culminating in the massive levels of urbanization that characterize the world today (see Chapters 1–3). With every technological advance, human society has learnt to turn its surroundings into an increasing range of things. Over a hundred years ago Karl Marx tried to understand this

process, suggesting that *production* is the critical force driving human interaction with nature:

> Not only do the objective conditions change in the act of reproduction, e.g. the village becomes a town, the wilderness a cleared field etc., but the producers change, too, in that they bring out new qualities in themselves, develop themselves in production, transform themselves, develop new powers and ideas, new modes of intercourse, new needs and new language.
>
> (Karl Marx 1973, originally 1861, *Notebook V*)

This quote captures one of the fundamental arguments of Marxist thought – that social development unfolds through the transformation of nature. From the basic work of a peasant clearing a field to the industrial exploitation of oil resources, economic production simultaneously transforms nature and society.

Perhaps nature existed *before* modern humans arrived on the scene, then? Conservationists talk about 'restoring' landscapes that have been spoilt by human activities to their natural condition. Environmentalists hark back to the natural conditions that existed before humans altered the levels of CO_2 in the atmosphere and biologists talk about how many more species there would be in the absence of humans. But *when* was this pre-human condition? For example, the dominant vegetation in pre-industrial Britain was temperate forest, and what is generally considered to be Britain's 'natural' landscape is constituted by the species that typified this landscape, like oak trees. But if one goes back to the last glaciation some ten thousand years ago (when human impact was even less apparent, and thus the landscape it could be argued was even more 'natural') there was no forest, and no temperate species existed in Britain (Birks 1997). Quintessentially British species like oaks did not arrive until the end of the last ice age. Depending on what timescale is used they can be seen as either native to Britain, or as an invasive species. When we go back in time no historical cut-off is more valid than any other as the supposed point at which the environment was in a *natural* state.

12.1.1 The social construction of nature

So, if nature does not exist as something separate, and never really did, then what is it? Many academics have argued that instead of being a pre-given reality, nature is actually an idea, or a *social construct,* which varies for different people in different times and places. The phrase 'social construction' was coined by sociologists Peter Berger and Thomas Luckmann (1966), who argued that humans establish meanings and truths through habits and institutions. So, for example, things like laws are social constructions because they have no essential existence outside of the society in which people have decided to agree and act as if they exist – they are 'social facts'.

Social constructionism recognizes that individuals and groups participate in the creation of their realities through the way in which they perceive things and events. Critically, it follows from this idea that various groups of people will interpret things and events differently, because their perception is coloured by their previous experiences and differing social norms. David Harvey gives a classic example of how this problem of perception can work in relation to nature:

> What's nature? What if I give you a chemical formula and say, how do you feel about your relationship to that? If I then take that chemical formula and represent it to you as a tree, then you relate to it in a different way. Now, what's nature? Is it all the molecules that make up the tree, or is it the tree? The point is that if you see a tree, you'll react to it differently than you would if you saw a bunch of molecules, and if you see a tree in a habitat in a forest with a spotted owl sitting in it you would react very differently than if you just saw a tree.
>
> (Harvey, in Banrffalo, 1996)

Plate 12.1 shows four representations of the same 'thing'. It is easy to argue that all are natural, and yet each prompts very different reactions and will be interpreted variously by different types of people. For example, a trained chemist will recognize the carbon molecule as constitutive of wood, while a conservationist will recognize that the forest depicted is not just any forest, but a rainforest – the most biodiverse habitat on the planet. Indigenous inhabitants

Plate 12.1 Carbon, tree, forest, wood.

(Photodisc, Inc.; P. Lunnon)

of that rainforest will react with a whole different set of emotions, although they may not recognize their home from an aerial photo. The pile of timber is a resource to be burned, while the solitary tree is not. The way in which each representation is perceived depends on the experience and knowledge that the viewer has, which in turn depends on the society and culture in which he or she lives. Viewing nature as a social construct helps us to understand why people from different places and times behave so differently towards the environment. To return to the example of trees, early US settlers were fearful of the impenetrable dark forests that dominated the Eastern seaboard, and believed that clearing trees for agriculture was 'God's work'. In stark contrast to the attitudes of today, chopping trees down was actually seen as morally good. The idea of nature is subjective – it means

something slightly different to you than it does to me. This matters because it helps us to understand why conflicts occur over the environment and how they might be averted.

That said, nature is often presented as if it *is* something fixed and objective, because it provides a powerful basis for how society should behave. As the American environmental historian William Cronon puts it, people tend to 'appeal to nature as a stable external source of non-human values against which human actions can be judged' (1996: 26). Constructions of nature are thus highly political, because specific versions of nature can be used to privilege certain social behaviours and actions over others. Geographers have sought to understand who benefits and suffers under different social constructions of nature (see Case study 12.1).

Thematic case study 12.1

The limits to growth: a radical critique

■ The idea of natural limits comes from the writings of Thomas Malthus in the late eighteenth century. Malthus was a churchman who deplored the overcrowding and unsavoury conditions in which the newly-created working classes of the British industrial revolution lived. In his *Principle of Population* he attributed this problem to a disjuncture between the geometric rate of population increase (1,2,4,8,16) and the arithmetic rate of food supply (1,2,3,4,5). Overstepping the limits of natural resources (most notably at that time, food) would inevitably result in famine and death. Malthus deplored the moral depravity of the working classes and sought to control their behaviour by invoking the idea of *natural limits*. Of course Malthus has been proved wrong. With the advent of mechanized production and better-yielding crop varieties, food supply has far outstripped population growth. Despite an almost ten-fold increase in the number of people living on this planet since Malthus's time, a higher percentage enjoy clean water than ever before.

■ However, debates over food, climate change and oil still revolve around the idea of natural limits. But are these resources really 'natural'? For example, in what way is it 'natural' to burn oil? Oil did not become a resource until the invention of the internal combustion engine, and with the proliferation of hydrogen cells and bio-fuels it will cease to be the most fought-over global resource. Oil is only impor-

tant to human existence in the relatively recent era of industrialization, and, even then, only if you are one of the fortunate few who either buys it or owns it. Far from being a 'natural' resource, the demand for oil depends entirely on the existence of a socio-economic system that values and uses it.

■ David Harvey (1974) argues that the real cause of scarcity is the unequal distribution of resources. For example, the richest 1 per cent of the world's adult population owned 40 per cent of global assets in the year 2000. By contrast, the poorest half of the world adult population owned only 1 per cent of global wealth (Davies *et al.* 2006). As Harvey notes, the earth could not support the current population if everyone enjoyed a Western lifestyle. The idea that resource shortages are due to overstepping 'natural' limits conceals these grotesque imbalances. When we look at famines on a global scale, the problem is not food production, but the distribution of the food that we produce. Blaming problems on overpopulation shifts the blame to the poorest people, while protecting the monopoly rich nations enjoy over resources. Radical geographers are interested in how constructions of nature serve different political interests. The notion of natural limits clearly serves the interests of those who control the lion's share of the world's wealth.

Q Can you find other environmental debates that use the language of natural limits?

Social constructionists seek to understand how ideas of nature reproduce the cultural, political and spiritual beliefs of a society, and the ways in which ideas of nature are used to support specific political viewpoints. If certain types of behaviour or ways of organizing society can be shown to be 'natural', then it makes it hard to justify any other way of acting. The authority to speak for nature becomes the authority to speak for society. Geography has traditionally tried to understand the interface between society and the environment, and the ways in which nature has been constructed and contested have particular relevance to this task.

12.2 Cultural constructions of nature

The great enemy of truth is very often not the lie –
deliberate, contrived and dishonest – but the myth –
persistent, persuasive and unrealistic.

(J.F. Kennedy)

David Demeritt (1998) identifies the main use of social constructionism within geography as *refuting* established claims about nature. This is perhaps not surprising as social constructions can be understood as a form of 'myth' that may become widely accepted in the absence of any counter evidence. The majority of work in geography on the construction of nature has sought to reveal myths concerning nature.

12.2.1 Wilderness

One of the most cherished notions of nature in Western societies is that of *wilderness*. Images of wilderness are used widely, to sell holidays, relax us and inspire us. Motivational posters showing 'man conquering the great outdoors', with captions like 'Success' and 'Risk', will be familiar to you if you have ever set foot in a gym. The idea of wilderness as a form of nature untouched by humans has a long history. In the USA it has been argued that the idea of wilderness appealed to settlers, who, as they moved west across the country, preferred to believe that the lands they were claiming were devoid of humans and waiting to be inhabited. Exploding what he calls the 'Pristine Myth', Bill Denevan (1992) uses extensive archaeological and historical research to argue that native Amerindians actually managed their landscapes through cutting, burning, terracing and building. These forms of management created landscapes that European settlers assumed were untouched because they were unable to recognize forms of management that they were unfamiliar with. America was only a 'discovery' for the white European and, similarly, the plains of the USA only appeared to be uninhabited wildernesses to white European settlers. While the idea of wilderness is obviously related to the inability of the settler to 'see' the human impacts upon unfamiliar landscapes, it had drastic effects. When indigenous people inconveniently did appear in these landscapes, they were often exterminated.

The idea of wilderness was fundamental to the birth of the modern environmental movement in North America. The first National Park was created by Congress in 1864 in Yosemite Valley, quickly becoming a national symbol of the spiritual home of the modern United States in the aftermath of the Civil War. In 1892 the so-called 'wilderness prophet' John Muir founded the Sierra Club in California, which went on to establish a national wilderness preservation system. Of course, national parks are exactly that – parks – and great effort was required to preserve these areas as wildernesses. People had to be kept out, animals had to be kept in. Vistas had to be opened up for tourists while simultaneously not compromising the appearance of the area as untouched. So painstaking were the efforts to create wilderness that Muir actually praised Yosemite for its resemblance to an artificial landscape garden (Schama 1995: 191).

The artificiality of wilderness is taken to its extreme in the landscapes of Frederick Olmsted, who is perhaps most famous for designing New York's Central Park. In 1887 Olmsted became involved in the preservation of another symbolic American landmark of natural vigour, Niagara Falls. As William Irwin notes (1996: 77), Olmsted 'believed that parks and nature retreats relieved the festering distress of the poor and working classes'. The area had become overrun with disorganized tourist developments and industry, and Olmsted sought to transform the area into a pure pastoral park in which the visitor might enjoy the splendour of nature in peace. In its return to nature, however, the minutest details were planned, including widespread landscaping and the installation of viewing points to maximize the vistas over the falls. The landscape that now surrounds the falls represents an idealized version of how Olmsted thought the Falls *should* look, rather than a reproduction of the original landscape, which was renowned for hindering the efforts of visitors to see the falls.

Wilderness appeals to the American psyche because it is reminiscent of the birth of their country through settlement, and it remains a powerful and emotive cultural construction of nature in contemporary North America. However, while the idea of wilderness is aesthetic on one level today, the politics involved in valuing landscapes that are free of humans are still influential (see Case study 12.2).

In Britain the excesses of industrialization and urbanization in the eighteenth and nineteenth centuries generated a cultural backlash that began to value environments that were seen as more natural. The Romantic movement which emerged at this time was typified by Wordsworth's love of the Lake District and an increasingly negative attitude towards cities. This was in part an aesthetic reaction to the physical degradation of the environment associated with the Industrial Revolution, but it was also a moral reaction to the squalor and poverty of the cities. In contrast, the countryside became seen as the repository of moral purity. Interestingly, this represented a U-turn in British attitudes to wild landscapes. In the sixteenth and seventeenth centuries mountainous and uninhabited regions were actually regarded with fear and disregard. We have reports of travellers drawing the curtains of their carriages to block out the 'Satanic' view as they travelled through mountainous regions. Less than one hundred years before Wordsworth claimed of the Lake District 'Who comes not hither ne'er shall know how beautiful the world below', the great diarist Daniel Defoe called the area a 'barren and frightful place'. So if you hate hiking up mountains, simply tell people you have an early Georgian attitude to nature!

As with the US environmental movement, the British Romantic movement established the idea that nature has *spiritual* value. In the United Kingdom this happened primarily as a reaction to industrialization, while in the United States it emerged from the pioneer/

Thematic case study 12.2

Contesting wilderness

- Clayoquot Sound is a small ocean inlet on Vancouver Island in British Columbia. The surrounding area contains 'old growth' temperate rainforest, one of the rarest habitats in the world. While such forests once covered large parts of Europe and Asia they now comprise just 0.01 per cent of the earth's land area. In the early 1990s the logging firm MacMillan Bloedel applied for a logging licence in the area, but was vigorously opposed by the Western Canada Wilderness Society. The geographer Bruce Braun (2002) explored how the resulting dispute played out through a series of representations and counter-representations of the area.
- MacMillan Bloedel produced a brochure titled *Beyond the Cut*, which represented Clayoquot Sound as a natural resource that was needed by the Canadian people. Constructing the forest as a resource, they presented their credentials as responsible and experienced managers who would use scientific techniques to create a sustainable industry and create much-needed jobs.
- By contrast, the Western Canada Wilderness Society produced a popular and glossily illustrated coffee-table book called *On the Wild Side*. In it they championed Clayoquot Sound as a pristine wilderness, evoking emotive spiritual ideas of the forest as a paradise unspoilt by humans.
- On one level, the representations of the forest as economic resource and ecological wilderness reproduced familiar opposing Western constructions of nature. But Braun's real insight is that both the representations of the logging company and the environmental group fail to include the indigenous Indians who live in the forest. He argues that this erasure of the indigenous population is a common feature of both the wilderness and resource constructions of nature, and can be traced back to the tendency of settlers to appropriate nature from the indigenous peoples. The importance of constructing nature as separate from humans (in this case as uninhabited) allows powerful (if opposing) claims to be made over the area by both the logging company *and* the environmental group. The real political import of this process is that the people who live there are excluded from their own home.

Q How does the idea of wilderness exert an influence over global conservation priorities?

settler mentality. In each case the idea of nature has been carefully constructed by societies seeking to preserve a world free of human development.

12.2.2 Landscape

This new relation to nature was expressed through the genre of landscape painting, which emerged in the seventeenth and eighteenth centuries. As a way of seeing, landscape takes a three-dimensional world and represents a single view of it in two dimensions. In doing so it establishes a highly *visual* relation between the viewer and nature that emphasizes aesthetic detachment and mastery over nature.

A classic example of how a landscape painting conveys social power is found in John Berger's (1990) analysis of Gainsborough's *Mr and Mrs Andrews*, which was completed *c.* 1750 (Plate 12.2). The painting depicts an upper-class land-owning couple in the foreground, with their estate stretching into the distance behind them. While the painting shows a highly cultivated patchwork of fields, there are no workers in this rural landscape. Their effacement represents the new balance of power in the countryside after the Enclosures Acts seized common land from peasants and parcelled it up to landowners. The landscape is familiar to us as an archetypically *English* landscape, Constable Country if you like, and the nature in it is archetypically English. There are other things going on in this representation too, such as the dominance of the male figure standing over his seated wife, and the phallic symbol of the gun. The symbolism of masculine

control over nature could not be more obvious. Landscape painting invented an English aesthetic of landscape, but it also established a new era of control, both of humans over nature and of the landed upper classes over the countryside in Britain.

Representations of landscape are central to national identities. Indeed, the word 'nature' derives from the Latin *natura*, which comes in turn from *nasci* – to be born. Thus *nature* is linked to other words from the same root, such as nascent, innate, native and nation. Just as wilderness plays an important symbolic role in the pioneer mentality of the USA, so the 'countryside' landscape is quintessentially English. Authors have explored the importance of the oak tree to British culture, as a sign of strength derived from their use to build the ships that allowed the British navy to rule the seas for 200 years.

These constructions of nature also have profound political effects. The historical geographer David Lowenthal (1994) argues that in France the legacy of peasant agriculture has left a national landscape of diverse smallholdings, which is cherished for the variety of cheese, wines and foods that it produces. Staunch French opposition to proposed reforms to the European Common Agricultural Policy, which would increase the exposure of its small producers to market forces, suddenly becomes intelligible as an attempt to preserve the French landscape and culture rather than simply blunt economic protectionism. The relationship between landscape and national identity was carried to its ideological extreme by the Nazis, who were probably the most ecologically aware government ever to exist.

Plate 12.2 *Mr and Mrs Andrews* by Thomas Gainsborough.

(*Mr and Mrs Andrews, c.* 1748–9 (oil on canvas), Gainsborough, Thomas (1727–88)/National Gallery, London, UK/The Bridgeman Art Library)

Plate 12.3 Painting by Banksy.
(Banksy)

Fond of being photographed in forest settings, they viewed the Black Forest of southern Germany as their spiritual home. Non-native plants were eradicated in exactly the same way as the non-Aryan human population. As Anne Whiston Spirn notes (1997: 253–4) 'the use of "native" plants and "natural" gardens to represent the Nazi political agenda should dispel forever the illusion of innocence surrounding the words *nature, natural,* and *native . . .*'.

The British association of nature with idealized rural landscapes has impacts that are felt beyond the art gallery. One is to divide the affairs of the city from the affairs of the country. The graffiti artist Banksy produced a series of works exploring the politics of this division, in which he placed typically 'urban' objects such as parking tickets and graffiti into traditional landscape paintings (see Plate 12.3). This juxtaposition disturbs constructions of the countryside as a somehow 'natural' place untroubled by the problems of cities, and in doing so highlights how exclusive this vision of the countryside actually is. As rural geographers have noted, the idea of the rural idyll is not particularly helpful as it conceals real problems of poverty, drug-use and deprivation in rural areas (see Chapter 11). Conversely, urban areas are assumed to be bereft of nature, despite often being more biodiverse than massive swathes of agricultural land, and supporting vast infrastructures to control elements such as water

and waste (Gandy 2002). In many ways elements of nature are *more* important in the city, but have simply been ignored until recently, as they do not correspond with unspoken cultural constructions of what nature should look like.

12.3 Environmental myths

12.3.1 The power of science

What we observe is not nature itself, but nature exposed to our method of questioning.

(Heisenberg 1958)

Science adheres to a very detailed and rigorous set of rules for building knowledge that accurately reflects the natural world, but in practice science is not immune to social constructions of nature. As the biologist Richard Lewontin says (1993: 3), 'Scientists do not begin life as scientists, after all, but as social beings immersed in a family, a state, a productive structure, and they view nature through a lens that has been moulded by their social experience'. However sanitized lab work is made to appear, experiments require human interaction and judgements to work. Science is a particularly influential realm in which nature is constructed and contested, because it has the capacity to make very strong truth claims about reality, and hence how we should live and behave. But its findings and models remain influenced by cultural factors.

12.3.2 Ecology and politics

In the past 30 years, geographers have become increasingly interested in how dominant models of ecology rely on social constructions of nature. The field of political ecology recognizes that ecological knowledge is not always neutral, but reflects the specific aims of those involved in producing it. This can have profound political consequences upon how different landscapes are managed, and for whom. In the words of Paul Robbins (2004: 12), the goal of political ecology is to 'take the hatchet' to environmental myths, using both scientific and social scientific studies to expose the false assumptions and unsuitability of certain ecological models. Bill Deneven's work on the 'Pristine Myth' described in the previous section is an example of this, but the hatchet has been taken to other scientific myths too (see Case study 12.3).

Thematic case study 12.3

Desertification

Desertification became an important international issue in the 1970s when successive years of drought in the Sahel (the area bordering the southern edge of the Sahara) caused widespread famine. The United Nations (UN) responded with the 1977 conference on desertification, and a subsequent Convention to Combat Desertification, which focused on land degradation in arid, semi-arid and dry sub-humid areas. This conference popularized and publicized the term, but was based on little actual science. David Thomas and Nick Middleton (1994) have identified four 'myths' of desertification in dryland areas:

1. That it is a 'voracious process' affecting one-third of the world's land area.
2. That drylands are fragile ecosystems.
3. That desertification is the primary cause of human suffering and misery.
4. That the UN is central to its understanding and solution.

They argue that the idea of desertification forms part of a longer history of reports from Westerners in the Sahara concerning advancing deserts. For example, early accounts of desertification originated in the eighteenth and nineteenth centuries when scholars believed the Sahara to have been created by the Romans and Phoenicians through deforestation, overgrazing and over-cultivation. In the twentieth century, colonial land managers took on this idea that deserts advance because of the mismanagement and over-exploitation of land. In 1935 E.P. Stebbing, a forester, published his views widely on the causes of the 'encroaching Sahara' in British West Africa, blaming

shortened agricultural fallow periods, shifting agriculture and overgrazing. It was concluded that the Sahara had grown, and was still growing, owing to poor land management, which had worsened under the colonial regime. The spectre of sand dunes encroaching upon fertile land remained an enthralling one, and a report in 1975 suggested that the Sahara was advancing at the astonishing rate of 5.5 km per year.

Thomas and Middleton question both the supposed pace and causes of this process. They suggest that biophysical processes have been the primary cause of Saharan advance over the twentieth century, because of the progressive desiccation (drying out) of North Africa since the end of the Pleistocene ice-age 10,000 years ago. Advances and retreats within this overall pattern reflect climatic variations in rainfall over tens of years. They also attack the idea that humans are to blame for desertification, arguing that policies to reduce the grazing of dryland areas are ineffective. 'Expert' knowledges about dryland fragility, they argue, actually undermine adaptive strategies that have evolved over centuries. For example, lowering livestock densities damages the ability of local farmers to resist drought. The focus on mismanagement obscures both the biophysical causes of dryland degradation and specific social problems, such as firewood scarcity. As Batterbury and Warren note (2001), degradation is usually localized and ephemeral, and it is very hard to overgraze in a dynamic non-equilibrium system dominated by annual grasses, such as rangelands, where external forces like drought far outweigh human factors like over-stocking.

The myth of desertification is bound up with the failure of Western scientists to understand the dynamics of dryland areas, simply falling back on received colonial stereotypes of overgrazing and advancing deserts. As Paul Robbins states (2004: 109), the notion of soil erosion was 'a social construction that helped to secure colonial power', and while the term is generally considered unhelpful amongst scientists today, it perseveres within the policies of international agencies such as the UN and NGOs.

Once specific constructions of nature become established, they assume a life of their own. Funding streams for research, policies and the vested interests of those that work within the field act to keep the concern alive. As they circulate through networks of scientists, institutions, funding bodies, the media and policies, such constructions of nature become established as seemingly incontrovertible facts.

Q Why is it hard to produce definite scientific answers to large-scale environmental questions?

These myths have real influence over the way in which environmental problems are tackled. For example, the assumption that degradation is caused by mismanagement generally leads to the imposition of solutions upon people, rather than developing solutions that use their local knowledge and understanding (see Robbins 2004, Chapter 8, for a wealth of examples). This resonates with Bruce Braun's work on Clayoquot Sound discussed in the previous section, where both economic and environmental representations of the forest actively excluded the indigenous population living there.

There are aspects of ecological science that clearly reflect the places and periods in which they developed. Directly parroting the language of US settlement, ecologists in the early twentieth century talked about the succession of plant 'communities' from *pioneers*, who come in first and settle an area, through to the stable 'climax' forest. There is some circularity to this process. Models of ecology reflect the social context in which they are produced, but then, having become established as scientific 'facts', become models for human society. Some authors have argued that the appeal of nature as a stable community goes back to the Judaeo-Christian myth of the Garden of Eden, when humans supposedly lived in perfect harmony with nature before their fall from grace. This yearning for a return to innocence often finds expression in the kinds of stories that are told about indigenous peoples, such as Amazonian Indians, who are romantically depicted living at one with nature, almost as part of nature rather than humanity (Slater 1995). The exclusion of humans from ecological models distorts conservation practices, which lack the ability to capture the value of landscapes influenced by humans. These oversights are unfortunate; not only do humans influence almost all habitats today, but, in order to live more sustainably, we require ecological models that can understand humans as part of ecosystems rather than just undesirable disturbances to them.

12.4 Constructing human nature

There is a long tradition of looking directly to the animal world in order to establish how humans should behave, and science has been integral to such attempts. If science can show certain behaviours or modes of social organization to be 'natural', then it is assumed that they are unquestionably 'right'. As evolutionary biologist and social commentator Richard Lewontin (1993: 87) suggests, 'the problem for political philosophers has always been to try to justify their particular view of human nature'. This is a game of high stakes; think about how often you hear political commentators talk about things like war being 'a part of human nature', or capitalists justifying the free market on the grounds that competition and the need to own things are 'in our nature'. It does not take much to spot the flaws in this logic. Animals do not cook, but no one would say cooking food is wrong because it is unnatural (apart from Raw Food advocates, of course). Putting it another way, James Weinrich says, 'when animals do something that we like, we say it is natural. When they do something that we don't like, we call it animalistic' (1982: in Bagemihl 1999: 77).

The attempt to draw parallels between animal and human behaviour is one of the most obvious ways in which nature acts as a mirror for the values of society.

Donna Haraway's (1989) work on primatology (the study of apes) re-examines evidence from major research projects on ape behaviour to show that they are biased by the gender of the researchers conducting the study. Male research teams demonstrate a tendency to explain ape behaviour in terms of sexual competition between aggressive males to impregnate passive females. By contrast, female primatologists focus on communication and basic survival activities of the apes, generating very different interpretations of their social behaviour. Depending on who is doing the research, different aspects of behaviour will be measured and different inferences drawn from them. Gender and politics become tangled up in the process of scientific enquiry, so that the ape becomes the site of legitimization for what is and is not 'natural'. This in turn provides a proxy moral framework for the social behaviour of humans. A more recent debate to hit the media involves the naturalness of homosexuality (see Case study 12.4).

Thematic case study 12.4

Gay animals

WHAT DOES A DOG HAVE TO DO ROUND HERE TO PROVE HE'S GAY????!!!!!   (Ingilby 2002)

In 1999 a biologist called Bruce Bagemihl published a book called *Biological Exuberance: Animal Homosexuality and Natural Diversity* (Bagemihl 1999). Drawing upon a lifetime of research, Bagemihl works his way through the animal world, from mammals to birds to cockroaches, describing a profusion of homosexuality, bisexuality, orgies, transvestism and transgenderism. The book includes field sketches of female bonobos engaging in cunnilingus and photographs of male whales entwining their penises in sexual communion, all of which provoked a considerable degree of outrage. Being gay is seen as an exclusively human behaviour, something that by definition is *not* natural. This is a longstanding Christian idea (although there is very little about it in the bible). The Church Council of Nablus in AD 1120 wrote the first law condemning homosexuality, which was subsequently preserved in the Vatican library. By the time of the Renaissance this idea had found its way into the laws of many countries, and today many groups in society claim that homosexuality is wrong because it is 'unnatural'.

Bruce Bagemihl's book threw this myth into doubt. In some penguin colonies, as many as one in 10 pairs of penguins may be same-sex, while in bonobo chimpanzees the whole species is bisexual. If homosexuality is found throughout the animal kingdom then how can it possibly be unnatural? As Bagemihl (1999: 78) says, 'what is remarkable about the entire debate about the naturalness of homosexuality is the frequent absence of any reference to concrete facts or accurate, comprehensive information about animal homosexuality'. Bagemihl also asks a more fundamental question: how has science overlooked this considerable subset of animal behaviour? His answer is that zoology is essentially a conservative profession, dominated by white male heterosexuals who impose social assumptions about homosexuality as perverse and unnatural on to the things they study. He gives examples of how the term 'homosexual' is avoided in favour of terms like 'male-only social interactions', 'multi-female associations', unisexuality, isosexuality or intrasexuality. As one newspaper reported at the time,

> A female ape wraps her legs around another female, rubbing her own clitoris against her partner's while emitting screams of enjoyment. The researcher explains: It's a form of greeting behavior. Or reconciliation. Possibly food-exchange behavior. It's certainly not sex. Not lesbian sex. Not hot lesbian sex . . .

(McCarthy 1999)

It is not only science that is to blame either. How common is it for parents to tell small children who enquire why two male dogs are mounting each other that they are just 'confused'? In the USA pet owners have been acquitted for shooting their dogs on the grounds that they were exhibiting 'gay' behaviour. The topic of gay animals is laced with contradiction, which may explain why it is a topic so beloved of comedians, from Ricky Gervais to South Park's 'Big Gay Al's Big Gay Animal Sanctuary' and Sparky the Gay Dog (famously played by George Clooney).

Prudishness is only half the story though. Bagemihl's work is scientifically controversial because it suggests that sex is not solely practised for reproduction. This seems to contradict the most important theory in biology – that of evolution. Researching animal homosexuality is not the road to a successful career: applications for research grants that appear ignorant of the theory of evolution do not tend to stand a good chance of receiving funding. However, sex between animals – as between humans – is often a matter of enjoyment, rather than procreation, and this applies to animals of the same sex as well as opposite sexes. Non-heterosexual practices can also be beneficial. For example, pairs of male birds may rear eggs 'donated' by a female, and two males can command a larger territory than a heterosexual pair, improving the chances of survival for chicks.

In 2006 the Oslo Natural History Museum ran an exhibition on gay animals, tellingly called 'Against Nature' (BBC 2006a). There was considerable hostility to the exhibition, with organisers reporting criticism and of being told by one opponent they would 'burn in hell'. An American commentator said it was an example of 'propaganda invading science'. More recently still, scientists in America have claimed that they can 'cure' gay sheep by manipulating hormone levels in their brains to make them attracted to ewes (Oakeshott and Gourlay 2006). The implication that gayness might be 'curable' in humans has caused outrage. Competing constructions of nature are not only highly charged politically, but cut across science and religion, parts of modern society that are often seen as unrelated.

Q Do you think there is any such thing as human nature?

12.4.1 Our place in nature

The biological sciences have also attempted to interrogate our genes to find out what makes humans distinctive from other organisms. Attempting to place humans *in* nature, while simultaneously maintaining their position *above* nature, is an old trick. The 'great chain of being' devised by medieval cosmologists represents a clear ordering of different things and beings in the world, with humans second only to God. Maintaining a scientific basis for human superiority has become progressively harder since Darwin placed us squarely among the apes, but has generally been based upon the argument that, while made of the same stuff, humans are much more complex than other animals.

While our attempts to benchmark human behaviour against 'natural' standards have tended to fail, even our own biology is not quite what we were hoping. The Human Genome Project (HGP) was a major international scientific collaboration to map the entire human genome (the entire set of human genes that determines the sequence of our DNA), begun in 1990 and completed in 2003. The project aimed to advance our understanding of human genes in order to support biotechnology research. Tori Holmberg (2005) has argued that the HGP represented a modern attempt to discover some inherent physical basis for what makes 'human-ness' distinct from nature. By counting the number of genes of other organisms as well,

the HGP promised a potential verification of human superiority.

As the HGP progressed the issue of how many genes human beings possessed became increasingly vexed. Estimates varied wildly from 40,000 and 200,000. When the preliminary results were published in 2001 it turned out that humans only have about 32,000 genes, far fewer than expected. The most disturbing element of this discovery was how close other organisms were, with the popular press jumping on the fact that the weed thale cress has 25,000 gene pairs, worms 19,000 and banana flies 13,600. Suddenly we were not as special as we had thought.

Within the scientific world the idea of human uniqueness, although momentarily shaken, was rescued by the explanation that our genes interact in complex ways that allow them to produce greater numbers of proteins than in other organisms. It appeared that individual genes did not control individual elements of an organism's phenotype (appearance and behaviour) after all. The idea of natural human superiority was transposed from gene numbers to protein numbers, and the things being compared shifted from genes to base pairs. So while reporters could write that we share 99 per cent of our genes with apes, they could simultaneously state that we do not share 3.5 million base pairs, thus upholding the distinction between human and animal. While science places us firmly in nature, our cultural beliefs require reaffirmation of our uniqueness.

12.5 Nature and the media

Did the planet betray us? Or did we betray the planet?

(Trailer for the film *An Inconvenient Truth*)

In each of the examples discussed so far, the idea of what is and is not natural forms a critical bone of contention. The way in which different positions are represented is important in understanding how certain ideas become dominant. For example, the idea of wilderness is inseparable from its representation in landscape painting and, more recently, photography. But how does this process work out in the mainstream media?

It is possible to identify two dominant constructions of nature that animate media discussions of environmental change: fragility and violence. The first is closely related to the emergence of environmentalism, which is concerned with the health of the planet that humankind depends upon for its survival. While the idea of caring for the planet seems quite normal to us now, this idea only emerged in the 1960s and 1970s. Denis Cosgrove (1994) argues that the environmental mindset was shaped by the release of photographs of the earth taken during the Apollo Space missions in 1968 and 1972. Splashed across newspapers and television, pictures of the earth from space had a huge cultural impact. Never before had the planet upon which we live been viewed as a single object, and the image of an orb hanging in space seemed to show the planet at its most vulnerable, as a finite, isolated island in a sea of nothingness. Books were written likening the earth to a 'spaceship', replete with life-support systems that require managing and conserving. The image of the earth became the leitmotif for Western environmentalism, a fragile blue image that transcends national borders and political divisions, and on which human features are invisible.

The second construction of nature beloved of the media is that of an uncontrollable, violent force. This version of nature is older, speaking to a pre-modern period in which human fortunes were largely determined by climate, natural disasters and disease. Press reports of floods and hurricanes often speak of nature's wrath, and the unstoppable forces of nature. While the constructions of fragility and violence appear to be contradictory, they both reveal insecurities in society's relation to the environment. Our original fear of nature's capricious power to destroy drove us to strive to control and dominate our surroundings, but our technological mastery has now made us the potential authors of our own destruction.

12.5.1 Natural disasters

Today, these versions of nature come together in media descriptions of natural disasters, which hold both a fascination with the frailty of human life, and highlight the potential for future ecological destruction. In the environmental era, the spectre of global catastrophe is constantly invoked, from the impending ice age predicted in the 1970s, to the threat of global warming that emerged in the 1990s. In *The Ecology of Fear*, Mike Davis (1998) explores this fascination with disaster through the city of Los Angeles, which he claims has been destroyed 145 times on film and in books by everything ranging from earthquakes and nuclear weapons to Bermuda grass and the Devil. Crisis fascinates humans; it sells papers and makes people watch TV and films. The film *The Day After Tomorrow* suggests what may happen if the Gulf Stream shuts down because of the melting of the Canadian and Greenland ice caps. While there is evidence from Quaternary studies that this has possibly happened in the past, the film suggests that it could freeze the North Atlantic overnight. In reality the process would take at least 50 (and probably more like 500) years. As the poster for the film shows (Plate 12.4), nature's capacity for destruction is used to sensationalize the film. Representations of crisis can also be politically useful, for example to international aid agencies seeking to attract funding.

But what is 'natural' about a natural disaster? The Cambridge dictionary defines a disaster as 'an event causing great damage, injury or loss of life', which means that it cannot be a disaster unless there are people involved. Yet equally a *natural* disaster would be a massive event not harming humans in any way. But the things commonly described as natural disasters *do* harm humans. Increasingly disasters are seen as events that exacerbate pre-existing social and economic problems. The flooding of New Orleans in 2005 in the wake of Hurricane Katrina demonstrates this point well – the people who were harmed most were the poorest because they were unable to leave. Media representations of the disaster tended to fall back on the traditional construction of nature as a destructive force. A *New York Times* editorial on 30 August 2005 led with the headline 'Nature's Revenge', even though it went on to criticize the *government* policies that exacerbated the disaster (notably the loss of protective coastal wetlands and the disrepair of the levees). The idea that nature can do things to us often results in its personification as a subjective force (in this case harbouring a whim for revenge). The quote from Al Gore's film on climate

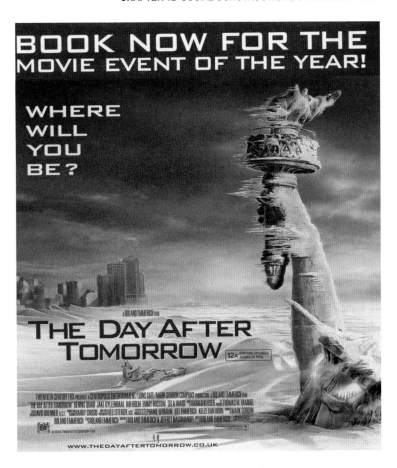

Plate 12.4 Film poster for *The Day After Tomorrow*.

(Ronald Grant Archive)

change, which opens this section, asks whether we betrayed the planet, or the planet betrayed us. While one can ask whether it is possible for a lump of rock to betray anything, let alone 'us', the personification of nature he uses evokes both our power over nature (the capacity for us to betray the planet) and fear of nature (the possibility that it is going to betray us).

Parallels with religious stories of God punishing humans are hard to ignore. But before we laugh too much at the polytheistic beliefs of ancient civilizations that saw nature as a world inhabited by gods, we should remember that the US National Hurricane Center continues an old tradition of giving *names* to hurricanes. Such disasters are also commonly referred to as 'acts of God' by politicians and insurance companies. The BBC website offers some clues as to what nature may have been taking revenge for, claiming that the levees surrounding New Orleans were a 'snub to nature'. Statements like this beg the question of whether it is actually possible to offend water and wind, and if so, how water and wind would be able to seek revenge. Representations of these disasters as somehow being the 'fault' of nature distract attention from the political, economic and social causes of human loss.

12.5.2 Environmental knowledge and the media

Climate change and natural disasters are pervasive, invading our TV screens and monopolizing newspaper column inches. And everyone seems concerned, from movie stars like Angelina Jolie to trendy politicians like David Cameron. Never has nature been so hot a topic.

The media is a critical sphere in which knowledges about nature are produced and consumed. The wildlife documentaries that are broadcast into our homes reproduce notions of Africa as a great wilderness, going to great lengths to capture scenes in which people and evidence of people are absent (Davies 2000). This is an increasingly difficult task given the level of tourist traffic and numbers of indigenous peoples that are found across many African savannah areas. Wildlife films have been edited for years to present only heterosexual animals, and TV shows like 'Meercat Manor' go out of their way to liken animal behavior to that found in typical human families. The Internet represents a huge source of scientific knowledge but is largely unregulated, producing a range of conflicting claims about the state of nature. For example, popular environmental websites vary wildly in their estimates of

---

Spotlight box 12.1

Climate porn

In August 2006 the Institute for Public Policy Research released a report claiming that the representations of climate change in the media are unhelpfully alarmist, focusing on thrilling headlines that prevent the public from understanding what they can do about the problem (Ereaut and Segnit 2006). The report is interesting because it shows that the government is increasingly aware that the media plays a critical role in shaping how people understand climate change. Based on research that analysed more than 600 newspaper articles, and 90 TV and radio excerpts, the report identified the dominant construction of nature driving the climate change debate as 'alarmist'. Such articles focus on the overwhelming size of the problem, and its potentially disastrous effects.

By focusing on the terrible, unstoppable forces of nature, alarmist constructions represent a form of climate pornography that seeks to thrill rather than increase understanding or engender action. This construction distracts us from what we can actually *do* to tackle environmental change. It is particularly destructive to the campaign of 'small actions', which emphasizes how small changes to behaviour can help combat climate change (and save money in the process). The report suggests that the small actions campaign is rendered impotent by the sheer scale of the alarmist claims.

Q How important do you consider the media to be in tackling climate change?

---

biodiversity loss (Ashlin and Ladle 2006 provide a salutary tale concerning the unreliability of information from the Internet).

The media has assumed increased importance as policy makers attempt to encourage people to shift towards more sustainable lifestyles. You may have heard of a UK government report branding the sensationalization of the climate debate a form of 'climate porn'; a scaremongering approach to environmental change that excites, but prevents meaningful action (Spotlight box 12.1). As the primary interface between the majority of the population and environmental issues, the way in which ideas of nature are communicated has a decisive impact on how people understand and respond to environmental challenges.

12.6 Conclusions

12.6.1 Nature: a dangerous idea?

This chapter started by asking what nature is. Geographers have sought to answer this question in a range of ways, suggesting that nature is a social construction that varies between places and periods. Rather than lessen its importance, the idea that nature is an idea reveals its power to dictate how we should behave, how we should live and how we should value and use the environment. It justifies the exploitation of certain people while validating the luxuries of others. It has been used to persecute homosexuals and legitimize genocide. Nature is indeed a dangerous idea.

At the beginning of the chapter it was argued that social constructionists seek to demonstrate two aspects of nature: first how ideas of nature do not represent some external 'truth' but, rather, come to reflect the cultural, political and spiritual beliefs of a society; and second, the ways in which ideas of nature are used to support specific political viewpoints. In light of the case studies, this picture has become more complex. As dominant cultural norms and preferences influence ideas about nature, so these arguments about nature are then used to support dominant cultural norms. Geographers seek to reveal the circularity of this process in order to expose repressive ideologies and open up new political possibilities for existence.

12.6.2 The challenge

The idea of nature remains central to the challenges facing society in the twenty-first century. The growing priority attached to climate change by leading political figures means that debates concerning how we should live, who should make sacrifices and how change should be managed are becoming ever more pressing. The increasing influence of biotechnology as a major global industry raises questions concerning how far we should manipulate biology to our benefit, and whether or not knowledge about our genes will reveal some essential truths of human existence. The idea of nature looms large in all these debates, and will continue to do so. You will see people talking about nature every day on the TV, in newspapers and around you in everyday life. Think about what people mean when they use the word. What characteristics are they attributing to nature? What are they excluding? What political and social arguments are they supporting through their appeal to nature? Be critical of how nature is used; the power to refute dominant presentations of the world and our place in it is the power to create a different world.

Learning outcomes

Having read this chapter, you should be able to:

- Understand nature as a social construction.
- Recognize that perceptions of nature vary across time and space.
- Identify basic constructions of nature in cultural representations.
- Understand the political power of different representations of nature.
- Critique ideas and concepts such as the 'limits to growth' and desertification.
- Appreciate how science is used to make arguments about human 'nature'.
- Understand how the media uses ideas of nature to report environmental issues.

Further reading

Castree, N. (2005) *Nature*, Routledge, London. Covers the major theoretical approaches to nature that geographers have used through time.

Castree, N. (2011) *Making Sense of Nature*, Routledge, London. Focuses on the ways in which nature is represented and the implications of these representations.

Demeritt, D. (2002) What is the social construction of nature? *Progress in Human Geography, 26*, 767–90. Reviews the social construction of nature literature in human geography.

Hinchliffe, S. (2007) *Space for Nature,* Sage, London. An introduction to current geographical thinking beyond nature.

Macnaghten, P. and Urry, J. (1998) *Contested Natures*, Sage, London. Covers a range of approaches to the sociology of nature.

Robbins, P. (2004) *Political Ecology*, Blackwell, Oxford. Accessible introduction to major work on the myths of environmental science, with lots of case studies.

Useful websites

www.sierraclub.org/john_muir_exhibit Interesting information and quotes concerning how wilderness became a valued idea in the USA.

www.nhm.uio.no/againstnature/index.html The website of the Natural History Museum of Norway for the 'Against nature?' exhibition.

www.funtrivia.com/ubbthreads/showthreaded.php?Cat=0& Number=34605&page=6 Great blog about gay animals.

www.youtube.com/watch?v=L6C_sADgVZo First quarter of a documentary film titled *The Truth about Gay Animals*. The other three quarters can also be found at this link.

www.earthobservatory.nasa.gov/Study/Desertification NASA website with an interesting take on the desertification myth in the era of GIS.

www.genome.gov Information on the Human Genome Project. Lots of outlandish rhetoric about how the HGP will reveal the secrets of our inner universe.

www.youtube.com/watch?v=wnjx6KETmi4 The trailer for *An Inconvenient Truth*. Climate porn at its best! Notice the images and wording. Does it '*shake you to your core*'?

For annotated, clickable weblinks and useful tutorials full of practical advice on how to improve your study skills, visit this book's website at **www.pearsoned.co.uk/daniel**

GEOGRAPHY, CULTURE AND GLOBAL CHANGE

Cheryl McEwan and Shari Daya

Topics covered

- A definition of culture
- The 'cultural' and 'spatial turns' in the social sciences
- An evaluation of the extent of cultural globalization
- The relationships between place and cultural identity
- Cultural production and consumption
- Progressive ways for geographers to think about culture

Culture is a word on everybody's lips these days. Hardly a moment seems to pass when we do not hear on the radio or television, see in newspapers and magazines, or read in academic texts some account of the world of the cultural. Governments at all levels announce cultural policies and provide funding for cultural activities; intellectuals announce the need for cultural initiatives or bemoan the loss of traditional cultural values; famous cultural icons – musicians, artists, novelists – themselves culture makers, are increasingly sought out for their opinions on the state of the world.
(Kahn 1995: ix)

13.1 What is culture?

This chapter explores some of the challenges posed by and for culture in the twenty-first century. First, however, it is necessary to define what is meant by culture. This itself is a complex and difficult task. By the 1950s, for example, there were over 150 different academic definitions of culture. As Mike Crang (1998: 1) argues, despite sounding like the most airy of concepts, culture 'can only be approached as embedded in real-life situations, in temporally and spatially specific ways'. Cultures are part of everyday life. They are systems of shared meanings often based around such things as religion, language and ethnicity that can exist on a number of different spatial scales (local, regional, national, global, among communities, groups or nations). They are embodied in the material and social world and are dynamic rather than static, transforming through processes of cultural mixing or transculturation (discussed below).

Cultures are also socially determined and defined and, therefore, not divorced from power relations. Dominant groups in society attempt to impose their ideas about culture and these are challenged by other groups, or **subcultures**. The latter might include various types of youth culture, gang culture, and different ethnicities or sexualities, where identities are organized around different sets of practices and operate in different spaces from dominant cultures (Crang 1998; Skelton and Valentine 1998). Culture makes the world meaningful and significant. Geographers suggest that we should think of culture 'not as a thing, but as a relationship' (Mitchell 2000: xviii) or as a process in which we are all involved (see Spotlight box 13.1).

13.1.1 The 'cultural turn'

As Kahn suggests, culture has generated a great deal of interest in recent years, for academics, policy makers, and at the popular level. Geographers have turned their attention towards cultural explanations of global, national and local phenomena, exploring issues such as the cultural embeddedness of economic processes (e.g. Amin and Thrift 2004), the relationship between cultures, identities and consumption (see Chapter 19), and cultural constructions of social relations of gender, ethnicity and class that shape people's lives (e.g. Nelson and Seager 2004). However, the current popularity of culture is not simply a trend in academe, but is reflective of a broader cultural turn in (Western) society as a whole.

The world has changed fundamentally in the past three decades and these changes are deeply cultural in character. For example, enormous changes have occurred in 'advanced' economies since the early 1980s (the decline in manufacturing, the growth of services, the feminization of the workforce, increased flexibility – all characteristic of 'post-Fordism' as discussed in Chapter 3, pp. 77–8). However, as Stuart Hall (1996: 233) argues, if 'post-Fordism' exists, it is as much a description of cultural as of economic change. Florida (2002) characterizes this as a shift from an industrial to a creative age, with 40 per cent of people in the US and UK economies now working in 'creative' sectors of science, technology, culture, arts and entertainment, and creative economies burgeoning in cities as diverse as New York, Shanghai and Bangalore. Culture has increasingly been brought into governmental and economic spheres, with creative economy initiatives developed for a number of years for small towns and small- and

---

Spotlight box 13.1

Cultures: a summary

- The systems of shared meanings which people who belong to the same community, group or nation use to help them interpret and make sense of the world, and to reproduce themselves.
- These systems of meanings include language, religion, ethnicity, custom and tradition, and ideas about 'place'.
- Cultures cannot be fixed, but shift and change historically through diffusion and transculturation.

- Culture is a process rather than a thing, but is embodied in the material and social world.
- Cultures include those social practices which produce meaning, as well as those practices which are shaped by those shared meanings.
- Cultures are one of the principal means by which identities are constructed; they give us a sense of 'who we are', 'where we belong' – a sense of our own identity and identity with others.
- Cultures are not divorced from power relations.

Source: adapted from Hall (1995)

Plate 13.1 Image marketing is central to popular consumption: the world of commodities is profoundly cultural.
(© Benjamin Rondel/CORBIS)

medium-sized cities as well as rural regions in the UK, Australia, New Zealand and Canada (Christopherson and Clark 2007; Christopherson 2008). These initiatives hinge on the arts and community cultural activities, including festivals and special events, which are believed to contribute to community identity, increase community pride and foster participation in local economies. Creative industries boost the local economy by attracting tourists, employers and a workforce who appreciate the sense of community that they generate.

Modern consumption also depends overwhelmingly on image (for example, the marketing of food and drink products and fashion clothing) (see Plate 13.1). Movements around the world of images, symbols, modes of thought and communications are unparalleled in terms of their volume, speed and complexity. As computer technology, video imagery and electronic music demonstrate, the material world of commodities and technologies is profoundly cultural. In addition, culture has become increasingly commodified; in other words culture is being translated into material goods that can be marketed and sold. We can see this in the rise of 'cultural tourism' (Gibson 2009). In southern Africa, for example, many 'cultural villages' have sprung up in rural areas to showcase local traditions and sell crafts.

These villages are intended to change the relationship between consumers in the global North and peoples in global South, being less exploitative of the latter in economic terms and supposedly fostering greater cultural exchange. However, as Saarinen (2007) points out, these cultural villages have developed largely in response to what tourists want; communities package and sell a version of their culture that fits with Western ideas of what African culture should be.

13.1.2 The 'spatial turn'

There has also been a 'spatial turn' in explanations in cultural and social theory. The world is changing fast, and the rate of change is probably greater than ever before. Fast-developing technologies such as the Internet and satellite communications mean that the world is becoming more global and more interconnected. The increased speed of transport and communications, the increasing intersections between economies and cultures, the growth of international migration and the power of global financial markets are among the factors that have changed everyday lives in recent decades. There is no historical equivalent of the global reach and volume of 'cultural traffic' (Held *et al.* 1999) through contemporary telecommunications, broadcasting and transport technologies. The challenge for geographers is to find ways of understanding and interpreting these changes.

Culture can be said to operate at three spatial scales: local, national and global. Two main interpretations have dominated discussion. The first highlights the global aspects of change. At its simplest, this approach suggests that it is possible to identify processes of cultural homogenization – the idea that everywhere is becoming the same – dominated by the USA and most easily recognized in terms such as 'Coca-Colonization', 'McDonaldization' and 'Hollywoodization' (Cochrane 1995: 250) (see Chapter 19, pp. 400–402). This *cultural globalization* involves the movement of people, objects and images around the world through telecommunications, language, the media industries, radio and music, cinema, television and tourism. The second interpretation places emphasis on the local and the localization of people's everyday lives and experiences. Instead of homogenization, emphasis is placed on the diversity of culture, on the ways in which global brands such as Coca-Cola or McDonald's are reinterpreted locally so that they take on different meanings in different places. The emphasis here is on the *interconnectedness* of global and local processes. For example, although the same event can be witnessed simultaneously around the world (e.g. an incident broadcast in a CNN news report, or an international sporting event), this event will be interpreted differently in different places. Furthermore, locality does not necessarily refer to the opposite of globality. For example, some environmentalists imagine the world as a locality, a 'global village'. Cultural theorists have a growing interest in how increasing globalization, especially of cultural production and consumption, affects people's sense of identity and place at both local and national levels (Goss 2006). Thus a geographic or spatial perspective has become central to studies of culture more widely. These are some of the concerns that form the focus of this chapter. Subsequent sections explore in more detail ideas about a global culture, examine ways of rethinking local cultures, and explore progressive ways of thinking about cultures in contemporary contexts.

13.2 Towards a global culture?

13.2.1 Imagining a global culture

Processes of cultural globalization have a very long history and are not peculiar to contemporary times.

Through global patterns of trade and migrations, and through the spread of religions and empires, people, objects and ideas have been circulating for centuries (see Chapters 1–3). However, contemporary globalization is distinctive in extent, form, rapidity of change, intensity and impact. Some commentators suggest that the idea of a global culture is becoming as meaningful as the idea of national or local cultures, with different places and cultural practices around the world converging and becoming ever similar. As Shurmer-Smith and Hannam (1994: 76) argue, a global culture might be the product of two very different processes:

1. The export or diffusion of supposedly 'superior' cultural traits (e.g. Western time-frames – the 24-hour day and the Gregorian calendar) and products (e.g. the motor car, television) from advanced countries, and their worldwide adoption ('Westernization', 'Americanization', 'modernization'). This is believed to create global cultural convergence – people around the world are becoming increasingly similar in terms of consumption, lifestyle, behaviour and aspirations (see Case study 13.1). It can be perceived positively (as 'modernization' or 'development') or negatively (as 'cultural imperialism', where 'we' assume that others in the world should aspire to be like 'us').

2. The mixing, or hybridization, of cultures through greater interconnections and time–space compression (the shrinking of the world through transport links and technological innovation), leading to a new universal cultural practice. This challenges the notion of unidirectional 'Westernization' and allows us to consider how Western cultures have influenced and are also being influenced by this mixing of cultures. Flows of music, food, ideas, beliefs and literature continue to percolate from around the world into the cultures of the West.

In reality, both these processes are flawed explanations for what is happening today. If a global culture exists, it is far from a product of unidirectional 'Westernization'. However, alternative ideas about cultures mixing to produce a universal global culture are also problematic. Cultures are mixing, but this does not necessarily mean we are all becoming the same.

13.2.2 Debunking global culture?

A different departure point for discussing global culture is that there is no such thing. Ideas about a singular global economy, politics and culture imply some

Thematic case study 13.1

The globalization of culture: some examples

Fashion

Clothing is globalized both in terms of the cultural identities to which different items appeal ('African' prints, 'Indian' embroidery) and in terms of the global commodity chains through which they are produced and sold. Through multiple media including magazines, television, blogs and websites, trends travel rapidly across the world. The production, distribution and consumption of clothing have a particular geography. Fashion design happens mainly in 'global cities' in the West, while the labour of cutting and sewing is concentrated in the developing world, particularly in Asia. The global geographies of fashion are also strikingly gendered; although men dominate the design industry, women are the main targets and consumers of fashion and also make up the majority of clothing factory workers (see Dwyer 2006).

Food

Food has perhaps the longest history of globalization of any cultural artefact. Spices, tea, and sugar are just three of the commodities whose trade have shaped our modern global economies. The geographies of what we eat are increasingly under discussion as consumers become more conscious of the environmental impacts of food production and travel (food miles), and the potential health risks of mass agriculture and genetic modification. As a result of these anxieties (Jackson 2010), movements promoting locally produced, organic food and 'slow' food have emerged in many places. Similarly, consumer recognition of the often exploitative nature of food production in developing countries has boosted sales of fair trade and other 'ethical' products in Western supermarkets. While these movements aim to break down unjust social divisions, we should also pay attention to the ways in which they may reinforce or re-shape identities of class (organic, 'slow' and fairly traded foods typically cost more) and understandings of Third World producers as 'other' (Cook *et al.* 2010).

Tourism

Tourism is one of the most obvious forms of globalization. The geography of tourism is skewed, since it is dominated by people of all classes from developed countries (North America, Western Europe and, increasingly, Japan and Australia). It can also be exploitative, particularly through the growth of international sex tourism and the dependency of some developing economies on the exploitation of women. However, it is a form of international cultural exchange that allows vast numbers of people to experience other cultures and places. It also locks specific places (tourist destinations) into wider international cultural patterns, such as in the case of the Lake District, which 'only really became part of England when many visitors, especially artists and writers, travelled to it from the metropolitan centre at the end of the eighteenth century onwards' (Urry 2005: 80). Many key English writers became known as 'Lake poets' even when they were not from the area, and poets such as Wordsworth and Coleridge became tourist attractions, 'locking' the Lake District into a relationship with broader ideas of landscape, literature and romance, which persists today as people from all over the world travel to the area on literary tours.

sort of worldwide commonality that does not exist. First, the image of rampant cultural imperialism by the West, and especially the USA, is flawed since apparent cultural sameness is limited in scope, located only in the consumption of products and media images. The possibilities of this eroding centuries of local histories, languages, traditions and religions are far-fetched and people in different parts of the world respond to these images and products in different ways. Many millions of people are not able even to access the Internet. For example, in 2007, there were on average only about 3 Internet users per 1000 people in Bangladesh and about 10 per 1000 people in Malawi, compared with 661 per 1000 in the USA and 905 per 1000 in the Netherlands (*CIA World Factbook* 2008). There is no single global culture and this is because of the unevenness of globalization.

Second, some theorists would argue that national cultures remain stronger than global cultures. This is borne out when we consider the many conflicts occurring throughout the world along the geopolitical fault-lines of national cultures (the ongoing conflicts between Pakistan and India are but one example of this). For the past 200 years, nation-states and national cultures have monopolized cultural power (state television is one example of a national institution influencing national cultures within national territorial boundaries). At the end of the twentieth century, this balance began to change, with international telecommunications and media corporations challenging the centrality and importance of national cultures. However, it could be argued that despite these changes a great deal of cultural life is still organized along national and territorial lines.

Third, if culture is a system of shared meanings, then looking at the world today there are clearly many systems of shared meanings and many different cultures. People in different places use different techniques and technologies to reproduce culture, such as oral histories, literature or television and cinema. These techniques have different patterns of dispersion, penetration and scale. Therefore, some cultures are more likely to become globalized than others – those reproduced through television and cinema have a greater range and speed of dispersion than those reproduced

through oral histories. However, this does not mean that globalized cultures completely erode localized cultures; the ways in which these different cultures intersect is important. Those 'things' (products, symbols, corporate entities) that have become global signifiers are clearly globalised (they are recognized the world over), but the ways in which people around the world make cultural responses to them are complex and multiple (see Case study 13.2). Globalization of products and symbols does not necessarily equal Westernization. For example, Japanese consumer goods do not sell on the back of exporting Japanese culture but on a market strategy based around the concept of *dochaku* ('glocalism'). This involves a global strategy not of imposing standardized products but of tailoring Japanese consumer products to specific local markets. These goods are, therefore, both globalized and localized. Consequently, how intersections between cultures are played out at local levels are of significance, and this suggests that imagining a universal global culture is quite problematic.

Thematic case study 13.2

Culture and social networking

Since their introduction in the early twenty-first century, social network sites (SNSs) such as MySpace, Facebook, Bebo and Cyworld have attracted millions of users, many of whom have integrated these sites into their daily practices. At one level, we might think of the globalization of SNSs as evidence of an emerging global culture. However, while the technological phenomenon has a global reach (with the caveat outlined previously that millions of people still do not have access to the Internet), the ways in which people around the world make cultural responses to SNSs are complex and multiple. As Boyd and Ellison (2007) argue, while their key technological features are fairly consistent, the cultures that emerge around SNSs are varied. For example, most sites help maintain pre-existing social networks, but others enable strangers to connect with each other based on shared interests, political views or activities. Some sites deliberately cater to diverse and heterogeneous audiences, while others attract people based on common language or shared identities, such as ethnic, sexual, religious or nationality-based identities. Sites also vary in the extent to which they incorporate new information and

communication tools, such as mobile connectivity, blogging and photo/video-sharing.

The question of whether and how much cultural differences impact upon the way people respond to and interact with social networks is an important one. It could be argued, for example, that networks such as Facebook mainly reflect and accommodate values and norms prevalent in Western cultures, which explains why they were at first more successful in countries such as the UK, USA and Canada than they are elsewhere. However, the design and use of SNSs also varies in different locations and are often adapted to local cultural norms, tastes and preferences. For example, the page design of SNSs varies from place to place, with greater use of pastel or muted colours and emoticons on South East Asian sites, in contrast to the bolder, darker colours used by sites in Europe and North America, for example. Different cultural expectations around privacy and personal modesty also shape the use of SNSs. For example, some users are often uncomfortable with posting pictures of themselves, preferring to use avatars. Therefore, while technologies such as SNSs become increasingly global, local and national cultural norms continue to shape how people use social networking.

13.2.3 Rethinking global culture

Instead of imagining a global culture that is erasing local and national cultures, we can think of local, national and global as three important, interconnected spatial scales at which culture operates. Those aspects of culture that operate at the global level are 'third cultures' (Featherstone 1995: 114). National institutions are no longer in complete control of cultural globalization and 'third cultures' (sets of practices, bodies of knowledge, conventions and lifestyles) have developed in ways that have become increasingly global and independent of nation-states. Phenomena such as patterns of consumption, technological diffusion and media empires are part of these third cultures, and transnational and multinational corporations are the institutions that make them global. In this sense, global cultures exist but only as third cultures, outside national and local cultures, yet intersecting on both these scales in different ways around the world.

Acknowledging that 'our' global view might be very different from that of people elsewhere, living in very different contexts, is also important. It is clear that multiple global cultural networks exist, such as those connecting the overseas Chinese with their homeland, or those linking Islamic groups around the world. These networks disrupt any notion of a singular global culture. Power and inequality bring into question the idea of global cultures. As Massey (2009) argues, a **power geometry** exists, which gives people with different access to power different notions of what global means. New institutions (like global media corporations) for the production, transmission and reception of cultural products are creating infrastructures supporting cultural globalization, including electronic infrastructure (radio, television, music, telecommunications), linguistic infrastructure (the spread of bi- and multilingualism, particularly the dominance of English), and corporate infrastructure (producers and distribution networks). As we have seen, these new institutions often operate at scales beyond the nation state, and they are sites of power in the production of culture. The ownership, control and use of these institutions remain uneven across and within countries (Held and McGrew 2007), thus creating 'power geometries' that are centred overwhelmingly on the West. People have very different experiences of culture because of their different locations in the world and their relationship to these sites of power. Mapping this power geometry, identifying sites of power and revealing the marginalization of some peoples around the world by cultural

globalization are increasingly significant. Of equal significance are forms of resistance (e.g. **culture jamming** (Dery 1993) – media hacking, information warfare, 'terror-art' and graffiti – which aims to invest advertisements, newscasts and other media artefacts with subversive meaning) and the ways in which marginalized peoples might be empowered by engaging with, and perhaps transforming, the new institutions driving cultural globalization.

In summary, global processes are occurring, but they do not produce a universal global culture, they are not distributed evenly around the world and are not uncontested. Global cultural processes are not simply a result of a unidirectional 'Westernization', since culture flows transnationally. A number of different global cultures exist as 'third cultures' – in patterns of consumption, flows of knowledge, the diffusion of technologies and media empires that operate beyond, but connecting with, the local and national scales.

13.3 Reinventing local cultures?

13.3.1 Locality and culture

It has often been assumed that there is a simple relationship between local place and local culture. Places were thought of as having a distinct physical, economic and cultural character; they were unique, with their own traditions and local cultures that made them different from other places. It is clear, however, that processes of globalization are also posing serious challenges to the meaning of place. Places and cultures are being restructured. According to Massey and Jess (1995a: 1), 'on the one hand, previous coherences are being disrupted, old notions of the local place are being interrupted by new connections with a world beyond'. The appearance of 7–11 stores in the rainforests of northern Thailand is one example of how even the remotest of places are becoming increasingly internationalized, in this case through tourism. 'On the other hand', Massey and Jess continue, 'new claims to the – usually exclusive – character of places, and who belongs there, are being made.' We see this in contemporary Western Europe and North America, where 'asylum seekers, migrants, Muslims, militant youths, pan-handlers, carriers of transmissible diseases' (Amin 2010: 10) are increasingly racialized and viewed as threats to established value systems and ways of life. Similar problems have also emerged in countries such as South Africa, which experienced a dramatic rise in xenophobic violence in 2008 that persists today (Dodson 2010). Therefore, modern life is characterized both by decentralization and globalization of culture *and* by the resurgence of place-bound traditions. Following this, the impact of the new global context on local cultures has two, possibly contradictory, outcomes.

13.3.2 Negative sense of culture

Where global processes are perceived to pose a threat to local culture, there might be an attempt to return to some notion of the exclusivity of culture. At the extreme, this might take the form of exclusivist nationalism or even 'ethnic cleansing'. Reactions to the perceived threat to local cultures include nationalistic, ethnic and fundamentalist responses, which also entail a strong assertion of local cultures, such as reviving or inventing local traditions and ceremonies. These can create a level of local fragmentation, with a parochial, nostalgic, inward-looking sense of local attachment and cultural identity. In this sense, cultures are thought of as **bounded**, with very clear definitions of 'insiders' and 'outsiders' in the creation of a sense of belonging, and producing **geographies of exclusion** (Sibley 1995).

For example, English rural areas are often conceptualized as the preserve of culture and identity (see Chapters 11 and 12). This idea is mobilized through the myth of the 'rural idyll', where rural cultures are thought of as timeless, unchanging and unaffected by global processes. This cosy vision of a peaceful countryside excludes many people who live in rural areas but do not fit this stereotype. Travellers, environmental protesters, hunt saboteurs, people with alternative lifestyles and people from minority ethnic communities are deemed to be 'outsiders' and a threat to local cultures (Garland and Chakraborti 2004). Ingrid Pollard's photography challenges ideas of Englishness as embodied in the countryside. Some of her most famous images depict black people in quintessentially English rural landscapes, challenging the unspoken understanding of these as white, upper- and middle-class spaces. The caption to one of her photographs reads 'it's as if the black experience is only lived in an urban environment . . . A visit to the countryside is always accompanied by a feeling of unease, dread' (Pollard 1984). As Halfacree (1996) argues, the 'rural idyll' is a selective representation. It is exclusive in its class, race and status connotations, is profoundly conservative and demands conformity. It is based on a very inward-looking sense of place and culture (see Plate 13.2).

Conservative reactions to change can be thought of as a kind of cultural fundamentalism through which the process of cultural change is often bitterly contested. Gender plays an important role in this. Women are often considered as guardians of the borders of culture (Yuval-Davis 1997). They not only bear children for the collective, but also reproduce it culturally. In closed cultures, the control of women's sexuality is seen as imperative to the maintenance of the purity of the cultural unit; women are discouraged from marrying outside their cultural and ethnic group. Ethnicity and culture, therefore, are seen to be one and the same. In addition, symbols of gender play an important role in articulating difference between cultural groups. Women's distinctive ways of dressing and

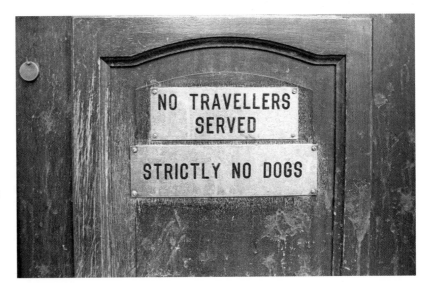

Plate 13.2 Exclusion in the countryside: travellers and other minority groups are often excluded from shops and public houses, marking their position as 'outsiders'.

(Joanne O'Brien Photography)

behaving very often come to symbolize the group's cultural identity and its boundaries. Women are often the intergenerational transmitters of cultural traditions, customs, songs, cuisine and the 'mother' tongue, primarily through their role as mothers. This is especially true in minority situations where the school and the public sphere present different and dominant cultural models from that of the home. Recent controversies over the Islamic veil, for example, have seen countries across Europe wrestling with issues of religious freedom, civil rights, women's equality, secular traditions and escalating fears of terrorism. France banned Muslim headscarves and other 'conspicuous' religious symbols at state schools in 2004, and in 2010 made it illegal in public places to wear any clothes designed to hide the face. Belgium passed a similar law in 2010, banning any clothing that hides the wearer's identity in public places. Turkey, a secular Islamic country, banned the wearing of head-scarves in all civic spaces in 2005 but amended the ruling in 2008 to allow women at universities to wear scarves tied under the chin. The Italian parliament approved anti-terrorist laws in July 2005, which make covering one's features in public – including through wearing the burqa – an offence. The Dutch cabinet backed a proposal in 2006 to ban the few dozen Muslim women who choose to wear the burqa from doing so in public places on grounds that it disturbs public order, citizens and safety. In the United Kingdom in 2010, the Conservative MP Philip Hollobone proposed a law to regulate wearing in public garments that cover the face, calling the burqa offensive. The debates over Islamic dress reveal some of the issues that can arise when marginal, minority cultures are seen to clash with the norms and expectations of the majority population, which in turn raises questions about the possibilities and challenges of multiculturalism as opposed to integration and/or assimilation (discussed below in Section 13.4).

Societies do not evolve smoothly from closed, bounded perceptions of culture to more open, dynamic notions. The question of cultural power, identity and resistance also needs to be considered. For many groups, cultural survival is seen to depend on a closed idea of culture, with strongly marked boundaries separating it from 'others'. The controversies over Islamic dress and the continuing sectarianism and possible threats of terrorism which still exist in Northern Ireland despite recent progress are examples of different cultures, religions and national identities colliding with each other. Elsewhere the mixing of cultures under the impact of globalization is often seen as threatening and

as weakening the sense of cultural identity. Immigration is seen as a particular threat, creating a revival of ethnicity that cuts across the political spectrum. Examples include the 'little England' reaction to closer European integration and the rise of neo-fascism in Europe (characterized by anti-immigration and racism). Perceived threats to religious identities have also witnessed the strengthening of Islamic and Christian fundamentalism around the world. These phenomena are not all the same, but they do share a response to globalization that involves a closed, fixed, bounded and often place-specific definition of culture, and a strong resistance to changes heralded by cultural globalization.

13.3.3 Positive sense of culture

A more positive response to global processes would be to imagine cultures as fluid, ever-changing, unbounded, overlapping and outward-looking – akin to Massey's (1994: 151) 'progressive sense of place'. This involves people being more cosmopolitan (free of prejudice and tolerant of difference). Increasing interconnectedness means the boundaries of local cultures are seen to be more permeable, susceptible to change, and difficult to maintain than in the past. Rather than everywhere becoming the same, some nation states have reconstituted their collective identities along pluralistic and multicultural lines, which take into account regional and ethnic differences and diversity. In Europe, this involves re-creation and invention of local, regional and sub-state or new 'national' cultures (for example, the cultural renaissance of the Scots, Basques and Catalans, or the cultural assertiveness of minority ethnic communities in cities such as London, Paris and Berlin). Thus, what can be perceived as destruction of local cultures by globalization might in fact be the means of creating new senses of locality. This still involves notions of local identity, but recognizes both the differences between cultures and their interconnectedness, taking account of the positive aspects of cultural mixing and increased cosmopolitanism. This new sense of local identity is based on notions of inclusion rather than exclusion.

This is not to say, of course, that all people within the same place will share the same culture and the same sense of locality. Within these more culturally pluralistic and cosmopolitan locales, different class factions, ages, genders, ethnicities and religious groupings mingle together in the same sites, consuming the same television programmes and products, but in highly uneven ways. These groups often possess different senses of affiliation to places and localities, possess different cultural identities and belong to different cultural groupings (Featherstone 1995: 97). A progressive sense of culture does not foresee the locale as a 'melting-pot', where everything becomes the same, but rather recognizes the different experiences of people, and that increasing interconnections might create new, dynamic and exciting cultural forms. An understanding of this is crucial to the creation of a progressive notion of place and culture, which recognizes cultures as fluid and open and interconnected, and accepts that older local cultures might decline as new ones emerge.

In summary, localities are important in maintaining cultural difference, but can also be sites of cultural mixing and transformation. Ideas about culture can be negative (bounded, fixed, inward-looking) or positive (progressive, dynamic, outward-looking). Bounded, fixed notions of culture can lead to localized resistance, racism, nationalism, and even 'ethnic cleansing'. Progressive ideas about culture involve the recognition of differences between cultures, the interconnectedness of cultures and their constant evolution.

13.4 Multi- and hybrid cultures?

13.4.1 Hybridity

One of the major contemporary challenges concerns what we do with the concept of culture in the changing global scene, where nation states are forced to tolerate greater diversity within their boundaries. Some want to see national identity as homogeneous and assimilatory – in other words, different cultures are subsumed into the dominant culture (the 'melting pot' idea) (see Case study 13.3). Denmark, for example, is highly assimilationist and has one of the toughest policies on immigration in Europe. Others call for the acceptance of ethnic pluralism and the preservation of minority ethnic cultures as a legitimate part of the national project. This is the politics of **multiculturalism**, which instead of thinking of different cultures as being absorbed by dominant cultures, relies more on a notion of a cultural mosaic, or a 'patchwork quilt' of cultures. Each culture is recognized as different and distinct, but these differences are understood and valued. Sweden, for example, rejects Denmark's assimilationist model in favour of multiculturalism.

Multiculturalism might seem more progressive, but it can sometimes reinforce difference because culture is seen as essentially connected to **race**, and ra-

cial difference as rooted in biological difference. (These ideas are no longer considered acceptable; anti-racists have demonstrated that 'race' is socially constructed and has little basis in biology (Price 2010). We could just as easily have 'races' of blue-eyed and brown-eyed people.) Multiculturalism, therefore, still relies on a negative notion of bounded cultures. It might suggest tolerance, but often results in segregation and ghettoization. The United Kingdom and Netherlands, for example, have tended towards multiculturalism but have increasingly witnessed tensions surrounding the lack of integration of some Muslim communities. The murder of Dutch film-maker Theo Van Gogh in 2004 and the London bombings in July 2005 – both by apparent 'home-grown radical Islamists' – are seen by some to represent the failures of multiculturalism. Similarly, the 2006 riots in the suburbs of Paris were seen as a product of the deep alienation of poor, largely immigrant communities facing high levels of unemployment, discrimination in housing and jobs markets, and police harassment.

A similar example is the appearance of 'gay spaces' in cities such as San Francisco, London, Manchester and Cape Town, which might appear to signal greater tolerance and cultural diversity, but points to segregation rather than integration. Indeed, in the USA politically correct accommodation agencies now ask applicants to tick a box for 'gay' or 'straight' so that they can be housed suitably. As Reeves (1999) argues, 'Homophobia, overt and covert, is herding lesbians and gay men into ghettos – which are then celebrated as signs of a tolerant society.' It is OK to be gay in these segregated spaces; it is still not OK to be openly gay in other spaces of the city. As Chapter 21 specifies, they become territorialized.

In contrast, a more progressive idea of culture (and its manifestations in ethnicity, gender and sexuality) might be developed through the concept of **hybridity**. Hybridity breaks down barriers, adhering to neither the 'melting pot' nor 'mosaic' idea of cultural mixing, but rather seeing different cultures coming together and informing each other in different ways to produce something entirely new. This process has a long historical trajectory. Indeed, some argue that cultures have always been hybrid forms – they have never existed in isolation from other cultures, and thus have always been subject to change and influences from elsewhere (Werbner 1997: 15).

One of the most obvious places we can observe hybridity is in popular mass culture, an immediate example being popular music. Innovations in music have always involved the fusion of different styles to create new sounds and rhythms. Rock-and-roll,

Thematic case study 13.3

Bounded or hybrid national culture?

Nationalists around the world cling to a notion of bounded cultures that make them distinct from others. One effect of globalization has been resistance in the form of increased nationalism to what is perceived to be the erosion of national cultures. Increased mixing of cultures is seen to pose a threat to the survival of national cultures. Nationalists seek to preserve the symbols of nationhood, such as language, lifestyles and cultural forms in the face of what are perceived to be sweeping changes. However, cultures are not unchanging; a fundamental flaw in nationalist ideology is the adherence to a notion of static culture, and its reliance on a mythical history of the origin of the nation. For example, English nationalists define Englishness as distinct, which is used to justify anti-immigrationist ideas, anti-Europeanism and, in some cases, racism. But who are 'the English'? After the last ice age many communities settled Britain and Ireland from all over Europe. They lived and fought with each other and in a short space of time produced a mixed group of people who eventually called themselves English. The islands have been subject to waves of invasion and settlement (e.g. Celts, Romans, Anglo-Saxons, Vikings, Normans). England has always been hybrid; peoples and cultures have mixed and evolved together. Some English nationalists avoid thinking about this point by arguing that the final invasion (by William the Conqueror in 1066) marks the origin of England and Englishness (Anderson 1983). This myth is also flawed. William the Conqueror spoke no English. Whom did he conquer? He conquered 'the English'. For many nationalists the founding father of England is French! It is also no small irony that one of the symbols of Englishness, the monarchy, changed its official name to Windsor in 1917 from Saxe-Coburg-Gotha, thus hiding its German origins.

Even protectionist policies towards language are flawed; as with culture more generally, language is always hybridised and evolving. Most sentences in 'English' contain words that derive from German, French, Spanish, Latin, Nordic and Celtic languages. One legacy of British imperialism has been the incorporation of elements of the languages of the colonized into everyday English. Words from Indian languages, for example, punctuate English, including Hindi (bangle, bazaar, caravan, cot, jungle, juggernaut, pajama, pundit, shampoo, thug), Sanskrit (atoll, aubergine, avatar, bandana, candy, cash, dinghy, karma, sugar), Tamil (catamaran, curry, mantra, pariah), and Urdu (bungalow, khaki). Such hybridity within national languages provides evidence for the ways in which cultures have always been dynamic and mutually influential. Similar myths of origin and notions of bounded cultures exist elsewhere in the world. In some places protectionist policies emanate from deeply contradictory ideas (anti-immigrationist views in the USA and Australia, for example). In other places notions of bounded culture have led to conflict (for example, between Georgia and South Ossetia in 2008, between the Tamil separatists and the Sri Lankan state in 2009, and the continuing struggle by Kurdish nationalist organizations, some of whom seek to create an independent nation state of Kurdistan, consisting of some or all of the Kurdish areas in eastern Turkey, northern Iraq, north-western Iran and northern Syria, while others campaign for greater Kurdish autonomy within these existing national boundaries). The idea of nationhood, based on fixed, bounded and unchanging cultures, is an ideological creation that masks profound cultural divisions of gender, race, class and religion within a nation state (see also Chapter 22), and ignores the fact that, in reality, all cultures are hybrid and dynamic.

rhythm-and-blues and Latin jazz are obvious examples. We might also think of recent and contemporary forms of music that fuse different styles, such as 'trip hop', which emerged in the 1990s; 'post-trip hop', which has since integrated trip hop with other genres, such as ambient, R&B, breakbeat, drum 'n' bass, acid jazz and new age; 'nu metal', which from the late 1990s until around 2005 combined heavy metal with other genres such as grunge and hip hop; and electro

house, which fuses house music with several other electronic dance music subgenres and came into prominence between 2000 and the present. Authors such as Barthes (1972), Bourdieu (1984) and Bakhtin (1984) see popular hybridity as an exciting challenge to, or subversion of, dominant cultures and the exclusive lifestyles of dominant elites. Such popular mixings and inversions, like the subversive elements of youth cultures (Hebdige 1979; see also Hammett

2009; Carr 2010), are hybrid in the sense that they bring together and mix languages and practices from different and normally separated domains. They have the potential to disrupt dominant cultures by their 'out-of-placeness'.

In many ways hybridity is related to the notion of **transculturation**. Transculturation describes one of the key cultural processes that operate between hitherto sharply differentiated cultures and peoples who are forced (usually by the processes of imperialism or globalization, and primarily through migration) to interact. This interaction often takes place in profoundly asymmetrical ways in terms of relative power between different groups. Despite the illusion of boundedness, cultures evolve historically through borrowings, appropriations, exchanges and inventions. Cross-fertilization of cultures is endemic to all movements of people throughout history (see Case study 13.3). However, for those who aspire to bounded notions of culture and refuse this idea of perpetual hybridity, cultural mixing is felt to be threatening and a deliberate challenge to social order. In reality there are no fixed cultures in modern nation states, but some people cling to ideas of pure or impure cultures. For others, however, hybridity remains the site of revitalization, resistance and fun (see Case study 13.4).

Thematic case study 13.4

Hybridity/diaspora – some examples

British Bhangra

British Bhangra developed out of traditional forms of folk dances and music of the people of the Punjab, which arrived in Britain in the late 1960s with the migration of people from South Asia and East Africa. It emerged in the mid-1980s among second-generation South-Asian British musicians, primarily in Birmingham and west London, combining the robust and energetic punctuated rhythms of double-sided drums (*dhol*), modern technology and urban sounds with Punjabi poetry, traditional Punjabi lyrics, acrobatic dance skills and celebrations of the Punjabi harvest festival (*bhaisakhi*). It played an important role in the proclaiming of a cultural identity for groups of youths defining themselves as South-Asian British. Although it appeals to the Punjab as a specific region of origin, Bhangra, with its fusion of instruments and musical styles from South Asia and urban Britain, developed a large following among diverse groups of South-Asian British youths. Throughout the 1990s, it was characterized by ongoing processes of fusion with broader ranges of musical styles (ragga, reggae, soul, jazzfunk, rock, hip-hop, pop), both to reflect the lived experiences of South-Asian Britons, and as an attempt to 'cross over' into mainstream charts. It is the antecedent to newer and more fashionable forms of post-Bhangra and 'Asian Kool' which have been less marginalized by mainstream culture, and are part of an emergent and vibrant British-Asian urban youth culture.

Source: adapted from Dudrah (2002)

Cuban Santería

Santería is an example of a syncretic (hybrid, mixed) religion. It is based on West African religions brought by slaves imported from what are now Nigeria and Benin to the Caribbean to work the sugar plantations. These religions were suppressed by the European plantation owners and in Cuba slaves were forced to convert to Catholicism. However, they were able to preserve some of their traditions by fusing together various West African beliefs and rituals and syncretizing these with elements from Catholicism. One factor enabling this process was that many of the *orishas* (primary gods) shared many of the same characteristics of Catholic saints. This enabled slaves to appear to be practising Catholicism while practising their own religions. This has evolved into what we know today as Santería, the Way of the Saints, whose traditions are transmitted orally from generation to generation. Despite suppression by Fidel Castro's Socialist Revolution since 1961, its influence is pervasive in Cuban life. Devotees are found in most households, Yoruba proverbs litter Cuban Spanish, and high priests (*babalawos*) offer guidance based on ancient systems of lore. Today, with less religious persecution, Santería is experiencing a rise in popularity and is part of an emerging Cuban youth culture. Similar syncretic religions are found in Haiti, Puerto Rico and other Caribbean and Latin American countries.

Source: adapted from Betts (2002)

13.4.2 Diaspora

Related to the idea of hybridity is the notion of **diaspora**. This term was originally used to refer to the dispersal of Jewish peoples, but is now used in reference to the long-term settlement of peoples in 'foreign' places that follows their scattering or dispersal from their original homeland. It refers to a modern condition where a sense of belonging is not derived from attachment to territory, and where different peoples mix together through the processes of migration (forced or free). European imperialism and associated processes of globalization have set many of these migrations in motion. Diasporas are classic **contact zones** – spaces in which two cultures come together and influence each other – where transculturation or hybridization takes place. Diasporic identities are at once local and global and based on transnational identifications encompassing both 'imagined' and 'encountered' communities (e.g. Irish-Americans belong to an imagined international community of people who have 'Irishness' in common, but whose identities are also informed by the communities in which they live in the USA). In other words, diasporas are a direct challenge to the idea that there is a simple relationship between place and culture. They transgress the boundaries of the nation state and provide alternative resources for constructing identity and fashioning culture.

The concept of diaspora space allows us to think of 'culture as a site of travel' (Clifford 1992), which seriously problematizes the idea of a person being a 'native' or an 'insider'. Diaspora space is the point at which boundaries of inclusion and exclusion, of belonging and otherness, of 'us' and 'them', are contested. As Brah

(1996: 209) argues, diaspora space is 'inhabited' not only by those who have migrated and their descendants, but equally by those who are constructed and represented as indigenous or 'native'. In the diaspora space called 'England', for example, African-Caribbean, Irish, Asian, Jewish and other diasporas intersect among themselves as well as with the entity constructed as 'Englishness', thoroughly reinscribing it in the process. Like notions of hybridity, the concept of diaspora is important since it allows for the recognition of new political and cultural formations that continually challenge the marginalizing impulses of dominant cultures.

13.5.3 Selling hybridity and the commodification of culture

In today's world, culture sells. Hybrid cultures, in particular, sell. Cities are now constructing themselves as cosmopolitan, and hybridity has become a form of 'boosterism' – where city authorities create marketing images to attract investment in the form of business and tourism. Hybrid culture is perceived as creating economic advantage. With increasing deindustrialization in Europe and North America, cultural strategies have become key to the survival of cities. Examples include the international marketing of cultural/religious festivals such as Mardi Gras in Sydney or New Orleans, or the importance of 'Chinatowns' and other 'ethnic' districts to tourism in cities throughout the developed world (see Plate 13.3). Ironically, hybridity is in danger of becoming just another marketable commodity. For example, treating the political work of some British-Asian bands as marketable, hybridity trivializes

Plate 13.3 Chinatown in San Francisco: 'ethnic' districts are promoted as major tourist attractions in many 'global' cities.

(© Dave G. Houser/Corbis)

black political activity and leaves problems of class exploitation and racial oppression unresolved. As Hutnyk (1997: 134) suggests, to focus on hybridity while ignoring (or as an excuse for ignoring) the conditions in which this phenomenon exists (the commodity system, global economic inequality, inequitable political relations), is problematic in that it maintains the status quo. Hybridity and difference sell, but in the meantime the market remains intact, power relations remain unequal, and marginalized peoples remain marginalized. Moreover, as culture is subsumed into capitalism, those marginalized peoples who might be capable of oppositional politics are also subsumed under the rubric of hybridity.

The notion of hybridity, therefore, can be problematic. In some Latin American countries, cultural elites and nation states have appropriated the hybrid mestizo (mixed) identity, making it dominant. This has been seen to be oppressive of 'Indian' populations, who have in turn been accused of ethnic essentialism (or emphasizing their racial difference) because of their desire to protect their cultures (Radcliffe and Westwood 1996). In the West, ideas of hybridity are currently popular with highly educated cultural elites, but ideas about culture, ethnicity and identity that develop in poverty-stricken underclass neighbourhoods are likely to be of a different nature (Friedman 1997: 83–4). Evidence of racial tensions in many North American and European cities points to the fact that class and local ghetto identities tend to prevail, with little room for the mixing pleaded for by cultural elites. The global, cultural hybrid, elite sphere is occupied by individuals who share a very different kind of experience of the world, connected to international politics, academia, the media and the arts. In the meantime, the world becomes more polarized in terms of wealth, and heads towards increasing balkanization where regional, national and ethnic identities are perceived as bounded, threatened, and in need of protection. As Bhabha (1994) reminds us, hybridity is an insufficient means through which to create new forms of collective identity that can overcome ethnic, racial, religious and class-based antagonisms – it sounds nice in theory, but does not necessarily exist outside the realms of the privileged.

Stuart Hall (1996: 233) argues, however, that we should not view the current fashionability of hybridity in a wholly negative light. Even as cultures are increasingly commodified, we should not forget the potential for the democratization of culture in this process, the increased recognition of difference and the diversification of the social worlds in which women and men now operate. This pluralization of social and cultural life expands the identities available to ordinary people (at least in the advanced economies) in their everyday working, social, familial and sexual lives. For Bhabha (1994: 9), the interconnections of different cultural spaces and the overlapping of different cultural forms create vitality and hold out the possibility of a progressive notion of culture.

We thus need to think about the place and meaning of cultural hybridity in the context of growing global uncertainty, xenophobia (fear of foreigners) and racism. Why is cultural hybridity still experienced as an empowering, dangerous or transformative force? Why, on the one hand, is difference celebrated through a consumer market that offers a seemingly endless choice of identities, sub-cultures and styles yet, on the other hand, hybridity continues to threaten and shock? Conversely, why do borders, boundaries and 'pure' identities remain important, producing defensive and exclusionary actions and attitudes, and why are the latter so difficult to transcend? Is the sheer pace of change in cultural globalization producing these reactions?

To summarize, hybridity and diaspora are examples of more progressive ways of thinking about culture. It could be argued that all cultures are always already hybrid; they are never pure, have always evolved and changed through time and through contact with other cultures, and they continue to evolve. In today's world, hybridity is being commodified, which might make it less radical. However, despite this, it has the potential to democratize culture and to allow us to rethink culture in ways that are more tolerant of difference. Finally, cultural hybridity needs to be understood in the context of growing global uncertainty, xenophobia and racism.

13.5 Conclusion

In this chapter it is suggested that there are two apparently contradictory tendencies in thinking about cultures – the attempt to secure the purity of a culture by conceptualizing it as strong, fixed, bounded, permanent and homogeneous, and the hybridity of most cultures. Culture is thus a contested concept. A progressive way of thinking about culture is to reject the idea of boundedness and internal cohesion. In the modern world especially, culture is a meeting point where different influences, traditions and forces intersect. There is, therefore, a continual process of change in cultural practices and meanings. Globalization is undermining closed, fixed ideas of culture and leading to new ways of conceptualizing cultures (transculturation, contact zones, hybridity and diaspora). However, the fact that cultures are not fixed or homogeneous does not mean that we will stop thinking of them in this way. As Stuart Hall (1995: 188–9) argues, this is because some people need 'belongingness' and the security that closed conceptions of culture provide.

Despite this, recent years have witnessed a decentring of culture, with nation states increasingly superseded by transnational institutions that are producing cultural globalization and greater cultural diversity. There has also been a shift in the awareness of the cultural capital of the West, and an understanding of the cultural dominance of developed countries. At the same time, there are now more voices talking back, reflecting the cultural assertiveness of marginalized groups and making us aware of new levels of diversity. Even though most people remain physically, ideologically and spiritually attached to a local or a national culture and a local place, complex cultural flows and networks ensure that it is becoming increasingly impossible for people to live in places that are completely isolated and disconnected culturally from the wider world. Thus,

> if there is a global culture it would be better to conceive of it not as a common culture, but as a field in which differences, power struggles and cultural prestige contests are played out. . . . Hence globalization makes us aware of the sheer volume, diversity and many-sidedness of culture. (Featherstone 1995: 14)

This points to a 'more positive evaluation by the West of otherness and differences' (ibid.: 89). For Massey and Jess (1995b: 134), globalization is not simply a threat to existing notions of culture, but a 'stimulus to a positive new response'.

People around the world have different cultures and systems of meaning, but we cannot avoid reading the world from within our own cultures and interpreting it through our own systems of meaning. Understandings of global culture for the majority of the readers of this book are filtered through the logic of the West (Spivak 1985). Western ideas and cultural forms are still considered superior and have become hegemonic, or dominant. Similarly, one's own cultural positioning (on the basis of gender, ethnicity, class, location, sexuality, stage in life cycle, ability) also influences understandings of local cultures. The same processes operate at local levels; dominant cultures marginalize others on the basis of ethnicity, sexuality, gender and religion. However, as we have seen, those dominant cultures also produce resistances that have the potential to create new ways of thinking about culture.

The challenge is to confront the limits of 'our' knowledge, to recognize other worlds, to acknowledge the legitimacy of other cultures, other identities and other ways of life. Accepting 'cultural translation' (Bhabha 1994) involves understanding the hybrid nature of culture, the influence of marginal cultures on dominant cultures, and that people in marginal cultural systems at local, national and international levels are also active in creating their own systems of meaning. They do not simply absorb ideas from, or become absorbed into, more dominant cultures. It is possible to develop cosmopolitanism in the twenty-first century that is global, sensitive to cultural difference, and dynamic.

Learning outcomes

Having read this chapter, you should be able to:

- Understand the complexities of culture; it is a process, rather than a thing, and subject to change over time.
- Demonstrate ways in which cultures operate at local, national and global levels.
- Discuss examples of global cultural processes and how these are filtered through localities to contest notions of a singular global culture.
- Discuss examples of negative (closed, bounded), or positive (progressive, hybridized) local cultures.

- Critique and utilize concepts such as hybridity and diaspora as a progressive way of thinking about culture.
- Reflect on your own cultures of knowledge, and the ways in which these condition your ideas about global, national and local cultures.

Further reading

Anderson, J. (2010) *Understanding Cultural Geography*, Routledge, London. This book provides a broad-based overview of cultural geography, arguing that its essential focus is *place*. The book presents specific chapters outlining the history of cultural geography, as well as the methods and techniques of doing cultural geography and focuses on topics such as corporate capitalism, nationalism, ethnicity, youth culture and the place of the body.

Anderson, K., Domosh, M., Pile, S. and Thrift, N. (eds) (2003) *Handbook of Cultural Geography*, Sage, London. This book offers an assessment of the key questions informing cultural geography and contains over 30 essays. It is an invaluable resource for students looking for an assessment of major issues and debates and the breadth, scope and vitality of contemporary cultural geography.

Cook, I., Crouch, D., Naylor, S. and Ryan, J. (eds) (2000) *Cultural Turns/Geographical Turns*, Prentice Hall, London. Aimed at advanced undergraduates and postgraduates, this book addresses the impact, significance and characteristics of the 'cultural turn' in contemporary geography and the intersections between space and culture.

Crang, M. (1998) *Cultural Geography*, Routledge, London. A useful introduction to the ways in which geographers think about culture, exploring the global and the local, cultures of production and consumption, and ideas about place, belonging and nationhood.

Duncan, J.S., Johnson, N.C. and Schein, R. (eds) (2004) *A Companion to Cultural Geography*, Blackwell, London. A comprehensive introduction to cultural geography, with 32 chapters written by leading authorities in the field and covering debates about cultural theories, nature/culture, culture and identity, landscapes and colonial/post-colonial geographies.

Shurmer-Smith, P. (ed.) (2002) *Doing Cultural Geography*, Sage, London. This is a good book for those students thinking of undertaking projects or writing dissertations in cultural geography. It explains the theory informing cultural geography and encourages students to engage directly with theory in practice.

Useful websites

www.culture.gov.uk The official website of the UK government's Department for Culture, Media and Sport (DCMS). An animated site that includes information about the role of the DCMS, and government policy towards media and arts, heritage, libraries and museums, sport, the National Lottery and tourism. Also has links to other useful sites in each category.

www.alt.culture.com An encyclopaedia of modern culture spanning modern musical styles, film, cyberpunk and street fashion, extreme sports and political correctness. A useful site for those interested in global cultures of consumption.

www.adbusters.org/home An example of 'culture jamming'.

www.un.org/womenwatch The official United Nations Internet Gateway on the Advancement and Empowerment of Women, and part of the global phenomenon of cyber-feminism, which might be considered an example of a 'third culture'. Has useful links to other sites advancing women's rights through new technologies.

For an introduction to Cuba's hybrid cultures see **http:// folkcuba.com** and **www.afrocubaweb.com**.

For annotated, clickable weblinks and useful tutorials full of practical advice on how to improve your study skills, visit this book's website at **www.pearsoned.co.uk/daniels**

PRODUCTION, EXCHANGE AND CONSUMPTION

Section 4

Edited by Peter Daniels

The idea of going 'global' is often associated with the economic aspects of human geography. This section of the book demonstrates how 'the economic' shapes many aspects of human and organizational behaviour and, thereby, the spatial patterns of growth and development. Many economic processes and their geographical outcome are increasingly a function of links and dependencies that extend far beyond the immediately local or national; they involve diverse and dynamic interactions with places and people distributed around the world. Witness the recent shifts in the global economic centre of gravity towards the fast-emerging economies of China or India. Many of the symptoms of this 'globalization' are most obvious in the sphere of consumption (Chapter 19); Coca-Cola (USA), Nokia (Finland), BMW (Germany) and Louis Vuitton (France) are brands that are almost universally known, if not equally accessible, worldwide. They also arise, although perhaps in a less obvious or visible way, in the context of food production and consumption (Chapter 15, see also Chapter 19). Such goods and services are often slightly customized to conform with local tastes, cultures or regulations (often reflected, for example, in the way that they are advertised or packaged for different markets). In the case of fashion goods the objective is to offer a uniform, high-quality, experience or convey a distinctive fashion statement wherever they are consumed in the world (see Chapter 19).

Consumption is both a consequence, and a driver, of the globalization of production. The shift from production systems largely focused on national markets to production facilities organized in complex ways and distributed around the globe has been accelerating and increasingly involves sophisticated global production chains and networks (see Chapter 16). These include detailed analyses of food chains or networks (Chapter 15) that vary in complexity and geographical coverage for different products. Such networks also arise in part from the search for new sources of suppliers or new markets as national sales opportunities have become saturated; but it also reflects the demand created by a growing band of global consumers, boosted by significant economic growth in newly industrializing countries such as China or Brazil, that can easily access common sources of information via the Internet, cable and satellite services that have, metaphorically, made the world smaller. Fordist methods of production helped both to fulfil and to create demand for high-volume, low-cost, standardized products at a time of steadily increasing incomes per capita during the first quarter of the twentieth century, but since then there has been a shift towards post-Fordist production systems with an emphasis on greater flexibility, specialization, customization, different firms joining together to undertake research while remaining competitors (strategic alliances) (see Chapter 16), even more fragmented divisions of labour as well as expertise (see Chapter 17) and horizontal rather than vertical integration of production. These have helped, on the one hand, to continue the trend towards globalization, with the first global shift led by manufacturing and the second global shift by service industries (see Chapter 17), while on the other, reinstating the role and importance of localities. The latter have been reasserting their role as the focus for production in conjunction with global-level influences while also contributing to an increasing disconnection between, for example, food producers and consumers that also give rise to food security challenges (Chapter 15) . The massive production plants employing thousands of workers that were symbols of the Fordist era have been replaced by industrial districts made up of networks of firms, small and large, each contributing particular skills and knowledges relating to particular products. Together they form distinctive nodes of production such as the Third Italy (small-scale skilled production in the North East, Emilia, Tuscany and Marches), Bangalore (Silicon Valley of India), or the City of London.

These changes should not lead us to conclude that large firms are no longer important in the globalization of economies. On the contrary, they are as influential as ever in shaping the geographies of production and consumption or the distribution of spaces that are strongly engaged with the global and those that are much less engaged (see Chapters 14, 16 and 17). Typical are the transnational corporations (TNCs) that often have more financial resources at their disposal than many national governments. They therefore exercise significant influence on decisions relating to the economy, investment in infrastructure, or regulation by nation-states and by international organizations such as the Organization for Economic Cooperation and Development (OECD) or the World Trade Organization (WTO).

Consumption and production are mediated or enabled by the circulation of finance capital (Chapter 18). National governments, companies large and small, or individuals require finance to allow them to execute investment strategies in the case of companies or purchasing decisions in the case of individual consumers. Advances in information technology and political shifts have greatly increased the global circulation of finance and encouraged innovation in the financial instruments traded daily on stock, commodity, currency and other exchanges in global financial centres such as London, New York, Chicago and increasingly Shanghai, Hong Kong or Singapore. Yet again, however, as with the other themes explored in this section of the book, the flows of finance capital and access to finance mediated by financial centres are spatially uneven. This operates to the significant disadvantage of marginalized nations, regions and groups within society while reinforcing the advantage of those individuals, companies and nation-states that already have very good access to finance capital.

This section demonstrates how the geography of the economy simultaneously incorporates global and local economic perspectives. They are inseparable even though there have been some important shifts in modes of production and mechanisms of consumption over time and across space. Some of the important causes and the consequences of the uneven economic development that results are highlighted in this section. An overview of all the contributions suggests that there are numerous important issues, questions and challenges that economic geographers can usefully address as the twenty-first century unfolds.

GEOGRAPHIES OF THE ECONOMY

Chapter 14

Peter Daniels and Andrew Jones

Topics covered

- The changing nature of economic geography
- The rise of the global economy
- A geographical approach to economic processes
- Places and localities in the global economy
- A new global informational economy?

The economy is everywhere. Wherever we happen to live, work or play our daily existence invariably involves making decisions that have an economic basis. Yet, we tend to take for granted the ways in which society determines how the scarce resources at its disposal are used to provide for its material needs and to produce wealth. Few days pass without the news media reporting on the economy: the balance of exports over imports, the unemployment rate, consumer spending, announcements of factory closures, trends in house prices, the relocation of jobs to other countries. Separate sections on business and the economy in the 'quality' daily newspapers, specialist weekly business publications, or 24-hour satellite television channels devoted exclusively to news and information about economic affairs are numerous. In summary, the economy is a set of human activities and institutions linked together in the production, distribution, exchange and consumption of goods and services.

You may have noticed that 'economies' or 'economy' are preceded by various adjectives such as: global, local, market, command (or redistributive), capitalist, informal, subsistence, mixed, Internet, information, new, or space economy. There are clearly many different kinds of 'economy' depending on the starting point for analysis; for example, an informal economy signifies how it is organized while a digital economy signifies something about the medium used to undertake economic transactions. Economic geographers are interested in the space economy; the way in which the organization of geographical space reflects the behaviour, values, and actions of actors (individuals, firms, institutions) that together make up the economy. Each actor operates from or occupies some unit of space: a field for a crop, a room used by a self-employed person working from home, a building of several thousand square metres for a corporate call centre, a site of several hundred hectares used for vehicle production in Korea or Japan, or an area of hundreds of square kilometers occupied by a sheep station in the Australian outback. The households that provide the labour required for production also require space in which to live and are, in turn, a major source for consumption goods and services that result from production; these are accessed via informal and formal transactions and trading. The challenge for the economic geographer is how to analyse and explain the geographical patterns of economic activity at different scales, how they change over time and the relationships between them.

14.1 The changing nature of economic geography

The uneven distribution of economic activities is plain enough to see and can be explained by the way in which different parts of our planet are endowed with different resources, climates, or different levels of accessibility. But there is much to understand and to explain and geographers who are interested in the consequences of these spatial variations comprise the major sub-discipline of economic geography which 'is concerned with the spatial organization and distribution of economic activity, the use of the world's resources, and the distribution and expansion of the world economy' (Stutz and de Souza 1998: 41). While the distribution of economic activities is influenced by historical, political, social, and environmental factors the outcomes are a function of spatial interaction. It is not possible for all the actors in the economy to be located at one point in geographical space. In its early guise in the early twentieth century economic geography was termed 'commercial geography': a largely factual compilation of the circumstances under which different commodities were produced and exchanged around the world. This kind of economic geography sought, for example, to analyse the factors that affected where particular commodities such as coal or iron ore were produced. Gradually the emphasis shifted towards explaining observed geographical variations in the location patterns of commodity production and trade. The transformation from commercial to economic geography was consolidated when explanations for the location of economic activity were derived from neoclassical economics (Case study 14.1). This is an early example of the way in which economic geography has used ways of thinking and modes of analysis used by other disciplines, in this case economics, to inform its own agenda. Neoclassical economics is the study of the allocation of scarce resources amongst alternative ends when there are several alternative outcomes. Interestingly, one of these is the maximization of social welfare which is based on the idea that the economic basis of society determines the form of its social institutions. We will see later that the social basis of the economy is very much part of contemporary analyses of geographies of the economy.

In the meantime, we need to consider the development of economic geography as a sub-discipline and the various transformations it has undergone since

Thematic case study 14.1

Models of economic location

There has been much interest in economic geography in theories or models that help to explain the development of economic activities within a spatial context. *Neoclassical economic theory* was first used by Von Thunen in 1826 (Hall 1966) to model patterns of agricultural land use and later applied to industrial location (Weber 1929) and for explaining the distribution of services and of settlements (Christaller 1966). In each of these and numerous other cases the objective is to generalize about patterns of economic activity. However, such is the complexity of the real world that it is necessary to make some assumptions: decision makers behave in a sensible (or rational) fashion; they possess complete and correct knowledge (often referred to as perfect knowledge); everyone is attempting to maximize profits; competition is unconstrained; economic activity takes place on a uniform land surface. The key assumption is that distance is the main influence on decision-making by households and businesses with the resulting spatial patterns of economic activity explained by examining the relationship between distance and transport costs. The outcome of this approach is an ideal or optimal pattern of land use or industrial location.

The main criticisms are the unrealistic nature of the assumptions and the oversight of many other factors that have an impact on the geography of economic activity such as:

- changes in the technology of production and consumption;
- the variety of ways in which businesses are organized or make decisions on where to locate;
- the role of social and cultural factors;
- the effects of international and national political and regulatory environments.

The limitations of neoclassical models encouraged the development of alternatives. The *behavioural model* also attempts to arrive at generalizations but the focus is shifted to the role of the individual as the principal explanation for spatial patterns. The motives, opinions, preferences and perceptions of the individuals making location decisions are incorporated and it is not unexpected that the outcomes are suboptimal (i.e. decision makers do not have perfect information and are not assumed to maximize profits). Examples of the behavioural approach include studies of consumer behaviour (Potter 1979), office location (Edwards 1983), and location change by manufacturing industries (Hayter and Watts 1983). The limitations of the behavioural approach are that it is too descriptive and only highlights variations in behaviour rather than why, for example, an entrepreneur starts a certain business at a particular location. Too much attention is given to factors specific to the firm or households, at the expense of more general processes operating in the wider environment.

The *structuralist approach* is a more holistic way to explain the location of economic activity and how it changes. It is based on the premise that behaviour is shaped or constrained by wider processes in the social, political and economic spheres. Notions of culture and of class rather than individual ideas determine the spatial structure of economic activity. The structuralist approach holds that firms work within a capitalist society in which production is essentially a social process structured by the relationship between capital and labour. It is not enough to explain patterns of economic activity by studying the various sectors such as manufacturing or services for their own sake; rather, it is necessary to examine the 'hidden' mechanisms or processes (e.g. social, political) that underpin economic patterns. These cannot be measured in the way typical of the neoclassical or behavioural approaches, so that much of the structuralist approach involves developing theories about, for example, the way in which changes in economic conditions affect the requirements for production that in turn change the requirements of an economic activity at a given location and that might ultimately cause it to move elsewhere (see Massey 1981).

---

the mid-twentieth century (see Scott 2004). During the 1950s and 1960s economic geographers led the way in the testing of theories and ideas, using numerically-based quantitative methods and modelling that were seen to be rigorous. However, by the 1970s a different political economy approach emerged within the sub-discipline that made extensive use of Marxist-based interpretations of economic change and its socio-economic consequences (Harvey 1982; Swyngedouw 1982). Analysis of economic crises such as the global oil shortage of 1973 or the debt crisis of the early 1980s focused on issues of regional inequality, manufacturing decline and social justice (Bluestone and Harrison 1982). In the 1980s the growth of political economic interests in the sub-discipline were also accompanied by empirical and theoretical analysis of some new and surprising economic growth in places previously regarded as peripheral or outside the mainstream. In Europe, for example, these included the proliferation of small-scale, skilled production units to form new industrial districts in North-east Italy, Emilia and Central Italy collectively known as the Third Italy (Brusco 1990). Economic geographers sought to analyse the changing nature of production in manufacturing in the world economy, particularly with the rise of new flexible kinds of post-Fordism production system (Amin 1992) in North America and Japan (Gertler 1992). This period also saw a marked increase in the international division of labour, led by multinational corporations (Fröbel *et al.* 1980), which became but one of the issues symbolic of globalization and international economic interconnectedness. This has become an increasingly central concern for economic geographers (Dicken 2010) and we will return to this theme later.

However, the 1990s saw two parallel developments that have diversified both the nature and disciplinary reach of economic geography. The first is the dramatic revival of quantitative and modelling approaches to economic geography within another social science discipline – namely, economics. At the forefront of this was the Nobel Prize winning economist Paul Krugman who advocated what became known as 'the new economic geography' (NEG) (Krugman 2000; Fujita and Krugman 2004). However, geographers have responded to this challenge and argued whether NEG is really economic geography as opposed to 'geographical economics' (Martin 1999a). Economists claim that, methodologically, NEG belongs to them. Economic geographers claim that economists have finally acknowledged the role of space (geography) and moved it from the edges of their discipline into the mainstream of economic theory; away from the 'wonderland of no dimensions' (Shepherd and Barnes 2000: 3). To avoid confusion, the rest of this chapter will use 'geographical economics' to refer to this strand within economics.

The second development is the impact on economic geography of the 'cultural turn' within human geography (Barnett 1998; Cook *et al.* 2000). This 'new eco-

nomic geography' as practiced by human geographers increasingly recognized the importance of 'social' and 'cultural' factors when interpreting the 'economic world' (see Chapter 10 for a discussion of the links between culture and the consumption of goods and services). It is argued that lifestyles, beliefs, languages, ideas, imaginations and representations interact with the economic to produce culturalization of the economy rather than the economization of culture. Thus, goods and services incorporate cultural attributes in, for example, their design, marketing, packaging and potential benefits to users. Material possession of a good (car, camera, mobile phone) or consumption of a service (tourism, fast food, financial transaction) is only part of the experience (a real as well as an imagined event) of using or being seen to use or to consume. Consumption of goods and services involves beliefs about what it says about us as individuals or groups: social or job status, image or wealth (see Chapter 19). Economic geographers have previously recognized cultural factors when using terms such as 'socialist economy', 'Chinese family production networks', or the various 'corporate cultures' encountered in transnational corporations (TNCs); but only recently have they considered the economic and the cultural to be intertwined rather than worthy only of separate study.

We could conclude the overview of NEG here but it is worth loitering a little longer to consider some further issues. For example, is it too simplistic to talk of 'economic geography' rather than 'economic geographies'? If the economy is shaped, at least in part, by social relations (that are by their very nature complex) and exchanges that reflect multiple variations in the value associated with the production, consumption, or circulation of a good or service, then 'are the geographies constituted through peoples' struggle to construct circuits of value sustainable across space and time'? (Lee 2006: 417) As Lee points out, if societies cannot do this in ways that allow them to make a living they can only materially reproduce themselves with great difficulty. In this sense, there is more than one economic geography.

Another important issue is the way that the NEG has spawned new strands of work around a range of different concepts for understanding the spatial nature of the economy and economic activity. In the 1990s, for example, economic geographers developed an increasing interest in the (spatial) nature of relations between different economic actors (Bathelt and Glückler 2003). Yeung (2005) suggests that economic geography underwent something of a 'relational turn' that is 'concerned

Spotlight box 14.1

Perfect competition

The intensity of competition among firms encourages efficiency and helps to keep prices low. The ultimate expression of this is *perfect competition*. In these purely theoretical circumstances the actions of any one individual buyer or seller have a very limited impact on market prices. This is because everybody has access to all the information they need, all products are the same, and firms earn only the base minimum profit. If firms earn excess profits (more than the base minimum) other firms will enter the market. This will continue to happen until profits are driven down so that only normal profits are made.

primarily with the ways in which socio-spatial relations of actors are intertwined with broader structures and processes of economic change at various geographical scales' (ibid.: 37). Central to this shift is a realization that economic geographers need to better understand how power shapes economic action. A second thread to the NEG has followed on from this, focusing broadly on the nature of economic practices and how their spatiality affects economies (Jones and Murphy 2010). This sociological perspective in economic geography has also been concerned with the significance of power in relation, for example, to how transnational firms are managed or how knowledge practices spread innovations through firms and industries (Amin and Cohendet 2004). This is also related to a third strand of NEG that has looked beyond the boundaries of geography to work in economic sociology, cultural and technology studies in seeking to develop what has been termed a 'cultural economy' approach (Amin and Thrift 2004). Economic geographers in this area have ranged into new industries and areas of the economy that have been previously ignored – notably creative sectors such as fashion design, music, or computer games software. The cultural economy approach has also brought a fresh perspective to the traditional concerns of the economic geographer, such as the growing body of work on financial and other kinds of markets that seeks to understand them as produced through the interactions between networks of social actors, their performances, power relations and technological devices (Berndt and Boeckler 2009).

NEG is therefore also about how to theorize economies, economic actors and their geographies. The scope that this provides for a healthy, ongoing debate can be illustrated using a set of basic propositions that are assumed to be true when we conceptualize a capitalist economy (Hudson 2004). First, there should be a variety of concepts of a capitalist economy that reflect the diversity of the flows of people or knowledge, for example, in space and time that make up economies. Second, the concepts used should not be constructed on the basis that there are separate economies; rather, they should necessarily be treated as interrelated. Third, economic behaviour and practice is undertaken by individuals or subjects possessing knowledge and skills, although not in the sense that everyone has complete knowledge and skills, as in perfect competition (see Spotlight box 14.1). Therefore, and fourth, economies are a social construct incorporating the full range from the informal habits of individuals to the formal institutions of the state. Fifth, individual and collective behaviour in a capitalist economy is influenced by, and its structures and institutions based on, long-term social relationships. Finally, capitalist economies are reproduced using various governance institutions that exist because capitalist economies are formed via social relations and practices that are competitive and not natural.

You may recall from the beginning of this chapter that there are many different kinds of economy; the capitalist economy is but one among many and you should consider whether conceptualizations such as the one above are likely to be the same, or different, to those applied to an informal, alternative, digital, or knowledge economy?

14.2 What is the economic problem?

Each of us has a variety of needs and wants that, when they are combined, comprise demand for goods and services. The resources available to fulfill this demand are finite. The economic problem (Heilbronner 1972) is how to devise a system that combines the physical and human resources needed to produce and distribute the output of goods and services to attain given ends, such as the maximization of social welfare. Possible solutions are numerous but they can be distilled to just three mechanisms (Dicken and Lloyd 1990). The first of these is **tradition** in which the allocation of resources to production relies on a set of 'rules' based on convention or past practice, such as the handing down of land from father to eldest son. Given the contemporary interest in the relationship between the economic and the social, it is notable that traditional solutions to the economic problem incorporate socially determined norms.

A second mechanism for resolving the economic problem is a **command economy** in which a key objective is redistribution of wealth based on public ownership of the factors of production. Prior to its demise during the late 1980s countries such as the former Soviet Union, Poland, Hungary and East Germany used central political authority to prepare directives setting production targets for a defined period (say three or five years) for the types, and quantities of goods and services to be produced. In order to achieve the targets and to distribute the output the centralized direction of national economic development is combined with control over the allocation of human and physical resources. Some countries such as Cuba and North Korea continue to use this approach for solving the economic problem. Capitalist countries may use this mechanism during national emergencies (e.g. wartime) in order to mobilize resources quickly.

The market is the third approach to solving the economic problem. In contrast to command economies, market economies rely on decentralized decisions by consumers and by firms about the quantities of goods to produce, the prices to charge, and how, when and where exchange transactions take place. Generalizations about the modern market system and its outcomes are actually quite difficult because decentralized decision-making creates complexity; for example, professional economists can rarely agree about the significance of the latest unemployment figures, the impact of a change in bank interest rates, or the likely impact of a steep rise in crude oil prices on the stability of the global economy.

Economies that operate by voluntary exchange in a free market that is not in some way managed or controlled by a central authority do not in practice exist. Most developed nations are mixed economies in that they allow markets to drive most of their economic activities, using regulations and government intervention to ensure stability and efficient economic operation. Market/mixed economies are now in the majority, with fast emerging economies such as China making big strides in the same direction.

Within such **capitalist** economies most of the resources and the means of production are controlled by a relatively small proportion of individuals and firms who are seeking to improve their economic well-being through competition (see Spotlight box 14.1). If we assume that there is more than one producer of a given good or service, each is seeking to create the most favourable value (such as lower price, better quality, superior design, or efficient after-sales support) so as to retain or enhance its share of the market.

The principle of freedom of action may be paramount but, in practice, market economies incorporate regulation by national governments and/or international institutions. In the interests of equity, society needs to control the behaviour of those who own the means of production; the market is not necessarily the most effective and equitable way to allocate its rewards. Thus, most countries have a national organization that monitors and adjudicates company acquisitions and mergers in an attempt to avoid the development of monopolies, i.e. a market situation in which there is only one provider of a particular product or service. The lack of economic competition will allow a monopoly to sell a lower quantity of a product at a higher price than firms in a purely competitive market, leading to monopolistic profits. Because of the increasing integration of national economies and firms (see below in this chapter), the actions of individual national governments are complemented by groupings of nation-states such as the European Union (EU) and the North American Free Trade Area (NAFTA) or international institutions such as the World Trade Organization (WTO) that regulate and facilitate trade flows of goods and services as well as the movement of labour.

One of the key targets for cross-border control and regulation are **transnational corporations** (TNCs) which serve markets and operate production plants or distribution facilities that extend well beyond their home countries (see Spotlight box 14.2) (see also Chapters 16 and 17). The distribution of TNCs varies by industry sector and by country; the majority are small relative to a limited number of very large global TNCs. But they generate significant flows of capital, knowledge, information, expertise, products, raw materials and components amongst their own establishments and between countries. There are few parts of the world that are not in some way affected by the activities and decisions of TNCs and, for some countries, it makes the difference between inclusion as opposed to exclusion from the economic mainstream.

Spotlight box 14.2

Transnational corporations (TNCs)

The definition of a TNC is that it:

- controls establishments/economic activities in at least one other country apart from its home country;
- possesses an ability as a result of its size or the ownership of particular knowledge or skills to move its operations and its resources quickly between international locations, i.e. it is relatively footloose;
- can exploit or take advantage of differences between countries, regions or cities around the globe in factor (land, labour, capital) and non-factor (information, knowledge, regulation) endowments;
- owns and controls overseas activities, although this is not a prerequisite for TNC status since there are

many other ways, such as franchising, licensing or joint ventures, of achieving a presence in markets outside the home country;

- organises aspects of its operation across national borders at the global-scale, rather than duplicating its activities in each national economy where it is present.

You may well encounter a related term: multinational corporation (MNC). This suggests a firm that has premises or production plants in several countries. TNC is therefore a much more all-embracing term than MNC; there are many more TNCs than there are MNCs and they provide the basis for a more realistic assessment of the scale of international investment.

14.3 What are economies?

So far we have considered the economy in more or less abstract terms, although we have made reference to various 'actors' such as individuals, households, companies, government departments, national governments, and so on. In order to help economic geographers to measure and interpret the processes and interactions that help to shape spatial patterns of economic activity, it is necessary to simplify the complexities of real economies into manageable categories or groupings.

One of the most common approaches to generalizing the structure of economies is to divide them into four broad economic sectors:

1. Activities engaged in the exploitation of natural resources, such as agriculture, fishing, mining and oil extraction, form the **primary sector**. Although still a significant share of activity in the less developed economies, this sector is declining overall. Some of the output from the primary sector has limited use and value until it has been transformed in some way to become part of usable goods.

2. Whether the transformation takes place near the source of the primary commodity or after transfer to a location some distance away it requires a **secondary sector** (or manufacturing). This sector is still expanding in some less developed economies and the emerging economies but contracting (employment or GDP share) in most developed countries. The outputs from the manufacturing

sector may be immediately suitable for final use by consumers, or they may be components for incorporation in other final products.

3. The outputs of the secondary sector require distribution to the places and markets where they can be assembled, consumed, or purchased (Plate 14.1). The **tertiary sector**, which includes wholesale and retail trade, transportation, entertainment and personal services, fulfils this role. Improvements in transport and telecommunications and their integration following major advances in computing technology since the early 1980s have transformed the operation and reach of firms in both the secondary and tertiary sector. The growth of international purchasing by firms and individuals via the Internet is symbolic of these changes.

4. A fourth, **quaternary**, sector has increasingly been identified as a separate grouping which includes banking, finance, business and professional services, the media, insurance, administration, education, and research and development. These intellectual services assemble, transmit and process the information, knowledge and expertise used by activities in the other three sectors to enable them to adjust effectively and efficiently to the changing geographic, economic, social and cultural parameters of doing business in the twenty-first century. The share of the tertiary and quaternary sectors in the total economy has expanded steadily for at least a hundred years and they now account for four out five jobs in countries such as the USA, Canada, Britain, Hong Kong or Australia.

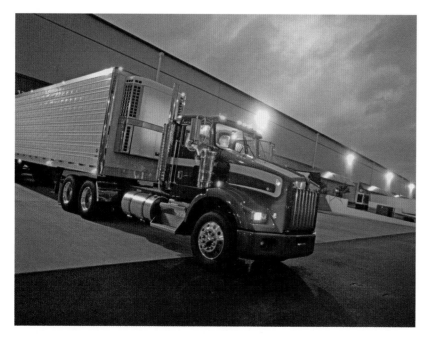

Plate 14.1 Large-scale distribution facilities at strategic locations with easy access to motorway networks serving national or global markets are a vital part of modern production.

(© Walter Hodges/CORBIS)

Another useful distinction for tracking change in economic structures is that between white-collar and blue-collar occupations. The former are salaried professionals (such as doctors, lawyers or airline pilots) and employees engaged in clerical or administrative occupations performing tasks that are mentally rather than physically demanding. Blue-collar workers, on the other hand, tend to earn hourly wages for performing skilled or unskilled tasks in factories, construction, or technical installation work. An intermediate category, the pink-collar worker, is also sometimes used to distinguish women mainly engaged in white-collar occupations that do not require as much professional training or who perform tasks to which lower prestige is attached. In line with the increasing share of the tertiary and quaternary sectors, the proportion of white-collar workers tripled during the twentieth century to around 60 per cent of total occupations.

14.4 A geographical approach to economic processes

One of the benefits of analyzing the structure of economies is that they enable economic geographers to measure the dynamics of change and whether, and how, this impacts at different spatial scales. These dynamics occur at all levels of analysis: from small rural localities to the largest metropolitan areas, from peripheral to core regions, from least developed to

most developed economies. This is not new or surprising but during the closing decades of the twentieth century it was dominated by changes which will shape the economic agenda well into the present century (such as more flexible production systems, see Chapter 16; or the changing nature of work, see Chapter 17).

As we have already noted, the share of country employment in the secondary sector has been declining steadily; by 2007 it directly supported only one in four jobs in the OECD countries (Table 14.1). The primary sector, already a very small part of the economy in developed countries, is contracting further as farm productivity is enhanced by improved crop disease resistance and better fertilizers. Meanwhile, the tertiary and quaternary sectors have expanded and diversified as rising standards of living and disposable incomes have boosted demand for tourism, travel or private health services. The production of goods and services also now incorporates more knowledge-intensive services that are heavily weighted towards the employment of engineers and scientists (see Chapter 17). These activities are growing most rapidly in the newly industrialized countries such as China that are 'catching up' with economies where services already provide more than 70 per cent of total employment (see Table 14.1).

There are numerous historical, political, social and institutional explanations for structural and functional changes in economies. However, from an economic geographers' perspective it is useful to think about the geographical nature of the processes of change that are by sector (%), selected OECD countries, 1997 and 2007

			Diff.	Secondary sector 2007	1997	Diff.	Tertiary sector 2007	1997	Diff.
			−1.7	21.5	22.0	−0.5	75.1	72.9	2.1
			2.2	40.3	41.6	−1.3	56.1	52.6	3.7
			.6	30	34.8	−4.8	67.7	62.3	5.4
			?	22.4	22.5	−0.1	66.1	57.7	8.4
				27.9	33.1	−5.2	67.9	61.6	6.3
				25.9	31.4	−5.5	66.7	57.8	8.9
				30.7	31.9	−1.2	54.5	47.5	7
			−0.6	21.6	26.0	−4.4	76.1	71.3	4.8
United Kingdom	1.4	1.9	−0.5	22.4	26.7	−4.3	76.3	71.4	4.9
United States	1.4	2.7	−1.3	19.8	23.9	−4.1	78.8	73.4	5.4
G7	2.4	3.6	−1.2	23.7	27.9	−4.2	73.9	68.5	5.4
EU-15	3.5	4.9	−1.4	26.3	29.5	−3.2	70.2	65.7	4.5
OECD Total	5.1	8.1	−3.0	24.9	27.6	−2.7	70.1	64.2	5.9

Note: G7: Group of seven industrialized nations – Canada, France, Germany, Japan, Italy, United Kingdom, United States.

Source: *Organization for Economic Co-operation and Development in Figures 2009*

occurring in the contemporary global space economy. At least five interconnected and overlapping processes are worth identifying when thinking about how the world economy (and within that, individual national economies) is evolving in the twenty-first century:

■ **Globalization** refers (in economic terms at least) to the integration across national boundaries of markets, finance, technologies and nation-states in a way not witnessed in the past. It has enabled nation-states, TNCs, as well as individuals, to extend their reach (markets, travel) across the globe faster, further, deeper and at lower cost than could ever have been imagined, even ten years ago (Plate 14.2). Economic globalization, however, cannot be separated from the social, political and cultural context in which it is embedded which means it is not really

Plate 14.2 Globalization has been accompanied by increasing geographical concentration of corporate control as symbolised by the density of office development and the skyline of Manhattan, New York City.

(T. Paul Daniels, Bromsgrove)

'a unified phenomenon', more a 'syndrome of processes and activities' (Dicken 2007: 8). We shall return to consider this in more depth shortly.

- **Informationalization** denotes the increasingly central role of information in production and consumption in the world economy. The sociologist Manuel Castells (2009) argues that we now live in an era of a global informational economy, where information (or knowledge) is the key input to economic activity whatever the industry. The rise of a whole range of new information and communications technologies (ICT) and the development of the Internet are, of course, key enablers of this process.
- **Neoliberalization** is used as a shorthand term to describe the spread of a complex set of political and policy trajectories that include the liberalization of international trade and financial markets.
- **Tertiarization/quaternerization** is used to describe the ongoing development of the global service economy. An increasingly proportion of GDP and employment in advanced economies is now in service-sector activities, and this trend is also evident in less developed economies (see Chapter 17).
- **Financialization** refers to the growing importance and centrality of financial capital in all geographical regions and industry sectors in the global economy; it is also used to describe the growing power of finance (see Chapter 18).

These processes impact and affect different parts of the world economy in different ways, and with different consequences. They form, however, the basis for developing an economic geographical understanding of why economic development is uneven at the global scale, and how it is likely to proceed in the coming decades. We now need to consider in more depth how these processes have led to the emergence of an increasingly globalized world economy that needs to be analysed as a whole rather than as a collection of many smaller economies restricted to nation-states or regions.

14.5 The rise of a global economy

Earlier conceptions of globalization suggested a convergence of economic, social and cultural values across the localities, regions and nations that come together as the 'global economy'. However, as Dicken (2007) argues, whilst 'there are undoubtedly globalizing forces at work, we do not have a fully globalized world economy'. Globalization is 'reflected in' and 'influenced by multiple geographies, rather than a single global geography' (Dicken 2007: 8). As all the other chapters in this section of the book show, we should beware of regarding globalization as producing convergence or uniformity. Rather, 'globalization' is a convenient shorthand that captures ideas of an integrating global economy, but it should not divert us from the reality of persistent variations in the levels of participation of different parts of the world as revealed, for example, by unemployment rates (Figure 14.1). Low-income countries are home to 12.4 per cent of the world's population but produced only 0.7 per cent of world gross national income (GNI) in 2007, equivalent to just US$457 per capita (Table 14.2). The high-income countries have 16.6 per cent of the world's population but almost 72.3 per cent of total GNI in 2007, at a rate per capita more than 12 times greater than for low-income countries generally but many times higher than for countries such as Ethiopia or Nicaragua where severe poverty (more than 50 per cent of the population living below US$1 a day) is a major challenge.

Globalization may be far from smoothing economic inequalities but if we assume for the moment that 'every country may not feel part of the global[ization] system, but every country is being globalized and shaped by it' (see Friedman 1999; 2005), then what are the symptoms? The first is the pivotal role of *knowledge*. The informationalization of economies (Castells 2009) means that production processes themselves are dependent on knowledge as the key input which creates value. For manufactured goods, for example, it is the design, marketing and customer support experience that makes products successful and makes firms' profits. And because the creation and exchange of knowledge tends to be embodied in people (see also Chapter 17) rather than machines, national economic health relies more than ever on 'producing' an educated and skilled workforce.

Second, *technology has become transnationalized*, especially amongst knowledge-intensive economic activities such as financial services or information and communications technologies (ICT). This does not necessarily mean better access to the factors of production; the increased complexity of the opportunities created by advanced technology means that only those individuals, companies and institutions with the resources to manage high technology can really take advantage.

Third, *there has been a marked increase in the power and influence of finance* through a range of financialization processes. Finance capital now takes many forms and moves almost seamlessly and with great speed, especially between the world's stock, currency, commodity and futures exchanges located in 'global cities' such as Tokyo, New York, London and Hong Kong

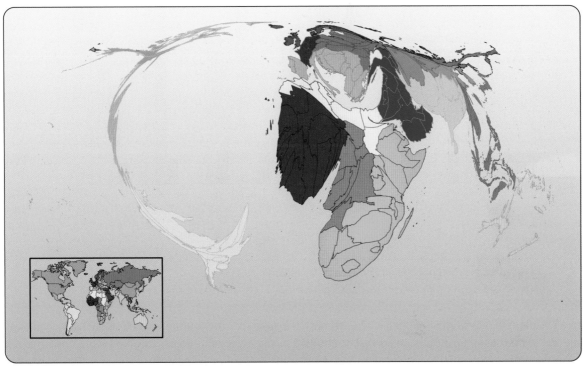

Figure 14.1 Cartogram of unemployment rates, by country, 2008–9.

Source: Compiled using data from the *CIA World Factbook* (2010)

Table 14.2 GNI-based and GDP-based measures for the relative performance of low-, middle- and high-income countries, 2007 (all data in US dollars)

Country groups	Population (millions)	GNI[1] ($ billions)	GNI per capita[2] ($)	Agriculture[5] value added[4] (% of GDP)[3]	Industry[6] value added (% of GDP)	Services[7] value added (% of GDP)
Low income[8]	810	995	403	27	26	47
Middle income[9]	4,708	13,852	2,682	10	37	53
High income[10]	1,102	41,487	36,953	2	26	73
World	6,620	55,631	8,100	3	28	69

Notes:

1. GNI (formerly GNP) is the sum of value added by all resident producers plus any product taxes (less subsidies) not included in the valuation of output plus net receipts of primary income (compensation of employees and property income) from abroad. Data are in current US dollars.
2. GNI per capita (formerly GNP per capita) is the gross national income, converted to US dollars using the World Bank Atlas method, divided by the mid-year population.
3. GDP (in current US$) is the sum of gross value added by all resident producers in the economy plus any product taxes and minus any subsidies not included in the value of the products.
4. Value added is the net output of a sector after adding up all outputs and subtracting intermediate inputs.
5. Agriculture corresponds to ISIC divisions 1–5 and includes forestry, hunting and fishing, as well as cultivation of crops and livestock production.
6. Industry comprises value added in mining, manufacturing (also reported as a separate sub-group), construction, electricity, water and gas.
7. Services include value added in wholesale and retail trade (including hotels and restaurants), transport and government, financial, professional and personal services such as education, health care, and real estate services.
8. Low-income economies are those in which 2009 GNI per capita was $995 or less.
9. Middle-income economies are those in which 2009 GNI per capita was between $996 and $12,195.
10. High-income economies are those in which 2009 GNI per capita was $12,196 or more.

Source: extracted from data in World Bank (2010) *World Development Indicators* database

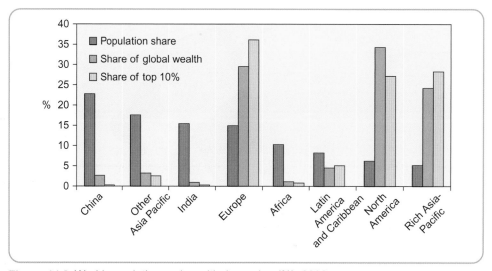

Figure 14.2 World population and wealth, by region (%), 2000.

Source: Davies *et al.* (2006), derived from figure showing population and wealth shares by region

(see Chapter 18). Electronic trading has ensured volatile and fast-changing financial markets that can transform the economic prospects of companies and, more importantly, of national or regional economies very quickly indeed. Equally, over the last couple of decades many new kinds of financial capital and financial firms have appeared replacing older and more traditional forms of finance for economic activity. This has not always been regarded as a positive development, however, with financialization processes seen by many within and beyond economic geography as creating greater instability and risk in an increasingly interdependent global economy (c.f. Harvey 2009).

The most recent example of this is, of course, the global financial crisis and economic downturn that began in 2007. The recent recession was the deepest economic downturn in the world economy since the Second World War, and was transmitted initially through increasingly globalized financial markets. It began in the United States where banks and other financial firms got into trouble after developing new forms of finance around the housing industry. The financial industry sold on household mortgages as new kinds of financial 'debt' products (known as mortgage-backed securities) that were sold on in global financial markets. Banks across the globe bought these new products, and banks in other countries (the UK, for example) also sold their mortgage debt on in a similar fashion. This had had the effect of making mortgages cheap in countries like the USA and the UK, but when their economies began to slow, it became clear that many of the original mortgage borrowers were unable to repay the loans. This

triggered a financial crisis with many banks suddenly being exposed to huge losses, or even going bankrupt in the case of Lehman Brothers and Bear Stearns in the USA. In Europe, governments had to intervene to prevent even large banks like Halifax Bank of Scotland (HBOS) and former building societies like Northern Rock from going bust. The 'contagion' that this financial crisis triggered, and the subsequent economic recession, penetrated every corner of the world economy and demonstrates how finance has bound the world economy ever more closely together.

A fourth symptom of globalization is the emergence of *global oligopolies*, that is, markets dominated by a small number of suppliers of a product or service (Case study 14.2). For a market sector in which there are very few buyers the term *oligopsony* has been coined. This is the converse of an oligopoly; for example, just six companies in the USA own the majority of movie theatres so that a film distribution company has very few negotiating alternatives. A similar story exists in many industries; the majority of drugs are increasingly made by a diminishing number of very large pharmaceutical companies like GlaxoSmithKline and Pfizer, and in the airline industry, mergers continue apace with even large airlines like British Airways, American Airlines and Iberia seeking to merge. There is now a sense in which corporate and economic survival requires 'going global'. The fifth symptom of globalization is that the power of individual nations to regulate their own economic development or to exercise a strong influence on the outcome, for example, of trade agreements has been diluted by *the rise of transnational institutions*

Thematic case study 14.2

An oligopoly: Nestlé

Nestlé is a Swiss-based company founded in 1866 that is now the top-ranking food and beverage producer in the world with sales of over US$100 billion in 2007. It employs 265,000 and has factories and operations in almost every country in the world. Nestlé has achieved this position by acquiring other related companies and brands; in 2001 it purchased Ralston Purina, making it the leading pet food manufacturer in the world; it purchased the leading French bottled water producer (Perrier) in the late 1990s to become number one for that product. It now owns eight further French waters including Vittel, as well as regional brands like Buxton in the UK, that are sold in 37 different countries in 2010. Nestlé is the leading producer of instant coffee (Nescafé was the first instant coffee on the market), and a major manufacturer of sweets and chocolates, and has increased

its dominance of the ice-cream sector by merging the Nestlé Ice Cream Company with the leading US company (Dreyers) in 2003. In addition to food products, Nestlé is involved in cosmetics (it has a large stake in L'Oréal), nutritional supplements and eye care. Its long-term strategy is to continue expanding worldwide, especially into emerging markets such as China, India, Latin America and Russia where, for example, in 2007 it invested US$120 million in a new coffee-processing plant.

Other oligopolies include Coca-Cola, Pepsico, Pearson, Interbrew and Gillette. For these leading oligopolies the key objective is to protect or to acquire world-class brands. There are currently some 40 consumer brands worldwide that sell more than one billion dollars annually (Coke, for example, sells over US$15 billion). The companies with these exceptional products are growing and expanding internationally at more than 10 per cent per annum.

---

such as the International Monetary Fund (IMF) or the World Trade Organization (WTO) that coordinate, steer and even regulate aspects of the world economy. Nation-states have also increasingly organized themselves into regional trading blocs to facilitate easier and greater volumes of trade. The European Union (EU) is one, but such regional economic arrangements have grown in importance in the last 20 years elsewhere. Examples include the North American Free Trade Agreement Area (NAFTA) and the Association of South East Asian Nations (ASEAN).

14.6 Global uneven development: the examples of trade and foreign direct investment

In light of the many processes of integration and interconnectedness in the world economy, it thus makes sense to talk of a 'global economy'. However, it is also important to realize that economic globalization has been and continues to be highly uneven. It produces a range of *global economic geographies* whether you are considering the complex spatial form of the global city networks that are key

in controlling global economic operations, or new kinds of global labour markets that shape patterns of employment and migration. In order to understand better the uneven consequences of the emergence of an increasingly globalized world economy, we consider, in this section, two aspects to this unevenness: trade and foreign direct investment.

14.6.1 Uneven trade

Trade in the world economy refers simply to the buying and selling of goods and services between actors in different localities. As the world economy has globalized, total trade has grown enormously but trade benefits some localities and not others depending on the nature of their economies. Whilst growth in total world trade stalled during the 2007–9 economic downturn, the long-term trend has been one of expansion. In 2008, world trade measured in terms of goods exports amounted to US$15.8 trillion, while exports of commercial services were worth US$3.7 trillion (WTO 2009). Much of this trade at the level of nations is concentrated between the wealthier countries in the global economy, although in the last decade developing countries like China and India have experienced huge trade growth, with China's trade surplus becoming an increasing source of tension in international politics (see Section 14.9).

Globalization processes have made understanding the idea of trade increasingly difficult, however. Conventionally, trade was measured at the national level with nation-states counting how many goods and services they exported and imported. However, globalization processes have complicated this in a number of ways. For one thing, a growing proportion of world trade is different parts of the same large transnational firm 'trading' with another part – this undermines the historical assumption that trade ended in the consumption of a product or service. Another issue is the nature of what is traded, with not only services but also new digitized products (e.g. software, music, film) hard to measure because they are sold and bought in different parts of the global economy.

The uneven impacts of international trade in certain markets have over several decades led to the emergence of a growing civil society movement to promote 'fair trade' amongst, in particular, the rich countries of the global North and developing countries of the global South. The argument is that the power of oligopolies in the global economy has forced down prices for basic commodities such as tea, coffee, sugar, cocoa and cotton that are often produced in poorer countries with the result that the producers (often small farmers) are penalized by low prices. Fair trade initiatives and campaigns for 'trade justice' thus seek to create a range of market-based mechanism to deliver better terms of trade to producers in low income countries (see Spotlight box 14.3) (see also Chapter 16).

We can illustrate the growth and development of world trade in the contemporary global economy with the example of the worlds' largest industry – namely, tourism. The United Nations World Tourism Organization (UNWTO) divided country arrivals in 2005 into leisure, recreation holidays (402 million), business travel (125 million), and visiting friends and relatives, religious purposes, health treatment and so on (212 million). Apart from demand for land, air and sea transport services, such large-scale flows also represent tourism receipts at a large number of destinations. Whether travelling on business or for leisure, visitors spend on accommodation, food and drink, local transport, shopping, entertainment and so on. In many cases this trade in tourism creates much-needed employment and other economic development opportunities, often in those less developed regions that have been largely bypassed by some of the other symbols of participation in the global space of flows (Table 14.3). Tourist destinations such as the Caribbean islands, Thailand, Bali, the Maldives or Fiji are benefiting from **comparative advantage** (Spotlight box 14.4) as well as the concept of **competitive advantage**, which is the advantage that a country or a business has over its

---

Spotlight box 14.3

Fair trade?

The idea of fair trade has its root in the 1940s and 1950s, but it was in the 1960s that a number of non-governmental organizations based in developed countries began to argue that the rise of large multinational firms and free market capitalism was not necessarily beneficial for the poorest people in the world, most of whom lived in the global South and worked in agriculture. MNCs (now TNCs), along with the tendency for oligopolies to form in many industries, meant that small farmers and commodity producers in poor countries received a very low price for their goods. The fair trade movement thus developed with the goal of rebalancing these trading relations and giving poor producers in developing countries a fair price for their products and produce. The fair trade movement started with fairly priced 'handicraft goods' in the 1960s, but by the 1980s the emphasis had shifted to commodities like coffee, cocoa, sugar and cotton.

In 1998, four organizations (Fair Trade Labelling International, the International World Fair Trade Association, the Network of European Workshops and the European Fair Trade Association) that had all been set up to achieve a more equitable trading system for producers in developing countries formed an organization called 'FINE'. Their aim was to come together as one international organization promoting fair trade in a coherent way. FINE produced a common definition of fair trade as 'a trading partnership, based on dialogue, transparency and respect, that seeks greater equity in international trade'. It should also contribute 'to sustainable development by offering better trading conditions to, and securing the rights of, marginalized producers and workers – especially in the South'.

Fair trade products have now achieved more market penetration in richer countries than at any point previously with, for example, major supermarket chains like Tesco and coffee retailers like Starbucks selling fair trade products.

Table 14.3 International tourist arrivals and receipts, world regions 1990 and 2009

Region	International tourist arrivals (million)				Tourist receipts (US$billion)	
	1990	2009	Share (%)	Change (%)	2009	Share (%)
Europe	266	460	52.2	72	413	48.5
Asia and the Pacific	56	181	20.6	223	204	23.9
Americas	93	141	15.9	51	165	19.4
Africa	15	46	5.2	206	28.9	3.4
Middle East	10	53	6.0	430	41.1	4.8
World	439	880	100.0	104	852	100.0

Source: United Nation World Tourism Organization (2010) *World Tourism Highlights, 2010 Edition,* UNWTO, Madrid

competitors because of the quality or superiority of its products or services which will persuade other countries or customers to buy from it rather than from competitors (Porter 1990). The latter is a more useful way of explaining how individual nations participate in the international tourist market in that it incorporates differences in values, cultures, histories and institutions as well as variations in endowments of the factors of production that make some tourist destinations more appealing than others.

14.7.2 Foreign direct investment

Another indicator of the integration of the global economy is *foreign direct investment* (FDI). In the context of our example of tourism, the development of many international tourist destinations relies on FDI, which is defined as the acquisition by an individual or enterprise resident in one country of assets (such as hotels, restaurants and clubs) located in another. World FDI inflows and outflows for developed and developing economies fluctuate from year to year in line with the performance of world, regional or individual national economies (Table 14.4). Overall, flows have been increasing in real terms since 1980, with the developed economies generating a net outflow of FDI in excess of US$820 billion in 2009, or almost 80 per cent of the world total. FDI fluctuates year on year and is particularly vulnerable to economic shocks such as the global economic crisis between 2007 and 2010; outflows from the developed economies

Spotlight box 14.4

Comparative advantage

Comparative advantage explains the tendency for countries (or regions/localities within countries) to specialize in certain goods and/or services even if they have the ability to fulfil their needs from domestic production. As long as countries or regions specialize in those products and/or services in which they have comparative advantage, they will gain from trade. Advantages can stem from spatial variations in, for example, mineral or land resource endowments, from variations in the educational levels of the labour force, in access to markets, or differences in levels of technology. For comparative advantage to work effectively it is necessary to assume a system of free trade, hence the significance of the trade liberalization that has been

high on the agenda of many countries since the 1980s. There is a nagging concern, however, that comparative advantage is not reflected in actual patterns of world trade. We would expect the biggest flows to be between the countries with the largest cost differences. It seems that consumer tastes and geographical proximity are actually more important than cost differences. This explains the fact that the vast majority of trade is between countries with relatively small cost differences, often involves similar rather than different goods (such as cars, electrical goods of all kinds, certain kinds of business services such as management consulting), and occurs on a 'nearest neighbour' basis. Well over half of EU goods and services trade takes place between the member states while Canada and Mexico are the major trading partners of the USA.

Table 14.4 FDI inflows and outflows, developed and developing economies, 1980–2009

$US (million)		1980	1990	2000	2009
Developed economies	Inflows	46,629	172,067	1,134,293	565,892
	Outflows	50,407	225,965	1,092,747	820,665
	Balance (I-O)	−3,777	−53,898	41,546	−254,774
Developing economies	Inflows	8,455	35,736	253,179	478,349
	Outflows	3,336	12,701	143,226	229,159
	Balance (I-O)	5,119	23,036	109,952	249,190
World	Inflows	55,108	207,878	1,396,539	1,114,189
	Outflows	53,743	238,681	1,239,149	1,100,993

Note: FDI inflows and outflows comprise capital provided (either directly or through other related enterprises) by a foreign direct investor to a FDI enterprise, or capital received by a foreign direct investor from a FDI enterprise. FDI includes the following components: equity capital, reinvested earnings and intra-company loans.

Source: United Nations Conference on Trade and Development (UNCTAD) *Major FDI Indicators* (extract from table at http://stats.unctad.org/FDI/TableViewer/tableView.aspx?ReportId=5 [accessed 25 October 2010]

exceeded US$1.1 trillion a year prior to 2007 but had fallen back by 2009. An earlier slowdown of the world economy in 2000–1 caused a decline in corporate cross-border mergers and acquisitions (M&As) which make up a large part of FDI activity, but by 2006 M&A activity once again reached a record high with deals worth $US4 billion before a significant decline again in the 2008–9 period (OECD 2009). FDI between developed countries is more important than FDI into developing countries. However, some of the biggest inflows in recent years, although still a relatively small proportion of the total, have been recorded by the newly emerging economies of Central and Eastern Europe and by the BRICs (Brazil, Russia, India and China).

TNCs contribute significantly to FDI activity. Although the number and size of developing country TNCs is now increasing, most FDI inflows result from the activities of developed country TNCs and over 60 per cent of their investments are made in other developed economies. Outflows from developing and transition economies (particularly BRICs) are now increasing as TNCs from these countries evolve into major regional and even global players. For those countries able to attract it, FDI helps to secure access to capital, technologies and organizational expertise. The net effect is modernization of infrastructure, an increase in industrial capability and an improvement in the quality and breadth of much-needed financial, business and professional services. China, a vast potential market comprising

more than one-fifth of the world's population, was closed to FDI until the mid-1980s but is now one of the leading recipients. Prior to an economic downturn during the late 1990s, almost two-thirds of the FDI flows to developing countries went to Asia (excluding Japan). These flows to Asia have subsequently recovered, but whilst there has been some increase in the last decade, much of the African continent continues to be bypassed for any form of FDI, even though many of its countries are resource-rich. With the exception of South Africa and a few West African countries which have seen some growth (UNCTAD 2009), consumer-purchasing power is low and average incomes per capita are at best growing slowly and not in a way that will raise the consumption of major consumer goods to a level that justifies the investment attention of most TNCs.

To summarize, the globalization of economies has strengthened the role of market forces while, as a result of advances in telecommunications and transportation technology, it has eased some of the constraints on interaction imposed by space and time. With the real cost of international telecommunications declining steadily over the past 20 years, global financial integration has strengthened and has been accompanied by diversified opportunities for new kinds of international trade, especially in services, and enabled easier transfer of the technology and innovation that encourages economic development and participation. The importance of non-government organizations, TNCs, and regional trading blocs such as the European Union and NAFTA for shaping the economic geography of globalization has been greatly enhanced and will continue to increase.

14.7 Places and localities in an uneven global economy

While the geography of economies accommodates a global dimension, this is not at the expense of individual localities. Indeed 'it is the combination of national and intensely local conditions that fosters competitive advantage' (Porter 1990: 158). Localities are subdivisions within nations, such as cities or regions or places that often have a particular economic identity because of the kinds of activities that take place there. Rural districts or subdivisions that specialize in the production of very particular kinds of wine (France), dairy products (Denmark) or woollen goods (Wales), for example, are also included. Many are identified as named territorial units while others are industrial districts or agglomerations whose identities are derived from specific economic activities, such as Motorsport Valley (Pinch and Henry 1999) in southern England, Silicon Valley in California or the light industrial districts of the Third Italy. Although these are often highly specialized and self-contained localities with systems of governance and regulation that fit their particular needs, they are inextricably linked with the wider national and global economic system.

These local economies and 'industrial clusters' thrive on the dynamism, innovation and 'untraded interdependencies' (knowledge and information that circulates through the transfer of key workers between firms or via social and other networks) that are made possible by proximity. Even TNCs, that are now often portrayed as being 'placeless' because their operations are so extensive and relatively mobile, started from businesses that were nurtured in a particular locality with its own economic and other characteristics to such an extent that the locally shaped attributes of such firms are carried through into their organization and transformation into TNCs (Dicken 2007). Economic geographers have, for example, sought to understand how the proximity of firms in clusters produces 'regional innovation systems' where the concentration of firms or sectors into one locality is argued to be instrumental in producing competitive success through innovation in the global economy (Asheim *et al.* 2007).

An outstanding example of the synergy between local and global processes is the City of London. This one-square-mile (259 hectares) district and the areas that fringe it in central London provided employment for approximately one million workers in 2008, with 30 per cent of these crammed into the City. The special characteristic of the City in particular is that the majority of its business activities are knowledge-intensive services. These include commercial banks and insurance companies, international banks, sophisticated private and corporate banking, firms operating in the foreign exchange, securities, commodities, shipping and derivatives markets, fund management, corporate finance, professional advisory services (legal, accountancy) that are associated with the City's financial-services complex, as well as advertising firms and other highly specialized activities. Furthermore, a large proportion of these activities are owned or managed by non-UK enterprises to such an extent that it is sometimes said that the City has stronger links with Europe and the rest of the world than it has with the rest of the United Kingdom. In 2008 over one-third of all global foreign currency transactions were undertaken in London; it was involved in 45 per cent of the total global market turnover in company equities (shares) that took place in exchanges other than the companies' domestic (or home market) exchanges (IFLS 2009).

These are just a few of the indicators of the unique localization of highly dynamic entrepreneurial, innovative and international economic activities in one very small space. There is clearly something about the environment in such localities that facilitates their competitiveness. A study by the City of London (2005) shows that local factors such as the availability of skilled personnel, access to suppliers of professional services, or the relatively open regulatory environment places the City ahead of its global competitors (Tokyo, New York, Frankfurt) (Table 14.5). It is not just about economic advantages; quality of life, cosmopolitan cultures, linguistic diversity, and a fair and just business environment are some of social/cultural factors that help us to understand how such localities work. On the basis of its prominent international status, firms operating from the City of London are able to attract the most skilled foreign workers who meet its needs both for highly educated staff offering very specialized knowledge as well as for more routine occupations in hotels and catering, transport services, or office servicing.

It has been argued that increased localization and agglomeration effects are thus a crucial converse outcome of wider globalization. Whilst London (and

Table 14.5 Rank of competitive factors, leading world financial centres, 2005

Factor of competitiveness	Rank	Average score	London rank[1]
Availability of skilled personnel	1	5.37	1
Regulatory environment	2	5.16	1
Access to international financial markets	3	5.08	1
Availability of business infrastructure	4	5.01	2
Access to customers	5	4.90	1
A fair and just business environment	6	4.67	1
Government responsiveness	7	4.61	1
Corporate tax regime	8	4.47	2
Operational costs	9	4.38	2
Access to suppliers of professional services	10	4.33	1=
Quality of life	11	4.30	3
Cultural and language	12	4.28	1
Quality/availability of commercial property	13	4.04	2
Personal tax regime	14	3.89	2

[1] Compared with three other leading financial centres: New York, Paris, Frankfurt.

Source: City of London Corporation (2005)

not just the City of London) is a leading example, work on global cities argues that to some extent all cities in the global economy are experiencing similar trends and that economic success in localities is increasingly affected by the complex mixture of place-based factors. National governments have responded to these ideas with policies aimed at giving localities greater autonomy over the institutions and instruments that they put in place, using the insights gained from detailed knowledge of local networks, to capitalize on their economic advantages or to create economic and social environments that will attract new investment. One very influential idea has been that increasingly the economic success of cities and other localities is dependent on skilled knowledge workers who undertake (in a broad way) the creative aspects of economic activity in the global information economy (Florida 2002). Attracting this new 'creative class' has thus been argued to be crucial in policy terms when local government or other institutions are seeking to develop strategies for strengthening regional economic development (see Spotlight box 14.5).

14.8 The rise of a new global digital economy?

In many respects the interaction between localities and the global economy is shaped by the so-called digital economy. It is estimated that there were two billion Internet users worldwide (approximately a third of the total population) at the end of 2010 (ITU 2010), up twentyfold from 100 million in 1997 (Table 14.6). E-commerce was already expanding at the end of the 1990s but it is now growing and diversifying even more rapidly, led by business-to-business (B2B) and business-to-customer (B2C) transactions by service sector activities such as telecommunications, information technology, publishing and the media, travel and tourism, retailing, transportation and professional services like management consultancy, industrial design and engineering. Although consumer sales attract the publicity (such as the phenomenal growth of eBay or Amazon), much of the expansion of e-commerce is between and within businesses. Global e-commerce sales (B2B, B2C, etc.) were estimated to be worth

Spotlight box 14.5

The importance of a 'new creative class'?

The American social scientist, Richard Florida, suggested the idea of a new creative class in his book *The Rise of the Creative Class* published in 2002. Florida's initial concern is the role in the US economy of what he identifies as a new class of creative workers that comprises some 40 million people (about a third of the workforce). Using the standard occupational classification, Florida divides this creative class into three groups, the two major ones are the 'super-creative core', which covers a wide range of occupations (science, engineering, education, computer programmers, research, arts, design) that creates products and consumer goods, and the 'creative professionals', who are essentially more conventional knowledge-based workers in finance, legal services, healthcare and education. A third, much smaller, group comprises the bohemians, including artists, writers and dramatists. The key argument Florida makes is that it is these segments of the workforce who are the major drivers of growth in today's global economy, and American (and hence other) cities and regions need to focus on attracting and supporting these groups.

The concept of the creative class has had enormous influence amongst policy practitioners beyond the USA since it was proposed, but it is not uncontroversial. Economic geographers (amongst others) have been critical of the idea in a range of ways, including the view that the concept has no causal mechanism (it is more about description than about any process) (Peck 2005) or that empirical research has found little sense of identity amongst those people supposedly in these groups with membership being based more on educational achievement than any measure of creativity (Markusen 2006).

US$438 billion in 2008, with the USA alone accounting for US$133.6 billion. The potential for growth in direct cross-border sales to consumers is substantial as Internet access (see Table 14.6), user confidence, payment systems, Internet/Web security, and mechanisms for tracking transactions for levying taxes and duties, continue to improve. Furthermore, the development of new so-called 'Web 2.0' technologies have again transformed the nature of Internet usage itself, with the rise of 'real-time' streamed data and increased storage capacity online. Social networking businesses such as Facebook offer new but as yet undeveloped e-commerce opportunities, and the rapid proliferation of mobile web devices (Apple's iPhone, for example)

Table 14.6 Internet usage, by world region, 2010

Region	Internet usage (000s)[1]	Penetration (% of total population)	Usage (% of world)	Population (% of world)	Usage growth 2000–2010 (%)
Africa	110,931	10.9	5.6	14.8	2,357
Asia	825,094	21.5	42.0	56.1	622
Europe	475,069	58.4	24.2	11.9	352
Middle East	63,241	29.8	3.2	3.1	1,825
North America	266,225	77.4	13.5	5.0	146
Latin America/Caribbean	204,690	34.5	10.4	8.6	1,033
Oceania/Australia	21,264	61.3	1.1	0.5	179
World total	1,966,515	28.7	100.0	100.0	449

[1] 30 December 2010

Source: http://www.internetworldstats.com/stats.htm [accessed 26 October 2010]

Thematic case study 14.3

Facebook versus Google?

One of the most dramatic developments in the global economy since the turn of the century is the continued rise of new media and web industries. During the late 1990s, the so-called 'tech boom' led to the rapid rise (and equally rapid demise) of many new firms in web-based software whose success was built on the emergence of e-commerce and new software for the World Wide Web that had increasing technical capacity. However, since the turn of the century, this new era of 'Web 2.0' has produced another phase comprising a mixture of new and existing firms enjoying meteoric success. No companies exemplify this more than California-based rivals Facebook and Google.

Founded in 1998 by two Stanford PhD students, Larry Page and Sergey Brin, Google is a software company that has become the dominant web search engine. Now based in California's Silicon Valley, and at more than a decade old in today's digital economy, it is a long-standing player. Google has developed its business beyond web-search into online advertising, and perhaps most importantly a range of software products that include the Android operating system for third-generation

cellphones. In this latter area it competes with Apple, but it is in the lucrative area of online web advertising where Google makes most of its money, it also competes with a relative newcomer, Facebook. Famously founded by Mark Zuckerburg in his Harvard dormitory in 2004, Facebook has grown spectacularly and had overtaken Google in terms of visitors to its website by mid-2010. Facebook's product is of course its social networking software which is based on Web 2.0 technology, giving users a personal profile they manage themselves. By mid-2010, over 500 million people globally had profiles and the company by this point had a staggering market valuation of over US$33 billion although not yet a fully listed public company (*Financial Times* 2010).

The very rapid growth of companies like Google and Facebook reflects the global penetration of the Internet in the contemporary global economy, and the potentially vast commercial revenues that new forms of web advertising and product sales may generate. The ongoing challenge for both companies is how to generate commercially viable business models in a digital world where technology moves at break-neck speed, and today's corporate success story fades into obscurity tomorrow as the next innovation emerges.

---

mean that online commerce is a rapidly changing sector both in terms of hardware and software (see Case study 14.3).

Apart from the numerous legal, regulatory, security and other challenges it presents, the global digital economy is likely to be accompanied by new geographies of the economy. Just as the Industrial Revolution generated significant economic and social changes (both positive and negative) in the form of new jobs, new industries and new industrial regions, so will the digital economy stimulate its own revolution. Perhaps the Internet will finally eliminate the effects of friction of distance on economic interactions. But this will require a truly global network of telecommunications and computer infrastructure that is accessible to all. As the data in Table 14.6 show, some national economies are yet to be 'plugged in' to the Internet. To be excluded is to widen the gap that already exists between those countries 'inside' and those 'outside' the global digital economy. A networked readiness index (Figure 14.3)

reveals wide differences in the geographical distribution (number, density and processing power, for example) of the computers, telecommunications networks and software required to participate in the global economy. However, Africa's digital economy deficit may be rectified much more rapidly than expected with the expansion of cellular (mobile) telephony. Cell phones not only offer voice services but also technologies that bypass the need for a computer to access the World Wide Web so that as early as 2001 Africa was the first world region where the number of mobile subscribers exceeded the number of fixed-line subscribers. Access to mobile phones opens up all kinds of economic and social benefits such as better access to information about crop prices, guidance and support from government institutions on how to run small businesses, accessing bank services without the need to travel (including money transfers – see Grameen Bank case study, Chapter 18), or keeping in touch with family and friends.

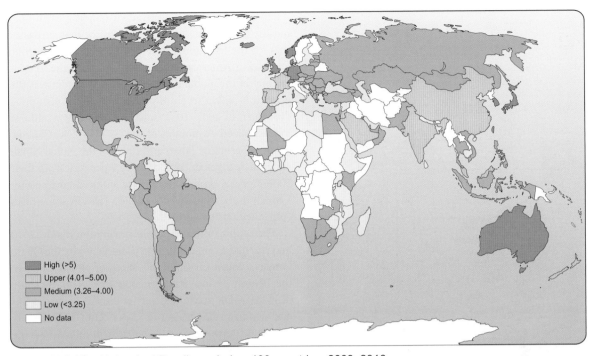

Figure 14.3 The Networked Readiness Index, 133 countries, 2009–2010.

Source: compiled from data at: www.weforum.org/issues/global-information-technology, accessed 15 October 2010.

A second challenge posed by the digital economy is the demand for human resources. More jobs will be created by the digital economy than will be lost but they are generally in higher-skilled and better-paid occupations. In order to fill new digital-economy jobs, economies everywhere need to retrain existing workforces and to train future workers. History suggests that the response to this challenge will be uneven; comparative advantage will enable some economies and some localities to respond more quickly and effectively than others. Perhaps less certain is how this will modify existing spatial patterns of economic development. A digital economy, for example, enables a wide range of purchasing and banking transactions to be undertaken from home via the telephone, laptop computers, televisions, or on the move using mobile phones that provide access to Internet services. An international door-to-door delivery time of 24 hours or much less is now commonplace, depending on the type of good or service, whether it is required just in time, or the location of the supplier. Moreover, cross-border transactions and payments are as easy as those made to the local supermarket or bookstore. With shopping sites on the Internet presenting themselves as 'virtual malls' and encouraging users to place their selections

in 'baskets' or 'carts', they are attempting to replicate offline retail outlets. More than 40 per cent of Internet users shopped online in the USA and in Western Europe in 2003, and by 2008 it has been estimated that 40 per cent of online buyers will purchase travel and 58 per cent will purchase books. This, combined with trends such as increased working from home or teleworking, whether self-employed or as an employee of a TNC, points to the potential for significant changes.

The challenge for economic geographers is how to understand the ways in which the digital economy impacts on the structure and pattern of economic activities at different scales, on the structure of localities and their relationship with the global economy, or on the relative importance of nations and TNCs in regulating or influencing the outcome.

14.9 Global re-balancing: the eastward shift

We end our overview of the geographies of economy with a consideration of probably the most pressing question that economic geographers can address in the

Plate 14.3 A street vendor walks past an old house marked with the Chinese character for 'demolish'. Parts of China, such as Shanghai or Guangzhou, are experiencing rapid economic transformation which is expected to continue well into the twenty-first century.

(AP Photo/Eugene Hoshiko)

Asian economies have enjoyed sustained growth during the first decade of the new century that surpasses that for the traditional developed countries in Europe and North America. Equally important to the 're-balancing' of the global economy towards the East, is the geographically uneven impact of the global economic downturn of 2007–10. Whilst the formerly wealthiest industrialized economies of the G7 (US, UK, Canada, France, Germany, Japan and Italy) in the global North suffered declines in GDP of up to 9 per cent from their 2006/7 peaks, economies in Asia suffered far less with a much more short-lived recession (and in some cases continued positive growth).

The result is a dramatic transformation that by 2010 had reached a critical turning point whereby the G7 no longer account for the lion's share of global GDP. In the 10-year period between 2000 and 2010, the share of global GDP accounted for by the G7 countries has declined from 72 per cent to 53 per cent (IMF estimate for 2011). Furthermore, annual growth rates across the globe seem set to continue this re-balancing. In 2010, countries like the UK will grow by less than 1 per cent, the US by around 2 per cent and even strong G7 economies like Germany are unlikely to breach 3 per cent. Compare this to the Chinese government's attempts to manage what is perceived to be too high (not too low) a growth rate that in late 2010 threatened to breach 10 per cent of an economy that is now the globe's second largest. China's economic power has thus not only increased over the last decade, but has in fact eclipsed almost all the formerly 'rich' countries of the G7. Whilst, because of its enormous population, GDP per capita still places it as a middle-income rather than a high-income country, China is no longer accurately described as a developing economy; many parts of its coastal regions are highly developed even if large areas of its inland provinces remain poor and agricultural.

The implications of this eastward shift in economic power for understanding the economic geography of the global economy are enormous. It seems likely that Asian economies will continue to grow more strongly than many economies of the global North over the next decade. The questions that face economic geographers are no longer so much about the impact of developing economies that become more developed, but rather about the effects of a fundamental further geographical shift in the centre of gravity of the global economy towards Asia.

early part of the twenty-first century: the re-balancing of the geography of the global economy away from the historical pattern of economic activity that has long been dominated by North America and Europe. In short, the key issue is the progressive 'shift eastwards' in terms of the centre of gravity for growth and output in the global economy.

Whilst there has been a longstanding trend since the 1980s for significant growth in the formerly 'less developed' economies of Asia, since the turn of the century this has become much more pronounced. China's rate of economic growth since 1995 has rarely fallen below 5 per cent per annum. Yet it is not just China; despite the economic crisis during the late 1990s other

Learning outcomes

Having read this chapter, you should be aware of:

■ The nature and changing concerns of economic geography as a sub-discipline.

■ The importance of economic restructuring and the shift from manufacturing to services in the contemporary economy.

■ The significance of globalization processes in the structure and spatial development of the world economy.

■ The major role performed by transnational corporations in trade, foreign direct investment and other indices of globalization.

■ The challenges posed to economic geographers by the rapid growth of the global informational economy and its potential for changing established patterns of economic activity.

Further reading

Dicken, P. (2010) *Global Shift: Mapping the Changing Contours of the World Economy*, 6th edition, Sage, London. A broad-ranging overview of the evolution of internationalization and the globalization of economic activities. Focuses on the complex interaction between transnational corporations, nation-states and the rapid developments in information and communications technology, drawing on fields outside geography such as political science and economics.

Knox, P., Agnew, J. and McCarthy, L. (2003) *The Geography of the World Economy: An Introduction to Economic Geography*, 4th edition, Hodder Arnold, London. A synthesis of the factors shaping the development of contemporary economic patterns in market-oriented and centrally planned economies. Strong emphasis on inter-dependence of economic development at different spatial scales, from local through national to international.

Mackinnon, D. and Cumbers, A. (2007*) Introduction to Economic Geography: Uneven Development, Globalization and Place*, Pearson, London. Explores the wide range of approaches and models that are debated

about and used by economic geographers. Focuses on globalization, uneven development and place. Covers conventional topics such as regional development and labour markets alongside an introduction to economic topics that are subject to rapid change such as consumption, information and communications technologies and tourist geographies.

Useful websites

www.ilo.org International Labour Organization. Provides documents and statistics on a wide variety of employment and labour market issues for individual countries and on a comparative basis for countries around the world.

www.un.org United Nations. Reflects the wide-ranging responsibilities of the United Nations but includes a useful section on global economic and social development, including an annual report on trends and issues in the world economy. Links to other websites.

www.wto.org World Trade Organization. Offers a wide range of documentation and statistical material about global trends in trade in goods and services, regional aspects of trade, electronic commerce and background research and analysis on all aspects of world trade. Links to related websites.

www.europa.eu.int/comm/indexren.htm The European Commission. Readers of this chapter will find the links to the various Directorates-General (DGs) sites within the Commission (Industry, Transport, etc.) useful starting points for monitoring economic development patterns, problems and policies across the European Union.

www.oecd.org Organization for Economic Cooperation and Development. Represents mainly the developed economies and provides free documents, summaries of OECD economic surveys and statistics.

www.unctad.org United Nations Conference on Trade and Development. Focuses in particular on the interests of less developed countries in relation to international trade and foreign direct investment. Publications, statistics and links to related websites.

For annotated, clickable weblinks and useful tutorials full of practical advice on how to improve your study skills, visit this book's website at www.pearsoned.co.uk/daniels

CHANGING GEOGRAPHIES OF FOOD PRODUCTION

Damian Maye and Brian Ilbery

Topics covered

- Thinking about food
- Farming, food chains and globalization
- Alternative geographies of food
- Food security
- The ethical foodscape

15.1 Thinking about food

This chapter introduces some debates about the changing nature of food provision and presents ideas and case studies to prompt critical reflection. Food has become a topic of great interest to human geographers in recent years (Atkins and Bowler 2001; Watts *et al.* 2005; Cook *et al.* 2006/2008; Morgan *et al.* 2006; Maye *et al.* 2007; Goodman *et al.* 2010) and the chapter begins by identifying its value as a tool for geographical analysis. The chapter reveals the changing nature of agri-food production systems and shows how the way food provisioning is conceptualized, framed and debated changes over time. A crude assessment of these debates indicates the cyclical nature of discussions about food provisioning. However, a comparison of past debates about industrial agriculture with more recent calls for the so-called 'sustainable intensification of global agriculture' (Beddington 2010) reveals important and significant differences. This more recent phase of debate and conceptualization is characterized by the need for an effective rather than exploitative use of resources, in response to growing concerns about the depletion and rising costs of energy supplies, population growth and climate change.

Such a 'mode of thinking' is significantly influenced by past events and a realization that energy-greedy production systems that propelled food systems in the 1980s and 1990s are no longer workable or sustainable (Harding 2010). Food has thus become a touchstone for wider environmental discourses. As Harding (2010: 4) puts it, 'feelings about eating and not eating are more immediate than thoughts about rainforests; like the energy or water embedded in the produce we buy, many fears, including fundamental ones about life and death, destruction and incorporation, are already embedded in food'. As we shall see later in this chapter, these feelings, fears and anxieties (see Jackson 2010) have been compounded by recent global price increases and the trends that food analysts suggest they reflect.

First though, it is useful to think about food more generally and to consider why we – as humans – relate to it in the ways Harding and others (see Belasco 2008, for example) suggest that we do. The food writer, Nigel Slater, in one of the first editions of the *Observer's Food Monthly* magazine, argued that:

> Never has there been more interest in what we eat and drink. But never have we needed to know more either. The bottom line is that the food we eat becomes part of us. Each blood-stained bite of meat, every cool sip of water, each spoonful of organic cabbage and every

additive-encrusted potato chip; the salads, the Mars bars, the wine and the beer, all play their small part in what we are. Our bodies take both the virtues and horrors of every forkful – for better or for worse – and turn it into Us. (Slater 2001: 1)

Food, then, extends well beyond its obvious biological significance as the protein for sustaining life. It is something that one relates to on a very human scale – it reflects who we are. According to Atkinson (1991) food is a 'liminal substance' that links humans and nature. It also reveals important social, cultural and economic geographies. On the social side, there are inequalities in terms of food access, good nutrition and the reported problems with obesity and diet, often tied to socio-spatial inequalities in terms of wealth and education. Places where access to affordable, healthy food is poor are described as '**food deserts**'; these are particularly prominent in urban areas (Wrigley 2002). Food also has cultural significance. Thus, 'food shapes us and expresses us even more definitively than our furniture or houses or utensils do' (Visser 1986: 12). Think, for example, about the different places associated with eating food (e.g. roadside café, Burger King, the Savoy Grill) and the different behaviours associated with them. As Bell and Valentine (1997) note, if 'we are what we eat' we can also say that 'we are *where* we eat'. There are also cultural associations between different cultural groups and ethnic or national cuisines (e.g. the favourite dish in the UK is Tikka Masala – this curry dish is thought to have originated in the first Indian restaurants in Soho, London in the 1970s and exemplifies the cross-cultural origin of some foods), as well as a range of cultural images used to market and sell food. Finally, the economic significance of food is enormous: for instance, one only has to think about the investment and the number of people involved in the production, processing and retailing of food. In the USA, for example, only 2–3 per cent of the population is directly employed in agriculture, but this figure rises significantly if the whole food chain is accounted for, and almost 10 per cent of US Gross Domestic Product is food-related (www.usda.gov [accessed 17/1/11]).

Thus it is quite easy to make a case as to why food matters as an object of study in human geography. However, tracing material and metaphorical associations with food is less obvious because they are connected together in complex ways and often 'hidden' as incidental parts of everyday life. For example, drinking a cup of coffee has numerous social and geographic connotations (Giddens 2001): it has symbolic value (as part of day-to-day life); it is a drug (caffeine provides 'extra lift', but coffee drinkers would not be seen as 'drug users'); it represents past social and economic relationships (e.g. in terms of colonization, mass consumption); and it is a symbol of globalization and world trade links (e.g. global brands like Nescafé and Starbucks). Another example of this global-local 'inter-connectedness' would be a visit to a local supermarket where one can identify 'the world on a shelf', with products such as apples from South Africa, spices from India and coffee from Colombia. For shoppers in Western society, it is easy to forget the complex array of economic and social ties that connect these local retail spaces to other parts of the world. Geographers argue that one cannot separate these localized, mundane acts from larger social settings that extend around the world. This realization about the need to trace *'connections'* between food production and consumption, sparked in part by on-going public debates about obesity, food risks and so on, has become central to agri-food studies and has been given further impetus more recently by debates about food security and the interdependence of food systems.

Geographical research in this area is increasingly focused on different types of food supply, which involves thinking about where and how food is produced, how it is retailed and how and where it is consumed. The routes traced by particular foodstuffs from 'farm to fork' are often referred to as a **'food chain'** (or 'network') and geographers have attempted to 'map' the system of connections for different products which may vary both in complexity and geographical coverage. Hartwick (1998: 425) defines food chains as 'significant production, distribution and consumption nodes, and the connecting links between them, together with social, cultural and natural conditions involved in commodity movements'. Geographers have thus adopted the supply chain metaphor to literally trace and follow the nature of 'connections' for particular commodities (Cook *et al.* 2006). Different food supply chains will be examined later in the chapter. The food chain is not a new concept but it has been recast as a 'food system', 'food circuit', 'food network' or 'food convention' (Jackson *et al.* 2006; Maye and Ilbery 2006). Despite the varied terminologies, the food chain continues to have symbolic and applied values.

Food, therefore, is a geographical topic. This includes its role in society and economy, and the food chain can be used as an overall organizing framework to explore changing geographies of food production. The rest of the chapter is divided into four parts: first, past and present modes of global food supply are examined, including the nature of global food chains and the resulting

'disconnection' between producers and consumers; second, some of the ways in which producers and consumers are beginning to establish alternative systems of food provision are outlined; third, new debates about food security are examined, including its definition and interpretations about how food should be produced now and in the future; and finally, the chapter summarizes some recent food chain trends and discussions about the ethics of food production, recognizing the need for geographers to continue to critically interrogate the conceptual foundations of food production and supply.

15.2 Farming, food chains and globalization

This section examines farming, food production and global food supply. It outlines some of the ways geographers and other food analysts have framed global food production until relatively recently. Two processes are significant: the **industrialization** of farming and the **globalization** of food supply. After introducing some concepts, case studies of two global food commodities are explored before identifying a series of questions that emerge in response to these particular food chain practices. Globalization is often defined in terms of the integration of systems among geographically dispersed places (see Dicken 2010 for details). Crucially, this process of global integration has been guided by powerful **transnational corporations** (TNCs), institutions and actors, which in this context led to a new **political economy** of agriculture, epitomised by the mass production of manufactured food (see also Chapter 14). Despite inequalities, developed and developing countries are reportedly linked together in 'highly industrialized and increasingly globalized networks of institutions and products, constituting an agri-food system' (Whatmore 1995: 37).

The global food economy is not new. For example, Friedmann and McMichael (1989) argued that relations between agriculture and industry have historically been more global than generally thought. Using the concept of **food regimes**, they linked international patterns of food production and consumption to the development of the capitalist system since the 1870s. They identified three food regimes, each one representing the modern food system of its time (Table 15.1). Thus, there has been a global dimension to the geography of food supply for some time (for useful reviews of food regimes see Robinson 2004 and McMichael 2009).

Table 15.1 Global food regimes since the 1870s

First regime: pre-industrial (1870s–1920s)
This involved settler colonies supplying unprocessed and semi-processed foods and materials to the metropolitan core of North America and Western Europe. Characterized by *extensive* forms of capital accumulation, the main products were grains and meat. The regime slowly disintegrated when agricultural production in developed countries competed with cheap imports and trade barriers were erected.

Second regime: industrial (1920s–1970s)
This regime relates to the productivist phase of agricultural change, focused on North America and the development of agri-industrial complexes based around grain-fed livestock production. Characterized by *intensive* forms of capital accumulation, the second regime incorporated developed and developing nations into commodity production systems. Agricultural surpluses and environmental disbenefits undermined this phase of production in the 1970s.

Third regime: post-industrial (1980s onwards)
This regime refers to the crisis surrounding industrialized farming systems and involves the production of fresh fruit and vegetables for the global market, the continued reconstitution of food, and the supply of inputs for 'elite' consumption in developed countries. Characterized by a *flexible* form of capital accumulation, this regime is dominated by the restructuring activities of agribusiness TNCs and corporate retailers.

Source: based on Robinson (2004), as derived from Friedmann and McMichael (1989)

The main criticism of this conceptualization is its view of the globalization of agriculture as a logical progression (similar in this respect to modernization theory). In reality, it is much more unstable. Take the fast food chain McDonalds, for example, often heralded as symbolic of global mass food consumption. Its global presence is in fact atypical of the complex and highly uneven process of globalization that has reshaped food production since the post-war period. Much less contentious in these debates is identifying the key agents involved. FitzSimmons (1997), among others, identifies TNCs as the primary agents of globalization in the agri-food sector; to her mind, they sit at the centre of webs of relations that link farming, processing and marketing. Food retailing in Europe and America is concentrated in the hands of a few supermarkets and these agents also play an important role in developing regulatory systems that ensure their dominance over the supply of key food products. It is also agreed that TNCs and corporate retailers played significant roles in the 'industrialization' of farming and food production activities, which, initially at least, were mostly concentrated in the developed world.

15.2.1 Industrialized agriculture in the developed world

Millstone and Lang (2003: 38) note that, 'encouraged by governments as a response to food shortages and the need to raise productivity after World War II, industrial farming has dominated . . . agriculture in the EU and North America since the 1950s'. The industrial model of farming, key features of which include a specialization of labour, product specialization and intensification, and assembly-line type production, led to three important food production trends in the developed world: first, the concentration of agricultural production on a limited number of large-scale farms; second, an increase in capital expenditure on major agricultural inputs like chemicals; and third, a growth in the processing and manufacturing of food. These developments explain why it is now vital to view farming in the wider context of an **agri-food system**, where the production sector itself is inextricably linked to various 'upstream' (input supplies) and 'downstream' (processing, distribution and marketing) industries. In many cases, the food supply system is also dominated by large **agribusinesses** which, according to Davis and Goldberg (1957), are the sum of all operations involved in the manufacture and distribution of farm supplies, the production operation of the farm, storage, processing and distribution of farm commodities and items made from them. These agribusinesses often develop commodity chains beyond national boundaries (Wallace 1985).

A number of papers have charted the **industrialization of agriculture**, especially in the United States. Two brief examples are presented here. The first is the Salinas Valley in California where lettuce production was transformed between 1950 and 1980 by two key things: *intensification*, as a result of a shift to more investment in intensive crops, increased labour productivity and intensified planting; and *restructuring*, whereby large farms dominated sales (Fitzsimmons

1986). By 1978, the ten largest grower–shippers in the region sold 65 per cent of all lettuce. Over time, therefore, power and control shifted towards the largest firms. Fitzsimmons also identified a process of **vertical disintegration**, whereby segments of the production process were subcontracted out to smaller-scale growers. The second example is derived from an examination of some well-known US agribusinesses, including Con Agra, one of the three largest flour millers in North America (Hendrickson and Heffernan 2002). The supply chain for Con Agra (Figure 15.1) shows that the company produces its own livestock feed,

Figure 15.1 The ConAgra food chain cluster (situation in 1999).

Source: Hendrickson and Heffernan (2002)

slaughters cattle, is involved in pork and broiler production and processing, crop protection and fertilizers (upstream), and the retail of foods (downstream). Foods produced through this vertically integrated chain are sold under a number of well-known US food labels, including Healthy Choice and Peter Pan Peanut Butter.

In these and other examples, it is often the non-farm sectors of the agri-food system that have become most industrialized and dominated by TNCs. This has occurred through two processes (Goodman *et al.* 1987): **appropriationism**, where certain agricultural inputs are replaced by 'industrial' alternatives (e.g. synthetic chemicals replacing manure); and **substitutionism**, which focuses on outputs rather than inputs and is concerned with the increased utilization of non-agricultural raw materials and the creation of industrial substitutes for food and fibre (e.g. sweeteners for sugar). Agri-food systems are different from other production systems because agriculture is bound by biological processes and cycles, and the essential aim of appropriationism and substitutionism is to 'replace' nature. Agribusiness TNCs also attempt to increase their influence over farming indirectly through a process of formal **subsumption**, where arrangements or contracts are made with farmers to provide 'raw materials' for their value-adding food manufacturing activities.

The overall key pattern of agricultural industrialization is *concentration* in particular sectors and regions characterized by large farm businesses that have adopted intensive farming methods and become integrated into global food networks. In the European Union, 80 per cent of agricultural production is concentrated on less than 20 per cent of farms in particular 'hot spots', including East Anglia, the Paris Basin, southern Netherlands, and Emilia Romagna in north-east Italy. A similar pattern of concentration is evident in the USA, with dairy farming now concentrated in the west, for example, and milk production highest in California. Agricultural industrialization is also increasingly global in scale, where food commodity chains lengthen and producers become 'distanced' from consumers. In response to various issues in the developed world market, another key feature of TNCs and major retailers has been their attempt to relocate the production function of the agri-food system to new agricultural spaces, often in **less developed countries** (LDCs) and usually in association with the intensive production of high-quality, high-value food commodities.

15.2.2 Developing world agriculture

Goodman and Watts (1997) argue that the classic export commodities (e.g. coffee, tea, sugar, cocoa) associated with the LDCs have been complemented by high-value foods, including fruits, vegetables and shellfish. By the 1990s, 24 low-middle income countries annually exported over US$500 million of **high-value foods**; just five countries (Brazil, Mexico, China, Argentina and Kenya) accounted for 40 per cent of such exports from the LDCs. The growth in high-value food exports reflects various things, including: technical changes in the food industry, the liberalization of world trade, and dietary changes in the developed world, with high-value foods produced to satisfy consumer tastes. Agribusiness TNCs are also attracted to high-value food production in newly agriculturalizing countries because of low-cost labour, government support (via **structural adjustment programmes**), good global communication links and the ability to produce high quality/value products for the developed world market. There are a number of detailed accounts of the export trade in high-value foods and cut flowers, especially for some African countries (see, for example, Hughes 2000; Freidberg 2004). However, the example used here is taken from Chile in South America.

15.2.3 Salmon farming in Chile

Salmon farming in Chile is mostly located in Chiloe, a large island that runs along the southern third of the Chilean coast and is characterized by a mild, rainy climate and protected inlets. Since the 1990s, the salmon farming industry has been growing rapidly in Chile. In 1987, Chile produced 2,000 tonnes of Pacific salmon, but did not produce any Atlantic salmon (Phyne and Mansilla 2003). By 1992, Chile produced less than 10 per cent of the world's Atlantic salmon, but by 2000 this had increased to 17 per cent (second to Norway). Chile also dominates the global production of Pacific salmon (1992: 37 per cent and 2000: 76 per cent). Figure 15.2 shows a simplified model for the salmon aquaculture supply chain, which starts when eggs are hatched and the fingerlings raised near the hatchery; once fingerlings become smolts, they are raised in smolt-rearing facilities in fresh-water cages; after a period of 12 months they are shipped to sea-water cages and raised for a further 12–18 months (thanks to new feeds, this grow-out stage is reducing rapidly); harvested salmon enter processing

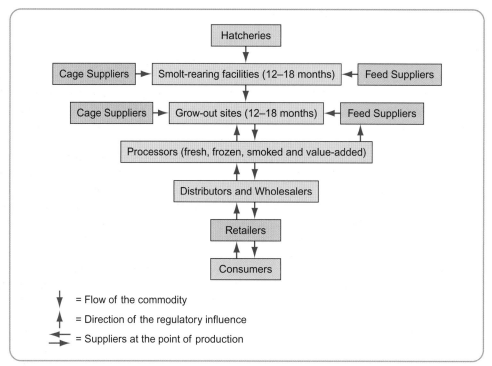

Figure 15.2 A simplified model of the salmon aquaculture supply chain.
Source: Phyne and Mansilla (2003)

plants (with automated feeding systems and where labour is most concentrated) and are prepared to standards set in Japan and America. Over 80 per cent of the product is sold to Japanese and American retail markets.

This case study shows how the Chilean salmon farming industry is inserted into a **'buyer driven' commodity chain**, where major distributors and retailers in export markets shape the nature of production. Power is at the retail end of the chain rather than the production end. Essentially, the industry is characterized by governance from lead firms and external authorities. Similar conclusions were noted in Cook's (2004) analysis of papaya in Jamaica (see Spotlight box 15.1 for details). In the middle part of the salmon supply chain, there is also clear evidence of concentration and squeezing, with forward integration by feed giants and increasing concentration of production by foreign and domestic firms. At the production end of the chain, benefits are skewed in favour of those in management positions. Grow-out sites and processing plants are where most local people work. Wages are low and the working conditions poor, with 80 per cent of plant workers being female. In this example, as in the case of papaya

in Jamaica, global food production raises significant questions, especially in terms of who truly benefits.

15.2.4 Problems associated with industrialized farming

Agricultural industrialization is thus extending its global reach; it is also clear that large-scale agribusinesses and corporate retailers have successfully linked regional economies and food sectors of some LDCs to a global system of food production and consumption. However, these systems of provision also raise questions about their suitability for LDCs, especially their contribution to sharp inequalities in income, productivity and technology compared to the sector producing staple foods (i.e. foods that are eaten regularly and provide a large proportion of a population's energy and nutrients, such as cereals or tubers), as well as the unsustainable nature of agricultural practices often favoured by agribusiness TNCs. In developed world economies, questions are also being asked about the logic and impacts of industrialized farming. Two areas of concern are notable. The first relates to food production. Questions about the merits of intensive

Spotlight box 15.1

Follow the thing: Papaya

Cook *et al.* (2006: 655–6) pick up on a recent trend in food geography to quite literally 'follow' things as a method to unpack those hidden food chain stories which foods might tell if they could talk. One of the best examples of this is an earlier paper by Cook (2004) which followed the supply chain for papaya (a large green fruit from the West Indies). It provides detailed accounts from multiple sites along this particular global fruit chain, including the papaya buyer (based just outside London), the papaya importer, the papaya farmer, the farm foreman, the papaya packer (all based in Jamaica) and the papaya consumer (in North London). The fruit itself is no passive agent in this story, secreting an enzyme that eats into the skin of pickers and packers; changing the sex of its flowers according to the weather – with only hermaphrodites producing exportable fruits; requiring specialist, intensive labour if fruits of the right shape, size and quality are to be produced for market; and dying once harvested, thus increasing pressures to get the fruit packed and on a plane to supermarket shelves in the United Kingdom and United States.

The research sets out to 'de-fetishize' commodities, but is unusual in that it chooses not to impose any specific theoretical framework, offering instead an apparently descriptive ethnographic account of individuals at different stages in the supply chain, interweaved with historical and contemporary contexts about the place, plant and its trade routes. The challenge Cook sets is for us to read and draw our own conclusions. Various important things emerge for consideration. First, the paper shows the complex and entangled nature of global food supply chains, linking in this example farmers and labourers in the Caribbean, fruit buyers, supermarkets and technicians in the United Kingdom, and First World consumers. Second, the role and evidence of power in these fruit chains are manifested in different ways and at different stages along the chain. Most notable are the pressures on growers and buyers to supply products that meet specific retail standards and are delivered on time. This confirms the buyer-driven rhetoric of most global food chains, but is equally disturbing in the way power games trickle down to the farm level, with the farm owner and foreman using surveillance to monitor workers' performance. Third, disconnections are highlighted between the peoples and places involved with papaya, including the farm workers who only know the 'consumer' as one vague entity that dictates market pressures, as well as the product itself which changes from plant to agricultural commodity to exotic fruit to beer by-product. Above all, it makes a clear case for alternative food systems that can better the livelihoods of workers and pickers at the bottom of the supply chain.

---

agricultural production entered public imaginations in large part due to mad cow disease, which emerged in the UK in the 1980s. This disease is caused by a protein called a prion, which entered the cattle-feed chain when animal carcasses were rendered down into meat and bone meal and incorporated into cattle feed without being adequately decontaminated. The disease is a direct consequence of the large-scale industrialization of food production and also represents a direct threat to human life in a novel form of the disease called 'variant Creuzfeld–Jakob Disease' or vCJD. Similar questions were being asked about the industrialization of agriculture after the outbreak of foot-and-mouth disease in the UK in 2001 and again after the spate of bird flu incidents reported in south-east Asia in 2003 and, more recently, in East Anglia in the UK as well.

The second area of concern relates to fair trade and environmental responsibility. International trade in food has expanded significantly in recent decades.

Between 1961 and 1999, there was a fourfold increase in the amount of food exported. In 2000, eight of the wealthiest countries accounted for almost half of all agricultural exports (Millstone and Lang 2003). One of the drivers of this process has been the **World Trade Organization** (WTO). The WTO's Agreement on Agriculture promoted trade liberalisation through reductions in agricultural subsidies, tariffs and import quotas. However, concerns have been voiced about the unfairness of global trade because of the continued use of national subsidies and tariffs. As well as unfair trade, the environmental impact of a major increase in world food trade is also being felt around the world. For example, trade-related transportation is one of the fastest growing sources of greenhouse gas emissions. The distance food travels is expressed in terms of **food miles**. As well as distance, the mode of transport (e.g. air freight vs. sea freight) is also an equally important, but sometimes forgotten, variable. Milk provides a useful example of unnecessary food trade. Until the 1960s, most people consumed milk produced locally, but between 1961 and 1999 there was a fivefold increase in milk exports, with many countries importing and exporting large quantities, resulting in millions of extra food miles.

15.3 Alternative geographies of food: concepts and case studies

One important way to respond to some of the above questions is to establish **alternative food networks** (AFNs). There has been a huge resurgence of interest within many developed market economies in foods of local and regional provenance (Parrott *et al.* 2002; Allen *et al.* 2003; Goodman 2003; Morgan *et al.* 2006; Harris 2009; Maye and Kirwan 2010a). These include consumer initiatives like the Slow Food movement and fair trade, as well as a growth in food purchases from 'alternative' supply chains rather than supermarket outlets, including farmers' markets (FMs), box schemes, community supported agriculture, buying-groups and food cooperatives, and home deliveries (Cone and Myhre 2000; Renting *et al.* 2003; Goodman 2004; Ilbery and Maye 2005; Renard 2005; Kirwan 2006; Kneafsey *et al.* 2008; Little *et al.* 2010). This interest is often seen as a response to the environmental and socio-economic disbenefits associated with agricultural industrialization and global food supply and is reflected, for example, by changes in consumer confidence in Austria, Italy and the UK (Sassatelli and Scott 2001).

Many authors have referred to the distinctions drawn between 'conventional' and 'alternative' agri-food systems (see Table 15.2). Binary opposites such as 'quality', 'embedded', 'sustainable', 'traditional' and 'natural' characterise alternative food production systems (Ilbery and Maye 2005). In reality, these binary opposites are not as simple and clear-cut as this. For example, while organic food may be regarded as 'alternative', most organic sales still occur through 'conventional' supermarkets. Interest in the alternative food economy has also led some geographers to proclaim the emergence of 'alternative geographies of food', which revolve around changing production and consumption relations that give rise to new regional and local food 'complexes'. It has been argued, for example, that alternative geographies of food may be associated

Table 15.2 Distinctions between 'conventional' and 'alternative' food supply systems

Conventional	Alternative
Modern	Post-modern
Manufactured/processed	Natural/fresh
Mass (large-scale) production	Craft/artisanal (small-scale) production
Long food supply chains	Short food supply chains
Costs externalized	Costs internalized
Rationalized	Traditional
Standardized	Difference/diversity
Intensification	Extensification
Monoculture	Biodiversity
Homogenization of foods	Regional palates
Hypermarkets	Local markets
Agrochemicals	Organic/sustainable farming
Non-renewable energy	Reusable energy
Fast food	Slow food
Quantity	Quality
Disembedded	Embedded

Source: based on Ilbery and Maye (2005)

with agriculturally peripheral regions because such regions have, 'for a variety of reasons, failed to fully engage with the productivist conventions that have predominated in the agri-food system in the second half of the twentieth century' (Parrott *et al.* 2002: 243). One can equally talk about alternative geographies of food in a global sense, notably through the international **fair trade** and organic food movements which challenge exploitative relations in agri-food systems (see also Spotlight box 15.2).

Goodman (2003) draws a useful comparison between American and European alternative food practices, conceptualizing US alternatives more as social and oppositional movements, in contrast to the European Union's more endogenous interventions, linked to historical and cultural traditions of product and place. These distinctions will now be examined a little further, via some empirical case studies; hopefully, they will also help you to think critically about the blurring between 'conventional' and 'alternative' systems of food provision. Before doing so, a summary of four concepts central to an understanding of the

Spotlight box 15.2

Towards conventionalization? Fair trade and organic foods

Conversations about alternative food networks will often make reference to fair trade and organic foods because they offer ethical and ecological possibilities that counter some of the negative externalities associated with conventional food supply. International trade often appears as a remote concern, but when commodity prices fall it has devastating impacts on the livelihoods of millions of small producers. The prices paid for coffee, for example, have not increased in real terms in the past forty years, whereas the costs of inputs like fertilizers and machinery have. Low coffee prices in the early 1990s had catastrophic impacts on the lives of millions of small farmers, mostly in LDCs, who were producing coffee at a loss. However, despite the positive benefits of fair trade and organic foods, both sectors have been subjected to recent debates about 'mainstreaming' and 'conventionalization'.

Fair trade offers an alternative to overcome the injustices of free trade, guaranteeing producers a fair price (http://www.fairtrade.org.uk/about_history.htm). It was started some fifty years ago by development agencies and charities like Oxfam and Traidcraft who realized the important role consumers could play in improving the lives of impoverished producers. Charities bought direct from the farmers at better prices and so were able to offer consumers the chance to buy products that were bought on fairly traded terms. However, to generate greater sales it became necessary to involve commercial manufacturers and to get these products into supermarkets where most people shop. Some twenty countries now have fair trade labeling initiatives, mostly in Europe and North America, and products include coffee, drinking chocolate, chocolate bars, orange juice, honey, tea, sugar and bananas.

Fair trade is now available in most European supermarket chains, with some products achieving 15 per cent of the national market share. It is, quite rightly, a global success story and fair trade consumerism is celebrated as enabling better everyday ethical practice. The success of these networks depends upon the ability and willingness of Northern consumers to pay redistributive premiums for such commodities. However, Renard (2005) suggests that the rapid growth of alternative commodities may be levelling off, with relatively few consumers willing to pay premiums and then only for products of seemingly better quality and taste. Some fair trade organizations have responded by competing on price, lowering premiums paid to producers and aggressively pursuing 'mainstreaming' strategies. This parallels developments in organic farming where some have argued that organic supply chains have become 'conventionalized' (especially in terms of rent structures, the size of businesses controlling production, conventional patterns of marketing and distribution), suggesting that it is often very difficult for alternative economies to maintain their differences from the global capitalist economy (Guthman 2004). As Mike Goodman (2004) suggests, such networks unveil the commodity fetish and simultaneously construct fetishes of their own.

---

development of so-called 'alternative geographies of food' is provided:

1. **Short food supply chains** (SFSCs): The key characteristic of SFSCs is that foods reach the final consumer having been transmitted through a supply chain 'embedded' with value-laden information concerning the mode of production, provenance and distinctive quality assets of the product. In many cases, the number of nodes between the primary producer and the final consumer will also be minimized (Renting *et al.* 2003). While this 'reconnection process' is best demonstrated through forms of direct marketing and thus face-to-face contact between producer and consumer, Marsden *et al.* (2000) identified two further types of SFSC: spatial proximity and spatially extended. The former is where products are sold through local outlets in the region, locality or place of production, so that the consumer is immediately aware of the locally embedded nature of the product at the point of retail. In contrast, the latter occur when products are sold to consumers (e.g. via the Internet) who are located outside the region of production and/or have no personal knowledge of the area.

2. **Social embeddedness**: This propagates the idea that economic behaviour is embedded in, and mediated by, a complex and extensive web of local social relations. In the case of local ('alternative') foods, both economic (e.g. price, markets) and social (e.g. local

ties, trust) relations are vital for success (for details see Hinrichs 2000). In stressing the role of social relations in generating the trust necessary for economic transactions to take place, it is easy to make the false assumption that social embeddedness relates just to alternative food systems. In reality, all economic relations are socially embedded in a range of contrasting ways and so there are different degrees of embeddedness in all food supply systems (Winter 2003). Nevertheless, social interaction between producer and consumer can make the difference between success and failure for local food businesses. This can take the form of acknowledgement, attention, respect, friendship and sociability, often

subsumed within the concept of 'regard' (Sage 2003; Kirwan 2006; Maye and Kirwan 2010a).

3. **The 'turn' to quality**: Closely linked to the ideas of SFSCs, embeddedness and regard is the notion that local (alternative) foods are of higher quality than products produced under more conventional farming systems (see Plate 15.1, for example). In other words, quality and locality are inextricably linked. Yet, 'quality' is a social construction and thus means different things to different people (Ilbery and Kneafsey 2000a). Product labelling is an important dimension of marketing quality and, in the case of local foods, labels tend to focus on different combinations of the three 'Ps': product (e.g. local cheese),

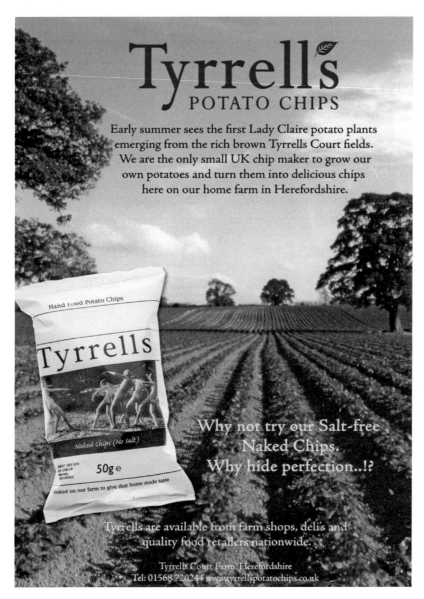

Plate 15.1 Speciality potato crisps in the UK.

(Image courtesy of The Advertising Archives)

process (e.g. organic) and place (e.g. Stilton) (Ilbery *et al.* 2005). The turn to quality has led to a growth in quality assurance schemes (QAS). However, the recent proliferation of QAS has been dominated by supermarkets, thus further blurring the distinction between 'conventional' and 'alternative' food supply systems.

4. **Defensive localism**: It is beyond the scope of this chapter to enter into debates surrounding the definition of terms such as 'local', 'locality' and 'regional' (see Maye and Kirwan 2010a for a full review); nevertheless, it is clear that the literature is divided about whether 'quality' or 'local' is more important in local food production. In fact, Winter (2003) used the concept of defensive localism to suggest that the turn to local is more important than a turn to quality based on, for example, organic or ecological principles. In his survey of consumers in different rural districts of England and Wales, he noted a clear preference for the purchase of local over organic food. This reflected support for local farmers and the local economy, irrespective of whether the food was produced using conventional or alternative methods. The turn to local is not just about alternative food systems; instead, it can cover different forms of agriculture and a range of consumer motivations.

It should be clear that the processes affecting 'alternative' food chains are complex and that there is considerable blurring between these and 'conventional' food production systems. The literature also suggests that a simple 'alternative geography of food' does not exist. Spotlight box 15.2 introduces one way of critically examining these processes, referring to the **conventionalization** of two well-known 'alternative' food examples: fair trade and organic foods. We now explore some of these issues in more detail using three case studies that attempt to 'reconnect' food production and consumption and make a statement about the alternative food economy, especially in terms of the nature of the food chain.

15.3.1 Quality food

One of the dominant features of AFNs, particularly in Europe, has been the attempt to link 'product and place' in order to add value to agricultural products (Watts *et al.* 2005; Ilbery and Maye 2010). This is often defined as a process of **relocalization**, in which locally distinctive quality food products (for an example, see Plate 15.1) are transferred to regional and

national markets as a mechanism to provide a valuable economic stimulant and reduce the deleterious impact of national and EU subsidy reforms and increasing trade liberalization. In particular, regional speciality food products have been linked to particular places in this way, especially via a formal system of quality food labels and the establishment of regional speciality food groups. The most notable example of this is the Protected Designation of Origin (PDO) and the Protected Geographical Indication (PGI) quality labels introduced by the European Union in the early 1990s to 'protect' and 'promote' food and drink products with a recognizable geographical origin (Ilbery and Kneafsey 2000b; Parrott *et al.* 2002). Groups of producers in Europe can thus apply for either a PDO label (product originates from a specific place and is linked to its natural environment) or PGI label (linked to place, but not necessarily in terms of raw materials). These labels protect producers from attempted copies and act as a marketing device. These changes have been interpreted as symbolizing a **quality turn** (Goodman 2003), with particular emphasis on specialist food production, especially in 'marginal economies'.

In terms of geography, the UK already has over 30 PDO/PGI labels. Specialist cheeses dominate the PDOs, indicating that these are made exclusively from local raw materials (e.g. Stilton cheeses). In contrast, drinks dominate the PGIs. There are now more than 350 PDO and 200 PGI designations in the EU, with a southern concentration in places like France and Italy (over 100 each, including cheeses, wines and meats, such as Parmigiano-Reggiano, Roquefort and Champagne) and Greece and Portugal (over 80 each), compared with limited numbers in northern member countries (e.g. Finland, Denmark and Ireland). This implies that regional/speciality food has more cultural significance in some European societies than in others (Parrott *et al.* 2002).

15.3.2 Direct food

Studies of the local food sector in the USA, UK and other parts of Europe typically include some of the archetypal examples noted earlier, especially farmers' markets, box schemes, farm shops and on-farm butchers (Watts *et al.* 2005; Maye and Kirwan 2010a). Significant in all these examples is the direct nature of the supply chain and the important social and economic benefits which accrue from these types of food transaction. This is illustrated in Plate 15.2, which shows the *direct interaction* between producers and

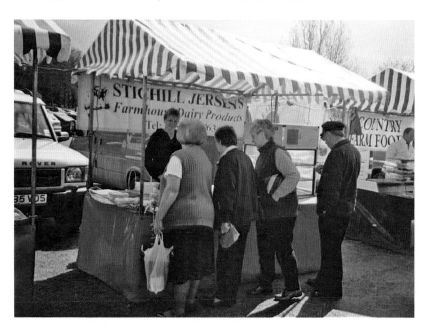

Plate 15.2 Farmers' market in Hawick, The Borders (Scotland).
(Authors' survey)

consumers at a small, rural farmers' market in Hawick in the Scottish Borders. To help explore some of these processes in more detail, we can focus on an organic meat producer business examined as part of a wider research project on the nature of direct marketing channels (Ilbery and Maye 2005).

The organic farm and on-farm butchery is located in the Northumberland uplands. Established in 1998, the business produces and retails a range of organically reared meat products, using local branding and traceability. The business supply chain is reproduced in Figure 15.3 and, crucially, records both the upstream inputs coming into the business and the downstream links in terms of how and where final products are sold. This 'whole chain' approach extends previous conceptualizations of SFSCs, which tended to focus only on downstream elements of the business supply chain (cf. Marsden et al. 2000).

Most of the inputs coming into the farm are from local suppliers. Constructed to add and retain value from primary production, the business sources some cattle, sheep, pigs and poultry from other organic farms in the region. All livestock are slaughtered at the (organically accredited) abattoir in Whitley Bay in North Tyneside, Tyne and Wear (about 20 miles from the farm), and then delivered to the on-farm butchery to be processed and packaged for retail. Businesses like this have to 'dip in' and 'dip out' of 'conventional' supply chains because of the ways in which the dominant agri-food system is currently structured. The meat products are sold through various SFSCs, for example

direct sales, local/regional specialist food shops and caterers, and mail order. Meat sales over the Internet are minimal. Over time, the business reduced sales output at farmers' markets and stopped supplying (specialist) butchers in the region in favour of more 'stable' alternatives (e.g. independent retail and catering). Thus the producer adjusts the business supply chain in a bid to establish as much control of it as possible.

15.3.3 Community food

Attempts to relocalize food production and establish 'alternative' supply chains also extend beyond farm-based production. This is an important point. The final case study of the Kansas City Food Circle (KCFC) is an example of a community food scheme, and was examined by Hendrickson and Heffernan (2002) who also looked at the ConAgra food chain. This is significant because the KCFC is presented as a (local) site of resistance within an industrialized food system dominated by global corporations like Con Agra, Cargill and Monsanto whose food chain arrangements are typified by joint ventures and strategic alliances. The KCFC organization began in November 1994, seeking to educate the public about the consequences of the industrial agricultural system and to persuade more people to participate in local, sustainable alternatives. It argued that industrial agriculture had eroded soils, made water unfit to drink, and increased pesticide resistance in insects; the food system had become so centralized that citizens lacked control over their food choices.

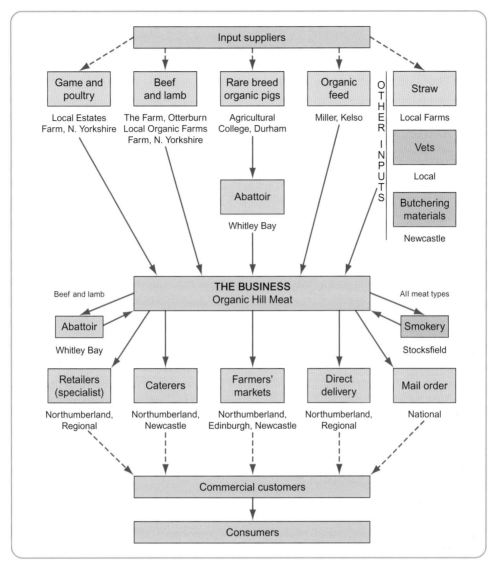

Figure 15.3 Business supply chain diagram for an organic hill meat producer.
Source: Ilbery and Maye (2005)

For KCFC, the industrialized food system was unhealthy, unjust, unethical and economically unviable.

The explicit political alternative that emerged from this critique was an attempt to create a local, organic food system where consumers can get seasonal, fresh food at a price that supports farmers who use sustainable practices. The KCFC thus connects all actors in the food system in a way that sustains and returns control to local communities. Effectively, the KCFC is about creating a new kind of community that recognizes the interconnectedness of people through the production and consumption of food. The Circle continues to grow, with 30 organic growers in the region surrounding Kansas City signing up to supply food, and roughly 600 consumers. The trust-based relationships that the KCFC is trying to cultivate between farmers and consumers mirror those of the Slow Food movement.

15.3.4 Critiquing AFNs

Given the range and diversity of work on 'alternative geographies of food', these mechanisms of food supply can also be critiqued, especially in terms of the privilege assigned to terms like 'local' and 'alternative'. Geographers and other agri-food researchers are thus beginning to ask what is alternative about the alternative food economy (see Whatmore *et al.* 2003; Maye *et al.* 2007)? This critique is also evident in the critical tone of the 'conventionalization thesis' introduced in Spotlight box 15.2, which raises important questions about the alternative status of some fair trade and organic food supply chains. Watts *et al.* (2005) distinguish between 'weaker' and 'stronger' alternative systems of food provision. The former place emphasis on quality and the labelling features of locality *food* networks (i.e. the product is key), whereas the latter focus on the revalorized and embedded characteristics of local food *networks* (i.e. the supply chain/network and nature of relations are key). In the three examples of AFNs, the first case study falls within this weaker classification, whereas the other two are stronger because of their emphasis on establishing alternative food chains. It may well be then that 'alternative food networks can be classified as weaker or stronger on the basis of their engagement with, and potential for subordination by, conventional food supply chains operating in a global, neoliberal polity' (Watts *et al.* 2005: 34). As we will see in Section 15.4, these debates have been further unsettled by recent increases in, and volatility of, global food prices alongside concerns about food availability. The scale of the discussion has therefore shifted from the regional and the local back to the global.

15.4 Food security: questions of scale, definition and interpretation

Some of the challenges associated with food security have been usefully mapped earlier in this book (see Chapter 7). It merits attention here too because it is significantly reshaping how we think about food. Food security has commonly been associated with developing countries in recent decades. Given the choice and daily availability of foods in supermarkets and restaurants, it is easy to see why the thought of food shortages or supply chain disruptions might not be uppermost in peoples' minds in most developed market economies. However, there is now a recognition that the food system and issues associated with securing food supplies need urgent attention.

The reason why political interest in food worldwide has suddenly arisen is easy to explain: global food prices have increased significantly in recent years (House of Commons 2009). The significant price spike was in late 2007, when the price for basic food staples rose sharply on international commodity markets (the price for wheat rose by 50 per cent and for rice by 20 per cent, for example). The price spike was unusual in one important way in that it 'applies to almost all major food and feed commodities, rather than just a few of them' (Chatham House 2008: 2). High prices in agricultural markets are not uncommon, but increases across such a range of food commodities are.

The escalation in food prices had some well-documented impacts, notably a series of violent protests and demonstrations that were witnessed in different parts of the developing world. Some countries (e.g. Argentina) also imposed trade restrictions to limit the amount of food being exported to protect national food supply. The price spike affected households in developed nations such as the UK in less dramatic ways, via food inflation. Prices for oils and fats in the UK rose by 29 per cent in the year to July 2008, meat by 16.3 per cent, bread and cereals by 16 per cent, vegetables by 11.1 per cent and fruit by 10.7 per cent (Barling *et al.* 2008, quoted in House of Commons 2009). Although less dramatic, these impacts served to underline 'our global interdependency and demonstrated the political and social importance of affordable food' (Chatham House 2009: 5).

The factors that contributed to the increase in food prices at this time were widely reported and included poor harvests, the use of food crops for biofuels, especially maize in the USA, rising energy prices, which led to high prices for fertilizers and fuel, changes in demand for certain foods (e.g. growing demand for more meat in China), export bans and speculation. There is still on-going debate about whether food price rises will be sustained in the longer term, or whether what was experienced in 2007 was merely a blip. Food prices have fallen again in international commodity markets, but they remain volatile. Industry analysts suggest higher and more volatile food prices are set to stay (see von Braun 2009).

Addressing the consequences of higher food prices is only part of the challenge. Constraints on global food supply are also influenced by longer-term trends, notably changing global patterns in diet and world population pressures. Population expansion is a particularly significant driver of the long-term increase in food demand, with predictions that the world may need to feed over nine billion people from 2050. Projections suggest that 95 per cent of this growth will occur in the developing world. At the World Food Security Conference in Rome in June 2008, it was estimated that food production needed to increase by 50 per cent by 2030 and to double by 2050.

Undercutting all of this discussion is an increased awareness of the risks posed by climate change and increasingly scarce natural resources. Various environmental and resource challenges require urgent consideration, notably the need to reduce greenhouse gas emissions produced by the food system, the need to reduce dependency on fossil fuels, and the need to reduce the depletion of natural resources and ecosystem services on which food depends (especially soil and water) (House of Commons 2009: 13). In other words, any response to the above crisis is not simply about producing more food; it is about doing it in a way that is sustainable and uses resources less exploitatively. Increases in food production will thus need to be achieved, making an effective, rather than exploitative, use of resources (Ilbery and Maye 2010). This has been described elsewhere as 'an unprecedented double challenge' (House of Commons 2009: 13). The following sub-sections consider two aspects of this debate: first, how the term food security is defined and for whom; and second, some ways that food analysts suggest we respond to the challenge.

15.4.1 Food security for whom?

The emphasis in much of the discussion on the current, perceived 'global food security crisis' is about the need to produce more food (i.e. food availability). There is recognition that increases in production must be consistent with systems of production that are sustainable and that we must produce food that consumers want. However, this policy rhetoric places too little emphasis on broader understandings of food security, which include questions about access and utilization as well as availability (Ericksen 2008). The standard definition of food security is that given by the United Nations Food and Agriculture Organization in its 1996 World Food Summit Plan of Action:

> Food security exists when all people, at all times, have physical and economic access to sufficient, safe and nutritious food to meet their dietary needs and food preferences for an active and healthy life.

(Quoted in Ericksen 2008)

Food security has multiple meanings for different people and organizations and varied scales of interpretation (global, national, regional, local, the household). It can be viewed as a 'master frame' with several distinct claims to ownership or, in other words, there is 'contested ownership behind the apparent consensus on food security' (Mooney and Hunt 2009: 470). Three collective action frames, which encompass food

security as a master frame, have been suggested by Mooney and Hunt (2009):

- Food security associated with hunger and malnutrition.
- Food security as a component of a community's developmental whole.
- Food security as minimizing risks in industrialized agricultural production in terms of the risk of 'normal accidents' and 'intentional accidents' associated with agri-terrorism.

These collective action frames are grounded in the experience of the United States and identify the different perspectives and interests on one seemingly consensual social problem. It also distinguishes between relatively tame institutional responses and much more critical viewpoints and positions. This resonates with earlier material in this chapter in which weaker and stronger variants of alternative food networks are identified as part of a critique of neoliberalism (Watts *et al.* 2005). Crucially, Mooney and Hunt (2009) do more than simply identify a plurality of framings by locating 'this process within an ordered, yet contentious, multi-organizational political field of differential power wielded by various insiders and outsiders' (ibid.: 493).

15.4.2 Three food security 'action frames'

Following Maye and Kirwan (2010b), this sub-section outlines three illustrative action frames that are emerging outwith the experience of the United States:

1. *Sustainable intensification*: There has been a strong science-orientated or techno-centric discourse underpinning the response to the current 'food crisis'. This is described by Godfray *et al.* (2010) as 'sustainable intensification'. The essence of this idea is that we need to achieve higher yields from the same acreage without damaging the environment. In a similar vein, the UK government's Chief Scientist, John Beddington (2010: 61), argues that we need 'a new, "green*er* revolution"' (original emphasis). Advocates of this approach claim that science and technology have time and again provided huge increases in yield growth when required. Important areas to focus upon in this context include, for example, crop improvement, smarter use of water and fertilizers, the introduction of novel non-chemical approaches to crop protection, the reduction of post-harvest losses, and more sustainable livestock and marine production. Projects and initiatives are also emerging to meet the challenge to produce more

food (or they are at least marketed in this food crisis context). One example is the development of modern hydroponic schemes. A state-of-the-art version now exists in the UK, called Thanet Earth (see Spotlight box 15.3), and such schemes are well known in the Netherlands and California. Another highly controversial example is the proposal to build the UK's first 'super dairy'. Plans were originally submitted for an 8,100-cow dairy herd on a single farm in Lincolnshire. Due to local resident and national resistance from animal welfare groups and others, the proposal has now been revised for a 3,770-cow herd. It is still the subject of a public consultation.

2. *Waste*: This example is about making better use of what we already have (i.e. we need to be less wasteful). There is growing realization that we are eating beyond our means and that food chain dependency on imports creates vulnerability. This approach is captured well in Tristram Stuart's (2009: xxii) excellent text, *Waste*, where he argues 'industrialized nations need to learn what it means to live in scarcity – because the appearance of infinite abundance is an illusion'. The essence of his argument is that we can increase food supply and reduce environmental consequences by reducing waste within the food supply chain. In the USA, for example, around 50 per cent of all food is wasted. In Britain, up to 20 million tonnes of food waste is created every year. The Japanese dispose of food worth ¥11 trillion (US $101.6 billion) annually. Food is treated as a 'disposable commodity' throughout the developed world (Stuart 2009: xvi). We waste food at all stages of the food chain, from production and harvesting through to post-purchase by the consumer. Waste may arise due to poor handling, poor storage, cultural perceptions, retailer demands, consumer ignorance or sheer laziness. Stuart (2009: xix) argues that we can turn this wastefulness to our advantage in that 'the world's mountain of surplus food is currently an environmental liability – but it is also a great opportunity'. Efficiency measures could create savings and help the fight against hunger. Salvaging food would also help tackle global warming. In Europe, for example, more than 30 per cent of Europe's greenhouse gas emissions come from food. If food waste was halved, emissions would be reduced by 5 per cent or more.

3. *The urban foodscape*: Burgeoning prices for basic foodstuffs and concerns about the security and sustainability of the agri-food system are also raising

Spotlight box 15.3

Is the future hydroponic? Thanet Earth

Hydroponic schemes are a well-established feature of the California and the Netherlands agro-foodscape. A state-of-the-art version also now exists in Kent, called 'Thanet Earth'. It is owned by a consortium of Dutch growers and Fresca, a fresh-produce agglomerate. The consortium bought 90 hectares of land and has assembled three gigantic glasshouses, with planning permission for another four.

In 2009, Thanet Earth's first year in production, 2.5 million tomatoes left the glasshouses for the sorting area in an average week and the cucumber and pepper harvests peaked at half a million and three-quarters of a million a week (Harding 2010). This amounts to about 2 per cent of UK demand and the target is to supply 4 per cent of all tomatoes, cucumbers and peppers consumed in the UK.

Thanet Earth is heralded as an example of low-carbon horticulture on a grand scale. As Hardy notes, it scores well in terms of certain facets of sustainability. In terms of water, for example, produce is grown on blocks of rock wool about a meter off the ground and surplus water drains into long trays underneath the blocks and from there to lagoons, along with rainwater off the roofs, to be pumped around again. Thanet Earth is also clever in terms of its energy strategy, using a system called combined heat and power. On other issues it is also virtuous – minimal pesticide use and preserving biodiversity – but it scores less well in terms of labour, with very few indigenous British workers among the workforce who harvest the crop, although such labour issues are not uncommon in other, smaller-scale, horticultural enterprises.

So could Thanet Earth-like growing models potentially bring Britain to self-sufficiency in 'salad'? The paradox is that it is not just about growing more food; it is also about getting people to eat the right type and balance of foods. As Harding (2010: 7) puts it: 'A larger query hanging over hydroponic growing in the UK is quite what it solves until we all start eating many more tomatoes, cucumbers and peppers.'

awareness of the pressures facing urban areas and their need to grow as well as consume food. For example, in a study of the evolution of food strategies in London and New York it is argued that both of these world cities are being forced to think anew about food security so that 'despite being highly developed sites of global capitalism, [they] have not managed to banish the spectre of hunger from their streets' (Morgan and Sonnino 2010: 222). Urban food security policy in both cities is focusing on two key dimensions of the problem: food production and food access. For the former new planning dispensation for urban agriculture allows the city to feed itself from within (as in London) or from neighbouring areas (as in New York). For the latter, both cities have used the school meals service as the main food access scheme for children, with New York actively promoting the consumption of fresh food in poor districts via schemes like *Green Cart*.

The last of these three indicative examples articulates a wider definition and framing of food security beyond that conceived in the pages of agricultural trade journals and the sustainable intensification mantra (i.e. not just about availability of food but also concerned with access and utilization – see Ericksen 2008). As Mooney and Hunt (2009) suggest, these framings exist side-by-side as part of a multi-organizational political field of differential power networks that constitutes food security.

15.5 Conclusion: the ethical foodscape

This chapter has reviewed the changing nature of food production, especially in the context of developed market economies. Food has become an object of much geographical attention. It provides an amenable reference point to examine society–economy relations. The important contributions that geographers have made to debates concerning the long-term sustainability of agri-food systems have been highlighted. Such a broad assessment is always partial in its coverage, but it shows how geographers have examined different systems of food production in different spatial and temporal contexts. At a very general level, it is clear that different types of production system have attracted particular research attention at certain times. A good deal of the research in the late 1970s, 1980s and early 1990s examined industrial agriculture and the industrialization of the food system. The capitalist mode of production still

prevails and large agri-food companies and corporate retailers have developed considerable power and effectively exert control over the entire food supply system. Geographers and others are continuing to provide valuable insights, as evidenced through the case study of salmon farming in Chile. Since the late 1990s, much attention has been placed on alternative and local food systems. Farmers and other producers have developed more sustainable and 'alternative' forms of agriculture, based on territorial association and direct consumer contact, and geographers continue to actively examine these modes of production. In some cases, these systems of provision directly challenge 'mainstream' forms of food provision, a point emphasized using the example of the Kansas City Food Circle (see also Little *et al.* 2010, who examine food buying-groups and cooperatives).

It is, of course, too simplistic to see these systems as diametrically opposed. They represent ways of growing and procuring food that are different in terms of scale and operation, but they constitute and exist within the one food system. At the time of writing, debates about food security are attracting a good deal of attention from food policy analysts, geographers and other agri-food researchers. This threat to food security is not new and many food analysts have in fact predicted it for some time. It has always been an issue in the countries of the developing world, but the new emphasis is forcing a rethink about the way we produce food, where and how we produce food, and how we value (good) food. Much of this is still to be worked out, but we are already seeing, for example, how the 'crisis' is forcing different actors in the food landscape to reposition and to reshape their alliances (Brunori and Guarino 2010; Maye and Kirwan 2010b).

This debate is also forcing us to reconceptualize food systems and the way in which we assess whether they are a sustainable and ethical way of food provisioning. Perhaps the food regime concept (see earlier for a definition) is a useful way to examine the 'world food crisis' because it not only reveals the structured moments and transitions in the history of capitalist food relations, but also of capitalism itself (McMichael 2009). It is a consequence of industrial capitalism's long-term over-dependence on fossil fuel, combined with the inflation-producing effects of biofuels offsets, financial speculation activities, and the concentration and centralisation practices of agribusiness capital. Rising costs have thus combined with monopoly pricing by agribusiness to inflate prices which have spread due to the liberalised nature of trade and finance.

Another example relates to 'ethical foodscapes', where 'morality is a key and growing currency in the provisioning of food in much of the post-industrial North and beyond' (Goodman *et al.* 2010: 1783). While all food has ethical implications, some food has taken on the connotation of being, often in quite particular ways, more ethical. This applies particularly to foods labelled as alternative. It has been argued that these foods are typically constructed as 'good', 'better' and/ or 'alternative', often described as 'local', 'organic' or 'fair trade' and working against the social, ecological and economic excesses of conventional food systems, but we need to be cautious in our understanding of the ethical foodscape (Goodman *et al.* 2010). Many of these more ethical foods are equally a part of conventional food systems and many foods that we label 'conventional' or 'ordinary' have their own implicit moralities and ethical relationships and meanings so that 'the ethics of ethical foodscapes can thus be ambiguous, slippery and consist of a number of interwoven layers' (Goodman *et al.* 2010: 1783).

This last example means that we need to break down dualistic characterizations of 'alternative' foods as uniquely ethical and conventional foods as unethical. We also need greater critical analysis of our assessments of 'goodness' and 'badness', whether in terms of practical indicators, material practices or acceptable food politics. The current 'food crisis' offers an excellent opportunity to carry out closer food system examinations of what constitutes 'goodness' and 'badness'. How and where should we grow food? Do we want our cities to be places where we grow (more) food? Should we challenge methods of growing food that promote so-called 'sustainable intensification'? What about the role of local food systems in the future? Recent Life Cycle Analysis assessments (see Coley *et al.* 2010, for example) suggest that, in environmental terms, local food systems may not always be the most sustainable way to buy food. However, this research misses any broader analysis of social and welfare aspects, for example. It nevertheless gives an indication of the more specific work that needs to be done to examine ethicality.

As a more general concluding point, it is suggested that an emphasis on food ethics in a context of growing uncertainty about food supplies is a highly fruitful means by which to examine food production geographies. A better understanding of food production geographies and capacities will also hopefully help to ameliorate social and ecological inequalities. The material presented in this chapter suggests the era of endless food may be winding down. Our habits will have to change because, as the food writer Felicity Lawrence (2008; quoted in Harding 2010: 3) puts it, 'they simply cannot go on. We are now entering a period of rapid transition'. Human geographers are well placed to provide analytical insight into how we respond to this transition.

Learning outcomes

After reading this chapter, you should have:

- An understanding of the importance of food to modern economies and societies and what is meant by a food chain and its application in different contexts.

- An awareness of how and why the agri-food system has become globalised – especially the dominance and spatially uneven penetration by transnational agribusinesses.

- An awareness of the emergence of quality-based commodity food production in some parts of the developing world and the reasons why this has happened.

- An understanding of so-called 'alternative food networks', including some of the main reasons why they have arisen, how they are conceptualized and case study examples to explain their diverse character.

- An appreciation of recent debates about food security and ethics and a sense of how these issues disturb and unsettle notions of 'goodness' and 'badness' in relation to food production and provisioning more generally.

- An ability to begin to think critically about changing geographies of food production.

Further reading

Atkins, P. and Bowler, I. (2001) *Food in Society: Economy, Culture and Geography*, Arnold, London. A collection of fairly short chapters on many of the global food production issues raised in this chapter, as well some useful introductory chapters on food consumption.

Friedberg, S. (2004) *French Beans and Food Scares: Culture and Commerce in an Anxious Age*, Oxford University Press, Oxford. A comparative case study of global food commodity networks, tracing the supply of green beans for two culturally specific trade links – France/Burkina Faso and England/Zambia.

Maye, D., Holloway, L. and Kneafsey, M. (eds.) (2007) *Alternative Food Geographies: Representation and Practice*, Elsevier, Oxford. This book examines debates and practices surrounding efforts to establish 'alternative' systems of food provision, with studies from Europe, North and South America, Australia and Africa.

Morgan, K., Marsden, T. and Murdoch, J. (2006) *Worlds of Food: Place, Power and Provenance in the Food Chain*, Oxford University Press, Oxford. This book provides a review of agri-food networks, including so-called 'conventional' and 'alternative' food networks, with case studies from Italy, the UK and the US.

Millstone, E. and Lang, T. (2008) *The Atlas of Food*, 2nd edition, Earthscan, London. A great introductory text that reveals the often bizarre facts about the way we make, process, ship, trade and eat foods. If you want to know the basic facts about food, this is the place to start. The graphics are particularly well-done and easy to digest.

Robinson, G. (2004) *Geographies of Agriculture*, Pearson, London. This textbook provides a comprehensive summary of the geographical aspects of agriculture and food supply systems. It is particularly good on the production side of the food chain equation.

Stuart, T. (2009) *Waste: Uncovering the Global Food Scandal*, Penguin Books, London. An excellent text on the way we treat food as a disposable commodity and have, in effect, created our own global food crisis. It also offers solutions to fix it.

Useful websites

http://ec.europa.eu/agriculture/index_en.htm Home page of the Directorate General (DG) responsible for Agriculture and Rural Development, and is part of the official (and much bigger) website of the European Union. This DG site contains information on the EU's Common Agricultural Policy, key speeches and policy papers, statistics and links to member state sites.

www.defra.gov.uk The site of the UK government's Department of Environment, Food and Rural Affairs. The site contains information on UK agricultural policy (including the most recent round of mid-term reforms), statistics and links to other government departments, non-ministerial departments and NGOs. Similar sites exist for other national governments (see, for example, www.usda.gov, home of the US Department of Agriculture, including useful details about the US Farm Bill).

www.sustainweb.org The site for Sustain – the alliance for better food and farming, an NGO umbrella organization which campaigns for more sustainable food chain practices. Contains information about various projects Sustain are working on and a raft of challenging position papers and responses to various food-related topics.

www.soilassociation.org This site is home to the UK organization dedicated to promoting organic food and farming. The site contains an online library with useful papers on various topics, including local/regional food schemes, food and farming policy, GMs and animals and food security, as well as a 'links' page which lists other useful websites related to food and farming. You can also visit their consumer website, www.why.organic.org, which contains useful information about the nutritional benefits of organic food.

www.fairtrade.org.uk Site for the Fairtrade Foundation. Contains information about the fair trade movement, products sold, suppliers, etc.; it also has a useful resources section with links to downloadable position papers and case studies. For those interested in ethical consumerism more broadly, see also www.ethicalconsumer.org, which examines the ethical credentials of individual products/organizations.

www.fcrn.org.uk Site for the Food Climate Research Network. Its aim is to better understand how the food system contributes to greenhouse gas emissions, which includes researching and promoting ways of reducing emissions. It is a fantastic resource for those interested in questions about food security and sustainability, with resources, case studies, interviews and a link to an email list.

www.bis.gov.uk/foresight/our-work/projects/current-projects/global-food-and-farming-futures Site for the Global Food and Farming Foresight study and contains various reviews and outputs from the study, which is due to report formally in early 2011. It also contains a link to other relevant Foresight studies, including one about the future of land use in the UK.

THE GEOGRAPHIES OF GLOBAL PRODUCTION NETWORKS

Neil M. Coe

Topics covered

- Defining and identifying production networks
- Spatial divisions of labour
- The governance of production networks
- Production networks in their institutional contexts
- Reshaping production networks through 'standards' and 'codes'

16.1 Engaging with global production networks

Imagine the scene. We could be in Birmingham, Boston, Berlin, Beijing or Brisbane, or many other places besides. A student – having just been to a lecture and en route to her evening job in a restaurant – stops for a drink in a café. After choosing from the extensive menu of coffee options, she takes a window seat and boots up her laptop computer – the café has recently installed free wireless Internet for its customers. Essay assignment in hand, the student starts surfing the Web, looking for the reading materials and background information she will need. Every so often, she pauses, takes a sip of coffee – a decaf, skinny latte or some such like – and gazes out at the street . . .

On the one hand, what is being described here is a unique consumption event, an individual lost in her thoughts as she consumes her coffee and uses her laptop at a particular time and place (see Chapter 19). And yet, on the other hand, the two central non-human artefacts in this particular story – the coffee and the laptop – may have remarkably similar economic geographies and histories, wherever the event is taking place.

For example, the coffee bar in question might be a branch of Starbucks, with revenues of US$10 billion making it the world's largest chain, serving millions of customers every day in some 16,500 cafés across more than 50 countries (as of December 2009). It is one of many other similar chains offering their own standardized take on the continental European coffee house experience: for example, Caffè Nero, Costa Coffee or Coffee Republic in the United Kingdom; Caribou Coffee or Coffee Bean and Tea Leaf in the United States; Barista in India; and Gloria Jean or Dome in Australia. There is a strong chance that the coffee beans were roasted and processed by one of just five American and European firms – Kraft (Maxwell House brand), Nestlé (Nescafé), Sara Lee (Douwe Egberts), Proctor and Gamble (Folgers) and Tchibo – that handle almost half of the world's coffee each year. Equally, the laptop used by our student may well be made by either Hewlett Packard (HP) or Acer, who between them, at the end of 2009, accounted for 39 per cent of the global market share. Indeed, just five companies – HP and Dell (US), Acer and Asus (Taiwan) and Toshiba (Japan) – account for almost 70 per cent of laptop sales worldwide.

Why are these observations interesting? In short, because both the cup of coffee being consumed and the laptop being used are the end result of far-reaching global production networks that are in turn dominated by powerful corporate interests of various kinds. Almost all commodities that we consume have complicated histories and geographies and yet, as a system, capitalism seems to conceal these. The purchase of commodities such as coffee or computers with money serves to *disconnect* producers and consumers, meaning many consumers are unaware of the nature of the production system that has enabled those commodities to be

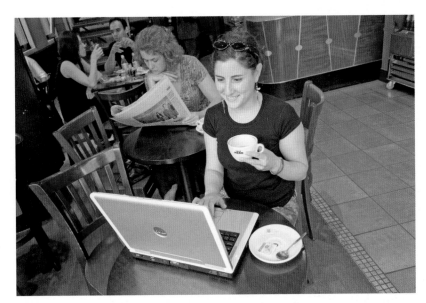

Plate 16.1 A common scene – but what stories lie behind these everyday products?

(Alex Segre/Alamy)

available to them. This poses profound challenges to both conscientious consumers actively curious about the history of the commodities they consume (see Chapter 19), and economic geographers who want to understand connections and interdependencies within the global economy. In reality, even drinking just one cup of coffee links the consumer – albeit unknowingly in most cases – to hundreds of thousands of workers involved directly or indirectly in its production through complex global webs of connections.

This chapter is therefore about how we, as economic geographers, can explore and understand the inherent variability of global production networks. In what follows, we will look in turn at four generic dimensions of all global production networks, namely their organization, geography, power relations or 'governance', and institutional context. We will use the contrasting examples of the coffee and laptop global production networks as illustrations of these various dimensions. In the final section of the chapter we will look at ongoing attempts to reshape production networks through the implementation of different kinds of standards and codes of conduct.

16.2 Production chains, production networks ...

Every economic activity can be thought of as a production chain – a linked series of value-adding activities (see Figure 16.1). In very simple terms, material and non-material inputs are combined and transformed through some kind of production process, leading to a new good or service that needs to be delivered to the customer who then consumes it. Consumption, in turn, is not just a single act of purchase, but an ongoing process that may include maintenance, repair, waste disposal, recycling and the like. This basic model holds whether it is a physical good – such as a television or bicycle – being produced, or a service such as a haircut or insurance policy, with the difference coming in the relative balance of tangible and intangible elements in the production chain (see Chapter 17). While simple chains can be thought of in linear terms, in reality, as Figure 16.1 shows, they are enmeshed within much wider *networks* of

Figure 16.1 A generic production network.

Source: Dicken (2007), Figure 1.4c

Spotlight box 16.1

From Fordism to post-Fordism?

Some commentators have characterized recent changes in production systems as heralding a shift from *Fordism* to *post-Fordism*. In its narrowest sense, Fordism signifies a range of industrial practices emerging from Henry Ford's innovations in the American automobile industry during the early decades of the twentieth century. The Fordist production system has four important elements:

1. It is characterized by a distinctive division of labour – that is, the separation of different work tasks between different groups of workers – in which unskilled workers execute simple, repetitive tasks and skilled technical and managerial workers undertake functions related to research, quality control, finance, etc.
2. It is a system in which the manufacture of parts and components is highly standardized.
3. It is organized not around groups of similar machinery, but machines arranged in the correct sequence required to manufacture a product.
4. The various parts of the production process are linked together by a moving conveyor belt – the assembly line – to facilitate the quick and efficient fulfilment of tasks.

Together, these four attributes can provide *economies of scale* whereby the large volume of production reduces the cost of producing a single product. In turn, prices can then be reduced, leading to increased sales and the potential development of a *mass market* for the commodity in question. In geographical terms, Fordist production is often associated with the formation of *branch plants* for undertaking the production process which can be relocated either nationally or internationally to benefit from variations in labour wages, skills, and levels of unionization (see Spotlight box 16.2).

Fordism has been contrasted with the emergence of so-called 'post'-Fordist production systems from the mid-1970s onwards. The chief characteristic of this new mode of production is *flexibility*, whereby the use of information technologies in machines and their operation allows for more sophisticated control over the production process, enabling firms to make a wide variety of products for different market niches. For some initial observers, this heralded the rise of an entirely new era of capitalism based on vertically disintegrated networks of nimble, innovative, small and medium-sized firms.

Any ideas of a simple shift from Fordism to post-Fordism, however, are highly problematic. For one thing, Fordism was never dominant across all sectors and geographies in the first place. Equally, no one production system predominates; it is more realistic to think in terms of, at least, three different kinds of production system that are evident today (Dicken 2007: see Table 16.1 for more details). First, variants of *Fordist* production are alive and well in certain sectors where simple scale economies remain crucial, e.g. agriculture and food processing, routinized service activities, and some electronic component industries. Second, in other instances, flexible production technologies have led to the resurgence of craft-based production driven by innovative, small independent firms. At the heart of this so-called *flexibly specialized* production system are skilled craft workers using flexible machinery to produce small volumes of customized goods such as shoes, jewellery, clothing and furniture, ceramics, and the like. Third, there are forms of *flexible production* emanating originally from Japan – for example, through the auto industry – that are distinctive from both the Fordist and flexibly specialized systems. Key here is the combination of information technologies with flexible ways of organizing workers that allows the highly efficient production of large volumes of customized goods. Intriguingly for economic geographers, the flexible specialization and Japanese flexible production systems both seem to exhibit a newfound tendency for firms to agglomerate – or group together – in particular places in what are commonly known now as *clusters* (see Spotlight box 16.3).

relationships involving a broad range of other functions that are necessary for economic activities to take place: for example, research and development, market research, technological inputs, logistics services, advertising, legal services, accounting, personnel management, software, security and so on. In turn, these inter-firm relations are embedded in broader financial and regulatory systems that may bring non-firm entities such as the state and its many institutions into view. In this chapter we use the term *production network* to refer to the full mesh of relationships that lies behind any economic activity.

Table 16.1 Alternative post-Fordist production systems

Characteristic	Flexible specialisation	Fordist mass production	Japanese flexible production
Technology	Simple, flexible tools/machinery; non-standardised components	Complex, rigid single-purpose machinery; standardised components; difficult to switch products	Highly flexible (modular) methods of production; relatively easy to switch products
Labour force	Mostly highly skilled workers	Narrowly skilled professionals 'conceptualise' the product; semi/unskilled workers 'execute' production in simple, repetitive, highly controlled sequences	Multi-skilled, flexible workers, with some responsibilities, operate in teams and switch between tasks
Supplier relationships	Very close contact between customer and supplier; suppliers in physical proximity	Arms-length supplier relationships; stocks held at assembly plant as buffer against disruption of supply	Very close supplier relationships in a tiered system; 'just-in-time' delivery of stocks requires 'close' supplier network
Production volume- and variety	Low volume and wide (customised) variety	Very high volume of standardised products with minor 'tweaks'	Very high volume; total partially attained through the production of range of differentiated products

Source: adapted from Dicken (2007), Table 3.2

Production networks vary greatly in their complexity. At one end of the scale is the single farmer who grows one particular kind of vegetable and transports it to a market for sale at harvest time. Conversely, a complicated manufactured product such as a car mobilizes a wide range of skills and technologies to bring together the thousands of components that comprise the finished product. The next step is to think about *who* undertakes the different activities in a given production network. Hypothetically, every single function might be carried out by one huge firm, which would coordinate the production network through its internal management hierarchy (a vertically integrated system). Equally, every single function might be undertaken by separate firms, meaning that the production network would take the form of a series of external, inter-firm relationships (a vertically disintegrated system). In reality, all production networks fall somewhere between these extremes. In a context of increased global competition, however, the dominant trend has been for firms in many sectors to focus on their core activity or 'competency', while seeking non-core inputs via external

relationships. For some commentators, this trend is part of a broader shift from *Fordist* to more flexible *post-Fordist* production systems (see Spotlight box 16.1).

These increasingly important external networks can take on many forms:

- *Markets*: some inputs – usually of low value and standardized – will simply be purchased by firms on the open market. In this case, there is no long-term relationship between the two parties and firms can readily switch between suppliers.
- *Subcontracting*: this involves firms buying inputs that have been made, under contract, to meet their own specific requirements. The stability of the relationship will vary according to the formality and length of the contract. Subcontracting may involve the entire manufacture of a particular good or service, known as *commercial* subcontracting, or it may take the form of a firm buying in particular inputs that it does not have the skills or capacity to produce cost-effectively 'in-house'– *industrial* subcontracting.

- *Strategic alliances/joint ventures*: this is where firms come together to create a new corporate entity in order to undertake a particular task, for example costly joint research. In many other respects the participating firms may still remain competitors.
- *Franchising and licensing*: here, firms allow a company (the franchisee or licensee) to sell their product or service in a particular territory under given terms and conditions, and in return for a set fee. Franchising is very common in the service sector (including fast food, retailing, coffee shops, etc.), allowing rapid geographical expansion.

In sum, every production network constitutes a unique constellation of activities performed via complex combinations of internal and external network connections. The first step in understanding any production system, therefore, is to map out the key participants and the *nature* of the relationships that connect them.

The coffee production network is relatively straightforward and is represented schematically in Figure 16.2. Even so, coffee travels a long way and changes hands several times on the journey from bean to cup. The major participants in the network are depicted by boxes, and the transactions that move coffee in its different forms between participants are shown by arrows. Coffee flows through the network from the growers, who in effect begin the system, to the consumers who represent the end point. Coffee – which comes in two main types, Arabica and Robusta – is generally grown on small farms or estates in the developing world. Once basic processing has extracted the 'green' coffee beans from picked coffee cherries, they will pass in 60-kg bags from an exporter to a developed country importer, trader or broker, then on to a roaster or instant coffee manufacturer, and then finally to a consumer via either a supermarket shelf or café of some kind. These relationships are essentially 'arms-length' *market* connections, with prices

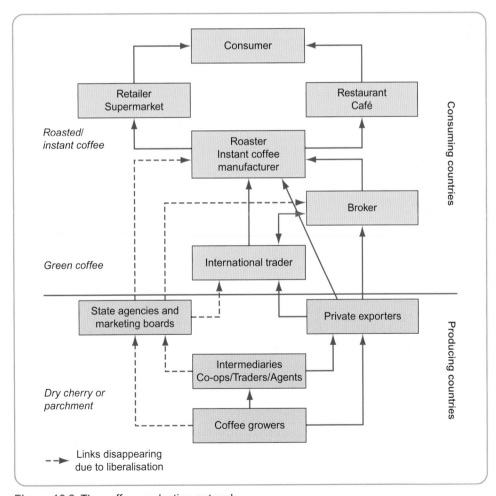

Figure 16.2 The coffee production network.

Source: adapted from Ponte (2002), Figure 1 and Talbot (2004), Figure 2.1

being set by international commodity markets such as the New York Board of Trade (Arabica) and the London International Financial Futures and Options Exchange (Robusta). We will see later, however, how the roles of these various participants have evolved over time.

In contrast, laptop computers are the outcome of a much more complicated production network that brings together hundreds of different components into the finished product. It is estimated, for example, that a HP laptop may contain around 2,200 separate parts (Dedrick *et al.* 2010). The personal computer (PC) industry is therefore a complex network of firms involved in a wide range of different industry segments – from microprocessors and other electronic components to applications and systems software providers – and covering a wide range of activities: R&D and design, manufacturing, assembly, logistics, distribution, sales, marketing, service and support. How these various functions are split between different companies has changed over time. Historically in the computer industry, vertically integrated giants such as IBM, HP and Siemens operated in all the industry segments and carried out the key functions of product innovation, manufacturing and customer relations internally.

Since the advent of the PC, however, a much more complex 'tiered' network has evolved in which most companies concentrate on one particular market segment, for example, assembling PCs, or making circuit boards or disk drives (see Figure 16.3). The PC can now be described as a *modular* product, whereby 10 to

Figure 16.3 The laptop PC production network.

Source: adapted from Foster *et al.* (2006), Figure 2

15 relatively self-contained sub-components (e.g. keyboard, monitor, hard drive, etc.) are brought together and assembled, an attribute that facilitates the disintegration of the production network across separate firms. Branded PC companies now focus primarily on design and customer relations, 'outsourcing' the remainder of the production process to other firms. This system reflects the nature of the PC as a standardized product assembled from components that can be produced in a wide variety of locations by a broad range of firms. Only limited value is added by assembling a PC: in most cases PC firms add value through customer relationships, either directly through their own direct sales and service relationships (e.g. help desks, repairs, etc.) or indirectly through their branding, marketing and quality assurance practices. They also extract value from the network through coordinating the logistics operations that turn components into finished products on customer doorsteps. As Curry and Kenney (2004: 114) suggest, 'a PC assembler is, in many ways, more a logistics coordinator than a manufacturer'.

Most laptop assembly operations, therefore, are subcontracted to contract manufacturers (CM), and increasingly, to a particular form of subcontractor known as the original design manufacturer (ODM) that also contribute to the design process. In the case of contract manufacturers, a branded PC firm is likely to employ its own, on-the-ground design and development teams throughout the production process to supervise subcontractors: this is the model preferred by Toshiba and Lenovo (the leading Chinese manufacturer), for example. With an ODM, however, the PC maker may take primary responsibility for design, but it will then pass a product specification on to the ODM for final development and manufacturing; HP and Dell operate in this way and this is currently the most common model in the industry. In yet another model, smaller PC vendors without the scale to undertake their own design activity may simply purchase generic 'off-the-shelf' products from ODMs to be labelled and sold under their own name. Backwards relationships from branded PC firms to ODMs may also extend forwards to the customer through the delivery and servicing of laptops on behalf of PC vendors. Overall, the result is a highly responsive and efficient production system in which an order via a branded PC company's website can be despatched by an ODM within 48 hours.

It is important to bear in mind that Figure 16.3 only provides a limited window on what is, in reality, a highly complex network connecting together hundreds of firms and tens of thousands of workers. For example, the hard disk drive industry (see 'Core Tier 3 suppliers' box in Figure 16.3) is itself a significant sector involving several tiers of firms and assemblers spread across East and South-east Asia. It is to these complex, on-the-ground geographies of production networks that we now turn.

16.3 Geographies of production networks: spatial divisions of labour

It goes without saying, of course, that the production networks described above do not exist on the head of a pin, but rather connect together, and indeed constitute, real places within the global economy. Every production network requires a *spatial division of labour*, which refers to the way in which certain elements of the production process are concentrated in particular places. The concept of the spatial division of labour is elucidated particularly well by Massey (1984) who explored how these spatial variations are both *created* and *exploited* by the ongoing restructuring of capitalist firms in their pursuit of profit. After examining the spatial restructuring of the UK economy in the 1960s and 1970s, her key argument was that under conditions of increased global competition, corporations were increasingly looking to separate spatially the *control* functions performed by managerial workers from the *execution* functions undertaken by manual workers. In its most simple form, low-skilled manual tasks were tending to locate in peripheral areas, while managerial and R&D tasks were concentrating in core cities and their surrounding regions. Massey argued that, over time, different *layers* of investment fan out across the economic landscape, redefining the nature of relationships between places as they gain or lose different kinds of activities.

Different spatial divisions of labour arise from variations in corporate structures which, at a simple level, can be divided into *cloning* and *part-process* structures. In the cloning structure, the production apparatus is simply replicated in different localities, with ultimate control residing at a headquarters, usually sited in the firm's initial place of origin. In the part-process format, there is a technical division of labour between branch plants, with components being made in one location and passed on to another for final assembly. The different stages of the production process will have varying requirements, and hence tend towards different kinds of location.

Massey's conceptual apparatus is still extremely powerful today, although her study – which was primarily about *intra-firm* and *intra-national* divisions of labour – needs to be extended in two important ways. First, as we have already noted, spatial divisions of labour can be constructed through combinations of intra-firm (internal), and increasingly, inter-firm (external) networks. Second, we need to apply these ideas at the *international* scale. One of the defining characteristics of the world economy over the past two to three decades has been the dramatic increase in the number of transnational corporations (TNCs) organizing their spatial divisions of labour at the international scale. It is now difficult to think of a production network that does not have at least some international elements, even if it is just seen in the sourcing of one or two inputs, or a limited export market for the final good/service. *Global* production networks, as we will call them here, have become one of the most important organizational features of the contemporary global economy. Spotlight box 16.2 considers different attempts to conceptualize the nature of international divisions of labour at different points in the global economy's evolution.

We can make three further arguments about the geographies of global production networks:

■ First, their *geographical complexity* is increasing, enabled by a range of developments in transport, communication and process technologies. As we shall see

---

Spotlight box 16.2

Conceptualizing global divisions of labour?

It is helpful to distinguish between three separate attempts to conceptualize international spatial divisions of labour, which broadly coincide with different phases of development of the global economy.

■ First, it is possible to identify the traditional **international division of labour** (IDL) that took shape by the nineteenth century and prevailed largely unaltered until the 1950s. The IDL essentially depicted a *trading* system – shaped initially by global trading empires, and later by the rise of the USA as an economic power – in which the developing world or 'periphery' was largely relegated to providing raw materials and agricultural plantation products (e.g. coffee from Brazil, copper from Chile, gold and diamonds from South Africa) for the industrialized economies of the 'core' (Western Europe and the USA). High-value manufactured goods were exchanged between the industrialized countries, and some were exported back to developing countries.

■ Second, from the 1960s onwards, a **new international division of labour** (NIDL) started to emerge in which European, North American and Japanese TNCs created labour-intensive export platforms in so-called 'newly industrializing economies' (NIEs) (especially in East Asia and, to a lesser extent, Latin America) in response to falling profit rates in the core countries. Crucially, the system depended on new technologies that allowed production fragmentation, thereby creating tasks that could use, often young and female, semi-skilled or unskilled workers in the periphery. Two kinds of technology were important: *process* technologies that allowed the subdivision of the manufacturing process into simple and self-contained tasks, and *transportation* technologies such as jet aircraft and containerized shipping that allowed the efficient shipment of both components and finished goods.

■ Third, it has become clear that in recent decades a *new global division of labour* has emerged that is far more complex than the system depicted in the NIDL model, which does not capture, for example: how the range of NIE economies has broadened and deepened considerably, particularly in Asia; increasing investment into NIEs, such as China and India, to access their domestic markets; significant outward investment by NIE transnational corporations, e.g. from South Korean giants such as LG and Samsung; continued high levels of investment between developed economies and the emergence of complex international divisions of labour in the service sector. The contemporary global economy is clearly not characterized by one single type of IDL but rather many different forms. Rather than disappearing, the traditional IDL and the NIDL remain important in a range of industries (e.g. natural resources and clothing/toys, respectively) and have subsequently been overlain by, and have interacted with, newer and more complex international divisions of labour. For more on these debates, see Coe (2011).

shortly, the assembly of a laptop requires components manufactured at places all across East Asia.

■ Second, the geographic configurations of global production networks are becoming *more dynamic and liable to rapid change*. This flexibility arises, first, from the use of certain 'space-shrinking' information and communication technologies and, second, from organizational forms that enable the fast spatial switching of productive capacity (Dicken 2007). In particular, the increased use of subcontracting and strategic alliance relationships (noted above) allows firms to switch contracts between different firms and places without incurring the costs of moving production themselves.

■ Third, we need to connect these ideas about the geographical extensiveness and complexity of global production networks with notions concerning the geographical *clustering* of economic activity. Not all the connections within the system can be 'stretched out' across the global economy; some interactions will need to take place within the same locality because of the sheer intensity of transactions or because of the importance of place-specific knowledge to the activity concerned. From this perspective, global production networks need to be seen as the organizational forms that connect clusters together.

Let us now move on to apply these ideas to our two products. The global map of coffee production and consumption is not very complicated. Growing coffee involves several million farmers, who ultimately support the consumption of an estimated 2 billion cups of coffee every day. The vast majority of coffee production takes place in developing countries while most of the consumption occurs in the wealthy markets of North America, Western Europe and East Asia. Some 7.2 million tonnes of coffee were produced in 2009, of which 5.7 million tonnes, or 80 per cent, were exported. Although over 50 countries currently produce coffee, the top 12 exporters account for almost 90 per cent of the global trade (Figure 16.4). The geography of production is shaped by the ecology of the coffee plant which requires a consistently warm and wet climate, making it most suitable for growing in tropical highland areas. The production geography is not static, however. Most notably, exports of Robusta coffee grown in the central highlands of Vietnam have expanded dramatically over the last decade – rising from just 100,000 tonnes in 1990 to 1 million tonnes by 2005 and remaining at or around that level ever since – creating an over-supply in the global market and putting severe downward pressure on prices. As a result, coffee growing is an increasingly marginal practice for growers across Central

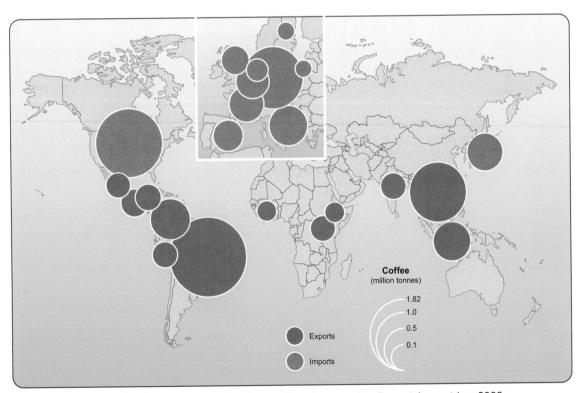

Figure 16.4 The top twelve coffee producing (exports) and consuming (imports) countries, 2009.

Source: data from www.ico.org, accessed 21/10/09

and Latin America and Africa. The consumption of coffee has a long history linked to colonial expansion by Europeans, which served to both spread the taste for coffee, particularly from the sixteenth century onwards, and spread coffee cultivation from its supposed origins in present-day Ethiopia to the band of tropical countries shown in Figure 16.4 (Topik 2009). The key commodity markets and international traders and roasters (see Figure 16.2) are, however, all based in the USA and Western Europe.

Turning to the geography of the laptop production network, the majority of the world's laptops are produced by Taiwanese firms. All branded manufacturers – such as HP, Acer, Apple and Toshiba – rely on Taiwanese ODM firms for manufacturing and product development. In 1998, Taiwanese companies accounted for 40 per cent of a world market of 15 million units, rising to 72 per cent of the 46 million laptops produced in 2004. By 2009, the figure had reached 90 per cent, with the top eight Taiwanese manufacturers (see Table 16.2) alone producing 144 million (or 85 per cent) of the 169 million laptops shipped worldwide that year. Production is actually growing faster than the PC market as a whole: in 2005, one in three PCs sold globally was a laptop, and by 2009 this figure had reached 55 per cent and was expected to reach 70 per cent by 2012. The laptop market is also becoming more differentiated, with the netbook and CULV (consumer ultra low voltage) segments growing particularly strongly. As Table 16.2 suggests, each leading laptop vendor tends to contract with two or three Taiwanese firms, arrangements which in the cases of HP

and Dell account for upwards of 90 per cent of their total global laptop production; for Japanese companies such as Sony, Sharp and Toshiba, the proportion tends to be somewhat lower (50–70 per cent). Taipei – and the nearby region of Hsinchu where many Taiwanese high-tech firms are based – is therefore a critical node in the laptop production network.

However, these geographies are also far from static. Taiwanese electronics firms have been moving production 'offshore' – to Southeast Asia, Europe and, most importantly, China – since the early 1990s. There are now two key PC clusters in China: the Shenzhen area of Guangdong province (in the south of the country), specializing in desktop machines, and the Shanghai/Suzhou/Yangtze River delta area, home to the laptop PC industry (see Figure 16.5). Before 2001, the Taiwanese government prohibited its laptop manufacturers from undertaking final assembly in China. When this restriction was lifted, the ODMs moved collectively, and incredibly rapidly, to the Shanghai area: in 2001 only 5 per cent of Taiwanese laptops were produced in China but by 2004 the figure had shot up to 80 per cent, and it is now almost 100 per cent. The lower costs of land, labour and facilities on the mainland allowed the firms to build giant factories, expand output and reduce costs through scale economies. The production levels involved are quite staggering: in 2009 the leading Taiwanese producer Quanta was manufacturing around 3 million laptops per month at its Chinese plants. The offices of these firms in Taiwan remain responsible for product development, technology research, materials procurement, financial management and marketing

Table 16.2 Top eight Taiwanese laptop PC manufacturers, 2009

Company	2009 Shipment (thousands)	Largest client	Second largest client	Third largest client
Compal	37,900	Acer	HP	Dell
Quanta	35,900	HP	Dell	Acer
Wistron	25,600	Acer	Dell	HP
Inventec	21,600	Toshiba	HP	Dell
Pegatron	10,900	Asus	Apple	Sony
Hon Hai	7,200	Asus	Apple	Sony
Elitegroup	3,600	Lenovo*	Hasee*	Founder*
Clevo	1,600	Hasee*	Founder*	

*Lenovo, Hasee and Founder are leading Chinese PC brands.

Source: www.researchinchina.com/Htmls/Report/2010/5822.html [accessed 21 October 2010]

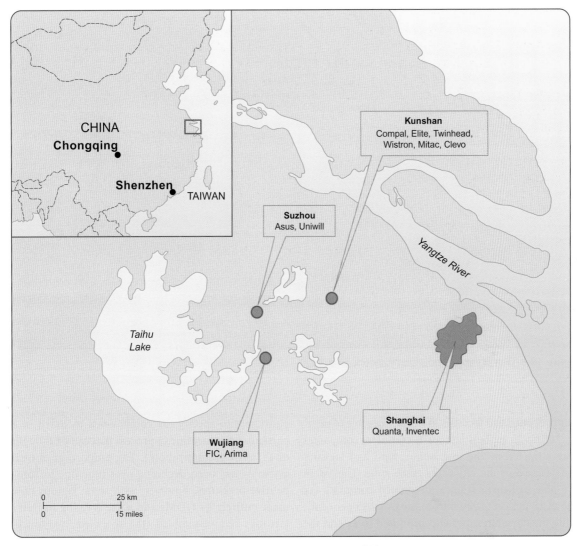

Figure 16.5 The laptop manufacturing cluster, Yangtze delta region, China.

Source: based on Yang (2006), Appendix 2

(Saxenian 2006). In this way, a new *intra-firm* spatial division of labour is added to the system depicted in Figure 16.3.

Many Taiwanese component suppliers had already relocated to the region by 2001, and now almost all the parts needed to make a laptop are manufactured in the Suzhou area. Initially, the highest-value components such as the central processor, hard drive and displays continued to be made overseas by Japanese, Korean and US companies, but over time many firms such as the US hard disk drive manufacturer Seagate and chip firm AMD have themselves established production facilities in the Shanghai region. Moving the entire supply chain to the Shanghai region in this way allowed the Taiwanese firms to overcome deficiencies in the Chinese logistics and distribution sectors. The scale

of the undertaking should not be underestimated: by 2005 some 3300 Taiwanese firms had already invested US$26 billion in Suzhou alone. Several of the transnational computer brand manufacturers (e.g. Toshiba, Samsung and Sony) also have bases in the region.

More recent developments suggest that these geographies are once again in flux, driven both by rising costs in the coastal provinces and strong growth in the Chinese domestic market for IT products and services of all kinds. In line with many other kinds of manufacturing activities in China, there are signs that production is starting to move to inland provinces. The massive city of Chongqing in Sichuan province, for example, aims to be home to laptop production of 80 million units by 2015 (*Global Times* 2010). In 2010 HP and the Taiwanese motherboard company Foxconn

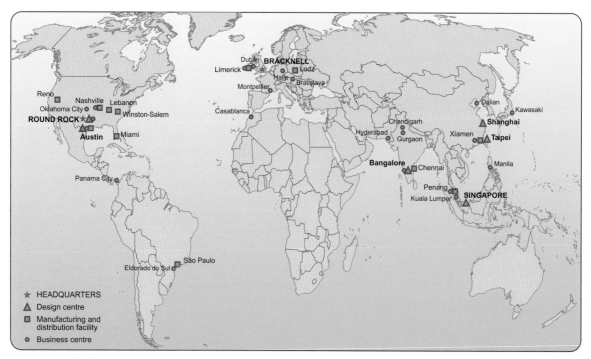

Figure 16.6 Dell's global production network, mid-2009.

Source: data from www.dell.com, accessed 22/6/09

jointly opened two facilities in the city capable of manufacturing 20 million laptops a year. The companies
were attracted by the low labour, land and logistics
costs as well as the city's free trade zone. They were
also able to benefit from a discounted corporate tax
rate of 15 per cent (down from the usual 25 per cent),
while the city authorities also promised to extend the
second runway of the local airport to meet HP's logisti-
cal requirements. Quanta, Inventec and Hon Hai have
likewise established manufacturing bases in the city,
with reports in mid-2010 suggesting that 73 suppliers
of laptop components were also already present in the
area. As these trends continue, the geography of global
laptop construction is once again being reworked.

While the laptop global production network has
a relatively concentrated geography around the key
nodes of Taipei/Hsinchu and Shanghai-Suzhou, the
PC industry generally has a somewhat more disparate geography. We can use the case of Dell as an example. There are three distinctive elements to Dell's
business model. First, the company only sells direct
to its customers, bypassing almost entirely distributors, resellers, and retailers (see Figure 16.3). Second,
unlike some PC manufacturers, Dell chooses to undertake a significant proportion of the final assembly
of both desktop and laptop computers itself, allowing

customers to specify the components included in a
particular product line (a process that can be thought
of as 'mass customization'). The proportion of final
assembly has been declining over time due to competitive pressures, however, with some laptops now
made entirely by ODMs and some arriving in Dell's
factories semi-assembled (the so-called 'two-touch'
approach). Third, it has aggressively used the Internet
to establish not only its system of direct sales, but also
the procurement and assembly operations, thereby
dramatically reducing its inventory of components.
Figure 16.6 illustrates how Dell services its global
market by using a series of regional production and
service clusters (Fields 2004). Such clusters are an integral part of the global space economy (see Spotlight
box 16.3).

16.4 The governance of production networks

So far we have exemplified the various actors within
production networks, the functions that they undertake, and where they are located. The next step is to
explain how such production networks function, and

Spotlight box 16.3

Types of clusters in the global economy

There is a wide variety of different kinds of clusters in the global economy, created through different historical processes, and bound together by varying combinations of both *traded* and *untraded* interdependencies. Traded interdependencies are created by firms co-locating in a cluster alongside suppliers, partners and customers with which they have formal trading relationships. Proximity to these firms reduces the costs of transportation, communication, information exchange, and searching and scanning for potential customers. Untraded interdependencies are the less tangible benefits of being located in the same place, for example the emergence of a particular pool of specialized workers, or the development of local institutions such as business associations, chambers of commerce, government offices, and universities with particular attributes and capabilities. In particular, clusters can facilitate patterns of intense and ongoing face-to-face communication – and thereby transfers of important forms of intangible knowledge –between people working in the same or closely related industries.

Recognizing the diversity of clusters in the world economy is important if economic geographers are to effectively understand the contemporary space economy. We can think of at least eight significant types, each performing different roles within the global system:

1. *Labour intensive craft production clusters*. These are often found in industries such as clothing where firms are involved in tight subcontracting networks and often use high levels of immigrant labour and homeworkers. Examples include the garment production districts of Los Angeles, New York and Paris.

2. *Design intensive craft production clusters*. These refer to dense agglomerations of small- and medium-sized firms specializing in a particular aspect of the high-quality production of a particular good or service. Examples include the renowned towns and districts of the Third Italy including Prato (textiles), Santa Croce (leather goods) and Sassuolo (ceramics).

3. *High-technology innovative clusters*. These clusters tend to have a large base of innovative small- and medium-sized firms and flexible, highly skilled labour markets in sectors such as computer software and biotechnology. Examples include Silicon Valley in the USA and Cambridge, United Kingdom.

4. *Flexible production hub-and-spoke clusters*. In these clusters, a single large firm, or small group of large firms, buys components from an extensive range of local suppliers to make products for markets external to the cluster. Examples include Boeing in Seattle, USA, and Toyota in Toyota City, Japan.

5. *Production satellite clusters*. These clusters represent congregations of externally owned production facilities. These range from relatively basic assembly activity, through to more advanced plants with research capacity. Examples are to be found across the export processing zones (EPZs) of the developing world – the Shanghai laptop cluster would also fall into this category.

6. *Business service clusters*. Business services activities such as financial services, advertising, law, accountancy are concentrated in leading cities – such as New York, London and Tokyo – and their hinterlands.

7. *State-anchored clusters*. Some clusters have developed because of the location of government facilities such as universities, defence industry research establishments, prisons, or government offices. Examples include agglomerations that have developed because of government research investment (Colorado Springs, USA; Taejon, South Korea) and universities (Oxford/Cambridge, United Kingdom).

8. *Consumption clusters*. There are also strong propensities to cluster – often in central urban areas – in a wide variety of consumer service activities including retailers, bars and restaurants, and cultural and leisure facilities. Tourist activities also tend to congregate around different kinds of amenities.

In reality, of course, applying any such typology is not an easy task. Many clusters are actually hybrid forms of the above categories, and a world city such as Los Angeles, for example, may contain examples of many, if not all, such cluster types.

more specifically, the way in which some firms use their power over other firms to control or 'drive' the overall system. These 'lead' firms are able to define which other firms can join the network, the roles they perform, and the financial and technological conditions under which they perform those roles – despite not directly owning them. This shaping of global production networks by powerful corporations through inter-firm relations is known as **governance**.

Governance can take different forms. A useful starting point is the distinction between networks that are *producer-driven* and those that are *buyer-driven* (Gereffi 1994; see Figure 16.7 and Table 16.3). Producer-driven networks tend to be found in sectors where large industrial corporations play the central role in controlling the production system, for example in capital- and technology-intensive industries such as aircraft, automobile and semiconductor manufacturing. Power in these networks is exercised through the headquarters operations of leading TNCs, and manifests itself in the ability to exert control over 'backward' linkages to raw material and component suppliers, and 'forward' linkages with distributors and retailers. High levels of profits are secured through the scale and volume of production in combination with the ability to drive technological developments within the production

system. The automobile industry provides an excellent example of producer-driven networks. Leading assemblers such as Toyota and Ford coordinate production systems involving literally thousands of subsidiaries and tiers of subcontractor firms around the world, as well as extensive global networks of distributors and dealers.

Buyer-driven networks, by comparison, are characteristic of industries where large retailers (e.g. Wal-Mart or Tesco) and brand-name merchandisers (e.g. Nike or Reebok) play the central role in establishing and shaping the network, with production activity itself often located in developing countries (see Figure 16.7). This form of global production network is common in labour-intensive consumer goods sectors, such as clothing, footwear, toys and handicrafts. Production is usually undertaken using tiered networks of subcontractors that supply finished goods subject to the specifications of the powerful buyers. These buyers in turn extract substantial profits from bringing together their design, sales, marketing and financial expertise with strong brand names and access to large consumer markets in developed countries.

The notions of producer- and buyer-driven networks are ideal types that allow us to get an overall impression of how a particular system functions, and

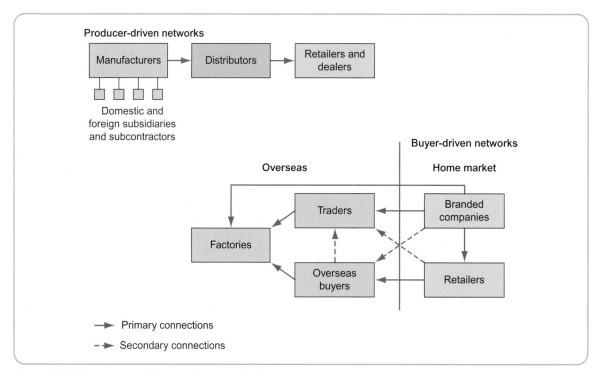

Figure 16.7 Producer- and buyer-driven production systems.

Table 16.3 Producer-driven and buyer-driven production networks compared

	Form of economic governance	
	Producer-driven	*Buyer*-driven
Controlling type of capital	Industrial	Commercial
Capital/technology intensity	High	Low
Labour characteristics	Skilled/high wage	Unskilled/low wage
Controlling firm	Manufacturer	Retailer
Production integration	Vertical/bureaucratic	Horizontal/networked
Control	Internalized/hierarchical	Externalized/market
Contracting/outsourcing	Moderate and increasing	High
Suppliers provide	Components	Finished goods
Examples	Automobiles, computers, aircraft, electrical machinery	Clothing, footwear, toys, consumer electronics

Source: adapted from Kessler and Appelbaum (1998)

in turn, who benefits most from the way it functions. In reality, however, the governance of global production networks is highly complex and variable both within, and between, different economic sectors. We can think about this complexity in three ways. First, there may be other forms of governance that seem to better characterize the overall nature of a particular production network. For example, in some contexts it might be useful to think of *relational* forms of governance that fall in between the producer- and buyer-driven models (Gereffi *et al.* 2005). These can be thought of as close inter-firm relationships that develop on a more even footing. In others, it might be trading firms – such as the Japanese giant trading firms, the so-called *soga sho-sha* –or intermediaries that drive the network. In the Internet era, for example, there is evidence that new forms of 'infomediaries' with extensive access to online consumer information (e.g. Google, AOL, etc.) may play increasingly important roles in production networks (Gereffi 2001). Second, when we look at the detail of a particular production network, it is important to recognize that there may be a variety of governance types in operation between the various actors within the overall system. A study comparing the PC production networks in the Yangtze and Pearl River delta regions of China, for example, found that in reality governance relations are extremely complicated and, indeed, varied significantly within the same industry (Yang and Coe 2009). Third, governance regimes are not static, but need to be seen in *dynamic* terms.

The nature of a production network – and indeed the constituent relationships within it – may change over time because of a wide range of factors, both internal and external to the system. In particular, firms and/or particular regions may seek to improve their position within the global production network through pursuing *upgrading* strategies.

At first glance, comparing coffee and laptops would seem to provide a nice contrast between buyer- and producer-driven networks. The coffee network is 'clearly buyer-driven and coffee roasters play the lead role in determining the functional division of labour along the chain. In other words, roasters define the key terms of participation directly for their immediate suppliers and indirectly for other actors further upstream' (Gibbon and Ponte 2005: 83). Roasters capture about 30 per cent of the value that is added in the coffee production network, but retailers (22 per cent) and international traders – such as Neumann, Volcafe, Cargill and others – also do well (8 per cent). The result is that at least 60 per cent of the economic value derived from coffee goes to developed country firms, although some estimates put the figure as high as 80 per cent (Fitter and Kaplinsky 2001). The domination by these actors is in large part a simple story of concentrated market power and economies of scale: the top four international coffee traders account for 40 per cent of the global trade; as noted in the introduction to this chapter, the top five roasters handle about 50 per cent of the world's coffee; and the top 30 global grocery retailers account for about one-third of global coffee sales. Roasters, for example, reap the economic benefits of buying the beans in bulk and having captive suppliers that depend on their purchases. Even so, the coffee network is not *as* buyer-driven as networks driven directly by supermarket chains (e.g. fresh fruit and vegetables) or by retailers and branded marketers (e.g. clothing and footwear). To a degree, the strong coffee roasters/brands are able to counteract the huge buying power of the retailers, meaning that influence is shared. Ideal-type buyer- and producer-driven chains, then, should perhaps be seen as the ends of a spectrum rather than discrete categories.

Typically, as noted in Table 16.3, the computer industry has been seen as a capital-intensive producer-driven network dominated by vertically integrated giants such as IBM. However, the computer industry has changed beyond recognition over the past 20 years as the PC has increasingly become a mass-market standardized product. The branded PC companies continue to drive the global production network, taking the decisions that drive the whole system and coordinating the activities of the other main players. But their role has changed over time. Leading manufacturers like HP and Dell have increasingly focused on final assembly, design and branding, with the vast majority of production outsourced to Asian subcontractors, thereby sharing the characteristics of the lead firms in buyer-driven systems such as clothing and shoe production. Over time, there has clearly been a shift in the governance characteristics of the PC industry.

However, it is possible to make another argument relevant to the PC sector. The branded PC marketplace has become increasingly competitive over recent years as the standardization process has progressively reduced profit margins for even the leading firms. The result has being ongoing consolidation and rationalization, with several large brands being swallowed up by competitors (e.g. Compaq by HP, IBM by Lenovo). O'Riain (2004) has suggested that we might think of the PC network (and others like it) as falling into another governance category – the technology-driven network, in which it is control over technical standards which is critical for exercising power over the system. In that sense, it is the firms that set the dominant architectural standards for PCs in terms of software (Microsoft's operating systems) and hardware (Intel's computer chips) that benefit most from the production network and secure profits rates significantly higher than any other participants in the industry. 'Power . . . resides, therefore, in what is getting made and mobilizing networks

of support for those products rather than in how that product is itself produced – emphasizing the place of intellectual property holders in securing the rewards . . . within the system' (O'Riain 2004: 645). These arguments are supported by detailed research into the gains associated with manufacturing a HP laptop computer in 2005 (Dedrick *et al.* 2010). The study found that while Microsoft and Intel secured net profits of 36 and 31 per cent respectively, and key component manufacturers in the range of 4 to 12 per cent, HP themselves only made a profit of 4 per cent, with the assembler (i.e. ODM) making an even lower profit rate of 2.4 per cent.

Returning to coffee, we can start to appreciate the complexity and dynamism of global production networks. The coffee production network is not static in organizational and governance terms. It can no longer simply be understood as a sequence of market transactions. Three changes illustrate this point. First, the roasters are increasingly concentrating on their roasting, blending and branding activities, employing coffee traders to source coffee and manage their supply networks for them, along with undertaking the steaming of certain kinds of beans. These are particularly important roles in a coffee network which is increasingly concerned with coffee quality testing and assurance, and reflect the establishment of more long-standing, relational connections between roasters and traders. Second, international traders are vertically integrating their operations by buying export firms in producing countries in order to ensure secure and high-quality supplies. This serves to consolidate the position of traders within the system and potentially increase further the value captured by developed-country firms. In turn, the potential for developing countries' producers and processors to upgrade their activity is restricted by the fact that roasting and final processing ideally needs to be undertaken near final markets. Third, growing consumer demand for new kinds of speciality, organic and fair-trade coffees has opened up some direct connections between producers and smaller roasting houses, potentially at least improving the financial returns for the coffee farmers connected to these new networks. We shall return to this issue in the last section of the chapter.

In terms of the laptop production network, at the same time as branded PC firms have retreated from manufacturing activity, Taiwanese manufacturers have evolved or upgraded to take on new roles and, as a result, now occupy much more significant positions within the production network. Many Taiwanese firms started out as simple assemblers of components for

PCs. Over time, they moved – through what is known as *process* upgrading – into the manufacture and assembly of PCs sold under the brand names of other firms (i.e. they became OEMs – original equipment manufacturers). The next step was to develop design capacity, as we have already noted, becoming ODMs (original design manufacturers) through a step known as *product* upgrading. Next, some firms have taken things further and started to manufacture and sell PCs under their own brand, thereby becoming OBMs (original brand manufacturers) through *functional* upgrading. Acer, for example, now the world's second largest PC brand, started out as one of Taiwan's main electronics suppliers, producing various computer products for international buyers. Over time, it separated off its OEM activity, leaving it to focus on the marketing and R&D activities of its own brand laptops. Acer's rise to the very top of the industry was confirmed by the purchase of two large US PC brands, Gateway and Packard Bell, in 2007 and 2008 respectively. Asus – the world's fifth largest brand – followed a similar path, spinning off much of its manufacturing activity into the now independent Pegatron. In reality, however, these categories may not map neatly onto individual firms, many of which will combine both OEM/ODM and OBM activities. In some cases, firms may be able to use their expertise to shift into entirely new production networks (e.g. from laptops to WAP phones) in a shift termed *chain* upgrading (Kaplinsky 2005).

Two further points should be noted here. First, there is no automatic or simple progression along this upgrading path. These shifts reflect changing *strategies* on the part of Taiwanese firms in response to changing market conditions. As PCs have become ever more standardized commodities, however, OEM and ODM producers have increasingly been squeezed for cost savings by branded PC firms. They have tended to respond in one or more of three ways: moving into original brand manufacturing themselves; making more components themselves as a way of cutting costs (i.e. vertically integrating their activity); or trying to cut costs through production relocation to China, as described earlier. Second, the successful ongoing upgrading of Taiwanese electronics firms is not only due to firm strategies and inter-firm networks. It also reflects a range of deliberate strategies pursued by the Taiwanese state to develop the industry since the early 1970s, for example: the establishment of the Hsinchu Science Park; direct investment in early key players in the sector including UMC (United Microelectronics Corporation) and TSMC (Taiwan Semiconductor Manufacturing Corporation); and steering developments in the industry through its agencies such as ERSO (Electronics Research Service Organization), ITRI (Industrial Technology Research Institute) and CETRA (China External Trade Development Council). These reveal the importance of also considering the institutional context of production networks, a topic to which we now turn.

16.5 The institutional context of production networks

So far we have focused almost entirely on the production network as a mesh of intra- and inter-firm connections. Production networks are also shaped by a wide range of *extra*-firm relationships that, as we saw in Figure 16.1, are also an integral part of the system and may incorporate a wide range of non-firm entities (e.g. supranational organizations, the state, labour unions, business associations, etc.). Expressed slightly differently, every relationship in a production network is shaped by its *institutional context*.

We can unpack the complexity of these institutional contexts in two ways. First, institutional context is significant at various spatial *scales*. At the subnational scale, local governments may seek to stimulate particular kinds of economic activity in their locality, for example by providing low-rent premises for small high-tech businesses. At the national scale, nation-states still wield a huge range of policy measures to try to promote, and steer, economic growth within their boundaries (Dicken 2007). Increasingly important in an era of globalization are measures designed to promote (or restrict) movements of traded products, investment and migrants across national boundaries. At the macro-regional scale, regional blocs such as the European Union or the North American Free Trade Agreement (NAFTA) have considerable influence on trade and investment flows within their jurisdiction. At the global scale, supra-national institutions like the World Trade Organization (WTO) and the International Monetary Fund (IMF) increasingly determine the regulatory frameworks for global financial and trading relationships. Even a relatively simple global production network will cross-cut and connect a wide range of institutional contexts. Second, it is important to distinguish between *formal* and *informal* institutional frameworks. The former relates to the rules and regulations that determine how economic activity is undertaken in particular places (e.g. trade policy, tax policy, environmental regulations and so on), while the latter refers to the rather less tangible, and often place-specific, ways of doing business that relate to the social, economic and political *cultures* of particular places.

Again, we can use the examples of coffee and laptop production to illustrate these arguments. First, the coffee production network demonstrates the importance of *formal* institutional frameworks at the national and global scales. From 1962 to 1989, the international trading of coffee was governed by a series of International Coffee Agreements (ICAs) managed by the International Coffee Organization (ICO), a supranational organization made up of representatives of a wide range of coffee importing and exporting countries. The ICAs combined price bands and export quotas to provide a coffee trading system that was widely credited with both raising and stabilizing coffee prices. During this period, many exporting countries established coffee marketing boards: government institutions that controlled markets, monitored quality and acted as a link to exporters and international traders (see Figure 16.2). Such boards provided an important protection for farmers and growers from the vagaries of the international coffee market. In 1989, however, the ICA was not renewed in the face of rising production levels and low-cost competition from non-member exporting countries.

The ending of the ICA regime has dramatically altered the balance of power in the coffee chain, as the now market-based coffee trading system has led to lower and more volatile coffee prices. Its demise has served to concentrate power in the hands of consuming-country firms, and in particular the small group of roasters introduced earlier. The collapse of the ICA has meant that coffee-producing countries are no longer cooperating, but are in fact competing with each other, resulting in overproduction and a drop in prices which the roasters have been able to benefit massively from, a process exacerbated by the fact that consumer coffee prices have not dropped in a similar way, but have in fact risen significantly. At the same time, the national coffee marketing boards in the exporting countries either have been eliminated, or have retreated into a restricted overseeing role that has left them marginalized within the production network (see dotted lines in Figure 16.2). As a result of these changes to the inter-linked international and national institutional contexts, millions of coffee farmers worldwide are exposed to price fluctuations on the global coffee market. On occasion, this can result in farmers receiving less for their coffee beans that it costs to grow them.

What this story shows is how changing institutional frameworks can impact on all the other dimensions of a global production network. In terms of its basic structure, the demise of the ICA has led to the bypassing of a previously important actor, namely the coffee marketing boards. In terms of changing geographies, the rapid growth of exports from Vietnam was both a cause of the ICA's demise and at the heart of the subsequent overproduction. The expansion of production in Vietnam – strongly supported and marshalled by the government – also shows the significance of national institutional factors. Overall, in governance terms, the post-ICA regime has enabled a further concentration of power in the hands of roasting firms.

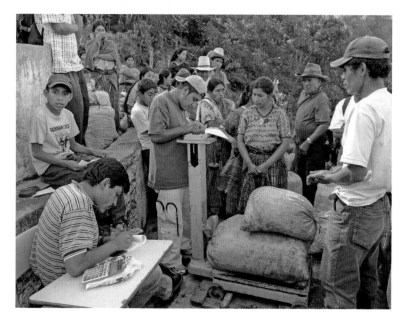

Plate 16.2 Coffee growers in Guatemala having their organic coffee beans weighed at the local cooperative.

(Sean Sprague/Panos Pictures)

The laptop industry is similarly affected by multi-scalar regimes of formal institutional relationships. For example, as we have seen, rules governing outward investment from Taiwan to China were crucial in determining the timing of shifts in laptop production, and the fact that, for geopolitical reasons, direct flights between Taipei and China only restarted in 2008 had a significant impact on the practicalities of investment. Moreover, the role played by municipalities such as Kunshan, Suzhou and Wujiang in setting up industrial parks and actively courting investments from Taiwanese ODMs was crucial in helping the Shanghai region steal leadership in the Chinese PC industry from the longer-established Guangdong cluster (Yang 2009). Here, however, we will use the laptop industry to open a window on a range of other less formal connections that are integral to the success of Taiwan's electronics industries. More specifically, there is now increasing recognition of the significance of migration flows of skilled engineers between Taiwan and California for driving the economic success of both high-tech regions. Hsu and Saxenian (2000) describe the development of a *transnational technical community* linking Hsinchu and Silicon Valley. For several decades now, Taiwanese migrants to Silicon Valley – in combination with those from India, China and other parts of Asia – have played an important part in the continued dynamism of the Californian region through their entrepreneurial activity and links back to Asian markets and supplier firms.

However, the more recent reversal of this standard 'brain drain' phenomenon has seen thousands of US-educated engineers returning to Taiwan – reaching a peak of 5,000 per year in the mid-1990s – and playing a pivotal role in transforming Taiwan into a high wage and skill economy. 'These returnees to Taiwan, many of whom had worked for at least a decade in the United States, brought with them not only technical skill but also organizational and managerial know-how, intimate knowledge of leading-edge IT markets, and networks of contacts in the United States technology sector' (Saxenian 2006: 149). Return migrants have taken on leading roles in Taiwan's technology sector: they are highly represented in the management cadre of leading firms, appear to be more entrepreneurial than non-migrants, and are active players in Taiwan's burgeoning venture capital markets. Not all are permanent returnees, however: some have instead become the 'new argonauts' of the global economy (Saxenian 2006), criss-crossing the Pacific on a weekly or monthly basis and facilitating connections between firms, suppliers, clients and investors in the two high-tech clusters. What these migrant flows serve to demonstrate is that global production networks in industries such as the PC sector are more than just formally regulated firm-to-firm connections – they are also constituted by broader personal and knowledge flows.

16.6 Reshaping global production networks?

We now have an appreciation of how to identify, analyse and explain global production networks. But what if various actors – and in particular, we as consumers – want to try to change how a production network is structured and operates? In an increasingly media-saturated and interconnected world, awareness has grown in consuming countries of the social, economic and environmental conditions under which commodities are produced, particularly in the developing

Plate 16.3 A worker at the Acer computer factory in Taiwan.

(Chris Stowers/Panos Pictures)

world. This has seen a rise of so-called *ethical* consumption, whereby interventions surrounding the consumption stage are designed to improve conditions at points 'upstream' in the production network (see also Chapter 19). There are a number of forms such intervention may take. Perhaps the simplest form of consumer campaign is to *boycott* – that is, not purchase – the products of a particular company. We can also think of *corporate campaigns* that seek to target highly visible corporations by mobilizing information regarding violations against workers or the environment, such as were targeted at the sweatshop operations of sportswear giants like Nike and Reebok in the late 1990s. Increasingly, however, the development of various kinds of *benchmarks* or *standards* – against which various end products and their production processes can be measured – are an increasingly important part of production network regulation within the global economy (see Figure 16.8).

The world of standards and codes has very rapidly become a broad and varied one. There are seven dimensions to this complexity:

1. They may be applied to different facets of the production system, for example environmental, social or labour conditions or rates of economic return.
2. They may take a variety of forms: a code of conduct, a label on a finished product, a tightly specified technical standard, a set of voluntary initiatives or a combination of some, or all, of these forms.
3. They may apply to a particular chain (e.g. beef), a sector (e.g. fresh meat), or be generic (e.g. all fresh foods).
4. They may be developed by firms, NGOs, trade unions or international organizations, and usually by a combination of some, or all, of these institutions.
5. The certification or accreditation of the standards –that is, judging whether they have been met – may

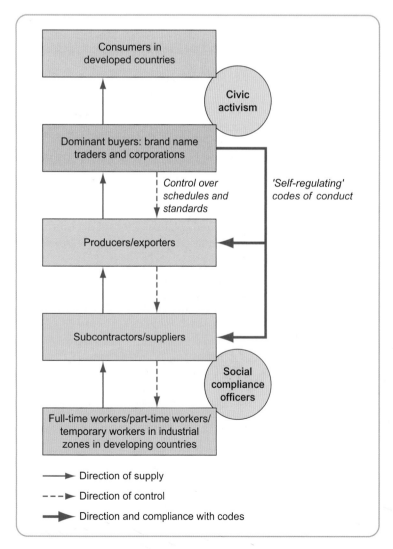

Figure 16.8 New modes of regulating production networks.

Source: adapted from Sum and Ngai (2005), Figure 1

be undertaken by public or private, and profit or not-for-profit organizations.

6. They will range from the voluntary (e.g. seeking 'Fair Trade' status for a product) to the mandatory (e.g. safety standards for plastic toys).

7. They are inherently geographical, in terms of both the territory in which they apply (often the place of consumption) and the places in which the effects are felt (usually places of production).

In terms of our two case studies in this chapter, we will focus exclusively on coffee, as these kinds of initiatives are more prevalent and developed in buyer-driven sectors characterized by labour-intensive production in the developing world (e.g. shoes, garments, food stuffs, etc.). A wide range of schemes seeking to promote sustainable coffee production currently exist (Table 16.4). The seven listed schemes clearly illustrate the degrees of variation just described. For example,

Table 16.4 Certification schemes for sustainable coffee

Name	Year established	Actors or organizations setting the standards	Characteristics	Geographic coverage
Fair Trade	1988 (in Netherlands)	Fair Trade Labelling Organizations International (FLO)	Minimum guaranteed price paid to registered small farmers' organizations that match standards on socio-economic development	Global
Rainforest Alliance certified (shade-grown)	1996	Rainforest Alliance	Certifies farms on the basis of sustainability standards; covers environmental protection, shade, basic labour and living conditions, and community relations	Primarily Latin America
Utz Kapeh	1997	Utz Kapeh Foundation	Includes standards on environmental protection and management, and labour and living conditions	Global
Bird-friendly coffee (shade-grown)	2000	Smithsonian Migratory Bird Center (SMBC)	Minimum standards on vegetation cover and species diversity needed to obtain use of label; also covers soil management	Latin America
Global-GAP Code for Green Coffee	2003 (Eurep-GAP until 2005)	Global-GAP Organisation	Designed to offer assurances about how coffee is grown on the farm with respect to environmental impacts, the use of chemical inputs and worker health and safety	Global
Starbucks CAFE (Coffee and Farmer Equity) practices	2004 (guidelines established in 2001)	Starbucks Corporation	Guidelines designed to ensure high-quality coffee, protect the environment, and promote fair relationships between farmers, workers and communities	Global
Common Code for the Coffee Community (4C)	2005	4C Association (Germany)	To achieve baseline levels of social, ecological and economic sustainability for all types of coffee production systems	Global

Source: adapted from Daviron and Ponte (2005), Table 5.11, and Neilson and Pritchard (2009), Table 6.1

they may apply primarily to environmental or economic conditions (Bird-friendly vs. Fair Trade, respectively); involve a label or be primarily a code of conduct (Fair Trade vs. Utz Kapeh); and be global or regional in their coverage (Global-GAP vs. Bird-friendly). Moreover, schemes may be initiated by single private companies (e.g. Starbuck's CAFE – Coffee and Farmers Equity – Practices), groups of companies (e.g. Global-GAP), civil society organizations (e.g. Fair Trade) or multi-stakeholders organizations (e.g. 4C) encompassing a range of domains. A growing proportion of global coffee falls under these schemes which are rapidly becoming mainstream: while the overall level was estimated at just 1 per cent of total production in 2003, by 2009 Utz Kapeh covered 5 per cent of the total, Rainforest Alliance 1.5 per cent, Bird-friendly 1.2 per cent and Fair Trade 1 per cent. In the same year, Starbuck's CAFE practices covered 81 per cent of their coffee purchases – equivalent to 136,000 tonnes of coffee – with a target of 100 per cent coverage by 2015.

Such schemes have certainly initiated improvements in the social, economic and environmental conditions of the coffee growers that are enrolled in them. Growers of certified Fair Trade coffee, for example, are guaranteed a premium above the market price, and their product may fetch significantly more in certain market conditions. And yet we need to add some significant notes of caution, such as the fact that the majority of the world's coffee farmers are still labouring outside the jurisdiction of these schemes which are voluntary in terms of participation. And we need to ask who pays for, and who benefits from, the certification process? The rise of standards-based schemes has fuelled the emergence of a new category of commodity chain participant, the independent auditor – such as Scientific Certification Systems (SCS), for example, who audit the Starbucks scheme – many of whom themselves are profit-seeking firms. The costs of auditing long and complicated production networks are significant, and firms will vary in their ability and willingness to meet these costs. In many instances, buying firms will expect suppliers to meet the extra production costs. Farmers involved in certified organic coffee production in Oaxaca State, Mexico, for example, may pay between 10 and 30 per cent of their gross receipts to certify their produce. This serves as a significant barrier to entry and only the best organized and most well-funded farmers can turn organic production into a profitable enterprise (Mutersbaugh 2005). Together, these points suggest that we need to look carefully and critically at initiatives to alter the nature of global production networks; there is always the risk that, in certain contexts, they will serve to enhance the very injustices and inequalities they seek to diminish.

16.7 Conclusion

Returning to where we began, we now understand far more about the global economic interconnections and exchanges that enable the student in the café to drink a coffee and surf the Internet on her laptop computer. We could tell similar, yet at the same time profoundly different, stories about the iPhone in her jacket pocket, about the jeans she is wearing, about the machine that brewed the coffee, the furniture in the café and so on. Whether the student in question is aware of the complex 'back stories' of the commodities around her is another matter: as economic geographers, however, it is beholden on us to unravel those stories and their implications for the people and places across the globe that they connect.

This chapter has provided us with a framework and a language for exploring and understanding these economic geographies. Sophisticated global production networks lie behind nearly all of the products and services that we consume on a daily business. These global production networks can be understood as the meshes of intra-, inter- and extra-firm relationships through which material and non-tangible inputs are transformed into consumable outputs of many kinds. They provide the organizational glue that connects together the disparate local clusters of economic activity that constitute the 'on the ground' reality of the global economy. As a result of recent shifts in the global economy associated with increased competition, technological change, and trade and foreign investment deregulation – for which globalization is often used as shorthand – global production networks in general have arguably become more geographically extensive, disintegrated and dominated by developed-country buyers. Ultimately, however, each global production network has its own unique organizational structure, geographical configuration, governance regime and institutional context. In this chapter we have used just two examples, namely the coffee and laptop computer production networks, to explore some of this inherent variability in the configuration of global production networks.

What this chapter has also demonstrated is the futility of seeking to understand the contemporary global economy and its workings without adopting a geographical perspective. Lead firms in global production networks both *use* uneven geographies – taking advantages in differences in labour costs and skill levels, for example – and in turn *reshape* those uneven geographies through their investment and disinvestment decisions. Today's global economy is undoubtedly more complicated and interdependent than ever before: it is the task of the economic geographer to unravel and explain this complexity.

Learning outcomes

After reading this chapter, you should understand:

- How all economic activities can be conceptualized as a form of global production network.

- That global production networks have distinctive, yet changeable, geographical forms.

- That global production networks exhibit a range of governance regimes according to the types of firms involved.

- That the range, complexity and efficiency of global production networks is heavily shaped by multi-scalar institutional contexts of different types.

- How attempts can be made to reconfigure production networks by introducing various kinds of standards and codes of conduct.

- The importance of a geographical perspective for understanding the organization of the contemporary global economy.

Further reading

Coe, N.M., Dicken, P. and Hess, M. (2008) Global production networks: realizing the potential, *Journal of Economic Geography*, 8, 271–95. Provides a useful overview of the current state of, and challenges facing, geographical research into global production networks.

Coe, N.M., Kelly, P.F. and Yeung, H. (2007) *Economic Geography: A Contemporary Introduction*, Blackwell, Oxford. A clear, engaging and student-friendly introduction to contemporary economic geography.

Dicken, P. (2011) *Global Shift: Mapping the Changing Contours of the World Economy*, 6th edition, Sage, London. Now in its sixth edition, this remains far and away the best textbook on the structure, evolution and geography of the global economy.

Massey, D. (1984) *Spatial Divisions of Labour*, Macmillan, London. This landmark book theorizes, and exemplifies, the concept of spatial divisions of labour. This is one of the most definitive accounts of why economic geography matters.

Neilson, J. and Pritchard, B. (2009) *Value Chain Struggles: Institutions and Governance in the Plantation Districts of South India*, Wiley-Blackwell, Oxford. A fascinating account of changes within the South Indian coffee and tea industries. A second edition was published in 1995 with a new concluding chapter. Particularly strong at revealing the institutional dimensions of the production networks.

Yang, C. (2009) Strategic coupling of regional development in global production networks: redistribution of Taiwanese personal computer investment from the Pearl River Delta to the Yangtze River Delta, China, *Regional Studies, 43*, 385–407. A fascinating, detailed study of the development of the Taiwanese laptop industry in and around Shanghai.

Useful websites

www.globalvaluechains.org   This site contains a wealth of conceptual and empirical material on global value chains.

www.unido.org   The United Nations Industrial Development Organization (UNIDO) website offers a wide range of data and reports on different commodity chains and the potential they offer for economic development in different localities.

goodelectronics.org/about   Excellent website about labour conditions in the electronics industry.

www.sourcemap.org   An open source site that allows users to map commodity chains for different products.

www.ico.org/index.asp   The website of the International Coffee Organization provides a range of information on the coffee industry, and in particular, its evolving regulatory structures.

pcic.merage.uci.edu   The Personal Computer Industry Centre at University of California, Irvine, offers a wide range of up-to-date materials on the industry.

www.ids.ac.uk/go/research-teams/globalisation-team Provides access to lots of globalization research at the Institute of Development Studies, University of Sussex.

www.fairtrade.org.uk The UK's Fairtrade Foundation is one of the most well-known attempts to improve the economic returns offered to commodity producers.

www.fairtracing.org Website of a research project seeking to support ethical trade.

For annotated, clickable weblinks and useful tutorials full of practical advice on how to improve your study skills, visit this book's website at **www.pearsoned.co.uk/daniels**

SERVICE ECONOMIES, SPATIAL DIVISIONS OF EXPERTISE AND THE SECOND GLOBAL SHIFT

Chapter 17

John R. Bryson

Topics covered

- Defining services
- Services capitalism
- Service employment versus output
- The body, the personality market and emotional labour
- The division of labour
- Spatial and gender divisions of labour
- Spatial divisions of expertise
- The second global shift

Other chapters in this book explore the environmental implications and inequalities of wealth creation at a variety of spatial scales. This chapter is about the large and complex capitalist economic system and the ongoing shift from manufacturing to services that is being experienced by a range of economies (Plate 17.1). Put another way, it is a shift towards various forms of expertise- or knowledge-intensive employment. The focus of this chapter is on understanding some of the dynamics of expertise-intensive service work but with the important caveat that the complexity of the modern economy makes it problematical to view service activities as a distinct and separate category from manufacturing. It follows that the other chapters in this section that include references to manufacturing, finance, knowledge-intensive services, and even consumption, must be seen as an integrated whole. Isolating finance, manufacturing or food production processes from a range of service functions overlooks the multiple interconnections that form a complex and integrated production system. The flows of money and finance that are transferred around the world in the twinkling of an eye are manipulated and coordinated by service workers, or more specifically knowledge workers who convert information into knowledge that then informs investment decisions of all kinds, but enabled by manufactured technologies. Many manufacturing firms now produce hybrid products that blend service and manufacturing tasks (Bryson and Rusten 2011). This implies that a post-industrial society is not inevitably a post-manufacturing society. This is to emphasize the relationship between production, manufacturing and services; manufacturing and service activities are just part of a much more complex production process. Geographers should explore production rather than separating manufacturing and service activities and treating them in relative isolation.

This chapter begins by defining services and placing them within the context of the overall production system at large. The continued development of this system over time (and space) reflects the process of economic specialization. This is driven by a continuous extension of the division of labour and the development of a new spatial division of expertise. The latter is a key concept that will be examined in more detail in the second part of this chapter, which explores the ways in which service expertise is integrated into complex production systems. The chapter concludes with an introduction to some of the characteristics of services offshoring, global sourcing or the rise of the second global shift (Bryson 2007).

17.1 Defining services

In the economically developed world, the vast majority, often more than 75 per cent, of all jobs involve some form of service work (Bryson *et al.* 2004). Furthermore, in Europe, North America, Japan, Australia, New Zealand as well as parts of the developing world, around 90 per cent of new jobs are created in services. As a result the economies and societies of these countries appear to revolve predominantly around service activities and the experience of service work. It is generally accepted by scholars and policy makers that developed market economies are dominated by various forms of service work, ranging from extremely well-paid lawyers and merchant bankers to less well-paid hotel and

Plate 17.1 An open plan office typical of the kind of space used by many service industries.
Source: © Shutterstock

retail workers. Services contribute to economic growth in a variety of ways; they are traded locally, regionally, nationally and internationally. They are also heavily wrapped within and around the production processes of manufactured goods as well as other services; they add value by smoothing the relationship between production and consumption, for example via market research, product design, development and testing and advertising (Bryson and Rusten 2011) (Spotlight box 17.1). Services can be exported either directly via transfer across borders or direct representation of the provider in another country or indirectly through the incorporation of a service into a product or another service which is then exported.

Spotlight box 17.1

Services and the production process

Producing any product or service requires service expertise to be embedded in different parts of the production process – pre-production, during production and post-production or consumption (Figure 17.1). Pre-production involves understanding the design process, including market research and the ability to innovate. This may involve the design of a production process or of an actual product or service (Bryson and Rusten 2011). The development of services may require the creation or modification of a process, for example the systems that support a financial services transaction or the check-in process at an airport. Services incorporated into production are concerned with the efficient management of the production process, for example queue management at leisure parks or call monitoring in call centres. Post-production involves marketing and related services or supporting services, for example finance packages to encourage

purchase, servicing agreements for software or content (music, text, pictures).

The production process can be divided into five parts with each part requiring different forms of service knowledge and expertise:

1. *Pre-manufacturing* – product development, research & development, design, product testing, market research, finance.
2. *During manufacturing* – finance, quality control, stock control, purchasing, safety, management, continuity/contingency planning, etc.
3. *Selling* – logistics, distribution networks, marketing, finance.
4. *During product and system utilization* – maintenance, leasing, finance, etc.
5. *After product and system utilization* – waste management, recycling, etc.

Source: Bryson *et al.* (2004)

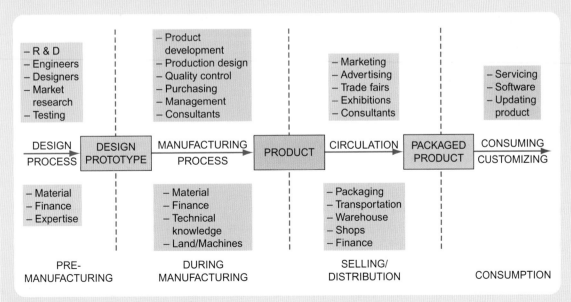

Figure 17.1 The production process.
Source: Bryson *et al.* (2004), p. 52

During the 1970s economists tended to consider services as 'immaterial goods or simply as goods' (Hill 1977: 315). But services are not goods; they exhibit a range of characteristics that fundamentally distinguish them from goods. Hill (1977) identified some of the primary differences between goods and services beginning with the observation that a diverse selection of services, the shipment of goods by a transport firm, vehicle repair, painting and decorating, the cleaning of a house by servants, hairdressing and dentistry, actually share a common set of characteristics. For Hill the two defining characteristics of services were the way in which they brought about:

1. A change or transformation in the condition of a good or person.
2. A change that is the result of the activity of another individual or firm.

These principles were used by Hill to construct what is considered the classic definition of services: 'A service is defined as a change in the condition of a person, or of a good belonging to some economic unit, which is brought about as the result of the activity of some other economic unit, with the prior agreement of the former person or economic unit' (Hill 1977: 318). This definition highlights one of the primary problems with the category of services, namely that in many instances the consumer often only observes or experiences the production or performance of the service. This simultaneous production and consumption, and even co-production of a service, creates confusion between the process of producing it and the final end process of consumption. This debate may appear to be quite complex but fundamental to the distinction between goods and services should be the realization that the delivery of a service does not result in the exchange of ownership of a material product. Thus, a service cannot be thought of as an immaterial good because nothing is exchanged. The output of many services is ephemeral or non-material, for example attendance at a lecture or viewing a live theatrical performance. It is often suggested that software is a material service since it can be stored on a computer, but this overlooks the fact that the software provider does not transfer complete ownership of the software to the user: it is owned under licence from the service provider and does not mirror the way in which a product such as a laptop computer or a mobile phone is physically transferred from the seller to the purchaser.

It is important not to become too distracted by attempts to arrive at a precise definition of services. Nevertheless, the classification of service activities is an important activity for those engaged in measuring the economy, for example national statistical agencies. It is far too easy to become preoccupied with systems of classification at the expense of understanding the changing dynamics of capitalism. In any event, such is the pace of economic change that classifications have to be continually modified to incorporate the development of new types of work or the on-going 'extended division of labour' (Sayer and Walker 1992; Bryson 2009).

All attempts to classify services must accommodate the complexity and diversity of the business activities involved (Illeris 2007). For our purposes a simple classification of service activities incorporates five different types:

1. Consumer services that provide services for final end-users, for example retailers, opticians, hotels and retail banks.
2. Producer and business services that provide intermediate inputs into the activities of private- and public-sector organizations.
3. Public services provided directly by the state or indirectly by the private sector and not-for-profit organizations.
4. Not-for-profit organizations working beyond the confines of the state (Bryson *et al.* 2002).
5. Informal services or unpaid service work, which is usually predominantly undertaken by women, and which is a vital element of peoples' daily lives.

Each of these categories includes a heterogeneous collection of service functions and it is not proposed to cover them all in this chapter. Rather, the focus here is predominantly on knowledge-intensive services (second in the list above). This reflects the increasing importance of a group of distinctive activities that have exhibited dramatic growth rates in both the numbers of firms established and their contribution to employment creation. It incorporates key sectors such as legal services, accountancy, market research, management consultancy, design and technical consultancy. All these services make an important contribution to economic development. They contribute directly to the creation of added value; they contribute to a national economy's balance of payments through exports; and they have also experienced dramatic growth rates both in employment as well as new firm formation.

17.2 The body, services and emotional labour

The classic definition of services developed by Hill (1977) highlights the importance of people-based expertise in the creation of services. This implies that many services must be delivered by people rather than machines. Unlike manufacturing, it is difficult and in many instances impossible to replace service workers with machines and this means that productivity improvements are difficult for those services that rely on people-based expertise. This is an important point in that some service activities can be codified and replaced by relatively simple computer programs. A good example is an application for a mortgage that can be processed by a computer using simple rule-based logic. Each rule can have a simple 'yes' or 'no' answer, for example, related to the value of the property or the income of the applicant(s) (Levy and Murnane 2004: 16–17). Many service jobs cannot be replaced by simple rule-based software as they require complex pattern recognition that may involve reading body language in service encounters.

The importance of face-to-face contact in the simultaneous production and consumption of a range of services plays an important role in differentiating the new world of service work from that of manufacturing (McDowell 2009). At the centre of interactive service relationships are three important elements: client interaction, an individual's reputation, and embodied knowledge. In this service age, the workplace is increasingly conceptualized as a stage upon which employees must execute an aesthetically pleasing performance. Service employment is not simply the exchange of goods or services but is a complex skill in which presentation, communication and display are integral to success (Bryson and Wellington 2003: 60). The implication is that for some service jobs, appearance, the body and accent matter (McDowell 2007) and this implies that service economies have within them cohorts of people who do not possess the desired bodily attributes to meet the expectations of employers and perhaps customers (Bryson and Wellington 2003). McDowell argues that in the United Kingdom this group contains white working-class males who find it difficult to obtain well-paid service work.

The academic literature on service work is heavily influenced by Arlie Hochschild's (1983) important work on the commercialization of emotions in face-to-face service encounters. Hochschild was inspired by C. Wright Mills's (1959 [1951]) classic work, *White Collar*. Mills was the first sociologist to explore the complexity and diversity of service work in his analysis of the new American middle class. Mills developed the concept of the 'personality market' to describe the 'shift from skills with things to skills with people' (1959: 182); central to this shift is the enhanced importance of the psychological dimensions of service work. In this account of the new world of work, men and women 'are to be shaped' (1959: 183) and their personalities managed, by themselves and by others.

Horchschild developed Mills's work by undertaking theoretically grounded research into the commercialization of the body and the feelings of male and female flight attendants and debt collectors. Central to this story is 'emotional labour', a concept which describes the management of employees' feelings during social interaction in the work process (Hochschild 1983: 137). The best example is the emphasis placed on providing a 'service with a smile' to show customers that they are valued. Hochschild reveals that much face-to-face interactive service work (flight attendants, debt collectors, waitresses, secretaries, fast-food operations) involves having to present the 'right', managerially prescribed, emotional appearance or mask to the customer or client, and that this involves real labour. In these occupations workers are faced with the dilemma of how to identify with their work role without it becoming part of their identity.

Service employees have to depersonalize the work by 'surface acting' and 'deep acting'. In surface acting 'we deceive others about what we really feel, but we do not deceive ourselves' (Hochschild 1983: 33); the body not the soul is the main tool of the trade; the smile on the face of the worker is a false smile, but it is still a smile. In emotional labour a smile becomes attached to the feelings that a company wishes to project rather than being attached to its usual function – to show a personal feeling (Hochschild 1983: 127). In deep acting the 'act' is no longer an act but becomes part of the individual's persona. It is about persuading employees to be sincere, 'to go well beyond the smile that's just "painted on"' (Hochschild 1983: 33). Unprecedented efforts are being made by employers to control employees not simply in terms of what people say and do at work, but also how they feel and view themselves. In deep acting the disjunction between displayed emotions and private feeling is severe and potentially psychologically damaging. The danger is that deep acting becomes part of the worker's personality and is used beyond the workplace. If this occurs then the new 'emotional proletariat' (Macdonald and Sirianni 1996) might find it difficult to 'interpret and take appropriate action in response to bodily signals' (Shilling 1993: 119). Hochschild's account of male debt collectors highlights how the display of aggression required by debt collectors can spill over into personal relationships with wives and children. According to Bryson and Wellington (2003: 62), 'active involvement in "emotional labour" renders employees' appearance and personality a form of "adjudicated cultural capital" that can be recruited, managed, manipulated and utilized to buy the hearts and minds of consumers'. The commodification of image, not surprisingly, is highly visible in interactive business service occupations and, in particular, the 'professional managerial classes'.

The importance of the body and image in service economies has led to a new occupation, that of image consultancy or impression management. Image consultants are employed by individuals and firms to alter the surface appearance of the body. Image consultants provide seminars and one-to-one consultations directly related to the restructuring of professional employees' bodies. According to Wellington and Bryson (2001: 940):

> KPMG [the accountancy firm] employs image consultants on a monthly basis primarily to provide employees with a 'confidence boost', and also to provide staff with a 'bonding experience'. The accountants Coopers and Lybrand as well as Ernst and Young hire image consultants to provide seminars in personal presentation for their audit teams. These seminars examine ways of increasing credibility and projecting the right image when undertaking client audits and when pitching for new business. Price Waterhouse employed image consultants to instruct potential partners in dining etiquette and in the art of looking, acting and sounding like a partner of a major global accountancy company. Note the use of the terms act and art and the link to the literature on flight attendants.
>
> (Taylor and Tyler 2000)

The literature on emotional labour suggests that in the service economy employees have to develop skills in dealing directly with people. The implication is that extrovert personalities may have little difficulty in fitting into this new world of service work, but that introverts may experience some challenges. It is important to remember that in many instances facing-based work that is heavily involved in emotional labour will be supported by back office or out-of-client-sight supporting labour. This suggests that a new division of labour is developing that links front and back office workers together.

17.3 Two common misconceptions about service economies

It is far too easy to assume that manufacturing no longer matters in economies that are dominated by service employment. Many service jobs are highly visible within the economy. A visit to a local shopping centre is saturated with service experiences. In contrast, manufacturing employment remains largely invisible as the production process is isolated from the consumption of the product (Bryson *et al.* 2008). This separation of process from consumption is not evident in many service-based economic relationships. The dominance of service employment and the apparent recent shift from manufacturing has led to two common misconceptions about service employment and it is to these that we now turn our attention.

17.3.1 Services as old as the Industrial Revolution

It is a common mistake to assume that the transformation of economies towards services is a phenomenon of the twentieth century. The social sciences have paid too little attention to the role services played during the Industrial Revolution or earlier. A good example is the development of London in the nineteenth century as the command and control centre of the British Empire. A detailed analysis of London's fire office registers between the years 1775 and 1825 concluded by noting that the

> service industries made no less contribution to the British economy during the Industrial Revolution than manufacturing, and that nowhere was this more true than in London. Its service economy was on a very large scale, serving the nation as a whole as well as the capital . . . London's service industries underpinned both its own and the national manufacturing and commercial infrastructure and at the same time contributed to the new 'commercialisation of leisure'.
>
> (Barnett 1998: 183)

During the late eighteenth century London was already being transformed into an important world city; a process that has continued to the present and which has further enhanced its status as a global city (see Chapter 14). Deane and Cole (1962: 166, 175) calculated that in 1851 some 45.3 per cent of the United Kingdom's national income was derived from service activities (trade, transport, housing, the professions and the civil service). The structure of employment in the

Plate 17.2 Clerks at work in the offices of a London bus company.

(Topical Press Agency/Getty Images)

United Kingdom changed dramatically during the nineteenth century as a result of technological innovation and the increasing maturity and extension of the capitalist system. Growth occurred in occupations that facilitated the exchange of goods and services between producers and consumers (Plate 17.2). Between 1881 and 1901 the number of business clerks increased from 175,000 to 308,000; bank officials from 16,000 to 30,000; and insurance officials and clerks from 15,000 to 55,000 (Marsh 1977: 124). During the nineteenth century the expansion of international trade was restricted by financial problems until it was enabled and supported by the introduction of bill markets and banking facilities of the kind associated with the flow of tea and silk from China to Europe between 1860 and 1890 (Hyde 1973). As the United Kingdom and the United States were becoming industrialized societies they were simultaneously being transformed into service economies; the growth of manufacturing employment went hand in hand with the growth of service employment.

17.3.2 Productivity and services and the myth of service economies

Since the nineteenth century the employment structure of many countries has steadily shifted away from manufacturing to service employment. This shift should not be equated with the demise of manufacturing or the complete displacement of manufacturing with service work. Many national economies are still dominated by the development, design, manufacture and sale of goods; the absence of manufacturing within a national economy would present serious problems since everyday life as well as service work is supported by a complex array of manufactured products – clothes, toasters, cars, clocks, toothpaste, tissues, deodorant, laptops and toys (Bryson and Rusten 2011). Some of these products need to be manufactured or customized locally, for example products that are difficult or impossible to transport (perishable foods, commercial air-conditioning systems) or production processes that require close contact between producers and consumers (Bryson *et al.* 2008).

It is possible to argue that manufacturing no longer needs to be undertaken in service-dominated economies as products can be traded for services. Nevertheless, many services support and are supported by manufacturing activities, for example the complete removal of manufacturing from the West Midlands (United Kingdom) would reduce the client base of local service companies by 25 per cent (Daniels and Bryson 2005: 3). It is also becoming apparent that a national economy that is over-reliant on products that are manufactured elsewhere is increasingly exposed to many different types of risk. First, since the 2008 recession manufacturing firms in Germany and the UK have proved to be more resilient than many service firms with increases in employment, output and exports (Markit 2011). Over-reliance on products manufactured abroad also has implications for health and safety. In August 2007, for example, Mattel, the world's largest toy company, issued a product recall for 436,000 toys made in China that were painted with lead-based paint and 18.2 million toys that were designed with small magnets that could become detached. This type of product recall provides American toy manufacturers with a competitive advantage as they are able to market their toys as made in America (Rusten *et al.* 2007) and as '100 per cent kid-safe' (Martin 2007).

After the Second World War manufacturing employment encountered gales of creative destruction (Schumpeter 1942) that eventually culminated in, what some commentators have termed, the crisis of Fordism (Gaffikin and Nickson 1984). This led to an ongoing reduction in manufacturing employment and a shift towards a diverse collection of service jobs. The difficulty is that scholars and policy makers tend to equate the decline of (often) high-profile manufacturing firms with a contraction in manufacturing overall. In practice the shift towards service employment does not mean that countries in Europe and North America, for example, have been transformed into service economies. It is essential that a distinction is made between adjustments to the employment composition of an economy and changes in economic output (productivity). These two indicators of economic change are not necessarily related, or they are related in unexpected ways.

The types of productivity improvements that have been achieved by manufacturers through automation and process improvements are not necessarily achievable by service providers. The category 'services' includes an extremely diverse group of business activities: from very highly paid professional occupations (lawyers, surgeons, bankers) to very poorly paid occupations (janitors, waiters). The common denominator is that, unlike manufacturing, many of these occupations rely on people-based skills or what is commonly termed 'embodied labour' (Bryson *et al.* 2004). In these circumstances the challenge of improving productivity is considerable. Improvements to productivity in services is extremely high and increasing where there is interaction between work practices and new technologies (such as the introduction of people-facilitated, as well as automated, call centres) but either non-existent or extremely low in, for example medical services, teaching and cultural or creative industries. It is very difficult for a dance company to achieve productivity improvements; it involves putting on more performances, using fewer dancers, or increasing the speed of delivery to enable two performances to be given during the time previously devoted to a single performance. This is illustrative of the fact that many service occupations are labour-intensive and often involve face-to-face interactions between service providers and clients for which productivity improvements are difficult to achieve. As a result, overall productivity improvements in the service side of the economy have, as a general rule, lagged behind manufacturing. The implication is that this productivity differential partly explains the shift from manufacturing to service work.

The significance of this reasoning can be demonstrated with reference to UK manufacturing exports which amounted to £227 billion or 58.8 per cent of total exports in 2009. Yet manufacturing directly employed only 2.5 million workers, with a further 3 million jobs dependent on the sector. A major problem is that official definitions of manufacturing employment do not include jobs and activities which depend on, or are closely allied to manufacturing, for example design or marketing activities undertaken by specialist 'service' providers. Although the gross added value (GVA) created by manufacturing increased by 26 per cent over the period 1992 and 2003 (Figure 17.2), it actually increased at much faster rates in other sectors of the economy. Other sectors were therefore growing at a much faster rate than manufacturing even though in absolute, rather than relative, terms manufacturing's contribution to the overall economy increased over this time period. That said, it is important to recognize that a restructuring of economic activity is occurring; new types of economic activity such as computer services or information management are flourishing alongside the development of innovations in well-established services such as insurance, banking and retailing. Such restructuring arises from, first, the externalization of previously in-house services to independent providers and, second, the continuation of an extended division of labour. The latter arises from the growing complexity that accompanies many service innovations as service providers seek to refine product differentiation in the marketplace.

While most of the key manufacturing sectors have therefore declined in relative terms, and some in absolute terms (for example, footwear, knitted goods, leather goods, man-made fibres, wearing apparel), the service sector has expanded, whether measured by employment or GVA. Between 1992 and 2004, GVA increased in the United Kingdom by 381.5 per cent in computer services, 252.2 per cent in other business services and 266.2 per cent in market research and management

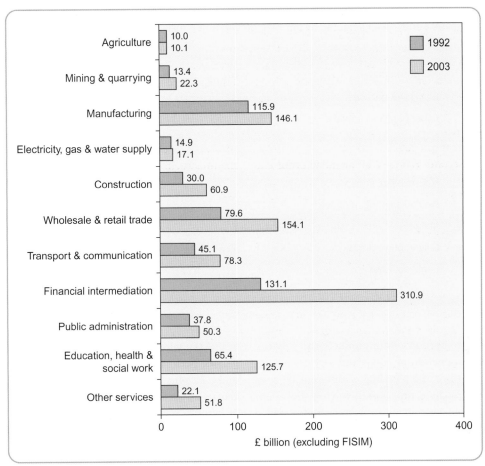

Figure 17.2 Comparison of gross value added in the UK at current basic prices by industry between 1992 and 2003.

Source: from Mahajan, S. (2005) *Input-Output Analysis: 2005*, p. 23, Office for National Statistics, London. Office for National Statistics licensed under the Open Government Licence v.1.0

consultancy (Figure 17.2 and Table 17.1). In 2004, computer services accounted for 2.9 per cent of total GVA (£30.0 billion). It is clear then that recent growth in the UK economy has been led by service industries which account for all of the top ten fastest growing industries between 1992 and 2004. An inexorable decline in manufacturing employment has therefore been offset by employment gains in services.

The economy is integrated in complex ways and it is important to begin to try to identify and explore some of the interactions between manufacturing and service activities. The question is 'How does this perspective affect our interpretation of employment gains in services at the expense of manufacturing?' A manufacturing company can be dominated by service workers employed in activities such as marketing, advertising, management, accounting and purchasing. Take the task of classifying and counting service activities that occur inside manufacturing companies. On the one hand, we can suggest that part of the shift towards

service employment reflects a transfer of service tasks that were previously undertaken within manufacturing firms to external service providers. On the other hand, we can suggest that 'although the objective of manufacturing industry is the production of commodities many people employed in manufacturing are not directly employed in the actual production process' (Crum and Gudgin 1977: 3). Questions follow about the wisdom of dividing the economy into service and manufacturing activities (Daniels and Bryson 2002; Bryson and Rusten 2011). In 1971 every two production workers needed to be supported by rather more than one non-production worker (Crum and Gudgin 1977: 5); the term non-production is used to identify people within a manufacturing firm who are not directly involved in the actual production process (managers, designers, sales team, research scientists). Since Crum and Gudgin's research, there have been two major organizational changes to production processes. First, knowledge has become more important as well as being increasingly

Table 17.1 The top ten fastest growing industry groups in the United Kingdom in terms of Gross Value Added (GVA) between 1992 and 2004 (weighted by per cent growth and £ million)

I-O no.	I-O group name	Change 1992–2004	
		%	£m
107	Computer services	381.5	24,265
114	Other business services	252.2	27,457
100	Banking and finance	139.1	35,506
103	Owning and dealing in real estate	196.3	15,185
121	Recreational services	169.0	19,153
92	Hotels, catering, pubs etc.	145.0	19,572
111	Market research, management consultancy	266.2	9,215
104	Letting of dwellings	119.8	45,266
109	Legal activities	174.4	9,488
117	Health & veterinary services	120.1	31,294

Source: Mahajan, S. (2006) *Input-Output Analysis: 2006*, p. 37, Office for National Statistics, London

combined or embedded within products and services. Second, the production of goods and services has increasingly required the blending of different types of expertise that include processes involving shop-floor workers (manufacturing as well as services) and a complex array of service and management expertise. All these different types of expertise can be organized within a manufacturing company or, alternatively, by combining the activities provided by many separate and independently owned service firms.

17.4 Services and the spatial division of expertise

The division of labour is a key concept for understanding the organization of work under capitalism and the shift towards service activities. In 1776 Adam Smith published *The Wealth of Nations*, in which he established the foundations of economic theory. For Smith,

> the greatest improvement in the productive powers of labour, and the greatest part of the skill, dexterity, and judgement with which it is anywhere directed, or applied, seem to have been the effects of the division of labour.
> (Smith 1977: 109)

To illustrate the importance of the division of labour, Smith explores pin-making, suggesting that an individual without knowledge of the production process would be unable to make even one pin in a day. The production process of pin-making can be divided into 18 distinct operations: from the drawing

out of the wire to the making of the pin head. If one person performs all of these tasks, then they might make 20 pins in a day. If, however, individuals specialize in particular tasks, by introducing a division of labour, Smith shows how ten people could make 48,000 pins in a day (Bryson and Henry 2005). Adam Smith's division of labour is based on his observations of specialization within factories. There is no reason, however, why its principles should not be applied to the services production process, not least to the ways in which new service activities are identified and firms developed to create a market for the provision of new forms of services. It is worth noting that the division of labour was extremely well known prior to Adam Smith's work. In 1678 Sir William Petty published an essay on political arithmetic in which he noted that one of the greatest English clockmakers of the seventeenth century, Thomas Tompion (1639–1713), produced watches via a complex division of labour in which 'one man [sic] shall make the *Wheels*, another the *Spring*, another shall engrave the Dial-place, and another shall make the *Cases*, then the *Watch* will be better and cheaper, then if the whole work be put upon one man' (Petty 2004, [1678]: 16).

Adam Smith's theory makes no explicit reference to geography. The obvious extension of the division of labour is to incorporate geography into the process, and this leads to a spatial division of labour (Massey 1984). Once a production process had been subdivided into its component parts it is a comparatively simple step to construct warehouses or offices dedicated to a particular part of the production process, such as consignment assembly and packing for despatch or a marketing function. The spatial division of labour began with manufacturers establishing specialist units that were contiguously located, but soon developed into a more dispersed spatial division of labour. Different parts of the production process were located close to sources of raw materials, the market or cheap or skilled labour. Very quickly, some regions and countries came to specialize in particular types of economic activity. For example, an (international) division of labour soon developed between the industrial countries (core) producing manufactured goods and the non-industrialized countries (semi-/periphery) supplying raw materials and agricultural goods as well as a market for manufactured products. This spatial division of labour can occur within service or manufacturing firms (Bryson *et al.* 2004) (Case study 17.1). One location (a major city such as London) is used for the headquarters (HQ), another for research and development (R&D), yet another (a peripheral region, South Wales, or a country, China) for the manufacturing branch plant, and sales and service centres will be distributed over the globe (Bryson and Rusten 2008).

Thematic case study 17.1

ICT and the establishment of data-processing factories in the Caribbean and call centres in India

Developments in ICT have made the transfer of service jobs to low-cost locations possible (Bryson *et al.* 2004; Bryson 2007). What is occurring is an intriguing ongoing international division of labour, but with a difference. The difference is that branch plants in developing or less developed countries used only to be associated with the assembly of products designed by and for the developed world: now, many such branch plants process data and/or interact with customers located thousands of miles away.

Data-processing factories, for example, in the Caribbean are linked via cable and satellite with service workers in Ireland, the Dominican Republic, Jamaica, Mauritius and the United States. To Freeman this development 'signals an intensification of transnational production and consumption – of labour, capital, goods, services and styles' (2000: 1). The Barbados Data Processing Centre, for example, employs 100 women who enter data from 300,000 ticket stubs for one airline's 2,000 daily flights, while on another floor women enter data from medical claims sent for processing by one of America's largest insurance companies (Freeman 2000: 1). This company employs nearly one thousand people, mostly women, who work eight-hour shifts and who are constantly monitored for productivity and accuracy.

Call-centre operations are labour-intensive, labour accounting for over 65 per cent of their running costs. In the 1990s, in the United Kingdom, American Express and British Airways (BA) transferred their customer services divisions from the United Kingdom to Delhi and

Plate 17.3 A call centre where business is outsourced from Western companies in New Delhi in India.
(Fredrik Renander/Alamy)

then Bombay. BA was attracted by India's large pool of English-speaking graduates who could be employed on starting salaries of between £1,500 and £2,500. India currently dominates the market for the provision of English-speaking call centres. Every year India produces 2 million graduates, mostly taught through English, and who are desperate for well-paid employment in a country where the national average wage is £300. In the United Kingdom, call centres are staffed by students and temporary workers and are considered to be twenty-first century sweatshops with high staff turnover rates and low salaries (£12,000–15,000). In India, they are desirable places, regarded as 'hip and funky places to work, somewhere to hang out with like-minded, outward-looking young people' (Spillius 2003: 44). In the 1990s the call-centre industry did not exist in India, but by 2003, 1,500 Indian call-centre providers employed 102,000 young people in the 'remote services' industry; the industry is growing by 70 per cent a year (Spillius 2003: 41). Over the next five to ten years, over 200,000 British call-centre jobs will be exported to South Africa, Malaysia, the Philippines and China, but India is currently becoming the world's back office.

Offshoring is a complex process that involves an on-going reorganization of the relationship between service functions, place and space. The first decade of this century was associated with the rise of service offshoring. The current decade has seen 'onshoring' or the return of previously offshored functions to relatively high cost locations. In July 2011 Santander, UK, announced that it had returned all its call centres in India to the UK because of customer complaints. The banks noted that customers were frustrated in dealing with offshore call centres and that such frustration rapidly turned into dissatisfaction. Quality issues have played an important role in the new process of 'return onshoring'. India has also experienced higher wage inflation whilst high unemployment rates in the UK have increased staff retention and reduced the cost differential related to training and wages.

Within geography, the division of labour concept is traditionally associated with deskilling, branch plants of manufacturing companies, and the ongoing development of an international division of labour, mostly associated with textiles and clothing, automotive, electronics and to a much lesser extent services (Dicken 2003). The services literature increasingly emphasizes the importance of emotional labour and embodied expertise in service relationships and production (Hochschild 1983; Fineman 2000; Warhurst *et al.* 2000; McDowell 2007, 2009). It follows that the concept of a spatial division of labour needs to accommodate the increasing centrality of embodied expertise in the economy (Wellington and Bryson 2001; Bryson and Wellington 2003). There has also been a shift in the nature of particular forms of work. Some forms of high-paid work, for example, are moving away from delivery by labour that is controlled by capital and regulated by company law to a situation in which control, or more correctly power, in the 'employment' relationship is transferred to the employed expert (from the employer). Such experts are the most important asset 'owned' or managed by expertise-intensive firms; these walking, highly mobile resources leave their firms' offices each evening and may, or may not, return next morning. This contrasts with manufacturing where owners of capital exercise most of the power in the relationship with their employees. The balance has shifted with the development of expertise-intensive occupations whereby power transfers from employers to employees. Indicative of this 'new order' are the difficulties experienced by firms trying to manage business service professionals and the role played by bonus payments and golden handcuffs for staff retention in,

for example, investment banking (Leicht and Fennell 1997; Alvesson 2000).

The *spatial division of labour* has continued to evolve and is centered on the expertise that is located within or outside client companies. This is not to suggest that the concept of a spatial division of 'production' labour is still not important, but it is now conceptually useful to distinguish between corporate strategies that affect production workers and those that involve expertise or forms of creative work. A spatial division of expertise distinguishes between the *spatial organization of expertise* (non-production activities, advanced business and professional services (BPS) and management functions) and the *spatial organization of production* (Bryson and Rusten 2005, 2006, 2008). It reflects the distinction between production and non-production activities within manufacturing (Crum and Gudgin 1977). The concept of a spatial division of expertise shifts the focus of analysis to the places where expertise is produced and consumed; it also exists within and between international business service firms and manufacturing firms. The building blocks of this 'expertise' economic system are distributed in a complex mosaic that reflects the ways in which a range of firms (from micro- to large transnational) try to maximize profitability. The spatial division of expertise is informed by the social division of labour, which gives considerable emphasis to relationships between people and to the joint supply and co-production of service knowledge/expertise.

The development of a spatial division of expertise has important implications for the geographies of service as well as manufacturing companies (Case study 17.2). Companies, and even governments,

Thematic case study 17.2

Boeing's new spatial division of expertise

It is becoming increasingly difficult to distinguish between manufacturing and service companies (Daniels and Bryson 2002). Many physical products contain embedded services, for example software, or are supported by service agreements. The increasing symbiosis that is developing between manufacturing and services is changing the geographies of production in complex ways, creating new business models that capitalize on service/manufacturing expertise and is also repositioning expertise as a key source of competitive advantage. There are many examples, but perhaps one of the most dramatic is found in the aerospace industry. Boeing, the American manufacturer, is currently building a new aircraft, the 787 Dreamliner, that will carry between 200 and 300 passengers with a range of 8,500 nautical miles (see Bryson and Rusten 2006, 2008, 2011). For Boeing the Dreamliner represents a new way of designing and manufacturing aeroplanes. An aeroplane is a complex product that is manufactured using an extended, complex supply chain. Before the Dreamliner programme, Boeing followed a build-to-print model. Companies in Boeing's supply chain were provided with detailed part specifications and they only became involved towards the end of the design and development process. In this production model, Boeing was responsible for the detailed design of the complete plane. For the Dreamliner, Boeing is responsible for the overall design of the

plane, for systems integration and final assembly, but detailed design work has been transferred to a network of global partners. Thus, Boeing has created a new spatial division of expertise that is constructed around capitalizing on its own internal expertise combined with that of its partners that are distributed around the globe. This expertise is a combination of research and development with industrial design inputs that have been developed in centres of excellence that support high-tech engineering activities.

The new model represents a radical alteration in the way Boeing designs and manufactures aeroplanes; most of the manufacturing will be undertaken at factories owned and managed by Boeing's partners and the only major part of the Dreamliner's airframe being manufactured by Boeing is the vertical tail. Boeing's major partners include three major Japanese manufacturing companies (Kawasaki Heavy Industries, Mitsubishi Heavy Industries and Fuji Heavy Industries) and Alenia Aeronautica, an Italian company. For the first time Boeing has outsourced the most important part of an aircraft – the design and production of the wings. The three Japanese companies are responsible for the wings. Initially, the Japanese wing design team worked closely with Boeing's design team and was based at Boeing's Everett facility in Washington State (USA). The Japanese design team returned to Japan to concentrate on the development of the detailed designs for the wing while the Boeing design team shifted its focus to concentrate on systems integration. The Dreamliner combines Boeing's 'service' expertise with that of its partners. Once assembled in the Kawasaki plant the completed wing box is transported by barge to Nagoya's new Centrair airport and from there flown to Charleston, South Carolina, in modified 747 cargo freighters (Figure 17.3). Charleston is the main fuselage hub in this spatial division of expertise; the fuselage sections manufactured in Japan, Italy and the USA will be assembled before being flown to Everett for a final assembly process that should take only three days. This is both a conventional spatial division of 'production' labour, with finished parts travelling from one supplier to another down the chain, and a spatial division of 'embodied' expertise. This is an expertise-rich production system in which conventional shop-floor manufacturing workers play a relatively minor role. A total of 3,600 engineers are directly employed by Boeing in the design of the Dreamliner, with the

Plate 17.4 The first Boeing 787 Dreamliner takes shape in the assembly plant in Everett, WA, USA.

(AP Photo/John Froschauer)

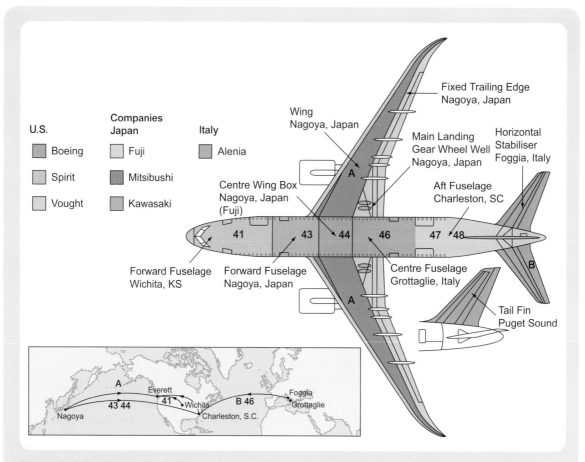

Figure 17.3 Spatial divisions of labour and expertise in the design and fabrication of Boeing's 787 Dreamliner aircraft.

Source: Bryson and Rusten (2008)

majority of these positions based at the company's Everett plant (Washington, USA) (Gates 2005). Another 670 engineers are employed by Spirit Aerosystems, Wichita (Kansas, USA) and 570 in Japan.

Unfortunately, for Boeing, this new production process based on global outsourcing has proved to be flawed. Partners in the Dreamliner's supply chain have experienced major capacity problems and there have been technical and quality difficulties. The Dreamliner made its maiden commercial flight in October 2011, three years behind schedule. Boeing has employed additional staff at Everett to overcome many of the difficulties the company has faced as it has tried to complete this new aircraft.

increasingly develop services (as well as products) using production processes designed to exploit differential comparative advantages. This is not an entirely new process; earlier international divisions of labour were based on core countries producing and exporting manufactured goods whilst peripheral countries exported raw materials. This was the first manufacturing division of labour, or the first global shift (Dicken 2003). The evolution into a spatial division of expertise came later. The comparative advantage for services is derived from a different form of raw material, that is, a highly educated and expert labour force and, importantly, language skills. It is to the geographies of the new spatial division of expertise that we now turn our attention.

17.5 The second global shift

Traditionally, services were considered to be produced and consumed locally; they were regarded as having limited export potential. As we noted earlier in this chapter this reflects the way in which services were conceptualized in traditional economic theory. Today, services are exported and companies are increasingly adopting business models that involve combining service expertise, activities and functions located in different parts of the world or, in other words, by developing competitive advantage through the development of an evolving spatial division of expertise. In the popular media or in political accounts this trend is known as 'service offshoring', a term developed in the Anglo-American context which should be used with considerable care. Services can indeed be 'offshored' but this is a less than helpful and even misleading term when applied to the transfer of work from, say, France to Eastern Europe. The concept of service outsourcing emerged during the 1990s, but only became of major concern during this century. Between January and May 2004, there were 2,634 reports in American newspapers that focused on service outsourcing (Amiti and Wei 2004: 4). Journalists tended to focus on the fear of job losses as service outsourcing had the potential to strip jobs out of the American labour market and relocate them to lower-cost production locations. Service offshoring is based on the concept of outsourcing, a term that describes the process by which a firm procures a material or service input from another organization. Outsourcing can occur at many different spatial scales – from the local to the international.

Service offshoring was made possible by developments in ICT. Initially, pioneer companies experimented with offshoring during the late 1980s and early 1990s. Companies began to offshore service functions on the basis of opportunistic searches to reduce costs and also in response to sequential learning (Lewin and Peeters 2006). In the early stages, companies offshored relatively simple business processes related to finance, accounting and information technologies. Experience of offshoring these functions encouraged firms to offshore more technically complex processes; offshoring continues to evolve as firms develop experience of offshoring more complex and higher-value added tasks. Offshoring was originally developed by business units within firms interested in cutting costs, but over the last decade was transformed into a standard business process. Offshoring became another management tool or approach that was sold by management consultancy firms to their clients and also was heavily marketed by providers of offshore services.

Service offshoring occurs when firms shift production to foreign locations. The objective may be to reduce costs, to service a foreign market, to reduce exposure to country risk, or to access skilled labour. An additional factor influencing the location of offshore service centres is the requirement to provide a 24-hour service to customers or an extended service beyond standard working hours. The cost of providing such services can be high as late shift workers must be attracted by higher wages or extended holidays. Advanced call-routing and networking technologies enables companies to get around this by implementing a 'follow-the-sun' geographical policy. Companies can link two or more call centres together with each open from between 8 to 12 hours per day. Country risk is removed when a company is able to shift the provision of a function between facilities located in different countries.

Service offshoring is not easily analysed because service tasks can be traded in four ways (United Nations 2002: 1):

Mode 1: *cross-border supply* occurs when suppliers of services in one country supply services to consumers in another country without either supplier or consumer moving into the territory of the other.

Mode 2: *consumption abroad* refers to the process by which a consumer resident in one country moves to another country to obtain a service.

Mode 3: *commercial presence* occurs when enterprises in an economy supply services internationally through the activities of foreign affiliates.

Mode 4: *presence of natural persons* describes the process by which an individual moves to the consumer's country to provide a service, whether on his or her own behalf or on behalf of his or her employer.

Three of these modes are concerned primarily with service transactions between residents and non-residents. Mode 1 involves the provision of services that require no direct contact with customers but procedures must be developed to overcome cultural barriers that exist between countries. Recently, there has been a particular interest in Mode 3, whereby enterprises supply services internationally through the activities of foreign affiliates (Bryson *et al.* 2004). For services, the Mode 3 'method of serving foreign markets is particularly important because it is often the only method that permits the close and continuing contact between service providers and their customers necessary to compete effectively with indigenous firms' (United Nations 2002: 54). In this instance the provision of services through foreign direct investment represents a type of *captive offshoring* or offshoring without outsourcing. Captive offshoring enables a firm to retain control over its assets, intellectual property and core business processes (Infosys 2006). Trade in services must address cultural differences between countries that restrict the ability of service providers to export standardized services. Modes 3 and 4 enable service providers to localize provision to take into consideration local cultures and client expectations. Modes 1, 3 and 4 involve what is commonly termed 'service offshoring' or more correctly 'service global sourcing'. This is encapsulated by the concept of a '**second global shift**' (Bryson 2007). The first global shift involved the relocation of manufacturing employment to low-cost production locations while the second implicates services in this process. There have been three distinct phases to the second global shift. First, during the early 1990s IT programming, testing and network support activities were outsourced and then globally sourced. Second, during the late 1990s global sourcing diversified into the provision of back-office and call-centre functions and also the development of computer applications. Third, during the early years of this century full service centres emerged that provide a wide range of administration, process, contact and support functions.

The second global shift highlights the potential for transferring service work from developed economies to low-cost locations, and it has provoked a major policy debate, or even panic, in the USA and Europe (Parker 2004; Blinder 2006). For example, Forrester (a consultancy company) has estimated that 1.1 million Western European jobs will move offshore during 2005–15 and that two-thirds of these jobs will originate from the United Kingdom (Parker 2004). Perhaps the best estimate comes from an analysis of occupational data for several OECD countries which suggests that around 20 per cent of total service employment has the potential to be geographically footloose as a result of rapid developments in ICT (van Welsum and Reif 2005: 6). Potentially, the second global shift will create unemployment in developed market economies. One observer has even gone as far as to argue that

> [The] number of service-sector jobs that will be vulnerable to competition from abroad will likely exceed the total number of manufacturing jobs. Thus coping with foreign competition, currently a concern for only a minority of workers in rich countries will become a major concern for many more. There is currently not even a vocabulary, much less any systematic data.
>
> (Blinder 2006: 114)

Perhaps the next round of creative destruction under capitalism will involve the destruction and relocation of a whole series of service jobs that, according to convention, were considered to be protected from foreign competition. Ultimately, the only types of service jobs that will be safe are those for which face-to-face interaction is absolutely essential (Levy and Murnane 2004).

The development of service offshoring represents a new type of international division of labour, but with a difference. It is various forms of service activity, ranging from call-centre-based work to back-office administration that is being relocated to low-cost locations rather than manufacturing or assembly activities. This new form of trade involves low-value call-centre-type activities as well as high-value services such as legal work, accountancy, design, business analysis and equity research (Table 17.2). Data scanned in Britain or America is transmitted to a low-cost offshore location along high-speed fibre-optic cables and undersea telephone lines to be processed in back offices or used in call centres.

A number of factors influence the decision to send a particular service activity offshore. First, it must be capable of some degree of standardization that does not require face-to-face interaction with consumers or clients. Second, the inputs and outputs required to deliver the service must be capable of being traded or transmitted with the assistance of ICT (OECD 2005b: 12). Third, some service activities are not fixed in space and can be provided either as a form of foreign trade or by the temporary relocation of a service worker to a client's premises, for example management consultancy or various forms of auditing. Fourth, specialist services can be provided from central locations with consumers travelling to avail themselves of the service. In many cases such services would be provided within the confines of a nation-state, but some of these services are being consumed by a form of service-based travel, for example education (secondary and tertiary), plastic surgery and a whole range of other surgical procedures.

During the 1980s India began to develop capability and capacity in the provision of offshore services. This capacity was based on three events. First, in the late 1980s Texas Instruments undertook some chip design in India. Some of their key chip designers in America were Indian and Texas let them return home to design chips for Texas using early forms of ICT. In 1989, Jack Welch, the then Chairman of the American conglomerate General Electric (GE), visited India and decided that it offered a pool of talented individuals who could be of benefit to GE. A team from GE was sent to India and GE set up a joint development project with Wipro.

Table 17.2 Jobs that have been sent offshore from the United Kingdom

Customer-facing – constant contact between producer and consumer	Back office – business process outsourcing
Call centres	Business services
Customer services	Accountancy
Out of hours claims (call centre)	Legal services
Back office administration	Finance
Back office processing	Human relations
E-commerce	IT provision
Credit control	Business systems
Internet services	Graphics and architectural services
Internet claims	Document production
IT helpdesk	Underwriting
Rail timetable enquiries	Technical lists
	Equity research
	Business analysts
	Legal secretaries
	Software development
	City analysts
	Transcription services (voice and shorthand)

Source: Bryson (2007)

Second, during the late 1990s the Y2 computer crisis, or the millennium bug, provided India with clients interested in computer remediation. When computers were originally designed, to save memory their clocks used six digits for dates – two for the day, two for the month and two for the year. The danger was that computers would not recognize the year 2000, but would think that it was 1900. Y2 upgrading was tedious work and many American and European companies employed Indian companies to deal with this problem. The Y2 problem allowed the Indian computer industry to enter the global marketplace for the provision of computer services. Third, the Y2 work was replaced with the e-commerce or dot-com boom. The dot-com boom led to investment in undersea fiber-optic cable, but the dot-com bust that followed meant that the cost of using this cable was virtually free. This new cable vastly increased the number of American companies wanting to outsource service activities to India (Friedman 2005: 106–110).

Outsourcing services to companies located in other countries comes with a number of risks attached to

language, culture and the quality of the provided service. Unlike the first 'global shift', the geography of the second global shift is determined by the educational and language abilities of service workers located in foreign locations that may also perhaps, but not always, be lower-cost locations (Bryson 2007). For the English-speaking world this means that potential suppliers must be able to provide English-speaking employees in other countries, for example France, Norway or Sweden require a pool of staff with fluent French, Norwegian or Swedish. Language and culture plays a much more important part in this global shift than they did during the development of an international division of manufacturing labour. This means that countries with relatively localized languages may be protected from the global sourcing of services whilst countries with more widely spoken 'global' languages (such as English, Spanish or French) will almost certainly participate in the second global shift. This means that the geography of the second global shift is also different from the first; it is more constrained by language as well as cultural nearness, that is, the ability of foreign service suppliers to relate to customers located in other countries (Figure 17.4).

The implication is that countries which developed extensive empires during the nineteenth century may have inadvertently laid the foundations for the emergence of foreign competitors who are able to exploit the benefits associated with their acquired or imposed non-native language. This is a form of 'post-colonial twist' in that the various trading and political empires encouraged the use of a common language and, in many cases, common legal and educational systems (Bryson 2007). It is therefore not surprising that countries such as the United Kingdom and the United States that are part of a globalized language grouping may be most at risk from companies choosing to deliver services from lower-cost locations either by establishing their own operations via a process of foreign direct investment or by subcontracting the activities to a third party and often foreign service provider. Whether this is a 'real' threat will depend on whether the availability of lower-cost services will actually enhance the competitiveness of other parts of the developed market economies, leading to future innovation and wealth-creation opportunities. It is also

important to remember that the second global shift does not just involve the transfer of jobs from high- to low-cost economies but it is also a two-way process in which service providers based in low-cost locations must also establish branch offices in developed market economies. Service offshoring is a complex process involving the development of firms that have the ability to provide service functions from a number of different locations. This type of transnational service firm can develop out of the activities of firms located, for example, in the United States or United Kingdom (developed market economies) or in India or China (developing or transition economies).

The evolving geography of service offshoring is complex. It is too easy to make generalizations based on media reports. It could be assumed that India is the most important provider of outsourced services. This is a mistake. An analysis of the trade date for 2002 identified that the United States was the most important outsourcer of business services (see Table 17.3). Germany and Japan were in second and third place (Amiti and Wei 2004; Bartells and Lederer 2009). It is worth noting that India and China are often considered to be major recipients of outsourcing, but they are also significant outsourcers of business services (with

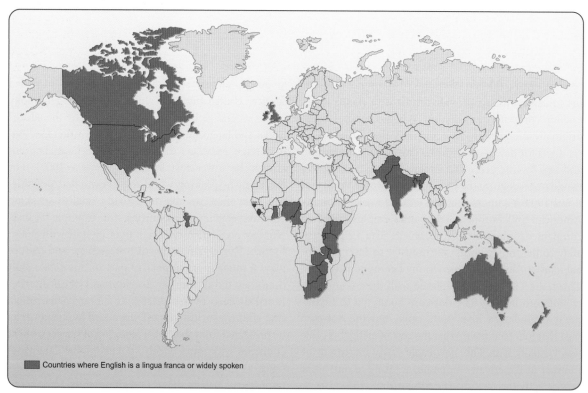

Countries where English is a lingua franca or widely spoken

Figure 17.4 Language as a factor in the geography of outsourcing countries where English is a *lingua franca* or relatively widely spoken.

Table 17.3 The largest absolute outsourcers and insourcers of business services, 2002

Outsourcers			Insourcers		
Rank	Country	US$ million	Rank	Country	US$ million
1	United States	40,929	1	United States	58,794
2	Germany	39,113	2	United Kingdom	36,740
3	Japan	24,714	3	Germany	27,907
4	Netherlands	21,038	4	France	20,864
5	Italy	20,370	5	Netherlands	20,074
6	France	19,111	6	India	18,630
9	United Kingdom	16,184	8	Japan	17,401
11	India	11,817	14	China	10,419
18	China	7,957	29	Russia	2,012
20	Russia	4,583			

Source: after Amiti and Wei (2004), pp. 22, 24

a value of US$11 billion for India and US$8 billion for China). The term insourcing is used to describe the amount of outsourcing that a country receives. The USA was the largest recipient of offshored services in 2002 and the top places in this list are dominated by developed market economies with India ranked in 6th place and China at 14th. It is worth emphasizing that the most important providers of offshored services are rich, developed market economies rather than poorer developing economies.

The global sourcing or offshoring of services does not have to entail the supply of services over large distances. It may occur as 'near-shoring' or the relocation or provision of services over short distances and often between locations on the same continental land mass (Gal 2009), for example the 'nearshoring' of American services to Canada. Services can now be supplied on-shore, near-shore and offshore. Presented in this manner these may appear as simple alternatives but in many instances firms have developed '*blended delivery systems*' that capitalize on the place-based advantages of coupling or blending activities located in a variety of different locations: home country, near, far (Bryson 2007).

17.6 Conclusion

The history of economic development is one of increased economic specialization combined with the development of new and often complex geographies. All this represents and reflects the constant working through of new divisions of labour and new ways of maximizing wealth creation. The development of service-dominated economies represents a working through of the division of labour but it is worth remembering that it is important to differentiate between employment and output. All services must be supported by manufactured products and all manufactured products are supported by services. The implication is that we have an integrated economy in which the worlds of production (services and production functions), finance and consumption combine to produce local, national and international economies.

The spatial division of expertise and the related second global shift continue to evolve and have the potential to rework the national and global dynamics of the capitalist economic system. Employment that was considered to be safe from foreign competition is suddenly exposed to business models that operate by blending together different forms of expertise related to differential comparative advantage. This may enhance productivity improvements, leading to new innovations, but it also has the potential 'to undermine the service-based solution to the employment crisis that is still being experienced by many developed regional economies' (Bryson 2007).

Learning outcomes

Having read this chapter, you should understand:

■ Different ways in which services have been defined and conceptualized by economists and geographers.

■ The concept of a 'personality market' or emotional labour highlights the importance of embodied labour, or skills with people, in the world of service work.

■ Appreciate that as the United Kingdom and the United States became industrialized societies they were simultaneously transformed into service economies.

■ Service economies should not just be identified by measuring employment, but should also take into consideration gross value added (GVA).

■ Economic production is a story of economic specialization and the continual reworking of the division of labour.

■ A spatial division of labour occurs with the geographical separation of production tasks, for example, writing, printing, binding and distributing a book.

■ The development of a concept of a spatial division of expertise draws attention to the development of new corporate strategies that are founded upon exploiting comparative advantage based on expertise.

■ Different economic geographies can be explained by distinctive combinations and patterns of spatial divisions of labour and of expertise.

■ Services were conventionally conceptualized as local untraded activities. The rise of the second global shift or service offshoring highlights one of the latest developments in the capitalist economic system.

Further reading

Bryson, J.R. (2007) A 'second' global shift? The offshoring or global sourcing of corporate services and the rise of distanciated emotional labour, *Geografiska Annaler*, 89B(1), 31–44.

Bryson, J.R. and Daniels, P.W. (eds) (2009) *The Handbook of Service Industries*, Edward Elgar, Cheltenham.

Bryson, J.R., Daniels, P.W. and Warf, B. (2004) *Service Worlds: People, Organizations, Technologies*, Routledge, London. The most recent account of the different divisions of labour amongst service activities and occupations as well as the role of ICT in altering the geographies of production systems.

Bryson, J.R. and Rusten, G. (2008) Transnational corporations and spatial divisions of 'service' expertise as a competitive strategy: the example of 3M and Boeing, *The Service Industries Journal*, 28(3), 307–23

Gál, A. (2009), 'Future Bangalores? The increasing role of Central and Eastern Europe in services offshoring', available at: http://papers.ssrn.com/sol3/papers.cfm?abstract_id=1334165.

Levy, F. and Murnane, R.J. (2004) *The New Division of Labour: How Computers are Creating the Next Job Market*, Princeton University Press, Princeton, NJ.

For annotated, clickable weblinks and useful tutorials full of practical advice on how to improve your study skills, visit this book's website at **www.pearsoned.co.uk/daniel**

THE GLOBAL FINANCIAL SYSTEM: WORLDS OF MONIES

Chapter 18

Jane Pollard

Topics covered

- Money in economic geography
- The social, cultural and political significance of money
- Geographies of monies
- Global monies, local monies and monetary networks
- Money, space, and power: geographies of inclusion and exclusion

Money is simultaneously everything and nothing, everywhere but nowhere in particular.

(David Harvey 1985b: 167)

This chapter explores the economic, social, cultural and political networks through which we establish value and manage exchange. Our attitudes to money and how we acquire and use it reveal not only its economic significance, but also its social, cultural, political and geographical qualities. As the opening quote from Harvey suggests, however, money can be a difficult subject. Novelists, film-makers and social theorists alike have grappled with its contradictory characteristics. Money can be everything and nothing, everywhere and nowhere; money can be the root of all evil and a source of independence and freedom. Even simple questions like 'what is money?' present some difficulties. Davies (2002: 29) suggests that money is 'anything that is widely used for making payments and accounting for debts and credits'. Historically, a wide range of items, including beads, shells, whales' teeth, cattle, salt, skins, tobacco, beer, gold and silver have been used as money. Such variation suggests that rather than asking what money is, we should perhaps ask what functions does money perform (see Spotlight box 18.1)? This question reveals the important economic functions of money and its status as a source of social power.

Spotlight box 18.1

The functions of money

- Money is a *unit of account*: it is the base of economic accounting systems.
- Money is a *measure of value*: it is the commodity against which the values of other commodities can be calibrated.
- Money is a *store of value*, in that you can sell a good for a certain amount of money and then use that money to buy something else at a later date. Money thus allows for the separation of the sale and purchase of commodities over space and time *if* it is a reliable store of value. Preserving this ability of money to store value is one of the reasons why governments are so concerned about the phenomenon of inflation.
- Money is a *medium of exchange and circulation*.
- Money is a *means of payment and a standard for deferred payments*.

These different functions of money often come into conflict with each other. As a store of value, governments, firms and individuals alike want monies to maintain their value and hence their purchasing power, between different times and spaces. As a medium of circulation, however, it is important for money to be extended as credit.

Banks create credit by extending loans to individuals, firms and governments. During economic growth, there is temptation for banks and other lending institutions to create large volumes of credit. Any sudden loss of confidence in the quality of credit, as happened in the Wall Street Crash of 1929 (see Plate 18.1) or, more recently, in the US–UK financial crisis of 2007–8, can trigger financial panics and devaluation as depositors, investors and financial institutions flee from credit moneys and seek out safer havens.

Plate 18.1 The Wall Street Crash 1929: the stock market collapse of October 1929 brought the 1920s boom to an end and led to widespread bank failures, bankruptcies and a drastic reduction in the availability of credit.

(Corbis UK/Bettman)

We are used to thinking of money 'oiling the wheels' of commerce and also being a measure of worth or value. Yet these two facets of money can be contradictory. As a store of value, it is desirable for money to be a stable, fixed representation of value. As a medium of circulation and as a form of capital – money that is thrown into circulation with the intent of generating more money – it is desirable to have money freely available as credit, divorced from its moorings in the 'real economy'. As Davies (2002:30) argues, the history (and geography) of the evolution of money reveals 'unceasing conflict' between borrowers, keen to expand the *quantity* of money in circulation, and lenders, keen to limit the supply of money and, at all costs, preserve the *quality* of monies in circulation. Managing this conflict or more accurately, reacting to financial crises resulting from it, has been and still is the mission of various regulatory authorities.

18.1 Money in economic geography

Many accounts of money and 'the global financial system' are written by economists or finance specialists, who pay little attention to its geographical anatomy. Is money something that we should think about geographically? As the Harvey quote at the start of the chapter suggests, money seems to be 'everywhere but nowhere in particular' and many contemporary debates about globalization stress the increasingly integrated, global machinations of the financial system in which 'virtual' monies can be moved around the globe at the touch of a button. One of the purposes of this chapter is to persuade you of the existence and importance – economically, culturally and politically – of different geographies of money and financial institutions that make up the so-called 'global financial system'.

Having said that, the organization of monies into different financial markets can appear dauntingly complex. In addition to the financial jargon describing different products and markets, the 'global financial system' can sometimes seem like something 'out there', a fast-moving, volatile 'thing', running out of control, increasingly uncoupled from the 'real' economy and operating beyond the regulatory reach of any one nation-state (Langley 2003). This global sense of money and finance can be difficult to grasp and sometimes it is difficult to understand just how its machinations affect us. For example, how did the South-east Asia financial crisis of the late 1990s change the lives

of people in Europe, Africa, the United States or elsewhere in the world? The US–UK crisis of 2007–8, however, triggered panic and recession across a wide range of countries and the high levels of public sector debt generated in order to save financial institutions and to stabilise financial markets will have significant consequences for employment, incomes and government spending for many years to come.

For all the difficulties, real and imagined, involved in understanding the workings of international, national and local finance, money and finance are now attracting a great deal of interest from geographers, other academics, governments and the popular press. Through the 1980s and 1990s, this was partly explained by various financial and political crises, including those in Pacific and South-east Asia, Russia and Brazil, and growing unease about levels of household indebtedness in the United Kingdom, United States and elsewhere. It is also the case that academic and popular commentators, most especially in the United States and the United Kingdom, have become aware of financialization: 'the increasing role of financial motives, financial markets, financial actors and financial institutions' (Epstein 2005: 3) in daily life. The most recent UK–US centred financial crisis, and the proliferation of complex financial instruments, has further intensified the pressing need to understand and manage volatility and instability in global financial markets and to reassess the role and power of various regulatory agencies. Also, as Leyshon (1995a) observed, money and finance have assumed a growing social and cultural prominence since the 1980s, related to the growth of financial services employment and the growing media and popular attention devoted to the rhythms and practices of financial services in London, New York and elsewhere.

A growing body of work on money and finance has grappled with a number of challenges including:

- the economic, social and cultural aspects of money;
- the links between our everyday experiences of money and the workings of 'the global financial system';
- the changing geographies of money in these increasingly global times;
- the creation of progressive, non-exclusionary forms of financial institutions in the contemporary world.

In dealing with these issues, geographers are not talking about 'the global financial system' as though it were a machine, but rather seeking to understand 'global finance' as a social, cultural and political phenomenon. Revealing the *social construction* of 'global finance' is important because its workings make a real difference to the well being of individuals and the choices/predicaments facing governments. The workings of the financial system shape who has access to credit, how, where and how much they pay for it and whether governments can afford welfare, education, healthcare and other services.

In the subsequent sections of this chapter, the above issues are explored in more detail. In the next section, we explore some of the different social and cultural aspects of money and how it is bound up with individual and group identities. The focus is broadened in the following section which considers the geographies of money that connect individuals and groups in distant places and at different times. By way of conclusion, the final section considers how some groups and regions are included or excluded from particular geographies of money, before exploring some examples of progressive, inclusive monetary geographies.

18.2 Everyday worlds of money

Although we may be used to thinking of money as primarily an economic phenomenon, geographers and other social scientists are now recognizing that money is saturated with cultural, social and political significance. In this section, we examine some of the social and cultural aspects of money in different societies. We focus initially on money and power, before moving on to consider money and identity, money as a transformer or leveller and money as a mediator of our experience of social relations.

18.2.1 Money and power

Many of the social and cultural aspects of money are related to its role as 'the very incarnation of social power' (Harvey 1982: 245). Again, however, we confront the contradictory qualities of money. As a means of expressing social power money has both desirable and less desirable qualities. Leyshon and Thrift (1997) talk of two discourses about money, one of suspicion and one of liberation.

To understand the discourse of suspicion, we can turn to the work of Georg Simmel. For Simmel (1978: 277), money is 'the most perfect representation' of the tendency to reduce quality to quantity. As a store of value and unit of account money transforms social relations – qualitatively different commodities and experiences – into an abstract quantity, namely their price. In contemporary capitalism, money has become

the mediator and regulator of economic relations between individuals, a measure of wealth and a means of expressing social power. All manner of social, political and cultural issues – from healthcare, housing, education, leisure and sports, the environment, and so forth – are debated not in terms of what they are worth, but in terms of *how much* they are worth in the sense of what can be afforded. Money 'affects our very ideals of what is good and beautiful and true' (Mitchell 1937: 371). Money, in this sense, has corroded the importance of other meanings and measures of 'value'.

Yet there is also a discourse of liberation that accompanies money. First, money provides a degree of individuality, freedom, security and independence for those that hold it. This is especially true in societies like Britain where laws regarding private property buttress ideologies of liberty, equality and freedom. The idea that money is empowering and liberating is central to the desire to possess money, to work for it, to accept it as payment and to save and invest it. Second, as Simmel (1991: 20) argues, money 'creates an extremely strong bond among members of an economic circle'; people have far more connections to other people in modern, monetized societies than was the case in feudal society. Through the **division of labour** in capitalist societies (see Chapter 17), money links people together in offices, factories, homes and shopping malls in different parts of the world. Money may in some senses be corrosive of social bonds, but it also creates a community that can forge new connections and hold people together.

18.2.2 Money and identities

Thus far, the discussion about money and power has been rather abstract. To bring this discussion closer to home, let's think about the role of money in the construction of identity. If we think of identity not as something that is fixed or innate but as a 'capacity to keep a particular narrative going' (Giddens 1990: 54), then money can be an extremely important part of telling and maintaining stories about ourselves. In the remainder of this section, we consider four brief examples.

First, money affects how individuals identify their class positions, what accommodation they live in, what type of schools they attend, and whether they can afford to stay on and pursue higher education and so forth. Being one of Britain's 'very rich' is associated with social and cultural networking practices which include particular (very expensive) rituals of conspicuous consumption, including balls and dinners and attendance at annual sporting events like Ascot and Henley. Such activities maintain relatively tight-knit networks, often linked by marriage. Within such moneyed networks, children can be sent to exclusive schools to mix with children from other wealthy families and be endowed with the social, if not financial, capital necessary to smooth their journey along their chosen future path.

Second, money stirs passions about nationhood, sovereignty and *national identities*. Consider the birth of the euro (€), which became legal tender on 1 January 2002. Swyngedouw (2000: 71) has described the new currency as 'arguably the most radical political-economic and cultural transformation of space in Western Europe since the French Revolution'. Each euro note has an image of an arch and a bridge, yet none of these are real bridges. Instead, they are all simulated bridges, symbolizing connection, interchange, communication; they hint of the economic, political and cultural aspirations of European integration (Pollard and Sidaway 2002).

Third, just as money can be a powerful symbol of class or national identity, it is also integral to the performance of *gender identities*. By way of example, we can think of how central the role of 'the male breadwinner' is to the construction of many masculinities. Films like *Roger and Me* (USA) and *The Full Monty* (UK) illustrate only too well that loss of employment in the car, steel and mining industries is not simply an economic catastrophe for the men involved. For those whose masculinity is rooted in 'industrial work' and the status of a 'breadwinner', the loss of regular employment and earnings can entail wrenching changes in social status, self-esteem and relations with partners and children.

Fourth, money is intricately linked to the *virtual financial identity* of individuals (and indeed countries), which are built up by credit-scoring companies that monitor credit histories and vet applications for credit. Those who seek credit cards, mortgages and other loans are the subject of credit checks by major credit-scoring companies that assess a wide range of factors including employment status, income, expenditure, outstanding debts, and repayment histories. Credit-scoring companies are not alone in constructing virtual identities for their customers. For example, grocery retailers in the USA, Canada, Britain, Australia and elsewhere use what are variously referred to as 'club cards', 'reward cards' or 'loyalty cards' as a means of tracking their customers' buying habits. The aim is to use customer information to target marketing initiatives towards those socio-demographic groups which will generate the greatest profits for the lowest marketing related expenses.

18.2.3 Money talks . . .

Money, as an incarnation of social power, can create new links between groups of people. As such, money can be a force for making people more tolerant of difference. For example, overseas investment in Birmingham, London, Manchester and other British cities is such that they embrace ethnic and racial diversity as part of their economic development strategies (Henry *et al.* 2002). In many British cities, 'Chinese Quarters' and other distinctive neighbourhoods associated with minority ethnicities have taken shape. Similarly, the 'pink pound' – the earning, investing and purchasing power of gay consumers – is courted by financial services providers, film and music media, travel firms, airlines, health and fitness firms, restaurants and bars. Surveys in Britain, the United States and Australia characterize gay consumers as groups that are relatively affluent, active, financially literate, and likely to travel overseas and eat out regularly (Anon. 1999). Although this stereotype is not representative of gay couples on low incomes or with children, the economic significance of pink pounds and dollars – already acknowledged in cities like London, San Francisco, Toronto and Sydney – is being recognized in a growing number of places.

With its ability to forge new associations between individuals and groups, money is also, potentially, a leveller. Money can disrupt traditional cleavages along the lines of class, gender, ethnicity and sexuality such that singers, actors, footballers and other high earners can join the ranks of the financial elite regardless of their backgrounds. The realization that 'money talks', that it can be transformative, a bridge to a different life (McLuhan 1964) is integral to the appeal of forms of gambling such as lotteries or casinos that are now very big business in Europe, North and South America, Australia, Hong Kong and, more recently, China.

18.2.4 Money as a mediator of social relations

The idea of money being a language that can connect and liberate different individuals and groups suggests that money can, in all sorts of ways, mediate our experience of social relations. By way of example, we can develop our previous discussion of gender identities and think about how money, its organization and uses are structured in ways that are gendered. Viviana Zelizer

(1989), for example, suggests that how monies are exchanged within households is usually subject to a very different, and usually gendered, set of rules that differ from those that govern the use of money in the market place. Zelizer (1989: 351) argues that social structures

> systematically constrain and shape (a) the *uses* of money, earmarking, for instance, certain monies for specified purposes; (b) the *users* of money, designating different people to handle specified monies; (c) the *allocation* system of each particular money; (d) the *control* of different monies; and (e) the *sources* of money, linking sources to specified uses.

Research in Britain has shown that men and women tend to be responsible for different forms of expenditure. Women are primarily responsible for expenditures on clothing, school-related expenses, food and charity donations and men responsible for expenditure on trips and holidays, repairs, redecorating, car maintenance and so forth (Pahl 1989).

Gendered assumptions about earnings and uses of money are also writ large in financial services industries. Financial spaces like the City of London and New York are renowned as elitist, masculinist environments (Lewis 1990; McDowell 1997) and many financial products target 'a male breadwinner' and are designed around gendered assumptions and stereotypes about work histories and sensitivity to risk. Pension products, for example, are designed to privilege those clients (most usually men) who work full-time and invest premiums regularly, and without interruption, over a 40-year period (Knights and Odih

1995). Similarly, Ravenhall (1999), citing a survey by *Money Magazine*, argues that male financial advisers treat men and women differently, with women being steered towards so-called 'widows and orphans investments', i.e. relatively conservative products with lower risks, and potential returns, than other investments. In Bangladesh, by contrast, it is often women who are the clients of banks, who receive credit and organize themselves and others into village groups to oversee the repayment of debt. Later in this chapter we will examine the establishment of Grameen Bank in Bangladesh. Its founder, Mohammad Yunus, established the bank believing that conventional banks were anti-poor and anti-women and insisted on collateral before making loans. Ninety-six per cent of Grameen Bank's clients are women. Women are regarded as better borrowers because 'they had most to lose, were more careful, had longer vision and were more concerned with the family than men' (Caulkin 1998: 2). In different ways in different societies, then, money can mediate the experience of, in this example, gender relations. Households and financial institutions can operate to reinforce or challenge prevailing assumptions and stereotypes concerning women and money.

In summary, money has cultural, political, and social as well as economic significance and this is reflected in the negative and positive discourses about the power of money. Money, as an 'incarnation of social power', is integral to the construction and performance of different identities. It is a 'leveller', a language for linking different individuals and groups, and it mediates our experiences of social relations.

Plate 18.2 Money changing lives? The possibility of a big win, however remote, helps to fuel various forms of gambling.

(Associated Press/Luke Palmisano)

18.3 Geographies of monies

How can we understand the links between some of our everyday experiences of money, described in Section 18.2, and the workings of so-called 'global finance'? This section broadens the focus to consider the geographies of money that connect individuals and groups in different places and times. In an increasingly globalized era, some would argue that geography is becoming relatively less important. Below, we examine some of the arguments concerning the globalization of money and finance, before outlining some of the problems with this view. The section concludes by introducing the concept of *monetary networks* as a way of thinking about money and its 'global' and 'local' qualities.

18.3.1 Global monies?

Many geographers, particularly those working in a political economy tradition, have argued that capitalism seems to be speeding up and spreading out (Harvey 1989; Massey 1994). Capitalism is seen to be spreading out in that more people in different countries and regions are becoming bound up with the logics of contemporary capitalism. It is also speeding up in that the pace of life seems to be increasing in different parts of the globe. In the second section, we noted the ability of money to transform qualitatively different commodities and experiences into an abstract quantity, a price. This has led many theorists to argue that money is a vehicle for the *homogenization of space,* for making different spaces more similar. While there are ongoing debates about the extent to which capitalism is generating a *global culture* (see Chapter 19), it is difficult to argue with the contention that financial markets in different countries have, since the mid-1970s, become more interconnected (Harvey 1989; Martin 1994) and homogeneous.

When we talk about the *globalization of finance*, however, we are talking about more than just the growth of international financial transactions, or the growing presence of multinational companies in domestic financial markets. Globalization implies

a strong degree of *integration* between the different national and multinational parts . . . the emergence of truly transnational banks and financial companies . . . that integrate their activities and transactions across different national markets. And above all, it [globalization] refers to the increasing freedom of movement, transfer and tradability of monies and finance capital across the globe, in effect integrating national markets into a new supranational system.

(Martin 1994: 256)

The emergence of this *supranational* system coincides with the increasing difficulties faced by *nationally* based regulatory authorities like the Bank of England or the US Federal Reserve. The 'discourse of suspicion' we mentioned earlier rears its head in many popular representations of 'global finance':

[T]he world is now ruled by a global financial casino staffed by faceless bankers and hedge fund speculators who operate with a herd mentality in the shadowy world of global finance. Each day they move more than two trillion dollars around the world in search of quick profits and safe havens, sending exchange rates and stock markets into wild gyrations wholly unrelated to any underlying economic reality. With abandon they make and break national economies, buy and sell corporations and hold the most powerful politicians hostage to their interests.

(Korten 1998: 4)

This kind of discourse was also prevalent in the United Kingdom in 2008; Alex Salmond, First Minister of Scotland, argued that 'spivs and speculators' were jeopardising 'the quality of life and jobs of hundreds, thousands and millions of people across the world at risk through 'short-trading' – where investors make money by betting that the price of shares in a company will fall (cited in McIntosh and Maddox 2008). Whether or not you agree with these views, it is true that the degree of integration of the financial system profoundly shapes and connects the lives of people thousands of miles away from each other. In this sense, Leyshon (1995b) talks of money being able to 'shrink' space and time, bringing some (but not other) parts of the globe relatively closer together through the working of financial markets. A good example of this is considered in Case study 18.1, which describes how the fates of property developers in Bangkok (Thailand) ultimately affected the fortunes of a bartender, Graham Jones, in Whitley Bay (United Kingdom). The fates of individuals and companies in Britain and Thailand are connected through the workings of foreign exchange markets that, in turn, link national currencies to the competitiveness of different nation-states.

This example highlights only some of the interconnections between changes in the value of the Thai baht and how these affected businesses across Asia and Europe.

Regional case study 18.1

From Bangkok to Whitley Bay . . .

1 Devaluation in Thailand, July 1997

In 1997, several property companies collapsed in Thailand; property prices and the stock market started to fall. Currency speculators, already nervous about slowing growth in the region, started to sell the baht (Thai currency) as they expected the currency to be devalued. In July, the baht was devalued. As a result, Thai exports became cheaper and, to stay competitive, Indonesia, Malaysia, South Korea and the Philippines allowed their currencies to fall sharply. Some Korean firms who needed foreign currency to pay off loans started dumping cheap microchips, forcing microchip prices down from US$10 a unit to US$1.50.

2 Factory closure in North Tyneside, United Kingdom, 31 July 1997

Rapidly falling semiconductor prices meant losses of £350 million for German electronics company Siemens. Siemens semiconductors were produced at plants in Tyneside (UK), Taiwan, Germany, France and the USA. Managers in Munich announced plans to close their £1.1 billion plant in North Tyneside that employed 1,100 people and had opened only in 1994.

3 The cleaning company

The Siemens factory provided 10 per cent of business in the north-east for Mitie, a cleaning company. One-third of the workforce of 90 people were facing redundancy and their boss had his salary bonus cut by 10 per cent, in line with the loss of work from Siemens.

4 The hotel

Siemens and its contractors, like Mitie, were the single biggest sources of business for the Stakis Hotel. The hotel responded to the closure by switching their market focus. As the flow of German executives and their UK contractors slowed and then ceased, the hotel sought to attract more families.

5 The taxi firm

Foxhunters, a taxi firm, had a contract with Siemens that generated more than a dozen runs a day, usually to the Stakis Hotel, the airport or the university. Since August, Siemens business had dried up and 85 drivers were chasing work for 50. Despite working longer hours, drivers' takings were down by between £100 and £200 a week. Drivers started to economize by bringing in their own lunches and by cutting down their trips to the local pub.

6 The local pub

Takings at Cameron's had fallen by £600 a week since August. The landlord, who blamed the Siemens shutdown for the reduced trade, cut his opening hours. Attempts to drum up more trade by price reductions had little effect. The landlord was hoping for more business around Christmas. He and his partner had less money for spending on their leisure activities, which included visiting places like Whitley Bay.

7 Graham Jones, bartender in Whitley Bay, October 1998

Graham was fired from his job at a pub in Whitley Bay; there was not enough business to occupy two bartenders in the public bar.

Source: adapted from Carroll (1998)

We could extend the example by thinking through how job losses in any other region/locality would also have been felt by those not in paid work. For those in households dependent on the wages of someone made redundant there would be further belt-tightening or a search for alternative sources of income. For the unemployed, job losses in the area might mean more competition for any new jobs that are created and so forth.

18.3.2 Forces for globalization

How has this globalization of money come about? There are a range of factors to consider here. First,

through the 1980s and 1990s, different governments and international institutions, like the International Monetary Fund (IMF), have pursued neoliberal, free market policies and encouraged the deregulation of financial markets (by eliminating exchange and capital controls) and the liberalization of flows of capital across national borders.

Second, there have been advances in telecommunications and computing technologies that can be summarized under three headings: computers, chips and satellites (Strange 1999). Computers have transformed payments systems. For many hundreds of years, payments and transfers of money were completed in cash – coins and notes – and written down in ledgers. Coins and notes were superseded by cheques which in turn have been superseded by electronic monies that can be moved around the world at the speed of light. Semiconductor chips in computers mean that consumers in many countries can use credit and debit cards to pay for goods. In some countries, computer chips are now embedded in plastic 'smart' cards to allow the use of 'digital money', or, more accurately, electronic representations of currencies, to pay for groceries and other goods. Finally, there are the systems of communication, using earth-orbiting satellites, that are integral to the operation of computing systems, email, the Internet and other forms of communication that are used by banks, governments and other financial players. These innovations have made 24-hour trading possible as stock markets in different countries are linked by computer.

The growth of the Internet through the 1990s has been closely linked to the rise of discourses about a 'new economy' (see Aglietta and Breton 2001; Thrift 2005). The Internet has not only altered how business can be done in financial markets – think, for example, about the rise of online banking in North America and parts of Europe – but also changed the cast of characters involved in undertaking that business. New communications and software technologies have spawned the proliferation of financial websites, information and intermediaries able to provide financial information to firms and consumers. Online providers, without the costs of maintaining the bricks and mortar associated with a high street presence, provide insurance, banking, pension and other financial products to firms and consumers with access to Internet technologies.

Third, and closely related to developments in technology, there have been innovations in tradable financial products that have made it easier and faster to move money around the globe. Derivatives are one example of this kind of innovation. Derivatives are contracts that specify rights/obligations based on (hence 'derived' from) the performance of some other currency, commodity or service. In the 1970s, financial derivatives were created to allow financial managers to deal with currency risk, but since that time they have become increasingly sophisticated and extended to more markets. Since 1997, for example, energy suppliers, transport agencies, construction companies, wine bar owners and other firms exposed to weather risk (the possibility that weather could have an adverse impact on their profits and cash flow) have been able to purchase weather derivatives contracts to protect themselves against this risk (Pollard *et al.* 2008).

With this extension, however, there has been the proliferation of purely speculative trading of such financial instruments – and the development of other much more complex contracts – as part of what Robin Blackburn (2006) terms the 'grey capitalism' of financialisation. Early warning signs of the kinds of losses that could be generated from derivatives contracts included the collapse of Barings Bank in London (1995); the US–UK banking crisis of 2007–8 generated losses running into trillions of dollars in the USA alone (www.pbs.org/wnet/need-to-know/economy/the-true-cost-of-the-bank-bailout/3309/) and was the worst financial crisis since the 1930s (see Blankenburg and De Palma 2009).

Innovations in computing and software technologies and in financial instruments are fundamental to arguments about the growing *financialization* of the economy alluded to earlier; this is an argument that asserts that the speculative accumulation of capital has become an end in itself in contemporary capitalism, that the financial system has come to feed on itself, so to speak, rather than supporting firms and industries and other elements of the 'real' economy (Strange 1999). Round the clock trading of financial instruments in different places and time zones opens up opportunities for speculation and arbitrage, which is the ability to profit from small differences in price when the same financial product is being traded on more than one market. One indicator of how growth in the international financial system is outpacing the growth of the 'real' economy is provided by foreign exchange trading data. According to data from the Bank for International Settlements (www.bis.org) the daily turnover of foreign exchange trading in 1973 was $10–$20 billion, roughly twice the amount necessitated by world trade. By 2004, daily trade in foreign exchange averaged $1.9 trillion, roughly 95 times that necessitated by world trade. By April 2010, daily trading volumes had reached $4 trillion (www.bis.org/publ/rpfx10.htm). For individual consumers with access to the appropriate technology there are now online financial bookmakers encouraging clients to enjoy spread betting (and possibly tax-free profits) on price movements of stocks, stock indices, currencies, interest rates, commodities and even house prices.

These, then, are some of the ways in which money has become more global since the 1970s. But what motivates such changes? What motivates the implementation of 'free market' policies, the development of new technologies and financial instruments like derivatives? For those working in a *political economy tradition*, like David Harvey (1989), the motivation for these changes is the search for profit. Capitalism is fundamentally

Spotlight box 18.2

The circulation of money as capital

In capitalist societies, the circulation of money as capital is as follows. From left to right in the equation, money (M) is invested by producers to purchase commodities (C), namely labour power (LP) and the means of production (MP), say pieces of wood and wood cutting machinery. Labour power and the means of production are combined in production (P) to make more commodities (C′), in this example, let us say chairs, which are then sold for more money (M′) than was originally invested (M):

$$M \rightarrow C \rightarrow \frac{(LP)}{(MP)} \rightarrow P \rightarrow C' \rightarrow M' \rightarrow \ldots$$

The purpose of production in capitalist societies is to produce profit and accumulate capital. The *turnover time of capital* is the amount of time it takes for money to complete this circuit. The shorter the turnover time, the more often money can be lent out and the more profit can be made. Producers therefore have a very strong incentive to, where possible, reduce the turnover time of capital; time *is* money.

about the accumulation of surplus. Competition drives capitalists to seek out new markets, new products and to reduce the *turnover time of capital* (see Spotlight box 18.2). In different parts of the globe, money is the language through which the imperatives of capitalism are being communicated.

So, there is a very strong economic rationale for the globalization of money. And it is difficult to argue with the contention that money has become more globalized since the 1970s, that it has, increasingly, connected people in distant places and homogenized financial space. For Richard O'Brien (1992) the growth of the international financial system is tantamount to 'the end of geography', the notion that geography is becoming relatively less important because money, in its different forms, is able to overcome the friction of distance and link distant places together. Others argue that this view is too simple, that geography remains critical to our understanding of global finance. In the next part of this section, we will consider these views in more detail.

18.3.3 Debunking 'global' monies

A rather different departure point for talking about the 'globalization of finance' is to argue that the financial system is not really global at all, that it is more like a *web of connections* between different financial systems, some of which are bound together more tightly than others. This argument has developed because the idea of a 'global financial system' is problematic for several reasons.

The first problem stems from the simple observation that 'global finance' is largely the province of North America, Europe and parts of Asia, most notably Japan, Hong Kong, Singapore and, more recently China; it is an idea centred on the experiences of the West (see

Table 18.1). Many parts of Africa, Asia and Latin America are not only not included in 'the global financial system', they are being actively excluded as banks refuse to lend money until outstanding debts have been repaid. There is what Massey (1993) describes as a *power geometry* at work when some commentators describe the financial system as 'global'. For some financial workers in New York, London and Tokyo, international finance may be regarded as 'global' in the sense that all countries and regions of the globe deemed creditworthy and capable of producing profits have been included; sub-Saharan Africa, however, ceases to exist in such a conception of 'the global'.

Second, and resulting from this geographical concentration of the management of 'global finance' in North America, Europe and parts of Asia, some currencies circulate more widely and have greater spatial reach than others. The Japanese yen, the euro and most especially the US dollar are very useful in international markets because they are accepted as forms of payment, unlike, for example, Indian rupees. Historically, the country that occupies a dominant economic and political position has underwritten the soundness of the international financial system and had its currency accepted internationally as the currency in which commodity prices are quoted and payments made. Before the Second World War, Britain and the pound sterling fulfilled this role; after the Bretton Woods conference in 1944, the US dollar became the key international currency. More recently, as the economic dominance of the US has declined, the yen, the euro and the Chinese renminbi have become relatively more important in international markets.

So, some currencies, like the US dollar, are truly international while some others are national. There are estimated to be over 4,000 *local currencies* in operation around the globe that facilitate exchange only within

Table 18.1 Average daily turnover of equities trading 2009 (millions of $US)

Region	Average daily turnover	Region as % of world total	Region	Average daily turnover	Region as % of world total
Americas		**60**	**Europe**		**15.4**
Bermuda SE	0.2		Athens Exchange	269.0	
BM&FBOVESPA	2,545.6		BME Spanish Exchanges	5,949.0	
Buenos Aires SE	12.3		Borsa Italiana	3,486.5	
Colombia SE	77.0		Budapest SE	101.2	
Lima SE	18.2		Cyprus SE	6.9	
Mexican Exchange	334.3		Deutsche Börse	8,608.0	
NASDAQ OMX	114,886.3		Egyptian Exchange	295.1	
NYSE Euronext (US)	70,573.8		Irish SE	138.6	
Santiago SE	152.4		Istanbul SE	1,195.0	
TSX Group	4,940.0		Ljubljana SE	4.8	
Asia-Pacific		**23.4**	London SE	13,403.6	
Australian SE	3,667.5		Luxembourg SE	1.1	
Bombay SE	1,083.8		Malta SE	0.1	
Bursa Malaysia	344.1		MICEX (Russia)	1,802.7	
Colombo SE	5.2		NASDAQ OMX Nordic Exchange	2,887.4	
Hong Kong Exchanges	6,030.7				
Indonesia SE	391.5		NYSE Euronext (Europe)	7,740.3	
Jasdaq	128.3		Oslo Børs	976.1	
Korea Exchange	6,162.2		SIX Swiss Exchange	3,025.4	
NSE India	3,237.4		Warsaw SE	226.2	
New Zealand Exchange	59.1		Wiener Börse	193.4	
Osaka SE	575.6		**Africa**		**0.4**
Philippine SE	86.0		Johannnesburg SE	1,343.5	
Shanghai SE	20,744.4		Mauritius SE	1.3	
Shenzhen SE	11,369.1		**Middle East**		**0.8**
Singapore Exchange	970.1		Amman SE	54.8	
Taiwan SE Corp.	3,606.1		Saudi Stock Market	1,348.1	
Thailand SE	518.9		Tehran SE	71.2	
Tokyo SE Group	16,423.5		Tel Aviv SE	349.0	

Source: adapted from World Federation of Exchanges www.world-exchanges.org/files/statistics/excel/EQUITY609.xls [accessed 28 October 2010]

very specific spatial and social contexts. For example, Local Exchange Trading Systems (LETS) are associations whose members list their services needed/offered in a directory and trade with each other in a local unit of currency, for example, 'bobbins' in Manchester and 'solents' in Southampton (Williams 1996). Since their establishment in Canada in 1983, LETS schemes have spread to Europe, Australia, New Zealand and North America, allowing members not only access to credit, but also the chance to engage in productive activity to earn such credit. Some local currency schemes are devised in times of hardship to allow local people to trade goods and services when they are unemployed and have little money. Other schemes are motivated by ecological concerns, the desire for community development and social cohesion, and the desire to construct alternative local economic geographies as a form of resistance against the global spread of capitalism (Lee 2000).

Third, as Martin (1999: 6) argues, for all the talk of 'global finance', there remain different *geographical circuits of money* that form the 'wiring' of an economy, along which 'currents' of wealth, consumption and power are conveyed. He identifies four geographies of money: locational, institutional, regulatory and public. The *locational geography* refers to the location of different financial institutions and markets. Financial institutions and specialized functions (like foreign exchange markets) tend to be agglomerated in large urban centres, with London, New York and Tokyo sitting at the top of the global hierarchy. Different countries also have different *institutional geographies*, in that they organize their financial institutions and markets in distinct ways. In Britain, for instance, banking is highly concentrated amongst a few large players; we are used to thinking of a few large retail banks like Lloyds-TSB, HSBC and Barclays, yet in the USA, by contrast, there were 7,830 federally insured commercial banks in 2010 (www2.fdic.gov/SDI/main4.asp). Meanwhile, in parts of Asia and west and South Africa, banks and other community-based financial institutions like Rotating Savings and Credit Associations (ROSCAs) (see Spotlight box 18.3)

are important in organizing and funding economic activity. ROSCAs take different forms in different regions and vary with the class, gender and ethnicity of their members (Ardener and Burman 1995).

These varied institutional geographies are products, in turn, of contrasting *regulatory geographies*. There are a wide range of supranational, national and regional regulatory agencies that govern the workings of financial institutions and markets. To return to the example of how the Asian financial crisis cost Graham Jones his job in Whitley Bay, a key player in how the crisis unfolded was the International Monetary Fund (IMF). The Chief Executive of Siemens argued that Korea was able to sustain a 'suicidal pricing strategy' for its memory chips (a strategy that led to losses of £350 million for Siemens) because it was receiving financial aid from the IMF (Milner 1998). To give another example, as Martin (1999: 9) observes,

[M]oney has had a habit of seeking out geographical discontinuities and gaps in these regulatory spaces, escaping to places where the movement of financial assets is less constrained, where official scrutiny into financial dealing and affairs is minimal.

Offshore financial centres like the Bahamas, Cayman Islands, Jersey and Guernsey are attractive because of their low tax rates and minimal regulation.

Finally, when Martin (1999) discusses *public geographies*, he refers to the role of states in distributing monies across regions in the form of goods and services, infrastructure, health, education and so forth, and in

Spotlight box 18.3

Rotating Savings and Credit Associations (ROSCAs)

A ROSCA is 'an association formed upon a core of participants who agree to make regular contributions to a fund which is given, in whole or in part, to each contributor in turn' (Ardener 1995: 1). ROSCAs are known by different names, depending on their form and scale, the social classes of their members and their location. In South India they are known as *kuris, chitties* or *chit funds*, in Cameroon as *njangis* or *tontines.*

ROSCAs were well developed in China, India, Vietnam and parts of West Africa by the end of the nineteenth century. Variants of ROSCAs also existed in Scotland and parts of northern England. Members,

usually ranging in number from a few to several hundred, make regular contributions, in cash or in kind, to a fund. The fund, or part of it, is then given to each member in turn, depending on age, kinship seniority, or by rules established by the organizer.

In addition to encouraging regular savings and providing small-scale capital and credit for their members, many ROSCAs have strong moral and social dimensions. Trust, social solidarity and responsibility are emphasized and members have an interest in ensuring that no member defaults on regular payment. Some ROSCAs are women-only and their potential to empower women, by giving them greater control over income and credit, has attracted the attention of anthropologists, sociologists and feminists (see Ardener and Burman 1995).

transferring monies in the form of various social and welfare programmes.

> What becomes clear from this discussion is that the world remains made up of a patchwork of different financial spaces or systems, and although there has been considerable integration of financial practices and processes across such systems in recent years, we are a long way from the seamless global financial space described by some commentators.
>
> (Leyshon 1996: 62–3)

18.4 From 'global monies' to monetary networks

The phrase 'global finance' therefore seems to be something of a misnomer. While it is true that different monies and markets have become more interconnected since the 1970s, 'the global financial system' seems to comprise some elements that are international in their reach, juxtaposed with other local and national currencies, regulations and institutional geographies. One way of juggling with this complexity is to think about the idea of *monetary networks.*

Recall that in the introduction to this chapter 'the global financial system' was described as a social, cultural and political *network of relations* in which individuals have different capacities to participate. When we talk about 'networks' what do we mean? Dodd (1994: xxiv) argues that monetary networks are, above all, *networks of information* that have five abstract properties (Spotlight box 18.4). These networks are formed by the

production, interpretation and circulation of information between actors through time and space. Thus we can think of foreign exchange traders in London, Tokyo, New York and elsewhere constituting a network of relations through which foreign exchange trading is performed, organized, regulated and understood.

How do networks work? Nigel Thrift (1996, 1998) has argued that the forms of information that constitute monetary networks are always ingrained in *practices.* To give an example, we can return to the workings of global telecommunications networks discussed earlier. Telecommunication networks rely on the workings of thousands of workers and machines. Similarly, the Internet consists of thousands of miles of fibre-optic cables, most usually laid alongside existing transport infrastructure, and computing hardware and software managed and sustained by programmers, engineers, users and machines (Bray and Kilian 1999). What we think of as the 'global financial system' is really an assemblage of consumers, workers, computers, software, telephones, office buildings, bits of paper, financial reports and so forth. There is no escaping the *materiality* (i.e. the bits of cable, bank branches, telephones, desks and so forth) and *practices* (using computers, telephones, paying for goods and so forth) that constitute these networks. More recently, research has started to highlight the significant and growing role of different groups of migrants in channelling capital to the global South through money transfers or remittances (see Datta 2007). Many of these groups – who are often excluded from holding bank accounts – now make use of mobile phone networks to send money overseas (see www.sendmoneyhome.org/foreign-currency-articles/mobile-money-transfers.php). These movements of

---

Spotlight box 18.4

Properties of monetary networks

- Networks contain a *standardized accounting system* into which each monetary form within the network is divisible; any goods, priced in terms of this standardized accounting system, are exchangeable.
- Networks rely on *information regarding expectations of the future*; 'money is accepted as payment almost solely on the assumption that it can be re-used later on' (Dodd 1994: xxiv).
- Networks rely on *information regarding their spatial extent*; institutional frameworks limit the

territories in which specific monetary forms may be used.
- Networks are based on *legal information*, usually in the form of rules that govern forms of contractual relations of network members.
- Networks presuppose *knowledges of the behaviour and expectations of others*. In order to trust money, those holding it need to anticipate, and trust, that it can be used and reused within the network.

Source: Dodd (1994: xxiv)

people and the financial networks and flows they generate all point not to the end of geography but to 'a more complex mosaic of states, regions, global city-regions and localities differentially integrated into the global economy' (Agnew 2009: 221).

When we talk of 'the global financial system', then, we are really talking about actor networks that combine human beings, technologies and documents which come together to sustain, in particular spaces and times, the flows of information that constitute different monetary networks. Note the use of the plural. Rather than thinking of one 'global financial system' we can think instead of there being a *multitude of monetary networks that overlap and intersect*. This is an important point, because different monetary networks, as we have seen, have very different geographies. Some monetary networks, like those of the foreign exchange markets, have the capacity to connect people and machines in London, New York and Tokyo at the touch of a button. Other monetary networks, like LETS schemes in Manchester, are much more compact and connect people over much shorter distances. Rather than thinking about networks as being either 'global' or 'local', Bruno Latour (1993: 122) argues that we should think of networks as being 'more or less long and more or less connected'.

Monetary networks differ not only in the *length of connections* they are capable of sustaining, but also in the *speed of connections* they are capable of sustaining. We have talked already of the computing networks that facilitate foreign exchange trading and allow banks to move monies around the world in seconds or the growing use of mobile phone networks to facilitate financial flows in parts of the global South. One of the attractions of ROSCAs (Spotlight box 18.3) for men and women in Ghana and elsewhere is the speed at which news of hardship or an emergency can be spread, and the order of rotation of the fund changed, to help out a member in trouble. ROSCAs may not have quite the speed of connection of foreign exchange trading (assuming, of course, that all the computers are working), but they can react with a speed rarely matched by local banks (Ardener 1995).

In summary, financial markets have become increasingly globalized since the 1970s, although truly 'global' markets are concentrated in a select number of major financial centres. There remain distinct regional, national and supranational geographies of money, such that the 'global financial system' can be conceptualized as a web of intersecting networks, some fast, some slow, some long, some short. These financial networks connect different geographies of employment, technology, regulation, government policies, leisure trends and so forth. Such a conceptualization allows us to understand how the fates of property developers in Thailand affected Graham Jones in Whitley Bay. It also alerts us to the dynamism of these networks; if the ninetheeth century belonged to the United Kingdom and the pound sterling and the twentieth century to the United States and the dollar, the early parts of the twenty-first century are witnessing significant geo-economic and political shifts towards Asian economies, most notably China. It is important to remember then that when we talk about 'global finance', we are not talking about a machine operating 'out there', but a socially, economically, culturally and politically constructed network of relations with which we engage, in some way, shape or form, on a daily basis.

18.5 Conclusion: money, space and power

From the preceding discussion, it should be clear that many monetary networks are very uneven and exclusive. Different monetary networks are not open to all who would like to participate and there are some very distinctive *geographies of financial inclusion and exclusion* that have already been mentioned. The benefits of some of the most extensive, fastest monetary networks are, for the most part, extended only to those areas where individuals, companies and governments can produce a profit for the major lenders in North America, Japan and Europe. Access to such networks is carefully screened and countries, firms, banks, public agencies and individuals are credit-rated. On the basis of their credit rating, decisions are made by major financial institutions on whether to offer credit or buy debt and, if so, for how long, and at what rate of interest.

So, who does this screening and what are the geographies of this credit scoring? The big three international agencies are Moody's Investment Services, Standard and Poor and Fitch Ratings, all US firms. Standard and Poor's so-called 'Investment grade' scales rate types of debt on a scale from 'AAA' ('Superior') through 'A' ('Good') to 'BBB'. Below this are the 'Speculative' grades which range from 'BB' ('Questionable') through to 'D' ('Default'). Although these are US-based firms, this is not to say that only the major capitalist economies are able to secure credit. In the late 1970s and early 1980s, for example, total lending to Africa, Asia and most especially Latin America increased sharply. Yet rising interest rates, falling commodity prices and a rapid appreciation of the US dollar left debtor countries unable to cope with repayments. Affairs came to a head in August 1982, when Mexico suspended its debt repayments, followed by Brazil, Argentina, Peru, Venezuela, Ecuador and more than 20 other countries by the end of 1983 (Porter and Sheppard 1998). The response of banks to the debt crisis was to drastically reduce their lending to these countries, to effectively exclude them from international bank lending networks.

More recently, investors in North America and Western Europe have been encouraged by fund managers and financial media to invest in 'emerging markets', which include parts of the Russian Federation and former Soviet Republics and other parts of Asia, Latin America and Africa or the so-called BRICS (Brazil, Russia, India and China). The lure for investors is the prospect, real or imagined, of higher rates of return than those that can be secured in what are represented as the more conservative, less volatile, confines of Western European and North American equity markets (see Sidaway and Pryke 2000). Again, the US-based credit ratings agencies are key players in the construction of these financial markets. Research in emerging economies demonstrates that a downgrade in, say, a country's bond rating not only affects the price of its bonds but also affects stock market returns and 'spillover' to neighbouring states as international investors anticipate further downgrades in the region (Kaminsky and Schmukler 2002).

The concentration of wealth and power in the major capitalist economies is such that they hold sway, economically and politically, over the networks through which international lending is conducted and its terms negotiated. Even supranational agencies like the World Bank and the IMF rely on funds from major capitalist countries that exert considerable influence over where the funds go and how they are spent. If South Africa, Russia and other post-socialist economies want to be included in the monetary networks of the World Bank and the IMF, then instituting reform packages that mimic the political and economic organization of major capitalist economies is the price of admission. Such **structural adjustment** packages have spread widely through east and west Africa, Latin America and Asia since 1982 (Porter and Sheppard 1998). For those that do not toe the line, life can be made very difficult. Money, as we saw in Section 18.2, is a very powerful tool of social and political control.

Such geographies of inclusion and exclusion operate within countries too as banks, private investors and other lenders seek out the most prosperous growth regions. For example, in Los Angeles major retail banks have closed bank branches in low-income neighbourhoods like South Central Los Angeles and opened new branches in prosperous areas like Beverly Hills (Pollard 1996); this pattern has been repeated in many other US cities. The financial institutions that tend to move into South Central Los Angeles are so-called 'alternative' financial institutions like pawnbrokers and cheque cashing establishments, which often charge very high fees for basic banking services. In similar fashion, the distribution of venture capital investments in the United Kingdom has traditionally favoured firms located in the prosperous south-east of England (Mason and Harrison 2002).

Financial exclusion tends to reinforce existing patterns of wealth, poverty and inequality. Those individuals and nations that already have wealth are deemed the most creditworthy and are offered the cheapest forms of credit to expand businesses, pay off debts, buy new goods and so forth. And, as discussed in Section 18.3.4, those consumers that have the financial, cultural and technical capacity to access the Internet will be the beneficiaries of new and sometimes cheaper financial products available online. For low-income groups, new migrants and countries reeling from crippling debt repayment schedules, credit is difficult to find and usually very expensive. For those groups suffering varying degrees of financial exclusion, what alternatives are available?

18.5.1 Some responses and alternatives to financial exclusion?

Although some alternative monetary networks, like those of doorstep lenders, may be expensive to use, there are plenty of examples of other monetary networks that are designed to be inexpensive and inclusive. One very famous example is the Grameen Bank, formed in 1983, which now employs over 23,000 people in 35,000 villages in Bangladesh (Case study 18.2). The successes of the Grameen Bank and other micro-credit institutions in Indonesia, Bolivia, Kenya and elsewhere have attracted a great deal of attention and spawned innumerable imitations in Africa, Asia, Latin America and elsewhere. In recognition of their impact in helping the poor, Yunus and Grameen Bank were awarded the 2006 Nobel Peace Prize. The prospect of linking access to credit for the poorest with job creation has

Regional case study 18.2

Grameen Bank

The basic principles of Grameen Banking are as follows:

- the bank lends only to the poorest landless villagers in Bangladesh;
- the bank lends, overwhelmingly, to women; 96 per cent of its 8.3 million members (2010) are women;
- loans are extended without borrowers needing collateral;
- the borrower, not the bank, decides what the loan will be used for;
- borrowers pay sufficient interest to keep the bank self-sustaining.

The average loan is for about US$100. Rather than insisting on collateral or security, women wanting a loan join a village group, attend weekly meetings and assume responsibility for the loans of other group members. Thus, there is strong peer pressure for individual women to maintain payments, otherwise the group is unable to secure further loans. A total of 97 per cent of Grameen Bank loans are repaid.

The success of Grameen Bank has spawned imitations in over 50 countries in many parts of Africa, Asia and Latin and North America. Grameen has now branched out into fisheries, mobile telephones, energy and the Internet.

Sources: Bornstein (1996), Caulkin (1998), www.grameen-info.org/index.php?option=com_content&task=view&id=779&Itemid=748

---

made these so called micro-finance schemes increasingly popular in Britain, the United States and Western Europe, and not just in those regions of the United Kingdom and the United States that are experiencing financial exclusion.

It is also important to note that some 'alternative' financial networks may be preferred by individuals and households who do not want to use financial institutions like traditional retail banks and instead want to participate in networks that encourage broader social, religious, environmental and other assessments of 'value'. In the last 15 years, for example, Islamic banking and finance (IBF) has grown very rapidly cross the globe. IBF practices are rooted in the rules and norms of Islam which prohibit the payment of *riba* (interest) (as it is viewed as exploitative and unfair) and investments deemed excessively risky or uncertain (*gharar*). Islamic institutions thus shun investments into, for example, gambling, alcohol and prostitution (Henry and Wilson 2004).

The growing popularity of various micro-credit schemes, like the LETS and ROSCA schemes is based not only on their economic successes, but also on their status as 'alternative' and, potentially, progressive schemes. One of the interesting features of micro-credit schemes, like the Grameen Bank, is their often strong emphasis on creating social bonds and a sense of shared responsibility between members. Another feature of many of these micro-finance experiments is

that it is often women who make up the bulk of the membership. As Goetz and Gupta (1996) point out, however, access to credit does not necessarily equate to control over money and, more generally, empowerment for women in environments where they frequently face extreme constraints on their lives. As we mentioned in Section 18.2, how monies are exchanged within households is usually subject to a rather different set of rules than those governing the use of money in the market place (Zelizer 1989).

To conclude, these examples of monetary networks illustrate only too well that money, how it is obtained, how it is used, how it is managed and by whom, is an economic, social, political, cultural and *geographical* phenomenon. Although many accounts of 'the global financial system' pay little regard to the geographies of money and finance, this chapter has argued that a host of monetary networks – some relatively slow, some relatively fast, some longer than others – make up the 'global financial system'. These monetary networks intersect most densely through institutions and practices in North America, Japan and Western Europe. This geopolitical reality profoundly shapes patterns of uneven development and structures which regions, industrial sectors, governments and individuals are deemed credit worthy and on what terms. 'Global finance' is thus a socially constructed, geographically rooted network of relations with which we engage, in some way, shape or form, on a daily basis.

Learning outcomes

After reading this chapter, you should be aware that:

- Money has geographical, economic, cultural, social and political significance and mediates our experiences of social relations.

- The 'global financial system' is a socially constructed web of intersecting networks operating at different speeds and over different distances.

- There are distinct regional, national and supra-national networks of money.

- There are many geographies of financial inclusion and exclusion.

- There are a range of financial networks which may be more inclusive and politically progressive than others.

Further reading

Leyshon, A. and Thrift, N. (1997) *Money Space: Geographies of Monetary Transformation*, Routledge, London. This is a compilation of Leyshon and Thrift's essays on money.

Martin, R.L. (ed.) (1999) *Money and the Space Economy*, Wiley, Chichester. A wide-ranging collection of essays covering geographies of banking, financial centres, money and the local economy, and money and the state.

Sassen, S. (2001). *The Global City: New York, London, Tokyo*. Princeton University Press, Princeton, NJ. This book looks at the growth and internationalization of the financial system and its effects on the economic base and social structure of its three 'command centres', London, New York and Tokyo.

Useful websites

www.ft.com The *Financial Times* website contains useful links to financial and other firm, sector and country data, as well as breaking news stories.

http://money.cnn.com The US-based CNN financial website, like the *Financial Times* site, contains numerous links to business, financial and market news, by country, sector and firm.

www.worldbank.org The World Bank website is home to a host of information on World Bank publications, research, data and related organizations.

www.bis.org/statistics This site, maintained by the Bank for International Settlement, provides sources of banking data and statistics.

www.imf.org The home page of the International Monetary Fund. This site provides a wealth of information about the IMF and its mission in addition to country information.

www.grameen-info.org The website of the Bangladesh-based Grameen Bank. It has information on the mission of Grameen Bank and how it works, together with material on other micro-credit schemes.

www.ecb.int/home/html/index.en.html A site dedicated to European central banking issues.

www.creativeinvest.com The site of Creative Investment Research, a US-based firm specializing in 'socially responsible' investment and providing many links on women and minority-owned financial institutions.

www.cdfa.org.uk The site of the CDFA, the trade association for Community Development Finance Institutions (CDFIs), which are designed to be sustainable, independent financial institutions that provide capital and support for individuals/organizations in disadvantaged communities or under-served markets.

http://blog.mint.com/blog/finance-core/a-visual-guide-to-the-financial-crisis/ A visual guide to the recent US–UK sub-prime financial crisis, illustrating the complex links between consumers, housing markets, credit ratings, financial instruments, banks and crisis.

For annotated, clickable weblinks and useful tutorials full of practical advice on how to improve your study skills, visit this book's website at **www.pearsoned.co.uk/daniel**

CONSUMPTION AND ITS GEOGRAPHIES

Chapter 19

Ian Cook and Philip Crang

Topics covered

- Economic geographies of consumption
- Global geographies of consumption
- Local geographies of consumption
- Consumption and geographies of (dis)connection

19.1 Economic geographies of consumption

If production refers to the economic process of making goods and services, then consumption refers to their utilization. It is vital to capitalist economies. At the simplest level, enterprises need the sales generated by consumption in order to realize value from their production processes. In turn, such sales necessitate huge organizational efforts; they do not simply happen. Thus, retail capital accounts for a growing proportion of paid employment and, in many sectors, exercises great organizational power within systems of commodity production (Wrigley and Lowe 1996). Marketing and advertising are also strategically central to contemporary economies, mediating between the worlds of producer and consumer and thereby helping to generate sales (Jackson and Taylor 1996; Nava *et al.* 1997; Nixon 1997; Hackley 2010). Discourses of consumption – framing thought and action in terms of a consumer focus – have come to dominate organizational theory (du Gay 1996), spreading out across not only private enterprise but on to marketised public services. Indeed, in parts of the economy production itself has been reconfigured away from material manufacture and towards the fashioning of 'signs and spaces' (Lash and Urry 1994) that are oriented to consumption.

According to Robert Goldman and Stephen Papson (1998), Nike are exemplary here, illustrating a wider shift towards a 'sign-economy' in which 'brands' are central to both production and consumption (c.f. Arvidsson 2006). Nike do not so much make and sell sportswear as make and sell the brand Nike, constituted through the sign and typography of the Nike name, the logo of the 'swoosh', and philosophies such as 'Just do it' (Plate 19.1). Sportswear, and other physical products, are the vehicles to mobilize the brand. So, Nike sub-contract manufacture; they do not make Nike stuff themselves. They are concerned about the physicality of Nike products, but less in terms of them as manufactured goods than in terms of them as outcomes of a 'design' process that materialises brand and sign value within the product. Nike invest far more in advertising and promotion (around 10 per cent of annual revenues) than in physical capital expenditure (around 3 to 5 per cent), and included in that capital expenditure is retail infrastructure, such as the 'Niketown' retail outlets (ibid.: 13). Nike's business and its corporate geographies are constructed to be market-led and market-leading. They are consumption-focused.

However, even as consumption is economically vital, it is not purely economic. If we reflect back on our opening definition of consumption – the utilization of the products of human labour – then we can see that consumption is the iceberg that lies below the tip of any purchase. Any sale has both a back story – the predispositions that shape our buying of this or that – and is but the start of an ongoing narrative concerned with the active use of the things we buy. Consumption is central to economics, but not reducible to it. Consumption is implicated in how we conduct our daily lives, in our management of household and family affairs, in our senses of self-identity, in our moral senses of how our lives ought to be and of how we ought to relate to others. In sum, consumption is cultural as well as economic.

This is one reason why consumption has been something of an enigma to mainstream economics (Douglas and Isherwood 1996). Think about something you have recently bought, about why you bought it, and how you plan for it to be used. In so doing, you can see how complicated the stories around a purchase can be. For our part, a recent purchase that comes to mind is an Apple iPhone 4. Philip and his partner bought this for their daughter, as a Christmas present.

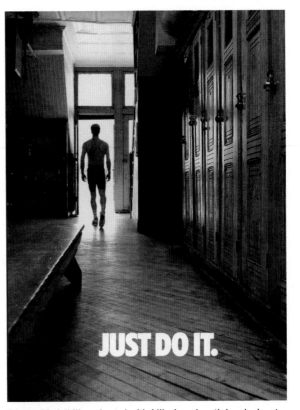

JUST DO IT.

Plate 19.1 'Nike: Just do it': Nike's advertising helps to establish the Nike brand's sign value.

Economics, in its simplest sense, certainly played a part in this purchase, as they weighed up whether they could afford it. But even that weighing up of price and value took them well beyond a narrow economic rationality. They considered, and argued over, what sort of price was appropriate for the main, perhaps only, Christmas present to give to a 14-year-old. They debated whether they were spoiling her. They reflected on her desire for an iPhone, moaned that she had not asked for anything else, and imagined how happy she might be to get one. They attempted to understand the teenage culture she was immersed in and the importance of digital and social media to it, batting back and forth views on her love of Facebook and Philip's (somewhat selfish!) concerns about her spending so much time chatting online to her friends rather than to him. They discussed the iPhone's touch-screen interface, and admitted they could see why someone would find it appealing. They laughed about their different attitudes to the Apple brand: Philip is a devotee, albeit one embarrassed at the predictability of his tastes in this regard; his partner sees brands as silly. In short, this single purchase was entangled in a complex economics of parental love and cultural taste (see also Miller 1998).

One might expect such acts of gift-giving to be invested with significance, but the same is true of purchases that we barely think about. For instance, consider Philip's most recent 'purchase': while writing and editing this part of the chapter in his study one morning, Philip got cold, and in response turned on the central heating. He is now buying additional gas. He is also, in small measure, further depleting the world of a non-renewable energy resource. He is doing so because his own sense of comfortable domestic ambient temperature was disturbed, and because he has become used to responding to a feeling of cold by turning the heating on or up. Both that embodied sense of comfort, and his response, are far from innate, or subject solely to crude economic forces of price. As Elizabeth Shove has demonstrated, our senses of things like comfort are not fixed, but change over time and between cultures (Shove 2003). Philip, for example, grew up in a house without central heating; if you were cold during the day, you moved to a room that had a heat source in it (an open fire or the cooking range) or put on more clothes. Now, like most households in the UK, he has got used to a normal domestic environment close to 20°C and the energy consumption that goes along with that. His habits and embodied senses of comfort have changed in conjunction with transformations in domestic infrastructure. This may seem pretty inconsequential, after all he only flicked a switch, but as Shove says, it 'acquires more sinister overtones when we recall the global environmental costs of maintaining what we now think of as comfortable conditions indoors' (ibid.: 28).

Such complexities to our purchases suggest, then, that consumption is a vital economic process, but not limited to a narrowly defined economic realm. It extends into our wider culture; and it does so in part because what is being produced and consumed has a distinctly cultural character. Images; meanings; feelings; experiences; habits and the technologies that underlie them; senses of identity and place: these are some of the sorts of stuff circumstances that consumption involves. Consumption is cultural as well as economic. One has to grasp consumption's status as a form of 'cultural economy' (Amin and Thrift 2004) if one is to understand its causes, characters and consequences.

In turn, this begs the question about the kinds of relationships between culture and economy generated through consumption. For some, the predominant trend is for economic, commercial imperatives to dominate and degrade our cultural lives: an 'economization of culture'. Famously, the philosopher Herbert Marcuse argued that in capitalist societies 'false needs' are produced (Marcuse 1964), a sense of wanting and requiring far more than is materially necessary and yet far less than might be culturally desirable (Plate 19.2). More specifically, numerous studies have debated the

Plate 19.2 'The iPhone. Not just a matter of economics.'

role of advertising in stimulating consumption, creating desires for products that would otherwise not exist (for a classic study in this vein see Packard 1977). In this portrait, then, it is economic forces that dominate. Indeed, in this instance consumer culture here is cast as a way of ensuring that mass production is accompanied by the appropriate levels of mass consumption. All in all, in such accounts consumers are portrayed as rather sad and pacified figures:

> irrational slave[s] to trivial, materialistic desires who can be manipulated into childish mass conformity by calculating mass producers. This consumer is a cultural dupe or dope, the mug seduced by advertising, the fashion victim … yuppies who would sell their birthright for a mass of designer labels. Ostensibly exercising free choice, this consumer actually offends against all the aspirations of modern Western citizens to be free, rational, autonomous and self-defining.
>
> (Slater 1997: 33)

On the other hand, it is possible to see consumer culture as less passive, as less controlled, as less subordinate. Here the emphasis is not on consumers being made to consume, but on consumers *using* the things they buy for their own ends. In this perspective, consumption is less the death of the commodity, the end of an economic chain that began with the commodity's production, and more its resurrection. Consumption involves, then, the 'culturalization of the economy'. In this spirit a range of research has stressed how consumers can creatively rework the products being sold to them, giving them new meanings in the process, and using them as launch pads for their own symbolic and practical creativity (Willis 1990). A classic example would be the Italian motor scooter and its histories of consumption in the UK (Hebdige 1987). Having been originally aimed in the post-war years at a market of young, urban women in Italy, in the United Kingdom the Lambretta and Vespa scooters became a central identifier of the 'Mod' subculture. Initially attractive to Mods for its Italianicity and design aesthetics, by the late 1960s the scooter was increasingly symbolically recast and materially customized to act as a subcultural icon opposed to the British, brute masculinity of 'Rockers' and their motorbikes (Plate 19.3). The scooter became part of a culture that its original designers had no knowledge of and certainly did not control. Designed as a form of transport for urban women in post-war Italy, it actually enabled a novel form of working-class culture and youthful masculinity to be created in the UK.

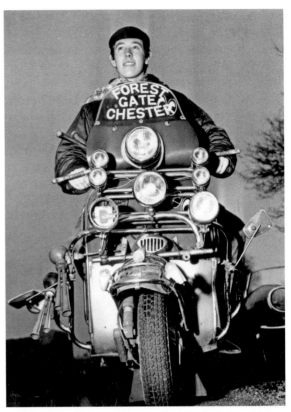

Plate 19.3 Originally designed to be a stylish mode of travel for Italian women, the Italian motor scooter became a crucial component of the British Mod subculture. The meaning of things we consume are not simply controlled by those who design them, but are remade through their consumption.

(Robert Stiggins/Express/Getty Images)

More generally, there have been a range of conceptual approaches that emphasize how consumption involves the use of commodities within at least partially autonomous cultural fields. Some emphasise their adoption within existing status structures and patterns of class distinction, so that consumption is regulated by and reproduces socially structured and inherited tastes (Bourdieu 1984). What we consume both reflects and creates what sort of class position we have. Others highlight the role of commodities within the private dream worlds of consumers, so that consumption is about fulfilling our fantasies and desires more than conforming to socially structured tastes (Campbell 1987). And still others, focusing on more mundane forms of consumption like household provisioning and food shopping, attend to how the things we buy enact domestic relations of love and care, so that consumption is about concretizing our relationships to our nearest and dearest (Miller 1998).

In weighing up these relations between economy and culture in consumption this chapter focuses on three geographies. First, focused in part on brands and their circulation, we highlight the global character of consumer culture, exploring its implications for cultural difference. Second, in examples ranging from shopping centres to nightclubs, we foreground the local contexts of consumption, reflecting on how they can both represent strategies of commercial interests, designed to promote and regulate consumption, and stage the remaking of places into forms that accord to different logics than that of commerce and economic rationality. Third, we turn to the connections between the local and global in consumption. The everyday acts of consumption we undertake in particular places are dependent upon wider networks of provision. The last part of the chapter therefore reflects on these networks and what we consumers know about them. In so doing, it offers a critical consideration of ideas of a (geo)ethical consumption and the kinds of worldly knowledge fostered through our acts of consumption.

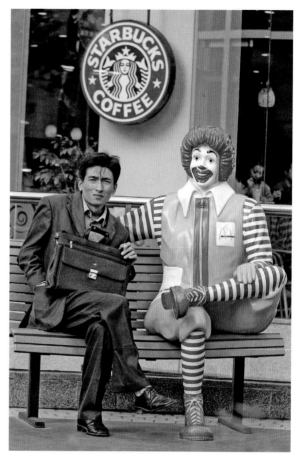

Plate 19.4 Man relaxes in the company of Ronald McDonald and Starbucks in Shanghai, China.

(© CLARO CORTES IV/Reuters/Corbis)

19.2 Global geographies of consumption

19.2.1 Consumption, cultural imperialism and the end of geography

Consumption is often cast as central to nothing less than the cultural fate of the world. At stake here is the extent to which modern consumption both produces and reflects a profound erosion of cultural traditions and differences, homogenizing the world into increasingly similar landscapes, people and social systems. Richard Peet, for example, argues that global capitalism and its consumer culture comprises 'a powerful culture which overwhelms local and regional experience . . . breaking down the old geography of society and culture' (Peet 1989: 156). Robert Sack concurs, detecting a 'trend toward a global economy and culture, which seems to require that places all over the world contain similar or functionally related activities and that geographical differences or variations that interfere with these interactions be reduced' (Sack 1992: 96).

More generally, it is hard to miss the worldwide spread of products and brands such as McDonald's, Coca-Cola or Starbucks. Indeed, such brands have entered academic language as bywords for a con-

sumerist homogenization – in the form of theses of the 'McDonaldization of society' (Ritzer 2000) or of 'Coca-colonization' (for a critique see Flusty 2004) (Plates 19.4 and 19.5). Some of these issues of cultural homogenization are discussed in Chapter 13. This section will consider them in more detail, focusing upon the issue of consumption.

For many, what modern consumption enacts is a form of 'cultural imperialism' (see Tomlinson 1991 for an excellent review). This thesis takes a number of forms. For some it is characterized by a 'dumping' of First World, and especially American, consumer products on to Third World markets: 'authentic, traditional and local culture in many parts of the world is being battered out of existence by the indiscriminate dumping of large quantities of slick, commercial and media products, mainly from the United States' (Tunstall 1977: 57). Of concern here is both the economic dominance of First World and American multinational

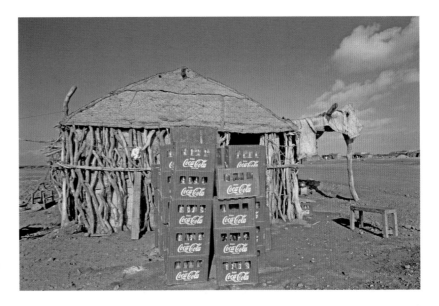

Plate 19.5 Coca-colonization?

(Stefan Boness/Panos Pictures)

companies (Mattelart 1979) and the predominance of products that promote First World and American cultural values (Dorfman and Mattelart 1975) (for a general review see Fejes 1981). This economic and cultural power results in the overwhelming of indigenous, 'authentic' local forms of life, so that the 'real' culture of an area and a people are lost.

For other critics, the problem is not just the conquest of some national and regional cultures by others, but also the ways in which consumerism destroys all culture as we have traditionally understood it. The suggestion here, then, is that what consumption represents is the contemporary dominance of a very peculiar cultural form – consumer culture – which in many respects is barely cultural at all. Consumption, some suggest, is all about the political economy of market shares, niches and profit, and as such fundamentally opposed to the intrinsic values of cultural meaning and communication. George Ritzer, in his recent account of what he calls 'the globalization of nothing' (2004), argues that consumption tends to produce empty forms and spaces, devoid of ties to particular times and places (Spotlight box 19.1). Consumer culture is marked by a 'nothingness'. What matters here is partly the loss of cultural diversity – the making of a duller world – but perhaps more importantly the loss of cultural alternatives – the making of a less free and creative world. Uniformity entails conformity. This results in a form of culture that is inadequate and unfulfilling for many of those it is imposed upon:

A cultural system which would be adequate for the poorest people in that system would mean a set of

---

Spotlight box 19.1

Consumerism as the end of culture?

'Culture' was one of the crucial terms through which anti-liberal forces counted the cost of modernity. '*Consumer culture*', in this perspective, is merely an ersatz, artificial, mass-manufactured and pretty poor substitute for the world we have lost in post-traditional society. In fact, it is the antithesis and enemy of culture. In it individual choice and desire triumph over abiding social values and obligations; the whims of the present take precedence over the truth embodied in history, tradition and continuity; needs, values and goods are manufactured and calculated in relation to profit rather than arising organically from authentic individual or communal life. Above all, consumerism represents the triumph of economic value over all other kinds and sources of social worth. Everything can be bought and sold. Everything has its price. 'Consumer culture', therefore, is a contradiction in terms for much of modern Western Thought.

(Slater 1997: 63)

The social world, particularly in the realm of consumption, is increasingly characterized by nothing. In this case, 'nothing' refers to a social form that is generally centrally conceived, controlled and comparatively devoid of distinctive substantive content . . . we are witnessing the globalization of nothing.

(Ritzer 2004: 3, 2)

instrumental, symbolic and social relations that would help them to survive in meeting such fundamental needs as food, clothing, housing, medical treatment and education. Such needs are not met if they are identified with the consumption of Kentucky Fried Chicken, Coca-Cola, Aspro, or Peter-Stuyvesant cigarettes.

(Hamelink 1983: 15)

In question here, then, are the values that underpin consumer culture, and the ability of consumption to satisfy the 'real' as opposed to the 'false' needs of consumers. Clearly these are fundamental issues. However, the accounts related above are partial and rest upon some problematic assumptions. First, they suggest that the authenticity of a culture is dependent upon its resistance to external forces and artefacts. This can produce a somewhat fossilised and bounded sense of cultural life. Second, they argue that non-Western consumers are duped or coerced into having consumption tastes that are inappropriate to their real needs, just as the mass of Western consumers have been too. There is a dangerous arrogance here; at times it seems that only the critics of consumer culture know what it is any of us 'really' need. And there is an empirically contestable assumption that consumers make exactly the same of global brands the world over. We shall now turn to some of these criticisms, through the lens of work that emphasizes how consumption involves processes not of 'homogenization' but of 'indigenization', of making global things 'local'.

19.2.2 Consumption and cultural indigenization

Let us start with someone who may seem an unlikely proponent of this idea of indigenization. Marshall McLuhan wrote one of the most famous social science books of the post-war period, as he sought to understand the character of a contemporary society saturated by the mass media (McLuhan 1964). His argument about the increasing linkages between different places, and phrases expressing this such as the notion of the 'global village', have passed into common parlance, echoed in numerous advertising campaigns promoting the unifying qualities of new communication technologies. However, his argument always went beyond suggesting that this increasing interconnection between people and places – or in the much-cited language of the Marxist geographer Harvey (1989), this 'time–space compression' of the world – meant simple homogenization. He elaborates on this in a later paper:

The question arises, then, whether the same figure, say Coca-Cola, can be considered as 'uniform' when it is set in interplay with totally divergent *grounds* from China to Peru? . . . The 'meaning' of anything . . . is the way in which it relates to the user . . . [For example] it might be suggested that the use of American stereotypes in the United Kingdom yields a widely different experience and 'meaning' from the use of the same stereotypes or *figures* in their American *ground*. The interplay between the *figure* and the *ground* will produce a wide range of different effects for the users.

(McLuhan 1975: 44; emphases in original)

McLuhan argues, then, that consumer culture does not float above the world in some abstract space of capital or modernity. Global products land and take root in particular places and times. In so doing, such forms of global consumer culture not only impact on their surroundings; they are also shaped by their locations. Anthropological studies in particular have developed this theme, exploring how globally distributed facets of consumer culture – whether particular goods, styles or cultural orientations – are locally incorporated.

Take, for example, one of the exemplars of homogenising American consumer culture: McDonald's. This brand is famed for its uniformity; the same decor, the same basic menu (with very small variations, such as McSpaghetti in the Philippines or the Maharajah Mac in India), and the same service style the world over. And yet, McDonald's may not be just a force for cultural homogenization. This was certainly Watson's (1997a) conclusion, based on studies of the consumption of McDonald's in East Asia (see Case study 19.1). These studies found that McDonald's has been localized, indigenized and incorporated into traditional cultural forms and practices, even as it has played a part in the ongoing dynamism of those traditions. As Watson poses the question:

Does the spread of fast food undermine the integrity of indigenous cuisines? Are food chains helping to create a homogeneous, global culture better suited to the needs of a capitalist world order? . . . But isn't another scenario possible? Have people in East Asia conspired to change McDonald's, modifying this seemingly monolithic institution to fit local conditions? . . . [Perhaps] the interaction process works both ways.

(Watson 1997a: 5–6)

And his conclusion is that there is indeed a two-way interaction:

McDonald's has effected small but influential changes in East Asian dietary patterns. Until the introduction of McDonald's, for example, Japanese consumers rarely, if

Regional case study 19.1

Consuming McDonald's in East Asia

McDonald's outlets have colonized much of East Asia, from their arrival in Hong Kong in the 1970s to their growth in metropolitan mainland China in the 1990s. However, just how McDonald's is consumed varies across the region. In Beijing, for example, McDonald's lost its American role as a place of fast and cheap food. Instead, Yan suggests, it became a middle-class consumption place, somewhere for a special family outing where 'customers linger . . . for hours, relaxing, chatting, reading, enjoying the music' (1997: 72). McDonald's here is seen as American, but Americana means something stylish, exotic and foreign, resulting in the meanings and experiences of McDonald's in Beijing being very un-American (Plate 19.6). In contrast, in Japan, whilst there is a similar leisurely use of McDonald's, it is not a place of exotic social prestige,

but a youth hangout, a place where someone in a business suit would be out of place (Ohnuki-Tierney 1997). In Hong Kong, McDonald's was likewise marketed to the youth market, and adopted as such initially, but here its American origins are of little contemporary consequence (Watson 1997b). Whilst initially patronized in the 1970s by adolescents seeking to escape products associated both with China and with Hong Kong's perceived provincialism, by the 1990s McDonald's has become routinely local, just another mundane part of the Hong Kong landscape: '[t]oday, McDonald's restaurants in Hong Kong are packed – wall to wall – with people of all ages, few of whom are seeking an American cultural experience. The chain has become a local institution in the sense that it has blended into the urban landscape . . . McDonald's is not perceived as an exotic or alien institution' (Watson 1997b: 87, 107).

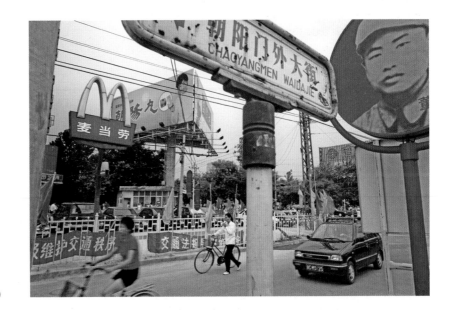

Plate 19.6 McDonald's in Beijing: American fast food becomes exotic, stylish Americana.

(National Geographic/Getty Images)

ever, ate with their hands . . . this is now an acceptable mode of dining . . . [However,] East Asian consumers have quietly, and in some cases stubbornly, transformed their neighborhood McDonald's into local institutions. [For example] In the United States fast food may indeed imply fast consumption, but this is certainly not the case . . . [in] Beijing, Seoul, and Taipei . . . [where] McDonald's restaurants are treated as leisure centers, where people can retreat from the stresses of urban life.
(ibid.: 6–7)

Hence the meanings and practices of consuming McDonald's – often cast as an icon of global homogenization – vary from place to place. Melissa Caldwell's work (2004) on McDonald's and consumer culture in post-Soviet Moscow reinforces the argument. In Russia too McDonald's has been indigenized or 'domesticated', and not just at the level of everyday practice. Counter-intuitively, McDonald's – which as outsiders we might read as an American export to Russia – has come to be understood by consumers in Moscow as an 'authentically Russian product', to the extent that eating there is seen as responding to 'nationalist-oriented consumer campaigns' (ibid.: 5). Whilst initially perceived and promoted as an exotic American import, more recently McDonald's has become Russian, explicitly marketed as 'Nash Makdonalds' (which Caldwell translates as 'our McDonald's'). Since the mid-1990s Russian consumer culture more generally has evidenced a patriotic enthusiasm for goods understood as 'Nash'. This Nash ideology entangles notions of national space with those of everyday familiarity and trust. McDonald's in Moscow has embraced this ideology and has thus been transformed from American exotica to both mundane familiarity and imagined Russianness.

Parallel accounts of indigenization have been produced through research on the consumption of a range of other global brands, products and commodity genres. The anthropologist Danny Miller, for example, has undertaken a series of studies in Trinidad, addressing how global phenomena ranging from the American television soap opera *The Young and the Restless* (Miller

1992), to 'Coke' (Miller 1997) and to the Internet (Miller and Slater 2001) are not just localized but made authentically Trinidadian. Jonathan Friedman's (1994) study of the subcultural consumption of European designer fashions in the Republic of Congo shows that not only do designer labels come to mean something different here to what they mean in Paris or London, but that this difference stems from distinctively Congolese relating of consumption and personhood (see Case study 19.2). The collective 'global denim project', convened by Danny Miller, has demonstrated how the ubiquity of this textile around the world has enabled, rather than eradicated, a range of local cultural expressions to be developed, whether that be through youthful Muslim fashions for decorated jeans in Kerala, India, or the use of stretch denim fabrics and figure hugging jeans in Brazil, or the wearing of 'carrot cut' jeans by hip-hop fans of Turkish extraction in Berlin (see Miller and Woodward 2010).

In sum, a body of research suggests that rather than there being one consumer culture that has spread across the world, there are many consumer cultures, with forms and meanings that are locally specific. At a

---

Regional case study 19.2

Western goods and cultural indigenization: the case of 'Les Sapeurs'

'Les Sapeurs' are a group of consumers in the Republic of the Congo (see Friedman 1994). They are men of the politically marginalized Bakongo ethnicity, drawn especially from the poor in large cities such as Brazzaville and Pointe Noire, who aim to move through a graded and spatialized consumption process of 'Western' fashions. Beginning with the conspicuous wearing of imported ready-to-wear items, Les Sapeurs look to progress to a stay in Paris (*l'aventure*). Here they accumulate *haute couture* from French and Italian design houses whilst usually living in extreme poverty, before returning to the Congo. There the sub-culture is supported by 'Sape clubs', where men perform competitively in '*la danse des griffes*', a catwalk display of one's dressed self.

All of this, especially at a first hearing, might be rather too easily dismissed as an extreme case of fashion victimhood, or as a bizarre oddity, thrown up by the maelstrom of globalized consumer culture as inflected in the Congolese context. However, for

Friedman, far from being a cultural oddity, 'la sape' is symptomatic of much more general local–global geographies of consumption, in which globally distributed commodities and materials – here Western European designer fashions – are encompassed within local cultural logics of identity practice. In the Congo, understandings of identity are very different to Western fashion culture's conceptions. For the Bakongo, senses of self-identity are framed by a conception of a cosmic 'life force', which can be tapped into through consumption in order to create a successful self-fashioning. For the socially marginalized, there is a concern to find sources of this life force that are external to the local political and cultural hierarchies from which they are excluded. As such, the conspicuous consumption of foreign fashion and designer labels becomes something very different from what one might first make of it. It is not any sort of dandyism, nor motivated by a desire to imitate Europeans. Rather it is, Friedman suggests, 'a life-and-death struggle for physical and social survival' (1994: 106) that accords to the existing, indigenous cultural logics of the Bakongo.

broader level this marks a conceptual shift away from seeing a modern capitalist world as a singular universal entity, and towards recognizing the many 'specific', 'prismatic', 'comparative' or 'global' modernities being forged through consumption practices worldwide (see for example Breckenridge 1995 on South Asia; Featherstone *et al.* 1995 generally; and Miller 1994 on Trinidad). Rather than necessarily destroying local, indigenous cultures, the consumption of non-local products may be irrelevant, indeed even helpful to, the ongoing production of authentic local cultures. Miller argues that in a world of globalized consumption 'authenticity has increasingly to be judged *a posteriori* not *a priori*, according to the local consequences not local origins' (Miller 1992: 181). It matters less where things come from than what they are used for. Indeed, it may be that the non-local origins of products may be positively beneficial, giving 'the periphery access to a wider cultural inventory' and providing new raw materials for local cultural and economic entrepreneurs (Hannerz 1992: 241; see also Edensor and Kothari 2006 on the acquisition and sale of locally produced branded goods in a market in Mauritius).

19.2.3 Consuming difference

Not only do consumer cultures differ from place to place, but ideas of cultural and geographical difference – what we might call 'imaginative geographies' (Driver 2005) – are produced and sold within consumer culture.

Exemplary here, it is often suggested, are the themed environments of consumption spaces such as amusement parks and shopping centres. In Disneyland and Disney World, for example, the amusements and other consumption opportunities are not just randomly arrayed, but organized into distinct lands: Frontierland themed on the American West; or Adventureland themed on landscapes of European colonialism and imperialism (see Gottdiener 1982; Wilson 1992; Bryman 1995). This echoes much earlier consumer festivals and fairs such as the Midway attractions at the 1893 Chicago exhibition (Rubin 1979) or Luna Park at Coney Island, New York (Kasson 1978), and is also a pattern to be found in many large shopping centres, most famously the huge West Edmonton Mall in Canada (Crawford 1992; see also Goss 1999a). Indeed, Gottdiener (1997) goes so far as to posit that consumer culture has produced a general 'theming of America', so that theme parks become models for what is happening across the urban landscapes outside them. Sack

(1992: 98) agrees on this production of 'pseudo-places' in consumer culture. Harvey (1989) sees the same general trend towards highly packaged, and indeed simulated, experiences of cultural and geographical difference, such that microcosms of the whole world are encountered in particular consumption spaces.

From a commercial perspective, geographical images are a resource to be used in projects of brand making and differentiation. (For a wider discussion of the geographical aspects of brands see Pike 2009.) Marianne Lien, for example, narrates a fascinating case study of frozen pizza marketing in Norway, as one company (given the pseudonym Viking Foods) differentiates its various product lines through geographical associations (Lien 2000). Some of these are predictable and none too subtle: one pizza line is 'made Italian' and more 'authentic' through packaging that references Il Tricolore's green, white and red, and uses Italian place names and language; another line is 'made American', more 'modern' and more oriented to a youthful market with a packaging that has stars and stripes, American football helmets, cowboys and the Statue of Liberty! But other positionings are more complicated, in particular in the case of a pizza product line that Viking Foods wanted to establish as meaningfully Norwegian to their Norwegian consumers, as what they term a 'folkepizzaen'.

More generally, you are probably familiar with a range of examples of such branded place association. Think, for example, about the bottled waters that appeal to place imagery to differentiate a somewhat generic commodity (on water branding see Wilk 2006; for the case of marketing Fijian water see Connell 2006 and Kaplan 2007). Or consider products that associate themselves in some way with imagined geographical areas such as the tropics or the rainforest (for a critical analysis of a 'rainforest' perfume see Slater 2004a; on tropical fruit see Cook *et al.* 2004). Or there are the brands that draw upon national identifications in framing their identities: whether that be the likes of Paul Smith and Burberry presenting their Britishness within the fashion economy (Goodrum 2005); or car brands like Alfa Romeo (Italian passion) or Mini (with its fascinating marriage of British quirkiness and Germanic efficiency). And of course there are famed examples of brands, like Benetton and Diesel, that deliberately deploy discourses of difference and multiculture to market themselves (Lury 2000).

There are a number of implications of these associations between the commodities we consume and ideas of cultural and geographic difference. For a start, it means we need to broaden the critique of consumer culture beyond a debate over its tendencies for homogenization or indigenization, to consider capitalist processes of 'heterogenization'. In Naomi Klein's terms, we need to recognize the emergence of a 'market masala', where differences become part of capitalist consumer culture (Klein 2000). For some commentators, such as Klein herself, it is the politics of incorporation that are key here. Capitalism, via consumer culture, increasingly absorbs forms of difference, whether cultural, geographical or political, that might seem in opposition to it. Oppositional voices need to find ways either to resist or to use this incorporation for their own ends (see also McGuigan 2009).

For others, this is less a process of incorporation than a more diverse range of ways in which difference becomes commodified and consumption acts as a space for cross-cultural encounters and translations (Jackson 1999). In this perspective a range of other issues come to the fore. How, for example, are consumer cultures implicated in present and past circulations of material culture, and with what effects on national cultures? What kinds of differences are fostered or occluded within consumption and what are the implications of this for the wider politics of identity? What options are there for both producers and consumers to express and develop their cultural identities through commercial cultures? Such questions have shaped a set of research projects by Philip Crang and colleagues on the presence of South Asian 'stuff' in British consumption cultures, especially of food and fashion. These studies range from conceptual reflections on the relations between commodity cultures and transnationality (Crang *et al.* 2003), to accounts of how 'Eastern' difference is presented and commodified in British food marketing (Jackson 2002) and fashion design and retail (Dwyer and Crang 2002; Dwyer and Jackson 2003; Jackson *et al.* 2007), to analyses of how contemporary fashions for South Asian styled goods relate to past colonial histories in which South Asian materials (such as cotton, or the boteh/Paisley pattern) were central to British cultures of design, fashion and dress (Crang and Ashmore 2009; Breward *et al.* 2010).

More generally, it is argued that consumer cultures can act as arenas of geographical knowledge, in which understandings of both the origins of what we consume and our own places in the world are played out. Think, for instance, of how food marketing presents geographically referenced cuisines (Indian food, Chinese food, Italian food, Mexican food and so on). Here, a mundane realm of consumption is also a

Plate 19.7 Eating into Britishness: by what process do these 'Indian' restaurants become 'British'?
(Photography courtesy of G.P. Dowling)

setting in which ideas of cultural difference and mixture are played out, as retailers and restaurants present consumers with 'the world on a plate' (Cook and Crang 1996; for a sample of related work on food geographies see Cook *et al.* 1999, 2008; Cook, Crang and Thorpe 2000; Duruz 2005; Monrreal 2008).

19.3 Local geographies of consumption

We have been suggesting that consumption involves the extension of the economy into realms of life that we might not usually think of as economic, into our wider 'culture'. This is also true at the more immediate scale of our own everyday economies. When we buy goods we take them home; we 'domesticate' them. In consuming things, we make them part of household relations; we live with them and their lives become shaped by ours (Gregson 2007). They become part of what we might call 'moral economies', insofar as: '[t]he household is a moral economy because the economic activities of its members are defined and informed by a wider set of cognitions, evaluations and aesthetics, which themselves are defined and informed by the biographies of the household and its members' (Silverstone *et al.* 1992: 11). The language here may be unfamiliar, but we all have experiences of how economic goods become part of something else, something more than economic, in their consumption: the toys that become childhood companions and then symbols of our childhood and its end (a narrative powerfully mined by the makers of Pixar's *Toy Story* movie trilogy); the little 'treat', bought by a loved one in a mundane shopping event, that says they were thinking about you (Miller 1998); and so on.

This entangling of economic and social relations through consumption is not limited to domestic space. It is also true of public spaces of consumption. Here commodity culture is implicated in our forms of public sociality and in how we relate to our immediate environments, our local places. By way of illustration, we start with the places where people go shopping.

19.3.1 Shopping and place

Shopping centres (or malls as they are more commonly termed in North America) are routinely cast as the cathedrals of our consumerist age, symbols of our worship of commodity culture. But what goes on inside a shopping centre, and what does this tell us about the kinds of places being produced in the world of consumption?

One response would be that in shopping centres we are manipulated into behaving in certain ways, and especially into buying things, by the power of the shopping centre as a place. It is certainly the case that shopping centres are in part designed according to a 'merchandise plan' (Maitland 1985: 8) that tries to maximize the exposure of consumers to goods. In older shopping centres and malls this plan tended to be understood in quite mechanical terms. Shopping centres were seen as functional 'machines for shopping'. 'Generator' stores are used to pull us to the shopping centre in the first place, and once we are there 'magnet' stores every 200 metres or so make sure we do not just pop in and out, but explore every part of it. Here, then, 'the public mall [is defined] as essentially the passive outcome of a merchandising plan, a channel for the manipulation of pedestrian flows' (Maitland 1985: 10). There are well-known tricks to make us behave as the shopping centre management and its shops want: escalators arranged so we have to walk past shop fronts or merchandise when going between floors; hard seats so we do not linger too long without getting up and seeing some more potential purchases; no water fountains so we have to buy expensive drinks (see Goss 1993).

In more recent shopping centres, however, the manipulation may be more subtle. Increasingly, retail planners and developers see their job as providing spectacular places that people will want to spend time in. This is because research suggests that the longer people stay, the more on average they spend, and because as competition between centres intensifies, each one needs more than just the usual shops to attract us in the first place – everything from fountains to funfairs (see, for example, Goss 1999b on The Mall of America). Here, then, the manipulation of shoppers is less mechanical and more affective. The hope is to excite, inspire, relax and please (see Case study 19.3). Of course, these are promotional intentions rather than shoppers' experiences. As Chaney (1990) puts it in his review of the MetroCentre in Gateshead, north-east England, the designer's utopian impulses to produce a place of perfection often result in rather underwhelming, 'subtopian' forms. Nonetheless, there is much here to be taken seriously. There is the attempt to produce a fabricated space in which the individual consumer can be made to feel like consuming (see Goss 1999a). There is the emphasis on creating an internal, closed-off, privately owned but partially public environment,

---

Regional case study 19.3

The shopping mall experience: Woodfield Mall, Schaumburg, Ilinois

Leaving the comfort of her air-conditioned car, the shopper hurries across the sun-broiled or rain-swept tarmac and thankfully passes through doors into an air-conditioned side mall. Drawn down this relatively dark, single-level tunnel, lined with secondary units, by the pool of light ahead, she reaches one of the squares ahead located at the end of each arm of the main mall system. She discovers this to be a two-storey space . . .

On one side of the cavern is a Magnet department store, but before deciding to enter it she turns to see that a second tunnel, much taller than the first with natural light spilling though holes punched in its roof, leads off into the heart of the building . . . This route now culminates in the central space, Grand Court . . . Its floor has been quarried and shaped to form an amphitheatre and pools . . . Emphasis is placed upon the creation of a plausible internal world, full of visual interest and variety, in which one is tempted not simply to pass through, but to stay.

(Maitland 1985: 26, 28)

divorced from the harsh exterior world, not only climatically but also socially, through the operation of security systems that ensure the absence of anyone who might threaten this consumer paradise or disrupt the pleasure of the shopper. Thus, these privately owned and managed spaces rework the older public spaces of the street or market:

> They reclaim, for the middle-class imagination, 'The Street' – an idealized social space free, by virtue of private property, planning and strict control, from the inconvenience of the weather and the danger and pollution of the automobile, but most important from the terror of crime associated with today's urban environment.
>
> (Goss 1993: 24)

Such shopping centres recognize that there is much more to going shopping than buying things (Glennie and Thrift 1992). Shopping is in part about experiencing an urban space, seeing and being seen by other shoppers. It is a social activity. But they seek to manage this social experience. The street gets recast as a purified space of leisurely consumption, cleansed of nuisances that inhabit the 'real' streets, like 'street people'.

However, this fashioning of the mall can be disturbed. Retail spaces can be reclaimed by those who use them. Leading the charge are a mixed assortment of senior citizens and adolescents. Take the case of Bert, 78 years old, and a regular visitor to a shopping mall in New England (USA):

> I come here at quarter to eleven and I leave at twenty minutes past one . . . I do this five days a week, except I missed one day this year . . . And I have my lunch here at noontime . . . I move every half-hour. I go from here down to where the clock is, I go down in front of Woolworth's, then I come back again and go up there and take my bus by the front of JC Penney.
>
> (Quoted in Lewis 1990: 126)

For Bert, the mall is not a place to shop, but somewhere sheltered to go outside his house, a chance to meet up with friends. For the mall management there is the option of evicting these low-spending visitors, but ejecting old people forcibly on to the pavement is hardly good public relations. Instead, and as part of wider moves to see malls as the new civic centres, Bert's mall has actually established its own walking club for the elderly, with over 300 members, for whom the doors open at 6 a.m., rather than 9 a.m. when shops open.

Adolescents also use the mall as a meeting place, somewhere to hang out. They also spend very little.

'Suburban kids come to malls to look around, meet and make friends, and hang out – because there is nowhere else to go' (Lewis 1990: 130). Or as Ed and Tammy, self-confessed 'mall rat' and 'mall bunny' respectively, put it:

> [Gesturing around himself] I met all these people here. I've met lots of other people, too. One place where you can always find someone . . .
>
> (Ed; quoted in Lewis 1990: 130)

> I used to come here every Saturday from 11 am to 9.30 pm and just walk around with my friends, like Gina here, and just walk around and check out the guys.
>
> (Tammy; quoted in Lewis 1990: 130)

Unlike the elderly, the mall is less indulgent of these teenagers. There are constant skirmishes with the security staff over how long ten people can share one Coke at the food court, and some mall rats and bunnies get banned altogether.

So what are we to make of our trip to the mall? What does it tell us about the local geographies of consumption? Well, it certainly tells us that place matters to consumption. But how? Here the answer is more ambivalent. In part what we see is a pseudo-public space symbolizing and enacting claims for consumption to be all there is to civic life, making citizens consumers in a very concrete way. In this light, shopping centres illustrate how place can produce both acts of consumption and people as consumers. Yet in shopping centres one also sees the small-scale, trivial resistances that people enact, their refusal to be just consumers in the sense of purchasers, their determination to use this space for their own ends. And that too has a wider symbolism, signalling the possibility that consumption is rarely just about what we are being sold, but also about fulfilling our own needs for social affiliation, participation and place making.

19.3.2 Economies of experience: skateboarding and dancing

Shopping is not just about buying goods, then; it is a placed experience. More generally, a range of research has framed consumption as experiential. For some, this involves understanding consumption as a resistant act in which we, as consumers, creatively use the places made for us. Writing from the discipline of architecture, Iain Borden (2001) pursues this approach in a study of skateboarding, space and the city. Paying particular attention to 'street skating', he argues that

Plate 19.8 The new Bullring shopping centre, Birmingham, United Kingdom was opened on 4 September 2003. The round tower (known as the Rotunda) that was part of the 'old' Bullring has been retained and redeveloped.
(Images of Birmingham/Alamy)

skateboarders remake urban spaces designed for other purposes: sometimes explicitly symbolic monumental space (e.g. town halls, national theatres, historical monuments, tourist attractions); but more often everyday spaces of neglect, the left-overs of rational, economically focused planning (e.g. mini-roundabouts, the spaces under bridges and urban highways, mini-malls). Thus spaces of public, official symbolism and of arid architectural functionality become the stages for energetic practices of place consumption. Handrails become tools for 'ollie nose grinds'; cement banks opportunities for 'shredding'. Skaters consume the city with their boards, recomposing it in the process. Skateboarding enacts a way of using the city, of consuming the city, that precisely emphasizes use over commercial or other values. The city becomes a site of play, of skillful embodied practices, of memorable performances. For Borden, this contests the commodified logic that underpins contemporary urban space.

However, these sorts of performative engagements with place can also be reclaimed as commodities. The business theorists Joseph Pine and James Gilmore talk more generally about the emergence of an 'experience economy' focused on the production and consumption of memorable experiences (Pine and Gilmore 1999). Place is again central here; this is a theatrical economy, in which places are the arenas in which certain kinds of social performances can be developed. Consumption is the undertaking of those performances. Clubs (in the sense of nightclubs) are exemplary.

In London alone there are over 500 club nights on offer every week. In Britain the industry has a turnover of £1.8 billion per annum, and it is estimated that 42 per cent of the population now visit a club once a year, 42 per cent of 15–19-year-olds at least once a month (Mintel 2004, 2006). Ben Malbon (1997, 1999) argues that clubs are spaces of play. This should really not need saying. It is patently obvious that people go to clubs to have fun. It is also patently obvious that this fun is achieved through a variety of means. Through enjoyment of the music; through an enjoyment of being part of a crowd of people with whom, temporarily, one can feel at one; through the feelings of competency and bodily expressiveness that dancing can give you; through the sharing of a special place and time with friends; through behaving in ways one cannot in other social situations, for example at work. Perversely, though, this fun needs to be taken seriously. We need to understand the playfulness of much consumption. We need to analyse the complicated microspatial practices through which play is enacted; the ways in which one gets to feel, and enjoy being, part of the clubbing crowd; the ways in which one dances so as to lose oneself in the music. We need to grasp the importance of place to playful consumption; the ways in which one has to be in the right place, with the right music and the right crowd, to have these feelings of exhilaration, communality and vitality; the way that in such consumption 'the emphasis is placed on the near and the affectual: that which unites one to a place, a place that is experienced among others' (Maffesoli 1996: 128). And we need to appreciate the wider significance of such playful consumption for our productions of local experience, such that consumer culture is not just made up of atomistic, avaristic individual consumers, but also facilitates temporary, some would say 'neo-tribal' gatherings, through which people come to feel part of something bigger than themselves (see Maffesoli 1996: 72–103).

Clubbing is, then, a place-specific practice of consumption. Whereas in the shopping centre or with street skateboarding senses of communal belonging tend to be forged by those occupying space for purposes other than those for which it was designed, in clubs that communal belonging is the very product that people are buying. Clubs may be distinctive places of consumption, characterized by an exceptional emotional and bodily intensity, but they also point to wider trends within the so-called 'experience economy' towards localized places of theatrical consumption.

19.4 Consumption and geographies of (dis)connection

So far in this chapter we have been considering the implications of consumption for both global dynamics and localized productions of place. In this final section, we bring these global and local scales together, to discuss the wider implications of our everyday practices of consumption. Our focus in particular is on how consumers are both connected to, but often feel disconnected from, the other people and places involved in providing the things we consume.

Consumption may often feel mundane, or perhaps not serious enough to matter that much, but it has profound political and economic implications. In the words of Danny Miller, the figure of 'the consumer' has become a sort of 'global dictator': 'Today, real power lies unequivocally with us, that is the waged consumers of the First World' (1995: 10). In consequence, opposing the study of consumption because so many in the world are not affluent consumers is profoundly misplaced. Quoting Miller again, 'The acknowledgement of consumption need not detract from the critique of inequality and exploitation, but this critique is foundering precisely because the enormous consequences and attractions of consumption are left out of the analysis' (ibid.: 21). A fundamental challenge for the twenty-first century is to fashion economic forms that fulfil those attractions of consumption, but deal too with its consequences.

19.4.1 Dis-connections

In Robert Sack's view, we consumers are profoundly ignorant of the geographies of provision that bring goods to us: 'A shop that sells Colombian coffee does not reveal the social structure that produces the coffee, the economic impact of coffee production on the Colombian economy, or the way coffee growing affects the Colombian environment' (Sack 1992: 200). Such ignorance, he argues, makes it impossible for consumers to behave responsibly in relation to those involved in commodity production and distribution. Unless we know about the origins of the goods we consume, then we lack the necessary understanding to assess the impacts of our actions as consumers; and without such understanding of consequences, responsible action is impossible. Thus, modern consumption's disconnected geographies make it deeply amoral.

One critical response to such ignorance and amorality is to see consumers as blinded by a veil that is draped over the things we consume, through a process of what Marx called 'commodity fetishism'. In this portrait the true history or biography of a product, and its basis in the work of others, is obscured to consumers. Instead, one is confronted by advertising imagery that seeks to associate the product with other fantasy worlds. The critical response, then, becomes to puncture these fantasies with some sharp shafts of reality, to 'defetishise' the commodity. Elaine Hartwick points to the startling incongruities and inequalities that such a puncturing can bring to light:

> Michael Jordan, basketball player extraordinaire, receives US$20 million a year to endorse Nike; Philip Knight, cofounder and chief executive officer of Nike, is worth US$5.4 billion; Indonesian workers are paid

US$2.40 a day and Vietnamese workers are paid US$10.00 a week to make the sneakers . . . The Walt Disney Company is publicly criticized for its treatment of Haitian workers making Pocahontas shirts for 28 cents an hour; Disney's chief executive officer makes US$78,000 a day.

(Hartwick 1998: 423)

Hartwick's suggestion is to approach consumption in terms of 'commodity chains'. Here she draws on a much wider body of scholarship (see Hughes and Reimer 2004) to argue for what she terms a 'geoknowledge' of consumption, one that traces back the things we consume through the links in a provision chain, eventually connecting consumers to producers (see also Chapter 16). Such a knowledge, she argues, allows a 'geoethics' of consumption, attuned to the politics of the economic systems upon which consumption depends.

Smith's account of Starbucks Coffee shops makes a similar 'defetishization' argument. He highlights how unfairly traded coffee sees only 25 per cent of the retail price returning to the producing country, with even less going to those who actually cultivate the crop (the smallholders, waged workers and seasonal pickers). He contrasts the stylish coffee-shop culture that a brand like Starbucks epitomizes with the lives of those who produce coffee. For the women who pick the coffee beans, their working day begins somewhere between three and four in the morning. Getting back home at six in the evening, they still have evening meals to prepare, water to fetch from the river, a family to feed and clear up after. Few live in houses with electricity or running water. They face a major health hazard from the pesticides used on the coffee plants (M.D. Smith 1996: 512–13). Other accounts profess a more open reading of the juxtapositions of different moments and

places in a commodity's life. Ian Cook's 'followings' of foods – particularly the papaya – are exemplary in that regard (Cook *et al.* 2004, 2006).

It is not just academic accounts that look to connect up spaces of consumption and production in this way. You have probably seen documentary films with a similar project; one of the best known is *Black Gold: Wake Up and Smell the Coffee*. Produced and directed by Marc and Nick Francis, and first released in 2006, *Black Gold* centres its narrative on Tadesse Meskela, the General Manager of the Oromia Coffee Farmers Co-operative Union in Addis Ababa, Ethiopia. It combines a strong message about the profound inequalities that structure the global coffee trade, with a representation of Ethiopian coffee growers not just as victims but as active political subjects. Tellingly, some of the film's most powerful scenes are created when Tadesse Meskela travels to London and Seattle, thereby disrupting the systemic distancing of coffee producers from First World consumers as he looks for 'his' coffee on a supermarket shelf. Artists too have made powerful works that recast our more customary geographies of production and consumption. Shelly Sacks' *Exchange Values* installation has been highly influential (see Cook *et al.* 2000 for a description and review). Initially as a gallery piece, but also as a website, this work presents the viewer with panels of dried banana skins, each accompanied by a reference number that designates the St Lucian farmer who grew those bananas and a sound recording of that same farmer speaking to the viewer (who is assumed to be a banana consumer) about their experiences of banana production. Melanie Jackson's 2006 animated film *Global Positioning System* (also available to view via a website) unpacks and traces out the constituent parts of a hand-held GPS device, itself

Plate 19.9 'Which one is it you're [looking for]?' 'Ethiopian': from a scene in the documentary film *Black Gold* in which Tadesse Meskela, General Manager of the Oromia Coffee Farmers Co-operative Union in Ethiopia asks a UK supermarket employee where 'his' coffee is on the shelf.

(www.blackgoldmovie.com/dvd)

mapping out the divergent global positionalities of those involved in this object's 'life'.

Other interventions are more explicitly activist in character. Naomi Klein (2000) notes how the power of global brands, combined with their concern for their sign values, has generated a political response, a 'brand boomerang', in which the meanings being promoted in marketing discourses are contested by other stories, often centred on the social relations of production. Sometimes generically referred to as 'culture jamming', these forms of activism rework the dominant commodity aesthetics of adverts, in ways that can be deeply problematic for corporations' brand values. One very famous example, circulated through viral marketing media itself, is Jonah Peretti's email exchange with Nike in 2001, concerning a pair of trainers that using Nike's iD platform he wanted customised with the word sweatshop (see Spotlight box 19.2).

19.4.2 Ethical consumption

These academic, artistic and activist attempts at reconnecting consumers and producers speak to a wider consumer movement, that of 'ethical consumption'. Here, consumption is explicitly cast in relation to wider provisioning systems and their human and

Spotlight box 19.2

Jonah Peretti's email exchange with Nike iD

Below is an email correspondence with customer service representatives at Nike iD, an online service that lets people buy personalized Nike shoes. The dialogue began when Nike cancelled an order for a pair of shoes customised with the word 'sweatshop'.

From: "Personalize, NIKE iD" <nikeid_personalize@nike.com>

To: " 'Jonah H. Peretti' " <peretti@media.mit.edu>

Subject: RE: Your NIKE iD order o16468000

Your NIKE iD order was cancelled for one or more of the following reasons.
1) Your Personal iD contains another party's trademark or other intellectual property.
2) Your Personal iD contains the name of an athlete or team we do not have the legal right to use.
3) Your Personal iD was left blank. Did you not want any personalization?
4) Your Personal iD contains profanity or inappropriate slang, and besides, your mother would slap us.

If you wish to reorder your NIKE iD product with a new personalization please visit us again at www.nike.com

Thank you,
NIKE iD

From: "Jonah H. Peretti" <peretti@media.mit.edu>

To: "Personalize, NIKE iD" <nikeid_personalize@nike.com>

Subject: RE: Your NIKE iD order o16468000

Greetings,

My order was canceled but my personal NIKE iD does not violate any of the criteria outlined in your message. The Personal iD on my custom ZOOM XC USA running shoes was the word "sweatshop." Sweatshop is not: 1) another's party's trademark, 2) the name of an athlete, 3) blank, or 4) profanity. I choose the iD because I wanted to remember the toil and labor of the children that made my shoes. Could you please ship them to me immediately.

Thanks and Happy New Year,

Jonah Peretti

From: "Personalize, NIKE iD" <nikeid_personalize@nike.com>

To: "'Jonah H. Peretti'" <peretti@media.mit.edu>

Subject: RE: Your NIKE iD order o16468000

Dear NIKE iD Customer,

Your NIKE iD order was cancelled because the iD you have chosen contains, as stated in the previous e-mail correspondence, "inappropriate slang".

If you wish to reorder your NIKE iD product with a new personalization please visit us again at www.nike.com

Thank you,
NIKE iD

From: "Jonah H. Peretti" <peretti@media.mit.edu>

To: "Personalize, NIKE iD" <nikeid_personalize@nike.com>

Subject: RE: Your NIKE iD order o16468000

Dear NIKE iD,

Thank you for your quick response to my inquiry about my custom ZOOM XC USA running shoes. Although I commend you for your prompt customer service, I disagree with the claim that my personal iD was inappropriate slang. After consulting Webster's Dictionary, I discovered that "sweatshop" is in fact part of standard English, and not slang. The word means:

"a shop or factory in which workers are employed for long hours at low wages and under unhealthy conditions" and its origin dates from 1892. So my personal iD does meet the criteria detailed in your first email.

Your web site advertises that the NIKE iD program is "about freedom to choose and freedom to express who you are." I share Nike's love of freedom and personal expression. The site also says that "If you want it done right ... build it yourself." I was thrilled to be able to build my own shoes, and my personal iD was offered as a small token of appreciation for the sweatshop workers poised to help me realize my vision.

I hope that you will value my freedom of expression and reconsider your decision to reject my order.

Thank you,
Jonah Peretti

From: "Personalize, NIKE iD" <nikeid_personalize@nike.com>
To: "'Jonah H. Peretti'" <peretti@media.mit.edu>
Subject: RE: Your NIKE iD order o16468000

Dear NIKE iD Customer,
Regarding the rules for personalization it also states on the NIKE iD web site that "Nike reserves the right to cancel any Personal iD up to 24 hours after it has been submitted".

In addition it further explains:
"While we honor most personal iDs, we cannot honor every one. Some may be (or contain) others' trademarks, or the names of certain professional sports teams, athletes or celebrities that Nike does not have the right to use. Others may contain material that we consider inappropriate or simply do not want to place on our products.

Unfortunately, at times this obliges us to decline personal iDs that may otherwise seem unobjectionable. In any event, we will let you know if we decline your personal iD, and we will offer you the chance to submit another."

With these rules in mind we cannot accept your order as submitted.

If you wish to reorder your NIKE iD product with a new personalization please visit us again at www.nike.com

Thank you, NIKE iD
From: "Jonah H. Peretti" <peretti@media.mit.edu>
To: "Personalize, NIKE iD" <nikeid_personalize@nike.com>
Subject: RE: Your NIKE iD order o16468000

Dear NIKE iD,
Thank you for the time and energy you have spent on my request. I have decided to order the shoes with a different iD, but I would like to make one small request. Could you please send me a color snapshot of the ten-year-old Vietnamese girl who makes my shoes?
Thanks,
Jonah Peretti
{no response}

Source: http://shey.net/niked.html [accessed 15 January 2011]

non-human participants, of which the consumer is a part (for a general introduction see Lewis and Potter 2011). Green consumerism, fair trade, vegetarianism and other animal welfare concerns would be principal strands.

In ethical consumption, being a consumer is not opposed to being a political actor or a citizen. Rather, consumption is seen as a realm for political agency. However, the motivations, forms and implications of such agency cannot be assumed. Ethical consumption is not as simple a matter as correcting the ignorances and amorality of most consumers through more accurate information about the origins of products. This is so in at least three ways.

First, we have been arguing that consumption is an arena for the production, circulation and reception of geographical knowledges. It is tempting, especially perhaps for geography academics and students, to respond by looking to export the factual knowledge of

our subject to this everyday arena, to correct the ignorances that we all operate under as consumers. But *that*, we suggest, is both too easy – offering a very simplistic account of both geographical knowledge and the politics of consumption – and impossibly difficult, given the complex histories and geographies of the multitude of things we consume (Cook *et al.* 2007). Just think, for example, about how many component parts your mobile phone has, how many hands those parts and their constituent materials have been through. Even the most committed ethical consumers therefore rely on shorthand rationales, whether that be guidance from those judged to be experts on appropriate ethical actions, recognized auditings of ethical standards (e.g. through certification schemes such as fair trade, organic or animal welfare), or advertisementss that present ethical products (with their carefully managed, and often romanticized, representations of producers; see Wright 2004 and Zick Varul 2008). Ethical consumption still involves the representation of commodity production and its relations to consumers (Goodman 2004).

Second, studies of ethical consumption also suggest that its existent forms stem as much from personal histories and collective organizations of political action as they do from the decisions of better informed consumers (Barnett *et al.* 2005; 2011; Clarke *et al.* 2007). In consequence, ethical consumption practice is as much a matter of political organization, and collective regulation, as it is of individual product choices. In turn, those forms of regulation operate within complex political fields where the interests of a range of interest groups have to be negotiated and may compete. Producers may and do find ethical regulation an imposition upon them by powerful consumer interests; corporate actors, such as supermarket retailers, may use ethical concerns to reinforce their role of speaking for consumers and dictating terms to producers (Freidberg 2003).

Third, a preoccupation with the dark side of consumption – to put it crudely with suffering, injustice, and our responsibility for it – is too partial. Rather than articulating worlds of consumption and production, it can end up dismissing worlds of consumption and their positive possibilities altogether. It is not a coincidence that Hartwick's commodity chain approach pays no attention to consumption per se. The danger is that consumption is framed as unequivocally bad. That is neither true, as hopefully the discussions earlier in this chapter have demonstrated, nor terribly attractive in shaping political action. Particularly if one's own worlds of production, as a worker, are thoroughly unrewarding – admittedly not usually the case for academic critics of consumption, but true for many others – then giving up all the pleasures of consumption is pretty hard to stomach. Certainly the last two decades in Western politics suggest that failing to recognize the genuine senses of autonomy and creativity that consumption can facilitate is a sure way to alienate those very same workers in whose name one wishes to re-centre the world of production (Miller 2001).

The pleasures and responsibilities of consumption do not have to be opposed in this way. There are connections to be made. Early empirical research on ethical consumers in London challenged any sense of ethical consumption as joyless moral prescription (Bedford 1999), instead emphasizing the positive pleasures gained through such consumer movements: the feelings of achievement, the senses of empowerment, the forms of sociality fostered (see also Barnett *et al.* 2011, pp. 153–80). What is more, much routine consumption has a moral core, centred on caring for one's nearest and dearest. Ethical consumption looks to extend the reach of such a consuming sensibility. We consumers are both global dictators and local freedom fighters. Understanding and inhabiting those two roles in conjunction remains a significant challenge.

Learning outcomes

After reading this chapter, you should have:

- An understanding of the importance of consumption to modern economies and cultures.

- An understanding of the global geographical impacts of modern consumption.

- A sensitivity to the continuing importance of cultural and geographical difference within contemporary consumer cultures.

- An understanding of the character of public space fashioned within the worlds of consumption.

- An understanding of the connections that exist between worlds of consumption and production, and of how ethical consumption looks to address these.

- An ability to think about the geographies of your own everyday consumption practices.

Further reading

Barnett, C., Cloke, P., Clarke, N. and Malpass, A. (2011) *Globalizing Responsibility: The Political Rationalities of Ethical Consumption*, Wiley-Blackwell, Oxford. This book argues for understanding ethical consumption as a political phenomenon, based on empirical research in the United Kingdom.

Bell, D. and Valentine, G. (1997) *Consuming Geographies: We Are Where We Eat*, Routledge, London. Focused on the example of food, this is a lively exploration of the geographies at stake in consumption, at every scale from the body to the home, the city, the nation and the world.

Cook, I. and Crang, P. (1996) The world on a plate: culinary culture, displacement and geographical knowledges, *Journal of Material Culture*, 1(2), 131–53. Modesty prevents us from saying how wonderful this article is, but we can say that it offers an account of the global array of foods found in the United Kingdom that emphasizes the global, local and knowledgeable geographies of consumption.

Cook I. (2004) Follow the thing: papaya, *Antipode*, 36(4), 642–64. An engaging narration of the lives connected through the travels of this tropical fruit from its Jamaican 'production' to its UK 'consumption'. This article attempts to show the complexity of 'defetishization' research.

Goss, J. (1993) The magic of the mall: form and function in the retail built environment, *Annals of the Association of American Geographers*, 83, 18–47. A brilliant, if somewhat sceptical, analysis of shopping malls and the consumer worlds they embody.

Hartwick, E. (1998) Geographies of consumption: a commodity chain approach, *Environment and Planning D: Society and Space*, 16, 423–37. A pithy comparison of the advertising imagery used to sell gold and the ways in which that gold is produced. Argues for a knowledgable and ethical connection of consumers and producers. Simplistic but powerful.

Miller, D. and Woodward, S. (eds) (2010) *Global Denim*, Berg, Oxford. A collection of studies that show how the seeming global uniformity of widespread denim wearing is in fact testament to the capacity of that textile to facilitate a wide range of local cultural expressions.

Useful websites

www.ucl.ac.uk/global-denim-project This academic website presents comparative findings from around the world by re-searchers analysing the wearing of denim. About 25 different projects are presented. An excellent site for recent research on the global geographies of consumer culture.

www.followthethings.com Launched in the Spring of 2011, this website is run by Ian Cook and collaborators, and presents work that seeks to understand consumption and its complex geographies using a 'follow the thing' approach.

www.pbs.org/independentlens/tshirttravels/track.html This website has some great materials about the director Shantha Bloemen's film *T-Shirt Travels*. The film follows a worn T-shirt donated to charity in New York through to its life as second-hand clothing, or what is known as 'salaula', in Zambia.

www.blackgoldmovie.com This is the website associated with the documentary film *Black Gold: Wake Up and Smell the Coffee*. The film explores the global political economies of coffee, by following one Ethiopian farmers' cooperative leader as he attempts to strike a better deal for his members.

www.adbusters.org This website presents examples of 'culture jamming' reworkings of well-known advertising campaigns, as well as magazine and blog materials devoted to contesting the cultural power of corporate promotional culture.

www.exchange-values.org This website presents Shelly Sacks' artwork *Exchange Values* in an online format. The artwork presents audio testimony from banana farmers in St Lucia, who speak to you, a banana consumer, about their hopes and concerns.

www.melaniejackson.net/projects/gps/index.php This website presents online materials for works from the artist Melanie Jackson. The link given here takes you to the pages associated with Jackson's Global Positioning System artwork, including background information, reviews and the animated film that is at the heart of the artwork.

POLITICAL GEOGRAPHIES:
GEOPOLITICS, TERRITORY, STATES AND GOVERNANCE

Section 5

Edited by James Sidaway

Those who have reached this section after reading the rest of the chapters (congratulations!) should already be aware that many elements of geography touch on political issues. Such 'politics' may not be confined to those things that are usually designated by the term: governments, elections (or the lack of them), political parties, revolutions and so on. Rather, readers will have noticed that debates about culture, economic change, history, changing gender relations and many other things described in this text are political. They are political in the sense of being about power, albeit at all kinds of levels and in lots of different ways. So, in an important way, all geographies are political geographies. Though focused on geopolitics territory, nation-states and global governance, the chapters that follow aim to bring that home. Chapter 20 examines the meanings of geopolitics. It will explore how geopolitics has been defined in different times and places: including South America in the second half of the twentieth century, as well as through the Cold War confrontation between the Soviet Union and the USA and the 'global war on terror' since; all building on the moment when geopolitics was first established

as a way of understanding (and indeed seeking to influence) world politics, in early twentieth-century Europe. As the chapter describes, when considering what geopolitics might mean today and in the future, these pasts are mobilized and reworked.

Chapter 21 examines 'territoriality'; the ways that human individuals and human social groups claim or are assigned to particular areas: in short, the human territorial strategies which regulate, demarcate and divide social and political spaces, and their uses. Whilst this is perhaps most clearly expressed in the form of the political geography of nations and states and their boundaries (the focus of Chapter 22), Chapter 21 stresses that territoriality operates in myriad ways and at local scales: for example, in the manner in which particular areas within many cities become associated with different ethnicities (as in the history of urban ghettos and city 'quarters' or other places that seen as being overwhelmingly white – such as many rural areas in Britain) or, more widely, in the ways that some spaces are seen as public and others private or as male and female. Readers will therefore (hopefully) find many connections with earlier sections of →

the book here. Chapter 22 introduces the historical and geographical variability of nations and states. It stresses that neither the nation nor the state is to be taken at face value. Or in other words, it examines how nation-states are complex symbolic systems that crucially depend upon particular visions and associations of territory, place and space.

The last chapter in the book should be read in tandem with the others in this section. Its consideration of the geographies of global governance also connects with many others in the book. In particular, it is easy to see how Chapter 23 relates to issues of population, resources, development and the environment that are considered in Section 2. In the relationships with culture, nature and cities, the links to Section 3 are evident, but so too are those with the economic and financial geographies described in Section 4. And it is not hard to appreciate how the present and future of all the issues considered in this section derive from the historical geographies examined in Section 1. Such interconnections are part of what makes a geographical approach so inviting and exciting.

GEOPOLITICAL TRADITIONS

Chapter 20

James Sidaway and Virginie Mamadouh

Topics covered

- Origins and history of geopolitics, critical perspectives and 'popular geopolitics'

- Diversity and dissemination of geopolitical discourses

- Changing 'World Orders'

Few modern ideologies are as whimsically all-encompassing, as romantically obscure, as intellectually sloppy, and as likely to start a third world war as the theory of 'geopolitics'. Popularized at the beginning of the twentieth century by an eccentric British geographer, Sir Halford Mackinder, geopolitics posits that the earth will forever be divided into two naturally antagonistic spheres: land and sea. In this model, the natural repository for global land power is the Eurasian 'heartland' – the territory of the former Russian empire. Whoever controls the heartland, wrote Mackinder, will forever seek to dominate the Eurasian landmass and ultimately the world.

(Charles Clover 1999)

20.1 Introducing the idea of a geopolitical tradition

Geopolitical rhetoric is more popular than ever. The sudden independence of the five former Soviet Socialist Republics of Central Asia and the fall of the communist regime in Afghanistan in the 1990s, the invasion of Afghanistan and later of Iraq by the United States and its allies in the 2000s, combined with the earlier establishment of an Islamic Republic in Iran and instability in Pakistan, have greatly boosted talks about geopolitics in that macro-region and more generally in world politics. Authors like Mackinder are being read and discussed again. Parallels are seen between the present competition in the region and The Great Game – the intense rivalry for the control of Central Asia between the British and Russian Empires in the nineteenth century.

Yet whilst the term geopolitics can and does refer to many things, and is in vogue, it is important for students of geography (a subject with which geopolitics has often been associated) to understand that geopolitics is commonly associated with particular ways of writing (and thinking) about space, states and the relations between them. Often this takes the form of mapping, emphasizing the strategic importance of particular places (as in Figure 20.1 or 20.3). Geopolitics in this sense often sees itself as a tradition: that is, something conscious of its unfolding historical development and with a sense of important founders (and certain key texts written by them). Many of those who write about this tradition or see themselves as working within or extending it usually trace its origins to the late-nineteenth-century writings of a conservative Swedish politician Rudolf Kjellén. Kjellén is reputedly the first person to have used the term 'geopolitics' in published writings (see Holdar 1992; Tunander 2001). But beyond this idea of a founding moment when the geopolitical 'tradition' begins with the first use of the term by Kjellén, things start to get complicated. The 'tradition' divides, fractures, multiplies and finds itself translated into many languages and cropping up in everything from the writings and speeches of American politicians to texts written by Brazilian generals and Russian journalists. All these reinvent and rework the 'tradition' as they go along. As the introduction to a critical collection on 'rethinking geopolitics' explains:

> the word 'geopolitics' has had a long and varied history in the twentieth century, moving well beyond its original meaning in Kjellén's work. . . . Coming up with a specific definition of geopolitics is notoriously difficult, for the meaning of concepts like geopolitics tends to change as historical periods and structures of world order change.

> (Ó Tuathail 1998: 1)

In other words, exactly what is meant by the term 'geopolitics' has changed in different historical and geographical contexts. However, a good way to understand something about this idea of a tradition of geopolitics is to look at some of the people, places and ideas associated with one of its most significant appearances, in the ideology of the right-wing dictatorships that ruled many South American countries in the 1960s, 1970s and 1980s. In most cases they had come to power through military coups and became more concerned with internal security (patrolling the towns and country as a kind of police force and running the wider government) than with fighting or preparing to fight wars with other countries.

Plate 20.1 General Pinochet: geopolitics in action.
(SIPA Press/Rex Features)

Consider the case of Chile. On 11 September 1973, the elected left-wing government of Chile (headed by the communist President Salvador Allende) was overthrown in a brutally violent coup led by the head of the armed forces General Augusto Pinochet (Plate 20.1). Pinochet and his generals ruled Chile with an iron fist. All opposition was crushed and thousands of people were rounded up by the armed forces. Many 'disappeared' into secret military jails. Hundreds of these prisoners were tortured and murdered. The use of torture became a routine instrument of state. For the next 17 years, Chile was ruled by Pinochet and his cronies who imposed a new 'neoliberal' economic and political model (of the form that is described in Chapter 8) and murdered any significant opposition. Pinochet finally stepped down as head of state in 1990, when democracy was restored, but not before he had entrenched himself as a senator (in the parliament) for life. He also guaranteed that the armed forces (which he continued to head) would continue to have a significant role in Chile's state structures and that they could not be prosecuted for any crimes committed during the coup and brutal years of military rule.

The ideology of the military regime in Chile (like those elsewhere in South America) was intensely nationalist. More importantly it was a particular conservative form of Chilean nationalism. The nation was held to be sacred and the military were rescuing this 'sacred body' from communists, subversives and so on (i.e. anyone who opposed the military vision of ultra-nationalism and order and the neoliberal economic and social model which was now imposed). It is in this nationalism and conception of the nation as a kind of sacred body that geopolitics enters the picture. A few years before the coup, Pinochet had published an army textbook on geopolitics. Intended mainly for use in Chilean military academies, the 1968 textbook indicated that Chile's future dictator took the subject deadly seriously. For Pinochet, geopolitics was a science of the state, a set of knowledges and programmes to perfect the art of statecraft, that is to strengthen the state in a continent and wider world in which it was held to be in competition with others. The starting point and the heart of Pinochet's textbook and of the wider bleak tradition of geopolitics that it presents is an organic theory of the state.

20.2 The organic theory of the state

This idea is at the core of most South American geopolitical writings and of the wider geopolitical tradition.

In summary, it holds that the state or country (and the sense of nation that goes with it) is best understood as being like a living being. Like any living organism, it therefore needs space to grow and it will be in competition with other living beings. An idea that the state needs 'living space' (a certain amount of resources and land) for the nation to thrive is particularly evident in South American geopolitics. A certain reading of Charles Darwin's idea of a struggle for the 'survival of the fittest' is therefore transferred to the realm of states. Drawing upon conservative German writings of the nineteenth and early twentieth centuries (which had been codified by the German academic Friedrich Ratzel and were later elaborated by another conservative German geographer and general Karl Haushofer) and with this organic notion of the state in mind, geopolitics claims to identify certain laws that govern state behaviour and that, once identified, can be a guide for those charged with furthering and protecting the 'national interest' (or at least a certain definition of the latter).

It is not difficult to see how this idea could appeal to South American military dictators, for it gives the military, equipped with the supposedly scientific study of geopolitics, a special mission. In other words it legitimizes their rule. Moreover, the organic idea was extended further to legitimize the extermination of those whom the military defined as enemies of the state. People labelled as communist or subversive, indeed all those who oppose the military dictatorship, can be compared with a disease or cancer which threatens the lifeblood of the state-organism and is best 'cut out', that is, eliminated. The otherwise unthinkable (the murder of thousands of people in the name of the wider welfare of the nation-state) is made to seem natural and a good thing, since it serves the longer-term interest of the health and power of the 'living being' that is the state (see Hepple 1992 for an exploration of this). With the 'cancer' of subversion and disloyalty eliminated, the 'state-body' can go on to grow and thrive.

20.3 Brazilian national integration

Aside from Chile, one of the most important expressions of geopolitics in South America has been in Brazil. The military in Brazil overthrew the democratic government in 1964 and stayed in power for the next 20 years. Although never as despotic as the regime led by Pinochet (or similar military governments in

Figure 20.1 National integration in Brazilian geopolitics: Amazon development poles. The shaded areas represent the 15 'development poles' where financial resources would be concentrated and concessions offered to mining, settlement, industrial, lumber and agricultural development resources.

Source: Hecht and Cockburn (1989: 127)

Argentina and Uruguay), the long years of military rule gave the Brazilian generals a chance to elaborate and impose a geopolitical vision on the country. One of the most evident aspects of Brazilian geopolitics is the idea that state security requires a measure of national integration. The vast scale of Brazil, the difficulty of travel across its Amazon 'heartland' (the world's largest tropical rainforest) and the fact that it shares borders with every other South American country except Chile and Ecuador gave the geopolitics of the Brazilian generals a special sense of its national mission and an obsession with the potential for the country to be a great power (known in Portuguese as *Grandeza*). The associated sense of the urgency of the integration of Brazil required the extension of a network of highways across Amazonia and the settlement of its lands by farmers and ranchers. It is this geopolitically inspired vision, combined with a highly corrupt system of patronage and favours to those close to the regime, that underlies the enclosure and division of Amazonia into private lands (some larger than European countries such as Belgium) and the accompanying transformation or

destruction of the tropical rainforest (see Foresta 1992; Hecht and Cockburn 1989) (Figure 20.1).

Although the Brazilian military have been back in the barracks for more than two decades and Brazil is again a lively democracy, the long-term consequences of this geopolitically motivated (and economically profitable) strategy have been disastrous for many of the indigenous peoples of Amazonia. They have found themselves forced off land that was traditionally theirs, sometimes murdered or attacked by ranchers, the state and settlers, and disoriented by the arrival of a frontier culture of violence, destruction and consumption. Poor peasants, particularly from the impoverished north-east of Brazil, have also moved into the Amazon region in search of land and freedom. They are fleeing from the oppression and landlessness they themselves face in the north-east of Brazil, where the vast majority of the land is still in the hands of an elite class. Of course, in the geopolitical visions of the Brazilian generals, this population movement was seen as overwhelmingly positive, reinforcing the Brazilian population of the Amazon and redistributing what they regard as marginal 'surplus' people in the process (in much the same way that the colonization of the 'wild west' was seen as the advance of civilization and America's 'manifest destiny' in the United States). At the same time this has defused some of the political pressures for land reform in north-east Brazil. In all this we can see how Brazilian geopolitics was implicated in a complex web of social and environmental transformations.

20.4 Antarctic obsessions

An interest, at times an obsession, with the last unexploited continent of Antarctica is evident in much South American geopolitics. Antarctica is also prominent in geopolitical writings from Argentina, Chile and Uruguay and also appears in Peruvian, Brazilian and Ecuadorian geopolitical texts (see Dodds, 1997). To understand what this amounts to and what forms it takes necessitates some understanding of Antarctica's exceptional political geography. As Chapter 23 (pp. 500–503) explains in more detail, Antarctica has a unique territorial status, in so far as it has no recognized state on its territory. Everywhere else in the world forms part of the patchwork of states that we learn to be familiar with and take for granted. Antarctica is a stark reminder that there is nothing natural or inevitable about this. The absence of recognized states in Antarctica reflects the inhospitability and remoteness of the continent, the only one without an indigenous human population. Although the plankton-rich waters around Antarctica were exploited for seal and whale hunting in the nineteenth century, it was not until the twentieth century that exploration of the interior began. Even today, Antarctica has only a non-permanent population (at any time) of a few hundred scientists. This presence, plus the relative proximity of the continent to South America and the possibility of exploitable mineral resources, has led to a series of territorial claims. Although Argentinean writers had already represented Antarctica as an extension of the southern Argentine area of Patagonia, the British made the first formal claim in 1908. In turn, parts of this claim were 'granted' to Australia and New Zealand. France and Norway made claims in the 1930s and 1940s, followed shortly by Argentina and Chile. After 1945, the USSR and the USA also established a wide network of bases, although without staking formal claims to territory. The claims made by Argentina, Chile and the United Kingdom overlapped – and

foreshadowing the British–Argentine conflict over the Falklands/Malvinas, British and Argentine forces exchanged fire in Antarctica in the late 1940s and early 1950s. In the context of this potential for conflict and growing possibility of Cold War confrontation (on the Cold War, see Section 20.7), a United Nations Treaty in 1959 agreed that all claims would be (forgive the pun) 'frozen' for at least 30 years, and the continent reserved for scientific (not commercial or military) use. The Treaty was extended in 1991. But prior to then, it was not clear what the future status of the continent would be. Even today, the Treaty merely defers the issue of claims.

Argentina is the South American country with the largest claim (Figure 20.2). The Argentine claim has also become central to geopolitical discourse there. As Child (1985: 140–1) explains in a study of geopolitics and conflict in South America:

> For Argentine geopolitical writers, the subject of Antarctica is not only linked to tricontinental Argentina [a power in the South American continent and the South Atlantic], but also to … national sovereignty, patriotism, and pride. This is a particularly touchy combination after the humiliating defeat of the Malvinas/Falklands conflict. The Argentine National Antarctic Directorate has professors of Antarctic geopolitics on its staff. Through the media, maps, and postage stamps and the centralized educational system, Argentines are constantly taught and reminded that there is an Argentine Antarctic just as much as there are Argentine Malvinas. The need to assert Argentine rights in the Argentine Sea, islands, and Antarctica is linked to dreams and national projects of Argentine greatness.

20.5 Heartland

Although Antarctica and the South Atlantic are significant components of Argentine (along with Brazilian and Chilean) geopolitics, the former also looks beyond other Argentine frontiers. In particular, Argentine geopolitical writers together with some Brazilians and Chileans have taken a particular interest in the security of Bolivia. They have scripted Bolivia (a relatively impoverished land-locked mountainous country, which has borders with Argentina, Brazil, Chile, Paraguay and Peru) and proximate areas of its neighbours as a key strategic continental **heartland** (see Kelly 1997). Control of Bolivia, in this vision, would be a vital key to a relative dominance in the

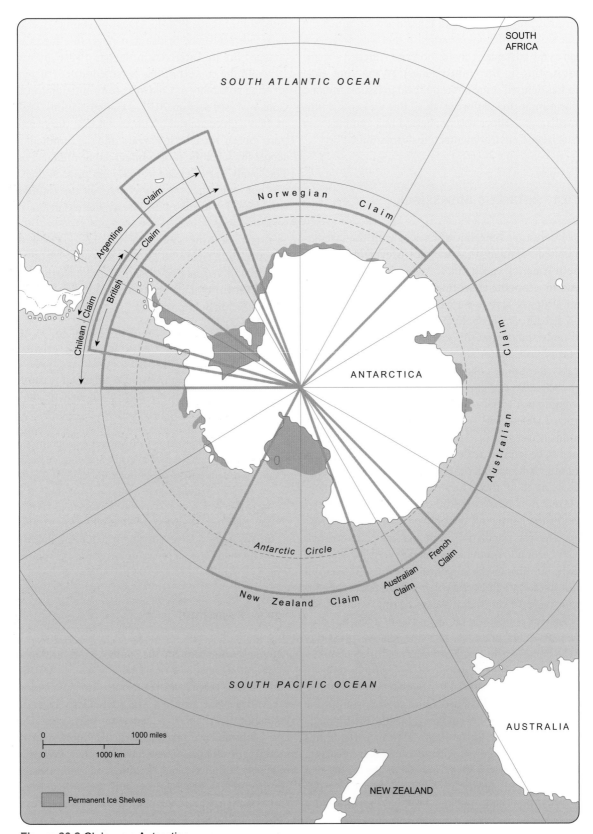

Figure 20.2 Claims on Antarctica.

Source: adapted from M.I. Glassner, *Political Geography*, John Wiley & Sons Inc., 1993, p. 498. Copyright © 1993 John Wiley & Sons Inc. This material is used by permission of John Wiley & Sons Inc.

South American continent. That Bolivia has a strong revolutionary tradition and was for many years characterized by chronic political instability has reinforced the tendency of the other South American countries to meddle in Bolivian politics. Indeed, during the years (1976–82) when Argentina was last ruled by a geopolitically obsessed military junta, the Argentine armed forces were actively involved in supporting a Bolivian military government. This activity took the form of the kinds of brutal suppression and frequent murder of those (trade union leaders, dissidents, opposition members and leaders) who opposed the military government and its economic and social strategies. The idea of Bolivia as a 'heartland', control of which would be a kind of magic (geopolitical) key to domination of South America, links back to one of the best-known genres in classical (European) geopolitical thought. For although the South American countries have seen some of the most significant expressions of geopolitical discourse of modern times, geopolitics originates in Europe and it is to some examples of European geopolitics, including the idea of heartland, that we now turn.

The designation of a heartland was first made by the British geographer (and strongly pro-imperialist conservative politician) Halford Mackinder. In what has since become a widely cited (if less often read) article first published in 1904 following its presentation to the Royal Geographical Society (RGS), Mackinder argued that the age of (European) geographical exploration was drawing to a close. This meant that there were hardly any unknown 'blank' spaces left on European maps of the world. According to Mackinder, the consequence of this closing of the map, this end of the centuries-long task of exploration and discovery, was that political events in one part of the world would invariably affect all others, to a much greater extent than hitherto. There would be no more frontiers for Europeans to explore and conquer. Instead, the great powers would now invariably collide against each other. Mackinder called this end of European exploration 'the post-Columbian age' and the closing of frontiers, the emergence of a 'closed political system'.

Given that this was the case, Mackinder claimed to identify the places of greatest world-strategic significance, control of which would give any great power a key to world power. In his 1904 paper, he termed this the 'pivot area'. With Mackinder's address to the RGS and his subsequent article came a series of maps, the most frequently reprinted one of which claims to describe 'The Natural Seats of Power' (Figure 20.3). As

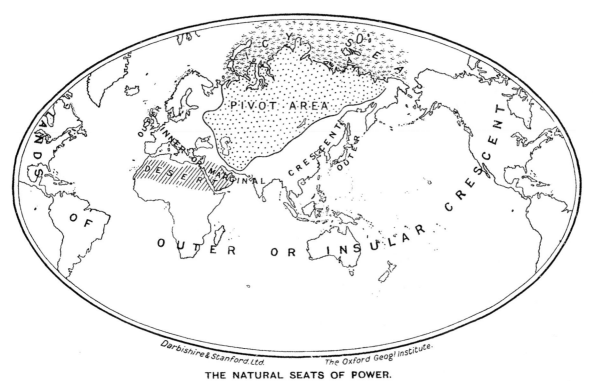

THE NATURAL SEATS OF POWER.

Pivot area—wholly continental.   Outer crescent—wholly oceanic.   Inner crescent—partly continental, partly oceanic.

Figure 20.3 'The Natural Seats of Power' according to Mackinder (1904).

Ó Tuathail's (1996a: 25) critical account of Mackinder argues:

> Mackinder's January 25 04, address to the Royal Geographical Society, 'The Geographical Pivot of History', is generally considered to be a defining moment in the history of geopolitics, a text to which histories of geopolitics invariably point.

A certain geographical determinism can be detected here, in which the physical environment ('essentially the same physical features' in Mackinder's words) provide the fixed, unchanging backdrop to and therefore something of the explanation of human history. This view, which ignores the social construction of resources (analysed in Chapters 5 and 12) and other features of nature, is common to virtually all geopolitics (something which it shares with a great deal of twentieth-century geography). One of the key formulas in Mackinder's notion of geographical determinism is his identification of a 'pivot area' or 'heartland zone' in east-central Europe, control of which would be a kind of magic key to world domination. Reworking the ideas of his 1904 article after the First World War, Mackinder refined this notion and penned a formula that encapsulated it:

> Who rules East Europe commands the Heartland.
> Who rules the Heartland commands the World-Island.
> Who rules the World-Island commands the world.
>
> (Mackinder 1919: 4)

According to this (simplistic) formula, 'rule' of the eastern portion of Europe offered the strategic path to that of the African–Asian–European continents (which together constitute what Mackinder terms the 'World-Island') and hence a dominant position on the world scene.

Mackinder's writings reflect his place and time. Mackinder, writing from the vantage point of imperial Britain, is concerned to identify threats and dangers to British power. At the time when Mackinder presented his paper to the RGS, Britain was the pre-eminent world power. It still seemed that way to Mackinder in 1919 when he wrote about the 'Heartland'. But British imperial politicians, like Mackinder, were aware of the growing power of America, Germany and Russia. In fact, potential British imperial competition with the latter in Asia provided a key context to Mackinder's work. As Peter Taylor (1994: 404) explained:

> Behind every general model there is a specific case from which it is derived. For the heartland model this is particularly easy to identify. Throughout the second half

of the nineteenth century Britain and Russia had been rivals in much of Asia. While Britain was consolidating its hold on India and the route to India, Russia had been expanding eastwards and southwards producing many zones of potential conflict from Turkey through Persia and Afghanistan to Tibet. But instead of war this became an arena of bluff and counter-bluff, known as the 'Great Game'. … Mackinder's presentation to an audience at the Royal Geographical Society would not have seemed so original as it appears to us reading this paper today. … Put simply, the heartland model is a codification and globalization of the Great Game: it brings a relatively obscure imperial contest on to centre stage.

Not only does this envisage the world in a particular way, as a 'stage', but it sees only select key actors as the significant figures at play. These are the European powers (plus Russia). Other peoples and places are merely the backdrop for action by White Men. The taken-for-granted racism of Mackinder's model, in which only Europeans make history, is also that of European imperialism and that of the bulk of wider European geographical and historical writings of the time (see Chapter 3).

Yet, although Mackinder's 1904 paper is very much a product of its time and Mackinder's own conservative world-view, it has proven durable and has been integrated into rather different contexts, which saved it from the relative obscurity that it deserves as a turn-of-the-century imperialist text. The transfer of the discourse of 'Heartland' to Bolivia by South American codifiers of geopolitics has already been noted. In addition, 'Heartland' was appropriated by German geopolitics in the 1930s and 1940s and formed part of the backdrop to Cold War American strategy from the late 1940s through to the last decade of the twentieth century. Taylor (1994: 405) notes:

> First, in the inter-war years the heartland theory became an integral part of German geopolitics. It fitted the needs of those who advocated lebensraum [living-space], the policy of expanding into eastern Europe [and enslaving and murdering its indigenous populations], coupled with accommodating the U.S.S.R., as a grand continental policy for making Germany a great power again. Second, with the onset of the Cold war after World War II, the heartland theory got another lease of life as the geostrategic basis of nuclear deterrence theory. The west's nuclear arsenal was originally justified in part as compensation for the U.S.S.R.'s 'natural' strategic advantage as the heartland power.

The next two sections of this chapter will examine aspects of Nazi, Fascist and Cold War geopolitics in greater detail.

20.6 Nazi and Fascist geopolitics

The formal tradition of writing about space and power under the title of 'geopolitics' also found fertile contexts in Italy, Portugal, Spain and Japan. Influenced and supported by Nazi Germany, all these countries (together with Hungary and Romania) saw the rise and victory of ultra-nationalist or Fascist governments (often through violent struggle or full-scale civil war with democratic or communist forces) (see Chapter 3). In each case geopolitical debates were crucially negotiated through other cultural and political debates about race, nationalism, the colonial pasts and futures, supposed national 'missions' and destinies and the European and global political contexts. In Portugal, for example, the right-wing dictatorship argued that it was bringing Christianity and civilization to the territories it had acquired in Africa and Asia, even projecting these onto a map of Europe to show that – if the African territories controlled by Portugal were taken into account (Figure 20.4), then Portugal was not a small country (Sidaway and Power 2005)! In similar terms, in Italy under the fascist dictator Benito Mussolini, geopolitical journals saw themselves as serving and expressing the aspirations of the fascist state to establish a new Roman Empire across the Mediterranean (Figure 20.5) (Antonsich, 2009). However, it is the German geopolitics of this epoch that has become the best-known. In Germany, organic notions of the state had already been popularized by conservative nineteenth-century academics. Moreover, Germany was characterized by extreme political and economic turbulence in the decades following its defeat in the 1914–18 World War. This combination provided a fertile environment for the elaboration and circulation of a distinctive geopolitical tradition. In Ó Tuathail's (1996a: 141) words:

> After the shock of military defeat and the humiliation of the dictated peace of Versailles, the Weimar Republic proved to be fertile ground for the growth of a distinct German geopolitics. Geopolitical writings, in the words of one critic, 'shot up like mushrooms after a summer rain'.

The main features of these writings (which they shared with a wider German nationalism, later codified in Nazism) were a critique of the established 'World Order', and of the injustices imposed on Germany by the victors. German claims were often presented graphically in maps that were widely circulated (see

Figure 20.4 Portugal is not a small country. Henrique Galvão, 1934.

Source: adapted from Power and Sidaway (2005)

Figure 20.5 Italian living space in the Mediterranean Region.

Source: *Geopolitica*; 3 (1939) p. 161. Reprinted in Antonisch (2009).

Herb 1989: 97). Like the variants of the geopolitical tradition that were developed amongst right-wing and military circles in Italy, Portugal, Spain and Japan, German geopolitics also asserted an imperial destiny. Indeed, as Agnew and Corbridge (1995: 58–9) explain: 'The Nazi geopoliticians of the 1930s came up with formalized schemes for combining imperial and colonized peoples within what they called "Pan-Regions".'

In this vision, notions of racial hierarchy were blended with conceptions of state 'vitality' to justify territorial expansion of the Axis powers (see O'Loughlin and van der Wusten 1990) (Figure 20.6). In Europe, related conceptions of the need for an expanded German living-space were used as justification for the mass murder of occupied peoples and those who did not fit into the grotesque plans of 'racial/territorial' purity. The practical expression of these was the construction of a system of racial 'purification' and mass extermination. At least 6 million Jewish people were murdered in concentration camps together with millions of others: disabled people, gays and lesbians, gypsies and political opponents. Historical debates about the role and relative significance of German geopolitics

within the Holocaust and within broader Nazi ideology and strategy continue (see Heske 1986: 87; Bassin 1987; Ó Tuathail 1996a; Natter 2003). Paul Gilroy (2000: 39), however, reiterates the connections between racial, early ecological and geopolitical thinking:

> connected in profound ways to the notions of Lebensraum (living-space) that figured in but were not created by the racist population policies and agricultural and scientific planning of the Nazi period . . . [and to] the geo-organic, biopolitical and governmental theories of the German geographers Friedrich Ratzel and Karl Haushofer and the early-twentieth century geopolitican Rudolf Kjellén. These writers supplied important conceptual resources to Nazi racial science, helping it to conceptualize the state as an organism and to specify the necessary connections between the nation and its dwelling area.

In the United States in the 1940s, German geopolitics became the subject of lurid tales and depictions, cropping up in media, military and government 'explanations' of Nazi danger. Popular magazines, such as *Readers Digest*, would inform Americans of the

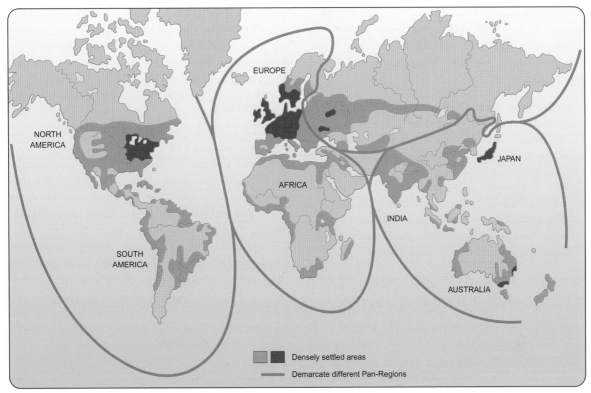

Figure 20.6 'Pan-Regions' as envisaged in Nazi geopolitics.

Source: O'Loughlin and Van der Wusten (1990) 'Political geography and the pan regions', *Geographical Review*, 80, adapted with permission of the American Geographical Society. Copyright © American Geographical Society, 1990

'scientists behind Hitler' describing German geopolitics as *the* key to Nazi strategy. And whilst the relative significance of the German geopolitical tradition in the wider genocidal ultra-nationalism of German Fascism was certainly overstated in such accounts, we should see geopolitics as a particular expression of *wider* academic and intellectual involvement and complicity in authoritarian state power, war-making and genocide. Much more widely, beyond the geopolitical tradition per se, academic geography was deeply implicated in these activities. Exploring this, Natter (2003: 188) notes how

> [T]he work of disciplinary historians of geography has demonstrated the extent to which the demarcation of geography seems inseparable from the history of war, imperialism and quests for national identity. . . . Geopolitics, thus, would mark a particular, but in no way separable (and hence containable) geopolitical deployment of geo-power.

An example of this wider complicity was the way in which the models of 'Central Place Theory' (an abstract model of the ideal spatial distribution of towns), which

were developed in Germany by Walter Christaller in the 1930s, were elaborated with the express purpose of providing planners with a model of German settlement to impose on conquered territories of Eastern Europe, once the minority Jewish and Gypsy populations there had been murdered and the majority Slavic population enslaved (see Rössler 1989).

The defeat of the Axis powers in 1945 (culminating in the use of American atomic weapons against the civilian populations of Hiroshima and Nagasaki in Japan) and the lurid wartime depictions of Nazi geopolitics in the United States dealt something of a blow to the formal tradition of geopolitics. In the 1990s, however, right-wing geopolitics returned – in Russia and Ukraine for example, where references to Nazi-era concepts (such as *Großraum*, 'larger space') and associated racist and anti-Semitic ideologies have become influential within some contemporary geopolitical thought (Inghram 2001; Wilson 2002; O Loughlin *et al.* 2005). Moreover, references to geopolitics had continued through the 1950s to the 1970s in both Spain and Portugal (see Sidaway 1999, 2000; Sidaway and Power 2005), where Fascist regimes remained in power to the mid-1970s, as well as in Turkey (especially when it too was under military rule through part of the 1960s, and again in the early 1970s and early 1980s). On the Turkish example, Pinar Bilgin (2007: 753) argues that:

> Constructed through texts authored by military geo-politicians, and disseminated through a variety of institutions including compulsory military service (with access to all males 18+ years of age), the National Security Academy (proving in service training to high level civil servants and journalists), and the compulsory high-school course 'National Security', Turkey's geopolitical discourse has allowed the military to play a central role in shaping domestic political processes but also make this role seem 'normal'.

In addition, as this chapter has detailed, geopolitics has been influential in recent decades in a number of South American countries, especially in the 1970s' and early 1980s' epoch of military rule through much of the continent. These regimes (and plenty of others) were integrated into the US-led anti-communist network of allies. In a policy that became known as containment, the USA aimed to encircle and block the potential expansion of Soviet power and influence beyond the immediate borders of the USSR and the pro-Soviet states installed by the USSR in the Eastern European territories that it had occupied during its Second World War battle for survival against Nazi Germany. In the allied countries (and as an expression of this strategy of containment), the functions of geopolitics, in particular its 'strategic vision' and claim of 'scientific' objectivity, continued to operate or were displaced into other disciplines and branches of knowledge, including geography and the expanding subject of International Relations. This displacement of strategic knowledge was particularly evident in the United States, which by 1945 was the greatest power that the world had ever seen, constituting over half of the world economy and (until the recovery of the Soviet Union and its own development of atomic weapons in 1949) possessing a virtually unrivalled military capacity.

Plate 20.2 The gateway to the Auschwitz concentration camp. The Nazi regime used concentration camps to execute their policy of racial/territorial 'purification'.

(© Ira Nowinski/CORBIS)

20.7 Cold War geopolitics and the logics of containment

Like a fast-moving glacier, the Cold War cut across the global terrain in the 1940s, scraping and grinding solid bases, carving and levelling valleys, changing the topography. For nearly half a century the Cold War competition between the United States and the Soviet Union dominated the international system. . . . Like all retreating glaciers, the receding Cold War left deposits of debris and exposed an altered, scarred landscape that leaders in the 1990s struggled to measure and fathom.

(Paterson 1992: 2)

The Cold War is used as a shorthand description of the conflict (which to many seemed to be a conflict for world domination) between the 'communist' East, led by the USSR, and the 'capitalist' West, led by the USA from around 1949 to around 1989. This was not simply a rift between two great powers, but a complex ideological conflict, which often appeared to be about different ways of life and contrasting social systems (the bureaucratic 'command economies' based on predominantly state ownership of the means of production, with a ruling class drawn from the bureaucracy; and the capitalist economies based on predominantly private ownership, with a ruling class drawn from the owners of capital). It was also a conflict in which *direct* military confrontation between the two great powers was avoided. Hence the metaphor of Cold, that is, 'poor, frozen or frosty' relationships, but short of all-out 'hot' war, involving direct exchanges of missiles, bombs and so on. In part this avoidance of direct conflict arose because both powers were conscious of the enormous stakes, armed with nuclear, chemical and biological weapons as well as vast arsenals of increasingly high-tech 'conventional' armaments. By the 1960s, they were each capable of destroying virtually all human life several times over. Moreover, neither could attack the other without being sure that the enemy would not still possess enough undamaged nuclear weapons to retaliate, a 'balance' of terror known, appropriately enough, as MAD (Mutually Assured Destruction). Instead, conflict took the form of the continual preparation for war (as in 'civil defense' preparations, see Plate 20.3) plus proxy wars in what became known as the (originally not pro-West, not pro-East, but contested) 'Third World' and many other political, economic and cultural forms of competition. That is, it was a conflict conducted by *every* means *except* direct military confrontation between the superpowers.

Of course there was no shortage of indirect confrontation and preparation for war. Europe was divided into two armed camps and split down the middle between Soviet and American zones of influence by a fortified military and ideological frontier that became known as the 'Iron Curtain'. Whilst Europe was characterized by an uneasy stability and balance of terror, by the 1970s the term 'geopolitics' had been revived (or rediscovered) by US national security advisers Henry Kissinger and Zbigniew Brzezinski (see Hepple 1986; Sidaway 1998) to refer to the strategic vision deemed necessary to circumvent the sense of growing Soviet power, particularly in the Third World, where a wave of successful revolutions had brought left-wing, pro-Soviet governments to power. As has already been noted, one of these was in Cuba, but by the end of the 1970s, there were also pro-Soviet governments in the countries of Indochina (Vietnam, Laos and Kampuchea) where the United States had been unable to defeat communist insurgencies despite a massive military effort, in the African countries of Angola and Mozambique, in the Arabian country of South Yemen and in Afghanistan as well as in the small Central American state of Nicaragua. Although China had been prised away from its close alliance with the Soviets by the early 1960s, communist China remained a force to be reckoned with and opposed American support for Taiwan (see Case study 20.1). Thousands of American troops stationed in South Korea were just a short distance from a hostile communist North Korea. By the early 1980s, American concern about the security of oil supplies from the Persian Gulf region had escalated to the extent that the then US President Carter declared that it was of vital importance to the USA (see Spotlight box 20.1). In this world of superpower competition, geopolitics again found its moment and expression in the visions of the national security advisers and American generals.

More widely, American leaders and policy makers declared that the Soviet Union had to be *encircled* and *contained*. The metaphor of disease (containment) was mixed with that of 'dominoes' – if one country (say Vietnam or Cuba) had 'fallen' to Soviet control or influence, then it could (like a chain reaction) 'infect' other proximate ones. The irony of this (which is equally present in the formal geopolitical tradition) is that in the name of security and strategy, the real complexity of *human* geographies in the places (such as Vietnam or Chile, or Afghanistan, Nicaragua or the Persian Gulf) that are deemed strategic is sometimes obscured or erased. Forget about the complex details of the people, culture and society. What matters is the 'strategic value' of the place, or the political identity of its government as an

Plate 20.3 The cold war comes to American cities

Source: © Alamy Images/Jim West

(a)

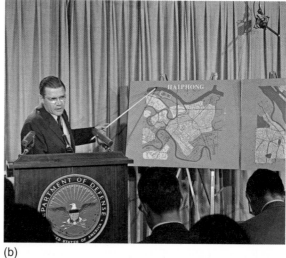

(b)

Plate 20.4 (left) Containment in action: the aftermath of napalm bombing on a Vietnamese village by US-backed (anti-communist) South Vietnamese forces. (right) 'Containment' explained: US Secretary of State for Defense briefing journalists during the Vietnam war.

(© Bettmann/CORBIS)

Spotlight box 20.1

The Carter Doctrine: contexts and consequences

In late 1979, the USSR invaded Afghanistan. For the Soviets, this meant supporting a communist government in a neighbouring country, who had come to power just a couple of years before (in a coup, after several years of political instability) and were facing rebellion and resistance in the Afghan countryside (already being fostered by the Pakistani and US intelligence agencies). For strategists in Moscow, the invasion would secure the USSR's southern flank, especially from the growing perceived threats from a new Islamic radicalism: the USSR had millions of Muslim citizens in its Central Asian territories, and feared that the Islamic revolution in Iran and the insurgency in Afghanistan might influence them and threaten Soviet power and stability. However, for the USA – at least to those in influential positions, like US National Security Advisor Zbigniew Brzezinski – this marked a new phase of Soviet expansion and a failure of containment. In tandem with the Islamic revolution in Iran which overthrew the Shah (who was an American ally and whose rise to power in the 1950s was orchestrated by the USA and Britain) was the second jump in oil prices in a decade, and American concern over the security of oil supplies from the Persian Gulf region grew. Brzezinski and the American media had already starting talking about the region as 'an arc [or crescent] of crisis' (see Sidaway 1998) (see Plate 20.5, which shows how the influential *Time* magazine represented these 'geopolitical dangers'). And soon after the Soviets invaded Afghanistan the US media started to describe it as the first Soviet step towards the Middle East, and President Carter announced that any threat by an outside force to the Persian Gulf was a matter of direct American strategic interest and would result in a military response. This commitment became known as the Carter Doctrine, taking its place amongst a succession of other US Presidential geopolitical doctrines: such as Truman in 1947,

who set out the overall commitment to containment, to (after Carter) the Reagan Doctrine in the 1980s (which sought to roll back Soviet influence in places like Afghanistan and elsewhere in the Third World where pro-Soviet regimes had come to power, rather than simply contain the USSR and its allies). Under the Reagan Doctrine, vast quantities of weapons were channelled to rebel forces fighting the pro-Soviet government in Afghanistan. Amongst those who took part in the anti-Soviet struggle in Afghanistan and through which the armaments were channelled was Osama Bin Laden – later to turn against the United States, whom he came to regard as intent on dominating the Muslim world (Coll 2004).

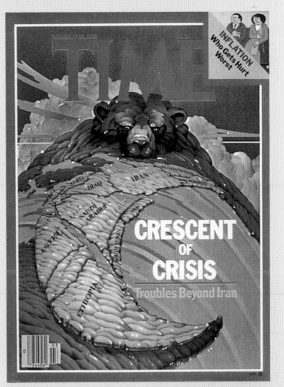

Plate 20.5 'The Crescent of Crisis' as depicted on the front cover of *Time* Magazine, 15 January 1979.

(*Time* Magazine)

ideological and strategic friend or foe. At times this could take quite extreme forms, as with US National Security Advisor Robert McNamara (previously head of Ford, and later head of the World Bank):

> Robert McNamara was, of course, the leading specimen of *homo mathematicus* – i.e., men who behave and

believe other men [and, we might suppose, women too] behave primarily in response to 'hard data', usually numbers (infiltration rates, 'kill ratios', bomb tonnage). Like the classic private eye on television, they are always looking for 'the facts', but usually the wrong ones. They miss reality, for they never get close enough or related enough to another society to do more than count things in it. If you relate to a country as a military target you do not need to know anything about it except details as are easily supplied by reconnaissance satellites, spy ships, secret agents, etcetera. You need never know who the victims of your attack were. Your task is merely to assess the results of what you have done and this is done by counting bodies, destroyed factories, enemy soldiers.

(Barnet 1973: 119)

Such logic is seemingly far removed from the prominence of maps in the formal geopolitical tradition. Yet we can see that, in *both* the kind of thought present in McNamara's mind and in classic geopolitics, there are moments when the myriad complexity of the world is reduced to a simple black and white representation, be it numerical or cartographic.

20.8 Cold War geopolitics in art and culture and 'popular geopolitics'

Few works of art explicitly refer to geopolitics. A famous one that does is *Geopoliticus Child Watching the Birth of the New Man* by the Catalan painter Salvador Dalí. It was painted in 1943 in New York, during his stay in the United States from 1940 to 1948 (Plate 20.6). The painting is generally read as

Plate 20.6 *Geopoliticus Child Watching the Birth of the New Man* by Salvador Dalí, 1943.

representing the rising of the USA as the new world power during the Second World War. The crushing of Europe is seen as a representation of its diminishing role in international politics and the oversize of Africa and South America as a growing importance of the (former and remaining) colonies. More widely, however, the art world and, in particular, the course of modern art in the second half of the twentieth century became intertwined with Cold War geopolitics. More generally, art and politics have frequently been intertwined. The Nazis (and their leader Adolf Hitler, who was a failed artist of minimal talents) detested modern art, condemning it as decadent. Other dictators, such as Joseph Stalin, the leader of the Soviet Union from the mid-1920s till his death in 1953, were also wary of modern art and favoured a garish realist style, depicting heroic smiling workers and peasants. Meanwhile, and especially in the years between the two World Wars, Paris became the centre of production of modern art. The American Central Intelligence Agency (CIA) stepped in after 1945 to fund and promote American modern art, bankrolling major exhibitions and tours, to demonstrate the creative artistic culture of the United States.

Of course, the Cold War was waged too in everything from science fiction comics and action movies and thrillers (in which good Americans, like Rambo, and the occasional heroic Brit, like James Bond (Plate 20.7), faced and defeated totalitarian, alien and frequently communist enemies) to chess (where 'Soviet Man' could demonstrate intellectual superiority and training) and Olympic sports (in which the Soviet, Chinese, East Germans and Bulgarians invested enormous resources in training and drug-enhanced muscles, to 'prove' the superiority of the communist system in the Olympic stadiums). Heroic men (and sometimes women) were invoked. Each one seeking to outdo the other, rather like men on sports fields or in street fights might (see Spotlight box 20.2). The 'Space Race', to build and launch artificial satellites and then to put men into space and reach the moon became a military, technical and a cultural expression of the Cold War – full of moments of national superpower pride, media spectacles and the basis of tales of heroic and daring American and Soviet men in space (Carter 1988). We might thus write of a 'popular geopolitics' to designate how geopolitical concepts are reflected in and take on meaning through popular culture: from art and literature, through film to comics, sport, computer games (Power 2007) or the Internet (Dodds

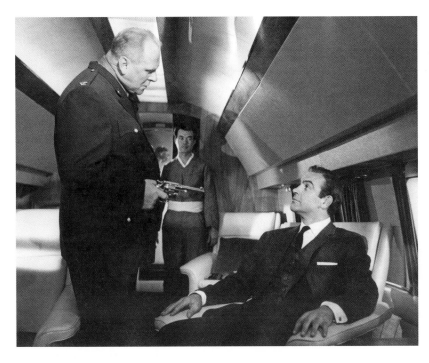

Plate 20.7 Cold War in cinema: James Bond (007) saves the West from communists and criminals.

(© 2005 Topfoto)

2006) and more traditional media like magazines, newspapers, radio and television. Cartoons and comic books have received ample attention in studies of such popular geopolitics – and continued to proliferate after the Cold War (Plate 20.8): see the article by Dittmar (2005) listed in the further roading for this chapter.

Spotlight box 20.2

Feminist Geopolitics

As in other academic fields, feminist scholars brought social struggles into academia to question taken-for-granted assumptions in geopolitical research and, more generally, in political geographical scholarship. Bringing issues of emancipation and empowerment, and their counterparts subordination and oppression, to the fore, feminist approaches problematized the state-centric approach of geopolitics, its elitism (as it is concerned with the visions and the behaviour of those in power), and its abstract, disembodied view of politics. They question, for example, the definition of security in terms of state security, undermine the idea of a national interest that would give clear direction for the conduct of foreign policy, and scrutinize by contrast how foreign and military policies affect the everyday life of people on the agenda. After all, if a state is more secure, does it automatically mean that the people it rules are? And whose security? Of ruling

elites, or the state? Or security from disease, detention or from violence in the home?

Fiona Smith (2001), for example, showed how the end of the Communist regime in Eastern Germany and German reunification were received with mixed feeling by people whose life was highly disrupted by the institutional changes (disorientation, massive unemployment, disappearance of many public services, etc). Addressing more acute situations, Jennifer Hyndman (2007, 2010) discusses the politics of body counts in the American intervention in Iraq and the fate of child soldiers in different geopolitical context. Another line of research has been the inquiry into the militarization of the lives of soldiers and their families, not only in wartime but also in military bases, for example as epitomized by the pioneering and prolific work of feminist scholar Cynthia Enloe (1983, 1989, 1993, 2000, 2000, 2004, 2007). As such, feminist geopolitics also examines environmental geographies of militarization (such as pollution by weapons in and around the battlefields and test sites – see Woodward 2004, 2005).

Plate 20.8 Captain America: geopolitics in action.

20.9 New World Order to the Long War

The notion of the Cold War faded with the collapse of the USSR and associated communist allies between 1989 and the early 1990s, leaving the United States as the sole effective global military superpower. Yet this Western and American 'victory' quickly produced a certain sense of disorientation. This had been evident before, as the so-called 'bipolar' world of the early years of Cold War confrontation in which Washington and Moscow were the main political, economic and ideological points of orientation became progressively more 'multipolar' from the 1960s onwards. That is, as Western Europe and Japan recovered from war-time devastation and as Communist China split on

Thematic case study 20.1

The continuing Cold War?

The Cold War in Asia was always a very complex mixture of ideological, territorial and national disputes. An extra dimension was added by the fact that China became a nuclear power in the early 1960s at the same time that the Chinese communists split with the USSR on the grounds of ideological differences.

Moreover, whilst it is now more than two decades since proclamations of the 'end of the Cold War' began to multiply in Europe, in the Asia-Pacific region remnants of the global Cold War confrontation remain in place. Despite substantial changes and accommodations, China, North Korea and Vietnam are still one-party 'communist' political systems. In addition, a series of divisions and territorial disputes whose history is deeply rooted in the Cold War remain unresolved. Most notably, the Korean peninsula remains divided between mutually antagonistic regimes and both the People's Republic of China and the offshore island of Taiwan have claimed to be the legitimate government of the whole of China and remain fundamentally opposed.

Arguably, therefore, there is a sense in which the narratives about 'the end of the Cold War' reflect the Euro-Atlantic experience rather than being globally valid. The 'end' may have happened in Europe and the Soviet Union may have ceased to exist years ago, but Cold War legacies and logics are active in East Asia (see Hara 1999) and beyond. However, as the main text details (see Sections 20.9 and 20.10), the extent to which new geopolitical narratives, notably 'the war on terror,' have replaced the Cold War as organizing frames for Western strategy is much debated – and contested.

Plate 20.9 South Korean soldiers patrol their side of the demilitarized zone between North and South Korea.

(© Michel Setboun/Corbis)

ideological grounds with the USSR, the sense that the world was divided into just two superpower points of orientation lessened. But the sensation of geopolitical complexity was to grow precipitously once one of the established 'poles' (Soviet communism and pro-Soviet regimes in Central and Eastern Europe) collapsed at the start of the 1990s. Having a clearly demarcated and identifiable enemy – as both sides had in the Cold War – offered contrasting but apparently solid ideas of identity, purpose and common cause that has faded in a post-Cold War world. If the Cold War embodied a relatively coherent geopolitical map (East against West, with the allegiance of the Third World as one of the prizes), the post-Cold War world is characterized by a diversity of maps and scripts. A variety of interpretations and geopolitical 'models' now attempt to offer explanation and impose meaning on contemporary events.

Amongst the most influential of these has been the discourse of **New World Order**, articulated by then US President George Bush (senior) in the early 1990s. In Bush's vision, what is called the 'international community' (the member states of the United Nations, led by the USA, but with financial and military support from key allies such as Germany, the United Kingdom and Japan) would act as a kind of global police force, intervening where and when they felt necessary to maintain or restore 'order'. Both the 1990–1 Gulf War and the 1999 conflict with Yugoslavia were justified and conducted (in part) in the name of building such a 'New World Order'. But it was his son, President George W. Bush, who was left to articulate a new American strategy (which became the Bush Doctrine, of pre-emptive attack on states that may – it was claimed – threaten the United States or harbour terrorists) in the aftermath of 9/11. And even this was, arguably, partially within a long American geopolitical tradition of overthrowing overseas regimes during and before the Cold War, which has its roots in the rise of American power in the Pacific and Americas in the late nineteenth century and has some deeper roots in the colonization of the continent (Kinzer 2006). Even well before 9/11 and the Bush administration's strategic reactions to it, critics had pointed out that, as Booth (1999: 49) put it, the New World Order 'means the New World [gives out the] Orders'. That is, the United States was seeking global military hegemony and with the Soviet resistance out of the way, at first there seemed little, save its own 'public opinion' (or perhaps, economic limits?) to stop it.

Others have stressed that such apparent American military superiority and rhetoric counts for less than might appear at first sight, and that what continues to be evident in the post-Cold War world is enhanced economic (and in some ways, cultural) competition between blocs or great powers. In this vision, American power is potentially contested by an integrating European Union and by Pacific powers, notably China. But the terms of the 'contest' are less in conventional military and strategic terms than in competition for markets, productivity and profit. A world of 'geopolitical' competition is, it was argued, being partially displaced by 'geoeconomic' competition (for an example, see Luttwak 1990). In the USA (and beyond), this was also sometimes interpreted as an impending 'supposed clash of civilizations' (e.g. Huntington 1993), between that of the 'West' and, for example, societies oriented to alternative belief systems such as (Chinese) 'Confucianism' or 'Islam'. Although superficially quite different, notions of geoeconomic competition and a clash of civilizations both betray a sense that the relative military dominance of the USA at the end of the Cold War may not in itself be enough to preserve America's sense of leadership, power and world-historical destiny in the context of a complex 'multipolar' world. It remains too early to tell how enduring the reassertion of American military power that unfolded under the Presidency of George W. Bush in the first decade of the twenty-first century will be. Certainly it quickly faced multiple challenges, as attempts to implement visions of 'A New American Century [of leadership]' (www.newamericancentury.org) mutated into an open commitment to what the US Department of Defense, in the form of National Security Strategy documents, has called (and codified in the February 2006 Defense Review as) 'The Long War'. Superficially this appears as a new Cold War, with the 'global war on terror' (GWoT) taking the centre stage once occupied by communism. However, as Buzan (2006: 1102) argued, despite the prominence of the global war on terror in these texts, it could not define contemporary US grand strategy overall:

> US grand strategy is much wider, involving more traditional concerns about rising powers, global energy supply, the spread of military technology and the enlargement of the democratic/capitalist sphere. US military expenditure remains largely aimed at meeting traditional challenges from other states, with only a small part specifically allocated for the GWoT. The significance of the GWoT is much more political . . . the main significance of the GWoT is as a political framing

that might justify and legitimize US primacy, leadership and unilateralism, both to Americans and to the rest of the world.

A few years later, Clark (2011: 20) noted how:

> President Obama's 2010 National Security Strategy . . . described 'a dynamic international environment in which different nations are exerting greater influence' and 'emerging powers in every region of the world are increasingly asserting themselves'. . . . Official US projections up to 2025 paint a broadly similar picture. These highlight the trend towards multipolarity, associated with a greater diffusion of power internationally.

Growing fears about cybersecurity, geoeconomic shifts, the assassinations of Bin Laden and other al-Qaeda leaders, and the 2011 war in Libya are currently at the fore of geopolitical debate in the United States. Who would have predicted these two decades ago? And what forms will such debates take in 20 years' time?

20.10 Conclusions: shifting hegemony?

Despite early post-Cold War proclamations about the great prospects for a more peaceful and secure New World Order, the world soon came to appear to American strategists (and many others) as more 'disordered' and unpredictable. Yet the USA is more conscious that the enemy is also configured as a loose and amorphous network (the form attributed to al-Qaeda, who are held responsible for the events of 9/11 and host of other terrorist attacks). At the same time, high-tech weapons spread, disseminating greater destructive capacity into areas of local confrontation and conflict, such as South and East Asia and the Middle East. Moreover, for the Western powers and for the USA in particular, no longer is the main 'enemy headquarters' so clearly demarcated and fixed behind a stable 'Iron Curtain' – as the USSR and its communist allies were. Following the hijacking of airliners and their use in attacks on highly symbolic economic and military targets in the United

Spotlight box 20.3

The EU as a geopolitical actor

The European Union (EU) incrementally developed a new type of geopolitical role. At the beginning of the 1950s, France and West Germany aimed at constructing new collaborative institutions between Western European states in an effort to prevent a fourth major conflict between themselves after the French Prussian war of 1870–1, the Great War of 1914–1918 and the Second World War of 1939–1945; these collaborative institutions later evolved into the European Union.

The first effort to collaborate more formally was the establishment in 1952 of the European Coal and Steel Community (ECSC) by six member states: France, West Germany, Italy, Belgium, Luxembourg and the Netherlands. This was a limited and technical form of integration, but economically and militarily crucial. Coal was still the major source of energy then and steel was key to manufacturing and weapons industry. The geopolitical idea behind the proposal of the French Minister of Foreign Affairs Robert Schuman was articulated by high-ranked civil servant Jean Monnet who had worked during the world wars and in the postwar reconstruction at collaborative efforts between the allies to improve their war and reconstruction logistics. The idea of a technical and technocratic approach was

successful and the role of the Community has spilled over to many other economic and social activities and enlarged to include 27 member states: including many former communist states that had been precluded from joining during the Cold War.

The EU remains, however, an extraordinary geopolitical actor as it does not have its own army and it has to coordinate its diplomatic services and its foreign policies with that of its member states.

The content of its geopolitical representations is worth underlining. The EU is primarily conceived as a civilian power: its weight in regional and global politics is derived from its economic, political and cultural influences, not from its military power (Bachmann and Sidaway, 2009). As the largest single market and the largest economy in the world (if the economy of the 27 member states is taken as one), with a large regulatory and normative power, it greatly influenced a wide array of countries with strong ties with it, economic or otherwise.

The civilian power of the EU is projected particularly strongly in the so-called European neighbourhood, the countries bordering the EU, some of them applicants to become EU members at some point, like the Balkan countries and Turkey, with others encouraged to become 'good' neighbours' like North Africa, Ukraine and the Caucasus.

States on 11 September 2001 ('9/11'), this has been felt even more deeply. Although there is no absence of traditional forms of geopolitical discourse, which define a clear boundary of 'our' space against 'them', some have argued that, today, Western strategy operates in the frame of a new kind of 'postmodern geopolitics'. If the twentieth century began with the modern age of geopolitics, announced by Kjellén and Mackinder, it might be argued that the twenty-first century will see new forms of geopolitical actor (see Spotlight box 20.3). The apparent simplicity of the great power and Cold War confrontations, let alone demarcated and fixed 'heartlands', have been replaced by the sense of a world of proliferating uncertainty and threats.

Scares, dangers and anxieties about terrorism since '9/11' have greatly reinforced this. The remaining superpower also faces long-term economic challenges (especially if the global role of the dollar diminishes) and must continue to reckon with more traditional state-centred powers, such as Russia and a rising China (even if they no longer pose the ideological competition of the Cold War years). There is no easy or predictable end to this uncertain situation and new challenges – climate change – complicate the global scene. New and old challenges and shifting patterns and forms of power, resistance and domination are expressed in space and time. Such geographies form the focus of Chapter 23.

Learning outcomes

Having read this chapter, you should understand:

- That the term 'geopolitics' has been associated with a wide variety of texts and contexts.

- However, geopolitics is frequently associated with the sense of a self-conscious tradition of writing about and mapping the relations between states, geography and power.

- This is usually traced to the late nineteenth-century writings of Rudolf Kjellén.

- One of the key features of this tradition is the idea that the state resembles a living organism which requires living space and will compete with others (the organic theory of the state).

- The oppressive military regimes in South America (notably Argentina, Brazil and Chile) during the 1960s and 1970s elaborated and applied a variety of the geopolitical tradition.

- However, it also proliferated in a number of European countries and Japan, particularly between the two world wars.

- The rise and eclipse of the Cold War have produced wide and deep geopolitical transformations.

- Since this proliferation of geopolitics was most evident in Fascist states, the defeat of the Axis powers in 1945 signified its relative decline as a formal tradition, although it continued in Fascist Spain and Portugal and in South American military circles.

- Beyond the self-conscious tradition of geopolitics, similar forms of thinking about territory, states and power have proliferated in many disciplines and countries. Indeed, we can also think of 'popular geopolitics' – whereby everyday culture (such as cinema) contains and expresses wider geopolitical narratives.

- This has been particularly evident during the Cold War, when broader geopolitical discourses suffused many aspects of art, science, culture and daily life.

- Since the decline of the Cold War, there has been a proliferation of geopolitical visions of the 'New World Order', amongst them a more complex cartography of perceptions of threats and dangers. However, the ability of any new single geopolitical 'big picture' (such as 'the war on terror') to inherit the mantle of the Cold War is radically open to question and contested.

Further reading

Agnew, J. (2003) *Geopolitics: Revisioning World Politics*, 2nd edition, Routledge, London and New York. A historical and contemporary guide which places geopolitics in comparative historical perspective.

Agnew, J. and **Corbridge, S.** (1995) *Mastering Space: Hegemony, Territory and International Political Economy*, Routledge, London and New York. Written in the aftermath of the Cold War and as debates proliferated about post-Cold War geopolitical and geoeconomic trends, Part II (Chapters 5, 6 and 7) remains useful on aspects of and critical reflections on the 'New World Order'.

Agnew, J., Mitchell, K. and Toal, G. (eds) (2003) *A Companion to Political Geography*, Blackwell, Oxford. Includes some useful accounts of geopolitical traditions, amongst other aspects of past and present political geographies.

Campbell, D. (1992) *Writing Security: United States Foreign Policy and the Politics of Identity*, University of Minnesota Press, Minneapolis, MN. A challenging but rewarding account of how foreign policy and the Cold War have been productive of America's sense of self-identity.

Dodds, K. (2004) *Global Geopolitics: A Critical Introduction*, Prentice Hall, Harlow.

Dodds, K. (2007) *Geopolitics: A Very Short Introduction*, Oxford University Press, Oxford.

Elden, S. (2009) *Terror and Territory: The Spatial Extent of Sovereignty*, University of Minnesota Press, Minneapolis, MN. This readable book links many of the themes considered in this chapter with those in Chapters 21 and 22 here.

Flint, C. (2006) *Introduction to Geopolitics*, Routledge, London. Three accessible introductions: all good starting points.

Halliday, F. (2002) *Two Hours that Shook the World. September 11, 2001: Causes and Consequences*, Saqi Books, London. Amongst the most nuanced of the hundreds of books that have appeared on this subject. Fred Halliday – a prolific and informative scholar of international relations – died in 2010. His last book was published posthumously:

Halliday, F. (2010) *Shocked and Awed: How The War on Terror and Jihad Changed the English Language*, I.B. Tauris, London.

Heffernan, M. (1998) *The Meaning of Europe: Geography and Geopolitics*, Arnold, London.

Loriaux, M. (2008) *European Union and the Deconstruction of the Rhineland Frontier*, Cambridge University Press, Cambridge.

McCormick, J. (2006) *The European Superpower*, Palgrave Macmillan, London.

Heffernan's book is a readable account of European geo-politics and the background to European integration; Loriaux investigates the geopolitics that engendered the European Union; McCormick claims that the European Union is a new form of 'superpower' (in geoeconomic and civilian rather than conventional geopolitical terms).

Two accounts by historians that offer startling insights into the geopolitics of the Soviet Union and Nazi Germany. Both are enabled by the opening of archives since the decline of the Cold War:

Jersild, A. (2011) The Soviet state as imperial scavenger: 'Catch Up and Surpass' in the transnational socialist bloc, 1950–1960, *The American Historical Review,* 116(1), 109–32.

Snyder, T. (2010) *Bloodlands: Europe between Hitler and Stalin*, Basic Books, New York.

An account of millions murdered in the name of geopolitical projects. Snyder's book is not a light read or good bedtime reading: it is the stuff of nightmares.

For a challenging collection on the work of a German political thinker (and one-time member of the Nazi party) whose work on space and politics became part of wider twentieth-century geopolitical writings, see:

Legg, S. (ed.) 2011 *Spatiality, Sovereignty and Carl Schmitt: Geographies of the Nomos,* Routledge, London and New York.

For more accounts seeking a critical (approach to) geopolitics or interrogating past geopolitical traditions, see:

Hyndman, J. (2003) Beyond either/or: a feminist analysis of September 11th, *ACME: An International E-Journal for Critical Geographies*, 2(1), 1–13 (available at: www.acme-journal.org). Indicates how feminist approaches might problematize and counter dominant geopolitical interpretations.

Larsen, H.G. (2011) 'The need and ability for expansion': conceptions of living space in the small-state geopolitics of Gudmund Hatt, *Political Geography,* 30(1) 38–48. The strange case of a Danish geographer who became entangled with Nazi geopolitics.

O'Loughlin, J. (ed.) (1994) *Dictionary of Geopolitics*, Greenwood Press, Westport, CT. Still a useful source.

Ó Tuathail, G. (1996) An anti-geopolitical eye: Maggie O'Kane in Bosnia, 1992–3, *Gender, Place and Culture*, 3(2), 177–85. Shows how commitment can produce alternative narratives amounting to an 'anti-geopolitics'. Suggestive and worth reflection.

Ó Tuathail, G. and Dalby, S. (eds) (1998) *Rethinking Geopolitics*, Routledge, London and New York. A collection of critical reflections on Cold War, popular and post-Cold War geopolitical narratives.

Ó Tuathail, G., Dalby, S. and Routledge, P. (eds) (2006) *The Geopolitics Reader,* 2nd edition, Routledge, London and New York. A collection of previously published short texts (readings) on different aspects of geopolitical discourse, including extracts from Mackinder, Haushofer and Kissinger and debates about the New World Order. Detailed editorial introductions contextualize the readings.

Sidaway J.D. (2010) 'One Island, One Team, One Mission': geopolitics, sovereignty, 'race' and rendition, *Geopolitics,* 15(4), 667–83. How colonial, Cold War and post Cold-War geopolitics interact in the Indian Ocean.

Three suggestive papers on aspects of popular geopolitics (cartoons and newspapers), worth comparing and contrasting:

Debrix, F. (2007) Tabloid imperialism: American geopolitical anxieties and the war on terror, *Geography Compass*, 1(4), 932–45.

Dittmer, J. (2005) Captain America's Empire: reflections on identity, popular culture and post-9/11 geopolitics, *Annals of the Association of American Geographers,* 95(3), 626–43.

Falah, G.-W., Flint C. and Mamadouh, V. (2006) Just war and extraterritoriality: the popular geopolitics of the United States war on Iraq as reflected in newspapers of the Arab world, *Annals of the Association of American Geographers*, 96(1), 142–64.

This account of French media's (TV, print and internet) geopolitical representations of an American presidential election is also rewarding:

Adams, P.C. (2007) *Atlantic Reverberations: French Representations of an American Presidential Election*, Ashgate, Aldershot.

But for a great introduction to popular geopolitics, (and with reflections on geopolitical traditions more widely):

Dittmer, J. (2010) *Popular Culture, Geopolitics and Identity*, Rownman and Littlefield, Lanham.

There are many other variants of 'popular geopolitics' in circulation and five minutes on the Internet will yield dozens of (mostly US-authored and hosted) religious 'end times' sites on the New World Order as well as 9/11 conspiracy theories. Read them critically with a view to understanding the basis on which each claims truth and the consequences of and basis of their *geopolitical* assumptions. For some primers, drawing on American popular culture, see:

Fenster, M. (2008) *Conspiracy Theories: Secrecy and Power in American Culture*, Minnesota University Press, Minneapolis, MN.

Jones, L. (2008) A geopolitical mapping of the post-9/11 world, *Aether: the Journal of Media Geography*, 3, June 2008. Freely downloadable at http://130.166.124.2/~aether/volume_03.html.

Finally, a century after its original publication, Mackinder's (1904) paper was reprinted in *The Geographical Journal* (Volume 170, Number 4, December 2004), along with a set of commentaries and reflections.

For annotated, clickable weblinks and useful tutorials full of practical advice on how to improve your study skills, visit this book's website at **www.pearsoned.co.uk/daniels**

TERRITORY, SPACE AND SOCIETY

Chapter 21

David Storey

Topics covered

- The creation of territories
- Territorial strategies and the concept of territoriality
- Social processes and spatial relations
- Territories and class, race, gender, sexuality
- Leisure, work and home space
- Contested spaces and geographies of resistance

Human spatial relations are the results of influence and power.
Territoriality is the primary spatial form power takes.
(Sack 1986: 26)

21.1 Territory and territoriality

We live in a world where we are regularly confronted with signs reminding us of where we are permitted to be or not to be, or how to behave in certain places. Messages that tell us 'authorized personnel only', 'no trespassing', 'no entry', 'keep off the grass' and so on are constant reminders that the world is highly territorialized (Plate 21.1). A glance at the global political map provides us with the most obvious formalized manifestation of this in the division of the earth into separate countries or states. However, this macro-scale territorialization is accompanied by a myriad of much more micro-scale, often less formal, variants. In everyday usage, territory is usually taken to refer to a portion of geographic space which is claimed or occupied by a person or group of persons or by an institution. In this way it can be seen as an area of 'bounded space'. Following from this, the process whereby individuals or groups lay claim to such territory can be referred to as 'territoriality'. However, as will be seen, these somewhat simplified definitions mask considerable complexity.

It is often assumed that humans have a natural tendency to behave in a territorial manner: to claim space and to prevent others from encroaching on 'our' territory. Considerable debate has occurred over the extent to which territorialization and territorial behaviour should be seen as a 'natural' or 'social' phenomenon, a debate echoing wider long-standing arguments over the relative influence of nature and nurture, a division that many see as somewhat blurred (see Whatmore 2002). In his landmark book *Human Territoriality: Its Theory and History*, published in 1986, Robert Sack cautions against determinist views of

Plate 21.1 No trespassing

human territoriality as a basic instinct and emphasizes instead its role as a geographic and political strategy. For Sack, territoriality is 'the attempt by an individual or group to affect, influence, or control people, phenomena, and relationships, by delimiting and asserting control over a geographic area' (Sack 1986: 19). He draws attention to the means through which territorial strategies may be used to achieve particular ends. In essence the control of geographic space can be used to assert or to maintain power, or, importantly, to resist the power of a dominant group:

> Territoriality, as a component of power, is not only a means of creating and maintaining order, but is a device to create and maintain much of the geographic context through which we experience the world and give it meaning.
>
> (Sack 1986: 219)

In this way territoriality is deeply embedded in social relations and territories, rather than being natural entities, result from social practices and processes (Delaney 2005). They are produced under particular conditions and serve specific ends.

Sack (1986) argues that territoriality involves a classification by area whereby geographic space is apportioned. However, territories are more than mere spatial containers; they link space and society and convey clear meanings relating to authority, power and rights (Sassen 2006; Delaney 2009). David Delaney has defined a territory as 'a bounded social space that inscribes a certain social meaning onto defined segments of the material world. A simple territory marks a differentiation between an "inside" and an "outside"' (2005: 14). These geographic spaces convey messages of political power and control which are communicated through various means, most notably through the creation and maintenance of *boundaries*. These and other mechanisms facilitate control over space and of those within it; they serve as divisions between those inside and those outside. In this way territory allows classification, communication and enforcement through boundary-making. The meanings conveyed may have important implications such as constraining, restricting or limiting mobility, for example.

While Sack's work focused usefully on territoriality as a political strategy, more recently Stuart Elden (2005, 2010) and Joe Painter (2010) have focused attention on the concept of territory itself. Elden suggests that the meaning of the term has been taken for granted but he argues that the manner in which the concept emerged is bound up with particular ways of thinking about geographic space; ways which reflect notions of power and control. From this, it follows that it is not just territorial strategies that need to be interrogated, but the broader ways in which we conceive of geographic space. Elden (2005) contends that developments in cartographic techniques, advancements in geometry and mathematical calculation led to our modern conceptualization of space. Territories are, therefore, more than simply land; instead territory can be viewed as a political technology, related to the measurement of land and the control of terrain. As such, it is suggested that both territories themselves and their boundaries (terms which we tend to take for granted) reflect a distinctive mode of social and spatial organization which is dependent on particular ways of calculating and thinking about space (see Elden 2007, 2010; Brighenti 2010a). Painter (2010) argues that territory is an effect or an outcome of a set of practices. Using the construction of the English regions as an example, he demonstrates how a set of techniques and processes, such as the issuing of regional statistics, the creation of development agencies and the devising of regional strategies, reproduce a sense of delimitation, contiguity and coherence to the idea of regional divisions. Yet these apparently solid territorial blocs are created and sustained through complex networks of agencies and processes that transcend the boundaries of those regions.

While being mindful of the complexity of ideas surrounding territory and territoriality, it is clear they serve useful political functions. Territoriality and the production of territories can be seen as devices that tend to reify power so that it appears to reside in the territory itself rather than in those who control it. Attention is thereby deflected away from the power relationships, ideologies and processes underpinning the maintenance of territories and their boundaries. In this way, as Delaney argues, 'territory does much of our thinking for us and closes off or obscures questions of power and meaning, ideology and legitimacy, authority and obligation' (2005: 18). Ultimately territoriality can be seen as 'a primary geographic expression of social power' (Sack 1986: 5). Once created, territories can become the spatial containers in which people are socialized through various social practices and discourses. As Paasi (2003) suggests, a number of important dimensions of social life and social power are brought together in territory. There is a material component such as land, there is a functional element associated with control of, or attempts to control, space and there is also a symbolic component associated with people's social identity. The social, cultural and political are brought

together so that people identify with territories in such a way that they can be seen 'to satisfy both the material requirements of life and the emotional requirements of belonging' (Penrose 2002: 282). Sack suggested that 'territoriality . . . forms the backcloth to human spatial relations and conceptions of space' (Sack 1986: 26). However, territory is more than a mere backdrop or the material manifestation of a set of social relations, it is also itself a form and effect of particular social relations bound into the intersections of power, space and society (Brighenti 2010a). The spatial is not simply the outcome of the social but the two are intrinsically bound up together (Delaney 2009).

The creation and imagining of territories and the utilization of territorial strategies can be observed at a variety of spatial scales. While some of the most obvious (and contested) expressions of territoriality are manifested at the level of the state (see Chapter 20), many more micro-level examples of territorial control and territorial strategies may be observed. Territorial configurations exist at a variety of spatial scales and are constantly reproduced, contested and transformed (Brighenti 2010a). Although these may be less obvious and may often seem more vaguely defined with less clear-cut boundaries, these 'informal' territories can still convey quite clear meanings to those concerned (Delaney 2005). They are bound up with strategies of enforcement and control. Elden (2007) has reminded us of the possible etymological links between terror and territory whereby territory might be interpreted as something which people are warned off or frightened away from. The bounding of space brings with it exclusion and control but human territories and their boundaries may be subject to periodic or continuous contestation, modification and destruction (Paasi 2003). Territory and territoriality run through a wide range of social issues including race, class, gender and sexuality so that spatial enclaves are created and sustained (Sidaway 2007). This chapter explores some of these examples, demonstrating how particular social practices are spatialized and the ways in which territories are constructed, contested and used to achieve particular outcomes.

The examples are arranged under three main headings with a focus on race and class; gender, sexuality and space; and work, rest and play. Despite the adoption of this structure, it should be abundantly clear that many of the issues raised are interrelated (and also overlap with earlier chapters in the book and others in this section). People have more than one single identity: gender, sexuality and ethnicity cross-cut each other in a system of overlapping identities. Ethnic groups and classes are not immutable but are social constructs. In other words, they vary in time and space and are not simply natural categories, but human (that is social) products. Rather than seeing identities as rigidly defined, it is more useful and realistic to regard them as unstable, relational and contingent. The examples used in the chapter raise obvious questions about identity, rights and belonging, thus touching on broader ideas of citizenship (Storey 2003; Low 2005). While traditional notions of citizenship are centred on the relationship between the individual and the state with a focus on rights and duties, more recently, ideas of citizenship have broadened to consider group rights

and issues below the level of the state (local issues) and beyond the state (responsibility to distant others). While people may appear to enjoy full citizenship rights, as we shall see, particular territorializations and the use of territorial strategies can result in some groups being excluded from participating fully within society. In effect some people may be made to feel like second-class citizens as a consequence of the creation and maintenance of informal territorial boundaries. This is one reason why various social movements and groupings have arisen to assert group identities and rights often through challenging particular territorializations (Case study 21.1). Consequently

Thematic case study 21.1

The spaces of political protest

Political protests such as marches, sit-ins or meetings take place in specific places. However, state authorities can delimit (or try to) where and when these may take place. Protest marches such as those against the wars in Iraq and Afghanistan have taken place periodically in London and many other centres. On these occasions permission to march through the streets (public space) is granted though routes must be pre-agreed. In Ireland in September 2010 part of central Dublin was closed to traffic as a group of protestors vented their anger against former British Prime Minister Tony Blair outside a bookshop where he was signing copies of his autobiography. The protestors viewed Blair as a war criminal who should be tried in connection with his role in the Afghan and Iraq wars. Such protest was permitted by Irish police but within strict territorial

limits. Protestors who tried to break through a cordon and throw objects at Blair's car were arrested. Their presence and actions were thus highly constrained.

In a related manner a political-spatial controversy surrounds Parliament Square in London. Its proximity to the UK parliament renders it a popular space in which to protest. Anti-war protestor Brian Haw set up camp there is 2001. He remained there until shortly before his death from cancer a decade later and various other individuals and groups have also occupied space there for periods of time. Despite attempts to have them removed, protestors remain, along with some homeless people. While some regard this as a spatial manifestation of a legitimate right to political protest, others tend to focus on the presumed illegality and the supposedly negative impact the 'residents' have on the appearance of the space (Plate 21.2)

Plate 21.2 Protestors 'camp' in Parliament Square, London.

the chapter is concerned with raising questions about who is 'allowed' to be in particular spaces and who is barred or discouraged from being there. In doing so it sheds light on the ways in which space is conceived, used and organized (Elden 2005).

21.2 Territoriality, race and class

21.2.1 Class and segregations

Society is seen to be stratified into different socio-economic groupings. An analysis of these social classes casts light on inequalities between different sections of society; between rich and poor, between property owners and those who are not; between those who own and control resources and those who are paid workers or are unable to obtain a job. These social divisions (together with race, ethnicity, gender and others) not only reflect inequality, they are also deeply spatialized. For example, most cities appear to have distinct residential neighbourhoods, colloquially defined as 'rich' or 'poor', 'working class' or 'middle class'. These spatial divisions are reflected in ideas of 'the wrong side of the tracks' and everyday discourses of 'good' and 'bad' areas within cities. These latter zones are often viewed as separate spaces inhabited by many people who are marginalized not just in social and economic terms, but also spatially. Sidaway (2007) has used the idea of enclaved space as a way of thinking about social (and other) divisions which serve both to reflect and reproduce inequality, especially marked in contexts of violence and war (Plate 21.3). One good example is the *banlieues* of Paris, which are both physically separate and socially distant from the heart of the city, cut

off by a ring road (the *périphérique*) from the centre and relatively isolated from each other (Morley 2000). Dikeç (2006a, 2007a) has highlighted the ways in which the *banlieues* have become synonymous with high unemployment, crime and related negative images. Their residents suffer discrimination and inequality, while being subjected to negative stereotyping and harsh policing. They have come to be characterized as dangerous neighbourhoods inhabited by dangerous people pathologised as an underclass. Urban riots in the poorer *banlieues* on the edge of Paris in 2005 (and smaller-scale events in 2007 and 2009) spread to other French cities. Young people there, including many from families of North African migrants, reacted to incidents of police violence in a wave of protests that rapidly spread. In these areas of high unemployment younger residents gave violent expression to their feelings of social and spatial alienation (see Plate 10.1). The unrest which has periodically arisen has deep structural roots embedded in the French urban planning system and in the wider French political structures (Dikeç 2006b, 2007b). As Hubbard suggests in Chapter 10, such stigmatizing of place (and people) in itself becomes part of the problem, serving to reinforce class (and often ethnic) divisions and to reproduce various forms of social exclusion. This can result in ghettoization with a whole series of negative consequences for both people and place so that the 'inhabitants come to wear the myths of that place' (Winchester *et al.* 2003: 176).

In general terms these, and similar, forms of segregation arise and are reinforced through various mechanisms, particularly the housing market which effectively determines who can afford to live where. The 'poor' cannot afford to live in 'rich' areas and the rich are generally unlikely to want to reside in close proximity

Plate 21.3 UN compound in Afghanistan: gated and guarded territory.

to poorer run down neighbourhoods. Nevertheless, attempts to regenerate older working class areas tend simply to reshape the geographies of class rather than eliminate them. Gentrification is where parts of the urban area experience regeneration or renewal, resulting in more affluent residents moving in and displacing the original predominantly working-class inhabitants. Driven in part by economic considerations and in part through consumer choice, it serves to reinforce economic divisions within society, thereby perpetuating the idea that some households do not belong in particular places (see Hubbard 2006). Linked to this, it is argued that the role of 'urban gatekeepers' (such as estate agents) may play a key role here in altering (or endeavouring to maintain) the social composition of particular areas (Knox and Pinch 2000; Shaw 2008). Gentrification reflects broader socio-economic processes and the resultant residential territorialization can be seen as an expression of both demand for housing (from homeowners) and supply of capital (from financial institutions). It serves to highlight how broader global processes play out alongside (and are implicated in) the destruction and reconstruction of local territorialized identities (Butler 2007). The major regeneration schemes undertaken in recent decades in older industrialised and dockland areas in cities throughout Europe and North America reflect this transformation from manufacturing and working-class residential spaces into service sector zones with middle-class residents. Dockland and waterfront areas, like other 'regenerated' urban zones, have been transformed into different places, with quite different uses and symbolic meanings (Plate 21.4). In South Wales,

for example, the regenerated Cardiff Bay area has been transformed from a place associated initially with the export of coal and subsequently with industrial decline and decay. Now, with its mix of residential and commercial developments, it is promoted as a symbol of a modern forward-looking city (see www.cardiffbay.co.uk) (Plate 21.4)). The scale and extent of contemporary developments, in global cities like New York and London, where a super-rich elite engages in intense investment and conspicuous consumption, has led to the coining of the term super-gentrification (Butler and Lees 2006). These attempts to purify urban space have led to the displacement of some urban residents while others are rendered homeless. In cities such as San Francisco fragile lives are lived out on the streets amidst an atmosphere in which the homeless are seen as a blot to be removed rather than a manifestation of a systemic problem. Homeless people are criminalized and medicalized so that urban space can be cleansed and put to more profitable uses (Gowan 2010).

Class-based segregation is rendered even more obvious through the long evident phenomenon of urban gating (see Davis 1990; Blakely and Snyder 1997; Atkinson and Blandy 2006; Glasze et al. 2006; Bagaeen and Uduku 2010). The apparent rise in the numbers and varied forms of gated communities within urban areas in various parts of the world in recent years could be interpreted as a very obvious manifestation of attempts to control and limit access to portions of geographic space. The creation of residential fortresses where security guards patrol the perimeter of walled residential zones in an effort to exclude those seen as 'undesirable' has been repeated and deepened in many other cities

Plate 21.4 Former Docklands in London.

Plate 21.5 Security gates on gated residential development.

in order to exclude those seen as not belonging there, so maintaining the 'undefiled' and 'exclusive' nature of the neighbourhood. The level of 'fortification' of these developments is quite varied ranging from perimeter walls, gates and barriers, through the limiting of non-residential access by intercoms and associated 'screening' devices, to more perceptual barriers or codes deterring access (Plate 21.5).

The idea of gating can be seen as a consequence of two inter-connected factors: security and prestige. Though not strictly the preserve of the better-off, discourses of safety and security serve as a useful rationale for developers to design, build and promote spatially exclusive housing often surrounded by security fences and with highly limited public access regulated by intercoms and other 'screening' technologies which have come to be regular features in contemporary landscapes of power (see Jones *et al.* 2004). These territorial strategies work in ways which ensure a particular residential mix and may well serve to link together both racial and class divisions. More broadly, Morley (2000) suggests social groups separate out from each other and, indeed, are encouraged to do so. He further suggests that these tendencies have been effectively reinforced in recent decades through geodemographics and the use of postcode data and associated marketing strategies of companies who are keen to identify particular types of consumer, and link these to geographic areas. In this way residential homogeneity is both reflected and reproduced. The social and the spatial are inextricably linked as 'gated minds' are translated into gated places (Landman 2010). While the forms it takes may vary

somewhat according to place-specific circumstances, gating in its various guises touches on vitally important questions of the privatization of previously public space, inclusion, exclusion and the territorialization of social life (Lemanski and Oldfield 2009; Rosen and Razin 2009).

The mixture of subtle and not so subtle processes of residential segregation are mirrored in other arenas. Taken alongside the privatization and 'gating' of residential zones, the proliferation of covered shopping malls can be seen as the erosion of shared urban street space and its replacement with privatized, more exclusionary spaces of consumption. Where once streets were open to a broad public, there are now privatized spaces whose owners (invariably resorting to security firms who take on some police powers) can evict those seen to behave inappropriately or who simply look 'out of place'. The shopping mall has become a privatized and highly regulated space in which people may be excluded by virtue of their appearance or behaviour (Staeheli and Mitchell 2006). In the United Kingdom, recurring debates and associated social panics over young people wearing 'hoodies' (hooded sweatshirts popular amongst teenagers) reflects particular hegemonic ideas about how teenagers and young adults should behave in socially acceptable ways in particular places. In 2005 a shopping mall in Kent banned young people from wearing 'hoodies' and baseball caps as part of a crackdown on what was deemed to be antisocial behaviour. The centre manager was reported as saying: 'We're very concerned that some of our guests don't feel at all comfortable in what really is a family environment' (see http://news.bbc.co.uk/1/hi/england/kent/4534903.stm). The centre is clearly considered private space into which people are 'invited' rather than having any automatic right to be there. These 'secure' shopping centres, office blocks and apartment buildings, complete with gates and intercom systems, exemplify a trend towards socio-spatial design whereby territorial strategies associated with crime prevention effectively exclude those not wanted. Some people are effectively barred from certain areas – a policy of territorial containment enforced through increased surveillance and architectural design features (Cozens *et al.* 1999). In part, these forms of exclusionary policing reflect broader geographies of fear and the perceived risk of crime associated both with specific groups and with specific geographic spaces (Pain and Smith 2008) (Case study 21.2).

In a different context the idea of spatial enclaves separate from the surrounding area is demonstrated in the existence of export processing zones (EPZs)

Thematic case study 21.2

Policing space

The geography of policing has begun to attract some research interest since Fyfe drew attention to it in 1991. One elementary way of seeing this is the division into police forces with responsibility for particular regions; West Mercia police in the UK and so on. Linked to this are ideas of particular spaces in which special units operate such as harbours and airports, where specific concerns linked to immigration, smuggling and security are seen as priorities. The manner in which policing is carried out also reflects ideas of territory and territorial control ranging from police having particular 'beats' or zones of responsibility to the spatial tactics employed by police, for example, funneling street marches and protests along agreed routes. Recent protests in London (in particular, student demonstrations in 2010) saw the police detain protestors in confined spaces – a controversial policing strategy known colloquially as 'kettling'. The increasing array of non-state actors engaged in forms of policing also brings with it a series of additional territorializations with private security firms patrolling office blocks and shopping malls. These create various tensions including attempts to prevent people taking photographs in certain spaces. The advent of citizen-based groups such as Neighbourhood Watch also rest on notions of local territories with people encouraged to report suspicious behaviour. Current UK government strategies appear designed to deepen trends towards greater community control and the broadening of policing partnership arrangements.

Policing is about more than enforcement, it is also about the regulation and endorsement of specific values and moral codes (Mawby and Yarwood 2011). In addition to territorialisations associated with the policing of space, there are those which endeavour to exclude individuals or particular forms of behaviour from certain spaces. In the United Kingdom, Anti-Social Behaviour Orders (ASBOs) and related instruments provide a controversial example of bringing a territorial approach to dealing with particular behaviours that are seen to constitute a nuisance. They were introduced in 1998 and they can be used to prevent specified individuals from being in particular places, streets, etc., or to prevent individuals from repeatedly engaging in specified behaviours such as aggressive begging, disturbing neighbours, drunkenness in public places. They have proved controversial, seen by some as the criminalisation of certain social behaviours (Squires 2008). What they clearly do is suggest the idea that certain forms of behaviour are not allowed in public space and they can be punished with territorial exclusion. What is also clear is that they tend to be imposed on people in certain types of place, mainly areas of public housing in UK cities (Painter 2006). Similar so-called 'bubble' laws in the USA also function to exclude people behaving in certain ways from specific places (Mitchell 2005). Regardless of our view of the rights and wrongs of such moves, and whether they might be deemed successful even on their own terms, they raise important geographical questions related to what is meant by 'public', who is allowed to be where and how people are expected to behave (Cameron 2007). For a good overview of geographies of policing see Yarwood (2007).

(and related creations such as Specal Economic Zones) in countries such as the Philippines and South Korea where 'normal' business regulations (such as environmental and labour laws and taxation regimes) are suspended or eased in order to encourage particular forms of production geared towards the export of goods. These have been subject to criticism for allowing large-multinational companies to enjoy lax regulations in zones within developing economies (see Klein 2000; Park 2005). In a similar vein, recreational spaces such as golf courses, wider tourist complexes in regions such as the West Indies or beach resorts in parts of Southeast Asia can be seen as further extensions of spatial enclaving designed to enclose the relatively privileged consumers from the excluded local 'other' (see Bunnell *et al.* 2006; Sidaway 2009). Viewed in a broader context the production of these spaces of exclusion can be interpreted as a long-standing and inherent tendency within capitalist social relations bound up with constant pressures towards the privatization of the public domain (see Harvey 2003; Vasudevan *et al.* 2008).

21.2.2 Ethnicity, 'race' and space

Just as class is mapped on to the urban fabric, so too is ethnicity (see also Chapters 10 and 13). While ideas of race appear firmly embedded in everyday discourses, racial and ethnic categories are social constructions rather than simple biological realities. Although race can be questioned as a problematic and often dubious form of social classification derived from past notions of hierarchy and domination (notably slavery and colonialism), there is no doubt that racism or 'race thinking' is a very real social phenomenon. The idea of a 'scientific' categorization of humanity into different 'races' has long ago been challenged and is now thoroughly discredited, but its legacy persists. Kobayashi and Peake argue that race is socially constructed but that 'racialization' is 'the process by which racialized groups are identified, given stereotypical characteristics, and coerced into specific living conditions, often involving social/spatial segregation and always constituting racialized places' (2000: 293). While issues linked to 'race' are clearly social phenomena, they are often manifested spatially. Amongst the most rigid examples of racialized space was that devised under the apartheid system in South Africa from the late 1940s through to the early 1990s: a territorial system that enhanced and entrenched the political, economic and social power of a minority white population over non-white populations. Both nationally and at the more localized level of individual urban areas, space was divided on racial lines. Non-white people were 'placed' in locations not of their own choosing in order to entrench minority white power. In this way, there was a legal transposition of inequality on to geographical space. This spatial arrangement was designed to ensure greater degrees of control over the majority black population and is a classic example of the utilization of a territorial strategy to attain political objectives. At localized levels, there was a racialization of space with buses, public toilets and other amenities reflecting this divide. A racial ideology was mapped on to the South African landscape. Although apartheid ended in the early 1990s, after a long struggle for democratic rights for all in South Africa, its legacy means that a division of space based on the racial and class lines reinforced during the apartheid era has left enduring marks on the social landscape.

Elsewhere, nationalist conflicts may give rise to attempts to create ethnically pure spaces through the forcible removal of ethnic 'others'. The phenomenon of ethnic cleansing in parts of the former Yugoslavia during the 1990s is a striking example of an attempt to 'purify' territory of those deemed to belong to other ethno-cultural groups in an apparent attempt to justify territorial control in the name of the nation (Case study 21.3). In a not unrelated manner, the generally negative stereotyping of gypsies in much of Europe has led to considerable discrimination, with gypsies seen as an undesirable 'other', as a consequence of which they are effectively de-territorialized; they are seen not to belong anywhere and active attempts are made to exclude them from certain spaces (Fonseca 1996). In 2010 the French government attempted to repatriate ethnic *Roma* (gypsy) people to Romania, in what appeared a contravention of EU policy on freedom of movement between member states, as well as a highly questionable attack on an ethnic group. Broader issues emerge here in relation to groups which pursue nomadic lifestyles and who, consequently, are subject to considerable public opprobrium. Attempts by gypsies and other traveller groups to use particular spaces are often resisted by settled residents. As a group they are partly de-territorialised, simultaneously belonging everywhere and nowhere, their mobility juxtaposed to the settled nature of place-based communities. The mobile lifestyles of some are seen as unnatural and those who practice them are often depicted as untrustworthy and are subject to a range of discriminatory practices (Shubin 2010). Yet travellers also have nuanced notions of community, territory and belonging amidst movement that may parallel as well as clash with those of more 'settled' communities.

Leaving aside these overt and coercive examples, it is clear that many cities exhibit high degrees of ethnic segregation. In most US cities, for example, the elites and middle class are disproportionately white. The spatialisation of class (discussed above) thereby contributes to ethnic segregation (Crump 2004). These patterns of exclusion and inclusion and attendant territorialities reflect the complex intersections of race, class and ideology. In considering the evident spatial concentrations of ethnic groups in urban areas, it might be argued that individuals choose to locate in such areas for a variety of reasons. In brief there are a combination of 'positive' and 'negative' factors; for some there are attractions such as 'being amongst one's own', while others may feel driven to seek sanctuary from a racist, hostile society. Following Knox and Pinch (2000), key reasons may be summarized as follows. Clustering affords *defence* against attack by the majority group. It also provides a degree of *mutual support* and bolsters a sense of belonging and community. This in turn may be a useful means through which group cultural norms and heritage may be preserved. Finally, clustering produces *spaces of resistance* whereby external threats, whether to cultural norms or of physical attack, may be reduced. However, these arguments should not detract from the

Thematic case study 21.3

'Ethnic' territorial cleansing

Following the disintegration of Yugoslavia in the early 1990s, violent conflicts erupted in Croatia, Bosnia-Herzegovina and, later, Kosovo. These ethno-national conflicts were characterized by attempts by various groups to eradicate other ethnic groups from 'their' territory. This strategy was built on an essentialist version of defining ethno-national identity and, quite literally, clearing the territory of those possessing a supposedly 'different' identity. Armed movements claiming to be representing Serbs, Croats and Bosnians (predominantly Muslims) tried to carve out spaces which they could call their own in the ruins of the former federal state. Viewed from the outside it is easy to argue that such a strategy is both dangerous and simplistic. However, for those directly involved the control of territory was seen as an essential element

in the conflict. The struggle over Bosnian territory was particularly violent. Despite the area's complex multi-cultural history, reductionist interpretations of identity led to attempts to assert territorial control through the elimination of 'others'. Nationalist rhetoric, and the associated desire to control particular portions of territory in the name of a specific group, hardened divisions which were of relatively minor significance only a few years previously when Yugoslavia was a federal (and communist) republic under one-party rule. Once the conflict was underway, however, it became impossible to reverse, as the conflict drew upon and *reinforced* group identities and competing territorial claims underpinned through reference to historical myths and forms of boundary-making. In order to achieve peace, areas were mapped and information gathered on the various ethno-national groupings living in different localities. In this way territory was designated 'Serb',

Figure 21.1 Southwest Europe on the eve of the First World War.

Figure 21.2 Majority ethnic groups within Yugoslavia in 1991.

'Muslim', 'Croat', and so on, with lines dividing towns and cities into different zones. After a number of failed alternatives, the Dayton Agreement of 1995 divided Bosnia-Herzegovina into two autonomous units: a Muslim-Croat Federation and a Bosnian Serb Republic (*Republika Srpska*). While this solution was lauded in many circles, critics have argued that it is inherently unstable and that it also reinforces ethnic divisions rather than rising above them (Figure 21.2). The partition of territories and 'transfer' of populations has occurred in many other situations of territorial, ethno-cultural, national and state conflict. Greece and Turkey in the early 1920s, India and Pakistan in 1947, Palestine and Israel in 1948 and Cyprus (whose northern part was occupied by Turkey in 1974) are all examples. Understanding them requires critical attention not only to the details of each case, but also to the concepts (territoriality, the political geography of nations and states and geopolitics) and background that are considered *both* in this *and* the subsequent chapter in this section. As Figure 21.1 indicates, they have long histories.

(See Campbell 1999; Storey 2002; Kolstø 2005; Jeffrey 2006; Robinson and Pobric 2006; and the books by Gagnon 2006 and Toal and Dahlman 2001 listed in the suggestions for further reading.)

fact that residential clustering may be more a function of necessity rather than free and unconstrained choice. The degree and extent of choice available to many may be extremely limited. The idea that people may choose to cluster is to ignore the fact that quite often no easy alternatives are available. Discriminatory ideologies of race work to exclude people from particular areas, thereby translating social exclusion into geographical exclusion. As Wacquant (1995) argues, 'ghetto' areas are institutionally produced; they are far from being simply the product of choice by those who come to inhabit them.

It is also clear that such clustering itself further contributes to future rounds of marginalization and exclusion so that the spatial divisions resulting from economic and social marginality have a tendency to lock people into a socio-spatial milieu which impedes economic advance or social mobility (Smith 2005). This further demonstrates the idea that territorial practices serve to reproduce particular social outcomes in a complex relationship whereby society does not simply impact on space, but spatial arrangements in turn impact on society. In considering so-called ghetto areas, there is a tendency to see them in quite negative terms. The ghetto is a territorial entity and it is one that evokes many negative connotations; the term is often seen as synonymous with 'slum', the juxtaposition leading to a stigmatizing of its residents. In such ways the hegemony of the dominant group is maintained and the 'other' (in this case, ethnic minority) remains geographically, as well as socially and economically, marginalized. In Western cities, 'ghettos' and other 'inner city' areas are frequently seen as the home of a so-called 'underclass' experiencing relative and absolute deprivation and disadvantage. Places such as South Central Los Angeles or the *banlieues* of Paris are represented as dangerous areas to avoid. In this way the 'ghetto' is portrayed *as a problem* rather than a place whose residents *experience* problems. It is also depicted in a somewhat monolithic way that ignores its internal social, economic and cultural diversity.

21.3 Geographies of resistance

The sorts of territories discussed above can be seen as manifestations of social inequality reflecting the uneven distribution of power in society. While there are many negative consequences of territorialization built on ideas of race, class or ethnicity it is danger-

ous to paint places as poor marginalized ghettos whose residents lie somehow passive and unable to resist the forces seemingly arrayed against them. The apparent consignment of the poor or of ethnic groups to certain areas can also of course provide the spatial framework for forms of resistance. The 'ghetto' can be used as a means of mobilizing residents, of providing the territorial frame with which people identify and within which they can operate with a view to improving their own conditions. Successful residents' groups often emerge and may become sufficiently well organized to be able to engage in lobbying, in forms of self-help, and so on. Such community action may well make a positive difference to the lives of ghetto residents. It may facilitate the election (at local level at least) of political representatives for residents of those areas, thereby providing them with a voice that might otherwise be denied them. It may facilitate the mobilization of people in support of, or in opposition to, issues of direct concern to them such as public transport, banking facilities, the provision (or non-provision) of a range of services and the preservation of open spaces (see Jones *et al.* 2004). In this way a territorial strategy can be utilized in order to defend the interests of those who, if more spatially scattered, would be unable to do so. Community groups in working-class areas in European cities or in the *barrios* and *favelas* of many Latin American cities exemplify this tendency. Resistance may also take more overtly subversive forms such as street gangs laying claim to their 'turf', the renaming of streets and the placing of territorial markers (see Case study 21.4). Ultimately the contestation of space through various forms reflects resistance and an assertion of rights. They indicate the ways in which the meanings of place are disputed and, hence, they could be said to reflect contested ideas of community.

Places associated with visible ethnic minority groups can seek and develop positive connotations. Chinatowns (and areas such as 'Banglatown', centred on Brick Lane in London, which is predominantly inhabited by people of Bangladeshi origin) in Western cities for example, have often become sites on tourist and gastronomic circuits. In the case of Brick Lane, an area with a long history of Jewish and Irish immigration has latterly come to be seen as epitomizing a certain cosmopolitanism, which can be mobilized as a resource (Dwyer 2005). However, while there may be many potentially positive outcomes to this territorial construction, it may also be contested and we need to be careful not to romanticize places through simplistic readings of ideas of 'local' and 'community' (Staeheli

Thematic case study 21.4

Marking territory

Within contested political spaces the names given to areas or to streets itself reflects power relations or challenges to existing power (see Berg and Vuolteenaho 2009). This can be seen in a variety of ways. At an official level, streets have often been named in honour of royalty or leading political or military figures or as a reminder of important battles. Significant political change is often accompanied by name changes. Thus, the collapse of communism and the disintegration of the Soviet Union saw Leningrad revert to its earlier name St Petersburg, while in Montenegro, Titograd (named after the former Yugoslav communist leader) became Podgorica once more. Elsewhere in eastern Europe street names were changed to reflect the changing political order (Jones *et al.* 2004). The creation of an independent Ireland in the 1920s saw Dublin's principal thoroughfare change from Sackville Street to O'Connell Street in honour of the nineteenth-century Irish nationalist Daniel O'Connell. The defeat of the United States and the unification of Vietnam in the 1970s saw Saigon renamed Ho Chi Minh City after the former leader of Vietnamese nationalism. In conflict zones the unofficial changing of such things as street names are important symbolic ways in which territory is reclaimed, and the phenomenon was well known in black townships during the apartheid era in South Africa and also in Northern Ireland. Unofficial Irish language street signs were often erected in republican parts of Belfast, for example, and one such street, Beechmount Avenue, was unofficially 're-named' as RPG (rocket propelled grenade) Avenue. In a more overtly political vein in the north of Ireland, red, white and blue kerbstones indicate loyalist areas, while Irish flags are highly visible symbols in nationalist areas. Wall murals reflecting political allegiances adorn walls on both sides of the political–religious divide (Plate 21.6). These markers serve both to remind residents and outsiders whose place this is. In these ways the authority of the dominant political or ethnonationalist group is either asserted or challenged (see Shirlow and Murtagh 2006).

A different form of street marking is often indulged in by graffiti artists and by members of urban gangs, the activities of which have received considerable media attention in recent years, particularly in the USA. In places such as South Central Los Angeles, a combination of high unemployment, deprivation and political powerlessness creates a situation where territoriality may be a means of expressing power, using the only resource available – the streets and neighbourhoods in which they live (Scott and Brown 1993). These gangs tend to lay down territorial markers to indicate to others their 'ownership' of particular places. Graffiti on walls, bridges and buildings is one very visual method of claiming space. Markers are quite literally placed on the landscape to signal control of territory or 'turf ownership'. Work in Philadelphia indicated that graffiti became denser closer to the core of that gang's territory. Ley and Cybriwsky (1974) were able to demarcate reasonably accurately the spatial extent of gang control in the city. In this way, aspects of popular culture are translated into a territorial frame. This claiming of space may be a means by which marginalized youth make their claim to existence; drawing becomes a territorial act (Brighenti 2010b). For young people living in gang areas there are multiple problems of exclusion, marginalization and victimization resulting from both the activities of gangs and the ways in which such places are policed and stigmatized (Ralphs *et al.* 2009).

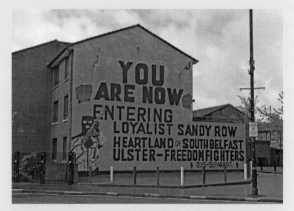

Plate 21.6 Loyalist wall murals in Northern Ireland.

2008; Storey 2010). Residents may not agree on the way in which the area is represented or portrayed. For example, there was a hostile reaction from some local residents to the fictional portrayal of the area by Monica Ali in her novel *Brick Lane* (2003). Another outcome has been the commercialization of the area and its 'culture', with tourist guidebooks promoting its restaurants. The area also hosts festivals and music events celebrating cultural diversity (see http://visitbricklane. com). While there are clearly positive dimensions to this, such strategies also run the risk of essentializing the identity of both places and their inhabitants. There is a risk of the portrayal of such 'ethnic spaces' as centres of exotica, to be exploited for commercial purposes. Areas like Banglatown become contested spaces and the subject of struggle over their identity and meaning.

We should also be mindful of the fact that much discussion surrounding issues of 'race' and ethnicity in Western societies tends to assume 'whiteness' as the norm. As McGuinness (2000) argues, much progressive research itself falls into this trap with a focus on non-white groups, tending to deflect attention away from white ethnicity. One consequence of the pursuit of this 'new exoticism', as McGuinness terms it, is that relatively little attention is given to 'white spaces'. Ideas of white flight to the suburbs (in response to the evolution of 'black' ghettos) and the creation of 'white' territories are themselves elements in the racialization of space. Similarly the construction of rural Britain as a relatively 'white space' reflects deeply embedded ideas associated with belonging, rurality and with national identity (Holloway 2005; *Journal of Rural Studies,* 25(4), 2009). Such constructions can have serious implications for those who do not (or are seen not to) 'fit in' with the dominant assumptions and ethos. Recent debates over immigration into the UK and other western European countries have tended to focus on a supposed 'invasion' of asylum seekers, 'illegals' and 'hordes' of eastern Europeans (Gilmartin 2008; Samers 2010). While much of this is inaccurate and misleading, the nature of the comments suggests an idea of the United Kingdom as a territorial entity that should increasingly seal itself off from invasions from 'outside' by those who do not 'belong' there. These arguments are often (misleadingly) bound into security discourses emphasizing the need to protect the country and its citizens. Territory, terrorism and identity become inextricably linked in debates calling into question who has the right to be in certain places and who has not (Elden 2009). Additionally, countries are seen to extend their border controls well beyond their own territory with immigration personnel screening passports in foreign airports in order to interdict those deemed 'illegal' (Mountz 2009).

This section has demonstrated the ways in which racist and exclusionary ideologies are transposed on to space. It has also indicated ways in which those racist and other social constructions are opposed. Just as particular power relations are refracted through a territorial frame, so those relations are contested through territorial strategies. The spaces to which people are consigned may provide the means through which they contest their marginalization. A territorial base may serve as a means through which an ethnic identity or a class identity is reinforced and reshaped, in part at least, in opposition to other identities. Other categorizations, such as religious affiliation, may also provide the basis for parallel territorialities. Of course we need to be mindful that such identities may be deeply contested and are far from monolithic, though there may often be attempts by some 'inside' or 'outside' to portray them as such (Yuval-Davis 2003).

21.4 Territoriality, gender and sexuality

21.4.1 Gendered spaces

The growth of what became known as the 'feminist movement' from the 1960s onwards, building on earlier attempts to achieve equality for women, has been influential in gaining recognition in many societies for the unequal status of women and men in all dimensions of life; in the home, in (paid) workplaces, in the broader political and social arena. Feminist geographic interventions have challenged certain assumptions and have opened up new avenues of enquiry and new modes of analysis (Kofman 2008). Feminist writers and activists have been instrumental in attempting to explain how patriarchal systems of power have tended to reinforce male dominance and how women have often been marginalized (Dixon and Jones 2006; England 2006). Feminist geographers have drawn attention to the manner in which space and place are heavily gendered and have challenged the relationships between gender divisions and spatial divisions (see, for example, McDowell 1999). Critical attention has been focused on divisions between the public and private domain and its spatial corollary of a separation between what is seen as public space and private space. It is argued that patriarchal systems of power have led to a division between predominantly 'male' public and mainly 'female' private space, resulting in social practices whereby certain activities and certain spaces are seen as male preserves. This duality reflects broader distinctions centred on the binary divide between masculinity and femininity (see Pratt 2005).

As with race and class, issues of gender are mapped on to space in various ways and the implications of gender are seen to be as important as other political, social and economic factors in the structuring of spaces and places. In its most simple form this is reflected in the sexist notion that 'a woman's place is in the home'. The home has tended to be seen as a space of reproduction juxtaposed to the workplace as a space of production (Laurie *et al.* 1999). Underpinning this are ideas that distinguish between sex as a biological fact and sex as gender, which refers to the socially constructed roles of both male and female identities. In emphasizing the role of social conditioning, the argument is that as individuals we are not biologically predetermined to be more suited to some roles rather than to others.

One reason for the relative absence of women in particular places is overt discrimination or active discouragement in the sense of certain activities or pursuits not being deemed suitable for women. Historically, women who transgressed these boundaries were often portrayed in a negative light, an idea reflective of notions of 'good' and 'bad' women. Women out alone at night might be seen as not conforming to what is expected of them. A crucial aspect of the relationship between women and place centres on the perception of some specific places as 'unsafe'. Many women do not feel safe in certain public places, most notably darkened streets. As Valentine (1989) suggested, women transfer a fear of male violence into a fear of certain spaces, which has profound implications for the ways in which men and women negotiate their way through urban areas (Fell 1991). Clearly, the various strands of feminist thought and practice have resulted in significant advances with regard to equal rights for women. While this can be seen within the arena of legislation in many countries associated with equal pay and related issues, it is also reflected in terms of spaces. Thus, the heightened visibility of women in public space reflects the changing status of women. Phenomena such as 're-claim the night' marches demonstrate the overt use of a spatial strategy to make a political and human point. While particular groups may find themselves excluded from certain spaces, those spaces can also be reclaimed (as an example see www.isis.aust.com/rtn) (Plate 21.7).

Historically, the gender division of labour tended to confine women to the private realm, leaving men to inhabit (much of) the public domain. This view of women as playing a subordinate role has in the past been reflected in discriminatory attitudes and practices, particularly in relation to women in the paid workforce, with active discouragement through lower wages, if not actual exclusion, from many jobs. These views are predicated on the undesirability of women going out to work. It can be argued that this ascribing of women's role, through delimiting the spaces in which women were encouraged to appear, is another spatial expression of power. In other words, the confining of women to domestic space, and their exclusion from

Plate 21.7 Reclaim the Night march.

male territories, was a key element in male control (see Little 2002). With increasing female participation in the workforce and a raft of equal opportunities legislation in many countries, such a generalization may appear to have lost some of its validity. Nevertheless, the division between a (largely) male public sphere and a (largely) female private sphere still has considerable resonance in many societies (although the extent of this is itself immensely geographically variable across the world).

Where women enter the workforce, they may still encounter territorial divisions in the workplace. Thus, Spain (1992) documents the 'closed door' jobs of managers (mainly men) and the 'open floor' jobs of manual workers (who may be predominantly women in certain countries/regions/sectors). Employers may locate in particular localities (or countries) in order to take advantage of what they see as an available (and exploitable) workforce based on prevailing wage levels or skills and assumptions about gender roles (Hanson and Pratt 1995). The presence of women in the armed forces periodically provokes debate over the supposed appropriateness of women performing such roles. In 2007 Iran's arrest of 15 British sailors, one of whom was female, led to media commentary on the appropriateness of a woman, particularly a mother (as in this instance) being engaged in military activity in a war zone, rather than being at 'home' playing the key role in bringing up her children. The deaths of female soldiers in recent years in the wars in Afghanistan and Iraq have been depicted in especially poignant terms in the media. Social processes reproduce attitudes that tend to 'naturalize' a gendered division of labour in which women perform certain functions which are acted out in specific spaces, e.g. 'home-making' and child-rearing in domestic space. Socially constructed gendered difference is inherently also spatialized. Within the arena of sport and leisure gender stereotyping remains prevalent. While there are undoubtedly marked changes, gendered ideas and practices about leisure activities and, hence, separate spaces for men and women, are still common. Such social practices are built upon ideas of *what* is or is not acceptable behaviour for men and women to engage in (built on socially or culturally constructed notions of masculinity and femininity) and *where*. Stereotypes of women spending leisure time shopping while men attend sporting events, or watch them on television at home or in pubs, have a self-perpetuating quality. Media interest surrounding the shopping habits of the so-called 'WAGS' (wives and girlfriends of English international football players) provides further reinforcement of this.

If women have sometimes been seen as 'belonging' in or closely associated with the home, it is not the case that the space of the home is undifferentiated. Even within the home, territorial divisions take place, the most obvious being the notion of the kitchen as a female preserve. The relative neglect of such territorial behaviour in all its manifestations is commented on by Sibley:

> In geography, interest in residential patterns wanes at the garden gate, as if the private province of the home, as distinct from the larger public spaces constituting residential areas, were beyond the scope of a subject concerned with maps of places.
>
> (Sibley 1995: 92)

Many communities and societies often had a clear spatial division between men's and women's spaces and roles (Spain 1992). Within most Western societies, deep-seated ideas about woman's role as homemaker, cook, cleaner, child-rearer, and so on, mean that women have often been historically presumed to 'belong' in some rooms and spaces more than others. While such ideas have been challenged (and in many cases transformed), they sometimes endure. All these reflect territorial expressions of power whereby the designation or apportionment of space within the domestic sphere reflects the relative status or roles of the individuals concerned. Spain therefore reminds us that 'houses are shaped not just by materials and tools, but by ideas, values and norms' (1992: 111).

21.4.2 Sexuality and space

The idea of places territorialized by particular groups on the basis of sexual orientation has begun to receive more academic attention in recent years (see *Political Geography*, 25(8), 2006; Browne *et al.* 2007). At its most elementary level this has seen the mapping of gay and lesbian 'zones' in selected cities. It is fair to point out that such spaces are not as easy to identify as, say, areas inhabited predominantly by a particular ethnic group. It is equally obvious that, in the main, these are not strictly demarcated areas. Rather, they are zones where gay people may feel more at ease through being accepted rather than rejected, scorned or ignored (or worse) by their neighbours. The mapping of 'gay territories' runs the risk of focusing on what some see as deviant behaviour as well as essentializing sexuality and reinforcing a gay/straight dichotomy (Knopp 1995; T. Davis 1995). Nevertheless, the fact that those who identify themselves as gay and lesbian do, in some instances, become associated with particular (usually urban) spaces, suggests that another form of territorial behaviour may be evident. A concentration of visibly gay restaurants, bars and clubs results in the creation of what Castells calls 'a space of freedom' (1997: 213). Places such as the Castro District of San Francisco (Plate 21.8) and the more spatially confined 'gay village' in Manchester serve as important examples, while the city of Brighton has acquired an image as the 'gay capital' of the UK (Browne and Lim 2010).

The construction of such zones may arise for reasons similar to those associated with ghettos and other forms of segregated space. Castells (1997) has argued that there are two key factors: protection and visibility. The first of these is fairly obvious. The idea of 'strength in numbers' may make people feel safer from homophobic 'gaybashers'. Visibility may have an emancipatory effect through which identification as gay or lesbian within a culture that is predominantly straight (heterosexual) and in which a straight discourse dominates, may work to nullify views that see homosexuality as deviant or abnormal. Gay neighbourhoods

Plate 21.8 Rainbow flag: a symbol of gay, lesbian, bisexual and transgender pride flying in the Castro District of San Francisco.

become a means of asserting identities. Harry Britt, a one-time key figure amongst San Francisco's gay community, once commented that 'when gays are spatially scattered, they are not gay, because they are invisible' (cited in Castells 1997: 213). The significance of San Francisco's gay area was reflected in that community's ability to gain political representation. In obtaining power over territory they also gained political representation and San Francisco has become recognized as something of a 'gay capital' of the United States with a somewhat more liberal attitude. In this way, the designation of 'gay territories' plays a crucial role in raising awareness of gay people and issues and also provides a means by which some degree of power and self-confidence can be attained. Celebratory events such as 'gay pride' marches can be seen as an assertion of citizenship rights through staking a claim to public space. Conversely, heightened visibility may render people targets for homophobic assault, both verbal and physical. In 1999 a gay pub in Soho, central London, was the target for a bomb, resulting in three deaths and over 80 people injured. The examples of gay and lesbian spaces suggest another important point, that of the temporality of territory. The longevity of these spaces may be quite brief as the 'scene' moves to somewhere else. Lesbian spaces may be very short-lived in time, whether caused by the transience of lesbian bars/clubs or the even more short-term phenomena of lesbian or gay evenings (Valentine 1995).

Of course, such places and events may be inextricably bound up with factors that extend well beyond the realm of identity. 'Gay spaces' have often been associated with an economic imperative as the importance of the 'pink pound (dollar, or euro)' in aiding urban regeneration has frequently added a strong commercial angle to these developments. One consequence may be the 'co-option' of the identity in order to present an 'acceptable' image of the group concerned. Not surprisingly, some activists have expressed disquiet over the appropriation of such events and their dislocation from their original social and cultural roots and from their original territorial base. The fact that Manchester's 'gay village' and San Francisco's Castro District are firmly on the tourist trails of their respective cities may be lauded as an acceptance of identities previously scorned but it can also be seen as a commercialization of that identity which may, to some extent, serve to further ghettoize it (see www.sfguide.com/sights/neighborhoods/castro.htm). The complex intersections between different identities can create specific tensions which play out in different ways in different places. For example, the ways in which gay and lesbian identities are interconnected with processes of gentrification through which they take on a spatial dimension is apparent in places such as the Marais district in Paris where tensions between 'old' and 'new' residents and within the 'gay community' emerge (Sibalis 2004). Similarly, the intersections of sexualized and racialized identities in post-apartheid Cape Town serve as another useful insight into how issues of identity play out in specific contexts (Tucker 2009). The relations between class, capital, sexual identity and place may be manifested through complex embodiments and manifestations of territoriality.

It might be argued that it is acceptable to be openly gay in an area so designated but in other spaces and societies the pressure to keep this identity hidden may well persist. While gay spaces may allow for more open expressions of sexual identity, homophobic assaults on sexual minorities reflect contested place meanings (Sumartojo 2004). Openly gay behaviour may be accepted or tolerated in some places but may remain decidedly unacceptable elsewhere. Here of course there are profound links between the small-scale (often neighbourhood or urban) territorialities and the wider policies of the city or state concerned. Many countries still criminalize gay sexuality and almost everywhere sexual mores and regulations are highly territorialized.

21.5 Work, rest and play

We can recognize two important social tendencies that bolster territoriality: the wish by people to have space of their own and the wish by others to exclude people from certain spaces. We have already seen ample evidence of the latter in this chapter. Even in what might appear to be very mundane or innocuous ways, the apparent claim to territory seems to manifest itself. At its most elementary level, the assertion of territoriality is reflected in claims to private property. Thus, people desire to mark their own home, to adorn it in their chosen style (influenced of course by social trends, technologies and fashions) and, in various ways, to mark it out as theirs. Homeowners are generally keen to stamp their personality on their home through the ways in which they choose to decorate it, alterations to layout, choice of colour schemes, furnishings and so on. This personalizing of space is further manifested through such things as the display of paintings, posters or photographs and the collection and arrangement of ornaments. The geographer

Jean Gottman suggested that people 'always partitioned the space around them carefully to set themselves apart from their neighbours' (1973: 1). This manifestation is commonly interpreted as being symptomatic of our inherently territorial nature. This emphasis on the centrality of the home also has a broader cultural and political significance. Symbolic connections are often made between the domestic home and the nation whereby images of the former are seen to give material meaning to the latter. The home is seen in some ways to be at the heart of the nation. In times of war, for example, people have been encouraged to fight for the 'homeland' and the defence of 'hearth and home'.

Private property is regarded by many as an outcome of human territorial behaviour and it represents a claim to space that is reinforced by the legal system of many countries. However, as Alland points out, it might well be the case that 'private property is the child of culture and develops into a major preoccupation only with the evolution of complex society' (1972: 64). It follows that we need to be careful to avoid the trap of translating a need for personal space into an ideological claim for the sanctity of private property. The centrality of the family home, encapsulated in such phrases as 'home, sweet home', glosses over the fact that the privacy which many of us associate with the home is comparatively recent and is specific to some societies. Where 'domestic' space is limited (or for those who find it constrains them), life may be lived in the street (or perhaps the mall or car), much more evidently than the 'behind four walls' lifestyle many take as 'natural'. We need to be mindful of social, ethnic and geographic differences in the ways in which the home is conceived. In the United Kingdom, the notion of private home space was initially quite a middle-class idea which has since permeated the rest of society (Morley 2000).

Hegemonic ideas of the home within Western societies, exemplified by such notions as the home being an 'Englishman's castle' or associated with 'The American Dream', can be argued to have led to an ignoring of internal tensions and, in particular, a consideration of the different positions, roles and experiences of men, women and younger people within this domestic space, as discussed earlier (McDowell 1999). While the home is commonly depicted as a refuge from the outside world it may also be a site for domestic violence and fear (Pain 1997; Squire and Gill 2011). Similarly teenagers may view the home as somewhere to escape from. In any event, the emphasis on the domestic idyll may have highly exclusionary consequences (Delaney 2005).

Even within buildings territorial behaviour can be recognized. As we have seen, the idea of the kitchen as a 'woman's place' is one example of this. The domestic home in many different cultural contexts is often spatially divided, not just in terms of gender but also in terms of age, with certain spaces being designated for women or for children (Spain 1992). The banning of children from some rooms and the proprietorial attitude towards one's own room in a house are other examples of this. In the home, space is even being claimed at the level of 'my chair', 'my place at the table' and so on. There are also distinctions amongst those allowed in, with differential access for close family and friends on the one hand and more casual acquaintances on the other. Even then, friends may be welcomed into the living room but are less likely to be invited into the more 'private' spaces such as bedrooms (Morley 2000). In any consideration of the home, we need to be mindful that it is not a straightforward and unambiguous entity. While for some it conjures up feelings of comfort and security, for others it may be a place of discomfort, alienation and tension (Blunt and Dowling 2006). For some the home may come to feel like a prison – quite literally so for those sentenced to home imprisonment of the type long imposed on political activist Aung San Suu Kyi in Burma or the many others subject to degrees of curfew, control and house arrest. Home is therefore a social construction in which social identities are (re)produced but which conveys different meanings to different people.

Equally, in workplaces some areas and rooms can only be entered by staff of a certain level and are out of bounds to more junior staff (Case study 21.5). These can be interpreted as managerial strategies designed

Thematic case study 21.5

Playgrounds and lecture halls

There has been a growing interest recently in the geographies of younger people (see Holloway and Valentine 2000; Leyshon 2008) and as an example of micro-scale territoriality it is instructive to consider the space of the school playground. Thomson (2005) suggests that while children's behaviour is clearly spatialised in a wide variety of ways, it might be thought that the playground is 'their' space in which school children can enjoy a degree of freedom to engage in a range of activities as distinct from the constraints imposed in the more formalized and controlled spaces of the classroom or other parts of the school buildings and grounds. However, while play spaces can be seen as emancipatory they are simultaneously regulating. Thomson's research indicates that on closer inspection the playground displays two key criteria associated with territories and territoriality. First, it remains heavily controlled by adults (teaching staff) who delimit the times it can be used and the activities which are permitted there. Second, it often contains internal territorial divisions in the sense of discrete spaces for different types of 'play' activity – sports, etc. What is also obvious of course is that children themselves will often endeavour to claim their own territory within the larger space of the playground with groups 'hanging out' in particular places. In addition it is equally apparent that these spatial practices are open to contestation. Rival groups of children may 'compete' for spaces within the playground. What is also apparent is the manner in which children may try to test or push the spatial boundaries imposed by teachers and supervisors through such actions as encroaching on grass playing pitches and so on. Another important dimension of this (in addition to classification, communication, enforcement and resistance) is the way in which the control of territory is masked through recourse to other discourses such as that of health and safety – prohibiting access to spaces deemed dangerous or hazardous (Thomson 2005).

In a similar vein we might take the example of the university teaching room. Here a lecturer may exert a strong degree of territorial control; occupying a space at the top of the room while controlling (or trying to) who else may speak and when they may do so. Even in highly interactive and more student-led sessions, a lecturer, by virtue of their status, continues to exert control over the space. She or he can 'command' the room and walk around it in ways which students cannot (or are discouraged from doing). However, such control has very obvious temporal constraints. What the students and lecturer may regard as 'their' room is usually limited to a regular time-tabled slot. Before and after that time, their right to be there is denied as others lecturers and students take over the space and

Plate 21.9 University lecture theatle.

control it. Your geography lecturer is unlikely to walk into a lecture theatre during a microbiology class and commence to speak to the assembled students. (Well, they might try to do so but they would be met by a bewildered reaction and, more than likely, a phone call to security!) (Plate 21.9).

to ensure a particular outcome; staff know their 'place' and can be more effectively controlled, sometimes through very obvious visual intrusion. Hanson and Pratt (1995) reveal how companies reproduce social segregation through spatial practices within the workplace whereby different sets of workers inhabit different parts of the factory and rarely, if ever, meet. Thus, office staff may be located downstairs in 'cubicles' separated by room dividers, with sales staff and management upstairs in individual or shared offices while production staff are located in an entirely separate part of the building. Socializing between workers tends to reflect their spatial segregation even to the extent of each department having separate annual parties. Work hierarchies are reflected in the spatial arrangements of the workplace. These practices have clear outcomes. They may render it difficult for workers to organize through physically keeping them separate and through engendering a sense of difference between different sections of the workforce.

Taking the idea of territory down to its most elementary level, the desire for personal space can be seen as a form of territorial behaviour. Humans like to have a pocket of space around them that is 'theirs' and they resent others 'invading' their space (unless invited!). This can be interpreted as a territorial claim to a portion of geographic space. While this might be taken as reflecting a natural tendency, it is worth noting that the amount of space needed appears to vary from one society to another, a fact noted long ago by Hall (1959). For many young people, their own room, apartment, etc., may seem like a 'natural' ambition, but for many of the world's inhabitants such a desire is completely unobtainable. For those living in overcrowded conditions, the amount of personal space available is extremely limited. For a homeless person, their 'own' space may be limited to a hostel bed in central London, a doorway in downtown Manhattan, or a small patch of pavement in the Tenderloin district in San Francisco (Gowan 2010). Nurture, culture, power and politics all need to be considered where the complexities and diversities of human territorialities are concerned.

21.6 Conclusions

The idea of territory reflects a way of thinking about geographic space while territorial strategies are utilized in conflicts concerned with social power and identity. Territorial thinking, the production of territories and the employment of territorial strategies are bound up with maintaining power or with resisting the imposition of power by a dominant group. Forms of exclusion can be consolidated and reinforced through territorial practices, yet they can also be resisted through similar means. The examples provided are evidence of the ways in which social relations are expressed through spatial patterns and they highlight how these geographies help in turn to shape social relations. Social phenomena such as class, racial or gendered identities invariably embody a territorial component. Territorial strategies are often used to control and police those who are defined as 'out of (their) place'. In this way particular ideologies are transposed on to space. People are confronted with wider practices through their use of space or through the ways in which they are allowed to use space. Power relationships take on a spatial dimension, even at the most mundane and everyday level. Issues of identity, particularly within multi-cultural societies, have a spatial expression as social divisions (associated with class, ethnic, religious, gender or other factors) are given material form through spatial divisions. The examples used here demonstrate the spatialization of wider ideas and they show how people are kept 'in their place' whether through overt mechanisms or more subtle means. Social boundaries are being communicated through space and 'the assignment of place within a socio-spatial structure indicates distinctive roles, capacities for action, and access to power within the social order' (Harvey 1990: 419). This also means that some groups of people are effectively treated as 'second-class (or indeed as non-) citizens' denied full rights in the society in which they find themselves, confined to spatial enclaves which become characterized as dangerous and threatening spaces. Social and spatial exclusion has the effect of denying full access to those rights often thought of as inalienable for all people everywhere. In this way many territorial strategies are invariably discriminatory and exclusionary and can be used to deny people effective participation in 'society': the latter being based on a categorization of those who 'belong' (by virtue of citizenship and income levels that enable a range of choices and possibilities to participate).

In all societies, those who are convicted (or even in some cases accused) of criminal offences may also be excluded by territorial confinement (this is a key function of prisons) or controls on their movement. The range of these territorial strategies will vary according to the crime and legal jurisdiction. But in the USA, the prison population has quadrupled in recent decades, so much so that a new extended geography of prisons and incarceration can be critically mapped (Gilmore 2007; Martin and Mitchelson 2009). In the current geopolitical climate, territory and territorial strategies are deployed in intriguing ways. In September 2010 it was reported that British military forces handed over responsibility for security in the Sangin area of Afghanistan to the United States. The fact that Afghanistan is neither British nor US territory (but has been under effective occupation since 2001) renders this a glaring example of the territorial exercise of political and military power. At a more micro-level, spaces of confinement and interrogation reflect complex territorial strategies. The US prison camp at Guantánamo Bay (on the island of Cuba) is *outside* the territorial United States and its ambiguous territorial status has proven an effective excuse in allowing the indefinite detention without trial of its inmates who are simultaneously subject to interrogation by US military authorities (see Reid-Henry 2007). This and other related practices renders the relationships between territory and sovereignty ever more complex and blurred (Gregory 2007; Elden 2009). Paralleling this is a geopolitical discourse of fear in which people are warned of the supposed risks of being in some places at certain times (Pain 2010). The everyday impact for people in particular places is highlighted by struggles for control of territory in such places as Colombia's Pacific coast region. Here a long-standing conflict between the government and left-leaning guerillas has resulted in the territorial displacement of many thousands of people (Oslender 2007). Meanwhile, in the Israeli–Palestinian dispute territory is 'made' Israeli through the construction of Jewish settlements in the Palestinian West Bank and the erection of a so-called 'security fence' excluding Palestinians from these places and restricting and impeding their movements (Yiftachel and Ghanem 2005) (Plate 21.10).

However, as indicated earlier, just as dominant ideologies can be reinforced through territorial practices, they can also be resisted. Territorial strategies are useful mechanisms in the assertion of identity. Spatial concentrations within particular geographic areas make visible people and issues that might otherwise remain unseen. They can be used to draw attention to exclusionary practices and to assert the right to be equal citizens. In doing so, this demonstrates the 'positive' and 'negative' dimensions to territoriality; it can be both a force for oppression and also one for liberation. Particular strategies can be used to assert an identity and

Plate 21.10 Israeli West Bank 'separation barrier'.

territorially transgressive acts can be employed to re-claim space and, hence, to assert basic rights. While many people do not necessarily freely choose their 'place', they may, nevertheless, identify with their immediate neighbourhood or locality. This sense of identity can in turn be converted into forms of action aimed at obtaining particular outcomes. The forma-tion of community or residence groups reflects feel-ings of belonging or attachment to a particular place. It follows that notions of territory are connected with ideas of social power. The claiming of space is a po-litical act whether it occurs in the 'public' or 'private'

arena (and the categorization and demarcation of these areas is a key expression of territoriality) and territories provide a material expression of the fusion of meaning, power and social space (Delaney 2009). In other words, *power* permeates society and this is something which is marked in and expressed through a range of human ge-ographies. And as the next chapter details, the nation-state has come to be a key container of power and ex-pression of human territoriality. As such, nation-states are frequently contested, not only by other states, but also by those acting in the name of other loyalties and other scales and ideologies of belonging.

Learning outcomes

Having read this chapter, you should understand:

■ How social issues linked to class, racism, ethnicity, identity, gender and sexuality are mapped on to geographic space and become the basis for territorial strategies.

■ Some of the ways in which social inequalities and differences are manifested spatially.

■ How territorial practices are used to exert or to resist control.

■ How territorial strategies can be mechanisms for groups to promote their identities, rights and claims to meaningful citizenship.

■ The ways in which many aspects of everyday life are reflected in spatial arrangements and territoriality.

Further reading

Sack, R. (1986) *Human Territoriality: Its Theory and History*, Cambridge University Press, Cambridge. As outlined in this chapter, Sack's book has become a classic geographic text on territory and territorial behaviour, setting it firmly within a 'social' rather than 'natural' framework. For commentaries revisiting Sack's book more than two decades after its original publication, see Classics in human geography revisited, *Progress in Human Geography*, 24(1), 91–9.

Delaney, D. (2005) *Territory: A Short Introduction*, Blackwell, Malden.

Storey, D. (2001) *Territory: The Claiming of Space*, Prentice Hall, Harlow.

These both build on Sack's work and deal with territory and territoriality in social and political contexts. Storey's book contains many more case studies of territoriality than can be considered here. Delaney's book also offers a good survey (and some critique) of Sack's ideas and lots of case studies.

The work of Stuart Elden extends earlier thinking on the nature of territory and the relationships between territories and geographic space. In particular, he stresses the necessity of thinking historically about how territory has evolved. See, in particular:

Elden, S. (2007) Governmentality, calculation, territory, *Environment and Planning D: Society and Space,* 25, 562–80.

Elden, S. (2009) *Terror and Territory: The Spatial Extent of Sovereignty*, University of Minnesota Press, Minneapolis, MN.

Elden, S. (2010) Land, terrain, territory, *Progress in Human Geography,* 34(6), 799–817.

See also the useful paper by **Joe Painter** (2010) Rethinking territory, *Antipode,* 42(5), 1098–1118.

For some reflections on how territorial strategies delimit and shape violence and terror, see:

Gregory, D. and Pred, A. (eds) (2006) *Violent Geographies: Fear, Terror, and Political Violence*, Routledge, London. See also **Ingram A. and Dodds, K.** (eds) (2009) *Spaces of Security and Insecurity: Geographies of the War on Terror*, Ashgate, Aldershot.

Some useful wide-ranging political geography texts include:

Blacksell, M. (2005) *Political Geography*, Routledge, London.

Jones, M., Jones, R. and Woods, M. (2004) *An Introduction to Political Geography: Space, Place and Politics*, Routledge, London.

Painter, J. and Jeffrey, A. (2009) *Political Geography,* 2nd edition, Sage, London.

These are all reader-friendly introductions to the intersections between politics and geography, providing very good examples of some of the themes covered here. Blacksell's book is a good basic starting point while the others take some of the arguments to a deeper level of analysis. They also contain lots of material relevant to other chapters in this section (for example, on the territorial state and geopolitics).

There are a range of useful social and cultural geography texts dealing with many of the concerns of this chapter. These can be read with a view to the complex operations and manifestations of territoriality:

Valentine, G. (2001) *Social Geographies: Space and Society*, Prentice Hall, Harlow.

Panelli, R. (2004) *Social Geographies: From Difference to Action*, Sage, London.

Del Casino, V.J. (2009) *Social Geography. A Critical Introduction*, Wiley-Blackwell, Chichester.

Anderson, J. (2010) *Understanding Cultural Geography. Places and Traces*, Routledge, London.

Sibley, D. (1995) *Geographies of Exclusion: Society and Difference in the West*, Routledge, London. This is a classic book focusing on issues of power and social and spatial marginalization.

For two contrasting cases (one on territoriality in a school playground, another on parades), both drawing on Sack, see:

Thompson, S. (2005) 'Territorialising' the school playground: deconstructing the geography of playtime, *Children's Geographies*, 3(1), 63–78 and **O'Reilly, K. and Crutcher, M. E.** (2006) Parallel politics: the spatial power of New Orleans Labor Day parades, *Social and Cultural Geography*, 7(2), 245–65.

Hubbard, P. (2006) *City,* Routledge, London. This is a very useful urban geography text touching on many of the issues dealt with here, such as gentrification.

For a more specific focus on urban gating there are a number of recent compilations, including **Bagaeen, S. and Uduku, O.** (eds) (2010) *Gated Communities. Social Sustainability and Historical Gated Developments*, Earthscan, London. This provides a useful range of perspectives drawn from a variety of geographical contexts.

Gagnon, V. P. (2006) *The Myth of Ethnic War: Serbia and Croatia in the 1990s*, Cornell University Press, Ithaca, NY. This provides an examination of the political mobilization of ethnicity, identity and territory in a conflict situation and a good way to follow up in detail some of the themes touched on here. It is worth reading in tandem with:

Toal, G. and Dahlman, C. T. (2001) *Bosnia Remade: Ethnic Cleansing and its Reversal*, Oxford University Press, Oxford.

Nelson, L. and Seager, J. (2004) *A Companion to Feminist Geography*, Blackwell, Oxford. A collection which illustrates the diverse contributions feminist geographic thought and practice have made to human geography. A good way to follow up some of the ideas in this chapter is to read Eleonore Kofman's chapter on feminist political geographies in *The Sage Handbook of Political Geography*, Sage, London, 73–86.

Browne, K., Lim, J. and Brown, G. (eds) (2007) *Geographies of Sexualities: Theory, Politics and Practice*, Ashgate, Aldershot. This is an interesting collection of chapters casting light on sexuality and space. See also the recent book by Andrew Tucker (2009) *Queer Visibilities: Space, Identity and Interaction in Cape Town*, Wiley-Blackwell, Chichester, and the commentaries on it by **Ward** *et al.* in *Political Geography,* 29(8), 454–62.

Blunt, A. and Dowling, R. (2006) *Home*, Routledge, London. A useful examination of the complex natures and meanings of home.

The law is a key aspect in regulating territorial behaviours and strategies (and related and overlapping economic, social and political geographies). See, for example, **Blomley, N.K., Delaney, D. and Ford, R.** (eds) (2001) *The Legal Geographies Reader: Law, Power, and Space*, Blackwell, Oxford. On geographies of policing **Nicholas Fyfe's** early paper in *Progress in Human Geography,* 15(3), 249–67, and **Richard Yarwood's** review in *Progress in Human Geography,* 31(4), 447–65 are useful starting points.

Finally, on the geographies of fear see the wide-ranging collection edited by **Rachel Pain and Susan Smith** (2008) *Fear: Critical Geopolitics and Everyday Life*, Ashgate, Aldershot.

For annotated, clickable weblinks and useful tutorials full of practical advice on how to improve your study skills, visit this book's website at **www.pearsoned.co.uk/daniels**

THE PLACE OF THE NATION-STATE

Chapter 22

James Sidaway and Carl Grundy-Warr

Topics covered

- The ubiquity and geographical diversity of nations and states
- Interpretations of nationalism as a territorial project
- Relationships between nations, states and territory
- Transnational identities and borderlands

The territory of a nation is not just a profane part of the earth's surface. It is a constitutive element of nationhood which generates plenty of other concepts and practices directly related to it: for example, the concepts of territorial integrity and political sovereignty; border control, conflict, invasion and war. It defines and has some control over many other national affairs, such as the national economy, products, industry, trade, education, administration, culture and so on. Unarguably, the territory of a nation is the most concrete feature of a nation for the management of nationhood as a whole. For a theoretical geographer, it is the territoriality of a nation . . . For people of a nation; it is a part of SELF, a collective self. It is a nation's *geo-body* . . . The geo-body, the territoriality of a nation as well as its attributes such as sovereignty and boundary, are not only political but also cultural constructs.
(Winichakul 1996: 67)

22.1 Historical and geographical variability of states

Although it is also other things, notably a provider of services, a system of regulations, ideologies, legal regulations and police powers ('law and order'), flows of capital (budgets, taxation and government spending) backed up by the threat of discipline and violence (for example, armed forces), the state can be interpreted as a form of *community*. As earlier chapters (in particular those in Section 1) have shown, forms of human community and their attendant territorialities (see Chapter 21) have been extremely variable historically and geographically.

The contemporary system of states, in which all of the land surface of the earth (with the partial exception of Antarctica as is detailed in Chapter 20) is divided into state units, whose outlines become familiar to us from maps and globes, is after all fairly new. In the early twentieth century, the borders between many of today's states were only vaguely defined, and more recently large areas of the world were ruled by colonial empires or dynastic realms (the Austro-Hungarian in Central Europe, Ottoman in the Balkans and Eastern Mediterranean, and Ch'ing empire in China, for example). Alternatively, the concept of nation-statehood was simply not in the political vocabulary. In the latter cases, ethnically, linguistically or religious specific groups owed allegiance to the imperial order rather than to a defined nation. That is, loyalty would be *foremost* to the empire and any sense of national or ethnic identity would be a local or 'private' matter.

Such imperial visions are no longer dominant and today few formal colonies and no large-scale dynastic empires remain (see Chapter 3). Therefore the territories that were once ruled as part of, for example, a Japanese, British, French, Ottoman (Turkish), Portuguese or Russian empire are today mostly divided into self-avowed and recognized *sovereign* states. They possess the same apparatus of statehood (leaders, flags, capital cities, administrations, postage stamps, seats at the United Nations and so on) that the former imperial powers have. Moreover, the global map of states continues to change. In recent decades, some have disappeared as separate states (like the former East Germany and South Yemen), whilst others have split into component parts (like the former USSR and Yugoslavia). All this reinforces the point that states, like other communities, particularly the 'nations' with which they are associated, are not to be taken at face value. The claims made for and on behalf of them deserve critical examination. However, this is not an easy task. For, as Benjamin Akzin (1964) pointed out nearly half a century ago, to discuss nations, states and nationalism is to enter a terminological maze in which one easily and soon becomes lost. This chapter will indicate some pathways into the maze. It begins with an account of 'nations' and nationalism before returning to the relationship of these to states.

Plate 22.1 The state as a system of organized violence: armed forces.

(GrahamGrieves/TRIP)

22.2 Nations as 'imagined' political communities

One of the most influential and suggestive critical studies of nations and nationalism is a book by Benedict Anderson (1983) entitled *Imagined Communities: Reflections on the Origin and Spread of Nationalism*. He begins with a reminder of the ubiquity of nations and nationalism:

> Almost every year the United Nations admits new members. And many 'old nations' once thought fully consolidated find themselves challenged by 'sub'-nationalisms within their borders – nationalisms which naturally dream of shedding this subness one happy day. The reality is quite plain: the 'end of the era of nationalism', so long prophesied, is not remotely in sight. Indeed, nationness is the most universally legitimate value in the political life of our time.
>
> (Anderson 1983: 3)

Although enormously variable between, say, Nepali, Israeli, Singaporean, Nicaraguan, Vietnamese, Eritrean, American, Greek, Turkish and Irish versions, the ideology of nationalism holds that everyone will have a primary identity with a particular 'nation'. Such communities should be able to express themselves in a state; that is, they should enjoy what is called 'sovereignty' within certain geographical boundaries. 'Sovereignty', which is a term of long vintage and was previously associated with royal dynasties (the sovereign monarch), shifts to the 'people' of a 'nation', and even if a royal figurehead is retained, she or he will have to become in some way a 'national' symbol. However, it is important to remember that territorial sovereignty as depicted on the world political map of today is of relatively recent origin (see Section 22.3), that more differentiated forms of sovereignty and other territorialities have existed in the past, and that sovereignty is continually being challenged in various ways.

Examples of quite different conceptions of sovereignty in the past are abundant in many parts of the world, from medieval Europe to most parts of the pre-colonial world. The labyrinthine world of medieval times in Europe was at some levels intensely 'local', involving much smaller communities and political units than today, although these were usually a 'part of a complex hierarchy of political or cultural entities, such as the Church of Rome, the Hanseatic League, or the dynastic Habsburg Empire' (J. Anderson 1986: 115). Sovereignty was not rigidly territorial as it mostly

is with modern nation-states (see Chapters 1 and 2). As James Anderson (1995: 70) explains:

> Political *sovereignty* in medieval Europe was shared between a wide variety of secular and religious institutions and different levels of authority – feudal knights and barons, kings and princes, guilds and cities, bishops, abbots, the papacy – rather than being based on territory *per se* as in modern times. Indeed the territories of medieval European states were often discontinuous, with ill-defined and fluid frontier zones rather than precise or fixed borders. Then the term 'nation' meant something very different and *non*-political, generally referring simply to people born in the same locality. Furthermore, the different levels of overlapping sovereignty typically constituted *nested* hierarchies, for example parish, bishopric, archbishopric for spiritual matters; manor, lordship, barony, duchy, kingdom for secular matters. People were members of higher-level collectivities not directly but only by virtue of their membership of lower-level bodies.

In many parts of the pre-colonial world, sovereignty was not based upon fixed boundaries and territorial control per se. For instance, much of North Africa and the Middle East had very different forms of political sovereignty in pre-modern times. According to George Joffé (1987: 27):

> political authority was expressed through communal links and was of varying intensity, depending on a series of factors involving, *inter alia*, tradition, geographic location and political relationships. The underlying consideration, however, was common throughout the region and involved a concept of political sovereignty that derived from Islamic practice. The essential condition was that ruler and ruled were bound together through a conditional social contract in which the ruler could expect loyalty in return for enduring the conditions in civil society for the correct practice of Islam.

Not surprisingly, many of the colonially inspired geometrical boundaries that define the modern states of the Middle East and North Africa have limited relation to pre-colonial political landscapes. This is also true of other parts of the world. In many pre-colonial Asian states, for example, the emphasis of sovereignty was not on the territorial limits of control 'but on pomp, ceremony and the sacred architecture of the symbolic centre' (Clarke 1996: 217). In the classical Indianized states of South and South-east Asia, sovereignty was often focused on rulers who claimed divinity, and further eastwards, emperors held the 'mandate of Heaven', and the mandarins' right to exercise their authority was derived from their being 'superior men' (Sino-Vietnamese, *quan-tu*; Chinese *chun-tzu*) 'who acted according to Confucian ideals' (Keyes 1995: 195). But it would be too simplistic to think of sovereignty purely in terms of emperors, kings, queens, chiefs and so on, as ruler–ruled/state–society relations in pre-modern societies were often complex, hierarchical, shifting and not based on strict territoriality. O.W. Wolters (1982: 16–7) described the scheme of power relations in South-east Asia as *mandala* (a Sanskrit word that defies easy translation, but which refers to a political apparatus that was without fixed boundaries, but which rested on the authority of a central court):

> [The] *mandala* represented a particular and often unstable political situation in a vaguely definable geographical area without fixed boundaries and where smaller centers tended to look in all directions for security. Mandalas would expand and contract in concertina-like fashion. Each one contained several tributary rulers, some of whom would repudiate their vassal status when the opportunity arose and try to build up their own networks of vassals.

Territories of pre-modern kingdoms were often discontinuous, with outer tributaries and chiefdoms often paying tribute to more than one authority at the same time, and with areas of sovereign ambiguity, particularly in heavily forested zones, high mountain areas, and remote regions. In a sense, the vast mountain 'frontiers' of the mainland Southeast Asian massif, Southwest China, Himalaya, and Afghanistan, have variously retained certain 'frontier' 'resistance' and 'refuge' characteristics well into the era of modern nation-states (refer to Case study 22.2).

This system of tributary relationships carried its own forms of obligations, sanctions and allegiance. Mandalas created complex geographies of power, 'a polycentric landscape–seascape' (Friend 2003: 18), including smaller chiefdoms paying tribute to more than one 'overlord' at the same time. Initially, the multiple sovereignties of the region were very confusing to the European imperial powers in the region who were eagerly trying to carve out their own spheres of unambiguous control. As Theodore Friend (2003: 21) notes in relation to the making of Indonesia:

> The Netherlands required centuries to unify Indonesia in their own fashion: first for mercantilist advancement of trade and then for nineteenth-century motives of geographic empire . . . How could so few succeed over so many? The answer: because only a handful of mandalas had to be overcome, each caring little about the others or knowing nothing of them. The Dutch brought

a layer of assiduous modernity to political vacuums strung throughout a vast archipelago. Geographically disconnected and culturally discordant but now administratively centralized, the Netherlands East Indies was for the length of one human generation the first comprehensive empire that region had ever known.

Although significant vestiges of such territorial structures remain, today they have mostly been displaced by nation-states with fixed (though sometimes disputed) boundaries. Sometimes the new geography was imposed by the colonial powers, but indigenous polities were also active in transforming the political map, such as China's long quest to have political control over numerous stubborn non-Han ethnic polities in the vast south-western frontier (Giersch, 2001), and in Siam in the late nineteenth century, where the Siamese Court began to employ 'modern' political cartography backed up by military power to determine the sovereign limits of the kingdom (Winichakul 1996). Through the twentieth century, the ideology of nationalism (and the associated idea of nation-states) became one of the dominant and most widespread influences on politics across the world, arguably becoming a key (or the key) manifestation of modern territoriality.

Occasionally nationalisms recognize that the state itself may be multinational (as in British nationalism, which contains English, Scottish and other affiliations), but in so doing, the wish is usually expressed that somehow a more inclusive national identity will evolve or has evolved which coincides with the boundaries of the state. Britain, China, Switzerland, South Africa, Nigeria and the United States of America are all cases where different versions of the claim and goal of an inclusive 'national identity' that supposedly unites disparate 'sub-nations' or communities have been asserted. Often, the assertion of a particular dominant nationalism in a territory has required the suppression of or conflict with other national claims on the same territory (the emergence of the state of Israel is a clear example of this; and in turn the project of a national home for Jewish people in the land of Palestine is in part a reaction to the genocidal extremes associated with German and other European nationalisms earlier in the twentieth century). As a result many 'national' communities that assert a claim to statehood are denied this (see Case study 22.1, for example). Others remain contested. Burma, for example, has witnessed over five decades of protracted ethno-nationalist struggles, particularly between successive military regimes holding power in Rangoon and the predominantly ethnic-Burman heartlands, and various movements in

Regional case study 22.1

Nations without states: the Kurdish case

Gerard Chailand (1993: 4) noted that: 'the Kurdish people have the unfortunate distinction of being probably the only community of over 15 million persons which has not achieved some form of national statehood, despite a struggle extending back several decades'.

The lands predominantly inhabited by Kurds are in fact divided between four states (see Figure 22.1): Iran, Iraq, Turkey and Syria (there are also smaller Kurdish communities in the former Soviet Union – Armenia, Azerbaijan and Georgia – and significant Kurdish migrant populations in Europe, especially in Germany).

Yet frequent attempts on the part of the states in the region to eradicate Kurdish identity and nationalism have not been able to quash either the widespread sense of Kurdishness or the often violent struggle in pursuit of a Kurdish nation-state. After the invasion of Iraq by the USA and its allies in 2003 and overthrow of Saddam Hussein's centralized dictatorship there, the northern part of the country – which is predominantly Kurdish – became an autonomous region. However, neither the occupying power, nor Iraq's neighbours have wanted to see this become the basis of a separate Kurdish state.

Figure 22.1 Regions historically inhabited by the Kurds.

Source: after Chailand (1993: ix)

the provinces that are seeking either greater political autonomy within a federal structure or complete independence from the fragile 'Union of Myanmar' (as Burma is named by the ruling military junta). Indeed, large tracts of northern and eastern Burma/Myanmar are patched together by a series of fragile ceasefires and natural resource joint ventures between the ruling military junta and various ethno-political parties, companies and warlords. Specific armed political groups, such as the Wa United State Party, have effective authority over sections of the Burma–China and Burma–Lao border areas, which provide examples of *de facto* territorial power by 'shadow state' organizations (Grundy-Warr and Dean, forthcoming). Thus, the political–legal integrity of 'the Union of Myanmar' remains contentious. Globally, there a numerous other examples of *de facto* states (Pegg 1998; Bartmann and Srebrnik 2004), such as Somaliland (Srebrnik 2004), Abkhazia (Walker 2007) and the Turkish Republic of Northern Cyprus (Navaro-Yashin 2005), as well as bi- or multi-'national' territories, such as Bosnia-Herzegovina (Dalhman and Ó Tuathail 2006). Such cases reveal the truly complex nature of the historical, cultural and political geographies that complicate and sometimes contradict territorial nation-statehood.

Two 'multinational' communist countries, namely Yugoslavia and the Soviet Union, sought to regulate nationalism by assigning citizens to one of a number of constituent national identities. This sometimes involved inventing nationalities to rationalize and simplify more complex tribal and religious identities, whilst asserting that these should be subservient to an overarching sense of Soviet or Yugoslav identity which coincided with the boundaries of the USSR or of Yugoslavia. So, for example, in the USSR, people could be declared on their identity documents as having one of a number of officially recognized nationalities (Uzbek, Latvian or Russian, for example), but they would also and supposedly above all be citizens of the USSR. In due course, it was in part because of local nationalist challenges to wider Soviet and Yugoslav affiliations that the USSR and Yugoslavia collapsed in the early 1990s.

Many of those self-identified nations without their own state (the Kurds are an example, see Case study 22.1; as are Palestinians, Basques, Karen and Tibetans, to name a few) claim either the right to one or at least a high level of self-rule or autonomy. And frequently, smaller recognized national identities within a multinational state (such as Scotland in the United Kingdom or Quebec in Canada) become the basis for

claims that they should enjoy full statehood. In many cases too, either the central state or some other community with another affiliation resists. There are numerous examples of this, including the complex case of Northern Ireland, where most Catholics (Republicans) would wish to see the province united with the rest of the Irish Republic (which itself successfully broke away from the British colonial empire earlier in the twentieth century). Most Protestants (Unionists), who claim descent from settler populations from England and Scotland, wish to remain part of a 'United Kingdom' (see Case study 21.4).

The Irish case is just one of dozens of situations where conflicting nationalist and confessional (religious–cultural) logics collide, often with violent consequences. There is clearly something very powerful going on, whereby nationalist visions are linked with particular territories and conceptions of state. Yet, as Anderson (1983: 3) recognizes:

> But if the facts [of the existence of many and sometimes conflicting nationalisms] are clear, their explanation remains a matter of long-standing dispute. Nation, nationality, nationalism – all have proved notoriously difficult to define, let alone analyse. In contrast to the immense influence that nationalism has exerted on the modern world, plausible theory about it is conspicuously meagre. Hugh Seton-Watson, author of by far the best and most comprehensive English-language text on nationalism, and heir to a vast tradition of liberal historiography [theories of history writing] and social science, sadly observes: 'Thus I am *driven* to the conclusion that no "scientific definition has existed and exists".'

Readers may wish to 'prove' Seton-Watson's observation for themselves, by trying to come up with a universally valid definition of a nation. Faced with this task, students will often work through a long list of characteristics ascribed to nationality. But it seems that exceptions can always be found. Language is a favoured criterion, but then many languages are spoken by more than one 'nation' (English is an example) and some 'nation-states' (Switzerland, Belgium or Mozambique, for example) contain substantial communities speaking different languages. Religion is sometimes chosen, but the same objections apply. Ethnicity and 'race' turn out to be problematic criteria, favoured by racists of all stripes, and often part of the basis for national identities, but susceptible to the obvious and undeniable point that everywhere is much too mixed up historically and genetically for such categorizations to

Regional case study 22.2

Zomia: highland resistance to nation-building and state-making

The map refers to a vast mountainous region of approximately 2.5 million square kilometers, spanning the boundaries of ten recognized nation-states, and sprawling across the designated 'world regional area' limits of 'South Asia', 'Central Asia', 'East Asia' and mainland 'South-east Asia'. This transnational highland zone is called 'Zomia' (based on a Tibeto-Burman term 'zomi') a term invented and proposed by historian Willem van Schendel (2005: 284–5), who argues that 'much of Zomia resisted the projects of nation-building and state-making of the countries to which it belonged'. Part of the reason for this are the great many dispersed 'national minority' groupings, such as Naga, Akha, Lahu, Hmong, Lisu, Shan and Kachin, as well as numerous other groups, each with their own peculiar socio-ecological adaptations to highland life, and cultural, linguistic and religious affinities that are distinct from dominant 'national' languages and religions. Similar communities (the four towns indicated on the map)–just a short distance apart–fall into four states and macro-regions.

James C. Scott (2009) has produced a wonderfully rich 'anarchist history of upland Southeast Asia', revealing the connected but often antagonistic co-evolution of hill and valley across much of Zomia. He (2008: 12) argues that 'Zomia coheres as a region not by political

unity, which it utterly lacks, but by comparable patterns of diverse agriculture, dispersal, mobility and egalitarianism'. Scott (2009) reveals that, over long periods of time, many groups of people viewed the uplands as economic, cultural and political 'refuges' from interfering and hierarchical lowland state authorities. Thus, historically, many people migrated to these areas in order to escape, evade or resist state interference. These spaces of state 'evasion', 'resistance' and 'refuge' were ultimately partitioned into various national geo-bodies following the nineteenth and early twentieth century rivalries and boundary-making activities of imperial powers, and the adoption of modern political geography by dominant indigenous polities, and subsequent post-colonial nationalist elites (Winichakul, 1996).

The arbitrary and abrupt nature of geopolitical boundaries generated tensions between central states and the upland groups, who have frequently been regarded as suspicious, backward, primitive and uncooperative 'minorities', and, occasionally, as security threats by lowland state authorities (van Schendel 2005; Michaud, 2009; Scott, 2009). In the modern history of these nation-states, Zomia has been characterized by 'a high incidence of regionalist and separatist movements, "non-state spaces", and discursive battles around concepts such as "tribe" and "indigenous people"' (van Schendel 2005: 285). Central territorial states have sought to extend administrative control

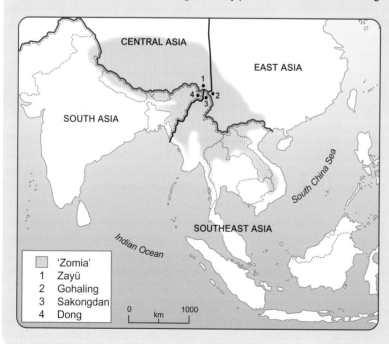

Figure 22.2 A map of Zomia.

Source: adapted from van Schendel (2002)

into the uplands of Zomia through various 'development projects', forced resettlement of highland groups, campaigns aimed to extend settled farming into higher lands, religious conversions, spatial extensions of central bureaucratic administration and infrastructure such as roads, making these 'spaces of evasion' more accessible over time.

By the early twenty-first century, the accessibility of 'non-state spaces' has increased, but not entirely, and these spaces continue to represent considerable challenges to nation- and state-building. Indeed, we could argue that historically the mountainous zone that straddles the Afganistan–Pakistan border, described by President Barack Obama in a speech he gave on 27 March 2009 as 'the most dangerous place in the world' (see also Imtiaz Gul 2010) has served as such a zone, confounding the military–security ambitions of imperial rivals, Cold War superpower interventions, and numerous efforts of centralized states to control people, relations and territory for the past two centuries.

'Zomia' is part of a new political map invented and elaborated by scholars to address relative academic neglect in 'area imagination'. Whilst its regional dimensions are greatly complicated by 'a cultural mosaic with contrasting colours, rather than an integrating picture in harmonizing shades' (Michaud 2010: 206), the idea of 'Zomia' and other such transnational zones is important. 'Zomia' helps us to break free of our rigid national geographic imaginaries, particularly with regard to the histories, ethnographies and geographies of mobile groups (Ludden 2003), the politics of place and belonging of so-called 'minorities' (Michaud 2010), and the complex interrelated, transitional and transient dimensions of borderlands and borderscapes (Donnan and Wilson 1999; Rajaram and Grundy-Warr 2007). We need to be critical of reifying 'national geo-bodies' and pay attention to myriad pre-existing and contemporary transnational social and political formations that challenge the map of nation-states.

be watertight. Besides, some nationalisms have come to celebrate their multiracial and multicultural composition, as in the 'melting pot' United States or the 'rainbow nation' of South Africa. 'Culture' usually crops up as a criterion. Yet as Chapter 13 has indicated, cultures are always (though of course to varying degrees) heterogeneous and contested. Think, for example, of age, gender, class and other variations and the coexistence of multiple sub-cultures and identities (which, as Chapter 21 has shown, are often related to distinctive 'local' territorialities) that characterize every supposedly 'national society'.

All this leads Anderson to declare that nations are in a sense 'imagined communities'. This imaginary status is not to deny that they are not in a sense real to those who feel they belong to them. Indeed, Anderson (1983: 6–7) feels that:

In fact, all communities larger than primordial villages of face-to-face contact (and perhaps even those) are imagined. Communities are to be distinguished, not by their falsity/genuineness, but by the style in which they are imagined (. . .) Finally, it is imagined as *community*, because regardless of the actual inequality and exploitation that may prevail in each, the nation is always conceived as a deep, horizontal comradeship. Ultimately it is this fraternity that makes it possible, over the past two centuries, for so many millions of people, not so much to kill, as willingly to die for such limited imaginings.

Within all this is a profound territorial link between the nation and the state. The state claims to be a sovereign expression of the nation – bound to it and to a particular place. That is, it has territorial limits. The nation-state has a *geography*, which is charted, demarcated, mapped and represented to the 'national population' in their school atlases and geography lessons. Such a system of 'national' geographical representation is always combined with an historical vision, a grand narrative of 'national history', often assuming that the nation is ancient, even primeval.

Yet Anderson and other critical accounts of nationalism stress that it is very much a modern ideology. For what makes mass nationalism possible are certain socio-economic and technical transformations, notably the arrival of media and national educational systems. Schooling, newspapers, and later radio and television all help to promote and popularize the idea that people belong to and share in the nation. Whilst others have emphasized the longer historical roots of many nationalisms in pre-modern ethnic affiliations (for example, Smith 1988), for Anderson and most other critical observers, what is striking is nationalism's relative modernity. Even where nationalists imagine the nation as ancient, such an imagination is itself more often than not predominantly a nineteenth- and twentieth-century phenomenon. At a pinch, nationalism might be traced to the seventeenth century. But the ideology

Plate 22.2 'Nation' imagined as race.

(Jim West/Alamy)

of the nation-state was not anywhere very evident before then. So whilst someone might talk about a thousand years (or more) of, for example, English, Korean or British or Indian national history, they conveniently forget that English or British, Korean and Indian nationalisms are relatively modern concepts, much less than a thousand years old.

The tendency for nationalists to 'reinvent the past', to pick out selective moments from the past, or to manipulate history, is one of the main arguments of historian Eric J. Hobsbawm (1996: 255), who argues:

> For historians are to nationalism what poppy-growers in Pakistan are to heroin addicts: we supply the essential raw material for the market. Nations without a past are contradictions in terms. What makes a nation *is* the past; what justifies one nation against others is the past and historians are the people who produce it.

Whilst history is undoubtedly significant, the nationalist brew would be incomplete without ancestral connections to 'homeland' and to particular places. In

other words, *territory* is central to nationalism, and this is often why whenever there are divergent nationalist claims to one piece of land, extreme violence often follows. Indeed, the terrible forms of so-called 'ethnic cleansing' witnessed in the former Yugoslavia in the 1990s are illustrations of recent nationalist extremism and the significance of historical imaginations concerning territorial–political and cultural identity.

Nationalist narratives and claims are not always linked to such violent actions. At times they may be rather more rhetorical or theatrical. Graham E. Clarke (1996: 231–2) raises a particular Nepalese example:

> [I]n the 1960s in Kathmandu the sole national newspaper carried articles, no doubt read tongue in cheek by some educated Nepalese, arguing that since Lumbini, the birthplace of the historical Buddha (Gautama), was located some few five miles north of the current southern border, that Buddha was *therefore* Nepalese and not Indian.

All of this relates to the important issue of how we deal with things like space and identity in history. As Morris-Suzuki (1996: 42) writes:

> The nation . . . casts a long shadow backwards on our vision of the past, and channels our perceptions into a particular spatial framework. In my bookcase, I have a volume on the history of Thailand since the tenth century, which, considering the repeated political and cultural realignments within the space we now label 'Thailand', seems only a little more bizarre than its neighbour on the shelf, a history of the Soviet Union from palaeolithic times to World War II.

Claims to nationhood frequently involve the blending of 'tangibles' and 'intangibles' in a unique brew that may contain such volatile ingredients as blood, language, race and religion. A fundamental goal of nationhood is to generate a strong *sense* of belonging associated with a particular territory. In practice it may prove difficult to distinguish between 'the nation' and other human collectivities, but as Walker Connor (1994: 93) puts it, 'what ultimately matters is not *what is* but *what people believe is*'.

There has been a strong tendency to see nations and nationalisms as essentially being derived from American and European prototypes. It is important to challenge the idea that nationhood is a nationalist project originating in the West that was simply implanted on to the rest of the world without significant transformations in the various cultural, ethnic and religious contexts it touched. African, Middle Eastern or Asian variants of nationalism or constructions of statehood are far more than simply colourful (or failed) replicas of the European–American model(s). To be sure, imported idioms and motifs from Europe and America have been important; for example, the official state language in a fair number of former colonies is that of the old colonial power. And European (and less often United States) imperialism has provided an important backdrop to the trajectory of nationalisms in Africa and Asia, not least because the anti-colonial nationalists themselves were often educated in Western institutions and adopted concepts such as self-determination, national liberation and territorial sovereignty in their political struggles against the colonial powers. Yet whilst 'globalizing' capitalism and especially imperialism are part of the picture, African, Middle Eastern and Asian nationalisms are also (like those of Europe and the Americas) rooted in *local* historical trajectories. In many cases these are of much longer vintage, often based on sophisticated cultural, religious and ethnic tapestries, some of the main patterns of which have subsequently been appropriated or (re)invented as precolonial 'national' histories.

Thus, we should perhaps think in terms of multiple histories and geographies of nations and states, and not just of particular European or American modular forms. The influential Bengali intellectual Partha Chatterjee (1993) has persuasively argued against the imagining of Indian national identity through the lens of the colonial power and stresses the 'essential' inner or spiritual domains of culture that were never colonized, never European. Furthermore, there are different coexisting national voices or 'fragments' – among women, peasants, elite, castes, outcasts, and so on – each with its separate discourse. As Stein Tønnesson and Hans Antlöv (1996: 32) put it:

> When the national idea entered Asia it could not be implemented without mediation, hence transformation, by indigenous agencies in particular settings. There were existing and alternative ideas with which European-style nationalism interacted and intermixed . . . the nationalist ideas were invested with local qualities, meanings and nuances which could not be found in Europe. People had their own views on what constituted a legitimate social order, and such views could not be ignored by the modernizing leaders of the anti-colonial struggles.

As well as different cultural 'forms of the nation', we may also consider different gendered perspectives of nationalism and nation-states. Although Chatterjee's (1993) arguments are quite specific to the South Asian case, they do reveal the coexistence of different voices, men and women, and this relates to the point about nationalism and constructions of nation-statehood invoking different conceptions of 'manhood' and 'womanhood' more widely. Although there are immense historical and geographical variations of this, there is a general tendency for women to be seen as particularly important transmitters of 'national culture' (for example, in the idea of a 'mother-tongue' language) (Yuval-Davis 1997) and as somehow the embodiment of the nation. McClintock (1994: 352) thus notes that:

> Nations are contested systems of cultural representation that limit and legitimize people's access to the resources of the nation-state, but despite many nationalists' ideological investment in the idea of popular *unity*, nations have historically amounted to the sanctioned institutionalization of gender *difference*. No nation in the world grants women and men the same access to the rights and resources of the nation-state.

Moreover, particular ideas of the 'family' are central to most imagined communities:

> Nations are frequently figured through the iconography of familial and domestic space. The term nation derives from [the Latin] *natio*: to be born. We speak of nations as 'motherlands' and 'fatherlands'. Foreigners 'adopt' countries that are not their native homes and are naturalized into the national 'family'. We talk of the 'Family of Nations', of 'homelands' and 'native' lands. In Britain, immigration matters are dealt with at the Home Office; in the United States, the president and his wife are called the First Family. Winnie Mandela was, until her recent fall from grace, honoured as South Africa's 'Mother of the Nation'. In this way, despite their myriad differences, nations are symbolically figured as domestic genealogies.

In several European nationalisms (including 'British') one significance of this is that the sexist notion of women (usually grouped with children) supposedly as *naturally* inferior to men, or as requiring the protection of *naturally* more powerful men, frequently provided the prior backdrop to depicting hierarchies (between dominating and dominated classes, for example) *within* the nation as natural, like those of a family. A similar conception provided part of the racist justification for overseas colonies. Peoples subject to colonial domination were often represented in such racist terms as being equivalent to 'families' of black 'children' ruled by a 'benevolent' white father.

22.3 Constructing boundaries

States come in diverse shapes and sizes. But they all have boundaries (which, if they have a coastline, also extend into the sea and are governed by international protocols: see Stienberg 2001). The extract from Winichakul (1996) quoted at the start of this chapter noted how the 'territoriality' of the modern nation-state is of a bounded space: 'a certain portion of the earth's surface which can be easily identified'. Pre-modern dynastic realms and empires could make do with loose boundaries. But modern nation-states have felt it necessary to *demarcate* their boundaries, to iron out perceived irrationalities and anomalies, sometimes by going to war. In short, they have sought to nationalize and unify 'their' space and 'their' (national) populations and render them into known, surveyed and defensible sovereign territory and communities. As Chapter 20 explored, the rise of geopolitics in the late nineteenth and early twentieth centuries represents a particular expression of just such a rationality, but it has wider forms and deeper roots. The nineteenth-century invention and refinement of statistics (*state-istics* = the science of the state) and demography (*demos* = people, *graphy* = writing) and the establishment of geography (*geo* = earth, *graphy* = writing) as a university discipline and field of research are all part of this regime of knowledge and power where people are surveyed and made into subjects of nation-states. For states with land boundaries the occupation and official demarcation of the frontiers is an important part of this process.

Consider the example of the border that is often recognized as the *oldest*, more or less stable, still-existing border in the world, that between Spain and Portugal. The mutual recognition of the border between Spain and Portugal is usually traced in modern Spanish and (particularly) Portuguese history to the Treaty of Alcañices signed on 12 September 1297 (in fact, the date on the Treaty is 12 September 1335, but amongst other things, our conventions for counting years have altered since). The Treaty has acquired something of the status of a foundational text. In other words, it is interpreted as a kind of proof of the ancient historical basis of the nations concerned.

Yet the treaty was *not* signed in the name of states or 'nations'. Named after the Templar castle in which it was sealed and witnessed, the text of the treaty of Alcañices begins with the words 'In the name of God. amen.' and is signed in the presence of the Templars

and other holy orders by those who describe themselves as:

> by the grace of God the King of Portugal and the Algarve and by the grace of God the King of Castille and Leon,

and has as its subject:

> towns, castles and lands, town boundaries, divisions and orderings . . . [disputes over which] have caused many wars, homicides and excesses, as a consequence of which the lands of both [kingdoms] have been looted, burnt and ruined, weighing heavily on God . . . [and] because of our sins, risking the danger of losing them and them falling into the hands of our enemies in the faith and most gravely [causing] the violation of God's will and injury to the holy church of Rome and Christendom.
>
> (Cited in Martíñez 1997: 15; our translation)

Moreover, the actual demarcation of the Portuguese–Spanish borderline had to wait until the nineteenth and twentieth centuries. Two formal delimitation treaties set out an agreed frontier and established mechanisms for its physical demarcation on the ground. But gone are 'amens' and references to 'the will of God'. Instead, the treaties (the first of 1864 for the northern half of the border and the second of 1926 for the southern half of the border) make reference to the need to impose order and well-defined demarcations, to 'eliminate the anomalous situation in which, in the shadow of ancient feudal traditions' (Tratado de Limites 1866: 1; our translation), some frontier areas (including a number of villages) were recognized as *shared* or common lands with usage rights by communities of both states/'nationalities'. Following the treaties, such areas were divided and the border, where not demarcated by a river, was marked on the ground by rectangular boundary stones every few hundred metres (although it took a couple of decades to put in place all the boundary markers and resolve local conflicts and differences over the 'fine-scale' division of lands). In other words, the Portuguese–Spanish case indicates how the epoch of modern nationalism (i.e. the nineteenth and twentieth centuries) saw what James Anderson (1995: 71) calls:

> a territorialisation of politics, with a sharpening of differences at the borders of states and of nations between 'internal' and 'external', 'belonging' and 'not belonging', 'us' and 'them'.

By the same logic, or an extreme extension of it, minorities who do not 'fit in' sometimes have to be moved, murdered or deported – or, in the euphemisms of our times, 'cleansed' (see Sibley 1995). This has been associated with, for example, Nazi racial territorial visions of a Europe cleansed of Jews and Gypsies and an expanded racially pure Germany. The Nazi Holocaust represents an extreme case. But 'cleansing' or 'purification' of national space, amounting to the expulsion or murder of people defined as 'foreign' or 'alien', has been widespread in the twentieth century. There are examples from all continents: the creation of modern Turkey whose Armenian population were slaughtered in the second decade of the twentieth century (and which still has not resolved the status of its Kurdish minority); the creation of Israel in the 1940s, which was accompanied by the flight of indigenous Palestinian Arabs from territories deemed part of the new Jewish national state; the system of reservations for native Americans in the USA and Canada; the violent exchange of Muslim, Hindu and Sikh populations which accompanied the 1947 partition of India; the 1974 partition of Cyprus; the wars of the 1990s in the former Yugoslavia (see Case study 21.2); and the partition of Ireland. These logics have also produced calls for the modification of boundaries, sometimes resulting in violence and war. No continent has been without examples of such conflict and violence and it is probably impossible to find a state which has not at some time expelled people or murdered them in the name of some nation or other. Yet each nationalism must also be unique, establishing itself as different from others (even if the apparatuses of flags and anthems are superficially similar), constructing a sense of self against others who are defined as outside the imagined community. We will return to borders and borderlands in Section 22.5.

22.4 Nation-states as symbolic systems

As we have seen, the state is the bureaucratic expression of nationalism. This is not to say that a widespread sense of nationalism inevitably precedes the state. Indeed, what is often termed the state apparatus (everything from regulations governing schools, to tax inspectors and politicians) seeks to foster national subjectivities out of the frequently ambivalent and disparate array of identities contained by its boundaries. If, as has been argued by many, nationalism is akin to religion, then the state becomes its symbolic structure. When examining the superficially

quite different cases of Australia and Sri Lanka, Bruce Kapferer (1988) claims that nationalism is itself a religion, owing to the fact that, as with most religious-like beliefs, nationalism demands the recognition (his word is 'reification', which means something more than this) of an immanent and all-encompassing entity (the nation). The nation fulfils the role of a sacred cause, something greater than any individual and something which may be worth dying for. The nation is the God or deity of the religion of nationalism and the state is its theology or temple. Moreover, as Herzfeld (1992: 37) notes:

> Every bureaucratic action affirms the theology of the state. Just as nationalism can be viewed as religion, bureaucratic actions are its most commonplace rituals. There are other such everyday rituals: Hegel saw the reading of the morning newspaper as the secular replacement of prayer.

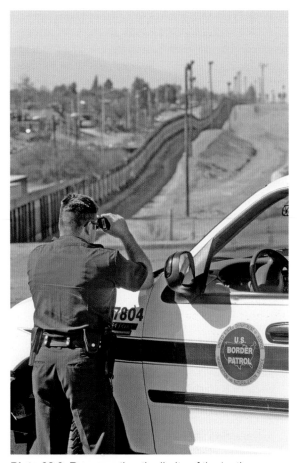

Plate 22.3 Representing the limits of the 'nation-state': boundary patrols.

(David R. Frazier Photolibrary, Inc./Alamy)

There are indeed many public rituals of nationality and statehood: coronations and remembrance days, military parades, national holidays, national prowess (or the lack of it) at football or the Olympics, swearing in of governments, state funerals. (Football in the United Kingdom is also revealing of the complexity of British nationalisms, and the survival of separate Scottish, English, Welsh and Northern Irish squads is a testimony to the limits to 'Britishness' and the survival or 'revival' of other national affiliations.) In all these rituals, the nation is reaffirmed and the state performed and made to seem omnipresent, historical and real. Dramatic cultural or political events (as well as notable national sporting occasions, such as soccer World Cup or Olympic successes) covered by the media reinforce senses of national community. The events of 11 September 2001 were thus narrated in the United States as an 'attack on America'. It is also said, for example, that virtually every adult *American* alive at the time can recall where they were on hearing the news of the assassination of US President John F. Kennedy in 1963. Whilst his funeral, like that of Diana Spencer in 1997, became a global media spectacle, it was represented and felt most acutely as a kind of national loss. 'Goodbye *England's* Rose . . .' as Elton John put it at Diana's funeral. (It should be noted that there was some controversy over the lyrics of Elton John's tribute song, in so far as Diana was not only the Princess of *Wales*, but also a former member of the avowedly *British* monarchy. Rather like the case of football, this bears out the complexities and unresolved tensions of British nationality.)

But the nation-state also demands (like all effective religions) personal commitment and more minor ritualistic acts, many of which rest on a geographical imagination of inside and outside, belonging and otherness (see Taussig 1997). The anthropologist Michael Herzfeld (1992: 109) examines this theme and is worth quoting at some length:

> Nationalist ideologies are systems of classification. Most of them are very clear about what it takes to be an insider. That, at least, is the theory. In practice, however, divergent interpretations give the lie to such essentialist claims, as to take one prominent and current example, in the debate currently waging in Israel about the definition of a Jew. Such taxonomic [naming] exercises . . . are central to the very existence of the nation-state. All other bureaucratic classifications are ultimately calibrated to the state's ability to distinguish between insiders and outsiders. Thus . . . one can see in bureaucratic encounters a ritualistic enactment of the

fundamental principles upon which the very apparatus of state rests. Seen in these terms, arguments about the number on a lost driver's license or an applicant's entitlement to social security do not simply challenge or reinforce the power of particular functionaries of state. They rehearse the logic of the state itself.

Not only that, but the whole exercise of state power gets taken for granted as the natural order of things. Only when many of those activities which are ascribed to the state are no longer carried out (in situations of war, for example) or when a person finds themselves on the wrong side of a state-sanctioned category (the wrong side of the boundary, the wrong side of the law) is the power (a power over life and death and thousands of lesser things) and the universality of the symbolic order of 'the state' revealed. The state claims the monopoly over these things. And for others to exercise judgment and to punish or to kill is 'to take the law into their own hands'. These things are reserved for the territorial state. But what if there is no state? Or as cultural anthropologist Bernhard Helander (2005: 193) asks: 'Who needs a State?' There are parts of the world where any semblance of effective state sovereignty has become so scattered, fragmented or *ineffective* in practice that other forms of rule and competing authority structures take over. For instance, Helander wrote about the complex power situations in Somalia affecting every single aspect of daily life, where 'any opinion expressed or action performed must always be positioned in the political landscape' of rival clans, warlords, and issues of ethnic, religious and cultural identity. In the contexts of atomistic and antagonistic power struggles we need to appreciate the local power configurations that enable communities to continue with everyday life in the absence of a meaningful state apparatus. Helander was critical of international organizations that tend to act too rashly in countries such as Somalia, often distributing massive sums of inappropriate aid according to the 'shifting agendas of foreign donors' and frequently without great familiarity with the nuances of the 'local' human, cultural and political landscape. Such perspectives would clearly be even more applicable in situations of *failed international interventions* creating uncertain and extremely dangerous regime changes, unpopular occupations, and an almost complete absence of any form of security for large segments of the population. This happened in much of Iraq after the US-led invasion to depose the government of Saddam Hussein in 2003. Arguably this was also the situation in Afghanistan in the 1980s, after the Soviet Union invaded to prop up a communist

Plate 22.4 Stansted Airport, UK – passport control signage.

(BAA Aviation Photo Library)

government in the capital city who faced rebellion in the more conservative countryside. Afghanistan then became a focus of Cold War geopolitical conflict (as was detailed in Chapter 20), laying the basis for many more years of confrontation, subsequent interventions (by the United States and its allies in the early twenty-first century) and enduring state failure. In such situations, we may actually have to ask, 'What kind of State? Where and who are the authorities?'

22.5 Sovereigntyscapes: 'shadows', 'borderlands' and 'transnationalisms'

We have seen that the geographies of 'nations' and states do not always coincide (consider again the Kurdish case, described in Case study 22.1), and that national territorial histories are insufficient to account for ethno-histories, transnational identities and human mobilities (as in illustrated in Case study 22.2). Thus sovereignty is rarely without contradictions or challenges from other states, potential states and nations. But how else might we think about the challenges to sovereignty and the historical and geographical complexity of nations? One way might be led to think of political landscapes as 'soveriegntyscapes' (Sidaway 2003; Sidaway *et al.* 2005) containing varieties of countervailing tendencies, fragmented state sovereignty, ambiguous forms of sovereign control, and situations (such as Somalia, as noted above) where state authority may have collapsed or been replaced by various contending de

facto forms of authority. Even when we consider the state as strong and central, we may view sovereignty as being 'graduated', whereby, as Aiwha Ong (1999: 217) observed, 'states make different subject populations, privileging one gender over the other, and in certain kinds of human skills, talents and ethnicities; it thus subjects different sectors of the population to different regimes of valuation and control.'

In addition to highly uneven and complex 'sovereigntyscapes' there are economic, political, cultural and social forces that could be considered to be largely non-state 'shadow powers' that are able to operate across national, regional and international borders. These may have parasitic relations to conventional sovereignty. Thus Carolyn Nordstrom (2000, 2004, 2007) has examined examples of 'shadow powers and networks' operating within numerous 'war torn' countries with global connections. The notion of 'shadow' here enables us to think of 'extra-state' transactions that transcend territorial boundaries, but are not 'officially' sanctioned by states, yet may involve a wide amalgam of actors and agencies, including state functionaries within complex transactional processes and extended networks. Undoubtedly, such 'shadows' are often associated purely with powerful international crime syndicates such as the Italian mafia, Russian 'Mafiya', Japanese *yakuza*, Chinese *Snakeheads*, and so on (Lintner 2002). In practice, 'shadow networks' may also incorporate a great many people who are deeply immersed in society and civil life albeit within shades of legality ranging 'between legal, quasi-legal, gray markets and downright illegal activities' (Nordstrom 2000: 36). Whilst 'Zomia' (Case study 22.2) represents a 'zone of state evasion',

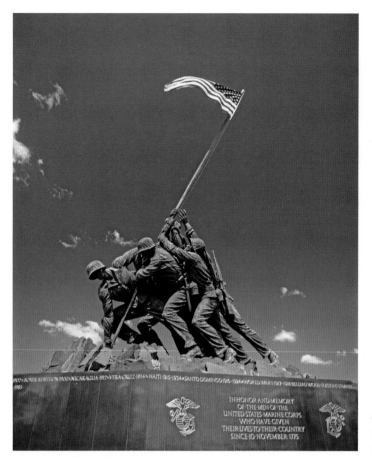

Plate 22.5 War-making and with it a sense of shared struggle can be productive of national identity.

(© Bill Ross/CORBIS)

shadow networks create exchanges 'outside the formal state institutions' and comprise both territorial and non-territorial dimensions operating across multiple scales. As Nordstrom (2004: 11) puts it: 'The relative freedom from controls found in warzones and the financial powerhouses found in cosmopolitan centers of the world combine in ways that tend to merge war and global profiteering.'

Furthermore, these networks are often far more significant economically and politically to the functioning of such societies than the malfunctioning apparatus of the official *national* 'sovereign states'. In such contexts, the official state-centred political map appears to be little more than an international juridical nicety without real *de facto* force (Jackson 1990). More than this, the research of Nordstrom and others (Strange 1996; Lintner and Black 2009) leads us to consider 'non-state', 'extra-state' and 'quasi-state' forms of spatial power alignments, as well as inter- and extra-territorial connections within the global political economy. They lead us to (re)consider questions such as *who* wields authority in the junctures and 'borders' of state/non-state transactions? *Who* and *what* constitutes the state when

'shadow networks' dominate the political economy? *How* does the existence of such networks undermine notions of national territorial integrity and formal international relations? *Why* and *how* do various communities, civil society groups and individuals become entangled in 'shadow networks'? Certainly, these questions relate to those situations of state 'failure' or 'collapse', which turns out to be much more complex than being a mere absence of authority.

A variation on the theme of 'shadow networks' is to consider myriad cross-border connections between people that somehow circumvent or subvert state-centred rules and regulations. In some contexts, such as the European Union, cross-border links are actively fostered and promoted (along with a common market and internal freedom of capital and personal mobility) with the aim of fostering a supranational (beyond nation) community (Kramsch and Hooper 2004). In other cases (such as the US–Mexican border) or places where there is large scale 'illegal' movement of people and goods, state power is defied or subverted. In this regard, Abraham and van Schendel (2005) draw upon a critical distinction between what states define as 'legal/illegal' and what ordinary people perceive as 'licit/illicit'. Thus many transnational movements of people, commodities and ideas are illegal because they defy the norms and rules of formal political authority, but they are quite acceptable, 'licit', in the eyes of the participants in these transactions and flows. Here the authors were not so concerned with the flows associated with big syndicates, but with the many 'micro-processes' and transactions that form 'everyday transnationality' in borderland spaces (van Schendel 2005: 55). Such transactions may involve kinship and family networks that were partitioned by superimposed boundaries, but revived subsequently in new forms. We should try to avoid overly dualistic perspectives of 'trans-border' or 'transnational' flows, such as 'domestic/foreign', 'internal/external', 'legal/illegal', because by doing so, we would be perceiving flows entirely from rigid state-centred and fixed territorial positions. Even if we just consider state practices such as the enormous efforts of the US federal authorities in trying to keep 'illegal aliens' 'out' (Plate 22.3), there are many more countervailing actions at official, quasi-official, unofficial levels, trying to make sure that as many people 'get in' because of their enormous contributions to economic sectors (Andreas 2000).

Why is such work of relevance to ideas about 'nation' and the 'nation-state'? It is precisely the very everydayness of many cross-border interactions, movements and connections that raises big questions about the impression of solidity of spatially fixed notions of nationhood. Despite attempts to fortify and police it, the southern border of the United States and the maritime and land frontiers of many southern and eastern member states of the European Union exhibit similar porosity. In other words, there are mini-'zomias' confounding and contradicting many parts of our political map of national territories (see Case study 22.2), with a great many transnational identities deserving of their histories, social studies and geographies just as much as recognized nation-states, such as the Kurds (Culcasi 2006), Kachin (Dean 2005) and Palestinians. Finally, global migrations have also produced significant transnational social networks, such as the Hispanics in the United States or Maghrebis in Europe. Arjun Appadurai (2003: 342–3) argued that there are new 'postnational cartographies' emerging, providing the geographical imaginary of a diasporic home beyond 'national space', such as 'Khalistan', an imagined home of some Sikhs in India and worldwide. Other 'translocal affiliations' are 'global or globalizing', such as those of Islamic, Christian and Hindu fundamentalisms, as well as 'racial and diasporic' ones

that 'cross-cut existing national boundaries' and intersect 'with other translocal formations'. The hypermobility of people, refugees, exiles, migrants and others, plus old and new identities, associations, networks and affiliations that are other than 'national', mean that the idea of the 'nation-state' is being challenged from all directions.

22.6 Conclusions: the place of the nation-state?

There have never been so many states as there are today. In the nineteenth century a wave of new states (countries such as Argentina, Mexico, Bolivia) emerged in the Americas in revolt against Spanish and Portuguese empires. These American prototypes and the USA itself (which dates from 1776) provided an example to nationalists elsewhere in the colonial world. But it was not until the twentieth century that most of Africa and Asia could escape direct colonial domination. This often required violent 'national-liberation' struggles against entrenched resistance from the colonial powers and white settler populations. In the decades between about 1945 and 1975, dozens of new 'sovereign' states emerged in place of the old colonial map. And if, for example, the Tamil nationalists of Sri Lanka, the Karen, Karenni and Shan of Burma, the Palestinian and Kurdish nationalists of the 'Middle East' or the Basque, Scottish and Corsican nationalists in Western Europe get their way, there will be even more.

Yet, for many years the *demise* or decline of the state has been discerned or predicted, and such claims have of late become even more common in discourses about 'globalization'. These usually argue that the growing scale and power of transnational flows, particularly of capital (but also of people, ideas and religious affiliations, technologies and so on) is *subverting* the capacity of the state and *weakening* national identities. The nation-state is often described as being 'hollowed-out' or 'eroded'. In this view, the state no longer has the power to command, for example, the society and economy inside its boundaries that was once attributed to it. And such a 'hollowing out' is sometimes seen to point the way to a post-national world (or some kind of *shared* global culture in which national cultures are replaced by a more hybrid, but common global mixture).

Yuval-Davis (1997, reprinted in Brenner *et al.* 2002: 322) has pointed to the lack of congruence between nations and states, arguing that there is often a lack of 'overlap between the boundaries of state citizens and "the nation"', which requires us to have a much more 'multi-layered' notion of people's citizenship needs, 'because people's membership in communities and polities is dynamic and multiple'. And, as we have seen, in recent years, the sovereignty of many states has also been questioned through imperial 'interventions' in the name of humanitarian intervention or the 'war on terror'.

Moreover, others object that transnational forces of 'globalization' are really nothing new and that capitalism in particular has shown itself to be able to coexist with states and nations and in symbiotic relationship with them. In other words, they reinforce each other. The same goes for many organized religious movements that coexist or support nation-states. (This is not to say that many movements with religious inspirations do not see themselves as anti-statist, even radically and violently so, as in the case of some fundamentalist Christian militias in the USA and Islamicist movements in Asia. But in so doing of course they ascribe to the state a certain power and centrality – thereby making it more real. A parallel to help readers understand is the atheist statement: 'There is no God', which cites something called God in order to deny its existence!) Moreover, 'globalization' is not only massively uneven (as we have seen in earlier chapters) but as likely to produce local backlashes as a universal culture of, say, the *same* fast food, drinks and soap operas and political orientations everywhere in the world. That is, the technologies of capitalism (particularly media) provide the preconditions for strengthening, rather than undermining, imagined communities of nationalism, while states still act to fine-tune the regulatory frameworks for continued capital accumulation (resorting where needed to force to suppress opposition). States still enact laws about business, trade unions, property rights and so on. Everywhere, buying a property or land or setting up a legal business requires some kind of registration with the state. No amount of globalization has ended this. However, instead of further entering the large and complex debates about the impacts of 'globalization' on nations and states, let us return to the conception of the state as a symbolic system, and as a complex of *representations*. The historian-philosopher and activist Michael Foucault (1979: 29) (see too Spotlight box 23.2) argued in one of his most famous essays that:

> We all know the fascination which the love, or horror, of the state exercises today; we know how much attention is paid to the genesis of the state, its history, its advance, its power and abuses, etc. . . . But the state, no more probably today than at any other time in its history, does not have this unity, this individuality, this rigorous functionality, nor, to speak frankly, this

importance; maybe after all the state is no more than a composite reality and a mythicised abstraction, whose importance is a lot more limited than many of us think.

Hence, to recognize the state as, in part at least, a symbolic system is also to recognize that, as Rose and Miller (1992: 172) argue: '"the state" itself emerges as an historically variable linguistic device for conceptualizing and articulating ways of ruling'. In other words, things like the 'nation' and the 'state' are *made* real mainly in certain words, texts (including maps) and deeds, that is, in language and action. The nation-state is an historically specific way of governance that links land, nation, population and polity. Think again of the staging of those *national* sporting, political and cultural occasions, which bring the nation home, and the more mundane or bureaucratic acts of state, such as the display of maps in schools and public buildings or the action of showing a passport or filling in a form with your *national* (insurance, social security, registration or identity) number. The political geographer Joe Painter (2006: 753) has called this 'prosaic stateness', 'the mundane *practices* through which something

Plate 22.6 Identifying citizenship.

which we label "the state" becomes present in everyday life'. According to Painter (2006: 753) therefore:

> Behind each of these registration numbers, licenses and certificates are yet more documents and records held in state archives tracking employment, earnings, criminal convictions, academic performance, visits to doctors and hospitals, ownership of vehicles and landed property and numerous other features of individuals' 'private' lives. If weighed down by anxieties about the scope of the state's knowledge of us, we repair to the local pub for a drink, we will find that the state decides when and where the pub can open, the possible sizes of our serving of beer, how much of its price goes in tax . . . how our drinks are labeled . . . hygiene required in the pub kitchen and the minimum wages paid to the staff.

So perhaps nation-states exist above all as systems of actions and beliefs – an 'imagi*nation*' if you like – which must be continually re-enacted, re-narrated and re-imagined as territorial sovereign spaces in order to seem important and real to us. Just as they have been here. And as they will be next time you go to the pub, or perhaps even to a café, library or classroom instead.

Learning outcomes

Having read this chapter, you should understand:

- Nations, nationalism and states are complex historically and geographically variable phenomena.

- Nationalism is an ideology (a system of beliefs) which holds that people have a primary identity to a particular nation and that such communities should be able to express themselves in a *geographically* defined state.

- Nations can be understood as a kind of imagined community. They are imagined because not all members of a national community can know each other.

- Such national imaginaries contain a geography, a mental map of national space and its boundaries.

- Nation-states embody particular ways of governing population, territory and economy: these modes of governance are variable in space and time (the role of the state has and continues to change) – and new modes and scales (such as the supranational project of the European Union) of governance have emerged.

- National imaginaries are also gendered, for example in the idea of mothers as key transmitters of 'national culture' to the next generation.

- States can be understood as complex symbolic systems.

Further reading

Agnew, J., Mitchell, K. and Toal, G. (eds) (2003) *A Companion to Political Geography*, Blackwell, Oxford.

Cox, K., Low, M. and Robinson, J. (eds) (2008) *The Sage Handbook of Political Geography*, Sage, London.

Both these edited collections contain several chapters relevant to the material covered here – and to other chapters in this section.

Anderson, B. (1983) *Imagined Communities: Reflections on the Origin and Spread of Nationalism*, Verso, London. A revised (third) edition (2006) is available. Readable and rewarding. It is worth comparing Anderson (who stresses the modernity of nations and nationalism) with another theorist who argues that many nations do have much deeper historical roots:

Smith, A. (1988) *The Ethnic Origin of Nations*, Blackwell, Oxford.

Anderson, J., Brook, C. and Cochrane, A. (eds) (1995) *A Global World? Re-ordering Political Space*, Oxford University Press and the Open University, Oxford. If you can track down a copy of this, then Chapter 2 on 'The exaggerated death of the nation-state' remains a good place to follow up debates about the impact of globalization mentioned above in Section 22.6.

McDowell, L. and Sharp, J.P. (eds) (1997) *Space, Gender and Knowledge: Feminist Readings*, Arnold, London. Section 7 on 'Gender, nation and international relations' contains extracts of some challenging articles and lots of suggestions for further reading.

McNeill, D. (2004) *New Europe: Imagined Spaces*, Arnold, London. Contains material on European nation-states and their relationships to alternative scales of politics (such as the European Union and the municipalities/cities). A good way to follow this up is to read the set of papers in: *Journal of Comparative European Politics*, 4(2/3), (July/September 2006) – Special Issue: Rethinking European Spaces: Territory, Borders, Governance.

Seton-Watson, H. (1977) *Nations and States: An Enquiry into the Origins of Nations and the Politics of Nationalism*, Westview Press, Boulder, CO. There are hundreds of case studies of particular nationalisms. Select a nation and you will usually find many to choose from: Seton-Watson reviewed them in the mid-1970s. Although long out of print, this book can often still be found in libraries (or second-hand) or browsed online via Google books. As well as tracking down a copy of or browsing extracts of Seton-Watson online, consult:

Delanty, G. and Kumar, K. (eds) (2006) *Handbook of Nations and Nationalism*, Sage, London.

Taylor, P.J. and Flint, C. (2007) *Political Geography: World-Economy, Nation State and Locality*, Prentice Hall, Harlow. Now in its fifth edition, this text may be consulted for further ideas on most political geography topics, including a treatment of nations and nationalism.

Yuval-Davis, N. (1997) *Gender and Nation*, Sage, London. An accessible and rewarding survey of its theme.

For accounts of how borders are changing, yet sometimes attitudes towards them as somehow 'natural' persist, compare:

Amoore, L. (2006) Biometric borders: Governing mobilities in the war on terror, *Political Geography*, 25, 336–51.

Fall, J.J. (2010) Artificial states? On the enduring geographical myth of natural borders, *Political Geography*, 29, 140–7.

It is worth reading both of these in tandem with the accounts of territory in Chapter 21 and geopolitics in Chapter 20. For a history of a nationalism that links these themes, we find this particularly rewarding and thought-provoking:

Wilson, A. (2002) *The Ukrainians: Unexpected Nation*, Yale Nota Benbe, New Haven, CT and London.

For different (but equally fascinating) studies of (mostly European) state boundaries and the (re)making of national identities around them:

Agnew, J. (2007) No borders, no nation: making Greece in Macedonia, *Annals of the Association of American Geographers*, 97(2), 398–422.

Ballinger, P. (2003) *History in Exile: Memory and Identity at the Borders of the Balkans*, Princeton University Press, Princeton, NJ.

Berdahl, D. (1998) *Where the World Ended: Re-unification and Identity in the German Borderland*, University of California Press, Berkeley, CA.

Dalakoglou, D. (2010) The road: an ethnography of the Albanian–Greek cross-border motorway, *American Ethnologist*, 37, 132–49.

Kramsch, O. and Hooper, B. (2004) (eds) *Cross-Border Governance in the European Union*, Routledge, London.

Megoran, N. (2006) For ethnography in political geography: experiencing and re-imagining Ferghana Valley boundary closures, *Political Geography*, 26(10), 622–40.

Paasi, A. (1996) *Territories, Boundaries and Consciousness: The Changing Geographies of the Finnish–Russian Frontier*, Wiley, Chichester.

Pelkmans, M. (2006) *Defending the Border: Identity, Religion and Modernity in the Republic of Georgia*, Cornell University Press, Ithaca, NY and London.

It is worth comparing any of these with a short but readable book on minorities, forced migration and boundary-making in the Balkans in the aftermath of the First World War (when the modern state of Turkey emerged from the collapsing Ottoman Empire) and/or a study of the aftermath of the partition of India (into the states of India and Pakistan) as the British Indian Empire ended in 1947:

Clark, B. (2007) *Twice a Stranger: How Mass Expulsion Forged Modern Greece and Turkey*, Granta, London.

Tan, T.Y. and Kudaisya, G. (2002) *The Aftermath of Partition in South Asia*, Routledge, London.

For two recent and engrossing studies of border-making in Southeast Asia, see:

Baird, I.G. (2010) Different views of history: shades of irredentism along the Laos–Cambodia border, *Journal of Southeast Asian Studies*, 41(2), 187–213.

Harris, I. (2010) Rethinking Cambodian political discourse on territory: genealogy of the Buddhist ritual boundary (sīmā), *Journal of Southeast Asian Studies*, 41(2), 215–39.

The second one of these is particularly suggestive on how deep-seated cultural and political practices interact with modern assumptions of sovereignty to produce a situation where: 'From Independence onwards the status of the national border has grown into a grand obsession and Cambodia's political elites have, usually as a means of whipping up nationalist sentiment and the fear of being swallowed up, chanted the same basic *mantra* to the effect that the border is continually being violated by neighbouring states.' (pp. 216–17).

For a more general review of recent work on borders:

Diener, A. C. and Hagen, J. (2009) Theorizing borders in a 'borderless world': globalization, territory and identity, *Geography Compass*, 3(3), 1196–216.

For accounts of death and suffering at the external borders of the EU, see:

Van Houtum, H. (2010) Human blacklisting: the global apartheid of the EU's external border regime, *Environment and Planning D: Society and Space*, 28, 957–76.

For annotated, clickable weblinks and useful tutorials full of practical advice on how to improve your study skills, visit this book's website at **www.pearsoned.co.uk/daniels**

GLOBAL GOVERNANCE

Chapter 23

Klaus Dodds and Chih Yuan Woon

Topics covered

- Defining and understanding governance

- The relationship between governance, territory and power

- The governance regimes for oceans, polar regions and the environment

23.1 What is governance about?

The term governance, derived from the Greek verb *kubernan*, suggests notions of steering and/or piloting. Governance as a term is, on the face of it, a source of reassurance. Actions involving steering and piloting connote a sense of being in control, of predictability and stability. If we cast our minds back to April 2010 and the volcanic eruption in Iceland (see Plate 23.1), we might consider what happens when the things we (in this case international travellers like ourselves) take for granted are suddenly disrupted (Donovan and Oppenheimer 2011). People, goods, and airmail were severely 'derailed' when air travel was grounded. The very things that partially govern our lives such as international regulations regarding air travel were powerless to cope with this natural event. We and countless other people were stranded abroad. Likewise, when severe disasters such as flooding and earthquakes affect communities around the world, we gain further insights into what a lack of governance looks and feels

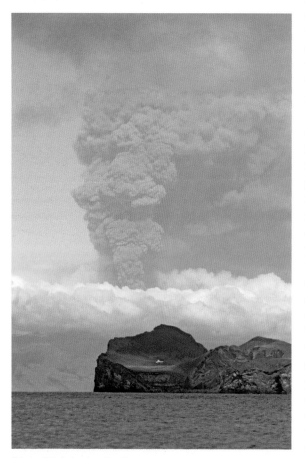

Plate 23.1 Iceland volcano.

like – it can be devastating and deadly as those seeking refuge in Australia, Europe, and North America have discovered (see Diagne 2007; de Genova and Peutz 2010; Neilson 2010).

But there are other ways in which governance manifests itself. For scholars working in fields like political geography and related disciplines such as International Relations (IR), governance refers to a system of rules, norms, codes, regulations and compliance mechanisms designed to regulate human activity, especially at the global level (Herod *et al.* 1998; Held and McGrew 2002; Sparke 2006). However, the mere identification of 'global governance' presents an important challenge to dominant strands of IR, notably realist depictions of the international arena. These stress the importance of nation-states and the absence of agreed rules between them (Sinclair 2004). Indeed, given realists stress the absence of agreed-on rules (they call this 'anarchy') and hence emphasize the role of state power as the ordering principle in the international system. For realists, the weakness of a formal world authority (such as the UN's inability to stop wars or the way that states ignore many of its rules) is a profound reminder that states, as the dominant actor, will have to rely on themselves to 'navigate' and 'steer' their way in the global arena in order to maximise what ruling elites define as the individual interests of their states (Donnelly 2000). Many of the assumptions present in geopolitics (those detailed in Chapter 20) are shared with such a perspective.

For the critics of realism, however, it is taken as axiomatic that our world is increasingly globalized, in so far as states, societies and markets are embedded and implicated with one another, across a range of geographical scales. As earlier chapters in this section have indicated, the state is complex, just one of many territorial actors and is frequently contested. It is apparent, moreover, that while the state remains the primary political actor there is not only a great deal of variation within the inter-state community but also a range of other organizations facilitating, respecting and obstructing 'global governance' (for earlier reflections on colonial governance, see Chapter 8, pp. 187–190). The list would include non-governmental organizations, inter-governmental organizations, corporations and/or criminal cartels. Likewise, it no longer seems plausible to adhere to a rigid distinction between the domestic and the foreign. As John Agnew (1994) rightly cautions, such a domestic/foreign polarity constitutes the 'territorial trap' whereby there is a neglect of the changing significance and meanings of the territorial states in different historical–geographical circumstances.

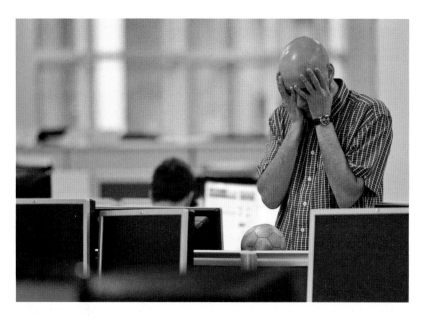

Plate 23.2 Financial crisis, 2008.

Indeed, states might well continue to enjoy formal sovereignty over their national territories, but they are also deeply affected by the contemporary transnational flows and global capital markets. The financial crisis of 2008–2010, for example (see Plate 23.2), duly demonstrated how the notion of the state as a territorial container was being challenged while ideas pertaining to the safeguarding of 'national economies' and 'national interests' were constantly (re)produced by (usually male) political leaders such as Gordon Brown and George W. Bush, desperate to convey the impression that they were in 'control' – raising interesting issues about 'manly states' (see Chapter 20, Spotlight box 20.2 and Chapter 21, pp. 455–7; see also Hooper 2001; Sidaway 2008).

We should, at this stage, make a fundamental distinction between two kinds of political–territorial world orders. First, there is an *idealized* order, in which the international system is composed of states, which enjoy *de jure* sovereignty over their national territories. This intricate link between sovereignty and territory is often traced back to the Treaty of Westphalia (1648) whereby there is an international codification of the basic principles of territorial integrity, border inviolability and the supremacy of the state. Alternatively, there is a *functional* order, which is rather more complicated and holds the de facto view that states, for example, do not enjoy complete and absolute authority over their territorial domains. This might manifest itself in all kinds of ways from an inability to impose law and order on parts of their national territory (e.g. Democratic Republic of Congo, Somalia and Afghanistan at certain historical junctures) to an inability to control flows of money,

people, goods and ideas. As John Agnew (2005; 2009) claims, sovereignty is never complete and/or absolute – rather than assuming sovereignty as fixed and naturally tied in to the nation-state and territory, attention should be given to how sovereignty can function on different geographical scales and can be shared by multiple actors (c.f. Jessop 2002; Brenner 2004).

In this chapter we explore a number of themes and issues germane to a discussion on global governance mindful of those *idealized* and *functional* senses of world order. But we will do so from some potentially unusual vantage points, namely the Polar Regions where there is no shortage of material pertaining to states and political leaders eager to perform, protect and exclude in the name of governance (see Chapter 20, p. 423). It might seem odd to go to the proverbial ends of the earth but in doing so we will be able to reflect on how governance has been understood and implemented in extreme environments. Initially, we consider why there has been an upsurge of reflection on global governance, with due consideration given to some relevant contextual and theoretical perspectives. Thereafter, the governance of the Arctic and Antarctic are used as examples of how the regulation of these apparently remote spaces has become embedded in an ever more complex matrix of states, non-governmental organizations and international governmental bodies (including the United Nations – see Spotlight box 23.1). Finally, the chapter concludes by looking at how governance questions remind us of the importance of accountability, access and authority in the contemporary world.

Spotlight box 23.1

United Nations

The United Nations (UN) has 193 member states, a figure that includes the vast majority of the world's sovereign states. Established in 1945, it was meant to succeed the League of Nations, to facilitate 'co-operation in international law, international security, economic progress, social progress, human rights and achievement of world peace' (UN website). The League of Nations (formed in the aftermath of the First World War) was an inter-governmental organization whose primary goals, as stated in its Covenant, included preventing war through collective security, disarmament and settling international disputes through negotiation and arbitration. However, the onset of the Second World War demonstrated that the League had failed in its primary purpose. There were many reasons for this failure. In particular, it has been pointed out that the structural weakness of the organization – many decisions required the unanimous consent of the entire Assembly, which made conclusive and swift action difficult – culminated in its eventual demise (see Gill 1996; Pollock 2003). The inspiration for the formation of the UN thus stemmed from the wish for a more effective organization for the managing of world affairs.

Currently based in New York, the UN has six major elements: the General Assembly (the debating chamber), the Security Council (addressing resolutions pertaining to peace and security), the Secretariat (providing support for the UN), the UN Trusteeship Council (inactive), the International Court of Justice (based in The Hague), and the Economic and Social Council (promoting international economic development and cooperation). There are also other major agencies associated with the UN, including the World Health Organization, the World Food Programme and UN Children's Fund. The current Secretary-General is the former South Korean diplomat, Ban Ki-Moon, who took over the post in 2007 (see Plate 23.3).

The organization is funded by voluntary and assessed donations from member states. The regular two-year budget of the UN and its specialized agencies are funded by assessments – the General Assembly approves the regular budget and determines how much each member state has to give. This is broadly based on the relative economic capabilities of various countries, as evaluated by their Gross National Income, taking into account adjustments for external debt and low per capita income (UN website). Given that the Assembly has established the principle that the UN should not be overly dependent on any one member to finance its operations, there is a 'ceiling' rate, setting the maximum amount a state can contribute to the budget.

Plate 23.3 UN Secretary-General.

23.2 Theorising global governance

In the last decade or so, terms like governance have become associated for many Western viewers with television reports of political leaders, global corporations and international financial organizations such as the World Bank (see Chapter 8, p. 193). G7/8/20 meetings around the world in places like Genoa, London, Seattle and Toronto further cement an impression of the rich and powerful gathering to manage the world (see Plate 23.4). In close proximity, it is common to find individuals and social movements taking to the streets of the same cities protesting against global

Plate 23.4 Recent G20 meeting.

economic and political integration, and highlighting the persistence of inequality. The June 2010 meeting of the G20 in Toronto was a timely reminder of how the governance of the world's economic system is a deeply contested affair, especially in the light of a financial crisis and public sector deficit reduction, which appear to be targeting the world's poorest and most vulnerable communities (Klein 2007).

In discussing global governance, we need to address two different constituencies. First, we should provide some contextual detail to these aforementioned activities. Why, for example, is it now so common for high-level meetings between the G7 and G20 to attract such protests, violence and even death? Second, how have theorists explored the nature of governance with reference to world order? Who, for example, benefits from good governance?

In an earlier phase, in the 1980s and 1990s, governance was a subject matter usually reserved for discussions of the global South (Sheppard and Nagar 2004; Slater 2004b). Indeed, as Williams *et al.* (2009: 3) rightly note, this period has seen the global South being constantly represented 'as a collection of place and peoples in need of external (i.e. Northern) intervention'. Under the euphemism of 'good governance', it was argued by international institutions like the World Bank and leading states such as the USA that public decision making would be better served if state bureaucracies were reduced, legal systems reformed, accountability improved, democracy and democratic institutions entrenched and market-led reforms implemented (see Chapter 8, p. 193). For the critics of good governance, these reforms were empowered by a neoliberal economic vision, which placed a premium on a reform package designed to make Southern economies more 'accessible' to foreign direct investment and corporations (Harvey 2005). As Roberts *et al.* (2003: 887) put it, the contemporary neoliberal idealism, bolstered by the 'simple master narrative about the inexorable force of economic globalization' upheld the idea that the universal extension of the virtues of openness, transparency and integration inherent in free markets will ultimately bring 'worldwide peace and prosperity'. Premised on a market-driven approach in shaping the political and economic priorities of the state, neoliberalism is unequivocally represented as a force that will lift the whole world out of poverty as more and more communities embrace and partake in the workings of the capitalist global economy (Tickell and Peck 2003). Thus, the critics complained that underlying claims to good governance was actually a disciplinary impulse (a form of governmentality, see Spotlight box 23.2) in which governments and their national populations were trained to operate in new ways, with the incentive being that they were more likely to attract additional aid and investment (Larner 2003; Watts 2003). In this sense, governance came with certain caveats and stringent conditions.

It was no accident that good governance debates and structural adjustment coincided with the ending of the Cold War and the intensification of global interdependence. To put it bluntly, Northern states (note the term, not *Western* states as we might have used during the Cold War) were still eager to control further aid packages, while at the same time take advantage of new economic opportunities in emerging markets,

Spotlight box 23.2

Governmentality

Developed by the French philosopher Michel Foucault (see Plate 23.5) in a series of essays delivered at the College de France in the late 1970s and early 1980s, governmentality addresses the ways in which subjects are governed via a series of techniques and rationalities. In so doing, it considers how governments attempt to manage their populations. Foucault himself described it as the 'art of government' and suggested that authorities might control populations in a manner whereby particular understandings of how to behave and act become simply accepted without contestation. Neoliberal governmentality, for instance, derives its power in part from the image of the absentee state and the acceptance of market mechanisms as the best way to allocate resources. However, as Peck (2004, p. 395) cautions, the *practical content* of neoliberal reform strategies is often quite 'interventionist', albeit in different ways. In his words, neoliberal imperatives are always melded with 'a range of paternalist, authoritarian, developmentalist and socio-democratic state form, together with the concomitant ascendancy of (appropriately elastic) notions of governmentality'. Hence beyond the clichés of more market/less state, the neoliberal script actually suggestively encompasses a wide range of proactive state strategies designed to refashion state–economy relations around a new constellation of elite, managerial and financial interests. In this formulation, there is a restructuring (rather than 'rolling back') of state responsibilities and spatialities – contemporary phenomena like the proliferation of NGOs, many of which are organized, financed or co-opted by national governments or international agencies, and deliver services and operationalise discourses that were once the privileged province of nation-states and their local outposts, signify that the clean lines once imagined to exist between the national state and the offshore world, or between bureaucracy and civil society, have become increasingly complicated (c.f. Ferguson and Gupta 2002; Brenner *et al.* 2003). What this means is that there is no one simple outcome to neoliberal governmentality, comprising a series of unified and fully integrated market-oriented policies; the focus is placed instead on how these 'neoliberal' processes are mediated and played out in different historical-geographical circumstances.

Plate 23.5 Michel Foucault.

especially with the demise of the Soviet Union and Cold War networks and relations (see Chapter 20). For Southern states, especially those with weak economies and limited control over their national territories, these kinds of demands were accepted as part of what was termed 'structural adjustment'. The end result was, as the IR theorist Robert Jackson (1990) claimed, to create a series of 'quasi-states' that were forced to embrace a series of World Bank and IMF policy conditions, as a consequence of their failure to govern their own populations and economic circumstances.

For instance, many Latin American nations, including Mexico, Argentina and Brazil, implemented both internal and external changes (notably economic privatization and deregulation and the reduction of trade barriers) in the 1980s so that they could qualify for loans from IMF and the World Bank to 'save' their debt-ridden economies. Furthermore, international institutions asserted further control in the late 1990s when the subject matter of corruption was addressed. Countries such as Kenya that were judged to be aberrant had funding support suspended for a period of time.

In the aftermath of the 11 September 2001 attacks on the United States, this good governance trend was supplemented by further financial and military pressures. Fearful of a repeat of 9/11, the USA led attempts by the international community to control more closely financial flows and, at the same, intervene in the global South, as these were concerns that some countries and regions might harbour terrorists and their terror networks. This concern for 'quasi-states' and 'weak states' manifested itself in the invasion of Afghanistan and Iraq and military initiatives elsewhere (e.g. South-east Asia and Africa). For example, the firm belief that the Philippines' susceptibility to transnational terrorist networks due to its government's weak control over peripheral islands with ongoing separatist insurgencies has led to the United States deploying armed forces in the South-east Asian state to counter such potential threats (Glassman 2005; Woon 2009). Thus, it can be seen that countering terrorism, promoting market-led reform and geopolitical realignment in the aftermath of the Cold War have all played their part in reshaping contemporary global governance. For the critics, including the crowds that gather at G20 meetings, these transformations have deepened market penetration and arguably reduced the capacity of countries in the global South to manage their own affairs.

Making sense of these kinds of changes has attracted a range of theoretical perspectives, and we can only briefly review the major ones. For disciplines such as International Relations (IR), a consideration of governance is never far removed from the central concern of order. For theorists such as Hedley Bull (1995), order in world politics, was epitomised by 'concerns [for] the pattern or disposition of international activity that sustains the elementary, primary or universal goals of the system and society of states'. These goals included: the preservation of the system; the maintenance of peace; and the upholding of the sovereign independence of states. For other theorists, order has also been understood as embracing both the production and preservation of stability and regular patterns of behaviour within the system of states and, as a consequence, the promotion of cooperative behaviour (Elster 1989).

This understanding of order remains influential but the nature of that order might, and indeed does, vary over time and space. For some theorists, the long-term stability of any world order depends on the role of a hegemonic party (Gilpin 1989; Cox 2001). In this conception, **hegemony**, denoting the political, economic, cultural and ideological dominance that one group, state or class has over others, is widely seen as the necessary condition for harnessing stability in the global arena. Hegemony is not simply domination, but relates to the ability to persuade others that the system as a whole is also in their interests. As such, interest is devoted to the role of the United States in the post-1945 era in shaping and maintaining a new international economic and political order based on ideas such as free trade and economic liberalization. While a hegemonic state might enjoy extraordinary military and other power capabilities, its influence lies in part in its capacity to persuade others that the prevailing system should be preserved. This approach has enjoyed renewed popularity in recent years because of interest in the growing importance of China and whether this might lead to a new hegemonic power replacing *Pax Americana*. Some have argued that the decline of American power and the rise of new challengers must be seen not simply as a transition from one hegemon to another, but a change in the very nature of hegemony, from one of states closely connected with their ruling classes to a more dispersed pattern (see Figure 23.1), whereby 'What binds these diverse regions and actors together is a shared commitment to an ideology of market economics and a growing recognition that territoriality alone is not a secure basis for economic or geopolitical power' (Agnew and Corbridge 1995: 205–7).

Others have explored world orders on the premise that stability is ensured by a so-called balance of power. This is largely based on a realist understanding of world politics, which posits that states are constantly in the process of negotiating their relations with one another, and that this occurs against a backdrop of anarchy. The international system is judged to be anarchical because there is no world government. So, as a consequence, states have to coexist with one another, sometimes uneasily. In so doing, if there is a stable world order then it is likely to be balanced by two or more states acting as restraining influences. So, according to this approach, the Cold War era was comparatively stable because the Soviet Union and the USA counterbalanced one another. What matters under this approach is a concern for power projection and the economic and military strengths of states. Of course these different forms of power and governance may coexist, albeit with the emphasis between them and their relative significance continuing to evolve.

You will note that the United Nations is not considered to be any kind of substitute for the absence of world government. Fundamentally, the UN is a collective of individual member states, which currently stands at 193. Realists tend to be sceptical about the nature and scope of the authority of this international body. This scepticism is in part rooted in the

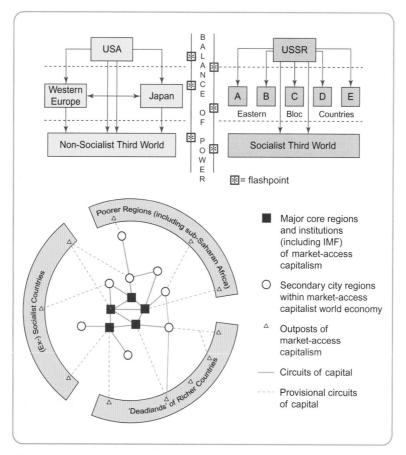

Figure 23.1 Diagrammatic representation of hegemony during (above) and after (below) the Cold War.

Source: based on Agnew and Corbridge (1995).

experiences of the pre-1945 League of Nations, which proved unable to prevent conflict in the inter-war period and the experience of the Cold War where the superpowers both routinely marginalized the UN when it did not suit their purposes. Critics of contemporary US foreign policy would also note that the Bush administration was eager to by-pass any restraints imposed by the UN General Assembly when it came to seeking authorization for the invasion of Iraq in 2003.

Finally, liberal scholars have considered how states work together to create something akin to an international society. Mindful of a lack of world government, and inspired by the writings of Immanuel Kant (Spotlight box 23.3) and his studies of 'perpetual peace', these analysts consider how governance is addressed through restraint and cooperation. How do sovereign states seek to avoid confrontation and exercise restraint? One answer is by building 'thick relationships' that emphasise mutual interdependence and universally accepted norms. For these authors, the creation of the United Nations and later regional blocs such as the European Union (EU) is emblematic of this trend of seeking to pro-

mote interdependence and cooperation. In particular, the EU has frequently been utilized as a case study to exemplify how multi-level governance has functioned to enable the formation of interlocking and collaborative networks between various nation states (Paasi 2003, 2009; Bialasiewicz *et al.* 2005). For instance, Sbargia's (2000) claim that the EU possesses a multi-level governance system is premised on the observation that overlapping competencies among multiple levels of government and the interaction of political actors across those levels exist within the regional bloc. Concurring that member state executives, while powerful, are only one set among a variety of actors in the European polity, Leitner *et al.* (2002), on the other hand, exemplify how transnational networks among cities and regions in the EU are good examples of new political relations emerging across national boundaries, but situated at subnational political scale. According to them, the 'horizontal collective action among cities and regions, forged around common agenda of mutual advantage [crosses] – and may challenge – the hierarchical relations between different scales of political governance

Spotlight box 23.3

Immanuel Kant

Immanuel Kant (see Plate 23.6) was an eighteenth-century philosopher who wrote an essay in 1795 entitled 'Perpetual peace: a philosophical sketch'. Kant outlined what he calls some 'preliminary articles' necessary for more pacific relations between states,

Plate 23.6 Immanuel Kant.

including 'No independent states, large or small, shall come under the dominion of another state by inheritance, exchange, purchase, or donation'. He then outlined what he called three definitive articles which would provide the necessary foundation for peace itself: every state should be republican, the 'law of nations' would function amongst a federation of free states and a law of world citizenship would be based on universal hospitality. Kant talked about republican not democratic states, which would be defined by representative government and that legislature will be separated from the executive branch of the government. He also does not regard that republican governments as sufficient by themselves to produce peace; rather he envisages that some kind of league of nations and freedom of emigration would be necessary to manage human affairs (Archibugi 1995; Kleingeld 2004). It is perhaps unsurprising then that observations have been made that the EU is in some ways Kantian, given how its regional structures and integrative (economic and immigration) policies are largely indicative of Kant's model of a 'federation' of states (Elden and Bialasiewicz 2006; Wolin 2010). This is contrary to the USA's (generally realist) view of the world, in which the absence of an effective world government signifies that the USA has to take the lead in maintaining peace and order in an otherwise anarchical geopolitical scene. Such a distinction, according to scholars like Balibar (2003), is most clearly seen in the post-9/11 era whereby (the majority of) Europe's attachment to international laws and norms to alleviate the 'terrorism' issue is contrasted to America's enhanced world role in this transformed security condition.

(local, regional, national and supranational)' (Leitner *et al.* 2002: 296–7). In this way, the EU can be said to mirror its precursor, the Treaty of Rome (1956), in attempting to engage the major European powers for the purpose of demarcating a more pacific era of cooperation by advancing cooperative measures in the political and economic spheres (see Chapter 20, p. 438).

Subsequent work informed by the Kantian tradition of IR have investigated further the manner in which like-minded states create international regimes for the purpose of sharing certain rules, values and norms in the apparent absence of a world authority. Indeed, the initial justification put forth by the Bush administration to invade Iraq in 2003 was premised

on the conviction that the country possessed weapons of mass destruction (WMDs). However, when these alleged WMDs were not found on Iraqi soil, the argument shifted to the notion that the Saddam regime was not democratic and abusive of human rights (see Chapter 20; see also Dalby 2003; Flint and Falah 2004). In both cases, while many members of that American administration were sceptical of the UN, they still wanted nonetheless the imprimatur of the UN Security Council. And the subsequent arguments about the legality of the US-led invasion of Iraq were rooted, on all sides, in competing arguments about international law and the scope of existing UN resolutions addressing Iraq's behaviour regarding weapons of mass destruction and human rights abuses. The debate was also mired in accusations that Iraq's substantial oil reserves were an important strategic consideration within the Bush administration (see Chapter 5; see also Le Billion and El Khatib, 2004).

Governance, in this sense, is a *shared* if imperfect project in which states (and the emphasis often remains on these actors) seek to embed and develop an international society. For Kantians, the existence of democratic/ republican states was considered to be critical to the long-term viability of such an international society not least because it was believed that democracies were less likely to wage war on one another. But in a world where not all states are democracies (and the democratic spectrum is a broad one), theorists disagree about what kinds of conditions improve stability and order.

At a systemic level, there is broad agreement that the development of multilateral institutions and institutionalized cooperation has been tremendously important in shaping contemporary global governance. Whether it is expressed through the United Nations and/or organizations such as the World Bank and IMF, states and non-state actors alike are embedded in a whole raft of obligations and restraints ranging from global human rights to international trade law and intellectual property rights. While the United States might be unrivalled as a military power, it is also deeply dependent on others for its economic and financial stability. Indeed, as noted earlier, the effects of the 'good governance' developments were to further entrench a particular global order – one that is perhaps best understood as a neoliberal one, empowered by shared values, norms and rules designed to ease global exchange. But the notion of 'shared' should not blind us to the fact this is an imposed project. Governance, as Richard Ashley (1987) once noted, is via imposition – imposing rules, silencing others and their knowledge, practices and projects.

Whatever understanding of governance we choose to adopt, it is never divorced from fundamental questions of power and knowledge. The manner in which the world appears to be organized and managed is not a natural condition of global life. The prevailing global economic order, based on neo-liberal capitalism and good governance strategies, is a human creation. The rules, norms and values selected to regulate that order are also imposed. This in turn shapes public debate about, for instance, feasible policy options. Some policies and practices appear more 'reasonable', such as cutting public spending in the current economic crisis, and those that seek to challenge those policies, such as activists, students and trade unionists, are often presented as iconoclastic and disruptive. The prevailing norms and mechanisms of governance work to promote particular agendas and geographers are increasingly mindful of how those 'structures' of governance impact, in highly differential ways, upon communities and places around the world.

Let's turn to the coldest places on earth to find some evidence of how institutions, networks and processes associated with governance shape parts of the world that are rarely considered by mainstream commentators. And in so doing, we explicitly want to move away from a sense in which the Arctic and Antarctic should be considered *exceptional*. Leaving aside their apparent remoteness and sparse population, there is plenty of evidence to be gathered about how their governance works, according to prevailing norms and values associated with the contemporary neoliberal economic and political order.

23.3 Governing the Arctic Ocean

In recent times, the changing physical state of the Arctic region (Spotlight box 23.4) has generated scores of commentaries (see Johnson 2010), and captured the

Spotlight box 23.4

Defining the Arctic

Definitions of the Arctic region vary from those based on biological, climatic, and/or geographical factors, such as the Arctic Circle, and/or political criteria. The Arctic Council's definition includes all oceans and territories north of the Arctic Circle and adjacent territories in Siberia and North America, and southern oceanic regions in the Atlantic and Bering Strait (see Figure 23.2).

All eight members of the Arctic Council are typically described as 'Arctic states', namely Canada, Denmark/ Greenland, Finland, Iceland, Norway, Russia, Sweden and the United States. Of those eight, five are described as Arctic Ocean coastal states (Canada, Denmark/ Greenland, Norway, Russia and the United States) because of their geographical proximity to the Arctic Ocean and the rights they acquire under the terms of the United Nations Convention on the Law of the Sea.

Maritime jurisdiction and boundaries in the Arctic region

	Internal waters		Norway claimed continental shelf beyond 200 nm (note 3)	———	Straight baselines
	Canada territorial sea and exclusive economic zone (EEZ)		Russia territorial sea and EEZ	———	Agreed boundary
	Potential Canada continental shelf beyond 200 nm (see note 1)		Russia claimed continental shelf beyond 200 nm (note 4)	– – –	Median line
	Denmark territorial sea and EEZ		Norway-Russia special area (note 5)		
	Denmark claimed continental shelf beyond 200 nm (note 2)		USA territorial sea and EEZ	·········	350 nm from baselines
	Potential Denmark continental shelf beyond 200 nm (note 1)		Potential USA continental shelf beyond 200 nm (note1)	·········	100 nm from 2500 m isobath (beyond 350 nm from baselines)
	Iceland EEZ		Overlapping Canada/USA EEZ		
	Iceland claimed continental shelf beyond 200 nm (note 2)		Eastern Special Area (note 7)	———	Svalbard treaty area
	Norway territorial sea and EEZ / Fishery zone (Jan Mayen)/Fishery protection zone (Svalbard)		Unclaimed or unclaimable continental shelf (note1)		

Figure 23.2 Map of the Arctic.

Source: based on map produced by the International Boundaries Research Group (IBRU) at Durham University. Original and notes are at: www.durham.ac.uk/ibru/resources/arctic

attention of states and other organizations including the European Union (EU). One event often cited as transformative in that regard was the 2007 planting of the Russian flag on the bottom of the central Arctic Ocean. Images of an underwater submersible gently dropping the titanium flagpole into place were widely circulated in national and global media networks (Dodds 2010). For Russian audiences, the flag planting exercise was widely interpreted as indicative of national prowess and technological sophistication even if the expedition itself was privately funded by a Swedish benefactor. For others, including senior Canadian political officials, the expedition and flag planting ceremony was judged to be provocative and unnecessary. While international lawyers noted that such an activity had little to no international legal significance, political scientists and journalists were penning evaluations of the resource and strategic significance of the Arctic in the post-Cold War era, while raising the spectre of un-governability (Anderson 2009; Emmerson 2010; and Chapter 5).

For our purposes, this flag planting episode brought to the fore two things: first, the changing geopolitical imagination of the Arctic and, second, the governance of the Arctic, and specifically the role of international cooperation between and beyond states. The two are closely interrelated because debates over polar governance have followed on from a sense in which the Arctic is in a state of physical and geopolitical flux (Johnson 2010). To put it simply, during the Cold War, while the Arctic was indisputably a frontline between the USA and the Soviet Union, it was a region in which the presence of thick sea ice acted as a physical barrier to movement – it was a space of 'thin governability'. The Arctic Ocean, while accessible, was not characterized by the kind of accessibility that might have been taken for granted in the other maritime spaces such as the Mediterranean. Both the Soviets and the US navies developed polar strategies predicated on the belief that each side would seek to contain the movement of their enemies' submarine fleets in particular. The dominant Cold War conceptualization of security was, therefore, based on containment (see Chapter 20, pp. 431–4). As part of this containment exercise, both sides conducted an assortment of underwater, drifting ice and aerial surveillance operations of the kind depicted in films such as *Ice Station Zebra* (1968) and *The Hunt for Red October* (1990) (see Plate 23.7). One immediate consequence of such activities was to ensure that the Arctic Ocean was a highly militarized space with little to no evidence of cooperation among the Soviet

Union and the four coastal states in particular, namely the USA, Canada, Denmark and Norway. With four out of the five Arctic Ocean coastal states being members of NATO, this sense of the Arctic as a geopolitical frontline was apparently well founded.

In the post-Cold War era, the dominant geopolitical imagination of the Arctic shifted. In part, this was due to the improvement of relations between the previously opposing sides. After Mikhail Gorbachev's noted speech in 1987 calling for the Arctic to be a 'zone of peace', concerted efforts were made to improve relations between the five Arctic Ocean coastal states and neighbouring countries such as Finland, Iceland and Sweden. One of manifestations of this rapprochement was the creation in 1996 of the Arctic Council (AC), which was designed to improve cooperation over matters of mutual concern including environmental management and shipping. Polar governance was entering a new phase but a period nonetheless dominated by states and their security-led agendas. Hence, rather than a straightforward 'rolling back' of the state, neoliberal governmentality functions through a restructuring of state's role and responsibilities – participation in a inter-state forum whereby negotiations of positions and agendas are made *vis-à-vis* other relevant actors.

The AC is an intergovernmental organization. It is intended to aid and abet Arctic states in areas of common interest. Notably, all parties agreed that any discussions involving military and security matters would be avoided even though the Cold War was considered to be finished. On a more progressive note, however, indigenous peoples and organizations such as the Sami Council were invited to be permanent participants of the Arctic Council. Underlying the creation of the Arctic Council was a sense in which the members involved recognized that the logic of containment was neither sufficient nor desirable in the changed geopolitical circumstances. Instead the Arctic has been understood as a weakly governed space needing effective spatial administration.

Apart from the apparently obvious point that the USA and Soviet Union were no longer threatening one another (with the Arctic as a major theatre of operations), a key driver of change has been the spectre of climate change (Johnson 2010). While serving as an indicator of global environmental change, melting sea ice was widely identified as indicative of greater accessibility in a political and material sense. Politically, the Arctic Ocean has attracted more interest from national governments, corporations, scientific communities,

Plate 23.7 Posters from the films *Hunt for Red October* (1990) and *Ice Station Zebra* (1968).

indigenous peoples and media networks. The 2007 flag planting episode, coinciding as it did with satellite images of an apparently open North West Passage, encouraged a maelstrom of media speculation about an apparent 'scramble for the pole'. Materially, a more accessible Arctic Ocean raises the spectre of further resource extraction in areas such as oil and gas (see Plate 23.8), alongside other activities such as fishing, tourism and trans-polar shipping routes, with their apparent promise of cutting thousands of nautical miles off the movement of freight from East Asia to Europe (Powell 2008).

This interest in spatial administration manifests itself in the interest shown by political actors including the European Union (EU), in examining a variety of governance options. At present, Arctic Ocean governance is shaped by three key factors – the Arctic Council, the Law of the Sea and a host of regional regimes such as fisheries agreements. The Arctic Council (AC) is a soft-law institution, meaning that it is largely advisory and lacking in a permanent secretariat, for example (McIver 1997; Bloom 1999). It does not have any legal competence. It cannot demand that member states restrain from, for example, drilling offshore for oil and gas. Organizationally, the Arctic Council has attached to its permanent membership so-called observer states such as Britain, and ad hoc observer states such as China. These 'observer states' cannot shape the agenda of the annual meetings but they are allowed to attend a certain amount of the formal business of the member states. Permanent participants such as indigenous peoples' groups (e.g. Inuit Circumpolar Council, Sami Council, Aleut International Association) also attend the meetings of the AC but do not have voting rights.

Since its creation in 1996, the AC arguably demonstrates something that is considered axiomatic by realist researchers, namely that states tend to veer towards self-interested behaviour. One of the central reasons why the AC is a soft-law institution is that the five Arctic Ocean coastal states in particular were reluctant to agree to a more substantial treaty-based international regime, as found in Antarctica (see below). Notwithstanding concerns about the growing accessibility of the Arctic Ocean, and the mounting interest from parties geographically remote from the region, the states concerned have sought to improve and indeed strengthen cooperation with one another in specific areas such as shipping, search and rescue, and environmental cooperation.

The United States and Russia have also been reluctant to strengthen the remit of the AC. Both countries have been reluctant to create a permanent secretariat, with increased funds, or to encourage the AC to become a more formal intergovernmental organization. However, smaller states such as Norway, Denmark and Sweden have shown a willingness to strengthen the AC. Between 2006–2012, the three Nordic countries have agreed to create a semi-permanent secretariat in Tromsø. Paradoxically, the reluctance of the two largest Arctic coastal states (in terms of military and political capabilities at least) has not dampened interest in the AC as the premier mechanism for inter-Arctic cooperation. One obvious manifestation is the publicly

Plate 23.8 Oil extraction in the Arctic.

stated desire for others such as the EU Commission and China to play a more active role in its proceedings.

The second major element in the governance of the Arctic Ocean lies in the provisions of the Law of the Sea Convention (LOSC). For the five Arctic Ocean coastal states, the LOSC regulates large sways of the Arctic Ocean by acknowledging a series of jurisdictions – ranging from the 12 nautical mile wide territorial sea, the 24 nautical mile wide contiguous zone and a 200 nautical mile wide exclusive economic zone in which coastal states enjoy sovereign rights when it comes to the exploitation of resources (Steinberg 2001; Dodds 2010). As Article 56 notes:

In the exclusive economic zone, the coastal State has:

(a) sovereign rights for the purpose of exploring and exploiting, conserving and managing the natural resources, whether living or non-living, of the waters superjacent to the seabed and of the seabed and its subsoil, and with regard to other activities for the economic exploitation and exploration of the zone, such as the production of energy from the water, currents and winds . . .

Moreover, under Articles 76 and 77, coastal states can seek to extend their sovereign rights over extended continental shelves (ECS). Interested parties, including all the Arctic Ocean coastal states barring the USA (which has not acceded to the LOSC), have (or will have) submitted geological and oceanographic materials to the Commission on the Limits of the Continental Shelf (CLCS) based within the United Nations in New York. If accepted by the CLCS, coastal states are allowed to identify ECS and extend their sovereign rights over potentially thousands of square miles of seabed. The implications for the Arctic Ocean are considerable – it is likely that the five coastal states will collectively enjoy sovereign rights to explore and exploit the natural resources of virtually the entire maritime region.

So the notion that there is a wild 'scramble for resources' in the Arctic Ocean is misplaced in one sense. There is no 'scramble' because there is no need to 'scramble'. In 2008, the five Arctic Ocean coastal states in the so-called Ilulissat Declaration reaffirmed their commitment to respect the Law of the Sea as the primary mechanism for ensuring cooperation in the Arctic Ocean and thus rejected calls for any kind of Arctic Treaty. Non-coastal states, under the terms of LOSC, enjoy rights of innocent passage in these particular maritime zones, subject to some regulatory rights of coastal states. Innocent passage, under Article 19 of the LOSC, is defined as:

Passage is innocent so long as it is not prejudicial to the peace, good order or security of the coastal State. Such passage shall take place in conformity with this Convention and with other rules of international law.

This right to innocent passage, alongside a growing commercial interest in trans-polar shipping routes, in large part explains the interests of the EU and China, for example. The EU Commission in 2008 outlined the EU's interests in the Arctic and the European Parliament called for the EU Commission to develop an Arctic policy in 2009. Shipping and fishing have been identified as core interests, and Denmark, Finland and Sweden are EU member states, and, at one stage, it was even possible that Iceland might have emerged as a fourth EU member state with distinct Arctic interests.

The third element shaping contemporary Arctic Ocean governance is a patchwork of national and regional fisheries management organizations such as the North East Atlantic Fisheries Commission, the North Atlantic Salmon Conservation Organization, the North Atlantic Conference and the North Atlantic Fisheries Ministers' Conference. With the 'opening up' of the Arctic Ocean in the coming decades, as a consequence of sea ice thinning, a number of Arctic states and others including the EU Commission have called for further multilateral mechanisms to be developed for the purpose of managing Arctic fisheries. One suggestion in particular has been to create an Arctic Regional Fisheries Management Organization for the purpose of coordinating such efforts across the Arctic Ocean.

Looking ahead, the future governance of the Arctic Ocean is likely to be largely determined by, on the one hand, the continued existence of the AC and, on the other hand, by the provisions of the LOSC. Neither Russia nor the United States appears willing to act in hegemonic ways. They and smaller Arctic states such as Norway and Denmark/Greenland benefit from this 'pooling' arrangement. These states have effectively agreed to abide by particular rules, customs, values and norms, while at the same developing their own national strategies for 'their' part of the Arctic. They have also been emphatic in their rejection of a proposal by the EU Commission in 2008 to create an Arctic Treaty, along the lines of the 1959 Antarctic Treaty. One reason for this initiative was to encourage the Arctic states to collaborate more closely with extra-territorial actors such as the EU and to work towards transforming the Arctic Ocean into a zone of peace and cooperation. For the five Arctic Ocean coastal states there is little or no apparent incentive to agree to any further treaty development. Each, in their different way, is eager to protect its sovereign rights in the Arctic Ocean without neglecting to cooperate in areas of mutual interest.

Contemporary governance in the Arctic Ocean reminds us of two things. First, states do tend, as realists would note, to articulate their interests in the Arctic in national terms. An emphasis on national security, resource control and maritime surveillance is common to all the major policy pronouncements of the five coastal states (e.g. Canada's Northern Strategy of 2007). Second, however, states and non-states are able to cooperate with one another and to create regimes and mechanisms (albeit with limited powers) to encourage information exchange, cooperation and negotiation. Indicative of neoliberal governmentality, the establishment of international norms and conventions help to arbitrate the different interests of relevant parties so that issues pertaining to the Arctic can be debated and resolved in a peaceful and orderly manner. Third, indigenous groups and organizations have become more active in demanding greater autonomy in terms of regional governance and future resource exploitation. In Greenland, for example, the Danish government agreed, after a referendum in 2008, to further autonomy and the awarding of resource rights to the population there. So oil and gas exploration off the coast of Greenland in the future might provide a new revenue stream for its eventual independence from Denmark, as might aluminium smelting – but this is controversial, with critics fearing that Greenland will become dependent on multinational corporations and their strategies, not to mention the whimsical nature of global markets (Nuttall 2008). Moreover, indigenous groups help to unite the native peoples who may be artificially divided by national territorial borders so that a platform is available for them to articulate their concerns. A good case in point is the Inuit Circumpolar Conference whereby Inuit living in the USA, Russia and Canada often come together to actively elicit responsibility on the part of the Arctic states to deal with the issue of climate change so that a sustainable living environment can be made available to the indigenous communities (Osofsky 2006, 2007).

23.4 Governing the frozen continent

The Antarctic is governed by a rather different politico-legal regime when compared to the Arctic (Dodds 1997). The physical geography of the Antarctic continent explains a great deal; it is a remote continent, far removed from major centres of population and lacking an indigenous human population (see Figure 20.2). Largely isolated from human affairs until the nineteenth century, it was only in the last century that Antarctica was partially settled through whaling stations, military outposts and scientific bases. Economically, the living resources polar continent and surrounding islands such as South Georgia have assumed considerable importance, especially in the field of whaling (Headland 1984). Products such as whaling oil were exported to European and North American markets and used for heating and in products including candle wax and margarine (Beck 1986). Later, in the 1950s and 1960s, commercial fishing occurred in the Southern Ocean and was dominated by Soviet and East European fishing fleets.

These economic activities occurred against a backdrop of political uncertainty. Between 1908 and 1940, seven states including the United Kingdom pressed claims to the Antarctic continent. Three claims (Argentina, Chile and the UK) overlapped with one another in the Antarctic Peninsula and the four others (Australia, France, New Zealand and Norway) occupied about four-fifths of the polar continent. One sector, the so-called Pacific Ocean segment, was never claimed by any state (see Figure 20.2, p. 424). In the aftermath of the Second World War, there were widely expressed fears that discord might arise, and the USA and the Soviet Union were mindful of the future status of the continent. Both superpowers, however, publicly rejected the seven claims and reserved the right to press a future claim to Antarctica. A sovereignty stalemate ensued. The seven countries were reluctant to drop their claims, and the Cold War superpowers were adamant about their position (Dodds 2002).

In the 1950s, the future shape of Antarctic governance was unclear. The United States had proposed a condominium composed of eight countries (the USA and the seven claimants) excluding the Soviet Union. This was rejected as unacceptable by the Soviets, who made it clear that they would not be debarred from any new polar order. What made it possible to proceed was something quite different, namely international science. The 1957–8 International Geophysical Year provided a *modus operandi* for polar cooperation: 12 countries committed to carrying out scientific research for 18 months and all parties signed up to the principle of open and free access to the polar continent. On the basis of this apparent success, the United States invited those 12 parties to attend a conference to discuss the future of Antarctica in October 1959 (see Chapter 21, p. 423).

Looking back on events some 50-plus years later, this conference was the polar cipher of the Bretton Woods agreement, where 44 nations gathered in 1944 to set up a system of rules, institutions (such as the World Bank and the International Monetary Fund) and procedures to regulate and govern the international monetary system. Admittedly smaller in scale, it nonetheless produced an Antarctic Treaty, which resolutely continues to shape the governance of this continent and surrounding ocean. As with Bretton Woods, the USA was pivotal in shaping the ethos of the treaty, especially in insisting upon 'access' and 'freedom of investigation' and thus overriding the territorial claims of others. And due to the relative isolation of Antarctica, it was possible to invite only 12 parties to the negotiating table, with only minimal opposition from other states (India raised the 'question of Antarctica' in the United Nations in 1956 and 1958). It would clearly be unthinkable now to have such a limited membership given the changing circumstances. But back in the late 1950s, Antarctica was considered marginal to world affairs.

After six weeks of negotiation, the Antarctic Treaty was signed and all parties agreed to the following: the region should be a zone of peace and cooperation; all claims to the continent are considered suspended for the duration of the Treaty and the rights of non-claimants are preserved; scientific cooperation should be encouraged; parties have the right to inspect one another's bases; and new parties can accede to the Treaty and its commitments provided they demonstrate their scientific and political credentials. The Treaty secured the support of all parties because the superpowers were able to maintain their policy of non-recognition of sovereignty claims while safe in the knowledge that both parties agreed to the non-militarization of the Antarctic. In addition, the seven claimant states could accede to the Treaty because no one was asking them to give up their claims and the others acknowledged the three other minor parties – Belgium, Japan and South Africa. For Japan as a defeated wartime state and for South Africa as an *apartheid* state, this was not insubstantial.

As a governance regime, the Antarctic Treaty is frequently trumpeted as a success story. Why has it endured? We think there are three fundamental reasons apart from all the original parties getting something tangible from the 1959 negotiations. First, the 12 parties produced an attractive creation myth: they used science and scientists to portray themselves as political visionaries seeking to introduce peace and harmony to a remote continent. While the experiences of the IGY and science in general were important, this was an emollient for a treaty created out of claims and counter-claims. The Treaty could not settle the disputed ownership issue because it was not solvable. But to the excluded wider world, it was important to 'sell' the Antarctic Treaty as progressive and attuned to the principles embodied in the United Nations Charter.

Second, the Treaty endured because there were provisions (and opportunities taken) to not only accommodate new members but also deal with issues as they arose, such as conservation, environmental protection and living resource exploitation. The Treaty with its associated conventions and protocols proved flexible. Institutionally, the parties agreed to create in 2005 an Antarctic Treaty secretariat based in Buenos Aires in order to better coordinate the activities of all parties. Even at moments of greatest political tension, such as the prospect of potential mineral exploitation in the 1980s, the Antarctic Treaty parties proved sufficiently adept at neutralising opposition as voiced in the United Nations General Assembly. They did so by inviting some of the most powerful critics such as India and China to join the Treaty and by signing a Protocol on Environmental Protection, which banned all forms of mining in the Antarctic Treaty region (defined as below of 60 degrees south latitude). The earliest any review conference can be called is 2048.

Third, the Treaty endured because the provisions of Article IV were largely respected. Article IV dealt with the potentially explosive issue of sovereignty claims. Getting agreement on this issue was helped, paradoxically perhaps, by the existence of the Cold War. It was clear that the Soviet Union and the United States did not want to make the Antarctic another area of strain given prevailing tensions over the Arctic Ocean. It also helped that no one was contemplating mining and resource extraction in the late 1950s. If anything, parties were more concerned about the frozen continent becoming a nuclear testing ground. So Article IV suited all concerned and, as noted earlier, there were only 12 parties to negotiate with. Compare such a situation to contemporary climate change summits and conferences where it is common to deal with over a hundred governments and political leaders and, something that did not exist in the 1950s, 24-hour news reporting.

But this is now changing. It has been pointed out that the Antarctic Treaty is facing a potential crisis in governance terms (Dodds 2011). One of the most important changes compared to 1959 are the range of activities now affecting the Antarctic. Commercial interests has brought into the Antarctic a phalanx of non-state actors engaged in tourism, biological prospecting and marine harvesting. Tellingly, there are now groups like the Coalition of Legal Toothfish Operators and the International Association of Antarctic Tour Operators involved in Antarctic affairs whereas in 1959 there were only states and their scientists alongside an advisory Scientific Committee for Antarctic Research. The hegemonic role of science and scientific interests within Antarctic Treaty system has disappeared. Joining those commercial operators over the last two decades or so has been an array of non-governmental organizations, especially in the field of environmental matters.

The scale and pace of change is noteworthy. Tourism, mainly ship-based, now attracts typically 30,000 people per year and these parties tend to visit a limited number of sites in the Antarctic Peninsula region. While tour operators and port states such as Argentina and Chile have benefited from this development, there are fears that the region is not well placed to cope with accidents such as the sinking of vessels. This happened in November 2007 when the vessel *Explorer* sank and 150 people had to be rescued. The polar tourism operators largely manage their own activities and while they liaise closely with the Antarctic Treaty parties, it is striking how the tourist market has grown, with larger cruise ships (sometimes carrying over 1000 passengers) searching for established and new sites to visit in the region. Controlling these kinds of commercial activities in a region where sovereignty is undecided is a difficult affair. A ship registered in a flag state such as Liberia, carrying a multinational crew and passenger list, traversing polar waters which are nominally claimed by Argentina, Chile and the UK, represents a jurisdictional challenge. Furthermore, there have been concerns that the increasing tourism numbers will add further stress to the already fragile ecosystem of the Antarctica – businesses have shown interest in

establishing various forms of tourism infrastructures including hotels and casinos to cash in on the tourism potential of the region. This makes it even more imperative that mechanisms and regimes are put into place to effectively deal with such contemporary circumstances.

So one area of major change is the range and role of other actors (i.e. non-states) in Antarctic affairs. Another notable alteration is in what we might term political and legal embroilment. In 1959, the Antarctic Treaty was the lodestar, which the parties followed, and indeed such was the state of affairs that they initially agreed to meet only once every two years. Fifty years later, the Antarctic is embedded in a raft of international conventions and treaties including the Law of the Sea Convention (LOSC). The Antarctic Treaty parties now meet annually and it is apparent that there are all kinds of overlapping commitments and obligations to be managed. One example involving the LOSC will provide a good example of how polar governance is now a more complex affair.

Under Article 76, as we noted in the case of the Arctic Ocean, coastal states are entitled to certain rights over the continental shelf beyond 200 nautical miles, if they can demonstrate its extent via a submission to the Commission on the Limits of the Continental Shelf (CLCS). While this might seem comparatively unproblematic in the case of the Arctic, it is a fraught issue in the Antarctic. Can there actually be something akin to a coastal state? If we understand the category of 'coastal state' as both a geographical category and legal-political creation then it becomes clearer. The seven claimant states consider themselves coastal states as understood under the terms of Article 76 and are seeking to create, perform and indeed reserve their rights. But under Article IV of the Antarctic Treaty, all parties agreed to suspend their territorial claims and counter-claims for the duration of the Treaty? So we have an apparent clash between the provisions of the LOSC and the Antarctic Treaty. It is worth recalling that when the Treaty was signed in 1959, the LOSC had not even been negotiated even though there was the 1958 Continental Shelf Convention in existence.

The end result over the last decade has been a delicate balancing act with claimant states making it clear that they consider themselves Antarctic coastal states while being careful to acknowledge the existence of the Antarctic Treaty. The Australians, for example, made a submission to the CLCS in November 2004 but asked

Plate 23.9 Indigenous inhabitants meet visitors to the Southern Ocean.

the UN body not to consider their materials pertaining to the Australian Antarctic Territory. Non-coastal states such as the USA and Russia have reiterated their non-recognition of territorial claims (and hence coastal state status). Such is the sensitivity of this issue, with attendant claims to sovereign rights over thousands of square miles of polar seabed, that the Antarctic Treaty parties have never formally discussed this in their final reports of their annual Antarctic Treaty Consultative Meetings.

What do we learn from the Antarctic in terms of governance? We think we can point to several things, all of which have geographical ramifications. First, the Antarctic Treaty endured because of the Antarctic's comparative remoteness from world affairs in the 1950s. By the time, the polar continent became more globalized, it had had 20 years to strengthen its regime dynamics and to socialize new members in terms of norms, values and practices. But since the early 1990s, the Treaty parties have been reluctant to regulate further fishing and tourism in particular. Second, the Treaty worked because it did not try to resolve the potentially divisive issue of ownership. Article IV was intended as part of a management strategy, seeking to keep all 12 parties in line. Third, the commercial development of Antarctica is arguably revealing the inadequacies of a governance structure that is still, notwithstanding all the changes, largely predicated on the dominant role of states and their organizations, mainly scientific. Tourism, fishing and biological prospecting (harvesting organisms for commercial gain) are all growing in scope and intensity, and some issues such

as illegal fishing in the Southern Ocean are revealing how difficult it is to manage activities far removed from individual countries (see Plate 23.9). Fourth, new legal regimes and conventions are creating ever-greater overlapping commitments and obligations. They are exposing tensions most notably over the sovereign rights of coastal states and other parties.

23.5 Conclusion

In both Polar Regions, the governance regimes are changing in large part because of their further integration into the contemporary neoliberal world order. But this is throwing up tensions between states and non-state organizations. In the Arctic, a limited number of coastal states are attempting to cement their sovereign rights, while others such as China are eager to 'open up' the Arctic Ocean to trans-polar shipping routes and resource exploitation. In the Antarctic, commercial activities such as fishing and tourism are placing renewed pressure on the Antarctic Treaty System and the delicate balance between claimant and non-claimant states alike. International law and protocols are emerging as increasingly important mechanisms for ensuring that these potential flashpoints are addressed in an orderly manner. And by orderly, we mean not only in a non-violent manner but also in ways that minimize disruptions to commercial activities and investment. Following Foucault, this is in line with neoliberal governmentality whereby international conventions help to arbitrate the different positions and interests of states so that consensual partnerships and collaborative efforts can be harnessed to deal with various issues pertaining to the Polar Region. Without suggesting the determinacy of such an outcome, sovereign states in the Arctic have also been eager to remind international audiences that they are determined to protect their interests, militarily if necessary.

But debates about governance should not be restricted to states alone and their sovereign rights. One danger of focusing on world order and global governance is that we give states a kind of privilege they don't merit. States and their claims to sovereignty don't just exist – they are actively produced. Likewise claims to governance are just claims – they will have to be constantly negotiated with other state and non-state actors in the global arena. This is especially pertinent given that many contemporary issues can no longer be confined within the limits of national territorial boundaries nor can they be adequately dealt with by a singular state authority. A good case in point is the regulation of 'global commons' (e.g. oceans and the atmosphere) whereby the role of the international community is essential in collectively governing the resources that are vital to the sustenance of life. Alternatively, concerns have also been raised with regards to the trans-national impacts arising from increased mobilities of people, goods and corporations in this globalizing era. It is then hardly surprising that inter-governmental organizations such as the UN and EU have functioned as governance regimes in facilitating dialogues and meaningful cooperation to tackle these 'problems'. For instance, the cross-border danger posed by infectious diseases (e.g. SARS and bird flu) and terrorism in recent years has led to an enhanced role by UN and other regional blocs such as the EU and Association of the South East Asian Nations (ASEAN) in coordinating joint efforts to deal with such imminent health threats. Not withstanding the contributions from these supranational entities, the increasing active involvement of NGOs in issues of governance also merits attention. In particular, these groups act as vigilant 'watchdogs' to scrutinize the activities and actions of states on the global stage. A good example would be the oppositional politics enacted by civil societies including Greenpeace to critique the lacklustre commitments of many developed nations in dealing with global warming/climate change. In this sense, governance is never emblematic of a natural state of affairs. It is a human intervention (however complex) that involves multiple actors, each with their own agendas. This makes governance a highly contested endeavour and closely bound up with issues of power – it is not only capable of being supported but can also be resisted, rejected or even overthrown (see Routledge 1996; Sharp *et al.* 2000 for ideas on power and resistance).

As the previous chapters reminded us, therefore, we are talking here in large part about how we, as human beings, choose to organize our affairs. Ideas about space, place, knowledge, power, international relations, governance and order are critical here and embedded within human cultures. Those ideas are then put to work and continue to inform and influence the ways in which we understand the world around us including our relationship with others. All of this is, of course, profoundly geographical.

Learning outcomes

Having read this chapter, you should understand:

■ The term governance and the manner in which it has been understood within geography and related disciplines such as International Relations.

■ The term international order and how we might make a distinction between functional and idealised orders.

■ That liberal and realist writers have different understandings of both governance and international order.

■ In the 1980s and 1990s governance was strongly associated with reforming governments in the global South, especially in the light of concerns over debt and corruption.

■ The driving logic behind this move towards 'good governance' was a form of neoliberal governmentality.

■ Governmentality is a term developed by Michel Foucault and addresses the 'art of government' and the way in which populations are organized and managed.

■ The Arctic is undergoing a fundamental 'state change' and provides an interesting example of evolving governance.

■ The changing physical properties of the Arctic (e.g. melting sea ice) are provoking different debates about future governance.

- The Antarctic has been governed by the 1959 Antarctic Treaty, which established the region as a zone of peace and international scientific cooperation.

- The growing commercialization of the Antarctic is leading to new fears that the existing governance arrangements are inadequate.

- In both the Arctic and the Antarctic, states are seeking to maximize their own advantages (as realists would expect) while seeking to develop collective governance structures and arrangements (as liberals would hope).

- Future governance will depend on future governmentalities and how ideas about space, power, knowledge and order are mobilized. These relate to changing forms and patterns of hegemony.

Further reading

Angew, J., Mitchell, K. and Toal, G. (eds) (2003) *A Companion to Political Geography,* Blackwell, Oxford. This provides a useful overview of political geography, with specific chapters relevant to governance such as territory and place – it is relevant to all the chapters in this section.

Agnew, J (2005) *Hegemony: The New Shape of Global Power,* Temple University Press, Philadelphia, PA. A very good overview of how the international system has evolved, especially in the post-1945 era.

See too this paper for an argument for the importance of thinking about power and authority beyond the conventional focus of many on the exclusive role of states:

Agnew, J. (1994) The territorial trap: the geographical assumptions of International Relations theory, *Review of International Political Economy,* 1(1), 1 53–80.

Fifteen years on from the original publication, a set of essays published in *Geopolitics* revisited Agnew's arguments. As the introduction to that set notes:

> Attempts to anchor states (both heuristically and practically) within readily definable territorial boundaries tend to obscure the intensity of the very trans-border flows which themselves secure the functioning of those states within their claimed territorial unit.
>
> (Reid-Henry, 2010: 753)

See: **Reid-Henry, S.** (2010) The territorial trap fifteen years on, *Geopolitics,* 15(4), 752–6, as well as other essays in the special section of *Geopolitics* that this introduces.

Anderson, A. (2009) *After the Ice: Life, Death and Politics in the New Arctic,* Virgin Books, London. Written by a former editor of *New Scientist,* this book explores likely changes to the Arctic in the next 20–30 years.

Beck, P. (1986) *The International Politics of the Antarctic,* St Martin's Press, London. A very useful guide to the political history of the Antarctic, especially in the first half of the twentieth century.

Borgerson, S. (2007) Arctic meltdown, *Foreign Affairs,* 87, 63–77. An influential essay, which set the tone of much of the alarmist debate about potential changes to the Arctic.

Cox, K., Low, M. and Robinson, J. (eds) (2008) *The Sage Handbook of Political Geography,* Sage, London. An accessible guide to the key concepts of political geography, including a section on governance and public policy.

Crampton, J. and Elden, S. (eds) (2007) *Space, Knowledge and Power: Foucault and Geography,* Ashgate, Aldershot. A compendium that highlights geographical readings of Foucault, with notions such as governmentality, power, knowledge and discourses duly interrogated.

Dodds, K. (2002) *Pink Ice: Britain and the South Atlantic Empire,* I B Tauris, London. Addresses the intrigue behind the creation of the 1959 Antarctic Treaty.

Emmerson, C. (2010) *The Future History of the Arctic,* Bodley Head, London. An accomplished account of the geopolitical dynamics at play in the contemporary Arctic.

Gallagher, C., Dahlman, C., Gilmartin, M., Mountz, A. and Shirley, P. (2009) *Key Concepts in Political Geography,* Sage, London. Useful chapters on governance and hegemony alongside others including neoliberalism and superpower.

Glassman, J. (2001), Structural adjustment in East and Southeast Asia: lessons from Latin America, *Geoforum,* 32, 77–90. A good article that addresses the consequences of structural adjustment and so-called 'good governance' policies on the continents of Asia and Latin America.

Herod, A., Ó Tuathail, G. and Roberts, S. (eds) (1998) *An Unruly World: Globalization, Governance and Geography,* Routledge, London. A classic edited volume that highlights the changing scales and meanings of governance in the context of globalization and the implications of such developments on a plethora of actors.

Jackson, R. (1990) *Quasi-states: Sovereignty, International Relations and the Third World,* Cambridge University Press, Cambridge. An important examination of the 'quasi states' and accompanying recognition that sovereignty is rarely absolute. This covers some of the ground that Chapter 22 here also considers.

Klein, N. (2007) *The Shock Doctrine: The Rise of Disaster Capitalism,* Metropolitan Books, New York. An argument about how economic and natural disasters are used to reshape states and societies in ways that favour the powerful.

Johnson, L. (2010) The fearful symmetry of Arctic climate change: accumulation by degradation, *Environment and Planning D: Society and Space*, 28(5), 828–47. A challenging paper on business and geopolitics in the Arctic.

Kjaer, A. (2004) *Governance: A Key Concept,* Sage, London. A short but accessible guide to some of the different meanings associated with the term governance.

Powell, R. (2008) Configuring an 'Arctic Commons'?, *Political Geography,* 27, 827–32. A short intervention, which sets out some of the competing arguments relating to the hydrocarbon potential of the Arctic.

Roberts, S., Secor, A. and Sparke, M. (2003) Neoliberal geopolitics, *Antipode,* 35, 886–97. An intervention piece that examines how discourses of neoliberalism have been closely imbricated with violence and militarism in contemporary geopolitics.

Rose-Redwood, R. (2008), Governmentality, geography and the geo-coded world, *Progress in Human Geography,* 30, 469–86. A useful paper that summarizes how Foucault's notion of governmentality has been engaged by geographers.

Sidaway, J. (2003) Sovereign excess? Portraying post-colonial sovereigntyscapes, *Political Geography,* 22, 157–78. A paper that relies on a post-colonial lens to rethink issues of sovereignty and the territorial state, with particular focus on the governance-related discourses pertaining to the supposedly 'weak' or 'failed' states in Africa.

Sinclair, T. (2004) *Governance: Critical Concepts in Political Science*, Routledge, London. A very clearly written account of the concept of governance by a political scientist.

Smith, B. (2007) *Good Governance and Development,* Macmillan, Basingstoke. A useful and clear account of the good governance agenda especially as it relates to the global South.

Sparke, M. (2006) Political geography: political geographies of globalization (2): governance, *Progress in Human Geography,* 30, 1–16. A good review article on how governance has been conceptualized within political geography, with particular focus on issues of neoliberalism, power and globalization.

Stokke, O. and Vidas, D. (eds) (1997) *Governing the Antarctic,* Cambridge University Press, Cambridge. This is an excellent collection of essays exploring the evolution of the Antarctic Treaty System.

Taylor, P.J. (2005) New political geographies: global civil society and global governance through world city networks, *Political Geography,* 24, 703–30. An analysis of the intersections between a variety of networks; state agencies, NGOs and cities.

Whitman, J. (ed.) (2009) *Palgrave Advances in Global Governance,* Palgrave, Basingstoke. An edited book which explores the different ways of understanding global governance depending in part on whether one addresses the role of the state, economic sectors, international organizations and non-state actors.

Useful websites

www.un.org The official website of the United Nations.

europa.eu The official website of the European Union.

www.lse.ac.uk/Depts/global LSE Global Governance.

www.globalgovernancewatch.org The right-wing American Enterprise Institute runs a Global Governance Watch.

www.globalgovernance.eu The home of the Global Governance Institute, an independent think-tank based in Brussels

www.igu-gog.org International Geographical Union Commission on Geography of Governance.

www.eui.eu/Projects/GGP/Home.aspx European University Institute Global Governance Programme.

www.arctic-council.org The home of the Arctic Council.

www.ats.aq/indexe.htm The Antarctic Treaty System and Secretariat.

www.tni.org The Transnational Institute.

For annotated, clickable weblinks and useful tutorials full of practical advice on how to improve your study skills, visit this book's website at **www.pearsoned.co.uk/daniels**

Glossary

A

AAA referring to the 'Anglo/American/Australasian' tradition in geography, which is sometimes also referred to (erroneously) as the Western tradition, where the latter should include the geography of scholars in Europe.

Abjection the process by which we seek to repress that which we regard as unclean, improper, impure or dangerous. In geographic terms, abject material tends to be located in marginal spaces which become repositories for those things which are regarded as threatening the social body.

Actor-network human actors in interaction through non-human intermediaries and power relations.

Agglomeration the concentration of productive activities in a particular region.

Agribusiness large-scale, capital-intensive, agricultural businesses incorporating supply, production and processing capacities.

Agri-food system the highly integrated system of agricultural production which involves both upstream (e.g. suppliers) and downstream (e.g. processing) industries.

Alternative food networks a reaction to conventional (e.g. supermarket-dominated) food chains, in which local and/or organic foods are marketed through alternative outlets such as farmers' markets, box schemes and home deliveries.

Anthropocentric a way of understanding the value of the environment which prioritizes humans as having the most value. In other words the environment is only useful as a resource for humans. In this approach humans do not have any responsibility to ensure the environmental sustainability of our actions.

Apartheid the policy of spatial separation on racial grounds employed in South Africa under National Party rule between the late 1940s and early 1990s.

Appropriationism the replacement of agricultural inputs with industrial alternatives. It forms a key process in the industrialization of the agri-food system.

B

Balance of payments (BOP) the difference in value between a country's inward and outward payments for goods, services and other transactions.

Balkanization progressive subdivision of a region into small political units.

Bands societies of hunter–gatherers typically numbering up to 500 individuals. See Spotlight box 1.1.

Billion thousand million.

Biodiversity the variability among living organisms from all sources including terrestrial, marine and other aquatic ecosystems and the ecological complexes of which they are part; this includes diversity within species, between species and of ecosystems.

Biomass plant and animal residue burnt to produce heat.

Biosphere that part of the earth's surface, extending from the uppermost part of the earth's rocky core (lithosphere or geosphere) to the lower atmosphere, including the hydrosphere. The biosphere corresponds to and sustains the domain of life.

Biotechnology a branch of technology concerned with the industrial production of living organisms and their biological processes.

Birth rate number of babies born per thousand population per year; known as the crude birth rate.

Bottom-up development economic and social changes brought about by activities of individuals and social groups in society rather than by the state and its agents.

Bounded applied to the behaviour of decision makers whose access to information, for example, is constrained by financial resources or time. They also have a limited capacity to process the information that they are able to obtain and they will also be constrained (bounded) by the environment within which their behaviour is taking place. Bounded cultures may be regarded as reactions to perceived threats to local cultures, entailing a strong assertion of the latter. These can create a level of local fragmentation, with a parochial, nostalgic, inward-looking sense of local attachment and cultural identity. Bounded cultures generally involve very clear definitions of 'insiders' and 'outsiders' in the creation of a sense of belonging.

Brundtland Report published in 1987, it adopted the position that it was possible to pursue economic growth without compromising the environment and introduced the first widely used definition of sustainable development: 'development which meets the needs of the present without compromising the ability of future generations to meet their own needs'.

Buyer-driven commodity chain a chain in which the nature of production (e.g. salmon farming) is shaped or driven by powerful 'downstream' actors such as distributors and retailers.

C

CAP Common Agricultural Policy of the European Union.

Capital money put into circulation or invested with the intent of generating more money. See Spotlight box 18.2.

Capitalism an historically specific economic system in which production and distribution are designed to accumulate capital and create profit. The system is characterized by the separation of those who own the means of production and those who work for them.

Chiefdom social formation based on societies characterized by the internal and unequal differentiation of both power and wealth and organized around the principle of kinship. See Spotlight box 1.3.

City-states urban regions having political jurisdiction and control over a specific territory. A form of social organization associated with some of the earliest states.

Civilization refers to an advanced form of social development characterized by such things as urban life, commercial activity, writing systems and philosophical thought.

Class social distinctions between groups of people linked to their material conditions and social status.

Clean development mechanism (CDM) a mechanism whereby industrialized countries can earn credits towards their own greenhouse gas reduction targets by investing in emission-reducing projects in the developing world. Part of the aim is to aid economic growth in developing countries without a commensurate increase in greenhouse gas emissions. At the same time this assistance is to the industrialized countries; such assistance is often cheaper than achieving their own emission reduction.

Climate change a change in climate that is attributed directly or indirectly to human activity that alters the composition of the global atmosphere and which is in addition to natural climate variability observed over comparable time periods. Includes temperature rise, sea-level rise, precipitation changes, droughts and floods.

Cluster a localized concentration of similar and interlinked economic activities.

Cold War a period extending from 1945 to the late 1980s during which two ideologically opposed blocs emerged within the world system, both with nuclear capabilities. The first was headed principally by the USA while the other was dominated by the Soviet Union.

Command economy an economic system characterized by the state-led central planning of economic activity combined with the simultaneous suppression of market-type relations.

Commodity a good or service produced through the use of waged labour and sold in exchange for money.

Commodity fetishism the process whereby the material and social origins of commodities are obscured from consumers.

Communism a political theory attributed to the works of Karl Marx and Friedrich Engels. Communism is characterized by the common ownership of the means of production. The foundations for full communism were supposed to be laid during a transitional period, known as socialism. The Soviet Union was the first state to be ruled by a communist government, from 1917 to 1991.

Communitarianism a moral and political philosophy based on the extension and preservation of community. It is the opposite of liberalism which is often regarded as insufficiently sensitive to the social/community sources of self-worth.

Community food scheme a site of resistance to the industrialization of agriculture, in which the local community retains control and where consumers get seasonal, fresh food at a price that supports farmers who use sustainable practices.

Comparative advantage in economic geography, refers to an advantage held by a nation or region in the production of particular set of goods or services. See Spotlight box 14.4.

Competitive advantage exists when a firm can deliver the same benefits as competitors but at lower cost (cost advantage). A firm may also deliver benefits that are greater than those of competing products (differentiation advantage).

Conditional reserve those (mineral or ore) deposits that have already been discovered but that are not economic to work at present-day price levels with currently available extraction and production technologies.

Contact zone the space in which transculturation takes place – where two different cultures meet and inform each other, often in highly asymmetrical ways.

Containment the Western strategy of encircling the Soviet Union and its allies during the Cold War.

Conventionalization also referred to as 'mainstreaming', this occurs when larger scale businesses (e.g. agribusinesses, supermarkets) increasingly control the production of 'alternative' products such as fair trade and organic through the adoption of conventional patterns of marketing and distribution.

Core according to Wallerstein, the core refers to those regions of the capitalist world economy characterized by the predominance of core processes associated with relatively high wages, advanced technology and diversified production. See Spotlight box 2.1.

Cosmology study of the universe as an ordered whole, and of the general laws or principles that govern it.

Counterurbanization population increases in rural areas beyond the commuting range of major urban areas.

Crude birth rate number of births in one year per 1,000 population.

Crude death rate number of deaths in one year per 1,000 population.

Cultural capital the possession of taste, style or attitude that can be converted, in many instances, into financial capital. In the post-industrial economy, where ideas are a global currency, cultural capital is a lucrative commodity.

Culture a system of shared meanings often based around such things as religion, language and ethnicity that can exist on a number of different spatial scales (local, regional, national, global, among communities, groups, or nations). Cultures are embodied in the material and social world, and are dynamic rather than static, transforming through processes of cultural mixing or transculturation. See Spotlight box 13.1.

Culture jamming tactics such as media hacking, information warfare, 'terror-art' and graffiti, which aim to invest advertisements, newscasts and other media artefacts produced by powerful agencies and organizations with subversive meaning.

D

DDT a chemical pesticide banned by many countries in the 1970s as a pollutant which is a probable human carcinogen and causes damage to internal organs.

Death rate number of deaths per year per thousand population; known as crude death rate.

Deep ecology An environmental approach which asks us to consider humans not only as part of nature but of equal value to other non-human entities. This is a holistic vision whereby all human and non-human entities are interconnected and interdependent. A holistic perspective enables us to understand that if we upset one element, it will have an impact on all other elements.

Defensive localism a process where consumers purchase local foods to support local farmers and the local economy, irrespective of their quality and whether they are produced organically or conventionally.

Deforestation the removal of forests from an area.

Deindustrialization refers to a relative decline in industrial employment. It may also refer to an absolute decline in industrial output as well as employment.

Demarginalization the process whereby a marginal or stigmatized space becomes 'normalized', and its population incorporated into the mainstream.

Demographic transition model traces the shift from high birth and death rates to low ones.

Dependency a viewpoint or theory of development which argues that global inequality is explained by the patterns of exploitation of the periphery by the capitalist core, established during the colonial period and perpetuated by neo-colonial economic relations in recent times. The dependency theorists' recommendation for poorer countries is to de-link from the global economy.

Depopulation the reduction of population in an area through out-migration or a reduction in the birth rate below the death rate.

Diaspora literally, the scattering of a population; originally used to refer to the dispersal of Jews in AD 70, now used to refer to other population dispersals, voluntary and non-voluntary. Evokes a sense of exile and homelessness.

Diaspora space the spaces inhabited not only by those who have migrated and their descendants, but also by those who are conceptualized as indigenous or 'native'. Similar to contact zone – a meeting point of different cultures.

Discourse social and cultural theorists understand discourse to be the dominant meanings that are attached to a linguistic term or utterance.

Displaced persons refugees who have no obvious homeland.

Division of Labour the separation of tasks in the production process and their allocation to different groups of workers.

Doubling time time taken for a population to double in size.

E

Earth Summit the UN Conference on Environment and Development held in Rio de Janeiro in 1992. A total of 178 countries negotiated a global strategy centrally concerned with sustainable development.

Ecocatastrophism the prediction of impending environmental disasters, often as a result of human actions.

Ecocentrism a way of understanding the value of the environment which critiques the priorities of anthropocentrism and suggests that non-human entities have intrinsic value.

Ecological democratization an approach that suggests that the only way to achieve environmentally sustainable practices is to value participation and justice in environmental decision-making. Thus, this approach requires extensive citizen participation and the development of democratic institutions to tackle environmental issues at all scales.

Ecological modernization an environmental management approach that argues that economic growth does not need to be slowed to ensure environmental protection. It does not undermine the limits to growth thesis entirely, but suggests that we can 'reorient' economic growth and use technological solutions for environmental problems, thus overcoming the environmental impact of growth. It is a very weak interpretation of sustainable development that allows for a reformist response where the current dominance of free-market capitalism is not challenged.

Economies of scale occur when mass production of a good (product) results in a lower average cost for each item. Economies of scale occur within a firm, such as using expensive equipment more intensively, or outside the firm as a result of its location, such as the availability of a local pool of skilled labour or good transport links.

Edge city (exopolis) a term referring to an area with city-like functions usually arising on the edge of an already urbanized area or conurbation and heavily dependent on fast communications systems. Edge cities are regarded as symptomatic of the most recent phase of urbanization.

Emerging markets refers to those markets that are perceived to have a substantial growth potential. The term is often used in relation to selected countries of the former Soviet Union, Asia, Latin America and Africa.

Emotional labour workers are expected to display certain emotions, often in line with the goals of their employer or organization, as part of their job.

Empire an extended territorial political unit or state built up, often by force, under a supreme authority. Empires usually involve rule over alien or subject peoples.

Energy mix the balance between various sources of energy in primary energy consumption.

Energy ratio the relationship between energy consumption and economic growth in an economy:

$$\text{Energy ratio} = \frac{\text{Rate of change in energy consumption}}{\text{Rate of change in economic growth (GDP)}}$$

A value greater than 1 indicates that the amount of energy required to create an additional unit of GDP is increasing; a value less than 1 suggests the reverse.

Enlightenment a philosophy or movement that emerged during the eighteenth century and was based on the idea of social progress through the application of reason.

Entitlements the set of all commodity bundles over which a person can establish command given the legal, political, economic and social arrangements of the community in which they live.

Environmentalism a broad term incorporating the concerns and actions in aid of the protection and preservation of the environment.

Epidemiological transition switch in the predominant causes of death noted in the movement from a pre-transition to a post-transition society, primarily from infections to degenerative diseases.

Ethnicity refers to the process through which groups are recognized as possessing a distinct collective cultural identity. See too **Race** and **Racism**.

EU European Union.

Exchange the process of interchange of goods and services between individuals, groups and/or organizations, whether involving money or not. Can also refer more broadly to social interactions.

Export processing zone a small closely defined area which possesses favourable trading and investment conditions created by a government to attract export-orientated industries.

F

Factors of production refers to those elements necessary for the effective functioning of the production process and typically includes land, labour and capital.

Fair trade attempts to overcome the injustices of free trade by guaranteeing producers a fair price and thus improving their lifestyles. Coffee, tea, bananas and chocolate are among the major fair trade products.

Fascism a term used particularly to describe the nationalistic and totalitarian regimes of Benito Mussolini (Italy, 1922–45), Adolf Hitler (Germany, 1933–45) and Francisco Franco (Spain, 1939–75).

Fast food a quick and accessible way to eat food; epitomized by McDonald's, it is usually eaten out of the home and out of our hands.

Fertility rates the number of like births per thousand women of child-bearing age.

Feudalism a hierarchical social and political system common in Europe during the medieval period. The majority of the population were engaged in subsistence agriculture while simultaneously having an obligation to fulfil certain duties for the landholder. At the same time the landholder owed various obligations (fealty) to his overlord.

Financial exclusion the processes that prevent disadvantaged social groups from gaining access to the financial system.

Financialization refers to the growing influence of capital markets, their intermediaries and processes in contemporary economic and political life.

Flexible specialization new methods of production such as multi-purpose machines and small-batch production. The term is frequently associated with agglomerations of firms, or industrial districts.

Food chain the route traced by particular foodstuffs from 'farm to fork'. Food chains involve a number of production, processing, distribution and consumption nodes, and the connecting links between them.

Food desert a place, usually in inner cities or remote rural areas, where access to fresh, affordable and healthy food is poor.

Food miles a term used to describe the distance that food travels from the point of production to the place of consumption. As well as distance, the mode of transport (e.g. air vs. ferry freight) is an important consideration.

Food regimes distinct relationships discerned between patterns of international food production and consumption and the developing capitalist system.

Food security when all people, at all times, have physical, social and economic access to sufficient, safe and nutritious food that meets their dietary needs and food preferences for an active and healthy life.

Fordism a regime of accumulation involving mass production and consumption. Named after Henry Ford (1863–1947), Fordism is known for a differentiated division of labour, assembly-line production and affordable mass-produced consumer goods.

Fordist or 'organized' capitalism a form of capitalism that reached its zenith during the period 1945 to 1973. It was characterized by the dominance of extractive and manufacturing industries, large-scale manufacture, mass-production and a significant level of state involvement. See Spotlight box 3.1.

Fourth World according to Castells (1998) a term referring to those regions effectively excluded from participation in the 'new' world economy. Importantly, it is of relevance not only to regions and countries of the recognized Third World but also to regions within the developed world. Also sometimes used to refer to indigenous or native peoples.

G

Gated communities residential developments protected by a range of mechanisms such as security gates, walls, private security guards and intercom systems.

Gemeinschaft a form of community said to be common in traditional societies (as distinct from industrial societies) and associated with notions of stability and informal personal contact.

Gender refers to socially constructed ideas of difference between men and women.

Gender division of labour a division of labour constructed around gender in which particular tasks and occupations are deemed to be male or female activities.

Gendered space the ways in which certain spaces are seen to be occupied exclusively or predominantly by either males or females.

Genetic modification (GM) human manipulation of genetic material (plant, animal and human) to create altered organisms.

Gentrification the process by which middle- and upper-class incomers displace established working-class communities. Often associated with new investment in the built environment, gentrification may be small-scale and incremental (i.e. instigated by individual incomers), or be associated with major redevelopment and regeneration schemes.

Geographies of exclusion the spatial processes by which a powerful grouping consciously seeks to distance itself from other less powerful groupings.

Geopolitics a term that has been used to refer to many things, including a tradition of representing space, states and the relations between them; also emphasizing the strategic importance of particular places.

Gesellschaft a form of association common in urban-based industrial societies (as distinct from traditional societies) and associated with non-permanent and utilitarian social relationships.

Ghetto refers to very high concentrations of people drawn from a particular ethnic or cultural background living in specific parts of an urban area. The term is now commonly associated with notions of deprivation, unemployment and social exclusion.

Global cities the term 'global city' was popularized by the sociologist Saskia Sassen in her 1991 book *The Global City* and is taken to refer to a small number of cities that serve as the command and control centres of the global economy. These cities possess a concentration of financial and business services linked to the multinational corporations that control the activities and organization of the global economy, and which are often conspicuously clustered in their downtown business districts. Global cities are also characterized by increasing social polarization as a result of a growing division in occupational and income opportunities, with growth in both high-skilled and low-skilled employment.

Global production networks the extensive webs of intra-, inter- and extra-firm connections through which commodities are produced, distributed, sold and consumed.

Global production system refers to the dominant form of production identifiable at a global scale. It has evolved over many centuries into its current form, referred to as post-Fordism.

Global triad this term describes how the world economy is essentially organized around a tripolar, macro-regional structure whose three pillars are North America, Europe, and East and South-east Asia.

Global warming an increase in the temperature of the Earth's surface caused by trapping infrared radiation in carbon dioxide, increased amounts of which are produced by burning fossil fuels.

Globalized or 'disorganized' capitalism a form of capitalism identifiable since the mid-1970s. It is characterized by a relative decline in the importance of extractive and manufacturing industries, together with a relative increase in the importance of services. There is also an increasing tendency for the production process to be dominated by small-scale and flexible forms of organization. See Spotlight box 3.3.

Globalization a contested term relating to transformation of spatial relations that involves a change in the relationship between space, economy and society.

Glocalization the interaction between the particular character of places or regions and the more general processes of change represented by globalization.

GM See **Genetic modification**.

Governance the way in which power operates through the relationships between different organizations.

Green revolution a large increase in crop production in developing countries achieved by the use of fertilizers, pesticides and high-yield crop varieties.

Gross domestic product (GDP) the value left after removing the profits from overseas investments and those profits from the economy that go to foreign investors.

Gross national product (GNP) a broad measure of an economy's performance; it is the value of the final output of goods and services produced by the residents of an economy plus primary income from non-residential sources.

H

Heartland identified by the British geographer Halford Mackinder (1904) as the zone in East-Central Europe and Siberia, control of which would be a key to world domination. The term has since been appropriated by Latin American geopolitics.

Hegemony term derived from the work of Antonio Gramsci which refers to the ability of a dominant group to exert or maintain control through a combination of overt and subtle mechanisms.

Heritage according to UNESCO, 'heritage is our legacy from the past, what we live with today, and what we pass on to future generations. Our cultural and natural heritage are both irreplaceable sources of life and inspiration'. In human geography the focus is often on rural and urban heritage landscapes, frequently regarded as playing a formative role in the maintenance of national identity.

High-value foods includes such foods as fruit, vegetables, poultry and shellfish. World trade in high-value foods has increased markedly during the last two decades.

Horizontal integration occurs when two companies, within the same industry and at the same stage of production, merge.

Hybridity refers to groups as a mixture of local and non-local influences; their character and cultural attributes are a product of contact with the world beyond a local place. See Case study 13.4.

Hyperincarceration the rolling out of the penal state that has led (in the USA at least) to a massive growth in prison populations, and the mass imprisonment in particular of young African-American men who suffer long-term judicial stigma as a consequence even when ex-convicts.

Hypersegregation a term used to describe extreme residential segregation, such as that experienced by African-Americans in the USA.

Hypothetical resources those resources that might be expected to be found in the future in areas that have only been partially surveyed and developed.

I

Imaginary geographies the ideas and representations that divide the world into spaces with particular meanings and associations. These exist on different scales (e.g. the imaginaries that divide the world into a developed core and less developed peripheries, or the imagined divide between the deprived inner city and the affluent suburbs).

Imperialism a relationship of political, and/or economic, and/or cultural domination and subordination between geographical areas.

Indigenous peoples those peoples native to a particular territory that was later colonized, particularly by Europeans.

Industrial revolution a term that is often taken to refer to the marked transformation of productive forces, initially within the British economic system, between the mid-eighteenth and mid-nineteenth centuries. It resulted in the movement of Britain from a largely rural-based economy to one that was dominated by manufacturing and industrial production. Such a transformation resulted in substantial social, political as well as economic changes.

Industrialization of agriculture (see also agribusiness) a process whereby methods commonly associated with the manufacturing industry are increasingly incorporated into farming, e.g. specialization of labour, assembly-line production systems.

Infant mortality rate (IMR) number of deaths of persons aged under 1 per 1,000 live births.

Informal economy those parts of the economy which operate beyond official recognition and outside formal systems of control and, often, of remuneration.

Informal settlements also known as shanty towns or squatter settlements, these are sections of a city where poor people have moved in, often illegally or unofficially, and have constructed improvised housing using informal means and scrap materials: often plywood corrugated metal and sheets of plastic. They are usually built on the periphery of cities and often do not have proper sanitation, electricity, or other services. Informal settlements are mostly found in the major cities of developing nations in the global South and result from a combination of intense rural to urban migration pressures coupled with a lack of provision of affordable housing for low-income urban households.

Inherent value the value something has for someone, but not as means to a further end.

Instrumental value the value which something has for someone as a means to an end.

International division of labour a term referring to the tendency for particular countries and regions of the globe to specialize in particular types of economic activity.

(International Monetary Fund) IMF international financial institution that originated from the 1944 Bretton Woods Conference. Its main roles include regulating international monetary exchange and controlling fluctuations in exchange rates in order to alleviate balance of payments problems.

Internationalization a process whereby products and services are adapted to specific local languages and cultures.

Intrinsic value simply the value something has. No appeal needs to be made to those for whom it has value.

Islam a monotheistic religion founded by the prophet Muhammad in the seventh century AD. Today there are two predominant divisions within Islam, the Sunni and Shi'i.

Islamophobia a prejudice against those who identify as Muslim, fuelled by post-9/11 discourses of Islamic fundamentalism and terrorist threat. Taking a variety of forms in the urban West, this can be interpreted as merely the latest inflection of a long-standing Orientalist discourse that contrasts Western democracy with Arab 'barbarism'.

K

Kondratieff cycle a term used to describe the cycles of boom and bust evident within the capitalist system since the mid-eighteenth century. Named after the Russian scholar N.D. Kondratieff.

Kyoto Protocol the agreed outcome of a meeting of 160 nations in Kyoto, Japan, in 1997 whereby developed nations agreed to limit their greenhouse gas emissions, relative to the levels emitted in 1990.

L

Legitimacy with regard to nation-states, a term meaning that the majority of people accept the rule of law of the governing political organizations.

Less developed countries (LDCs) countries at a disadvantage in today's global competitive environment because their comparative advantage in cheap labour or natural resource endowments has become subordinated to knowledge-based factors. LDCs suffer from poor productive capacities and competitiveness.

Liberalism a political philosophy distinguished by the importance it attaches to the civil and political rights of individuals. May also refer to belief in the importance of the free market.

Life expectancy (at birth) average number of years of life expected on the basis of age-specific mortality schedules for the specified year.

Limits to growth the belief that there are natural limits to possible growth which if exceeded will lead to environmental catastrophe.

Local Agenda 21 (LA21) the implementation by local administrations of sustainable development practices as defined in Agenda 21, an outcome of the Earth Summit, Rio de Janeiro, 1992.

Locality a place or region of sub-national spatial scale.

Low impact development (LID) a form of low impact living, a deep green vision where humans minimize their environmental impact in all aspects of their daily lives. LID is a radical form of housing and livelihood that works in harmony with the landscape and natural world.

M

Market-based states modern states where the market is the dominant means by which land, labour, capital and goods are exchanged and has a major influence over social and political organization.

Marxism a form of socialism and mode of analysis derived from the teachings of Karl Marx (1818–83). Marxism regards capitalism as an inherently unjust system with the capitalists (those who own the means of production) exploiting the proletariat (those who must sell their labour in order to live). It aims to replace capitalism with a fairer system, socialism maturing into communism.

Mega-cities giant metropolises with populations of at least 10 million people. Also used by Manuel Castells to describe large cities in which some people are connected to global information flows while others are disconnected and information poor. This use of the term serves to highlight the problems associated with mega-cities in that they are large, highly unequal and beset by problems such as slum housing, pollution and urban sprawl, but also often lack the economic and political power of global cities and

therefore the resources to adequately address their problems.

Merchant capitalism refers to an early phase of capitalist industrial development dominant in the larger urban regions of Europe from the late fifteenth century. Merchants were the principal actors engaged in both the provision of capital and the movement and trade of goods (predominantly bulky staples such as grain and manufactured goods).

Mergers occur when two firms agree to form a new company.

Modernism a term typically associated with the twentieth-century reaction against realism and romanticism within the arts. More generally, it is often used to refer to a twentieth-century belief in the virtues of science, technology and the planned management of social change.

Modernity refers to a period extending from the late sixteenth and early seventeenth centuries (in the case of Europe) to the mid to late twentieth century characterized by the growth and strengthening of a specific set of social practices and ways of doing things. It is often associated with capitalism and notions such as progress.

Monopoly in theory, exists in an industry when one firm produces all the output of a market; in practice varies between countries. In the United Kingdom, for example, any one firm that has 25 per cent of the market is considered to hold a monopoly.

Monopsony in theory exists in an economy when one firm or individual purchases all of the output in a given sector. An agribusiness company can gain a geographical monopsony over a given area through the use of contracts that prevent sale of produce to other parties.

Moral panic a term describing periodic episodes of concern about the threat of a particular group to the nation-state. Moral panics are normally fuelled by sensationalist media reporting, and are generally diffused by the state through policies which aim to counteract this imagined threat.

More developed countries (MDCs) countries with significant competitive advantages in today's globalizing economy. They have well-developed, increasingly knowledge-based and strongly interconnected manufacturing and service sectors that provide a significant proportion of employment and contribute to significant national and individual wealth. Indices such as literacy levels, incomes and quality of life are high and these countries exercise considerable political influence at the global scale. Examples are the United Kingdom, the United States, Germany and France.

Multiculturalism refers to a belief or policy that endorses the principle of cultural diversity and supports the right of different cultural and ethnic groups to retain distinctive cultural identities. It has often been criticized for being too symbolic and not politically radical enough in challenging racism.

Multiple deprivation usually refers to a situation in which an individual or group suffers a series of disadvantages, including poverty, poor health, exposure to criminality, inadequate housing and so on. Geographically often used to denote those denied easy spatial access to a range of services, such as housing, health care, education and also transport.

N

Nationalism the ideology and sentiment of belonging to a 'nation' and the claim that the 'nation' should be expressed in a 'state'.

Nation-state and state a symbolic system of institutions claiming sovereignty over a bounded territory.

Natural increase surplus of births over deaths in a population.

Neoliberalism an economic doctrine promoting market-led growth, deregulation and the privatization of state-owned enterprises.

Neo-Malthusian the belief that environmental problems are a consequence of population growth, following the arguments set out by Thomas Malthus in the late eighteenth century.

New economic geography (NEG) an economic geography that recognizes the importance of culture as an influence on economic processes and outcomes. In this way it draws attention to the culturalization of the economy in contrast to the economization of culture.

New industrial districts (NIDs) areas that specialize in a particular industry because of external economies which result from specialization.

New international division of labour (NIDL) the global shift of economic activity that occurs when the process of production is no longer constructed primarily around national economies.

New World Order the notion of a Western-led post-Cold War structure of global power, which sanctions intervention.

Newly industrialized countries (NICs) countries where there has been a relatively recent and significant shift away from primary activities towards manufacturing production. Examples are South Korea and Mexico.

NGO non-governmental organization.

Niche markets refer to the existence of consumer groups with identifiable tastes and lifestyles. A good example would be speciality food sales, often purchased by consumers as a symbol of social status or cultural capital.

Non-governmental organization (NGO) an organization formed by members of the public and one that has no government connections.

Non-renewable (stock) resources those resources, mainly mineral, that have taken millions of years to form. Their availability is therefore finite as there is no possibility of their stock being replenished on a timescale of relevance to human society.

O

Overpopulation used to suggest that the finite resources of a particular area will run out if the population expands beyond a given point. Similar to the idea of a carrying capacity, that there is a limit beyond which environmental degradation occurs.

P

Palaeolithic the stage in the development of human society when people obtained their food by hunting, fishing and gathering wild plants, as opposed to engaging in settled agriculture. Also referred to as the Early Stone Age.

Participatory democracy a form of governance that encourages involvement of all people in political decision-making, as opposed to representative democracy.

Pastoral nomadism a form of social organization that is based on livestock husbandry for largely subsistence purposes. Pastoral nomads are characterized by a high level of mobility which allows them to search continually for new pastures in order to maintain their herds of animals.

Patriarchy system of gendered power relations through which men exercise power over women.

Peasant the term has had a number of different interpretations over the years and in different parts of the world. In general it usually refers to those individuals whose livelihood is largely dependent on the land and on rural subsistence-type activities. The term 'peasant' is usually reserved for those living in organized states, thus distinguishing them from band and tribal members, and so on.

Periphery according to Wallerstein, refers to those regions of the world capitalist economy characterized by low wages, simple technology and limited production.

Permaculture an environmental approach inspired by nature's patterns to create sustainable human habitats. It is based around three core principles – earth care (working with nature and designing systems that draw upon natural systems for inspiration), people care (looking after ourselves on a community and individual level through cooperation and mutual support) and fair shares (ensuring we only consume our share of the earth's resources).

Personal space the apparent desire by humans to have a pocket of space around them and into which they tend to resent others intruding without invitation.

Political economy an approach to social study that emphasizes the political/social construction and consequences of economic activity.

Post-Fordism production system comprises a mix of different ways of organizing production at a number of spatial scales. Common to all these types of production is some form of flexible production.

Postmodernism a philosophy that holds that the traits associated with twentieth-century modernism, such as belief in the possibility of managing social change according to sets of agreed principles, are now in retreat in the face of increasing individualism, pluralism and eclecticism.

Post-productivist transition a term used to describe the movement away from productivist agricultural systems. This new phase of agricultural production is characterized by extensive and diversified patterns of farming and the growing importance of non-agricultural activities in the countryside, for example, recreation.

Poverty the condition of possessing an income insufficient to maintain a minimal standard of living. Definitions of poverty are culturally specific, and thus relative to the social norms and expectations endemic to a given nation-state. However, the condition of absolute poverty (i.e. lacking the income to maintain a minimum diet) is acknowledged worldwide.

Power geometry the ways in which different social groups/individuals are placed in relation to the forces of globalization, enabling some to benefit and others to be disadvantaged.

Prehistoric societies societies that have left no written records.

Primary energy the energy in the basic fuels or energy sources used, e.g. the energy in the fuel fed into conventional power stations.

Primary sector comprises economic activities that exploit naturally occurring resources.

Proto-industrialization refers to the early phase of capitalist industrial development in Europe. Characteristics of this period include a significant level of rural-based industrial activity and a low level of technological application in the production process.

Proven (proved) reserves those deposits of a resource that have already been discovered and are known to be economically extractable under current demand, price and technological conditions.

Q

Quality of life a composite measure that reflects individual preferences that include, for example, education, health, entertainment, living environment.

Quality turn the idea that local/alternative foods are of higher quality than products produced under conventional farming systems; in this way, quality is inextricably linked to locality.

Quaternary sector economic activities, mainly business enterprises, involved with the processing and exchange of information.

R

Race refers to the division of human beings into supposedly recognizable groups based predominantly on physical characteristics.

Racialized space the ways in which certain spaces are seen to be occupied exclusively or predominantly by a particular 'racial' or ethnic group.

Racism practices and attitudes that display dislike or antagonism towards people seen as belonging to particular ethnic groups. Social significance is attached to culturally constructed ideas of difference.

Ranked society society in which there is an unequal division of status and power between its members, where such divisions are based primarily on such factors as family and inherited social position.

Regulated states social formations organized on the basis of territory. The regulated, or pre-modern, state was not dominated by the market system with political, social and religious considerations usually being of greater importance.

Relational approaches a term referring to a range of approaches that are concerned more with the network as a unit of analysis than the spatial unit. Despite considerable diversity between approaches, relational approaches focus primarily on the connections that exist between places and things. Particularly influential in recent iterations of urban and economic geography.

Relocalization (of food) refers to the renewed interest in foods of local and regional provenance, in which the link between 'product' and 'place' is emphasized and often formalized through a system of quality assurance or quality labels. Such locally distinctive quality food products are sold in regional and national markets.

Renewable energy energy sources such as winds, waves and tides that are naturally replenished and cannot be used up.

Renewable (flow) resources those resources that are naturally renewed within a sufficiently short time-span to be of use to human society. The continued availability of such resources is increasingly dependent upon effective management.

Rent gap a term describing the difference between the current ground rental for a building and its potential ground rent. As such, gentrifiers seek to exploit the rent gap by investing in property and real estate in areas where they perceive rentals to be below their potential market value.

Replacement rate number of babies that an average woman should have to replace her generation, given prevailing levels of mortality (2.1 per woman in post-transition societies).

Representative democracy a form of governance that limits involvement of people in political decision-making to representatives often voted into positions of power by the broader public.

Reserves/production ratio (R/P) if the reserves remaining at the end of any year are divided by the production in that year, the result is the length of time those remaining reserves would last if production were to continue at that level.

Residential segregation the ways in which, most obviously in urban areas, housing patterns can be observed where people live in areas divided along class or ethnic lines. In some instances this is conscious policy, in others it results from the interaction of social and economic processes.

Resource a substance in the physical environment that has value or usefulness to human beings and is economically feasible and socially acceptable to use.

Right to Food This is a human right recognised in the Universal Declaration of Human Rights and the International Covenant on Economic, Social and Cultural Rights. It requires signatory governments to respect, protect and fulfil their citizens' rights to adequate and sufficient food consistent with cultural traditions.

Romantic movement artistic, literary and philosophical movement that originated in the eighteenth century as a backlash against industrialization.

Rural services social goods that comprise the fabric of non-productive life in rural areas (e.g. education, health, transport, shops).

Rurality functionalist, critical political economy and social representation approach for distinguishing between rural and urban economy and society.

Rural–urban fringe refers to the zone where rural and urban land uses meet and intermix around the edges of urban areas. While some regard the rural urban fringe as a transition zone between the city and the countryside, others argue that it represents a distinctive landscape type in its own right. This zone has long been regarded as problematic in planning and management terms. It is typically regarded as lacking aesthetic quality and being dominated by the functional landscape elements of the urban infrastructure, the transport and distribution networks of the city and low-grade rural land.

Rust Belt a region of the north-eastern USA that suffered substantial industrial decline, especially after the Second World War.

S

Second global shift offshoring of service functions; the first global shift involved the offshoring of manufacturing functions.

Second industrial revolution a term sometimes used to account for the profound technological and accompanying social changes that affected industrial capitalism from the late nineteenth century. During this period Britain's industrial might was challenged by Germany and the USA.

Secondary sector comprises economic activities that transform primary sector outputs into useable goods.

Semi-periphery according to Wallerstein, refers to those regions of the world capitalist economy which, while exploiting the periphery, are themselves exploited by the core countries. Furthermore, they are characterized by the importance of both core and peripheral processes.

Sense of place the feelings, emotions and attachments to a locality by residents (past or present), which may be articulated in art, literature, music or histories, or may become part of individual or group memory.

Services work done as an occupation or business for other individuals or businesses that brings about a change in the condition of a person or of a good belonging to some economic unit which does not produce or modify physical goods.

Sexuality refers to social differences linked to sexual identity and behaviour.

Short food supply chains (SFSCs) where the number of nodes in the food chain is reduced to as few as possible. An ideal SFSC involves the direct marketing of food between a producer and the final consumer. The key feature is that foods reach the final consumer 'embedded' with value-laden information about the production, provenance and distinctiveness of the product.

Simulacrum a copy without an original. An example might be Disneyland's 'Main Street', which represents an ideal American high street, but is not modelled on an original.

Slow food promoted as an anti- 'fast food' culture, slow food aims to decelerate the food consumption experience and celebrate the cultural connections surrounding local cuisines and traditional products. The slow-food movement aims to embed food in the local territory and culture.

Social construct a social concept or idea (such as race, class, gender or age) that is institutionalized and normalized within a culture to the extent that people behave as if it were a 'real' or a pre-social given.

Social constructionism a sociological theory of knowledge which claims that meanings and truths are established socially through habitual use and institutionalization. In opposition to realism, which regards truths as pre-given and objective, social constructionism recognises that individuals and groups participate in the creation of their realities through the way in which they perceive things and events.

Social embeddedness the idea that economic behaviour is embedded in, and mediated by, a complex and extensive web of social relations. Trust and regard are examples of such social relations.

Social exclusion the various ways in which people are excluded (economically, politically, socially, culturally) from the accepted norms within a society.

Social movement comprised of individuals, groups and organizations united by a common purpose or goal.

Social representation lay and academic discourses that describe an event or context.

Spatial divisions of labour the concentration of particular sectors or production tasks in specific geographic areas.

Spatial interaction a term used to indicate interdependence between geographical areas. Covers the movement of people, goods, information and money between places.

Spatial relations the ways in which people are connected across geographic space through economic, social, cultural or political processes.

Speculative resources those resources that might be found in unexplored areas which are thought to have favourable geological conditions.

State a political unit having recognizable control (claiming supreme power) over a given territory. Unlike earlier social formations like bands and tribes, states have always based their power on their ability to control a specified territory and its inhabitants.

Stigma a term describing the condition of possessing a 'spoiled' or discredited identity.

Stratified society society within which there is an unequal division of material wealth between its members.

Structural adjustment programmes loans designed to foster structural adjustment in LDCs by promoting market-led growth and a reduction in government intervention in the economy.

Structuralism a theory, evolved in Latin America in the 1940s and 1950s, which argues that the relative position of a given economy vis-à-vis the global economy is a function of the nature of the manner of insertion of that economy historically. Changing the fortunes of an economy therefore requires state intervention to alter its structure.

Studentification an increase in the student population in a residential area, principally associated with student renting in the private sector, though also associated with newly built off-campus accommodation developed by private investors. A widely noted phenomenon in those countries where students in higher education tend to live away from home, it is sometimes associated with gentrification, but has markedly different social and environmental outcomes.

Subculture a subdivision of a dominant culture or an enclave within it with a distinct integrated network of behaviour, beliefs and attitudes.

Substitutionism refers to the increased use of non-agricultural raw materials characterized by the replacement of food and fibre with industrial alternatives. It forms a key process in the industrialization of the agri-food system.

Subsumption a process whereby an agribusiness TNC increases its control over farming by offering contracts to farmers to provide 'raw materials' for its food-manufacturing activities.

Suburbanization, suburbs suburbanization describes the growth of areas on the fringes of major cities. Whilst residents of metropolitan regions have often tended to work within the central urban area, they have frequently chosen to live in satellite communities called suburbs and commute to work via car or mass transit systems. Suburbs are therefore most frequently viewed as residential areas, either being part of a city or separate residential communities within commuting distance of a city. In physical form they represent a mix of urban and rural elements and most have a lower population density than inner city neighbourhoods. Often considered dull and homogeneous in character, more recently research has highlighted the diverse and changing nature of suburbs with an increasing range of urban populations and activities evident in outer or edge of city locations.

Sun Belt refers to the major growth areas of the southern and western parts of the USA during recent years in contrast to the contracting and declining industrial base of the north-east. Can also be used in other parts of the developed world to describe dynamic regions, for example, the M4 corridor in England.

Sustainability, sustainable development terms whose usage and meanings have grown exponentially in recent years since their original discussion in the 1987 Brundtland Report. At their most basic they refer to the idea that development can continue in ways that respects the needs of future generations. There is considerable debate though concerning how sustainability and these needs can be defined and measured. Often sustainability has come to refer to economic or social sustainability rather than just environmental sustainability. As sustainability discourses have proliferated it is fair to say that the meaning of sustainability and sustainable development have become less clear and subject to political manipulation. The terms have come to invoke a degree of cynicism amongst many radical/critical geographers.

Sustainable intensification This term is attributed to concerns about global food security. It evokes the idea that there is a need to achieve higher yields from the same acreage without damaging the environment. Advocates argue this can be achieved through science and technology innovations.

T

Takeover occurs when one company purchases a controlling interest in a second company against the wishes of the latter's directors.

Territoriality a term used to describe an expression of ownership and control by an individual, group or state over a particular area of land in order to achieve particular ends.

Territory a recognizable region (area of land or sea) occupied and controlled by an individual, group or state.

Tertiary sector comprises economic activities engaged in enabling the exchange and consumption of goods and services (often referred to as the service sector).

Third World a rather vague term used to describe those regions of the world in which levels of development, as understood by such measures as GDP, are significantly below those of the economically more advanced regions. The term is increasingly seen as an inadequate description of the prevailing world situation since it disguises a significant amount of internal differentiation.

TNCs see transnational corporations.

Top-down development economic and social changes brought about by activities of the state and its agencies.

Total fertility rate (TFR) in general terms equal to the average number of live births per woman within a population.

Traceability in economic geography refers to the ability to determine the origin of agricultural foodstuffs.

Trade bloc a group of nation-states that act together in order to further their influence over world trade, for example, the European Union (EU) or the North American Free Trade Agreement (NAFTA).

Transculturation the ways in which subordinated or marginal groups select and invent from dominant cultures; although such groups cannot control what emanates from the dominant culture, they do determine to varying extents what they absorb into their own and what they use it for.

Transgression a term describing actions that breach social expectations of what is appropriate in a particular place. Deliberate transgression may thus constitute an act of resistance.

Transnational corporations (TNCs) major business organizations that have the power to coordinate and control operations in more than one country. They are the primary agents of globalization in the agri-food sector. See Spotlight box 14.2.

Transnationalism multiple ties and interactions linking people or institutions across the borders of nation-states and measured, for example, as flows of capital, people, information and images.

Travellers formerly referred to as 'gypsies', persons that are nomadic in lifestyle living in mobile homes of various types that are moved around from time to time.

Tribe a type of social formation usually said to be stimulated by the development of agriculture. Tribes tend to have a higher population density than bands and are also characterized by an ideology of common descent or ancestry. See Spotlight box 1.2.

U

Underclass a term referring to poorer, more marginalized groups in society who are seen to experience multiple deprivation.

Urban myths long-standing partial, at times stereotypical, images or impressions of the urban that are deeply ingrained within cultures and present throughout different historical periods. Often constructed in opposition to rural myths and implicated in moral debates about different ways of life associated with the rural and the urban.

Urban revanchism literally, the process of 'revenge' by which middle-class citizens take back the city from the marginal groups whom they portray as taking it from them. In practice, this is promoted by urban policies promoting law and order on the city streets and excluding those who do not fit in with the middle-class consumer ambience vital to the success of urban gentrification.

Urban village a residential sector of a city (often inner-city) believed to possess social characteristics typical of a village-type community.

Urbanization refers to the increasing importance of the urban relative to the total population. It is initially stimulated by the movement of people from rural to urban areas.

V

Value-adding activity sequential steps in a production process that enhances the saleable value of a commodity.

Vertical disintegration a process whereby segments of the production process (usually in a vertically integrated business) are subcontracted out to smaller-scale producers.

Vertical integration occurs when firms at different stages of the production chain merge together.

Vital trends changes in fertility and mortality.

W

World Bank international financial institution established in 1944. Its main role is to provide development funds to LDCs in the form of loans and technical assistance.

World systems theory Immanuel Wallerstein's conceptualization of the changing nature of the world socio-economic system into three distinct historical categories.

WTO World Trade Organization.

Bibliography

A

Aas, F.K. (2007) *Globalization and Crime*, Sage, London.

Abraham, I. and van Schendel, W. (2005) *Illicit Flows and Criminal Things, States, Borders, and the Other Side of Globalization*, Indiana University Press, Bloomington and Indianapolis, IN.

Abrahamsen, R. (2000) *Disciplining Democracy: Development Discourse and Good Governance in Africa*, Zed Books, London.

ActionAid (1995) *Listening to Smaller Voices: Children in an Environment of Change*, ActionAid, Chard, Somerset.

ActionAid (2006) *Real Aid II: Making Technical Assistance Work*, ActionAid International, Johannesburg, available at: www.actionaid.org.uk/doc_lib/real_aid2.pdf [accessed 7 October 2010].

ActionAid (2009) *Assessing the Alliance for Green Revolution in Africa*, ActionAid International, Johannesburg.

Aggarwal, R., Kaushal, M., Kaur, S. and Farmaha, B. (2009) Water resource management for sustainable agriculture in Punjab, India, *Water Science and Technology*, **60**(11), 2905–11.

Aglietta, M. and Breton, R. (2001) Financial systems, corporate control and capital accumulation, *Economy and Society*, **30**, 433–66.

Agnew, J. (1994) The territorial trap: the geographical assumptions of international relations theory, *Review of International Political Economy*, **1**, 53–80.

Agnew, J. (2003) *Geopolitics: Revisioning World Politics*, 2nd edition, Routledge, London and New York.

Agnew, J. (2005) *Hegemony: The New Shape of Global Power*, Temple University Press, Philadelphia, PA.

Agnew, J. (2007) No borders, no nation: making Greece in Macedonia, *Annals of the Association of American Geographers*, **97**(2), 398–422.

Agnew, J. (2009) *Globalization and Sovereignty*, Rowan and Littlefield, Lanham, MD.

Agnew, J. (2009) Money games: currencies and power in the contemporary world economy, *Antipode*, **41**, 214–38.

Agnew, J. and Corbridge, S. (1995) *Mastering Space: Hegemony, Territory and International Political Economy*, Routledge, London.

Agnew, J., Livingstone, D. and Rogers, A. (eds) (1996) *Human Geography: An Essential Anthology*, Blackwell, Oxford.

Agnew, J., Mitchell, K. and Toal, G. (eds) (2003) *A Companion to Political Geography*, Blackwell, Oxford.

AGRA *see* Alliance for a Green Revolution in Africa.

Ahluwalia, P. (2001) *Politics and Post-colonial Theory: African Inflections*, Routledge, London.

Akzin, B. (1964) *State and Nation*, Hutchinson, London.

Ali, M. (2003) *Brick Lane*, Doubleday, London.

Alland, A. (Jr) (1972) *The Human Imperative*, Columbia University Press, New York.

Allen, J. (1995) Global worlds, in Allen, J. and Massey, D. (eds) *Geographical Worlds*, Oxford University Press and Open University, Oxford, 105–44.

Allen, J. (1999) Cities of power and influence: settled formations, in Allen, J., Massey, D. and Pryke, M. (eds) *Unsettling Cities*, Routledge, London, 182–227.

Allen, P., FitzSimmons, M., Goodman, M. and Warner, K. (2003) Shifting plates in the agrifood landscape: the tectonics of alternative food initiatives in California, *Journal of Rural Studies*, **19**, 61–75.

Allen, T. and Thomas, A. (2000) *Poverty and Development into the Twenty-first Century*, Oxford University Press and Open University, Oxford.

Alliance for a Green Revolution in Africa (AGRA) (2010) Statement on plant breeding and genetic engineering, Alliance for a Green Revolution in Africa, available at: www.agra-alliance.org/section/about/genetic_engineering [accessed 12 October 2010].

Alvarez-Rivadulla, M.J. (2007) Golden ghettos: gated communities and class segregation in Montevideo, Uruguay, *Environment and Planning A*, **39**, 47–63.

Alvesson, M. (2000) Social identity and the problem of loyalty in knowledge-intensive companies, *Journal of Management Studies*, **37**(8), 1101–23.

Amin, A. (ed.) (1994) *Post-Fordism: A Reader*, Blackwell, Oxford.

Amin, A. (2002) *Ethnicity and the Multicultural City: Living with Diversity*, Report for Department of Transport, Local Government and the Regions, University of Durham.

Amin, A. (2004) Regions unbound: towards a new politics and place, *Geografiska Annaler B*, **86**(1), 31–42.

Amin, A. (2010) The remainders of race, *Theory, Culture and Society*, **27**(1), 1–23.

Amin, A. and Graham, S. (1997) The ordinary city, *Transactions of the Institute of British Geographers, NS,* **22,** 411–29.

Amin, A. and Graham, S. (1999) Cities of connection and disconnection, in Allen, J., Massey, D. and Pryke, M. (eds) *Unsettling Cities,* Routledge, London, 7–47.

Amin, A. and Thrift, N. (1994) Living in the global, in Amin, A. and Thrift, N. (eds) *Globalization, Institutions and Regional Development in Europe,* Oxford University Press, Oxford, 1–22.

Amin, A. and Thrift, N. (1994) Holding down the global, in Amin, A. and Thrift, N. (eds) *Globalization, Institutions and Regional Development in Europe,* Oxford University Press, Oxford.

Amin, A. and Thrift, N. (2002) *Cities: Reimagining the Urban,* Polity Press, Cambridge.

Amin, A. and Thrift, N. (2004) *The Blackwell Cultural Economy Reader,* Blackwell, Oxford.

Amiti, M. and Wei, S.J. (2004) *Fear of Service Outsourcing: Is it justified?,* International Monetary Fund Working Paper, WP/04/186.

Anderson, B. (1983) *Imagined Communities: Reflections on the Origins and Spread of Nationalism,* Verso, London.

Anderson, J. (1986) Nationalism and geography, in Anderson, J. (ed) *The Rise of the Modern State,* Harvester, Brighton, 113–26.

Anderson, J. (1995) The exaggerated death of the nation-state, in Anderson, J., Brook, C. and Cochrane, A. (eds) (1995) *A Global World? Re-ordering Political Space,* Oxford University Press and the Open University, Oxford, 65–112.

Anderson, A. (2009) *After the Ice: Life, Death and Politics in the New Arctic,* Virgin Books, London.

Anderson, J. (2010) *Understanding Cultural Geography: Places and Traces,* Routledge, London.

Anderson, J. (1997) Territorial sovereignty and political identity: national problems, transnational solutions? In Graham, C. (ed.) *In Search of Ireland: A Cultural Geography,* Routledge, London, 215–36.

Anderson, J., Brook, C. and Cochrane, A. (eds) (1995) *A Global World? Re-ordering Political Space,* Oxford University Press, Oxford.

Anderson, K. (1991) *Vancouver's Chinatown: Racial Discourse in Canada 1875–1980,* McGill-Queens University Press, Montreal.

Anderson, K. (2000) 'The beast within': race, humanity and animality, *Environment and Planning D: Society and Space,* **18,** 301–20.

Anderson, K., Domosh, M., Pile, S. and Thrift, N. (eds) (2003) *Handbook of Cultural Geography,* Sage, London.

Andreas, P. (2000) *Border Games: Policing the US–Mexico Divide,* Cornell University Press, Ithaca, NY.

Andreasson, S. (2005) Orientalism and African Development Studies: the 'reductive repetition' motif in theories of African underdevelopment, *Third World Quarterly,* **26**(6), 971–86.

Anon. (2001) *Text painted on a wall at Carters Road Community,* Margaret River, Western Australia.

Anonymous (1996) Ubuntu and democracy in South Africa, *Government Gazette,* Pretoria, No. 16943, 18.

Anonymous (1999) South Africa's middle-class pink currency, *Johannesburg Mail and Guardian,* 18 June, available at: www.globalarchive.ft.com

Antonisch, M. (2009) *Geopolitica*: the 'geographical and imperial consciousness' of fascist Italy, *Geopolitics,* **14,** 256–77.

Appadurai, A. (2003) Sovereignty without territoriality: notes for a postnational geography, in Low, S.M. and Lawrence-Zúñiga, D. (eds) *The Anthropology of Space and Place,* Blackwell, Oxford, 337–49.

Applebaum, A. (2003) *Gulag: A History,* Allen Lane, London.

Aranya, R. (2003) Globalization and urban restructuring of Bangalore, India: growth of the IT industry, its spatial dynamics and local planning responses. Unpublished Dr.ing thesis, Department of Urban Design and Planning, Norwegian University of Science and Technology, Trondheim.

Archibugi, D. (1995) Immanuel Kant, cosmopolitan law and peace, *European Journal of International Relations,* **1,** 429–56.

Ardener, S. (1995) Women making money go round: ROSCAs revisited, in Ardener, S. and Burman, S. (eds) *Money Go-Rounds: The Importance of Rotating Savings and Credit Associations for Women,* Berg, Oxford.

Ardener, S. and Burman, S. (eds) (1995) *Money Go-Rounds: The Importance of Rotating Savings and Credit Associations for Women,* Berg, Oxford.

Argent, N. (2002) From pillar to post? In search of the post-productivist countryside in Australia, *Australian Geographer,* **33,** 97–114.

Arvidsson, A. (2006) *Brands: Meaning and Value in Media Culture,* Routledge, London.

Ashley, R. (1987) Untying the sovereign state, *Millennium,* **17,** 227–62.

Ashlin, A. and Ladle, R. (2006) Environmental science adrift in the blogosphere, *Science,* **312,** 201.

Atkin, S. and Valentine, G. (eds) (2006) *Approaches to Human Geography,* Sage, London and Thousand Oaks, New Delhi.

Atkins, P. and Bowler, I. (2001) *Food in Society: Economy, Culture and Geography,* Arnold, London.

Atkinson, A. (1991) *Principles of Political Ecology,* Belhaven, London.

Atkinson, A. (2007a) Cities after oil – 1: 'sustainable development' and energy futures, *City*, **11**(2), 201–13.

Atkinson, A. (2007b) Cities after oil – 2: background to the collapse of 'modern' civilisation, *City*, **11**(3), 293–312.

Atkinson, A. (2008) Cities after oil – 3: collapse and the fate of cities, *City*, **12**(1), 79–106.

Atkinson, D. and Cosgrove, D. (1998) Urban rhetoric and embodied identities: city, nation and empire at the Vittorio Emanuele II monument in Rome, 1870–1945, *Annals of the Association of American Geographers*, **88**, 28–49.

Atkinson, R. and Blandy, S. (2006) *Gated Communities*, Routledge, London.

Atkinson, R. and Flint, J. (2004) Fortress UK? Gated communities, the spatial revolt of the elite and time-space trajectories of segregation, *Urban Studies*, **19**(6), 875–92.

Auty, R. (1979) World within worlds, *Area*, **11**, 232–5.

Auty, R. (1993) *Sustaining Development in Mineral Economies: The Resource Case Thesis*, Routledge, London.

Auty, R. (2001) *Resource Abundance and Economic Development*, Oxford University Press, Oxford.

B

Bachmann, V. and Sidaway, J.D. (2008) *Zivilmacht Europa*: a critical geopolitics of the European Union as a global power, *Transactions of the Institute of British Geographers*, NS, **34**, 94–109.

Bagaeen, S. and Uduku, O. (eds) (2010) *Gated Communities: Social Sustainability and Historical Gated Developments*, Earthscan, London.

Bagemihl, B. (1999) *Biological Exuberance: Animal Homosexuality and Natural Diversity*, Martin's Press, New York.

Bakhtin, M. (1984) *Rabelais and his World* (trans. Helen Iswolsky), Indiana University Press, Bloomington, IN.

Balibar, E. (2003) Which power? Whose weakness? On Robert Kagan's critique of European ideology, *Theory and Event*, **6**.

Ballinger, P. (2003) *History in Exile: Memory and Identity at the Borders of the Balkans*, Princeton University Press, Princeton, NJ.

Banrffalo, R. (1996) *Interview with David Harvey: The Politics of Social Justice*, available at: www.uky.edu/AS/SocTheo/DisClosure/Harveydi.htm.

Barkawi, T. and Laffey, M. (2002) Retrieving the imperial: empire and international relations, *Millennium*, **31**, 109–27.

Barnes, T. and Farish, M. (2006) Between regions: science, militarism, and American geography from World War to Cold War, *Annals of the Association of American Geographers*, **96**(4), 858–63.

Barnet, R.J. (1973) *Roots of War: The Men and Institutions behind US Foreign Policy*, Penguin, Harmondsworth.

Barnett, C., Cloke, P., Clarke, N. and Malpass, A. (2005) Consuming ethics: articulating the subjects and spaces of ethical consumption, *Antipode*, **37**(1), 23–45.

Barnett, C., Cloke, P., Clarke, N. and Malpass, A. (2011) *Globalizing Responsibility: The Political Rationalities of Ethical Consumption*, Wiley Blackwell, Oxford.

Barnett, D. (1998) *London: Hub of the Industrial Revolution: A Revisionary History: 1775–1825*, Tauris Academic Studies, London.

Barnett, J. (2007) The geopolitics of climate change, *Geography Compass*, **1**(6), 1361–75.

Bartells, F.L. and Lederer, S. (2009), *Outsourcing Markets in Services: International Business Trends, Patterns and Emerging Issues and the Role of China and India*, United Nations, Research and Statistics Branch, Working Paper 03/2009.

Barthes, R. (1972) *Mythologies*, Paladin, London.

Bassin, M. (1987) Race contra space: the conflict between German Geopolitik and National Socialism, *Political Geography Quarterly*, **6**, 115–34.

Bater, J.H. (1986) Some recent perspectives on the Soviet city, *Urban Geography*, **7**, 93–102.

Batterbury, S. and Warren, A. (2001) Desertification, in Smelser, N. and Baltes, P. (eds), *International Encyclopædia of the Social and Behavioral Sciences*, Elsevier Press, Amsterdam, 3526–29.

Bayley, C. (1989) *Atlas of the British Empire*, Hamlyn, London.

BBC *see* British Broadcasting Corporation.

Beaverstock, J.V., Hubbard, P. and Short, J.R. (2004) Getting away with it? Exposing the geographies of the super-rich, *Geoforum*, **35**, 401–7.

Beck, P. (1986) *The International Politics of the Antarctic*, St Martin's Press, London.

Beddington, J. (2009) *Food, Energy, Water and the Climate: A Perfect Storm of Global Events?*, Government Office for Science, London, available at: www.bis.gov.uk/assets/bispartners/goscience/docs/p/perfect-storm-paper.pdf [accessed 12 May 2011].

Beddington, J. (2010) Food security: contributions from science to a new and greener revolution, *Philosophical Transactions of the Royal Society B*, **365**, 61–71.

Bedford, T. (1999) Ethical consumerism, consumption, identity and ethics, unpublished PhD thesis, University of London, London.

Belasco, W. (2008) *Food: The Key Concepts*, Berg Publishers, Oxford.

Bell, D. and Jayne, M. (2006) *Small Cities: Urban Experience Beyond the Metropolis*, Routledge, Abingdon.

Bell, D. and Valentine, G. (eds) (1995) *Mapping Desire: Geographies of Sexualities*, Routledge, London.

Bell, D. and Valentine, G. (1997) *Consuming Geographies: We are Where we Eat*, Routledge, London.

Bell, M. (1994) Images, myths and alternative geographies of the Third World, in Gregory, D., Martin, R. and Smith, G. (eds) *Human Geography: Society, Space and Social Science*, Macmillan, London, 174–99.

Benko, G. and Strohmayer, U. (2004) *Human Geography: A History for the Twenty-First Century*, Arnold, London.

Benton, L. (1995) Will the real/reel Los Angeles please stand up? *Urban Geography*, **16**, 144–64.

Benton-Short, L. and Short, J.R. (2008) *Cities and Nature*, Routledge, London.

Berdahl, D. (1998) *Where the World Ended: Reunification and Identity in the German Borderland*, University of California Press, Berkeley, CA.

Beresford, M.W. (1988) *East End, West End: The Face of Leeds during Urbanisation 1684-1842*, Thoresby Society, Leeds.

Berg, L.D. and Vuolteenaho, J. (eds) (2009) *Critical Toponymies: The Contested Politics of Place Naming*, Ashgate, Farnham.

Berger, J. (1990) *Ways of Seeing*, Penguin, London.

Berger, M.T. (1994) The end of the 'Third World'?, *Third World Quarterly*, **15**, 257–75.

Berger, M.T. (2001) The post-cold war predicament: a conclusion, *Third World Quarterly*, **22**, 1079–85.

Berger, P. and Luckmann, T. (1966) *The Social Construction of Reality: A Treatise in the Sociology of Knowledge*, Anchor Books, Garden City, New York.

Betjeman, J. (1937) *Continual Dew*, John Murray, London.

Betts, V. (2002) Geographies of youth, religion and identity in (post-)socialist Cuba, unpublished PhD thesis, University of Birmingham.

Bhabha, H.K. (1994) *The Location of Culture*, Routledge, London.

Bialasiewicz, L., Elden, S., Painter, J. (2005) The constitution of EU territory, *Comparative European Politics*, **3**, 333–63.

Bilgin, P. (2007) Only strong states can survive in Turkey's geography: the uses of geopolitical truths in Turkey, *Political Geography*, **26**(7): 740–56.

Bingham, N., Blowers, A. and Belshaw, C. (2003) *Contested Environments*, Open University Press, Milton Keynes.

Birks, H. (1997) Environmental change in Britain: a long-term paleaoecological perspective, in Mackay, A. and Murlis, J. (eds) *Britain's Natural Environment: A State of the Nation Review*, Ensis, London, 23–8.

Blackburn, R. (2006) Finance and the fourth dimension, *New Left Review*, **39**, 39–70.

Blacksell, M. (1994) Irredentism, in Johnson, R.J., Gregory, D. and Smith, D.M. (eds) *The Dictionary of Human Geography*, 3rd edition, Blackwell, Oxford, 299.

Blacksell, M. (2005) *Political Geography*, Routledge, London.

Blakely, E.J. and Snyder, M.G. (1997) *Fortress America: Gated Communities in the United States*, The Brookings Institution, Washington, DC.

Blankenburg, S. and Palma, J. (2009) Introduction: the global financial crisis, *Cambridge Journal of Economics*, **33**, 531–8.

Blaut, J.M. (1993) *The Colonizer's Model of the World: Geographical Diffusionism and Eurocentric History*, Guilford Press, New York and London.

Blinder, A.S. (2006) Offshoring: the next industrial revolution, *Foreign Affairs*, Mar/Apr, 113–28.

Blomley, N.K. (1994) *Law, Space, and the Geographies of Power*, Guilford Press, New York.

Blomley, N.K. (2004) *Unsettling the City: Urban Land and the Politics of Property*, Routledge, New York.

Blomley, N.K., Delaney, D. and Ford, R. (eds) (2001) *The Legal Geographies Reader: Law, Power, and Space*, Blackwell, Oxford.

Bloom, E. (1999) Establishment of the Arctic Council, *The American Journal of International Law*, **93**, 712–22.

Bluestone, B. and Harrison, B. (1982) *The Deindustrialization of America*, Basic Books, New York.

Blunden, J. (1995) Sustainable resources?, in Sarre, P. and Blunden, J. (eds) *An Overcrowded World: Population, Resources and Environment*, Oxford University Press and Open University, Oxford, 161–213.

Blunt, A. and Dowling, R. (2006) *Home*, Routledge, London.

Blunt, A. and Wills, J. (2000) *Dissident Geographies: An Introduction to Radical Ideas and Practice*, Prentice Hall, Harlow.

Bobek, H. (1962) The main stages in socio-economic evolution from a geographical point of view, in Wagner, P.L. and Mikesell, M.W. (eds) *Readings in Cultural Geography*, University of Chicago Press, Chicago, IL, 218–47.

Bonnett, A. (2003) Geography as the world discipline: connecting popular and academic geographical imaginations, *Area*, **35**, 55–63.

Bonnett, A. (2008) *What is Geography?* Sage, London and Thousand Oaks, New Delhi.

Bonnett, A. and Nayak, A. (2003) Cultural geographies of racialization – the territory of race, in Anderson, K., Domosh, M., Pile, S. and Thrift, N. (eds) *Handbook of Cultural Geography*, Sage, London, 300–12.

Booth, K. (1999) Cold wars of the mind, in Booth, K. (ed.) *Statecraft and Security: The Cold War and Beyond*, Cambridge University Press, Cambridge, 29–55.

Borden, I. (2001) *Skateboarding, Space and the City: Architecture and the Body*, Berg, Oxford.

Borgelt, C., Ganssauge, K. and Keckstein, V. (1987*) Mietshaus im Wandel, Wohnungen der behutsamen Stadterneuerung*, S.T.E.R.N, Berlin.

Borgerson, S. (2007) Arctic meltdown, *Foreign Affairs*, **87**, 63–77.

Bornstein, D. (1996) *The Price of a Dream: The Story of the Grameen Bank and the Idea that is Helping the Poor to Change their Lives*, University of Chicago Press, Chicago, IL.

Borstelmann, T. (2001) *The Cold War and the Color Line: American Race Relations in the Global Arena*, Harvard University Press, Cambridge, MA.

Bosco, F. J. (2006) Actor-network theory, networks and relational approaches in human geography, in Aitken, S. and Valentine, G. (eds) *Approaches to Human Geography*, Sage, London, 136–46.

Boserüp, E. (1990) *Economic and Demographic Relationships in Development*, Johns Hopkins University Press, Baltimore, MD.

Bourdieu, P. (1984) *Distinction*, Routledge & Kegan Paul, London.

Boyd, D. and Ellison, N. (2007) *Social Network Sites: Definition, History, and Scholarship*, Guilford College, Greensboro, NC, available at: www.guilford.edu/about_guilford/services_and_administration/library/libguide_images/boyd.pdf.

BP (2010) *BP Statistical Review of World Energy June 2010*, BP, London, available at: www.bp.com.

Brah, A. (1996) *Cartographies of Diaspora: Contesting Identities*, Routledge, London.

Bradshaw, M.J. (2009) The geopolitics of global energy security, *Geography Compass*, **3**(5), 1920–37.

Bradshaw, M.J. (2010) Global energy dilemmas: a geographical perspective, *The Geographical Journal*, **176**(4), 275–90.

Bradshaw, M.J. (2012) *Global Energy Dilemmas, Energy Security, Globalization and Climate Change*, Polity Press, Cambridge. Forthcoming.

Braidotti, R., Charkiewicz, E., Hausler, S. and Wieringa, S. (eds) (1994) *Women, Environment and Sustainable Development: Towards a Theoretical Synthesis*, Zed, London.

Brandt, W. (1980) *North–South: A Programme for Survival*, Pan, London.

Braun, B. (2002) *The Intemperate Rainforest: Nature, Culture and Power on Canada's West Coast*, University of Minnesota Press, Minneapolis, MN.

Bray, K. and Kilian, T. (1999) God does not pay for it: the political economy of cyberspace. Paper presented at the 95th Annual Meeting of the Association of American Geographers, Honolulu.

Breckenridge, C.A. (ed.) (1995) *Consuming Modernity: Public Culture in a South Asian World*, University of Minnesota Press, Minneapolis, MN.

Breheny, M. (1995) The compact city and transport energy consumption, *Transactions of the Institute of British Geographers NS*, **20**(1), 81–101.

Brenner, N. (2004) *New State Spaces: Urban Governance and the Rescaling of Statehood*, Oxford University Press, New York.

Brenner, N., Jessop, B., Jones, M. and Macleod, G. (eds) (2002) *State/Space: A Reader*, Blackwell, Oxford.

Breward, C., Crang, P. and Crill, R. (eds) (2010) *British Asian Style: Fashion and Textiles/Past and Present*, V&A Publishing, London.

Bridge, G. and Watson, S. (eds) (2000) *A Companion to the City*, Blackwell, Oxford.

Brighenti, A.M. (2010a) On territorology: towards a general science of territory, *Theory, Culture and Society*, **27**(1), 52–72.

Brighenti, A.M. (2010b) At the wall: graffiti writers, urban territoriality, and the public domain, *Space and Culture*, **13**(3), 315–32.

British Broadcasting Corporation (BBC) (1999) *GM Foods: An Evaluation*. Online, available at: www.bbc.co.uk/science [accessed 12 December 2002].

British Broadcasting Corporation (BBC) (2005a) *The Changing Face of Poverty*. http://news.bbc.co.uk/1/hi/business/4070112.stm [accessed 23 May 2007].

British Broadcasting Corporation (BBC) (2005b) *Hurricane Aid Efforts Gather Pace*. http://news.bbc.co.uk/1/hi/world/americas/4210646.stm [accessed 23 May 2007].

British Broadcasting Corporation (BBC) (2006a) *Oslo Gay Animal Show Draws Crowds*, http://news.bbc.co.uk/1/hi/world/europe/6066606.stm.

British Broadcasting Corporation (BBC) (2006b) *Zero Carbon Homes Plan Unveiled*, http://news.bbc.co.uk/go/pr/fr/-/1/hi/sci/tech/6176229.stm [accessed 13 December 2006].

Brooks, R. (2005) *Our Homegrown Third World*, www.commondreams.org/views05/0907-24.htm [accessed 23 May 2007].

Broome, R. (1994) *Aboriginal Australians: Black Responses to White Dominance, 1788–1994*, 2nd edition, Allen & Unwin, St Leonards, NSW.

Brown, L. (2001) *Eco-economy: Building an Economy for the Earth*, Norton, New York.

Brown, L.R. (2011) *World on the Edge: How to Prevent Environmental and Economic Collapse*, W.W. Norton & Company, New York.

Browne, K. and Lim, J. (2010) Trans lives in the 'gay capital of the UK', *Gender, Place and Culture*, **17**(5), 615–33.

Browne, K., Lim, J. and Brown, G. (eds) (2007) *Geographies of Sexualities: Theory, Politics and Practice*, Ashgate, Aldershot.

Brunori, G. and Guarino, A. (2010) Security for whom? Changing discourses on food in Europe in times of a global food crisis, in Lawrence, G., Lyons, K. and Wallington, T. (eds) *Food Security, Nutrition and Sustainability*, Earthscan, London, 41–60.

Bryman, A. (1995) *Disney and his Worlds,* Routledge, London.

Bryson, J.R. (2000) Spreading the message: management consultants and the shaping of economic geographies in time and space, in Bryson, J.R., Daniels, P.W., Henry, N.D. and Pollard, J.S. (eds) *Knowledge Space, Economy*, Routledge, London, 157–75.

Bryson, J.R. (2007) A 'second' global shift? The offshoring or global sourcing of corporate services and the rise of distanciated emotional labour, *Geografiska Annaler*, **89B**(1), 31–44.

Bryson J.R. (2009) Economic geography: business services, in Kitchin R. and Thrift N. (eds) *International Encyclopedia of Human Geography*, Elsevier, Amsterdam, 368–74.

Bryson, J.R. and Daniels, P.W. (eds) (2007) *The Handbook of Service Industries*, Edward Elgar, Cheltenham.

Bryson, J.R. and Henry, N.D. (2005) The global production system: from Fordism to post Fordism, in Daniels, P.W. et al. *Human Geography: Issues for the 21st Century*, Prentice Hall, London, 313–36.

Bryson, J.R. and Rusten, G. (2005) Spatial divisions of expertise: knowledge intensive business service firms and regional development in Norway, *The Service Industries Journal*, **25**(8) 959–77.

Bryson, J.R. and Rusten G. (2006) Spatial divisions of expertise and transnational 'service' firms: aerospace and management consultancy, in Harrington, J.W. and Daniels, P.W. (eds) *Knowledge-based Services, Internationalisation and Regional Development*, Aldershot, Ashgate, 79–100.

Bryson, J.R. and Rusten, G. (2008) Transnational corporations and spatial divisions of 'service' expertise as a competitive strategy: the example of 3M and Boeing, *The Service Industries Journal*, **28**(3), 307–323.

Bryson, J.R. and Rusten, G. (2010), *Design Economies and the Changing World Economy: Innovation, Production and Competitiveness*, Routledge, London.

Bryson, J.R. and Taylor, P. (2006) *The Functioning Economic Geography of the West Midlands Region, Advantage West Midlands*, Birmingham, West Midlands Regional Observatory.

Bryson, J.R. and Wellington, C. (2003) Image consultancy in the United Kingdom: recipe knowledge and recreational employment, *The Service Industries Journal*, **23**(1) 59–76.

Bryson, J., Daniels, P.W. and Henry, N. (1996) From widgets to where? A region in economic transition, in Gerrard, A.J. and Slater, T.R. (eds) *Managing a Conurbation: Birmingham and its Region*, Brewin Books, Birmingham, 56–168.

Bryson, J.R., Daniels, P.W. and Warf, B. (2004) *Service Worlds: People, Organizations, Technologies*, Routledge, London.

Bryson, J.R., McGuinness, M. and Ford, R.G. (2002) Chasing a loose and baggy monster: almshouses and the geographies of charity, *Area*, **34**(1), 48–58.

Bryson, J.R., Taylor, M. and Cooper, R. (2008) Competing by design, specialization and customization: manufacturing locks in the West Midlands (UK), *Geografiska Annaler: Series B, Human Geography*, **90**(2), 173–86.

Bryson, J., Henry, N., Keeble, D. and Martin, R. (eds) (1999) *The Economic Geography Reader: Producing and Consuming Global Capitalism*, John Wiley, Chichester.

Buckingham-Hatfield, S. (2000) *Gender and Environment*, London, Routledge.

Bull, H. (1995) *The Anarchical Society*, Oxford University Press, Oxford.

Bullard, R.D. (1990) *Dumping in Dixie: Race, Class and Environmental Quality*, Westview Press, Boulder, Co.

Bundy, D. and Gotur, M. (2002) *Education and HIV/AIDS: A Window of Hope*, World Bank, Washington.

Bunnell, T., Muzaini, H. and Sidaway, J.D. (2006) Global city frontiers: Singapore's hinterland and the contested socio-political geographies of Bintan, Indonesia, *International Journal of Urban and Regional Research*, **30**(1), 3–22.

Bunyard, P. and Morgan-Grenville, F. (eds.) (1987) *The Green Alternative*, Methuen, London.

Burgess, E.W., Park, R.E. and McKenzie, R.D. (1925) *The City*, University of Chicago Press, Chicago, IL.

Burgess, J., Bedford, T., Hobson, K., Davies, G. and Harrison, C. (2003) (Un)sustainable consumption, in Berkhout, F., Leach, M. and Scoones, I. (eds) *Negotiating Environmental Change: New Perspectives from Social Science*, Edward Elgar, Cheltenham, 261–91.

Bush, G. (2007) *President Bush Delivers State of the Union Address*, 23 January 2007, www.whitehouse.gov/news/releases/2007/01/20070123-2.html [accessed 25 May 2007].

Butler, T. (2007) For gentrification?, *Environment and Planning A*, **39**(1), 162–81.

Butler, T. and Hamnett C. (2010) 'You take what you are given': the limits to parental choice in education in east London, *Environment and Planning A*, **42**(10), 2431–50.

Butler, T. and Lees, L. (2006) Super-gentrification in Barnsbury, London: globalization and gentrifying global elites at the neighbourhood level, *Transactions of the Institute of British Geographers NS*, **31**(4), 467–87.

Butlin, R.A. (1993) *Historical Geography, Through the Gates of Time and Space*, Edward Arnold, London.

Buzan, B. (2006) Will the 'global war on terrorism' be the new Cold War? *International Affairs*, **82**(6) 1101–18.

Byrne, D. (2001) *Understanding the Urban*, Palgrave, London.

C

Caldeira, T. (1996) Building up walls: the new pattern of social segregation in São Paulo, *International Social Science Journal*, **147**, 55–65.

Caldeira, T. (2001) *City of Walls: Crime, Segregation and Citizenship in São Paulo*, University of California Press, Berkeley, CA.

Caldwell, M.L. (2004) Domesticating the French fry: McDonald's and consumerism in Moscow, *Journal of Consumer Culture*, **4**, 5–26.

Calhoun, C. (2002) The class consciousness of frequent travellers: towards a critique of actually existing cosmopolitanism, in Vertovec, S. and Cohen, R. (eds) *Conceiving Cosmopolitanism: Theory, Context and Practice*, Oxford University Press, Oxford, 86–109.

Callicott, J.B. and Rocha, F.J.R. (eds) (1996) *Earth Summit Ethics: Toward a Reconstructive Postmodern Philosophy of Environmental Education*, State University of New York Press, New York.

Calvocoressi, P. (1991) *World Politics since 1945*, Longman, London.

Cameron, A. (2007) Geographies of welfare and exclusion: reconstituting the 'public', *Progress in Human Geography*, **31**(4), 519–26.

Campbell, B. (1993) *Goliath: Britain's Dangerous Places*, Methuen, London.

Campbell, C. (1987) *The Romantic Ethic and the Spirit of Modern Consumerism*, Blackwell, Oxford.

Campbell, D. (1992) *Writing Security: United States Foreign Policy and the Politics of Identity*, University of Minnesota Press, Minneapolis, MN.

Campbell, D. (1999) Apartheid cartography: the political anthropology and spatial effects of international diplomacy in Bosnia, *Political Geography*, **18**(4) 395–435.

Carapico, S. (1985) Yemeni agriculture in transition, in Beaumont, P. and McLachlan, K. (eds) *Agricultural Development in the Middle East*, John Wiley & Sons, Chichester, 241–54.

Carley (1999) Neighbourhoods – building blocks of national sustainability, *Town and Country Planning*, February, 58–60.

Carneiro, A. de M. (1990) *Descrição da Fortaleza de Sofala e das mais da Índia*, Fundação Oriente, Lisbon.

Carr, J. (2010) Legal geographies – skating around the edges of the law: urban skateboarding and the role of law in determining young peoples' place in the city, *Urban Geography*, **31**(7), 988–1003.

Carroll, R. (1998) The chill east wind at your doorstep, *The Guardian*, 28 October, 2–3.

Carson, R. (1962) *Silent Spring*, Hamish Hamilton, London.

Carter, D. (1988) *The Final Frontier: The Rise and Fall of the American Rocket State*, Verso, London and New York.

Carter, N. (2001) *The Politics of the Environment: Ideas, Activism, Policy*, Cambridge University Press, Cambridge.

Caspian Revenue Watch (2003) *Caspian Oil Windfalls: Who Will Benefit?*, The Open Society Institute, New York. Available at: www.revenuewatch.org/reports

Castells, M. (1989) *The Informational City: Information Technology, Economic Restructuring and the Urban and Regional Process*, Blackwell, Oxford.

Castells, M. (1996) *The Information Age: Economy, Society and Culture*, vol. 1, *The Rise of the Network Society*, Blackwell, Oxford.

Castells, M. (1997) *The Power of Identity*, Blackwell, Malden, MA.

Castells, M. (1998) *The Information Age: Economy, Society and Culture*, vol. 3, *End of Millennium*, Blackwell, Oxford.

Castles, S. and Miller, M.J. (2003) *The Age of Migration: International Population Movements in the Modern World*, Palgrave Macmillan, Basingstoke.

Castree, N. (2003) A post-environmental ethics? *Ethics, Place and Environment*, **6**(1) 3–12.

Castree, N. (2005) *Nature*, Routledge, London.

Castree, N (2011) Commentary: the future of geography in English universities, *The Geographical Journal*, **177**(4), 294–99.

Caulkin, S. (1998) Credit where credit's overdue, *Observer*, 15 November, 2.

Central Intelligence Agency (CIA) (2003) *The World Factbook*, CIA, Washington, DC.

Central Intelligence Agency (CIA) (2005) *The World Factbook: Uganda*, CIA, Washington, DC.

Central Intelligence Agency (2006) *The World Factbook* (access online at https://www.cia.gov/library/publications/the-world-factbook/index.html).

Chaliand, G. (1993) Introduction, in Chaliand, G. (ed.) *A People without a Country: The Kurds and Kurdistan*, 2nd edition, Zed, London, 1–10.

Chaliand, G. and Rageau, P. (1985) *Strategic Atlas: World Geopolitics*, Penguin, London.

Chambers, R. (1982) *Rural Development: Putting the Last First*, Longman, London.

Champion, A. (1999) Urbanisation and counterurbanisation, in Pacione, M. (ed.) *Applied Geography: Principles and Practice*, Routledge, London, 347–57.

Chandler, J. (ed.) (1993) *John Leland's Itinerary*, Sutton, Stroud.

Chaney, D. (1990) Subtopia in Gateshead, the Metro-Centre as a cultural form, *Theory, Culture and Society*, 7, 49–68.

Charlesworth, A., Stenning, A., Guzik, R. and Paszkowski, M. (2006) 'Out of place' in Auschwitz? Contested development in post-war and post-socialist Oświęcim, *Ethics, Place and Environment*, 9(2) 149–72.

Chatham House (2008) *Rising Food Prices: Drivers and Implications for Development*, A Chatham House Report, Chatham House, London.

Chatham House (2009) *Food Futures: Rethinking UK Strategy*, A Chatham House Report, Chatham House, London.

Chatterjee, P. (1993) *The Nation and its Fragments: Colonial and Postcolonial Histories*, Princeton University Press, Princeton, NJ.

Chatterton, P. (1999) University students and city centres: the formation of exclusive geographies, *Geoforum*, 30, 117–33.

Chiesura, A. (2003) The role of urban parks for the sustainable city, *Landscape and Urban Planning*, 68(1), 129–38.

Child, J.C. (1985) *Geopolitics and Conflict in South America: Quarrels among Neighbours*, Praeger, New York.

Chomsky, N. (1998) Power in the global arena, *New Left Review*, 230, 3–27.

Chossudovsky, M. (1997) *The Globalisation of Poverty*, Zed Books, London.

Christaller, W. (1966) *Central Places in Southern Germany* (trans. by C.W. Baskin), Prentice Hall, Englewood Cliffs, NJ.

Christopherson, S. (2008) Beyond the self-expressive creative worker: an industry perspective on entertainment media, *Theory, Culture and Society*, 25(7–8), 73–95.

Christopherson, S., and Clark, J. (2007) *Remaking the Region: Power, Labor, and Firm Strategies in the Knowledge Economy*, Routledge Press, New York.

Chung, R. (1970) Space–time diffusion of the demographic transition model: the twentieth-century patterns, in Demko, G.J., Rose, H.M. and Schnell, G.A. (eds) *Population Geography: A Reader*, McGraw-Hill, New York.

CIA *see* Central Intelligence Agency

City of London Corporation (2005a) *London's Place in the UK Economy, 2005–2006*, London, City of London Corporation.

City of London Corporation (2005b) *The Competitive Position of London as a Global Financial Centre*, London, City of London Corporation.

Clark, B. (2007) *Twice a Stranger: How Mass Expulsion Forged Modern Greece and Turkey*, Granta, London.

Clark, I. (2011) China and the United States: a succession of hegemonies?, *International Affairs*, 87(1), 13–28.

Clarke, G.E. (1996) Blood, territory and national identity in Himalayan states, in Tønnesson, S. and Antlöv, H. (eds) *Asian Forms of the Nation*, Curzon, Richmond, 205–36.

Clarke, N., Barnett, C., Cloke, P. and Malpass, A. (2007) Globalising the consumer: doing politics in an ethical register, *Geoforum*, 26(3), 231–49.

Claval, P. (1998) *An Introduction to Regional Geography*, Blackwell, Oxford.

Claval, P. (2009) National schools of geography, in Kitchin, R., Thrift, N. *et al.* (eds) *The International Encyclopedia of Human Geography*, Elsevier, Oxford, 236–41.

Clifford, J. (1992) Travelling cultures, in Grossberg, L., Nelson, C. and Treichler, P. (eds) *Cultural Studies*, Routledge, London.

Clifford, N. and Valentine, G. (eds) (2003) *Key Methods in Geography*, Sage, London.

Cloke, P. and Johnston, R. (eds) (2005) *Spaces of Geographic Thought: Deconstructing Human Geography's Binaries*, Sage, London.

Cloke, P., Crang, P. and Goodwin, M. (eds) (2004) *Envisioning Human Geographies*, Arnold, London.

Cloke, P., Crang, P. and Goodwin, M. (eds) (2005) *Introducing Human Geographies*, Arnold, London.

Cloke, P., Crang, P., Goodwin, M., Painter, J. and Philo, C. (2002) *Practicing Human Geography*, Sage, London.

Cloke, P., Philo, C. and Sadler, D. (1991) *Approaching Human Geography: An Introduction to Contemporary Human Geography*, Paul Chapman Publishing, London.

Cloke, P.J. (2000) Rural, in Johnston, R.J., Gregory, D., Pratt, G. and Watts, M. (eds) *The Dictionary of Human Geography*, Blackwell, Oxford, 718.

Cloke, P.J. (2005a) Conceptualising rurality, in Cloke, P., Marsden, T. and Mooney, P. (eds) *Handbook of Rural Studies*, Sage, London.

Cloke, P.J. (2005b) The country, in Cloke, P.J., Crang, P. and Goodwin, M. (eds), *Introducing Human Geographies*, 2nd edition, Arnold, London.

Cloke, P.J., Marsden, T. and Mooney, P. (eds) (2005) *Handbook of Rural Studies*, Sage, London.

Clover, C. (1999), Dreams of the Eurasian heartland: the reemergence of geopolitics, *Foreign Affairs*, 78(2), 9–13.

Coafee, J., Murakami Wood, D. and Rogers, P. (2009) *The Everyday Resilience of the City: How Cities Respond to Terrorism and Disaster*, Palgrave Macmillan, Basingstoke.

Cochrane, A. (1995) Global worlds and worlds of difference, in Anderson, J., Brook, C. and Cochrane, A. (eds) *A Global World*, Oxford University Press, Oxford, 249–80.

Cochrane, A. (2007) *Understanding Urban Policy: A Critical Approach*, Blackwell, Oxford.

Cockcroft, E. (1974) Abstract expressionism, weapon of the Cold War, *Artform*, **12**, 39–41.

Coe, N.M. (2011) Unpacking globalisation: changing geographies of the global economy, in Lee, R., Leyshon, A., McDowell, L. and Sunley, P. (eds) *The Compendium of Economic Geography*, Sage, London.

Coe, N.M., Dicken, P. and Hess, M. (2008) Global production networks: realizing the potential, *Journal of Economic Geography*, **8**, 271–95.

Coe, N.M., Kelly, P.F. and Yeung, H. (2007) *Economic Geography: A Contemporary Introduction*, Blackwell, Oxford.

Cohen, I. (1997) *Green Fire*, Harper Collins, Australia.

Coley, D., Howard, M. and Winter, M. (2009) Local food, food miles and carbon emissions: a comparison of farm shop and mass distribution approaches, *Food Policy*, **34**, 150–5.

Coll, S. (2004) *Ghost Wars: The Secret History of the CIA, Afghanistan and Bin Laden from Soviet Invasion to September 10, 2001*, Penguin, New York and London.

Collier, P. (2008) The politics of hunger: how illusion and greed fan the food crisis, *Foreign Affairs*, **87**(6), 67–79.

Commission of the European Communities (2007) *An Energy Policy for Europe*, Commission of the European Communities, Brussels.

Cone, C. and Myhre, A. (2000) Community supported agriculture: a sustainable alternative to industrial agriculture? *Human Organisation*, **59**, 187–97.

Connell, J. (2006) The taste of paradise: selling Fiji and Fiji water, *Asia Pacific Viewpoint*, **47**, 342–50.

Connelly, J. and Smith, G. (2003) *Politics and the Environment: From Theory to Practice*, 2nd edition, Routledge, London.

Connor, W. (1994) *Ethnonationalism: The Quest for Understanding*, Princeton University Press, Princeton, NJ.

Conzen, M.P. (1990) *The Making of the American Landscape*, Unwin Hyman, Boston, MA.

Cook, I. (2004) Follow the thing: papaya, *Antipode*, **36**(4), 642–64.

Cook, I. (2006) Geographies of food: following, *Progress in Human Geography*, **30**(5), 655–66.

Cook, I. *et al.* (2008) Geographies of food: mixing, *Progress in Human Geography*, **32**, 821–33.

Cook, I. *et al.*, (2000) Social sculpture and connective aesthetics: Shelley Sacks' 'exchange values', *Ecumene: A Journal of Cultural Geographies*, **7**, 337–44.

Cook, I. and Crang, P. (1996) The world on a plate: culinary culture, displacement and geographical knowledges, *Journal of Material Culture*, **1**(2), 131–53.

Cook, I., Crang, P. and Thorpe, M. (1999) Eating into Britishness: multicultural imaginaries and the identity politics of food, in Roseneil, S. and Seymour, J. (eds) *Practising Identities*, Macmillan, London, 223–48.

Cook, I., Crang, P. and Thorpe, M. (2000) Regions to be cheerful, culinary authenticity and its geographies, in Cook, I., Crouch, D., Naylor, S. and Ryan, J. (eds) *Cultural Turns/Geographical Turns*, Longman, London, 109–39.

Cook, I., Crang, P. and Thorpe, M. (2004) Tropics of consumption: getting with the fetish of 'exotic' fruit?, in Hughes, A. and Reimer, S. (eds) *Geographies of Commodities*, Routledge, London.

Cook, I., Crouch, D., Naylor, S. and Ryan, J. (eds) (2000) *Cultural Turns/Geographical Turns*, Prentice Hall, London.

Cook, I., Evans, J., Griffiths, H., Morris, B., Wrathmell, S. *et al.* (2007) 'It's more than just what it is': defetishising commodities, changing pedagogies, mobilising change, *Geoforum*, **38**(6), 1113–1126.

Cooke, B. and Kothari, U. (2001) *Participation: The New Tyranny?* Zed, London.

Cooper, D.E. (2001) Arne Naess, in Palmer, J.A. (ed.) *Fifty Key Thinkers on the Environment*, Routledge, London, 211–16.

Corbridge, S. (ed.) (1995) *Development Studies: A Reader*, Edward Arnold, London.

Corry, E. (2006) *The Nation Holds its Breath: Great Irish Soccer Quotations*, Hodder Headline Ireland, Dublin.

Cosgrove, D. (1994) Contested global visions, *Annals of the Association of American Geographers*, **84**, 270–94.

Cosgrove, D. (2001) *Apollo's Eye: A Cartographic Genealogy of the Earth in the Western Imagination*, Johns Hopkins University Press, Baltimore.

Cowen, M.P. and Shenton, R.W. (1996) *Doctrines of Development*, Routledge, London.

Cox, K., Low, M. and Robinson, J. (eds) (2007) *The Sage Handbook of Political Geography*, Sage, London.

Cox, M. (2001) The new liberal empire: US power in the twenty-first century, *Irish Studies in International Affairs*, **12**, 39–56.

Cozens, P.M., Hillier, D. and Prescott, G. (1999) Crime and the design of new-build housing, *Town and Country Planning*, **68**(7), 231–33.

Crampton, J. and Elden, S. (eds) (2007) *Space, Knowledge and Power: Foucault and Geography*, Ashgate, Aldershot.

Crang, M. (1998) *Cultural Geography*, Routledge, London.

Crang, P. (1997) Cultural turns and the (re)constitution of economic geography, in Lee, R. and Wills, J. (eds) *Geographies of Economies*, Arnold, London, 3–15.

Crang, P. and Ashmore, S. (2009) The transnational spaces of things: South Asian textiles in Britain and the grammar of ornament, *European Review of History*, **16**, 655–78.

Crang, P., Dwyer, C. and Jackson, P. (2003) Transnationalism and the spaces of commodity culture, *Progress in Human Geography*, **27**, 438–56.

Crawford, M. (1992) The world in a shopping mall, in Sorkin, M. (ed.) *Variations on a Theme Park: The New American City and the End of Public Space*, Noonday, New York.

Crawford, M. (2004) The world in a shopping mall, in Miles, M., Hall, T. and Borden, I. (eds) *The City Cultures Reader*, 2nd edition, Routledge, London.

Cresswell, T. (1996) *In Place/Out of Place: Geography, Ideology and Transgression*, University of Minnesota Press, Minneapolis, MN.

Cresswell, T. (2004) *Place: A Short Introduction,* Blackwell, Oxford.

Crone, P. (1986) The tribe and the state, in Hall, J.A. (ed.) *States in History*, Basil Blackwell, Oxford, 48–77.

Crone, P. (1989) *Pre-Industrial Societies*, Blackwell, Oxford.

Cronon, W. (1991) *Nature's Metropolis: Chicago and the Great West*, W.W. Norton, New York.

Cronon, W. (1996) *Uncommon Ground: Toward Reinventing Nature,* Norton, New York.

Crouch, D.P., Garr, D.J. and Mundigo, A.I. (1982) *Spanish City Planning in North America*, MIT Press, Cambridge, MA.

Crum, R.E. and Gudgin, G. (1977) *Non-production Activities in UK Manufacturing Industry*, Commission of the European Communities, Brussels.

Crump, J.R. (2004) Producing and enforcing the geography of hate: race, housing segregation, and housing-related hate crimes in the United States, in Flint, C. (ed.) *Spaces of Hate: Geographies of Discrimination and Intolerance in the USA*, Routledge, New York, 227–44.

Crush, J. (1995) Imagining development, in Crush, J. (ed.) *Power of Development*, Routledge, London, 1–26.

Cummins, S. and MacIntyre, S. (1999) Food deserts, evidence and assumptions in health policy-making, *British Medical Journal*, **325**, 436–8.

Cummins, S., Findlay, A., Higgins, C., Petticrew, M., Sparks, L. and Thomson, H. (2008) Reducing inequalities in health and diet: findings from a study on the impact of a food retail development, *Environment and Planning A*, **40**(2), 402–22.

Curry, J. and Kenney, M. (2004) The organizational and geographic configuration of the personal computer value chain, in Kenney, M. and Florida, R. (eds) *Locating Global Advantage: Industry Dynamics in the International Economy*, Stanford University Press, Stanford, 113–41.

Cybriwsky, R. and Ford, L. (2001) Jakarta, *Cities*, **18**, 199–210.

D

D'Silva, J. and Webster, J. (eds) (2010) *The Meat Crisis: Developing More Sustainable Production and Consumption*, Earthscan, London.

Daily Star (1995) Korma out for a meal, *Daily Star*, 5 December.

Dalby, S. (2003) Geopolitics, the Bush doctrine and war in Iraq, *Arab World Geographer*, **6**, 7–18.

Danforth, L.M. (1995) *The Macedonia Conflict: Ethnic Nationalism in a Transnational World*, Princeton University Press, Princeton, NJ.

Daniels, P.W. (1975) *Office Location: An Urban and Regional Study*, G. Bell, London.

Daniels, S. (2011) Geographical imagination, *Transactions of the Institute of British Geographers NS*, **36**(2), 182–7.

Daniels, P.W. and Bryson, J.R. (2002) Manufacturing services and servicing manufacturing: changing forms of production in advanced capitalist economies, *Urban Studies*, **39**(5–6), 977–91.

Daniels, P.W. and Bryson, J.R. (2005) Sustaining business and professional services in a second city region: the case of Birmingham, UK, *The Service Industries Journal*, **25**(4), 505–24.

Daniels, S. and Lee, R. (eds) (1996) *Exploring Human Geography: A Reader*, Arnold, London.

Dankelman, I. and Davidson, J. (1988) *Women and Environment in the Third World*, Earthscan, London.

Darwin, J. (2007) *After Tamerlane: The Rise and Fall of Global Empires*, Penguin, Harmondsworth.

Datta, K. (2007) *Money Matters: Exploring Financial Exclusion Among Low Paid Migrant Workers in London*, Queen Mary, University of London, London.

Davidson, M. and Lees, L. (2010) New-build gentrification: its histories, trajectories, and critical geographies, *Population, Space and Place*, **16**(5), 395–411.

Davies, G. (1994) *A History of Money: From Ancient Times to the Present Day*, University of Wales Press, Cardiff.

Davies, G. (2000) Narrating the natural history unit, *Geoforum*, **31**, 539–51.

Davies, J., Sandström, S., Shorrocks, A. and Wolff, E. (2006) *World Distribution of Household Wealth*. London:

Foreign Press Association and United Nations Secretariat, presentation 5 December.

Daviron, B. and Ponte, S. (2005) *The Coffee Paradox: Global Markets, Commodity Trade and the Elusive Promise of Development*, Zed Books, London.

Davis, A., Hirsch, D. and Smith, N. (2010) *A Minimum Income Standard for the UK in 2010*, Joseph Rowntree, York.

Davis, J. and Goldberg, R. (1957) *A Concept of Agribusiness*, Harvard University Press, Boston, MA.

Davis, M. (1990) *City of Quartz: Excavating the Future in Los Angeles,* Verso, London.

Davis, M. (1995) Fortress Los Angeles: The militarization of urban space, in Kasinitz, P. (ed.) *Metropolis: Centre and Symbol of our Times*, Macmillan, Basingstoke, 355–68.

Davis, M. (1998) *Ecology of Fear: Los Angeles and the Imagination of Disaster*, Metropolitan Books, New York.

Davis, T. (1995) The diversity of queer politics and the redefinition of sexual identity and community in urban spaces, in Bell, D. and Valentine, G. (eds) *Mapping Desire: Geographies of Sexualities*, Routledge, London, 284–303.

De Genova, N. and Peutz, N. (eds) (2010) *The Deportation Regime: Sovereignty, Space and the Freedom of Movement*, Duke University Press, Durham, NC.

de Ru, N. (2004) Hollocore, in Koolhas, R., Brown, S. and Link, J. (eds) *Content*, Taschen, Cologne, 336–49.

Dean, K. (2005) Spaces and territorialities on the Sino-Burmese boundary: China, Burma and the Kachin, *Political Geography*, **24**, 808–30.

Deane, P. and Cole, W. (1962) *British Economic Growth 1688–1959*, Cambridge University Press, Cambridge.

Dear, M. (2000) *The Postmodern Urban Condition,* Blackwell: Malden, MA.

Dear, M. (ed.) (2002) *From Chicago to L.A.: Making Sense of Urban Theory*, Sage, Thousand Oaks, CA.

Dear, M. and Flusty, S. (1998) Postmodern urbanism, *Annals of the Association of American Geographers*, **88**(1), 50–72.

Dear, M. and Flusty, S. (2002) The resistable rise of the LA school, in Dear, M. (ed.) (2002*) From Chicago to L.A.: Making Sense of Urban Theory*, Sage, Thousand Oaks, CA, 3–16.

Debrix, F. (2007) Tabloid imperialism: American geopolitical anxieties and the war on terror, *Geography Compass*, **1**(4), 932–945.

Dedrick, J. and Kraemer, K.L. (2006) Is production pulling knowledge work to China? A study of the notebook PC industry, *Computer*, **39**, 36–42.

Dedrick, J., Kraemer, K.L. and Linden, G. (2010) Who profits from innovation in global value chains? A study

of the iPod and notebook PCs, *Industrial and Corporate Change*, **19**(1), 81–116.

Deffeyes, K.S. (2001) *Hubbert's Peak: The Impending World Oil Shortage*, Princeton University Press, Princeton, NJ.

Dekmejian, R.H. and Simonian, H.H. (2003) *Troubled Waters: The Geopolitics of the Caspian Region*, I B Tauris, London.

Del Casino, V.J. (2009) *Social Geography: A Critical Introduction*, Wiley-Blackwell, Chichester.

Delaney, D. (2005) *Territory: A Short Introduction*, Blackwell, Malden.

Delaney, D. (2009) Territory and territoriality, in Kitchin, R. and Thrift, N. *et al.* (eds) *International Encyclopedia of Human Geography*, Elsevier, Oxford.

Delanty, G. and Kumar, K. (eds) (2006) *Handbook of Nations and Nationalism*, Sage, London.

Demeritt, D. (1998) Science, social constructivism and nature, in Castree, N. and Braun, B. (eds) *Remaking Reality: Nature at the Millennium*, Routledge, London, 173–93.

Demeritt, D. (2002) What is the social construction of nature? *Progress in Human Geography*, **26**, 767–90.

Denevan, W. (1992) The pristine myth: the landscapes of the Americas in 1492, *Annals of the Association of American Geographers*, **82**, 369–85.

Dennis, R. (1984) *English Industrial Cities of the Nineteenth Century*, Cambridge University Press, Cambridge.

Department for International Development (DFID) (2005) *Country Profile: Uganda,* copy available at: www .dfid.gov.uk/countries/africa/uganda.asp [accessed 23 May 2007].

Department of Energy and Climate Change (2010) *The Green Deal: Energy Savings for Homes and Business*, DECC, London, available at: www.decc.gov.uk.

Department of Work and Pensions (2010) *Households Below Average Income 2009/10*, National Statistics Office, London.

Derudder, B. and Wilcox, F. (2005) An appraisal of the use of airline data in assessing the world city network: a research note on data, *Urban Studies*, **42**(13), 2371–88.

Dery, M. (1993) *Culture Jamming: Hacking, Slashing and Sniping in the Empire of Signs*, Open Magazine Pamphlet Series, Open Magazine.

Desai, V. and Potter, R. (eds) (2001) *The Arnold Companion to Development Studies,* Arnold, London.

Desforges, L. (1998) 'Checking out the planet': global representations/local identities and youth travel, in Skelton, T. and Valentine, G. (eds) *Cool Places: Geographies of Youth Cultures*, Routledge, London, 174–91.

Destouni, G. and Frank, H. (2010) Renewable energy, *Ambio*, **39**, 18–21.

Devall, B. and Sessions, G. (1985) *Deep Ecology: Living as if Nature Mattered*, Peregrine Smith Books, Salt Lake City, UT.

DFID *see* Department for International Development.

Diagne, K. (2007) Governance and natural disasters: addressing flooding in Saint Louis, Senegal, *Environment and Urbanization*, **19**, 552–62.

Dicken, P. (1992) *Global Shift: The Internationalization of Economic Activity*, Paul Chapman, London.

Dicken, P. (2003) *Global Shift: Reshaping the Global Economic Map in the 21st Century*, 4th edition, Sage, London.

Dicken, P. (2007) *Global Shift: Mapping the Changing Contours of the World Economy,* 5th edition, Sage, London.

Dicken, P. (2011) *Global Shift*, 6th edition, Sage, London.

Dicken, P. and Lloyd, P. E. (1990) *Location in Space: Theoretical Perspectives in Economic Geography*, Harper & Row, New York.

Dicken, P., Kelly, P.F., Olds, K. and Yeung, H. W-C. (2001) Chains and networks, territories and scales: towards a relational framework for analysing the global economy, *Global Networks*, **1**(2), 89–112.

Dickenson, J., Gould, B., Clarke, C., Mather, S., Prothero, M., Siddle, D., Smith, C. and Thomas-Hope, E. (1996) *A Geography of the Third World*, Routledge, London.

Dikeç, M. (2006a) Two decades of French urban policy: from social development of neighborhoods to the republican penal state, *Antipode*, **38**(1), 59–81.

Dikeç, M. (2006b) The badlands of the republic: revolts, the French state and questions of the banlieues, *Environment and Planning D*, **24**, 159–63.

Dikeç, M. (2007a) Revolting geographies: urban unrest in France, *Geography Compass*, **1**(5), 1190–206.

Dikeç, M. (2007b) Space, governmentality, and the geographies of French urban policy, *European Urban and Regional Studies*, **14**(4), 277–89.

Dittmer, J. (2005) Captian America's empire: reflections on identity, popular culture and post-9/11 geopolitics, *Annals of the Association of American Geographers,* **95**(3), 626–43.

Dixon, D.P. and Jones, J.P. (2006) Feminist geographies of difference, relation and construction, in Aitken, S. and Valentine, V. (eds) (2006) *Approaches to Human Geography*, Sage, London, 42–56.

Dobb, M. (1946; 1963) *Studies in the Development of Capitalism*, Routledge & Kegan Paul, London.

Dobson, A. (2001) *Green Political Thought*, 3rd edition, Routledge, London.

Dodd, N. (1994) *The Sociology of Money: Economics, Reason and Contemporary Society*, Continuum Publishing Company, New York.

Dodds, K. (1997) *Geopolitics in Antarctica: Views from the Southern Ocean Rim*, Wiley, Chichester.

Dodds, K. (2002) *Pink Ice: Britain and the South Atlantic Empire*, I B Tauris, London.

Dodds, K. (2004) *Global Geopolitics: A Critical Introduction*, Prentice Hall, Harlow.

Dodds, K. (2006) Popular geopolitics and audience dispositions: James Bond and the Internet Movie Database, *Transactions of the Institute of British Geographers,* **31**(2), 116–30.

Dodds, K. (2007) *Geopolitics: A Very Short Introduction*, Oxford University Press, Oxford.

Dodds, K. (2010) Flag planting and finger pointing: the law of the sea, the Arctic and the political geographies of the outer continental shelf, *Political Geography*, **29**, 63–73.

Dodds, K. (2011) Sovereignty watch: resources, sovereignty and territory in contemporary Antarctica, *Polar Record*, in press.

Dodge, M. and Kitchin, R. (2000) *The Atlas of Cyberspace*, Continuum, London.

Dodgshon, R.A. (1987) *The European Past: Social Evolution and Spatial Order*, Macmillan, Basingstoke.

Dodgshon, R.A. (1998) *Society in Time and Space: A Geographical Perspective on Change*, Cambridge University Press, Cambridge.

Dodgshon, R.A. and Butlin, R.A. (eds) (1990) *An Historical Geography of England and Wales*, 2nd edition, Academic Press, London.

Dodson, B. (2010) Locating xenophobia: debate, discourse, and everyday experience in Cape Town, South Africa, *Africa Today*, **56**(3), 2–22.

Doherty, B. (2002) *Ideas and Actions in the Green Movement*, Routledge, London.

Donnelly, J. (2000) *Realism and International Relations*, Cambridge University Press, Cambridge.

Donovan, A. and Oppenheimer, C. (2011) The 2010 Eyjafjallajökull eruption and the reconstruction of geography, *Geographical Journal*, **177**, 4–11.

Dorfman, A. and Mattelart, A. (1975) *How to Read Donald Duck: Imperialist Ideology in the Disney Comic*, International General Editions, New York.

Dorling, D. (2004) *The Human Geography of the UK*, London, Sage.

Dorling, D. and Fairbairn, D. (1997) *Mapping: Ways of Representing the World*, Prentice Hall, Harlow.

Dorling, D. and Rees, P. (2003) A nation still dividing: the British census and social polarisation 1971–2001, *Environment and Planning A*, **35**, 1287–313.

Dorling, D. and Rees, P. (2004) A nation dividing? Some interpretations of the question, *Environment and Planning A*, **36**, 369–73.

Dorling, D. and Shaw, M. (2000) Life chances and lifestyles, in V. Gardiner and H. Matthews (eds) *The*

Changing Geography of the UK, London, Routledge, 230–60.

Douglas, M. and Isherwood, B. (1996) *The World of Goods: Towards an Anthropology of Consumption*, 2nd edition, Routledge, London.

Doyle, T. (2000) *Green Power: The Environment Movement in Australia*, University of New South Wales Press, Sydney.

Doyle, T. (2005) *Environmental Movements in Majority and Minority Worlds: A Global Perspective,* Rutgers University Press, London.

Doyle, T. and McEachern, D. (2001) *Environment and Politics,* 2nd edition, Routledge, London.

Doyle, T. and McEachern, D. (2007) *Environment and Politics,* 3rd edition, Routledge, London.

Driver, F. (2005) Imaginative geographies, in Cloke, P., Crang, P. and Goodwin, M. (eds) *Introducing Human Geographies,* 2nd edition, Arnold, London, 144–55.

Dryzek, J.S. and Schlosberg, D. (eds) (2004) *Debating the Earth: The Environmental Politics Reader,* Oxford, University Press, Oxford.

du Gay, P. (1996) *Consumption and Identity at Work,* Sage, London.

Dudrah, R.K. (2002) Drum n dhol: British Bhangra music and diasporic South Asian identity formation, *European Journal of Cultural Studies,* **5**(3), 363–83.

Duncan, J.S., Johnson, N.C. and Schein, R. (eds) (2004) *A Companion to Cultural Geography*, Blackwell, London.

Dunn, K.M. (2001) Representations of Islam in the politics of mosque development in Sydney, *Tijdschrift voor Economische en Sociale Geografi,* **92**, 291–308.

Dunning, J.H. (1981) *International Production and the Multinational Enterprise*, Allen & Unwin, London.

Duruz, J. (2005) Eating at the borders: culinary journeys, *Environment and Planning D: Society and Space,* **23**, 51–69.

Dwyer, C. (2005) Diasporas, in Cloke, P., Crang, P. and Goodwin, M. (eds) *Introducing Human Geographies,* 2nd edition, Hodder Arnold, London, 495–508.

Dwyer, C. and Crang, P. (2002) Fashioning ethnicities: the commercial spaces of multiculture, *Ethnicities,* **2**, 410–30.

Dwyer, C. and Jackson, P. (2003) Commodifying difference: selling EASTern fashion, *Environment and Planning D: Society and Space,* **21**, 269–91.

Dwyer, C.L. (2006) Fabrications of India: transnational fashion networks, in Breward, C. and Gilbert, D. (eds) *Fashion's World Cities*, Berg, Oxford, 217–33.

E

Earthscan (1998) *Internally Displaced People: A Global Survey*, Earthscan, Guildford.

Edensor, T. and Kothari, U. (2006) Extending networks and mediating brands: stallholder strategies in a Mauritian market, *Transactions of the Institute of British Geographers, NS,* **31**(3), 323–36.

Edwards, L.E. (1983) Towards a process model of office location decision making, *Environment and Planning A,* **15**, 1327–42.

Edwards, M. and Hulme, D. (1992) *Making a Difference: NGOs and Development in a Changing World*, Earthscan, Guildford.

EIA (Energy Information Administration) (2007) *International Energy Outlook 2007.* Washington DC: US Department of Energy.

EIA (Energy Information Administration) (2010) *International Energy Outlook 2010*, EIA, Washington, DC.

EJF (2008) *No Place Like Home: Where Next for Climate Refugees?*, EJF, London.

Elden, S. (2005a) Territorial integrity and the war on terror, *Environment and Planning A,* **37**, 2083–104.

Elden, S. (2005b) Missing the point: globalization, deterritorialization and the space of the world, *Transactions of the Institute of British Geographers,* **30**(1), 8–19.

Elden, S. (2007a) Governmentality, calculation, territory, *Environment and Planning D: Society and Space,* **25**(3), 562–80.

Elden, S. (2007b) Terror and territory, *Antipode,* **39**(5), 821–45.

Elden, S. (2009) *Terror and Territory: The Spatial Extent of Sovereignty*, University of Minnesota Press, Minneapolis, MN.

Elden, S. (2010) Land, terrain, territory, *Progress in Human Geography,* **34**(6), 799–817.

Elden, S. and Bialasiewicz, L. (2006) The new geopolitics of division and the problem of a Kantian Europe, *Review of International Studies,* **32**, 623–44.

Elliott, D. (2003) *Energy, Environment and Society,* 2nd edition, Routledge, London.

Elster, J. (1989) *Nuts and Bolts for the Social Sciences*, Cambridge University Press, Cambridge.

Emmerson, C. (2010) *The Future History of the Arctic*, Bodley Head, London.

Engels, F. (1845) *The Condition of the Working Class in England*, 1987 edition, Penguin, London and New York.

England, K. (2006) Producing feminist geographies: theory, methodologies and research strategies, in Aitken, S. and Valentine, V. (eds) (2006) *Approaches to Human Geography*, Sage, London, 286–97.

Enloe, C. (1983) *Does Khaki Become You? The Militarisation of Women's Lives*, Pluto Press, London.

Enloe, C. (1989), *Bananas, Beaches and Bases: Making Feminist Sense of International Politics*, University of California Press, Berkeley, CA.

Enloe, C. (1993) *The Morning After: Sexual Politics at the End of the Cold War*, University of California Press, Berkeley, CA.

Enloe, C. (2000) *Maneuvers: The International Politics of Militarizing Women's Lives*, University of California Press, Berkeley, CA.

Enloe, C. (2004) *The Curious Feminist: Searching for Women in a New Age of Empire*, University of California Press, Berkeley, CA.

Enloe, C. (2007) *Globalization and Militarism: Feminists Make the Link*, Rowan & Littlefield, Lanham, MD.

Epstein, G. (2005) *Financialization and the World Economy*, Edward Elgar, London.

Ereaut, G. and Segnit, N. (2006) Warm Words: How are we telling the climate story and can we tell it better? http://www.ippr.org.uk/members/download.asp?f=%2Fcomm%2Ffiles%2Fwarm%5Fwords%2Epdf.

Ericksen, P. (2008) Conceptualizing food systems for global environmental change research, *Global Environmental Change*, **18**, 234–45.

Escobar, A. (1995) *Encountering Development*, Princeton University Press, Princeton, NJ.

Esposito, J.L. (ed.) (1995) *The Oxford Encyclopedia of Islam in the Modern World*, Oxford University Press, New York.

Esteva, G. (1992) Development, in Sachs, W. (ed.) *The Development Dictionary*, Zed, London.

EU (2005) *Sustainable Financing of Social Policies in the European Union*, Commission Staff Working Document. SEC (2005) 1774, Brussels.

European Commission (1999) *Sustainable Urban Development in the European Union: A Framework for Action*, Office for Official Publications of the European Communities, Luxembourg.

Evans, R.L. (2007) *Fuelling our Future: An Introduction to Sustainable Energy*, Cambridge University Press, Cambridge.

Eyben, R., Moncrieffe, J. and Knowles, C. (2006) *The Power of Labelling in Development Practice*, Institute of Development Studies, IDS Policy Briefing 28, April.

F

Fairbairn, D. and Dorling, D. (1997) Mapping and map-making: new approaches to the teaching of cartography, Proceedings of the 18th International Cartography Conference, Stockholm, 23–27 June, 4, 1955–1962.

Fairlie, S. (1996) *Low Impact Development: Planning and People in a Sustainable Countryside,* Jon Carpenter, Oxfordshire.

Falah, G-W., Flint, C. and Mamadouh, V. (2006) Just war and extraterritoriality: the popular geopolitics of the United States war on Iraq as reflected in newspapers of the Arab world, *Annals of the Association of American Geographers*, **96**(1) 142–64.

FAO *see* Food and Agriculture Organization.

Featherstone, M. (1995) *Undoing Culture: Globalization, Postmodernism and Identity*, Sage, London.

Feeding America (2010) *Hunger in America, Feeding America*, Washington, DC, available at: http://feedingamerica.org/faces-of-hunger/hunger-in-america-2010/hunger-report-2010.aspx.

Featherstone, M., Lash, S. and Robertson, R. (eds) (1995) *Global Modernities*, Sage, London.

Fejes, F. (1981) Media imperialism, an assessment, *Media, Culture and Society*, **3**, 281–9.

Fell, A. (1991) Penthesilea, perhaps, in Fisher, M. and Owen, U. (eds) *Whose Cities?* Penguin, London, 73–84.

Ferguson, J. (1990) *The Anti-Politics Machine: 'Development', Depoliticization and Bureaucratic Power in Lesotho*, Cambridge University Press, Cambridge.

Ferguson, J. and Gupta, A. (2002) Spatializing states: towards an ethnography of neoliberal governmentality, *American Ethnologist*, **29**, 981–1002.

Fieldhouse, D.K. (1999) *The West and the Third World*, Blackwell, Oxford.

Fields, G. (2004) *Territories of Profit*, Stanford Business Books, Stanford.

Fineman, S. (ed) (2000) *Emotion in Organization*, Sage, London.

Fitter, R. and Kaplinsky, R. (2001) Who gains from product rents as the coffee market becomes more differentiated? A value chain analysis, *IDS Bulletin Paper*, IDS Sussex.

FitzSimmons, M. (1986) The new industrial agriculture: the regional differentiation of specialty crop production, *Economic Geography*, **62**, 334–53.

FitzSimmons, M. (1997) Restructuring, industry and regional dynamics, in Goodman, D. and Watts, M. (eds) *Globalising Food*, Routledge, London, 158–68.

Flint, C. (2006) *Introduction to Geopolitics*, Routledge, London.

Flint, C. and Falah, G.W. (2004) How the United States justified its war on terrorism: prime morality and the construction of a 'just war', *Third World Quarterly*, **25**, 1379–99.

Florida, R. (2002) *The Rise of the Creative Class and How It's Transforming Work, Leisure and Everyday Life*, Basic Books, New York.

Flowerdew, R. and Martin, D. (1997) *Methods in Human Geography: A Guide for Students Doing a Research Project*, Longman, Harlow.

Flusty, S. (2004) *De-Coca-Colonization: Making the Globe from the Inside Out*, Routledge, New York.

Fold, N. and Pritchard, B. (eds) (2005) *Cross-continental Food Chains*, Routledge, London.

Foley, R. (1995) *Humans Before Humanity: An Evolutionary Perspective*, Blackwell, Oxford.

Fonseca, I. (1996) *Bury Me Standing*, Vintage, London.

Food and Agriculture Organization (FAO) (1996) Rome declaration on world food security, *World Food Summit*, FAO, Rome.

Food and Agriculture Organization (FAO) (2005) *The State of Food Insecurity*, FAO, Rome.

Food and Agriculture Organization (FAO) (2006) *Food Security: Policy Brief*, FAO, Rome, available at: ftp://ftp.fao.org/es/ESA/policybriefs/pb_02.pdf.

Food and Agriculture Organization (FAO) (2010) *The State of Food Insecurity in the World: Addressing Food Insecurity in Protracted Crises*, FAO, Rome.

Forest, B. (1995) West Hollywood as symbol: the significance of place in the construction of a gay identity, *Environment and Planning D: Society and Space*, **13**, 133–57.

Foresta, R.A. (1992) Amazonia and the politics of geopolitics, *Geographical Review*, **8**(2), 128–42.

Foster, W., Cheng, Z., Dedrick, J. and Kraemer, K.L. (2006) *Technology and Organizational Factors in the Notebook Industry Supply Chain*, The Personal Computer Industry Center, University of California, Irvine, available at: www.pcic.merage.uci.edu/papers.asp [accessed 29 August 2011].

Fotheringham, S.A., Brunsdon, C. and Charlton, M. (2000) *Quantitative Geography: Perspectives on Spatial Data Analysis*, Sage, London and Thousand Oaks, New Delhi.

Foucault, M. (1979) On governmentality, *Ideology and Consciousness*, **6**, 5–29.

Frank, A. (1966) The development of underdevelopment, *Monthly Review*, **18**(4), 17–31.

Frank, A.G. (1969) *Capitalism and Underdevelopment in Latin America*, Monthly Review Press, New York.

Frank, A.G. (1997) The Cold War and me, *Bulletin of Concerned Asian Scholars*, **29**, available at http://www.rrojasdatabank.into/agfrank/syfrank.htm [accessed 7 December 2011].

Frank, A.G. (1998) *ReOrient: Global Economy in the Asian Age*, University of California Press, Berkeley, CA.

Freeman, C. (2000) *High Tech and High Heels in the Global Economy*, Duke University Press, Durham, NC.

Freeman, M. (1986) Transport, in Langton J. and Morris, R.J. (eds) *Atlas of Industrializing Britain, 1780–1914*, Methuen, London and New York.

Freidberg, S. (2003) Cleaning up down South: supermarkets, ethical trade and African horticulture, *Social and Cultural Geography*, **4**, 27–44.

Friedberg, S. (2004) *French Beans and Food Scares: Culture and Commerce in an Anxious Age*, Oxford University Press, Oxford.

Friedman, J. (1994) Globalization and localization, in Friedman, J. (ed.) *Cultural Identity and Global Process*, Sage, London, 102–16.

Friedman, J. (1997) Global crises, the struggle for cultural identity and intellectual porkbarrelling: cosmopolitans versus locals, ethnics and nationals in an era of dehegemonisation, in Werbner, P. and Modood, T. (eds) *Debating Cultural Hybridity. Multi-Cultural Identities and the Politics of Anti-Racism*, Zed, London, 70–89.

Friedman, J. and Rowlands, M.J. (1977) Notes towards an epigenetic model of the evolution of civilization, in Friedman, J. and Rowlands, M.J. (eds) *The Evolution of Social Systems*, Duckworth, London, 201–76.

Friedman, T.L. (1999) It's a small world, *Sunday Times*, 28 March, Section 5.

Friedman, T.L. (2005), *The World is Flat: A Brief History of the Globalized World in the Twenty-first Century*, Allen Lane, London

Friedmann, H. and McMichael, P. (1989) Agriculture and the state system: the rise and decline of national agricultures, 1870 to the present, *Sociologia Ruralis*, **29**, 93–117.

Friedmann, J. (1966) *Regional Development Policy: A Case Study of Venezuela*, MIT Press, Cambridge, MA.

Friedmann, J. (1986) The world city hypothesis, *Development and Change*, **17**(1), 69–83.

Friedmann, J. (1992) *Empowerment: The Politics of Alternative Development*, Blackwell, Oxford.

Friend, T. (2003) *Indonesian Destinies*, The Belknap Press of Harvard University Press, Cambridge, MA.

Friends of the Earth (2006) What kind of world do you want? *Friends of the Earth Limited*, June 2006.

Fröbel, F., Heinrichs, J. and Kreye, O. (1980) *The New International Division of Labour*, Cambridge University Press, Cambridge.

Fukuyama, F. (1992) *The End of History and the Last Man*, Penguin, London.

Furtado, C. (1964) *Development and Underdevelopment*, University of California Press, Berkeley, CA.

Fyfe, N. (1991) The police, space and society: the geography of policing, *Progress in Human Geography*, **15**(3), 249–67.

G

Gad, G. and Holdsworth, D.W. (1987) Corporate capitalism and the emergence of the high-rise office building, *Urban Geography*, **8**, 212–31.

Gaffikin, F. and Nickson, A. (1984) *Jobs Crisis and the Multi-Nationals*, Birmingham Trade Union Group for World Development, Birmingham.

Gagnon, V.P. (2006) *The Myth of Ethnic War: Serbia and Croatia in the 1990s*, Cornell University Press, Ithaca, NY.

Gallagher, C., Dahlman, C., Gilmartin, M., Mountz, A. and Shirley, P. (2009) *Key Concepts in Political Geography*, Sage, London.

Gallent, N. (2006) The rural-urban fringe: a new priority for planning policy?, *Planning Practice and Research*, **21**(3), 383–93.

Gallent, N. and Anderson, J. (2007) Representing England's rural-urban fringe, *Landscape Research*, **32**(1), 1–21.

Gandy, M. (2002) *Concrete and Clay: Re-working Nature in New York City*, MIT Press, New York.

Gans, H. (1968) *People and Problems: Essays on Urban Problems and Solutions*, Basic Books, New York.

Garland, J. and Chakraborti, N. (2004) England's green and pleasant land? Examining racist prejudice in a rural context, *Patterns of Prejudice*, **38**(4), 383–98.

Garner, R. (2000) *Environmental Politics: Britain, Europe and the Global Environment*, 2nd edition, Macmillan, London.

Garreau, J. (1991) *Edge City: Life on the New Frontier*, Doubleday, New York.

Gary, I. and Karl, T.L. (2003) *Bottom of the Barrel: Africa's Oil Boom and the Poor*, Catholic Relief Services. Available at: www.catholicrelief.org/africanoil.cfm

Gates, D. (2005) Boeing 787: parts from around world will be swiftly integrated, *Seattle Times,* 11 September.

Gautier, C. (2008) *Oil, Water, and Climate: An Introduction*, Cambridge University Press, Cambridge.

Gereffi, G. (1994) The organisation of buyer-driven global commodity chains: how US retailers shape overseas production networks, in Gereffi, G. and Korzeniewicz, M. (eds) *Commodity Chains and Global Development*, Praeger, Westport, 95–122.

Gereffi, G. (2001) Shifting governance structures in global commodity chains, with special reference to the Internet, *American Behavioural Scientist*, **44**, 1616–37.

Gereffi, G., Humphrey, J. and Sturgeon, T. (2005) The governance of global value chains, *Review of International Political Economy*, **12**, 78–104.

Ghazvinian, J. (2007) *Untapped: The Scramble for Africa's Oil*, Harcourt, London.

Gibbon, P. and Ponte, S. (2005) *Trading Down: Africa, Value Chains, and the Global Economy*, Temple University Press, Philadelphia.

Gibson, C. (2009) Geographies of tourism: critical research on capitalism and local livelihoods, *Progress in Human Geography*, 1–8.

Giddens, A. (1990) *The Consequences of Modernity*, Polity Press, Cambridge.

Giddens, A. (1991) *The Constitution of Society*, Polity Press, Cambridge.

Giddens, A. (2001) *Sociology,* Polity Press, Cambridge.

Giles, C. and Goodall, I.H. (1992) *Yorkshire Textile Mills, 1770–1930*, HMSO, London.

Gill, G. (1996) *The League of Nations from 1929 to 1946*, Avery Publishing Group, New York.

Gilmore, R.W. (2007) *Golden Gulag: Prisons, Surplus, Crisis and Opposition in Globalizing California,* University of California Press, Berkeley, CA.

Gilmartin, M. (2008) Migration, identity and belonging, *Geography Compass*, **2**(6), 1837–52.

Gilpin, R. (1989) *The Political Economy of International Relations*, Princeton University Press, Princeton, NJ.

Gilroy, P. (1993) *The Black Atlantic: Modernity and Double Consciousness*, Verso, London.

Gilroy, P. (2000) *Between Camps: Nations, Cultures and the Allure of Race*, Penguin, London.

Glassman, J. (2001), Structural adjustment in East and Southeast Asia: lessons from Latin America, *Geoforum*, **32**, 77–90.

Glassman, J. (2005) The 'war on terrorism' comes to Southeast Asia, *Journal of Contemporary Asia*, **35**, 3–28.

Glassner, M.I. (1993) *Political Geography*, John Wiley, New York.

Glasze, G., Webster, C. and Frantz, K. (2006) *Private Cities: Global and Local Perspectives*, Routledge, Abingdon.

Gleeson, B. (1998) The social space of disability in colonial Melbourne, in Fyfe, N. (ed.) *Images of the Street*, Routledge, London, 93–110.

Glennie, P. and Thrift, N. (1992) Modernity, urbanism and modern consumption, *Environment and Planning D: Society and Space*, **10**, 423–43.

Global Times (2010) Chongqing plans Asia's largest PC manufacturing base, *Global Times*, 14 January, available at: http://business.globaltimes.cn.

Godfray, C J., Crute, I., Haddad, L. *et al.* (2010) The future of the global food system, *Philosophical Transactions of the Royal Society B*, **365**, 2769–77.

Godlewska, A. and Smith, N. (eds) (1994) *Geography and Empire*, Blackwell, Oxford.

Goetz, A.M. and Gupta, R.S. (1996) Who takes the credit? Gender, power, and control over loan use in rural credit programs in Bangladesh, *World Development*, **24**, 45–63.

Gold, J.R. and Revill, G. (2004) *Representing the Environment*, Routledge, London.

Goldman, R. and Papson, S. (1998) *Nike Culture: The Sign of the Swoosh*, Sage, London.

Goodier, C. and Pan, W. (2010) The Future of UK Housebuilding, Royal Institution of Chartered Surveyors (RICS) Research Report, London.

Goodman, D. (2003) The quality 'turn' and alternative food practices: reflections and agenda, *Journal of Rural Studies*, **19**, 1–7.

Goodman, D. and Redclift, M. (1991) *Refashioning Nature: Food Ecology and Culture*, Routledge, London.

Goodman, D. and Watts, M. (eds) (1997) *Globalising Food*, Routledge, London.

Goodman, D., Sorj, B. and Wilkinson, J. (1987) *From Farming to Biotechnology: A Theory of Agro-industrial Development*, Blackwell, Oxford.

Goodman, M.K. (2004) Reading fair trade: political ecological imaginary and the moral economy of fair trade foods, *Political Geography*, **23**, 891–915.

Goodman, M.K., Maye, D. and Holloway, L. (2010) Ethical foodscapes: premises, promises and possibilities, *Environment and Planning A*, **42**, 1782–96.

Goodrum, A. (2005) *The National Fabric: Fashion, Britishness, Globalization*, Berg, Oxford.

Goodwin, M. (1995) Poverty in the city, you can raise your voice, but who is listening?, in Philo, C. (ed.) *Off the Map: The Social Geography of Poverty in the UK*, Child Poverty Action Group, London, 134–54.

Gore, A. (2006) *An Inconvenient Truth: The Planetary Emergency of Global Warming and What We Can Do About It*, Bloomsbury Publishing, London.

Goskomstat (2003) *Rossiiskii Statisticheskii Ezhegodnik (Russian Statistical Yearbook)*, State Committee of the Russian Federation for Statistics, Moscow.

Goskomstat (2006) *Rossiiskii Statisticheskii Ezhegodnik (Russian Statistical Yearbook)*, State Committee of the Russian Federation for Statistics, Moscow.

Goss, J. (1993b) The magic of the mall: form and function in the retail built environment, *Annals of the Association of American Geographers*, **83**, 18–47.

Goss, J. (1999a) Consumption, in Cloke, P., Crang, P. and Goodwin, M. (eds) *Introducing Human Geographies*, Arnold, London, 114–21.

Goss, J. (1999b) Once upon a time in the commodity world, an unofficial guide to the Mall of America, *Annals of the Association of American Geographers*, **89**, 45–75.

Goss, J. (2006) Geographies of consumption: the work of consumption, *Progress in Human Geography*, **30**(2), 237–49.

Gottdiener, M. (1982) Disneyland: a utopian urban space, *Urban Life*, **11**, 139–62.

Gottdiener, M. (1997) *The Theming of America: Dreams, Visions and Commercial Spaces*, Westview Press, Boulder, CO.

Gottman, J. (1973) *The Significance of Territory*, University Press of Virginia, Charlottesville.

Gould, P. and Pitts, F.R. (eds) (2002) *Geographical Voices: Fourteen Autobiographical Essays*, Syracuse University Press, Syracuse, NY.

Gourevitch, P., Bohn, R. and McKendrick, D. (2000) Globalization of production, *World Development*, **28**, 301–17.

Gowan, T. (2010) *Hobos, Hustlers and Backsliders: Homeless in San Francisco*, University of Minnesota Press, Minneapolis, MN.

Graham, B.J., Ashworth, G.J. and Tunbridge, J.E. (2000) *A Geography of Heritage: Power, Culture and Economy*, Arnold, London.

Graham, S. (ed.) (2004) *Cities, War, and Terrorism: Towards an Urban Geopolitics*, Blackwell, Oxford.

Graham, S. and Marvin, S. (1996) *Telecommunications and the City: Electronic Spaces, Urban Places*, Routledge, London.

Graham, S. and Marvin, S. (2001) *Splintering Urbanism: Networked Infrastructures, Technological Mobilities and the Urban Condition*, Routledge, London.

Green, A. (1997) Income and wealth, in Pacione, M. (ed.) *Britain's Cities: Geographies of Division in Urban Britain*, Routledge, London, 78–102.

Green, A. and Owen, D. (2006) *The Geography of Poor Skills and Access to Work*, Joseph Rowntree Foundation, York.

Green, F. and Sutcliffe, B. (1987) *The Profit System: The Economics of Capitalism*, Penguin, London.

Gregory, D. (1982) *Regional Transformation and Industrial Revolution: A Geography of the West Yorkshire Woollen Industry*, Macmillan, London.

Gregory, D. (1994) *Geographical Imaginations*, Blackwell, Oxford.

Gregory, D. (2006) The black flag: Guantánamo Bay and the space of exception, *Geografiska Annaler*, **88B**(4), 405–27.

Gregory, D. (2006) The death of the civilian?, *Environment and Planning D: Society and Space*, **24**, 633–8.

Gregory, D. (2007) Vanishing points: law, violence and exception in the global war prison, in Gregory, D. and Pred, A. (eds) (2007) *Violent Geographies: Fear, Terror, and Political Violence*, Routledge, London, 205–36.

Gregory, D. and Pred, A. (2006) (eds) *Violent Geographies: Fear, Terror, and Political Violence*, Routledge, London.

Gregson, N. (2007) *Living with Things: Ridding, Accommodation, Dwelling*, Sean Kingston Publishing, Wantage.

Grenville, K. (2006) *The Secret River*, Canongate Press, London.

Griffiths, I.L. (1994) *The Atlas of African Affairs*, 2nd edition, Routledge, London.

Groves, P.A. (1987) The northeast and regional integration, 1800–1860, in Mitchell, R.D. and Groves, P.A.

(eds) *North America: The Historical Geography of a Changing Continent*, Hutchinson, London.

Grundy-Warr, C. and Wong, S.Y. (2002) Geographies of displacement: the Karenni and the Shan across the Myanmar–Thailand border, *Singapore Journal of Tropical Geography*, **23**, 93–122.

Guthman, J. (2004) Back to the land: the paradox of organic food standards, *Environment and Planning A*, **36**, 511–28.

Gutkind, E.A. (1964) *Urban Development in Central Europe*, Macmillan, London and New York.

Gwynne, R.N., Klak, T. and Shaw, D.J.B. (2003) *Alternative Capitalisms: Geographies of Emerging Regions*, Arnold, London.

H

Hackley, C. (2010) *Advertising and Promotion: An Integrated Marketing Communications Approach*, 2nd edition, SAGE, London.

Hagen, J. (2004) The most German of towns: creating an ideal Nazi community in Rothenburg ob der Tauber, *Annals of the Association of American Geographers*, **94**(1), 207–27.

Halfacree, K. (1996) Out of place in the country: travelers and the 'rural idyll', *Antipode*, **18**, 42–72.

Hall, C., Tharakhan, P., Hallock, J., Cleveland, C. and Jefferson, M. (2003) Hydrocarbons and the evolution of human culture, *Nature*, **426**, 318–22.

Hall, C.M. (2005) *Tourism: Rethinking the Social Science of Mobility*, Prentice Hall, Harlow.

Hall, E.T. (1959) *The Silent Language*, Doubleday, Garden City, NY.

Hall, P. (ed.) (1966) *Von Thünen's Isolated State*, Pergamon, Oxford.

Hall, P. (2002) *Cities of Tomorrow: an Intellectual History of Urban Planning and Design in the Twentieth Century*, 3rd edition, Blackwell, Oxford.

Hall, P. and Pain, K. (2006) *The Polycentric Metropolis: Learning from Mega-city Regions in Europe*, Earthscan, London.

Hall, P. and Pfeiffer, U. (2000) *Urban Future 21: A Global Agenda for 21st Century Cities*, Spon, London.

Hall, R. (1995) Households, families and fertility, in Hall, R. and White, P. (eds) *Europe's Population: Towards the Next Century*, UCL Press, London.

Hall, S. (1992) The West and the rest: discourse and power, in Hall, S. and Gieben, B. (eds) *Formations of Modernity*, Open University Press, Milton Keynes, 275–331.

Hall, S. (1995) New cultures for old, in Massey, D. and Jess, P. (eds) *A Place in the World? Places, Cultures and Globalization*, Oxford University Press, Oxford, 175–213.

Hall, S. (1996) The meaning of new times, in Morley, D. and Chen, K.-H. (eds) *Stuart Hall: Critical Dialogues in Cultural Studies*, Routledge, London, 223–37.

Hall, T. (2003) Car-ceral cities: social geographies of everyday urban mobility, in Miles, M. and Hall, T. (eds) *Urban Futures: Critical Commentaries on Shaping the City*, Routledge, London, 92–105.

Hall, T. and Hubbard, P. (1996) The entrepreneurial city: new urban politics, new urban geographies?, *Progress in Human Geography*, **20**(2), 153–74.

Hall, T. and Hubbard, P. (eds) (1998) *The Entrepreneurial City: Geographies of Politics, Regime and Representation*, Chichester, Wiley.

Halliday, F. (2002) *Two Hours that Shook the World. September 11, 2001: Causes and Consequences*, Saqi Books, London.

Hamelink, C. (1983) *Cultural Autonomy in Global Communications*, Longman, New York.

Hamilton, N. and Chincilla, N.S. (2001) *Seeking Community in a Global City: Guatemalans and Salvadorans in Los Angeles*, Temple University Press, Philadelphia, PA.

Hammett, D. (2009) Local beats to global rhythms: coloured student identity and negotiations of global cultural imports in Cape Town, South Africa, *Social and Cultural Geography*, **10**(4), 403–19.

Hamnett, C. (2003) *Unequal City: London in the Global Arena*, Routledge, London.

Hannerz, U. (1992) The global ecumene, in Hannerz, U. (ed.) *Cultural Complexity: Studies in the Social Organization of Meaning*, Columbia University Press, New York, 217–67.

Hanrou, H. and Obrist, H.-U. (1999) Cities on the move, in Bradley, F. (ed.) *Cities on the Move: Urban Chaos and Global Change: East Asian Art, Architecture and Film Now*, Hayward Gallery, London, 10–15.

Hanson, S. and Pratt, G. (1995) *Gender, Work and Space*, Routledge, London.

Hara, K. (1999) Rethinking the 'Cold War' in the Asia-Pacific, *Pacific Review*, **12**, 515–36.

Haraway, D. (1989) *Primate Visions: Gender, Race and Nature in the World of Modern Science*, Routledge, New York.

Harding, J. (2010) What we're about to receive, *London Review of Books*, **32**(9), 13 May, 3–8.

Hardt, M. and Negri, A. (2000) *Empire*, Harvard University Press, Cambridge, MA.

Harper, S. (2005) *Ageing Societies*, Hodder Arnold, London.

Harris, C. and Ullman, E. (1945) The nature of cities, *Annals of the American Academy of Political Science*, **242**, 7–17.

Harris, D.R. (1989) An evolutionary continuum of plant–animal interaction, in Harris, D.R. and Hillman,

G.C. (eds) *Foraging and Farming: The Evolution of Plant Exploitation*, Unwin Hyman, London, 11–26.

Harris, D.R. (1996a) Introduction: themes and concepts in the study of early agriculture, in Harris, D.R. (ed.) *The Origins and Spread of Agriculture and Pastoralism in Eurasia*, UCL Press, London, 1–9.

Harris, D.R. (1996b) The origins and spread of agriculture and pastoralism in Eurasia: an overview, in Harris, D.R. (ed.) *The Origins and Spread of Agriculture and Pastoralism in Eurasia*, UCL Press, London, 552–70.

Harris, E. (2009) Neoliberal subjectivities or a politics of the possible? Reading the difference in alternative food networks, *Area*, **41**, 55–63.

Harris, F. (ed.) (2004) *Global Environmental Issues*, John Wiley and Sons, Chichester.

Harrison, S. (2009) Physical geography and human geography, in Kitchin, R., Thrift, N., *et al.* (eds) *The International Encyclopedia of Human Geography*, Elsevier, Oxford, 163–8.

Hart, G. (2001) Development critiques in the 1990s: culs de sac and promising paths, *Progress in Human Geography*, **25**, 649–58.

Hartwick, E. (1998) Geographies of consumption: a commodity-chain approach, *Environment and Planning D: Society and Space,* **16**, 423–37.

Harvey, D. (1973) *Social Justice and the City*, Arnold, London.

Harvey, D. (1974) Population, resources and the ideology of science, *Economic Geography*, **50**, 256–77.

Harvey, D. (1982) *The Limits to Capital*, Blackwell, Oxford.

Harvey, D. (1985a) *The Urbanization of Capital: Studies in the History and Theory of Capitalist Urbanization*, Blackwell, Oxford.

Harvey, D. (1985b) Money, time, space and the city, in Harvey, D. (ed.) *The Urban Experience*, Johns Hopkins University Press, Baltimore, MD.

Harvey, D. (1989) *The Condition of Postmodernity: An Enquiry into the Origins of Cultural Change*, Blackwell, Oxford.

Harvey, D. (1990) Between space and time: reflections on the geographical imagination, *Annals of the Association of American Geographers*, **80**, 418–34.

Harvey, D. (2003) *The New Imperialism*, Oxford: Oxford University Press.

Harvey, D. (2006) *Spaces of Global Capitalism: Towards a Theory of Uneven Capitalist Development*, Verso, London.

Haskell, T.L. (1985a) Capitalism and the origins of the humanitarian sensibility, part 1, *American Historical Review*, **90**, 339–61.

Haskell, T.L. (1985b) Capitalism and the origins of the humanitarian sensibility, part 2, *American Historical Review*, **90**, 547–66.

Hastings, A. and Dean, J. (2003) Challenging images: tackling stigma through estate regeneration, *Policy and Politics*, **31**, 171–84.

Haughton, G. and Hunter, C. (1994) *Sustainable Cities*, Regional Studies Association, London.

Hayter, R. and Watts, H.D. (1983) The geography of enterprise: a re-appraisal, *Progress in Human Geography*, **7**, 157–81.

Hayward, K. (2004) *City Limits: Crime, Consumer Culture and the Urban Experience*, Glasshouse Press, London.

Headland, R. (1984) *A History of South Georgia*, Cambridge University Press, Cambridge.

Hebdige, D. (1979) *Subculture: The Meaning of Style*, Methuen, London.

Hebdige, D. (1987) Object as image: the Italian scooter cycle, in *Hiding in the Light: On Images and Things,* Comedia/Routledge, London, 77–115.

Hecht, S. and Cockburn, A. (1989) *The Fate of the Forest: Developers, Destroyers and Defenders of the Amazon*, Verso, London and New York.

Hedberg, D, Kullander, S. and Frank, H. (2010) The world needs a new energy paradigm, *Ambio*, **39**, 1–10.

Heffernan, M. (1998) *The Meaning of Europe: Geography and Geopolitics*, Arnold, London.

Heilbronner, R.L. (1972) *The Economic Problem*, 3rd edition, Prentice Hall, Englewood Cliffs, NJ.

Heisenberg, W. (1958) *Physics and Philosophy: The Revolution in Modern Science*, Harper, New York.

Helander, B. (2005) Who needs a state? Civilians, security and social services in north-east Somalia, in Richards, P. (ed.) *No Peace, No War: An Anthropology of Contemporary Armed Conflict*, Ohio University Press, Athens, OH, 193–202.

Held, D. and McGrew, A. (eds) (2002) *Governing Globalization*, Polity Press, Cambridge.

Held, D. and McGrew, A. (2007) *Globalization/Antiglobalization: Beyond the Great Divide*, Polity Press, Cambridge.

Held, D., McGrew, D., Goldblatt, D. and Perraton, J. (1999) *Global Transformations: Politics, Economics and Culture*, Polity Press, Cambridge.

Helm, D. (ed.) (2007) *The New Energy Paradigm*, OUP, Oxford.

Hendrickson, M.K. and Heffernan, W.D. (2002) Opening spaces through relocalization: locating potential resistance in the weaknesses of the global food system, *Sociologia Ruralis*, **42**, 347–69.

Henry, C.M. and Wilson, R. (eds) (2004) *The Politics of Islamic Finance*, Edinburgh University Press, Edinburgh.

Henry, N., McEwan, C. and Pollard, J. S. (2002) Globalization from below: Birmingham – postcolonial workshop of the world?, *Area*, **34**(2) 118–27.

Hepple, L. (1986) The revival of geopolitics, *Political Geography Quarterly*, **5**, 521–36.

Hepple, L. (1992) Metaphor, geopolitical discourse and the military in South America, in Barnes, T.J. and Duncan, J.S. (eds) *Writing Worlds*, Routledge, London, 136–54.

Herb, G.H. (1989) Persuasive cartography in Geopolitik and national socialism, *Political Geography Quarterly*, **8**, 289–303.

Herf, J. (1984) *Reactionary Modernism: Technology, Culture and Politics in Weimar and the Third Reich*, Cambridge University Press, Cambridge.

Herod, A., Ó Tuathail, G. and Roberts, S. (eds) (1998) *An Unruly World: Globalization, Governance and Geography*, Routledge, London.

Herzfeld, M. (1992) *The Social Production of Indifference: Exploring the Symbolic Roots of Western Bureaucracy*, University of Chicago Press, Chicago, IL.

Heske, H. (1986) German geographical research in the Nazi period, *Political Geography Quarterly*, **5**, 267–81.

Hettne, B. (1995) *Development Theory and the Three Worlds,* 2nd edition, Longman, London.

Hildyard, N. (1997) *The World Bank and the State: A Recipe for Change?*, N. Hildyard, Bretton Woods Project, London.

Hill, T.P. (1977) On goods and services, *Review of Income and Wealth*, **23**, 315–38.

Hinchliffe, S. (2007a) *Space for Nature,* Sage, London.

Hinchliffe, S.J. (2007) *Geographies of Nature,* London, Sage.

Hinchliffe, S. and Belshaw, C. (2003) Who cares? Values, power and action in environmental contests, in Hinchliffe, S., Blowers, A. and Freeland, J. (eds) *Understanding Environmental Issues*, John Wiley and Sons, Chichester, 89–126.

Hinchliffe, S., Kearnes, M.B., Degen, M. and Whatmore, S. (2005) Urban wild things: a cosmopolitical experiment, *Environment and Planning D: Society and Space*, **23**(5), 643–58.

Hindess, B. and Hirst, P. (1975) *Pre-Capitalist Modes of Production*, Routledge & Kegan Paul, London.

Hinrichs, C.C. (2000) Embeddedness and local food systems: notes on two types of direct agricultural market, *Journal of Rural Studies*, **16**, 295–303.

Hirschmann, A.O. (1958) *The Strategy of Economic Development*, Yale University Press, New Haven, CT.

Hobsbawm, E. (1995), *The Age of Extremes: The Short Twentieth Century, 1914–1991*, Abacus, London.

Hobsbawm, E.J. (1996) Ethnicity and nationalism in Europe today, in Balakrishnan, G. (ed.) *Mapping the Nation*, Verso, London, 255–66.

Hochschild, A.R. (1983) *The Managed Heart: Commercialization of Human Feeling,* University of California Press, London.

Hoggart, R. (1957) *The Uses of Literacy: Aspects of Working Class Life*, Chatto and Windus, London.

Hoggart, K., Lees, L. and Davies, A. (2002) *Researching Human Geography*, Arnold, London.

Holdar, S. (1992) The ideal state and the power of geography: the life-work of Rudolf Kjellen, *Political Geography*, **11**, 307–23.

Holloway, L. and Hubbard, P. (2001*) People and Place: The Extraordinary Geographies of Everyday Life*, Prentice Hall, London.

Holloway, S. and Valentine, G. (eds) (2000) *Children's Geographies: Playing, Living, Learning*, Routledge, London.

Holloway, S.L. (2005) Articulating otherness? White rural residents talk about gypsy-travellers, *Transactions of the Institute of British Geographer, NS*, **30**, 351–67.

Holloway, S.L., Rice, S.P. and Valentine, G. (eds) (2003) *Key Concepts in Geography*, Sage, London.

Holmberg, T. (2005) Questioning 'the number of the beast': Constructions of humanness in a Human Genome Project (HGP) narrative, *Science as Culture*, **14**, 23–37.

Holt-Jensen, A. (1999) *Geography: History and Concepts*, 3rd edition, Sage, London and Thousand Oaks CA and New Delhi.

Homburger, E. (1994) *The Historical Atlas of New York City*, Henry Holt, New York.

Hoogevelt, A. (1997) *Globalisation and Postcolonialism*, Macmillan, London.

hooks, b. (1984) *Feminist Theory: From Margins to Centre*, South End Press, Boston, MA.

Hooper, C. (2001) *Manly States*, Columbia University Press, New York.

Hooper, J. and Connolly, K. (2001) Berlusconi breaks ranks over Islam, *Guardian*, 27 September.

Hosking, G. (1992) *A History of the Soviet Union, 1917–1991,* Third edition, Fontana Press, London.

House of Commons (2005) Hansard written answers for 18 October 2005 (pt 17). Available at: www.publications.parliament.uk/pa/cm200506/cmhansrd/vo051018/text/51018w17.htm

House of Commons Environment, Food and Rural Affairs Committee (2009) *Securing Food Supplies up to 2050: The Challenges Faced by the UK*, Fourth Report of Session 2009–09, Vol. 1, House of Commons, London.

Hoyt, H. (1939) *The Structure and Growth of Residential Neighbourhoods in American Cities*, Federal Housing Administration, Washington, DC.

Hsu, J-Y. and Saxenian, A. (2000) The limits of guanxi capitalism: transnational collaboration between Taiwan and the USA, *Environment and Planning A, 32,* 1991–2005.

Hubbard, P. (1999) *Sex and the City: Geographies of Prostitution in the Urban West,* London, Ashgate.

Hubbard, P. (2006) *City,* Routledge, London.

Hubbard, P. (2008) Regulating the social impacts of studentification: a Loughborough case study, *Environment and Planning A,* **40**(2), 323–41.

Hubbard, P. (2009) Geographies of studentification and purpose-built student accommodation: leading separate lives?, *Environment and Planning A,* **41**, 1903–23.

Hubbard, P., Kitchin, R., Bartley, B. and Fuller, D. (2002) *Thinking Geographically: Space, Theory and Contemporary Human Geography,* Continuum, London.

Hudson, R. (2004) Conceptualising geographies and their economies: spaces, flows and circuits, *Progress in Human Geography,* **28**(4) 447–71.

Hudson, R. (2005) *Economic Geographies: Circuits, Flows and Spaces,* Sage, London.

Hughes, A. (2000) Retailers, knowledges and changing commodity networks: the case of the cut flower trade, *Geoforum,* **31**, 175–90.

Hughes, A. (2005) Corporate strategy and the management of ethical trade: the case of the UK food and clothing retailers, *Environment and Planning A,* **37**(7), 1145–63.

Hughes A. (2007) Supermarkets and the ethical/fair trade movement: making space for alternatives in mainstream economies?, in Burch, D. and Lawrence, G. (ed.) *Supermarkets and Agri-food Supply Chains,* Edward Elgar, Cheltenham.

Hughes, A. and Reimer, S. (eds) (2004) *Geographies of Commodity Chains,* Routledge, London.

Hugo, G. (2007) Population geography, *Progress in Human Geography,* **31**(1), 77–88.

Hulme, M. (2004) A change in the weather? Coming to terms with climate change, in Harris, F (ed.) *Global Environmental Issues,* John Wiley and Sons, Chichester, 21–44.

Huntington, S.P. (1993) The clash of civilizations? *Foreign Affairs,* **72**, 22–49.

Huq, S., Reid, H. and Murray, L.A. (2006) *Climate Change and Development Links,* IIED Gatekeeper series 123, available at: www.iied.org/pubs/display.php?o=14516IIED [accessed 21 September 2010].

Hutnyk, J. (1997) Adorno at Womad: South Asian crossovers and the limits of hybridity-talk, in Werbner, P. and Modood, T. (eds) *Debating Cultural Hybridity: Multi-cultural Identities and the Politics of Anti-racism,* Zed, London, 106–36.

Hutton, D. and Connors, L. (1999) *A History of the Australian Environment Movement,* Cambridge University Press, Melbourne.

Hyde, F.E. (1973), *Far Eastern Trade: 1860–1914,* Adam and Charles Black, London.

Hyndman, J. (2003) Beyond either/or: a feminist analysis of September 11th, *ACME: An International E-Journal for Critical Geographies,* **2**, 1–13, www.acme-journal.org

Hyndman, J. (2007) Feminist geopolitics revisited: body counts in Iraq, *The Professional Geographer,* **59**, 35–46.

Hyndman, J. (2010) The question of 'the political' in critical geopolitics: querying the 'child soldier' in the 'war on terror', *Political Geography,* **29**, 247–56.

I

IEA (International Energy Agency) (2006) *World Energy Outlook 2006,* OECD/IEA, Paris.

IEA, UNDP and UNIDO (2010) *Energy Poverty: How to Make Modern Energy Access Universal?,* OECD/IEA, Paris.

Ilbery, B. and Kneafsey, M. (2000a) Producer constructions of quality in regional speciality food production: a case study from south west England, *Journal of Rural Studies,* **16**, 217–30.

Ilbery, B. and Kneafsey, M. (2000b) Registering regional speciality food and drink products in the United Kingdom: the case of PDOs and PGIs, *Area,* **32**, 317–25.

Ilbery, B. and Maye, D. (2005) Alternative (shorter) food supply chains and specialist livestock products in the Scottish–English borders, *Environment and Planning A,* **37**, 823–44.

Ilbery, B. and Maye, D. (2010) Agricultural restructuring and changing food networks in the UK, in Coe, N. and Jones, A. (eds) *Reading the Economy: The UK in the 21st Century,* Sage, London, 166–80.

Ilbery, B., Morris, C., Buller, H., Maye, D. and Kneafsey, M. (2005) Product, process and place: an examination of food marketing and labelling schemes in Europe and North America, *European Urban and Regional Studies,* **20**, 331–44.

Illeris, S. (2007) The nature of services, in Bryson, J.R. and Daniels, P.W. (eds) *The Handbook of Service Industries,* Edward Elgar, Cheltenham.

Ingilby (2002) Are animals gay? Available at: www.funtrivia.com/ubbthreads/showthreaded.php?Cat=0&Number=34605&page=6.

Ingram, A. (2001) Alexander Dugin: geopolitics and neofascism in post-Soviet Russia, *Political Geography,* **20**(8), 1029–51.

Ingram, A. and Dodds, K. (eds) (2009) *Spaces of Security and Insecurity: Geographies of the War on Terror,* Ashgate, Aldershot.

Institute of Development Studies (IDS) (2009) After 2015: rethinking pro-poor policy, *IDS 'In Focus' Policy Briefs*, Issue 9, available at: www.oecd.org/dataoecd/43/13/43148514.pdf.

International Communications Union (2006), www.internetworldstats.com/stats.htm [accessed 3 January 2007].

International Council for Local Environmental Initiatives (2002) *Second Local Agenda 21 Survey*. Published as UN Dept of Economic and Social Affairs, Background Paper 15 for the 2nd Preparatory Conference for the World Summit on Sustainable Development, DESA/DSD/PC2/BP15, 29pp. Available at: www.iclei.org/rioplusten/finalrdocument.pdf

IPCC (2007) Working Group III Contribution to the Intergovernmental Panel on Climate Change Fourth Assessment Report, *Climate Change 2007: Mitigation of Climate Change*, IPCC Secretariat, Geneva.

Irwin, W. (1996) *The New Niagara: Tourism, Technology, and the Landscape of Niagara Falls, 1776–1917*, Pennsylvania State University, College Park, PA.

Iyer, P. (1989) *Video Nights in Kathmandu: Reports from the Not-so-far East*, Black Swan, London.

J

Jackson, P. (1999) Commodity culture: the traffic in things, *Transactions of the Institute of British Geographers, NS*, **24**, 95–108.

Jackson, P. (2002) Commercial cultures: transcending the cultural and the economic, *Progress in Human Geography*, **26**, 3–18.

Jackson, P. (2010) Food stories: consumption in an age of anxiety, *Cultural Geographies*, **17**, 147–65.

Jackson, P. and Taylor, J. (1996) Geography and the cultural politics of advertising, *Progress in Human Geography*, **19**, 356–71.

Jackson, P., Dwyer, C. and Thomas, N. (2007) Consuming transnational fashion in London and Mumbai, *Geoforum*, **38**(5), 908–24.

Jackson, P., Ward, N. and Russell, P. (2006) Mobilising the commodity chain concept in the politics of food and farming, *Journal of Rural Studies*, **22**, 129–41.

Jackson, R.H. (1990) *Quasi-States: Sovereignty, International Relations and the Third World*, Cambridge University Press, Cambridge.

Jacobs, J. (1961) *The Death and Life of Great American Cities*, Penguin, Harmondsworth.

Jacobs, J. (1984) *Cities and the Wealth of Nations*, Random House, New York.

Japan Institute for Social and Economic Affairs (1996) *Japan 1997: An International Comparison*, Keizai Koho Centre, Tokyo.

Jarosz, L. (2009) Energy, climate change, meat and markets: mapping the coordinates of the current world food crisis, *Geography Compass*, **3**, 1–19.

Jeffrey, A. (2006) Building state capacity in post-conflict Bosnia and Herzegovina: the case of Brčko District, *Political Geography*, **25**(2), 203–27.

Jessop, B. (2002) *The Future of the Capitalist State*, Polity Press, Cambridge.

Joffé, G. (1987) Frontiers in North Africa, in Blake, G.H. and Schofield, R.N. (eds) *Boundaries and State Territory in the Middle East and North Africa*, MENAS Press, Berkhamsted.

John Muir Trust Flyer (2006) *Love the Wild? Help Keep it Wild*, John Muir Trust flyer, 2006.

Johnson, L. (2010) The fearful symmetry of Arctic climate change: accumulation by degradation, *Environment and Planning D: Society and Space*, **28**, 828–47.

Johnson, N. (2003) *Ireland, the Great War and the Geography of Remembrance*, Cambridge University Press, Cambridge.

Johnston, J. (2011) Creating human geography in the English-speaking world, in Agnew, J. and Duncan, J. (eds) *The Wiley-Blackwell Companion to Human Geography*, Wiley Blackwell, Oxford, 89–113.

Johnston, R. (2007) Commentary: on duplicitous battleground conspiracies, *Transactions of the Institute of British Geographers*, **32**, 435–8.

Johnston, R.J. (1984) *Residential Segregation, the State and Constitutional Conflict in American Urban Areas*, Academic Press, London.

Johnston, R.J. and Sidaway J.D. (2004) *Geography and Geographers: Anglo-American Human Geography since 1945*, 6th edition, Arnold, London.

Johnston, R.J., Gregory, D., Pratt, G. and Watts, M. (eds) (2000) *The Dictionary of Human Geography*, 4th edition, Blackwell, Oxford.

Johnston, R.J., Gregory, D., Pratt, G. and Watts, M. (eds) (2008) *The Dictionary of Human Geography*, 5th edition, Blackwell, Oxford.

Jones, G. and Hollier, G. (1997) *Resources, Society and Environmental Management*, Paul Chapman, London.

Jones, H. (1990) *Population Geography*, 2nd edition, Paul Chapman, London.

Jones III, J.P., Woodward, K., and Marston, A.A. (2007) Reply: situating flatness, *Transactions of the Institute of British Geographers NS*, **32**, 264–76.

Jones, M., Jones, R. and Woods, M. (2004) *An Introduction to Political Geography: Space, Place and Politics*, Routledge, London.

Jones, N., Holmes, R. and Espey, J. (2008) *Gender and the MDGs*, ODI Briefing Paper 42, ODI, London.

Jones, P.S. (2000) Why is it alright to do development 'over there' but not 'here'? Changing vocabularies and common strategies across the 'First' and 'Third Worlds', *Area*, **32**, 237–41.

Jones, R. and Fowler, C. (2007) Placing and scaling the nation, *Environment and Planning D: Society and Space*, **25**, 332–54.

Journal of Comparative European Politics (2006), **4**(2/3), Special Issue: Rethinking European Spaces: Territory, Borders, Governance.

Journal of Rural Studies (2009) De-centring White Ruralities: Ethnicity and Indigeneity, special issue of *Journal of Rural Studies*, **25**(4).

Jurgens, U. and Gnad, M. (2002) Gated communities in South Africa: experiences from Johannesburg, *Environment and Planning B: Planning and Design*, **29**, 337–53.

K

Kabeer, N. (1994) *Reversed Realities: Gender Hierarchies in Development Thought*, Verso, London.

Kahn, J.S. (1995) *Culture, Multiculture, Postculture*, Sage, London.

Kaminsky, G. and Schmukler, S. (2002) Emerging markets instability: do sovereign ratings affect country risk and stock returns?, *World Bank Economic Review*, **16**, 171–95.

Kapferer, B. (1988) *Legends of People, Myths of State: Violence, Intolerance and Political Culture in Sri Lanka and Australia*, Smithsonian Institution Press, Washington, DC.

Kaplan, M. (2007) Fijian water in Fiji and New York: local politics and a global commodity, *Cultural Anthropology*, **22**, 685–706.

Kaplan, R.D. (2009) The revenge of geography, *Foreign Policy*, May–June.

Kaplinsky, R. (2005) *Globalization, Poverty and Inequality*, Polity Press, Cambridge.

Kasson, J.F. (1978) *Amusing the Million: Coney Island at the Turn of the Century*, Hill & Wang, New York.

Kay, C. (2001) Reflections on rural violence in Latin America, *Third World Quarterly*, **22**, 5.

Kaye, H.J. (1984) *The British Marxist Historians*, Polity Press, Cambridge.

Keay, J. (1993) *The Honourable Company: A History of the English East India Company*, Harper Collins, London.

Keivani, R. and Werna, E. (2001) Refocusing the housing debate in developing countries from a pluralist perspective, *Habitat International*, **25**(2), 191–208.

Kelly, P. (1984) *Fighting for Hope*, Chatto and Windus, London.

Kelly, P. (1997) *Checkerboards and Shatterbelts: The Geopolitics of South America*, University of Texas Press, Austin.

Kenna, T. and Dunn, K. (2009) The virtuous discourse of private communities, *Geography Compass*, **3**(2), 797– 816.

Kennedy, J.F. (1962) Commencement address to Yale University, 11 June, New Haven, Connecticut.

Kershaw, I. (2000) *The Nazi Dictatorship: Problems and Perspectives of Interpretation*, 4th edition, Arnold, London.

Kessler, J. and Appelbaum, R. (1998) The growing power of retailers in producer-driven commodity chains: a 'retail revolution' in the US automobile industry?, unpublished manuscript, Department of Sociology, University of California at Santa Barbara, CA.

Keyes, C.F. (1995) *The Golden Peninsula: Culture and Adaptation in Mainland Southeast Asia*, University of Hawaii Press, Honolulu.

Kim, Y.-H. (2008) Global-local, in Hall, T., Hubbard, P. and Short, J.R. (eds) *The Sage Companion to the City*, Sage, London, 123–37.

Kinzer, S. (2006) *Overthrow: America's Century of Regime Change from Hawaii to Iraq*, Times Books, New York.

Kipling, R. (1912) *Rudyard Kipling's Verse: Definitive Edition*, Hodder & Stoughton, London.

Kirwan, J. (2004) Alternative strategies in the UK agrofood system: interrogating the alterity of farmers' markets, *Sociologia Ruralis*, **44**(4), 395–415.

Kirwan, J. (2006) The interpersonal world of direct marketing: examining conventions of quality at UK farmers' markets, *Journal of Rural Studies*, **22**, 301–12.

Kjaer, A. (2004) *Governance: A Key Concept*, Sage, London.

Klare, M. (2004) *Blood and Oil: How America's Thirst for Petrol is Killing Us*, London: Penguin.

Klein, N. (2000) *No Logo: Taking Aim at the Brand Bullies*, Flamingo, New York and London.

Klein, N. (2007) *The Shock Doctrine: The Rise of Disaster Capitalism*, Metropolitan Books, New York.

Kleingeld, P. (2004) Approaching perpetual peace: Kant's defence of a League of States and his ideal of a World Federation, *European Journal of Philosophy*, **12**, 304–25.

Kleveman, L. (2003) *The New Great Game: Blood and Oil in Central Asia*, Atlantic Books, London.

Kneafsey, M., Holloway, L., Cox, R., Dowler, L., Venn, L. and Tuomainen, L. (2008) *Reconnecting Consumers, Producers and Food: Exploring Alternatives*, Berg Publishers, Oxford.

Knights, D. and Odih, P. (1995) It's about time! The significance of gendered time for financial services consumption, *Time and Society*, **4**, 205–31.

Knopp, L. (1995) Sexuality and urban space: a framework for analysis, in Bell, D. and Knox, P. and Pinch, S.

(2000) *Urban Social Geography: An Introduction*, 4th edition, Prentice Hall, Harlow.

Knox, P. and Agnew, J. (1994) *The Geography of the World Economy*, 2nd edition, Edward Arnold, London.

Knox, P. and Agnew, J. (2003) *The Geography of the World Economy*, 4th edition, Edward Arnold, London.

Knox, P. and Marston, S.A. (2001) *Human Geography: Places and Regions in Global Context*, 2nd edition, Prentice Hall, Upper Saddle River, NJ.

Knox, P. and Marston, S.A. (2004) *Human Geography: Places and Regions in Global Context*, 3rd edition, Prentice Hall, Upper Saddle River, NJ.

Knox, P. and Marston, S.A. (2006) *Human Geography: Places and Regions in Global Context*, 4th edition, Prentice Hall, Upper Saddle River, NJ.

Knox, P. and Pinch, S. (2000) *Urban Social Geography: An Introduction*, 4th edition, Prentice Hall, Harlow.

Knox, P., Agnew, J. and McCarthy, L. (2003) *The Geography of the World Economy: An Introduction to Economic Geography*, 4th edition, Hodder Arnold, London.

Kobayashi, A. and Peake, L. (2000) Racism out of place: thoughts on whiteness and an antiracist geography in the new millennium, *Annals of the Association of American Geographers*, **90**, 392–403.

Koczberski, L. (1998) Women in development: a critical analysis, *Third World Quarterly*, **19**, 395–409.

Kofman, E. (2008) Feminist transformations of political geography, in Cox, K.R., Low, M. and Robinson, J. (eds) *The Sage Handbook of Political Geography*, Sage, London, 73–86.

Kolstø, P. (ed) (2005) *Myths and Boundaries in South-Eastern Europe*, Hurst & Company, London.

Kondratieff, N.D. (1925) The major economic cycles, *Voprosy Konjunktury*, **6**, 28–79; an English translation is to be found in *Lloyds Bank Review* (1978), **129**, 41–60.

Kong, L. (2007) The promises and prospects of geography in higher education, *Journal of Geography in Higher Education*, **31**(1), 13–7.

Korten, D.C. (1998) Your mortal enemy, *Guardian*, 21 October, 4–5.

Koter, M. (1990) The morphological evolution of a nineteenth-century city centre: Lódz, Poland, 1825–1973, in Slater, T.R. (ed.) *The Built Form of Western Cities*, Leicester University Press, Leicester and London.

Kothari, U. and Minogue, M. (eds) (2002) *Development Theory and Practice: Critical Perspectives*, Macmillan, London.

Kramsch, O. and Hooper, B. (2004) (eds) *Cross-Border Governance in the European Union*, Routledge, London.

Kuus, M. (2007) 'Love, peace and Nato': Imperial subject making in Central Europe, *Antipode*, **39**(2), 269–90.

L

Laïdi, Z. (1988) Introduction: What use is the Soviet Union?, in Laïdi, Z. (ed.), *The Third World and the Soviet Union*, Zed Books, London, 1–23.

Landes, D.S. (1969) *The Unbound Prometheus*, Cambridge University Press, Cambridge.

Landman (2010) Gated minds, gated places: the impact and meaning of hard boundaries in South Africa, in Bagaeen, S. and Uduku, O. (eds) (2010) *Gated Communities: Social Sustainability and Historical Gated Developments*, Earthscan, London, 49–61.

Lang, H.J. (2002) *Fear and Sanctuary: Burmese Refugees in Thailand*, Southeast Asia Program Publications, Cornell University, New York.

Langley, P. (2003) *The Everyday Life of Global Finance*, International Political Economy Group Working Paper No. 5, University of Manchester, Manchester.

Langton, J. (1984) The industrial revolution and the regional geography of England, *Transactions of the Institute of British Geographers*, NS, **9**, 145–67.

Langton, J. and Morris, R.J. (eds) (1986) *Atlas of Industrializing Britain, 1780–1914*, Methuen, London and New York.

Langton, M. (1998) *Burning Questions: Emerging Environmental Issues for Indigenous Peoples in Northern Australia*, Centre for Indigenous Natural and Cultural Resource Management, Northern Territory University, Darwin.

Larner, W. (2003) Neoliberalism?, *Environment and Planning D: Society and Space*, **21**, 509–12.

Lash, S. and Urry, J. (1987) *The End of Organized Capitalism*, Polity Press, Cambridge.

Lash, S. and Urry, J. (1994) *Economies of Signs and Space*, Sage, London.

Latham, A. (2008) Cities (2002): Ash Amin and Nigel Thrift, in Hubbard, P., Kitchin, R. and Valentine, G. (eds) *Key Texts in Human Geography*, London, Sage, 215–23.

Latour, B. (1993) *We Have Never Been Modern*, Harvester Wheatsheaf, Brighton.

Laulajainen, R. (1998) *Financial Geography*, School of Economics and Commercial Law, Göteborg University.

Laurence, J. (1997) The poor of Britain are going hungry, *The Independent*, 11 June, 7.

Laurie, N., Dwyer, C., Holloway, S. and Smith, F. (1999) *Geographies of New Femininities*, Longman, Harlow.

Lawrence, F. (2008) *Eat Your Heart Out*, Penguin, London.

Lawton, R. (1989) Introduction: aspects of the development and role of great cities in the Western world in the nineteenth century, in Lawton, R. (ed.) *The Rise and Fall of Great Cities*, Belhaven, London.

Laxton, P. (1986) Textiles, in Langton, J. and Morris, R.J. (eds) *Atlas of Industrializing Britain, 1780–1914*, Methuen, London and New York, 106–13.

Le Billon, P. (2001) The political ecology of war: natural resources and armed conflicts, *Political Geography*, **20**, 561–84.

Le Billon, P. (2007) Geographies of war: perspectives on 'resource wars', *Geography Compass*, **1**(2), 163–82.

Le Billon, P. and El Khatib, F. (2004) From free oil to 'freedom oil': terrorism, war and US geopolitics in the Persian Gulf, *Geopolitics*, **9**, 109–37.

Le Galès, P. (2002) *European Cities: Social Conflicts and Governance*, Oxford University Press, Oxford.

Le Heron, R.B. (1993) *Globalised Agriculture*, Pergamon Press, London.

Le Heron, R., Murphy, L., Forer, P. and Goldstone, M. (eds) (1999) *Explorations in Human Geography: Encountering Place*, Oxford University Press, Auckland.

Leach, E.R. (1960) The frontiers of 'Burma', *Comparative Studies in Society and History, **3**, 49–68.

Lee, R. (2000) Radical and postmodern? Power, social relations and regimes of truth in the social construction of alternative economic geographies, *Environment and Planning A*, **32**, 991–1009.

Lee, R. (2006) The ordinary economy: tangled up in values and geography, *Transactions of the Institute of British Geographers*, NS, **31**, 413–32.

Lee, R.B. (1968) What hunters do for a living, or how to make out on scarce resources, in Lee, R.B. and DeVore, I. (eds) *Man the Hunter*, Aldine Press, Chicago, IL, 30–48.

Lees, L. (2003) Super-gentrification: the case of Brooklyn Heights, New York City, *Urban Studies*, **40**(12), 2487–509.

Lees, L. (2008) Gentrification and social mixing: towards an inclusive urban renaissance?, *Urban Studies*, **45**, 2449–70.

Leicht, K.T. and Fennell, M.L. (1997) The changing organizational context of professional work, *Annual Review of Sociology*, **23**, 215–31.

Leitner, H., Pavlik, C., Sheppard, E. (2002) Networks, governance and the politics of scale: inter-urban networks and the European Union, in Herod, A. and Wright, M. (eds) *Geographies of Power: Placing Scale*, Blackwell Publishing, Malden, MA, 274–303.

Lemanski, C. and Oldfield, S. (2009) The parallel claims of gated communities and land invasions in a southern city: polarised state responses, *Environment and Planning A*, **41**(3), 634–48.

Lemon, J.T. (1987) Colonial America in the eighteenth century, in Mitchell, R.D. and Groves, P.A. (eds) *North America: The Historical Geography of a Changing Continent*, Hutchinson, London.

Lester, A. (1999) Historical geographies of imperialism, in Graham, B. and Nash, C. (eds) *Modern Historical Geographies*, Prentice Hall, Harlow.

Lester, A. (2002) Obtaining the 'due observance of justice': the geographies of colonial humanitarianism, *Environment and Planning D: Society and Space*, **20**, 277–93.

Levy, F. and Murnane, R.J. (2004) *The New Division of Labour: How Computers are Creating the Next Job Market*, Princeton University Press, Princeton, NJ.

Lewin, A.Y. and Peeters, C. (2006) Offshoring work: business hype or the onset of fundamental transformation?, *Long Range Planning*, **39**, 221–39.

Lewis, D., Rodgers, D. and Woolcock, M. (2005) The Fiction of Development: Knowledge, Authority and Representation, London School of Economics, Working Paper 05-61, Development Studies Institute, LSE, London.

Lewis, G. (1990) Community through exclusion and illusion: the creating of social worlds in an American shopping mall, *Journal of Popular Culture*, **24**, 121–36.

Lewis, M. (1990) *Liar's Poker: Rising Through the Wreckage on Wall Street*, Penguin Books, London.

Lewis, T. and Potter, E. (eds) (2011) *Ethical Consumption: A Critical Introduction*, Routledge, London.

Lewontin, R. (1993) *Biology as Ideology: The Doctrine of DNA*, Harper Perennial, New York.

Ley, D. and Cybriwsky, R. (1974) Urban graffiti as territorial markers, *Annals of the Association of American Geographers*, **64**(4), 491–505.

Ley, D. and Olds, K. (1999) World's fairs and the culture of consumption in the contemporary city, in Anderson, K. and Gale, F. (eds), *Cultural Geographies*, 2nd edition, Longman, Sydney, 221–40.

Leyshon, A. (1995a) Geographies of money and finance, *Progress in Human Geography*, **19**, 531–43.

Leyshon, A. (1995b) Annihilating space? The speed up of communications, in Allen, J. and Hamnett, C. (eds) *A Shrinking World?*, Oxford University Press, Oxford.

Leyshon, A. (1996) Dissolving difference? Money, disembedding and the creation of 'global financial space', in Daniels, P.W. and Lever, W.F. (eds) *The Global Economy in Transition*, Longman, London.

Leyshon, A. and Thrift, N. (1994) Access to financial services and financial infrastructural withdrawal: problems and policies, *Area*, **26**, 268–75.

Leyshon, A. and Thrift, N. (1997) *Money Space: Geographies of Monetary Transformation*, Routledge, London.

Leyshon, M. (2008) The betweenness of being a rural youth: inclusive and exclusive lifestyles, *Social and Cultural Geography*, **9**(1), 1–26.

Lien, M. (2000) Imagined cuisines: nation and market as organizing structures in Norwegian food marketing, in Jackson, P., Lowe, M., Miller, D. and Mort, F. (eds)

Commercial Cultures: Economies, Practices, Spaces, Berg, Oxford, 153–73.

Lintner, B. (2002) *Blood Brothers. Crime, Business and Politics in Asia,* Allen and Unwin, Crows Nest NSW.

Little, J. (2002) Rural geography: rural gender identity and the performance of masculinity and femininity in the countryside, *Progress in Human Geography,* **26**(5), 665–70.

Little, J. and Leyshon, M. (2003) Embodied rural geographies: developing research agendas, *Progress in Human Geography,* **27**(3), 257–72.

Little, R., Maye, D. and Ilbery, B. (2010) Collective purchase: loving local and organic foods beyond the niche market, *Environment and Planning A,* **42**, 1797–813.

Livingstone, D. (1992) *The Geographical Tradition: Issues in the History of a Contested Enterprise,* Blackwell, Oxford.

Low, M. (2005) States, citizenship and collective action, in Daniels, P., Bradshaw, M., Shaw, D. and Sidaway, J. (eds) *An Introduction to Human Geography. Issues for the 21st Century,* 2nd edition, Prentice Hall, Harlow, 442–66.

Lowenthal, D. (1994) European and English landscapes as national symbols, in Hooson, D. (ed.) *Geography and National Identity,* Blackwell, Oxford, 15–38.

Lury, C. (2000) The united colors of diversity: essential and inessential culture, in Franklin, S., Lury, C. and Stacey, J. (eds) *Global Nature, Global Culture,* Sage, London, 146–87.

Luttwak, E. (1990) From geopolitics to geoeconomics, *National Interest,* **20**, 17–24.

M

Macdonald, C.L. and Sirianni, C. (1996) *Working in the Service Economy,* Temple University Press, Philadelphia.

Mackinder, H. (1904) The geographical pivot of history, *Geographical Journal,* **23**, 421–44.

Mackinder, H. (1919) *Democratic Ideals and Reality: A Study in the Politics of Reconstruction,* Constable, London.

Mackinnon, D. and Cumbers, A. (2007) *Introduction to Economic Geography: Uneven Development, Globalization and Place,* Pearson, London.

Macnaghten, P. and Urry, J. (1998) *Contested Natures,* Sage, London.

Maffesoli, M. (1996) *The Time of the Tribes: The Decline of Individualism in Mass Society,* translated by Don Smith, Sage, London.

Mahajan, S. (ed.) (2006) *Input–Output Analysis: 2006,* Office for National Statistics, London.

Mahon, R. (1987) From Fordism to ?: new technology, labour markets and unions, *Economic and Industrial Democracy,* **8**, 5–60.

Maisels, C.K. (1993) *The Emergence of Civilization,* Routledge, London.

Maitland, B. (1985) *Shopping Malls, Planning and Design,* Construction Press, London.

Malbon, B. (1997) Clubbing, consumption, identity and the spatial practices of every-night life, in Skelton, T. and Valentine, G. (eds) *Cool Places,* Routledge, London, 266–86.

Malbon, B. (1999) *Clubbing,* Routledge, London.

Malpass, A., Cloke, P., Barnett, C. and Clarke, N. (2007) Fairtrade urbanism: the politics of place beyond place in the Bristol Fairtrade City campaign, *International Journal of Urban and Regional Research,* **31**(3), 239–49.

Mann, P. (1965) *An Approach to Urban Sociology,* Routledge, London.

Mansfield, B. (2005) Beyond rescaling: reintegrating the 'national' as a dimension of scalar relations, *Progress in Human Geography,* **29**(4), 458–73.

Marchand, M.H. and Parpart, J.L. (1995) *Feminism, Postmodernism, Development,* Routledge, London.

Marcuse, H. (1964) *One-Dimensional Man,* Abacus, London.

Marcuse, P. (2000) The language of globalization, *Monthly Review,* **52**(1–4), available at: www.monthlyreview.org/700marc.html

Markit (2011) *Markit/CIPS UK Manufacturing PMI™,* www.markit.com, 4 January 2011.

Marquez, P. (2005) *Dying Too Young: Addressing Premature Mortality and Ill Health Due to Non-Communicable Diseases and Injuries in the Russian Federation,* The World Bank, Europe and Central Asia Human Development Department.

Marsden, T., Banks, J. and Bristow, G. (2000) Food supply chain approaches: exploring their role in rural development, *Sociologia Ruralis,* **40**, 424–38.

Marsh, D.C. (1977) *The Changing Structure of England and Wales: 1871–1961,* Routledge and Kegan Paul, London.

Marsh, G.P. (1869) *Man and Nature; or, Physical Geography as Modified by Human Action,* C. Scribner & Co., New York.

Marston, S. (2000) The social construction of scale, *Progress in Human Geography,* **24**, 219–42.

Marston, S.A., Jones III, J.P. and Woodward, K. (2005) Human geography without scale, *Transactions of the Institute of British Geographers, NS,* **30**, 416–32.

Marston, S.A., Liverman, D.M., Del Casino, V. and Robbins, P. (2011) *World Regions in Global Context: People, Places and Environments,* 4th edition, Prentice Hall, Upper Saddle River, NJ.

Martin, A. (2007) Trouble in China is good news for American Toy Manufacturers, *New York Times,* 15 August.

Martin, L.L. and Mitchelson, M.L. (2009) Geographies of detention and imprisonment: Interrogating spatial practices of confinement, discipline, law, and state power, *Geography Compass*, **3**(1), 459–77.

Martin, R. (2010) The local geographies of the financial crisis: from the housing bubble to economic recession and beyond, *Journal of Economic Geography*, **11**(4), 587–618.

Martin, R.L. (1994) Stateless monies, global financial integration and national economic autonomy: the end of geography?, in Corbridge, S., Martin, R. and Thrift, N. (eds) *Money, Power and Space*, Blackwell, Oxford, 253–78.

Martin, R.L. (1999a) The new 'geographical turn' in economics: some critical reflections, *Cambridge Journal of Economics*, **23**, 65–91.

Martin, R.L. (1999b) The new economic geography of money, in Martin, R.L. (ed.) *Money and the Space Economy*, Wiley, Chichester, 3–27.

Martin, R.L. (ed.) (1999c) *Money and the Space Economy*, Wiley, Chichester.

Martínez, M. (1997) *Olivenza y el tratado de Alcanices*, Ayuntamiento de Olivenza.

Marx, K. (1867) *Das Kapital*; translated into English as *Capital*, 3 vols, International Publishers, New York (1967) and Penguin, London (1976).

Marx, K. (1973) *Gründrisse: Foundations of the Critique of Political Economy*, Penguin, London.

Mason, C. and R.T. Harrison (2002) The geography of venture capital investments in the UK, *Transactions of the Institute of British Geographers*, **27**, 427–51.

Massey, D. (1981) The geography of industrial change, in Potter, D., Anderson, J., Clarke, J., Coombes, P., Hall, S., Harris, L., Holloway, C. and Walton, T. (eds) *Society and the Social Sciences*, Open University Press, Milton Keynes, 302–13.

Massey, D. (1984) *Spatial Division of Labour: Social Structures and the Geography of Production*, Macmillan, London.

Massey, D. (1993) Power-geometry and a progressive sense of place, in Bird, J., Curtis, B., Putnam, T., Robertson, G. and Tickner, L. (eds) *Mapping the Futures: Local Cultures, Global Change*, Routledge, London, 59–69.

Massey, D. (1994) *Space, Place and Gender*, Polity, Cambridge.

Massey, D. (1996) Space/power, identity/difference, tensions in the city, in Merrifield, A. and Swyngedouw, E. (eds) *The Urbanisation of Injustice*, Blackwell, London, 190–205.

Massey, D. (1999) Cities in the world, in Massey, D., Allen, J. and Pile, S. (eds) *City Worlds*, Routledge in association with the Open University, London, 99–175.

Massey, D. (2009) Concepts of space and power in theory and in political practice, *Documents d'Anàlisi Geogràfica*, **55**, 15–26.

Massey, D. and Jess, P. (eds) (1995a), *A Place in the World? Places, Cultures and Globalization*, Oxford University Press, Oxford.

Massey, D. and Jess, P. (1995b) The contestation of place, in Massey, D. and Jess, P. (eds) *A Place in the World? Places, Cultures and Globalization*, Oxford University Press, Oxford, 133–74.

Massey, D., Allen, J. and Pile, S. (eds) (1999) *City Worlds*, Routledge, London.

Mather, A.S. and Chapman, K. (1995) *Environmental Resources*, Longman, London.

Mattelart, A. (1979) *Multinational Corporations and the Control of Culture*, Harvester Press, Brighton.

Mawby, R.I. and Yarwood, R. (eds) (2011) *Rural Policing and Policing the Rural: A Constable Countryside?*, Ashgate, Farnham.

Maye, D. and Kirwan, J. (2010a) Alternative food networks, sociology of agriculture and food, entry for *Sociopedia.isa*, Sage, London.

Maye, D. and Kirwan, J. (2010b) *The Future of Food*, CCRI Policy Conference, Cheltenham, 19 October 2010.

Maye, D., Holloway, L. and Kneafsey, M. (eds) (2007) *Alternative Food Geographies: Representation and Practice*, Elsevier, Oxford.

McCarthy, C., Rodriguez, A., Buendia, E., Meacham, S., David, S., Godina, H., Supriya, K. and Wilson-Brown, C. (1997) Danger in the safety zone: notes on race, resentment and the discourse of crime violence and suburban security, *Cultural Studies*, **11**, 274–95.

McCarthy, J. (2005) Rural geography: multifunctional rural geographies – reactionary or radical?, *Progress in Human Geography*, **29**, 773–82.

McCarthy, J. (2006) Rural geography: alternative rural economies: the search for alterity in forests, fisheries, food, and fair trade, *Progress in Human Geography*, **30**(6), 803–11.

McCarthy, S. (1999) The fabulous kingdom of gay animals, available at www.salon.com/it/feature/1999/03/cov_15featurea.html [accessed 18 December 2006].

McClintock, A. (1994) *Imperial Leather: Race, Gender and Sexuality in the Colonial Context*, Routledge, London.

McCormick, J. (1989) *The Global Environmental Movement*, Belhaven Press, London.

McCormick, J. (1991) *British Politics and the Environment*, Earthscan, London.

McCormick, J. (2006) *The European Superpower*, Palgrave Macmillan, London.

McDowell, L. (1997) *Capital Culture: Gender at Work in the City*, Blackwell, Oxford.

McDowell, L. (1999) *Gender, Identity and Place: Understanding Feminist Geographies*, Polity Press, Cambridge.

McDowell, L. (2007) Gender divisions of labour: sex, gender, sexuality and embodiment in the service sector, in Bryson, J. R. and Daniels, P. W. (eds) *The Handbook of Services*, Edward Elgar, Cheltenham, 395–408.

McDowell, L. (2009) *Working Bodies: Interactive Service Employment and Workplace Identities*, Wiley-Blackwell, Chichester.

McDowell, L. and Sharp, J. P. (eds) (1997) *Space, Gender and Knowledge: Feminist Readings*, Arnold, London.

McGreal, C. (2002) Thabo Mbeki's catastrophe, *Prospect*, **72**, available at: www.prospect-magazine.co.uk/article_details.php?id=4995

McGuigan, J. (2009) *Cool Capitalism*, Pluto Press, London.

McGuinness, M. (2000) Geography matters? Whiteness and contemporary geography, *Area*, **32**, 225–30.

McIntosh, L. and D. Maddox (2008) Salmond fury at 'bunch of spivs' behind bank crisis, *News.Scotsman.com*, available at: http://news.scotsman.com/latestnews/Salmond-fury-at-39bunch-of.4503240.jp [accessed 26 October 2010].

McIver, J. (1997) Environmental protection, indigenous rights and the Arctic Council: rock, paper, scissors on the ice?, *The Georgetown International Environmental Law Review*, **10**, 147–68.

McKibben, B. (1999) *The End of Nature*, Random House, New York.

McLaughlin, E. and Muncie, J. (eds) (1999) Walled cities: surveillance, regulation and segregation, in Pile, S., Brook, C. and Mooney, G. (eds) *Unruly Cities? Order/Disorder*, Routledge, London.

McLeod, G. and Ward, K. (2002) Spaces of utopia and dystopia: landscaping the contemporary city, *Geografiska Annaler*, **84B**(3–4), 153–70.

McLuhan, M. (1964) *Understanding Media: Extensions of Man*, Routledge & Kegan Paul, London.

McLuhan, M. (1975) The implications of cultural uniformity, in Bigsby, C.W.E. (ed.) *Superculture: American Popular Culture and Europe*, Bowling Green University Popular Press, Bowling Green, OH, 43–56.

McMichael, P. (2009) A food regime analysis of the 'world food crisis', *Agriculture and Human Values*, **26**, 281–95.

McNeill, J. (2000) *Something New Under the Sun: An Environmental History of the Twentieth Century*, Penguin Books, London.

McNeill, D. (2004) *New Europe: Imagined Spaces*, Arnold, London.

Meadows, D. H., Meadows, D. I., Randers, J. and Behrens III, W. W. (1972) *The Limits to Growth: A Report to the Club of Rome*, Universe Books, New York.

Megoran, N. (2006) For ethnography in political geography: experiencing and re-imagining Ferghana Valley boundary closures, *Political Geography*, **26**(10), 622–40.

Meinig, D. W. (1986) *The Shaping of America. A Geographical Perspective on 500 Years of History: Atlantic America, 1492–1800*, Yale University Press, New Haven and London.

Merchant, C. (2005) *Radical Ecology*, 2nd edition, Routledge, London.

Meyer, D. R. (1990) The new industrial order, in Conzen, M. P. (ed.) *The Making of the American Landscape*, Unwin Hyman, London and Winchester, MA.

Middleton, N. (2003) *The Global Casino: An Introduction to Environmental Issues*, 3rd edition, Arnold, London.

Miles, D. (2006) *The Tribes of Britain*, Phoenix, London.

Miller, D. (1992) The young and the restless in Trinidad: a case study of the local and the global in mass consumption, in Silverstone, R. and Hirsch, E. (eds) *Consuming Technologies: Media and Information in Domestic Spaces*, Routledge, London, 163–82.

Miller, D. (1994) *Modernity, an Ethnographic Approach: Dualism and Mass Consumption in Trinidad*, Berg, Oxford.

Miller, D. (1995) Consumption as the vanguard of history, a polemic by way of an introduction, in Miller, D. (ed.) *Acknowledging Consumption: A Review of New Studies*, Routledge, London, 1–57.

Miller, D. (1997) Coca-Cola, a black sweet drink from Trinidad, in Miller, D. (ed.) *Material Cultures*, UCL Press, London.

Miller, D. (1998) *A Theory of Shopping*, Polity, Cambridge.

Miller, D. (2001) The poverty of morality, *Journal of Consumer Culture*, **1**(2), 225–43.

Miller, D. and Slater, D. (2001) *The Internet: An Ethnographic Approach*, Berg, Oxford.

Miller, D. and Woodward, S. (eds) (2010) *Global Denim*, Berg, Oxford.

Millington, A. C., Mutiso, S. K., Kirkby, J. and O'Keefe, P. (1989) African soil erosion: nature undone and the limitations of technology, *Land Degradation and Rehabilitation*, **1**, 279–90.

Millstone, E. and Lang, T. (2003) *The Atlas of Food*, Earthscan, London.

Milner, M. (1998) IMF puts UK jobs at risk, *Guardian*, 17 July, 23.

Ministry of Defence (MOD) (2006) *Defense Spending*, copy available from: www.mod.uk/DefenceInternet/AboutDefence/Organisation/KeyFactsAboutDefence/DefenceSpending.htm [accessed 23 May 2007].

Mintel (2004) *Nightclubs*, Mintel Market Intelligence, London.

Mintel (2006) *Nightclubs*, Mintel Market Intelligence, London.

Mitchell, D. (1996) Public space and the city, *Urban Geography*, **17**, 127–31.

Mitchell, D. (2000) *Cultural Geography: A Critical Introduction*, Blackwell, Oxford.

Mitchell, D. (2003) *The Right to the City: Social Justice and the Fight for Public Space*, Guilford, New York.

Mitchell, D. (2005) The S.U.V. model of citizenship: floating bubbles, buffer zones, and the rise of the 'purely atomic' individual, *Political Geography*, **24**(1), 77–100.

Mitchell, J. B. (ed.) (1962) *Great Britain: Geographical Essays*, Cambridge University Press, Cambridge.

Mitchell, R. D. and Groves, P. A. (eds) (1987) *North America: The Historical Geography of a Changing Continent*, Hutchinson, London.

Mitchell, W. C. (1937) *The Backward Art of Spending Money and Other Essays*, McGraw-Hill, New York.

MOD *see* Ministry of Defence

Mohan, J. (2000) Geographies of welfare and social exclusion, *Progress in Human Geography*, **24**, 291–300.

Mohanty, C. T. (1991) Under Western eyes: feminist scholarship and colonial discourses, in Mohanty, C. T., Russo, A. and Torres, L. (eds) *Third World Women and the Politics of Feminism*, Indiana University Press, Bloomington, 51–80.

Momsen, J. (1991) *Women and Development in the Third World*, Routledge, London.

Moore, R. I. (ed.) (1981) *The Hamlyn Historical Atlas*, Hamlyn, London and New York.

Momsen, J. (2009) *Gender and Development*, 2nd edition, Routledge, London.

Monrreal, S. (2008) A novel, spicy delicacy: tamales, advertising and late nineteenth century imaginative geographies of Mexico, *Cultural Geographies*, **15**, 449–70.

Mooney, P.H. and Hunt, S.A. (2009) Food security: the elaboration of contested claims to a consensus frame, *Rural Sociology*, **74**, 469–97.

Moran, D. (2004). Exile in the Soviet forest: 'special settlers' in northern Perm Oblast, *Journal of Historical Geography*, **30**, 395–413.

Morgan, K. and Sonnino, R. (2010) The urban foodscape: world cities and the new food equation, *Cambridge Journal of Regions, Economy and Society*, **3**, 209–24.

Morgan, K., Marsden, T. and Murdoch, J. (2006) *Worlds of Food: Place, Power and Provenance in the Food Chain*, Oxford University Press, Oxford.

Morley, D. (2000) *Home Territories: Media, Mobility and Identity*, Routledge, London.

Morris-Suzuki, T. (1996) The frontiers of Japanese identity, in Tønnesson, S. and Antlöv, H. (eds) *Asian Forms of the Nation*, Curzon, Richmond, 41–66.

Moseley, W., Lanegran, D. and Pandit, K. (eds) (2007) *The Introductory Reader in Human Geography*, Blackwell, Oxford.

Moser, C. (1993) *Gender, Planning and Development: Theory, Practice and Training*, Routledge, London.

Mosse, G.L. (1975) *The Nationalization of the Masses*, H. Fertig, New York.

Mountjoy, A.B. (1976) Worlds without end, *Third World Quarterly*, **2**, 753–7.

Mountz, A. (2009) Border, in Gallaher, C., Dahlman, C.T., Gilmartin, M., Mountz, A. and Shirlow, P. (eds) *Key Concepts in Political Geography*, Sage, London, 198–209.

Murdoch, J. (2006) *Post-structuralist Geography: A Guide to Relational Space*, Sage, London.

Murray, W.E. (2001) The second wave of globalization and agrarian change in the Pacific Islands, *Journal of Rural Studies*, **17**(2), 135–48.

Murray, W.E. (2005) *Geographies of Globalization*, Routledge/Taylor and Francis, London and New York.

Murray, W.E. (2006) Neo-feudalism in Latin America? Globalisation, agribusiness, and land re-concentration in Chile, *Journal of Peasant Studies*, **33**, 646–77.

Murray, W.E. (2008) Neoliberalism, rural geography and the global south, *Human Geography*, **1**(1), 33–8.

Murray, W.E. and Terry, J.P. (2004) Niue's place in the Pacific, in Terry, J.P. and Murray, W.E. (eds), *Niue Island: Geography on the Rock of Polynesia*, INSULA, UNESCO, Paris.

Mutersbaugh, T. (2005) Fighting standards with standards: harmonization, rents, and social accountability in certified agrofood networks, *Environment and Planning A*, **37**, 2033–51.

N

Naess, A. (1973) The shallow and the deep: long-range ecological movement, *Inquiry*, **16**, 95–100.

Nagel, C. (2002) Geopolitics by another name: immigration and the politics of assimilation, *Political Geography*, **21**, 971–87.

National Research Council (2000) *Beyond Six Billion: Forecasting the World's Population*, National Academy Press, Washington, DC.

National Statistics Online (NSO) (2004) *Family Expenditure Survey HMSO*, London.

Natter, W. (2003) Geopolitics in Germany 1919–1945, in Agnew, J., Mitchell, K. and Toal, G. (eds) *A Companion to Political Geography*, Blackwell, Oxford, 197–203.

Nature (2010) How to feed a hungry world (Editorial), *Nature*, **466**, 531–2.

Nava, M., Blake, A., MacRury, I. and Richards, B. (eds) (1997) *Buy This Book: Studies in Advertising and Consumption*, Routledge, London.

Nayek, A. (2003) Last of the real Geordies? White masculinities and the subcultural response to deindustrialization, *Environment and Planning D: Society and Space*, **21**, 7–25.

Nayak, A. and Jeffrey A. (2011) *Geographical Thought: An Introduction to Ideas in Human Geography*, Pearson Education, Harlow.

Neilsen B. (2010) Between governance and sovereignty: remaking the borderscape to Australia's north, *RMIT Local-Global Journal*, **8**, available at: http://mams.rmit.edu.au/56k3qh2kfcx1.pdf.

Neilson, J. and Pritchard, B. (2009) *Value Chain Struggles: Institutions and Governance in the Plantation Districts of South India*, Wiley-Blackwell, Oxford.

Nellemann, C., MacDevette, M., Manders, T. *et al.* (eds) (2009) *The Environmental Food Crisis – The Environment's Role in Averting Future Food Crises: A UNEP Rapid Response Assessment*, United Nations Environment Programme, Arendal, Norway.

Nelson, L. and Seager, J. (2004) *A Companion to Feminist Geography*, Blackwell, Oxford.

Newman, O. (1972) *Defensible Space: People and Design in the Violent City*, Architectural Press, London.

Newman, P., Beatley, T. and Boyer, H. (2009) *Resilient Cities: Responding to Peak Oil and Climate Change*, Island Press, Washington, DC.

Nitkin, A. (2007) The end of 'post-Soviet space': the changing geopolitical orientations of the newly independent states, *Russia and Eurasia Briefing Paper 07/01*, Chatam House, London.

Nixon, S. (1997) Circulating culture, in du Gay P (ed.) *Production of Culture/Cultures of Production,* Sage, London, 177–234.

Nord, M., Andrews, M. and Carlson, S. (2009) *Household Food Security in the United States, 2008*, Economic Research Report 83, United States Department of Agriculture, Economic Research Service, Washington DC, available at: www.ers.usda.gov/Publications/ERR83/ERR83.pdf.

Nordstrom, C. (2000) Shadows and sovereigns, *Theory, Culture and Society*, **17**, 35–54.

Nordstrom, C. (2004) *Shadows of War: Violence, Power, and International Profiteering in the Twenty-first Century*, University of California Press, Berkeley, CA.

Nostrand, R. L. (1987) The Spanish borderlands, in Mitchell, R. D. and Groves, P. A. (eds) *North America: The Historical Geography of a Changing Continent*, Hutchinson, London.

Nove, A. (1987) *The Soviet Economic System*, 3rd edition, Allen & Unwin, London.

Nuttall, M. (2008) Climate change and the warming politics of autonomy, *Indigenous Affairs*, **1**, 44–51.

O

Oakeshott, I. and Gourlay, C. (2006) Science told: hands off gay sheep, *Sunday Times*, London.

O'Brien, R. (1992) *Global Financial Integration: The End of Geography*, Royal Institute of International Affairs, Pinter Publishers, London.

O'Connor, A. (1976) Third World or one world, *Area*, **8**, 269–71.

Odell, P.R. (1989) Draining the world of energy, in Johnston, R.J and Taylor, P.J. (eds) *A World in Crisis?*, 2nd edition, Blackwell, Oxford.

OECD *see* Organization for Economic Co-operation Development

Ogborn, M. (1999) Historical geographies of globalisation, *c.* 1500–1800, in Graham, B. and Nash, C. (eds) *Modern Historical Geographies*, Prentice Hall, Harlow.

Ó Gráda, C. (2007) Making famine history, *Journal of Economic Literature*, **45**(1), 5–38.

Ohmae, K. (1990) *The Borderless World: Power and Strategy in the Interlinked Economy*, Collins, London.

Ohnuki-Tierney, E. (1997) McDonalds in Japan: changing manners and etiquette, in Watson, J. L. (ed.) *Golden Arches East: McDonalds in East Asia*, Stanford University Press, Stanford, CA, 161–82.

O'Loughlin, J. (ed.) (1994) *Dictionary of Geopolitics*, Greenwood Press, Westport, CT.

O'Loughlin, J. and van der Wusten, H. (1990) Political geography and panregions, *Geographical Review*, **80**, 1–20.

O'Loughlin, J., Toal, G. and Kolossov, V. (2005) Russian geopolitical culture and public opinion: the masks of Porteus revisited, *Transactions of the Institute of British Geographers*, **30**(3), 322–35.

Ong, A. (1999) *Flexible Citizenship: The Cultural Logics of Transnationality*, Duke University Press, Durham, NC.

Oppenheimer, S. (2006) *The Origins of the British*, Constable, London.

O'Reilly, K. and Crutcher, M. E. (2006) Parallel politics: the spatial power of New Orleans Labor Day parades, *Social and Cultural Geography*, **7**(2), 245–65.

Organization for Economic Co-operation and Development (2005a) *OECD in Figures 2005*, OECD, Paris.

Organization for Economic Co-operation and Development (2005b) *Potential Offshoring of ICT-intensive Using Occupations, Working Party on the Information Economy*, DSTA/ICC/IE(2004)19, OECD, Paris.

Organization for Economic Co-operation and Development (2006) *Putting Climate Change Adaptation in the Development Mainstream*, OECD Policy Brief, March 2006, OECD, Paris, available at: www.oecd.org/dataoecd/57/55/36324726.pdf [accessed 21 September 2010].

Organization for Economic Co-operation and Development (2010) Development aid rose in 2009 and most donors will meet 2010 targets, report of the Development Co-operation Directorate, OECD, Paris, available at: www.oecd.org/document/11/0,3343,en_

2649_34447_44981579_1_1_1_1,00.html [accessed 21 September 2010].

Organization for Economic Co-operation and Development/IEA (2010) *Key World Energy Statistics 2010*, IEA, Paris.

O'Riain, S. (2004) The politics of mobility in technology driven commodity chains: developmental coalitions in the Irish software industry, *International Journal of Urban and Regional Research*, **28**, 642–63.

O'Riordan, T. (ed.) (2000)*Environmental Science for Environmental Management*, 3rd edition, Prentice Hall, Harlow.

Oslender, U. (2007) Spaces of terror and fear on Columbia's Pacific coast: the armed conflict and forced displacement among black communities, in Gregory, D. and Pred, A. (eds (2007) *Violent Geographies: Fear, Terror, and Political Violence*, Routledge, London, 111–32.

Osofsky, H. (2006/2007) The Inuit petition as a bridge? Beyond dialectics of climate change and indigenous people's rights, *American Indian Law Review*, **31**, 675–97.

Østergaard, L. (1992) *Gender and Development: A Practical Guide*, Routledge, London.

Oswalt, P. (ed.) (2004) *Schrumpfende Städte. Band 1: Internationale Untersuchung*, Hantje Cantz, Ostfildern-Ruit.

Ó Tuathail, G. (1996a) *Critical Geopolitics: The Politics of Writing Global Space*, Routledge, London.

Ó Tuathail, G. (1996b) An anti-geopolitical eye: Maggie O'Kane in Bosnia, 1992–3, *Gender, Place and Culture*, **3**(2), 177–85.

Ó Tuathail, G. (1998) Postmodern geopolitics? The modern geopolitical imagination and beyond, in Ó Tuathail, G. and Dalby, S. (eds) *Rethinking Geopolitics*, Routledge, London, 16–38.

Ó Tuathail, G. and Dalby, S. (eds) (1998) *Rethinking Geopolitics*, Routledge, London.

Ó Tuathail, G., Dalby, S. and Routledge, P. (eds) (2006) *The Geopolitics Reader*, 2nd edition, Routledge, London.

Overton, J.D. and Murray, W.E. (2011) Playing the scales: regional transformations and the differentiation of rural space in the Chilean wine industry, *Journal of Rural Studies*, **27**(1), 63–72.

P

Paasi, A. (1996) *Territories, Boundaries and Consciousness: The Changing Geographies of the Finnish–Russian Frontier*, Wiley, Chichester.

Paasi, A. (2003a) Territory, in Agnew, J., Mitchell, K. and Toal, G. (eds) *A Companion to Political Geography*, Blackwell, Malden, 109–22.

Paasi, A. (2003b) Region and place: regional identity in question, *Progress in Human Geography*, **27**, 475–85.

Paasi, A. (2009) The resurgence of the 'region' and 'regional identity': theoretical perspectives and empirical observations on regional dynamics in Europe, *Review of International Studies*, **35**, 121–46.

Pacione, M. (2009) *Urban Geography: A Global Perspective*, 3rd edition, Routledge, London.

Packard, V. (1977) *The Hidden Persuaders*, Penguin, Harmondsworth.

Paddison, R. (2001) Studying cities, in Paddison, R. (ed.) *Handbook of Urban Studies*, London, Sage, 1–9.

Pahl, J. (1989) *Money and Marriage*, Macmillan, London.

Pain, R. (2010) The new geopolitics of fear, *Geography Compass*, **4**(3), 226–40.

Pain, R. and Smith, S.J. (eds) (2008) *Fear: Critical Geopolitics and Everyday Life*, Ashgate, Aldershot.

Pain, R.H. (1997) Social geographies of women's fear of crime, *Transactions of the Institute of British Geographers, NS*, **22**, 231–44.

Painter, J. (2006) Prosaic geographies of stateness, *Political Geography*, **25**, 752–74.

Painter, J. (2010) Rethinking territory, *Antipode*, **42**(5), 1098–118.

Painter, J. and Jeffrey, A. (2009), *Political Geography*, 2nd edition, Sage, London.

Palin, M. (2010) Presidential address. Annual General Meeting of the Royal Geographical Society (with the Institute of British Geographers) 7 June 2010, *The Geographical Journal*, **176**(3), 235–55.

Pallot, J. (2005) Russia's penal peripheries: space, place and penalty in Soviet and post-Soviet Russia. *Transactions of the Institute of British Geographers,* **30**(1), 98–112.

Palmer, J.A. (ed.) (2001) *Fifty Key Thinkers on the Environment*, London, Routledge.

Panelli, R. (2004) *Social Geographies: From Difference to Action*, Sage, London.

Pantazis, C., Gordon, D. and Levitas, R.A. (eds) (2006) *Poverty and Social Exclusion in Britain: The Millennium Survey*, Policy Press, Bristol.

Park, B.-G. (2005) Spatially selective liberalization and graduated sovereignty: politics of neo-liberalism and 'special economic zones' in South Korea, *Political Geography*, **24**(7), 850–73.

Park, C. (1994) *Sacred Worlds: An Introduction to Geography and Religion*, Routledge, London.

Parker, A. (2004) *Two-Speed Europe: Why 1 Million Jobs will Move Offshore*, Forrester Research, Cambridge, MA.

Parker, J. (2006) Cold War II: the Eisenhower Administration, the Bandung Conference, and the reperiodization of the postwar era, *Diplomatic History*, **30**(5), 867–92.

Parks, B.C. and Roberts, J.T. (2008) Inequality and the global climate regime: breaking the north–south impasse, *Cambridge Review of International Affairs*, **21**(4), 621–48.

Parrott, N., Wilson, N. and Murdoch, J. (2002) Spatializing quality: regional protection and the alternative geography of food, *European Urban and Regional Studies*, **9**, 241–61.

Patel, R. (2000) Urban violence: an overview, in Miles, M., Hall, T. and Borden, I. (eds) *The City Cultures Reader*, Routledge, London, 261–5.

Paterson, T.G. (1992) *On Every Front. The Making and Unmaking of the Cold War*, revised edition, W.W. Norton, New York and London.

Peach, C. (1996) Does Britain have ghettos?, *Transactions of the Institute of British Geographers*, **21**, 216–35.

Peach, W.N. and Constantin, J.A. (1972) *Zimmerman's World Resources and Industries*, 3rd edition, Harper & Row, London.

Pearce, F. (2003) Saving the world, plan B, *The New Scientist*, 13 December, 6–7.

Peck, J. (2004) Geography and public policy: constructions of neoliberalism, *Progress in Human Geography*, **28**, 392–405.

Peck, J. (2010) *Constructions of Neoliberal Reason*, Oxford University Press, Oxford.

Peck, J.A. and Yeung, H.W. (eds) (2003) *Remaking the Global Economy*, Sage, London.

Peet, R. (1989) The destruction of regional cultures, in Johnston, R. J. and Taylor, P.J. (eds) *A World in Crisis?* 2nd edition, Blackwell, Oxford, 150–72.

Peet, R. with Hartwick, E. (1999) *Theories of Development*, Guilford Press, London.

Pelkmans, M. (2006) *Defending the Border: Identity, Religion and Modernity in the Republic of Georgia*, Cornell University Press, Ithaca, NY and London.

Peluso, N. L. and Watts, M. (eds) (2001) *Violent Environments*, Cornell University Press, Ithaca, NY.

Penrose, J. (2002) Nations, states and homelands: territory and territoriality in nationalist thought, *Nations and Nationalism*, **8**, 277–97.

Pepper, D. (1984) *The Roots of Modern Environmentalism*, Croom Helm, Beckenham.

Perkins, J.H. (1997) *Geopolitics and the Green Revolution: Wheat, Genes and the Cold War*, Oxford University Press, Oxford.

Perry, S. and Schneck, C. (eds) (2001) *Eye-to-eye: Women Practicing Development Across Cultures*, Zed Books, London.

Peterson Del Mar, D. (2006) *Environmentalism*, Longman, New York.

Petty, W. [1678] (2004) *Essays on Mankind and Political Arithmetic*, Kessinger Publishing, Whitefish, MT.

Philip's Great World Atlas (2003), George Philip in association with the RGS with the IBG, London.

Phillips, M. (2002) Distant bodies? Rural studies, political economy and post-structuralism, *Sociologia Ruralis*, **42**(2), 81–105.

Phyne, J. and Mansilla, J. (2003) Forging linkages in the commodity chain: the case of the Chilean salmon farming industry, 1987–2001, *Sociologia Ruralis*, **43**(2), 108–27.

Phillips, D. (2005) Parallel lives? Challenging discourses of British Muslim self-segregation in multicultural Britain, *Environment and Planning D: Society and Space*, **24**, 25–40.

Philo, C. (2008) Introduction, in Phlio, C. (ed.) *Theory and Methods in Human Geography*, Ashgate, Aldershot, xiii–xlix.

Pickerill, J. (2008) From wilderness to WildCountry: the power of language in environmental campaigns in Australia, *Environmental Politics*, **17**(1), 93–102.

Pickles, J. (2004) *A History of Spaces: Cartographic Reason, Mapping and the Geo-coded World*, Routledge, London and New York.

Pieterse, J.N. and Parekh, B. (eds) (1995) *Decolonization of the Imagination: Culture, Knowledge and Power*, St Martins Press, London.

Pike, A. (2009) Geographies of brands and branding, *Progress in Human Geography*, **33**, 619–45.

Pile, S. (1996) *The Body and the City: Psychoanalysis, Space and Subjectivity*, Routledge, London.

Pile, S. (1999) The heterogeneity of cities, in Pile, S., Brook, C. and Mooney, G. (eds) *Unruly Cities? Order/ Disorder*, Routledge, London, 7–52.

Pinch, S. (1994) Social polarization: a comparison of evidence from Britain and the United States, *Environment and Planning A*, **25**, 779–95.

Pinch, S. and Henry, N. (1999) Paul Krugman's geographical economics, industrial clustering and the British motor sport industry, *Regional Studies*, **33**, 815–27.

Pine, B.J. and Gilmore, J.H. (1999) *The Experience Economy: Work is Theatre and Every Business a Stage*, Harvard Business School Press, Cambridge, MA.

Pinstrup-Andersen, P. (2009) Food security: definition and measurement, *Food Security*, **1**(1), 5–7.

Pletsch, C. (1981) The three worlds or the division of social scientific labour 1950–1975, *Comparative Studies in Society and History*, **23**, 565–90.

Poguntke, T. (1993) Goodbye to movement politics? Organisational adaption of the German Green Party, *Environmental Politics*, **2**, 379–404.

Policy Commission on the Future of Farming and Food (2002) *Farming and Food: A Sustainable Future*, HMSO, London.

Political Geography (1998) Space, Place and Politics in Northern Ireland, special issue of *Political Geography*, **17**(2).

Political Geography (2006) Geographies as Sexual Citizenship, special issue of *Political Geography*, **25**(8).

Pollard, J.S. (1996) Banking at the margins: a geography of financial exclusion in Los Angeles, *Environment and Planning A*, **28**, 1209–32.

Pollard, J.S. and Sidaway J.D. (2002) Euroland: economic, political and cultural geographies, *Transactions of the Institute of British Geographers,* **27**, 7–10.

Pollard J.S., Oldfield, J., Randalls, S.C. and Thornes, J.E. (2008) Firm finances, weather derivatives and geography, *Geoforum*, **39**(2), 616–24.

Pollock, F. (2003) *The League of Nations*, The Lawbook Exchange, New Jersey.

Ponte, S. (2002) The 'latte revolution'?: Regulation, markets and consumption in the global coffee chain, *World Development*, **30**, 1099–122.

Popke, E.J. (2001) Modernity's abject space: the rise and fall of Durban's Cato Manor, *Environment and Planning A,* **33**, 737–52.

Population Division of the Department of Economic and Social Affairs of the United Nations Secretariat (2007) *World Population Prospects: The 2006 Revision, Highlights*, United Nations, New York.

Population Reference Bureau (2007) *Human Population: Fundamentals of Growth: Future Growth*, available at: www.prb.org/Educators/TeachersGuides/HumanPopulation/FutureGrowth/TeachersGuide.aspx?p=1

Porritt, J. (1984) *Seeing Green*, Blackwell, Oxford.

Porritt, J. (2009) Perfect storm of environmental and economic collapse are closer than you think, *guardian.co.uk*, 23 March, available at: www.guardian.co.uk/environment/2009/mar/23/jonathon-porritt-recession-climate-crisis [accessed 12 May 2011].

Porritt, J. and Winner, D. (1988) *The Coming of the Greens*, Fontana, London.

Porter, M.E. (1990) *The Competitive Advantage of Nations*, Free Press, New York.

Porter, P.W. and Sheppard, E.S. (1998) *A World of Difference: Society, Nature, Development*, Guilford Press, New York.

Potter, R. and Lloyd-Evans, S. (1998) *The City in the Developing World*, Longman, Harlow.

Potter, R.B. (1979) Perception of urban retailing facilities: an analysis of consumer information fields, *Geografiska Annaler*, **61B**, 19–27.

Potter, R., Binns, T., Elliott, J.A. and Smith, D. (1999) *Geographies of Development*, Longman, London.

Potter, R. B., Binns, T., Elliot, J.A. and Smith, D. (2004) *Geographies of Development*, 2nd edition, Prentice Hall, Harlow.

Potter, R.B., Binns, T., Elliott, J.A. and Smith, D. (2008) *Geographies of Development*, 3rd edition, Pearson Education, Harlow.

Poulsen, M. and Johnston, R. (2006) Ethnic residential segregation in England: getting the right message across, *Environment and Planning A*, **38**(12), 2195– 9.

Poulsen, M., Forrest, J. and Johnston, R. (2002) From modern to post-modern?: Contemporary ethnic residential segregation in four US metropolitan areas, *Cities*, **19**(3), 161–72.

Pounds, N.J.F. (1947) *An Historical and Political Geography of Europe*, Harrap, London.

Pounds, N.J.G. (1990) *An Historical Geography of Europe*, Cambridge University Press, Cambridge.

Powell, R. (2008) Configuring an 'Arctic Commons?', *Political Geography*, **27**, 827–32.

Power, A. and Tunstall, R. (1997) *Dangerous Disorder: Riots and Violent Disturbances in Thirteen Areas of Britain, 1991–1992*, Joseph Rowntree Foundation, York.

Power, M. (2003) *Rethinking Development Geographies*, Routledge, London.

Power, M. (2007) Digitized virtuosity: video war games and post 9/11 cyber-detterence, *Security Dialogue*, **38**(2), 271–88.

Pratt, G. (2005) Masculinity–feminity, in Cloke, P., Crang, P. and Goodwin, M. (eds) *Introducing Human Geographies*, 2nd edition, Hodder Arnold, London, 91–103.

Price, P.L. (2010) At the crossroads: critical race theory and critical geographies of race, *Progress in Human Geography,* **34**(2), 147–74.

Priestley, J.B. (1937) *English Journey,* William Heinemann, London.

Pritchard, B. (1998) The emergent contours of the third food regime: evidence from Australian dairy and wheat sectors, *Economic Geography*, **74**(1), 64–74.

Programme Evaluation Organization (2005) *Performance Evaluation of Targeted Public Distribution System*, Planning Commission, New Delhi.

Progress in Human Geography, **24**(1), – Classics in Human Geography revisited, 91–9.

R

Raban, J. (1990) *Hunting Mr Heartache*, Collins Harvill, London.

Radcliffe, S. and Westwood, S. (1996) *Remaking the Nation: Place, Identity and Politics in Latin America*, Routledge, London.

Radcliffe, S. A. (1999) Re-thinking development, in Cloke, P., Crang, P. and Goodwin, M. (eds) *Introducing Human Geographies*, Arnold, London, 84–92.

Radcliffe, S. A. (2005) Re-thinking development in Cloke, P., Crang, P. and Goodwin, M. (eds) *Introducing Human Geographies*, 2nd edition, Arnold, London.

Radical Routes (2006) *How to Work Out Your Ecological Footprint*, available at: www.radicalroutes.org.uk/pub.html

Rahnema, R. (1997) Introduction, in Rahnema, R. and Bawtree, V. (eds) *The Post-Development Reader*, Zed Books, London.

Ralphs, R., Medina, J. and Aldridge, J. (2009) Who needs enemies with friends like these?: The importance of place for young people living in known gang areas, *Journal of Youth Studies*, **12**(5), 483–500.

Ram, P., Mohanty, S.K. and Ram, U. (2009) Understanding the distribution of BPL cards: all-India and selected states, *Economic and Political Weekly*, **44**(7), 66–71.

Rathgeber, E. (1990) WID, WAD, GAD: trends in research and practice, *Journal of Developing Areas*, **24**, 489–502.

Ravenhall, Y. (1999) *Women's Investing*, Part 1, online, available at: www.fool.co.uk/personalfinance

Rawcliffe, P. (1998) *Environmental Pressure Groups in Transition*, Manchester University Press, Manchester.

Reddish, A. and Rand, M. (1996) The environmental effects of present energy policies, in Blunden, J. and Reddish, A. (eds) *Energy, Resources and Environment*, Open University and Hodder & Stoughton, London, 43–91.

Rees, J. (1985) *Natural Resources: Allocation, Economics and Policy*, 2nd edition, Routledge, London.

Rees, J. (1991) Resources and environment: scarcity and sustainability, in Bennett, R. and Estall, R. (eds) *Global Change and Challenge: Geography in the 1990s*, Routledge, London, 5–26.

Rees, W. and Wackernagel, M. (2008) Urban ecological footprints: why cities cannot be sustainable – and why they are a key to sustainability, *Urban Ecology*, **V**, 537–55.

Reeves, R. (1999) Do we really need gay florists?, *Observer*, 8 August.

Reid-Henry, S. (2010) The territorial trap fifteen years on, *Geopolitics*, **15**(4), 752–6.

Reid-Henry, S.M. (2007) Exceptional sovereignty? Guantanamo Bay and the re-colonial present, *Antipode*, **39**(4), 627–48.

Renard, M. (2005) Quality certification, regulation and power in fair trade, *Journal of Rural Studies*, **21**, 419–31.

Renner, M. (2002) The Anatomy of Resource Wars, Worldwatch Paper No. 162, Earthwatch Institute, Washington, DC.

Renting, H., Marsden, T.K. and Banks, J. (2003) Understanding alternative food networks: exploring the role of short food supply chains in rural development, *Environment and Planning A*, **35**, 393–411.

Rist, G. (1997) *The History of Development*, Zed, London.

Ritzer, G. (2000) *The McDonaldisation of Society*, 2nd edition, Pine Forge Press, Thousand Oaks, CA.

Ritzer, G. (2004) *The Globalization of Nothing*, Pine Forge/Sage, London.

Robbins, P. (2004) *Political Ecology*, Blackwell, Oxford.

Roberts, S., Secor, A. and Sparke, M. (2003) Neoliberal geopolitics, *Antipode*, **35**, 886–97.

Robinson, G. (2004) *Geographies of Agriculture*, Pearson, London.

Robinson, G.M. and Pobric, A. (2006) Nationalism and identity in post-Dayton Accords: Bosnia-Hercegovina, *Tijdschrift voor Economische en Sociale Geografie*, **97**(3), 237–52.

Robinson, J. (1997) The geopolitics of South African cities: states, citizens, territory, *Political Geography*, **16**, 365–86.

Robinson, J. (1999) Divisive cities, in Pile, S., Brook, C. and Mooney, G. (eds) *Unruly Cities? Order/Disorder*, Routledge, London, 149–200.

Robinson, J. (2002) Global and world cities: a view from off the map, *International Journal of Urban and Regional Research*, **26**(3), 531–54.

Robinson, J. (2005a) *Ordinary Cities, Between Modernity and Development*, Routledge, London.

Robinson, J. (2005b) Urban geography: world cities, or a world of cities, *Progress in Human Geography*, **29**(6), 757–65.

Robinson, J. (2006) *Ordinary Cities: Between Modernity and Development*, Routledge, Abingdon.

Roche, M. (2002) Rural geography: searching rural geographies, *Progress in Human Geography*, **26**(6), 823–29.

Roche, M. (2005) Rural geography: a borderland revisited, *Progress in Human Geography*, **29**, 299–303.

Rogers, A. and Viles, H. (eds) (2003) *The Student's Companion to Geography*, 2nd edition, Blackwell, Oxford.

Rogers, R. (1999) *Towards an Urban Renaissance, Final Report of the Urban Task Force*, chaired by Lord Rogers of Riverside, E. and F. N. Spon, London.

Rogers, R. and Power, A. (2000) *Cities for a Small Country*, Faber and Faber, London.

Rollins, W. H. (1995) Whose landscape?: Technology, Fascism and environmentalism on the National Socialist Autobahn, *Annals of the Association of American Geographers*, **85**, 494–520.

Ronald, L. (2003) The demographic transition: three centuries of fundamental change, *Journal of Economic Perspectives*, **17**(4), 167–90.

Rooum, D. (1995) *What is Anarchism?: An Introduction*, Freedom Press, London.

Rose, N. and Miller, P. (1992) Political power beyond the state: problematics of government, *British Journal of Sociology*, **43**, 173–205.

Rosen, G. and Razin, E. (2009) The rise of gated communities in Israel: reflections on changing urban governance in a neo-liberal era, *Urban Studies*, **46**(8), 1702–22.

Rose-Redwood, R. (2008), Governmentality, geography and the geo-coded world, *Progress in Human Geography*, **30**, 469–86.

Ross, M. (2001) *Extractive Sectors and the Poor: An Oxfam America Report*, Oxfam. Available at: www.oxfam america.org/cirexport/index.html

Rössler, M. (1989) Applied geography and area research in Nazi society: central place theory and planning, *Environment and Planning D: Society and Space*, **7**, 419–31.

Rostow, W. W. (1960) *The Stages of Economic Growth: A Non-Communist Manifesto*, Cambridge University Press, Cambridge.

Rostow, W. W. (1971) *The Stages of Economic Growth*, 2nd edition, Cambridge University Press, Cambridge.

Rothenberg, T. (1995) 'And she told two friends': lesbians creating urban social space, in Bell, D. and Valentine, G. (eds) *Mapping Desire: Geographies of Sexualities*, Routledge, London, 165–81.

Routledge, P. (1996) Critical geopolitics and terrains of resistance, *Political Geography*, **15**, 509–31.

Routledge, P. (2003) Anti-geopolitics, in Agnew, J., Mitchell, K. and Toal, G. (eds) *A Companion to Political Geography*, Blackwell, Oxford, 236–48.

Rowell, A., Marriott, J. and Stockman, L. (2005) *The Next Gulf: London, Washington and Oil Conflict in Nigeria*, Constable, London.

Rubin, B. (1979) Aesthetic ideology and urban design, *Annals of the Association of American Geographers*, **69**(3), 339–61.

Rucht, D. and Roose, J. (1999) The green environmental movement at a crossroads, *Environmental Politics*, **8**(1), 59–80.

Rusten, G., Bryson, J. R. and Aarflot, U. (2007) Places through product and products through places: industrial design and spatial symbols as sources of competitiveness, *Norwegian Journal of Geography*, **61**(3), 133–44.

Rutledge, I. (2006) *Addicted to Oil: America's Relentless Drive for Energy Security*, I. B. Tarus, London.

S

Saarinen, J. (2007) Cultural tourism, local communities and representations of authenticity: the case of Lesedi and Swazi cultural villages in Southern Africa, in Wishitemi, B., Spenceley, A. and Wels, H. (eds) *Culture and Community: Tourism Studies in Eastern and Southern Africa*, Rozenberg, Amsterdam.

Sachs, J.D. and Warner, A. M. (2001) Natural resources and economic development: the curse of natural resources, *European Economic Review*, **45**, 827–38.

Sachs, W. (1992) *The Development Dictionary*, Zed Books, London.

Sack, R. (1986) *Human Territoriality: Its Theory and History*, Cambridge University Press, Cambridge.

Sack, R. (1992) *Place, Modernity and the Consumers World*, Johns Hopkins University Press, Baltimore, MD.

Sage, C. (2003) Social embeddedness and relations of regard: alternative 'good food' networks in south-west Ireland, *Journal of Rural Studies*, **19**, 47–60.

Salway, P. (ed.) (1981) Roman Britain, in *Oxford History of England*, Oxford University Press, Oxford.

Samers, M. (2010) *Migration*, Routledge, London.

Sandercock, L. (1998) *Towards Cosmopolis: Planning for Multicultural Cities*, Wiley, Chichester.

Sandercock, L. (2003) *Cosmopolis 2: Mongrel Cities of the 21st Century*, Continuum, London.

Sandercock, L. (2006) Cosmopolitan urbanism: a love song to our mongrel cities, in Binnie, J., Holloway, J., Millington, S. and Young, C. (eds) *Cosmopolitan Urbanism*, Routledge, Abingdon, 37–52.

Santos, M. (1974) Geography, Marxism and underdevelopment, *Antipode*, **6**, 1–9.

Sassatelli, R. and Scott, A. (2001) Novel food, new markets and trust regimes: responses to the erosion of consumers' confidence in Austria, Italy and the UK, *European Societies*, **3**, 213–44.

Sassen, S. (1991) *The Global City: New York, London, Tokyo*, Princeton University Press, Princeton, NJ.

Sassen, S. (1994) *Cities in a World Economy*, Pine Forge Press, London.

Sassen, S. (2001) *The Global City: New York, London, Tokyo*, updated 2nd edition, Princeton University Press, Princeton, NJ.

Sassen, S. (2006) *Territory, Authority, Rights: From Medieval to Global Assemblages*, Princeton University Press, Princeton, NJ.

Saunders, D. (2010) *Arrival City: How the Largest Migration in History is Reshaping Our World*, William Heinemann, London.

Savage, M., Warde, A. and Ward, K. (2003) *Urban Sociology, Capitalism and Modernity*, Palgrave, Basingstoke.

Savitch, H. and Kantor, P. (2002) *Cities in the International Marketplace: The Political Economy of Urban Development in North America and Western Europe*, Princeton University Press, Princeton, NJ.

Saxenian, A. (2006) *The New Argonauts: Regional Advantage in a Global Economy*, Harvard University Press, Cambridge, MA.

Sayer, A. and Walker, R. (1992) *The New Social Economy: Reworking the Division of Labour*, Blackwell, Oxford.

Sbargia, A. (2000) The European Union as a coxswain: governance by steering, in Pierre, J. (ed.) *Debating Governance: Authority, Steering and Democracy*, Oxford University Press, Oxford, 219–40.

Schama, S. (1995) *Landscape and Memory*, HarperCollins, London.

Schifferes, S. (2007) Chrysler questions climate change, BBC News online, 10 January, http://news.bbc.co.uk/go/pr/fr/-/1/hi/business/6247371.stm

Schneiders, L. (2006) Is wilderness racist? *Chain Reaction*, Autumn, 25–7.

Schumpeter, J. (1939) *Business Cycles: A Theoretical, Historical and Statistical Analysis of the Capitalist Process*, McGraw-Hill, London.

Schumpeter, J.A. (1942) *Capitalism, Socialism and Democracy*, Harper, New York.

Schuurman, F.J. (2001) Globalization and development studies: introducing the challenges, in Schuurman, F.J. (ed.) *Globalization and Development: Challenges for the 21st Century*, Sage, London, 3–16.

Schuurman, N. (2004) *GIS: A Short Introduction*, Blackwell, Oxford.

Scott, A. (2000) *The Cultural Economy of Cities: Essays on the Geography of Image-Producing Industries*, London, Sage.

Scott, A. (2004) A perspective on economic geography, *Journal of Economic Geography*, **4**, 479–99.

Scott, A.J. and Brown, E. R. (eds) (1993) *South-Central Los Angeles: Anatomy of an Urban Crisis*, The Ralph and Goldy Lewis Center for Regional Policy Studies, Working Paper Series, 06. Available at: http://repositories.cdlib.org/lewis/wps/06

Scott, A. and Soja, E. (eds) (1996) *The City: Los Angeles and Urban Theory at the End of the Twentieth Century*, University of California Press, Berkeley, CA.

Seel, B. and Plows, A. (2000) Coming live and direct: strategies of Earth First! in Seel, B., Patterson, M. and Doherty, B. (eds) *Direct Action in British Environmentalism*, Routledge, London, 112–32.

Service, E.R. (1971) *Primitive Social Organization*, 2nd edition, Random House, New York.

Sen, A. (1981) *Poverty and Famines: An Essay on Entitlement and Deprivation*, Oxford University Press, Oxford.

Sennett, R. (1994) *Flesh and Stone: The Body And The City In Western Civilization*, Penguin, London.

Seton-Watson, H. (1977) *Nations and States: An Enquiry into the Origins of Nations and the Politics of Nationalism*, Westview Press, Boulder, CO.

Seyfang, G. (2002) Tackling social exclusion with community currencies: Learning from LETS to time banks, *International Journal of Community Currency Research*, **6**, available at: www.uea.ac.uk/env/ijccr.

Seyfang, G. (2006) Ecological citizenship and sustainable consumption: examining local organic food networks, *Journal of Rural Studies*, **22**(4), 383–95.

Seyfang, G. (2009) *The New Economics of Sustainable Consumption: Seeds of Change*, Palgrave Macmillan, Basingstoke.

Seyfang, G. and Smith, A. (2007) Grassroots innovation for sustainable development: towards a new research and policy agenda, *Environmental Politics*, **16**(4), 584–603.

Shabi, R. (2002) The e-waste land, *The Guardian*, 30 November.

Shakespeare, T. (1994) Cultural representations of disability: dustbins for disavowal, *Disability and Society*, **9**, 283–99.

Sharp, J. (1999) Critical geopolitics, in Cloke, D., Crang, P. and Goodwin, M. (eds) *Introducing Human Geographies*, Arnold, London, 179–88.

Sharp, J., Routledge, P., Philo, C. and Paddison, R. (eds) (2000) *Entanglements of Power: Geographies of Domination/Resistance*, Routledge, London.

Shaw, K. (2008) Gentrification: what it is, why it is, and what can be done about it, *Geography Compass*, **2**(5), 1697–728.

Shaw, M., Thomas, B., Davey Smith, G. and Dorling, D. (2008) *The Grim Reaper's Road Map: An Atlas of Mortality in Britain*, Policy Press, Bristol.

Sheehan, J.M. (2000) *The Greening of the World Bank: A Lesson in Bureaucratic Survival*, Cato Institute Foreign Policy Briefing No. 56, April 12. Available from the Cato Institute: www.cato.org/

Sheller, M. and Urry, J. (2006) The new mobilities paradigm, *Environment and Planning A*, **38**(2), 207–26.

Shepherd, A. (2008) *Achieving the MDGs: The Fundamentals*, ODI Briefing Paper 43, September, available at: www.odi.org.uk/resources/download/1933.pdf [accessed 21 September 2010].

Sheppard, E. (2002) The spaces and times of globalization: place, scale, networks and positionality, *Economic Geography*, **78** (3), 307–30.

Sheppard, E. and Barnes, T.J. (eds) (2000) *A Companion to Economic Geography*, Blackwell, Oxford.

Sheppard, E. and Nagar, R. (2004) From east–west to north–south, *Antipode*, **36**, 557–63.

Sherratt, A. (ed.) (1980) *The Cambridge Encyclopaedia of Archaeology*, Cambridge University Press, Cambridge.

Shevky, E. and Bell, W. (1955) *Social Area Analysis*, Stanford University Press, Stanford, CA.

Shields, R. (1991) *Places on the Margin*, London, Routledge.

Shilling, C. (1993) *The Body and Social Theory,* Sage, London.

Shirlow, P. and Murtagh, B. (2006) *Belfast: Segregation, Violence and the City*, Pluto Press, London.

Shiva, V. (1991) The Green Revolution in the Punjab, *The Ecologist*, **21**(2), 57–60.

Shiva, V. (1999) We will drive Monsanto out of the country, *Asian Age*, 6 March, 9.

Shiva, V. (2002) *Water Wars: Privatization, Pollution, and Profit*, Pluto Books, London.

Shiva, V. and Moser, I. (1995) *Biopolitics: A Feminist and Ecological Reader on Biotechnology*, Zed Books, London.

Short, J.R. (1989) Yuppies, yuffies and the new urban order, *Transactions of the Institute of British Geographers, NS,* **14**, 173–88.

Short, J.R. (1996) *The Urban Order: An Introduction to Cities, Culture and Power,* Blackwell, Oxford.

Short, J.R. (2004) *Global Metropolitan: Globalizing Cities in a Capitalist World*, Routledge, New York.

Short, J.R. (2006) *Imagined Country*, Syracuse University Press, Syracuse, NY.

Short, J.R., Kim, Y-H., Kuus, M. and Wells, H. (1996) The dirty little secret of world city research: data problems in comparative analysis, *International Journal of Urban and Regional Research*, **20**(4), 697–719.

Shove, E. (2003) *Comfort, Cleanliness and Convenience: The Social Organization of Normality*, Berg, Oxford.

Shubin, S. (2010) 'Where can a gypsy stop?' Rethinking mobility in Scotland, *Antipode,* **42**(5), 494–524.

Shurmer-Smith, P. (ed.) (2002) *Doing Cultural Geography*, Sage, London.

Shurmer-Smith, P. and Hannam, K. (1994) *Worlds of Desire, Realms of Power: A Cultural Geography*, Arnold, London.

Sibalis, M. (2004) Urban space and homosexuality: the example of the Marais, Paris 'Gay Ghetto', *Urban Studies,* **41**(9), 1739–58.

Sibley, D. (1995) *Geographies of Exclusion: Society and Difference in the West*, Routledge, London.

Sibley, D. (1999) Creating geographies of difference, in D. Massey, J. Allen and P. Sarre (eds) *Human Geography Today*, Polity, Cambridge, 115–28.

Sidaway, J.D. (1998) What is in a Gulf? From the 'arc of crisis' to the Gulf War, in Ó Tuathail, G. and Dalby, S. (eds) *Rethinking Geopolitics*, Routledge, London and New York, 224–39.

Sidaway, J.D. (1999) American power and the Portuguese Empire, in Slater, D. and Taylor, P.J. (eds) *The American Century*, Blackwell, Oxford, 195–209.

Sidaway, J.D. (2000) Iberian geopolitics, in Dodds, K. and Atkinson, D. (eds) *Geopolitical Traditions: A Century of Geopolitical Thought*, Routledge, London, 118–49.

Sidaway, J.D. (2003) Sovereign excesses? Portraying postcolonial sovereigntyscapes, *Political Geography*, **22**, 157–78.

Sidaway, J.D. (2007) Enclave space: a new metageography of development?, *Area*, **39**(3), 331–9.

Sidaway, J.D. (2008) Subprime crisis: American crisis or human crisis?, *Environment and Planning D: Society and Space*, **26**, 195–8.

Sidaway, J.D. (2009a) Shadows on the path: negotiating geopolitics on an urban section of Britain's South West Coast Path, *Environment and Planning D: Society and Space*, **27**(6), 1091–116.

Sidaway, J.D. (2009b) Accumulation by dispossession: the play of capital, *Human Geography*, **2**(3), 104–10.

Sidaway, J.D. and Power, M. (2005) The tears of Portugal: race and destiny in Portuguese geopolitical narratives, *Environment and Planning D: Society and Space*, **23**(4), 527–54.

Sidaway, J.D. and Pryke, M. (2000) The free and the unfree: 'emerging markets', the Heritage Foundation and the 'index of economic freedom', in Bryson, J. R., Daniels, P. W., Henry, N. D. and Pollard, J. S. (eds) *Knowledge, Space, Economy*, Routledge, London.

Sidaway, J.D., Grundy-Warr, C., Park, Bae Gyoon (2005) Asian sovereignty scapes, Editorial, *Political Geography*, **24**, 779–83.

Silverstone, R., Hirsch, E. and Morley, D. (1992) Information and communication technologies and the moral economy of the household, in Silverstone, R. and Hirsch, E. (eds) *Consuming Technologies: Media and Information in Domestic Spaces*, Routledge, London, 9–17.

Simmel, G. (1978) *The Philosophy of Money*, Routledge & Kegan Paul, London.

Simmel, G. (1991) Money in modern culture, *Theory, Culture and Society*, **8**, 17–31.

Simmel, G. (2004) Metropolis and mental life, in Miles, M., Hall, T. and Borden, I. (eds) *The City Cultures Reader*, 2nd edition, Routledge, London, 12–19.

Simmons, I.G. (1996) *Changing the Face of the Earth: Culture, Environment, History*, 2nd edition, Blackwell, Oxford.

Sinclair, T. 2004 *Global Governance*, Routledge, London.

Singer, P. (2009) *Animal Liberation: The Definitive Classic of the Animal Movement* (updated edition), Harper Collins, New York. (First published 1975.)

Six, C (2009) The rise of postcolonial states as donors: a challenge to the development paradigm?, *Third World Quarterly*, **30**(6), 1103–21.

Skelton, T. and Valentine, G. (eds) (1998) *Cool Places: Geographies of Youth Cultures*, Routledge, London.

Skidmore, T.E and Smith, P.H. (1988) *Modern Latin America*, 2nd edition, Oxford University Press, New York.

Slater, C. (1995) Amazonia as Edenic narrative, in Worster, D. (ed.) *Uncommon Ground: Reinventing Nature*, Norton, New York, 114–59.

Slater, C. (2004) Marketing the rain forest: raw vanilla fragrance and the ongoing transformation of the jungle, *Cultural Geographies*, **11**, 165–80.

Slater, D. (1993) The geopolitical imagination and the enframing of development theory, *Transactions of the Institute of British Geographers, NS*, **18**, 419–37.

Slater, D. (1997) *Consumer Culture and Modernity*, Polity Press, Cambridge.

Slater, D. (2003) Beyond Euro-Americanism: democracy and post-colonialism, in Anderson, K., Domosh, M., Pike, S. and Thrift, N. (eds) *Handbook of Cultural Geography*, Sage, London, 420–32.

Slater, D. (2004) *Geopolitics and the Post-Colonial: Rethinking North–South Relations*, Blackwell, Malden, MA.

Slater, N. (2001) Food and eating, *The Observer Food Monthly*, April 2001.

Smil, V. (1994) *Energy in World History*, Westview, Boulder, Co.

Smil, V. (2010) *Energy Transitions: History, Requirements, Prospects,* Praeger, Santa Barbara, CA.

Smith, A. (1776) *An Inquiry into the Nature and Causes of the Wealth of Nations*, edited by Skinner, A. (1974) Penguin, London.

Smith, A. (1977) *The Wealth of Nations,* edited by Skinner, A. Penguin, Harmondsworth.

Smith, A.D. (1988) *The Ethnic Origin of Nations*, Blackwell, Oxford.

Smith, B (2007) *Good Governance and Development*, Macmillan, Basingstoke.

Smith, D.M. (1995*) Geography and Social Justice*, Blackwell, Oxford.

Smith, D.A. and Tomberlake, M. (1995) Conceptualizing and mapping the structure of the world's city system, *Urban Studies*, **32**(2), 287–302.

Smith, D.P. and Holt, L. (2007) Studentification and 'apprentice' gentrifiers within Britain's provincial towns and cities: extending the meaning of gentrification, *Environment and Planning A*, **39**(1), 142–61.

Smith, F. (2001) Refiguring the geopolitical landscape: nation, 'transition' and gendered subjects in post-cold war Germany, *Space and Polity*, **5**, 213–35.

Smith, M.D. (1996) The empire filters back: consumption, production, and the politics of Starbucks coffee, *Urban Geography*, **17**, 502–24.

Smith, N. (1984) Isiah Bowman: political geography and geopolitics, *Political Geography Quarterly*, **3**, 69–76.

Smith, N. (1996) *The New Urban Frontier: Gentrification and the Revanchist City*, Routledge, London.

Smith, N. (2002) New globalism, new urbanism, gentrification as global urban strategy, *Antipode*, **34**, 434–57.

Smith, N. (2003) *American Empire: Roosevelt's Geographer and the Prelude to Globalization*, University of California Press, Berkeley, CA.

Smith, S. (1989) *The Politics of Race and Residence*, Polity, Cambridge.

Smith, S. J. (2005) Society–space, in Cloke, P., Crang, P. and Goodwin, M. (eds) *Introducing Human Geographies*, 2nd edition, Hodder Arnold, London, 18–33.

Sogge, D. (2002) *Give and Take: What's the Matter with Foreign Aid?*, Zed Books, London.

Soja, E. J. (1989) *Postmodern Geographies: The Reassertion of Space in Critical Social Theory*, Verso, London.

Soja, E. (1996) *Thirdspace: Journeys to Los Angeles and Other Real and Imagined Places*, Blackwell, Oxford.

Soja, E. (2000) *Postmetropolis: Critical Studies of Cities and Regions*, Blackwell, Oxford.

Sondheimer, S. (ed.) (1991) *Women and the Environment: A Reader*, Zed Books, London.

Soper, K. (1995) *What is Nature?* Blackwell, Oxford.

Spain, D. (1992) *Gendered Spaces*, University of North Carolina Press, Chapel Hill.

Sparke, M. (2006) Political geography: political geographies of globalization (2): governance, *Progress in Human Geography*, **30**, 1–16.

Speer, A. (1995) *Inside the Third Reich*, Phoenix, London.

Spicker, P. (2006) *The Idea of Poverty*, Polity, Bristol.

Spillius, A. (2003) Dialing tone, *Telegraph Magazine*, 19 April, 38–43.

Spirn, A. W. (1997) The authority of nature: conflict and confusion in landscape architecture, in Wolschke-Bulmahn, J. (ed.) *Nature and Ideology: Natural Garden Design in the Twentieth Century*, Dumbarton Oaks Research Library and Collection, Washington, 249–61.

Spivak, G. C. (1985) Three women's texts and a critique of imperialism, *Critical Inquiry*, **12**, 243–61.

Squire, G. and Gill, A. (2011) 'It's not all heartbeat you know': policing domestic violence in rural areas, in Mawby, R.I. and Yarwood, R. (eds) *Rural Policing and Policing the Rural: A Constable Countryside?*, Ashgate, Farnham, 159–68.

Squires, P (ed.) (2008) *ASBO Nation: The Criminalisation of Nuisance in Contemporary Britain*, Policy Press, Bristol.

Stamp, L.D. and Beaver, S. (1933) *The British Isles: A Geographic and Economic Survey*, Longman, London.

Stamp, L.D. and Beaver, S. (1963) *The British Isles: A Geographic and Economic Survey*, 5th edition, Longman, London.

Stannard, K. (2003) Earth to academia: on the need to reconnect university and school geography, *Area,* **25**, 316–22.

Staeheli, L. (2008) Citizenship and the problem of community, *Political Geography,* **27**(1), 5–21.

Staeheli, L.A. and Mitchell, D. (2006) USA's destiny?: regulating space and creating community in American shopping malls, *Urban Studies,* **43**(5–6), 977–92.

Steinfeld, H., Gerber, P., Wassenaar, T., Castel, V., Rosales, M. and de Haan, C. (2006) *Livestock's Long Shadow: Environmental Issues and Options,* FAO, Rome.

Stern Review (2006) *Stern Review: The Economics of Climate Change,* HM Treasury, London.

Stevens, P. (2003) *Resource Impact – Curse or Blessing? A Literature Survey,* Centre for Energy, Petroleum and Mineral Law and Policy, electronic journal, available at: www.cepmlp.org

Steinberg, P.E. (2001) *The Social Construction of the Ocean,* Cambridge University Press, Cambridge.

Stiglitz, J.E and Charlton, A. (2005) *Fair Trade for All: How Trade can Promote Development,* Oxford University Press, Oxford.

Stobart, J. and Raven, N. (eds) (2004) *Towns, Regions and Industries: Urban and Industrial Change in the Midlands, c. 1700–1840,* Manchester University Press, Manchester.

Stöhr, W. B. and Taylor, D. R. F. (1981) *Development from Above or Below? The Dialectics of Regional Planning in Developing Countries,* John Wiley, Chichester.

Stokke. O. and Vidas, D. (eds) (1997) *Governing the Antarctic,* Cambridge University Press, Cambridge.

Storey, D. (2001) *Territory: The Claiming of Space,* Prentice Hall, Harlow.

Storey, D. (2002) Territory and national identity: examples from the former Yugoslavia, *Geography,* **87**(2), 108–15.

Storey, D. (2003) *Citizen, State and Nation,* Geographical Association, Sheffield.

Storey, D. (2010) Partnerships, people and place: lauding the local in rural development, in Halseth, G., Markey, S. and Bruce, D. (eds) *The Next Rural Economies: Constructing Rural Place in Global Economies,* CABI, Wallingford, 155–65.

Storper, M. (1997) *The Regional World: Territorial Development in a Global Economy,* The Guilford Press, New York.

Strange, S. (1996) *Casino Capitalism,* Blackwell, Oxford.

Strange, S. (1996a) *The Retreat of the State: The Diffusions of Power in the World Economy,* Cambridge University Press, Cambridge.

Strange, S. (1999) *Mad Money: When Markets Outgrow Governments,* University of Michigan Press, Ann Arbor, MI.

Stuart, T. (2009) *Waste: Uncovering the Global Food Scandal,* Penguin, London.

Stutz, F. P. and de Souza, A. R. (1998) *The World Economy: Resources, Location, Trade and Development,* 3rd edition, Prentice-Hall, Upper Saddle River, NJ.

Subranabiab, M. and Ati, B. (2006) *Captives in India: A Research Study,* Infosys Technologies Ltd, Delhi.

Sugden, D. (1982) *Arctic and Antarctic,* Basil Blackwell, Oxford.

Sukarno, I. (1955) Modern History Sourcebook: President Sukarno of Indonesia: Speech at the opening of the Bandung conference, 18 April 1955, available at: www.fordham.edu/halsall/mod/1955sukarno-bandong.html [accessed 7 April 2003].

Sum, N-L. and Ngai, P. (2005) Globalization and paradoxes of ethical transnational production: code of conduct in a Chinese workplace, *Competition and Change,* **9**, 181–200.

Sumartojo, R. (2004) Contesting place: antigay and lesbian hate crime in Columbus, Ohio, in Flint, C. (ed.) *Spaces of Hate: Geographies of Discrimination and Intolerance in the USA,* Routledge, New York, 87–107.

Swanton D, (2010) Sorting bodies: race, affect, and everyday multiculture in a mill town in northern England, *Environment and Planning A,* **42**(10), 2332–50.

Sweetman, C. (ed.) (2000) *Gender in the Twenty-first Century,* Oxfam, Oxford.

Swyngedouw, E. (2000) Authoritarian governance, power, and the politics of rescaling, *Environment and Planning D: Society and Space,* **18**, 63–76.

Sylvester, C. (1996) Picturing the Cold War: an art graft/eye graft, *Alternatives,* **21**, 393–418.

T

Takahashi, L. M. and Dear, M. J. (1997) The changing dynamics of community opposition to human service facilities, *Journal of the American Planning Association,* **63**, 79–93.

Takeuchi, K., Katoh, K., Nan, Y. and Kou, Z. (1995) Vegetation change in desertified Kerqin sandy lands, *Inner Mongolia/Geographical Reports,* Tokyo Metropolitan University, **30**, 1–24.

Talbot, J. M. (2004) *Grounds for Agreement: The Political Economy of the Coffee Commodity Chain,* Rowman and Littlefield, Lanham, MD.

Tan, T. Y. and Kudaisya, G. (2002) *The Aftermath of Partition in South Asia,* Routledge, London.

Taussig, M. (1997) *The Magic of the State,* Routledge, London.

Taylor, I. (1997) Running on empty, *The Guardian,* 14 May, 2–6.

Taylor, P. (2004) *World City Network: A Global Urban Analysis*, Routledge, London.

Taylor, P.J. (1989) *Political Geography*, Longman, London.

Taylor, P.J. (1994) From heartland to hegemony: changing the world in political geography, *Geoforum*, **25**, 403–17.

Taylor, P.J. (2005) New political geographies: global civil society and global governance through world city networks, *Political Geography*, **24**, 703–30.

Taylor, P.J. and Flint, C. (2000) *Political Geography: World-Economy, Nation State and Locality*, Prentice Hall, Harlow.

Taylor, S. and Tyler, M. (2000) Emotional labour and sexual difference in the airline industry, *Work, Employment and Society*, **14**(1), 77–95.

Tesfahuney, M. (1998) Mobility, racism and geopolitics, *Political Geography*, **17**, 499–515.

The Economist (1999), The Next Shock, 4 March, 29. Available electronically from www.Economist.com.

The Economist (2006) The new titans: a survey of the world economy, 16 September.

The Independent (2004) Students no longer welcomed, *The Independent*, 21 October, 11.

The Times Archaeology of the World (1999) New edition, Times Books, London.

The Times Atlas of World History (1989) 3rd edition, Guild Publishing, London.

The Times Atlas of World History (1999) New edition, Times Books, London.

The Wilderness Society (2006) A Fringe of Green – Protecting Australia's Forests and Woodlands, available at: www.wilderness.org.au/campaigns/forests/fringe/#why-save

Thomas, D. and Middleton, N. (1994) *Desertification: Exploding the Myth*, Wiley, Chichester.

Thompson, E.P. (1963) *The Making of the English Working Class*, Penguin, London and New York.

Thompson, S. (2005) 'Territorialising' the school playground: deconstructing the geography of playtime, *Children's Geographies*, **3**(1), 63–78.

Thoreau, H.D. (1854) *Walden, Or: Life in the Woods*, Ticknor and Fields, Boston.

Thrift, N. (1996) New urban eras and old technological fears: reconfiguring the goodwill of electronic things, *Urban Studies*, **33**(8), 1463–94.

Thrift, N. (1998) Virtual capitalism: the globalisation of reflexive business knowledge, in Carrier, J. G. and Miller, D. (eds) *Virtualism: A New Political Economy*, Berg, Oxford.

Thrift, N. and Kitchin, R. (2008) *The International Encyclopedia of Human Geography*, Elsevier, London.

Thrift, N.J. (2005) *Knowing Capitalism*, Sage, London.

Tickell, A. (2003) Cultures of money, in Anderson, K., Domosh, M., Pile, S. and Thrift, N. (eds) *Handbook of Cultural Geography*, Sage, London, 116–30.

Tickell, A. and Peck, J. (2003) Making global rules: globalization or neoliberalization, in Peck, J. and Yeung, H. (eds) *Remaking the Global Economy*, Sage, London, 163–81.

Tiffin, M., Mortimore, M. and Gickuki, F. (1994) *More People, Less Erosion: Environmental Recovery in Kenya*, John Wiley, Chichester.

Time Out (1995) The world on a plate, *Time Out*, 16 August, 31–7.

Tissot, S. and Poupeau, F. (2005) La spatilisation des problems sociaux, *Actes de la recherché en Sciences socials*, **159** (September), 4–9.

Toal, G. (2003) Re-asserting the regional: political geography and geopolitics in world thinly known, *Political Geography*, **22**, 653–55.

Toennies, F. (1887) *Community and Society* (1957 edn), Michigan State University Press, East Lansing, MI.

Toje, A. (2005) The 2003 European Union security strategy: a critical appraisal, *European Foreign Affairs Review*, **10**, 117–33.

Tokar, B. (1994) *The Green Alternative*, 2nd edition, R. and E. Miles, San Pedro, CA.

Tomlinson, J. (1991) *Cultural Imperialism: A Critical Introduction*, Pinter, London.

Tønnesson, S. and Antlöv, H. (1996) Asia in theories of nationalism and national identity, in Tønnesson, S. and Antlöv, H. (eds) *Asian Forms of the Nation*, Curzon, Richmond, 1–40.

Topik, S. (2009) Historicizing commodity chains: five hundred years of the coffee global commodity chain, in J. Bair (ed.) *Frontiers of Commodity Chain Research*, Stanford University Press, Stanford, CA, 37–62.

Tratado de Limites (1866) *Tratado de Limites entre Portugal e Espanha assinado em Lisboa aos 29 de Setembro de 1864*, Imprensa Nacional, Lisbon.

Truman, H. (1949) *Public Papers of the President, 20 January*, US Government Printing Offices, Washington, DC.

Tuan, Yi-Fu (1999) *Who am I?: An Autobiography of Emotion, Mind and Spirit*, University of Wisconsin Press, Madison, WI.

Tuan, Yi-Fu (2002) *Dear Colleague: Common and Uncommon Observations*, University of Minnesota Press, Minneapolis, MN.

Tucker, A. (2009) *Queer Visibilities: Space, Identity and Interaction in Cape Town*, Wiley-Blackwell, Chichester.

Tunander, O. (2001) Swedish-German geopolitics for a new century: Rudolf Kjellén's 'The State as Living Organism', *Review of International Studies*, **27**, 451–63.

Tunstall, J. (1977) *The Media are American*, Constable, London.

Tunstall, R. and Coulter, A. (2006) *Twenty-Five Years on Twenty Estates: Turning the Tide*, Policy Press, Bristol.

Tyler May, E. (1989) Cold War – warm hearth: politics and the family in postwar America, in Fraser, S. and Gerstle, G. (eds) *The Rise and Fall of the New Deal Order, 1930*, Princeton University Press, Princeton, NJ, 153–81.

U

UK Universities (2005) *Studentification: a guide to challenges, opportunities and good practice*. UK Universities, London.

UN *see* United Nations.

UN Development Programme (2009) *Human Development Report 2009*, United Nations, New York.

UN HABITAT (2009) *Planning Sustainable Cities: Global Report on Human Settlements 2009*, Earthscan, London.

UN Millennium Development Goals (2000) Available at: www.millenniumgoals. org/

UNAIDS (2006) *Report on the global AIDS epidemic 2006*, available to download at: www.unaids.org/en/HIV_data/2006GlobalReport/default.asp

Understanding Cities (1999) DD304, Open University, Milton Keynes.

UNDP *see* United Nations Development Programme.

UNEP *see* United Nations Environment Programme.

UN-HABITAT (2006) The state of the world's cities 2006/7, *The Millennium Development Goals and Urban Sustainability*, Earthscan, London.

UNICEF *see* United Nations Children's Fund.

United Nations (UN) (2001) *The state of the world's cities Istanbul*, Habitat Publications Unit, available at: www.unhabitat.org/Istanbul+5/statereport.htm

United Nations (UN) (2002) *International Migration Report 2002*, Department of Economic and Social Affairs, United Nations, New York.

United Nations (UN) (2002a) *Manual on Statistics on International Trade in Services*, United Nations, New York.

United Nations (UN) (2003) *World Population Prospects: The 2002 Revision*, United Nations Population Division, New York.

United Nations (UN) (2006a) *Population Challenges and Development Goals*, available at: www.un.org/esa/population/publications/pop_challenges/Population_Challenges.pdf.

United Nations (UN) (2006b) *World Urbanisation Prospects – the 2005 update*, United Nations, New York.

United Nations (UN) (2010) *The Millennium Development Goals Report 2010*, United Nations, New York.

United Nations Children's Fund (UNICEF) (2000) *The Progress of Nations 2000*, online, available at: www.unicef.org [accessed 22 February 2002].

United Nations Conference on Trade and Development (UNCTAD) (2006) http://stats.unctad.org/FDI/TableViewer/tableView.aspx?ReportId=5 [accessed 31 December 2006].

United Nations Development Programme (UNDP) (1998) *Human Development Report 1998*, Oxford University Press, Oxford.

United Nations Development Program (UNDP) (2009) *Human Development Report – Overcoming Barriers: Human Mobility and Development*, UNDP, Geneva, available at: http://hdr.undp.org/en/media/HDR_2009_EN_Complete.pdf.

United Nations Environment Programme (UNEP) (1999) *Global Environmental Outlook 2000*, UNEP-Earthscan Publications, London.

United Nations Population Division (1998) *Briefing Packet*, 1998 Revision of World Population Prospects.

United Nations Population Division (2004) *World Population Prospects: The 2004 Revision*, available at: www.un.org/esa/population/publications/WPP2004/wpp2004.htm

United Nations Population Division (2006) *World Population Prospects: The 2006 Revision*, available at: http://esa.un.org/unpp/index.asp?panel=2.

United Nations World Tourism Organization (UNWTO) (2006) *World Tourism Highlights*, UNWTO, Madrid.

Urry, J. (2005) The place of emotions within place, in Davidson, J., Bondi, L. and Smith, M. (eds) *Emotional Geographies*, Ashgate, Aldershot, 77–86.

V

Valentine, G. (1989) The geography of women's fear, *Area*, **21**(4), 385–90.

Valentine, G. (1995) Out and about: geographies of lesbian landscapes, *International Journal of Urban and Regional Research*, **19**(1), 96–111.

Valentine, G. (2001) *Social Geographies: Space and Society*, Prentice Hall, Harlow.

van der Zee, H.I. (1982) *The Hunger Winter: Occupied Holland 1944–45*, Norman and Hobhouse, London.

van Schendel, W. (2005) Spaces of engagement: How borderlands, illegal flows, and territorial states interlock, in Abrahams, I. and van Schendel, W. (eds) *Illicit Flows and Criminal Things*, Indiana University Press, Bloomington and Indianapolis, IN, 38–68.

Van Welsum, D. and Reif, X. (2005) *Potential Offshoring: Evidence from Selected OECD Countries*, OECD, DSTI-ICCP.

Vasudevan, A., McFarlane, C and Jeffrey, A. (2008) Spaces of enclosure, *Geoforum*, **39**(5), 1641–6.

Visser, G. (2010) Gay men, leisure spaces and South African cities: the case of Cape Town, *Geoforum,* **34**(1), 123–37.

Visser, M. (1986) *Much Depends on Dinner,* Collier/Macmillan, New York.

Visvanathan, N., Duggan, L. and Nisonoff, L. (eds) (1997) *The Women, Gender and Development Reader,* Zed Books, London.

von Braun, J. (2009) The food crisis isn't over, *Nature*, **456**, 701.

W

Wackernagel, M. and Rees, W.E. (1998) *Our Ecological Footprint: Reducing Human Impact on the Earth*, New Society Publishers, Gabriola Island, BC, Canada.

Wacquant, L. (2007) *Urban Outcasts: A Comparative Sociology of Advanced Marginality*, Polity Press, Cambridge.

Wacquant, L. (2010) Class, race and hyperincarceration in revanchist America, *Daedulus*, **139**(3), 74–90.

Wacquant, L.J.D. (1993) Urban outcasts: stigma and division in the black American ghetto and the French urban periphery, *International Journal of Urban and Regional Research*, **17**, 366–83.

Wacquant, L.J.D. (1995) The ghetto, the state and the new capitalist economy, in Kasinitz, P. (ed.) *Metropolis Centre and Symbol of our Times*, Macmillan, London, 418–49.

Wade, R. (2001) Winners and losers, *The Economist*, April, 79–82.

Wait, G., McGuirk, P., Dunn, K., and Hartig, K. (eds) (2000) *Introducing Human Geography: Globalization, Difference and Inequality*, Longman, French's Forest, NSW.

Walker, R. (1996) Another round of globalization in San Francisco, *Urban Geography*, **17**, 60–94.

Wall, D. (1999) *Earth First! and the Anti-roads Movement*, Routledge, London.

Wallace, D. and Wallace, R. (1998) *A Plague on Your Houses: How New York was Burned Down and How National Public Health Crumbled*, Methuen, London.

Wallace, D. and Wallace, R. (2000) Life and death in Upper Manhattan and the Bronx: toward an evolutionary perspective on catastrophic social change, *Environment and Planning A*, **32**, 1245–66.

Wallace, I. (1985) Towards a geography of agribusiness, *Progress in Human Geography*, **9**, 491–514.

Wallerstein, I. (1979) *The Capitalist World Economy*, Cambridge University Press, Cambridge.

Wallerstein, I. (1980) Imperialism and development, in Bergeson, A. (ed.) *Studies of the Modern World System*, Academic Press, New York.

Wallerstein, I. (1991) *Geopolitics and Geoculture: Essays on the Changing World System*, Cambridge University Press, Cambridge.

Wallerstein, I. (1994) Development: lodestar or illusion?, in Sklair, L. (ed.) *Capitalism and Development*, Routledge, London, 3–20.

Wa Ngugi, M. (2005) *New Orleans and the Third World*, www.zmag.org/content/showarticle.cfm?ItemID=8694 [accessed 23 May 2007].

Ward, D. (1987) Population growth, migration, and urbanization, 1860–1920, in Mitchell R. D. and Groves, P.A. (eds) *North America: The Historical Geography of a Changing Continent*, Hutchinson, London.

Ward, D. (1989) *Poverty, Ethnicity and the American City, 1840–1925*, Cambridge University Press, Cambridge.

Ward, K., Binnie, J., Brown, G. *et al.* (2010) Reading Andrew Tucker's Queer Visibilities: space, identity and interaction in Cape Town, *Political Geography*, **29**(8), 454–62.

Ward, S. (2004) *Planning and Urban Change*, Sage, London.

Warhurst, C., Nickson, D., Witz, A. and Cullen, A. (2000) Aesthetic labour in interactive service work: some case study evidence from the 'new' Glasgow, *The Service Industries Journal*, **20**(3), 1–18.

Watney, S. (1987) *Policing Desire: Pornography, AIDS and the Media*, Methuen, London.

Watnick, M. (1952–3) The appeal of Communism to the peoples of underdeveloped areas, *Economic Development and Cultural Change*, **1**, 22–36.

Watson, J. L. (1997a) Introduction: transnationalism, localization, and fast foods in East Asia, in Watson J. L. (ed.) *Golden Arches East: McDonalds in East Asia*, Stanford University Press, Stanford, CA, 1–38.

Watson, J. L. (1997b) McDonalds in Hong Kong: consumerism, dietary change, and the rise of a children's culture, in Watson, J. L. (ed.) *Golden Arches East: McDonalds in East Asia*, Stanford University Press, Stanford, CA, 77–109.

Watson, S. and Gibson, K. (eds) (1995) *Postmodern Cities and Spaces*, Blackwell, Oxford.

Watts, D.C.H., Ilbery, B. and Maye, D. (2005) Making re-connections in agro-food geography: alternative systems of food provision, *Progress in Human Geography*, **29**(1), 22–40.

Watts, M. (2003) Development and governmentality, *Singapore Journal of Tropical Geography*, **24**, 6–34.

Weber, A. (1929) *Alfred Weber's Theory of the Location of Industries*, Chicago University Press, Chicago, IL.

Weigold, A. (1999) Famine management: the Bengal famine (1942–44) revisited, *South Asia*, **22**(1), 63–77.

Weinrich, J. (1982) Is homosexuality biologically natural?, in Paul, W., Weinrich, J., Gonsiorek, J. and Hotvedt, M. (eds) *Homosexuality: Social, Psychological and Biological Issues*, Sage, Beverley Hills, CA, 197–208.

Wellington, C. and Bryson, J. R. (2001) At face value?: image consultancy, emotional labour and professional work, *Sociology*, **35**(4), 933–46.

Werbner, P. (1997) Introduction: The dialectics of cultural hybridity, in Werbner, P. and Modood, T. (eds) *Debating Cultural Hybridity: Multi-Cultural Identities and the Politics of Anti-Racism*, Zed Books, London, 1–26.

Whatmore, S. (1995) From farming to agribusiness: the global agro-food system, in Johnston, R., Taylor, P. and Watts, M. (eds) *Geographies of Global Change*, Blackwell, Oxford, 36–49.

Whatmore, S. (2002) *Hybrid Geographies*, Routledge, London.

Whatmore, S., Stassart, P. and Renting, H. (2003) What's alternative about alternative food networks?, *Environment and Planning A*, **35**, 389–91.

Whitelegg, J. (1993) *Transport for a Sustainable Future*, Belhaven Press, London.

Whitman, J. (ed.) (2009) *Palgrave Advances in Global Governance*, Palgrave, Basingstoke.

Whittlesey, D. (1936) Major agricultural regions of the earth, *Annals of the Association of American Geographers*, **26**, 199–240.

Wiener, M. J. (1985) *English Culture and the Decline of the Industrial Spirit, 1850–1980*, Penguin, Harmondsworth.

Wiggins, S., Kirsten, J. and Llambi, L. (eds) (2010) The Future of Small Farms, special issue of *World Development*, **38**(10), 1341–526.

Wilk, R. (2006) Bottled water: the pure commodity in the age of branding, *Journal of Consumer Culture*, **6**, 303–25.

Williams, C.C. (1996) Local exchange and trading systems: a new source of work and credit for the poor and unemployed?, *Environment and Planning A*, **28**, 1395–415.

Williams, G., Meth, P. and Willis, K. (2009) *Geographies of Developing Areas: The Global South in a Changing World*, Routledge, Abingdon.

Williams, P. and Hubbard, P. (2001) Who is disadvantaged?: Retail change and social exclusion, *International Review of Retail, Distribution and Consumer Research*, **11**, 267–86.

Williams, R. (1973) *The City and the Country*, Chatto and Windus, London.

Willis, K. (2005) *Theories and Practices of Development*, Routledge, London.

Willis, P. (1990) *Uncommon Culture*, Open University Press, Milton Keynes.

Willis, R.P. (ed.) (2001) New Zealand in the 1990s, special edition of *Asia Pacific Viewpoint*, **42**, 1.

Wilson, A. (1992) Technological utopias, world's fairs and theme parks, in Wilson, A. (ed.) *The Culture of Nature: North American Landscape from Disney to the Exxon Valdez*, Blackwell, Oxford, 157–90.

Wilson, D. (2005) *Inventing Black-on-Black Violence: Discourse, Space, Representation*, Syracuse University Press, Syracuse.

Wilson, G. and Rigg, J. (2003) 'Post-productivist' agricultural regimes and the South: discordant concepts?, *Progress in Human Geography*, **27**(6), 681–707.

Wilson, G. A. (2001) From productivism to post-productivism and back again?: Exploring the (un) changed natural and mental landscapes of European agriculture, *Transactions of the Institute of British Geographers*, **26**(1), 77–102.

Wilton, R. (1996) Diminished worlds?: The geography of everyday life with HIV/AIDS, *Health and Place*, **2**, 69–83.

Winchester, H. and White, P. (1988) The location of marginalized groups in the inner city, *Environment and Planning D: Society and Space*, **6**, 37–54.

Winchester, H.P.M., Kong, L. and Dunn, K. (2003) *Landscapes: Ways of Imagining the World*, Prentice Hall, Harlow.

Winchester, S. (2003) *Krakatoa: The Day the World Exploded*, Penguin Books, London.

Winichakul, T. (1995) *Siam Mapped: A History of the Geobody of a Nation*, University of Hawaii Press, Hawaii.

Winichakul, T. (1996) Maps and the formation of the geobody of Siam, in Tønnesson, S. and Antlöv, H. (eds) *Asian Forms of the Nation*, Curzon Press, Richmond, 67–91.

Winter, M. (2003) Embeddedness, the new food economy and defensive localism, *Journal of Rural Studies*, **19**, 23–32.

Wirth, L. (1938) Urbanism as a way of life, *American Journal of Sociology*, **44**, 1–24, reprinted in Hart, P. K. and Reiss, A. J. (eds) (1957) *Cities and Society*, Free Press, Chicago, IL.

Wiston Spirn, A. (1997) The authority of nature: conflict and confusion in landscape architecture, in Wolschke-Bulmahn, J. (ed.) *Nature and Ideology: Natural Garden Design in the Twentieth Century*, Dumbarton Oaks Research Library and Collection, Washington, DC, 249–61.

Withers, C. W. J. and Mayhew, R. J. (2002) Rethinking 'disciplinary' history: geography in British universities, c. 1580–1887, *Transactions of the Institute of British Geographers*, **27**, 11–29.

Wolch, J. (2002) Anima urbis, *Progress in Human Geography*, **26**(6), 721–42.

Wolf, E. R. (1966) *Peasants*, Prentice-Hall, Englewood Cliffs, NJ.

Wolfe, T. (1988) *Bonfire of the Vanities*, New York, Bantam.

Wolfe-Phillips, L. (1987) Why Third World: origins, definitions and usage, *Third World Quarterly*, **9**, 1311–19.

Wolin, R. (2010) The idea of cosmopolitanism: from Kant to the Iraq war and beyond, *Ethics and Global Politics*, **3**, 143–53.

Wolters, O. W. (1982) *History, Culture, and Religion in Southeast Asian Perspectives*, Institute of Southeast Asian Studies, Singapore.

Women and Geography Study Group (1997) *Feminist Geographies: Explorations in Diversity and Difference*, Longman, Harlow.

Wood, A. and Welch, C. (1998) *Policing the Policemen: The Case for an Independent Evaluation Mechanism for the IMF*, Bretton Woods Project, Friends of the Earth US, London.

Wood, G. (1985) *Labelling in Development Policy: Essays in Honour of Bernard Schaffer*, Sage, London.

Woods, C. (1998) *Development Arrested: Race, Power and Blues in the Mississippi Delta*, Verso, London.

Woodward, K. (1997) Concepts of identity and difference, in Woodward, K. (ed) *Identity and Difference*, Sage, London, 8–59.

Woodward, R. (2004) *Military Geographies*, Blackwell, Oxford.

Woodward, R. (2005) From military geography to militarism's geographies: disciplinary engagements with the geographies of militarism and military activities, *Progress in Human Geography*, **29**, 718–40.

Woon, C.Y. (2009) A 'New Mecca for terrorism'?: Unveiling the 'second front' in Southeast Asia, in Ingram, A. and Dodds, K. (eds) *Spaces of Security and Insecurity: Geographies of the War on Terror*, Ashgate, 85–108.

World Bank (1999) *World Development Indicators*, World Bank, Washington, DC, also available as *World Development Indicators on CD-Rom*, World Bank, Washington, DC.

World Bank (2001a) *World Development Indicators*, World Bank, Washington, DC.

World Bank (2001b) *Attacking Poverty: World Development Report 2000/2001*, Oxford University Press and World Bank, Washington, DC.

World Bank (2002) *World Development Report 2002: Building Institutions for Markets*, World Bank, Washington, DC.

World Bank (2005) *World Development Indicators 2005*, World Bank, Washington, DC.

World Bank (2006) *World Development Report: Equity and Development*, World Bank, Washington, DC.

World Bank (2007) *The Little Green Data Book 2007*, World Bank, Washington DC.

World Bank (2009) *The World Bank Annual Report 2009*, World Bank, Washington, DC.

World Bank (2010a) *2010 World Development Indicators*, World Bank, Washington, DC.

World Bank (2010b) *The Little Green Data Book 2010*, World Bank, Washington, DC.

World Bank (2010c) *World Development Report 2010: Development and Climate Change*, World Bank, Washington, DC.

World Bank (2011) *Global Development Horizons 2011. Multipolarity: The New Global Economy*, The World Bank, Washington, DC.

World Federation of Exchanges (WFE) (2006) Data available at: www.world-exchanges.org/WFE/home.asp?menu=378&document=3557.

World Wind Energy Council (2010) *Global Wind Energy Outlook 2010*, World Wind Energy Council, Brussels.

Wright, C. (2004) Consuming lives, consuming landscapes: interpreting advertisements for CaféDirect coffee, *Journal of International Development*, **10**, 665–80.

Wright Mills, C. (1959) *White Collar Work: The American Middle Classes*, Oxford University Press, New York.

Wright, R. (1995) *The Color Curtain: A Report on the Bandung Conference*, Banner Books, New York.

Wrigley, N. (2002) 'Food deserts' in British cities: policy context and research priorities, *Urban Studies*, **39**, 2029–40.

Wrigley, N. and Lowe, M. (eds) (1996) *Retailing, Consumption and Capital: Towards the New Retail Geography*, Longman, Harlow.

Wrigley, N. and Lowe, M. (2002) *Reading Retail*, Arnold, London.

Wrigley, N., Warm, D. and Margetts, B. (2003) Deprivation, diet and food-retail access: findings from the Leeds 'food deserts' study, *Environment and Planning A*, **35**, 151–88.

Wylie, J. (2007) *Landscape*, Routledge, London.

Y

Yan, Y. (1997) McDonalds in Beijing: the localization of Americana, in Watson, J. L. (ed.) *Golden Arches East: McDonalds in East Asia*, Stanford University Press, Stanford, CA, 39–76.

Yang, C. (2009) Strategic coupling of regional development in global production networks: redistribution of Taiwanese personal computer investment from the Pearl River Delta to the Yangtze River Delta, China, *Regional Studies*, **43**, 385–407.

Yang, Y. (2006) *The Taiwanese Notebook Computer Production Network in China: Implications for Upgrading of the Chinese Electronics Industry,* The Personal Computer Industry Center, University of California, Irvine, CA, available at: www.pcic.merage.uci.edu/papers.asp [accessed 29 August 2011].

Yang, Y-R. and Coe, N.M. (2009) The governance of global production networks and regional development: a case study of Taiwanese PC production networks, *Growth and Change,* **40**, 30–53.

Yarwood, R. (2007) The geographies of policing, *Progress in Human Geography,* **31**(4), 447–65.

Yawnghwe, C. T. (1987) *The Shan of Burma: Memoirs of a Shan Exile,* Institute of Southeast Asian Studies, Singapore.

Yeates, N. (2004) Global care chains: critical reflections and lines of enquiry, *International Feminist Journal of Politics,* **6**(3), 369–91.

Yergin, D. (2006) Ensuring energy security, *Foreign Affairs,* **85**(2), 69–82.

Yiftachel, O. and Ghanem, A. (2005) Understanding ethnocratic regimes: the politics of seizing contested territories, *Political Geography,* **23**(4), 647–76.

Young, M. and Willmott, P. (1962) *Family and Kinship in East London,* Pelican, London.

Yuval-Davis, N. (1997) *Gender and Nation,* Sage, London.

Yuval-Davis, N. (2003) Citizenship, territoriality and the gendered construction of difference, in Brenner, N., Jessop, B., Jones, M. and MacLeod, G. (eds) *State/Space: A Reader,* Blackwell, Oxford, 309–25.

Z

Zelizer, V. (1989) The social meaning of money: 'special monies', *American Journal of Sociology,* **95**, 342–77.

Zick Varul, M. (2008) Consuming the campesino – fair-trade marketing: between recognition and romantic commodification, *Cultural Studies,* **22**, 654–79.

Zook, M.A. (2003) Underground globalization: mapping the space of flows of the internet adult industry, *Environment and Planning A,* **35**(7), 1261–86.

Zouza, L.B. (2002) Vital interests and budget deficits: US Foreign Aid after September 11, *Middle East Insight,* March–April, 25–9.

Zukin, S. (1989) *Loft Living: Culture and Capital in Urban Change,* Rutgers University Press, Newark, NJ.

Zukin, S. (1995) *The Cultures of Cities,* Blackwell, Oxford.

Index